MODERN
PHYSICAL GEOGRAPHY

MODERN PHYSICAL GEOGRAPHY
FOURTH EDITION

ALAN H. STRAHLER
ARTHUR N. STRAHLER

JOHN WILEY & SONS, INC.
New York · Chichester · Brisbane · Toronto · Singapore

ACQUISITIONS EDITOR	Barry Harmon
DEVELOPMENTAL EDITOR	Janice Haymes
PRODUCTION MANAGER	Katharine Rubin
DESIGNER	Ann Marie Renzi
PRODUCTION SUPERVISOR	Sandra Russell
COPY EDITOR	Marjorie Shustak
PHOTO RESEARCH MANAGER	Stella Kupferberg
ILLUSTRATION	Edward Starr

Library of Congress Cataloging in Publication Data:

Strahler, Alan H.
 Modern physical geography / Alan H. Strahler, Arthur N. Strahler.
 —4th ed.
 A.N. Strahler's name appeared first on the earlier edition.
 Includes bibliographical references and index.
 ISBN 0-471-53392-0 (cloth)
 1. Physical geography. I. Strahler, Arthur Newell, 1918–
II. Title.
GB54.5.S78 1992   91-27551
550—dc20   CIP

Printed and bound by Von Hoffmann Press, Inc.

10 9 8 7 6 5 4 3 2

ABOUT THE AUTHORS

ALAN H. STRAHLER   (b. 1943) received his B.A. degree in 1964 and his Ph.D. degree in 1969 from The Johns Hopkins University, Department of Geography and Environmental Engineering. His published research is in the fields of plant geography, forest ecology, quantitative methods, and remote sensing. He has held academic appointments at the University of Virginia, the University of California at Santa Barbara, and Hunter College of the City University of New York, and is now Professor of Geography at Boston University. He is a coauthor of several textbooks on physical geography and environmental science.

ARTHUR N. STRAHLER   (b. 1918) received his B.A. degree in 1938 from the College of Wooster, Ohio, and his Ph.D. degree in geology from Columbia University in 1944. He is a fellow of the Geological Society of America and the Association of American Geographers. He was appointed to the Columbia University faculty in 1941, serving as Professor of Geomorphology from 1958 to 1967 and as Chair of the Department of Geology from 1959 to 1962. He is the author of several widely used textbooks of physical geography, environmental science, and the earth sciences.

PREFACE

The first edition of Strahler's *Physical Geography*, published four decades ago by John Wiley & Sons, firmly established the tradition of a comprehensive physical geography text grounded in sound principles of natural science. Now, as then, these principles are carefully selected from atmospheric science (meteorology and climatology), hydrology, the geosciences (geology, geophysics, and geomorphology), soil science (pedology), ecology, and biogeography. Uniquely interwoven, this broad spectrum of topics comprises a physical geography that relates the planetary biota to the physical environment. From the outset, and continuing to the present, a thorough text has been complemented by a rich graphic presentation that features numerous original drawings, schematic diagrams, and maps, along with a large collection of excellent photographs.

Throughout the 1950s and 1960s, *Physical Geography* was expanded in its coverage of both soil science—adopting the modern international Soil Taxonomy—and relevant biogeographical concepts of ecosystem functioning and the global system of biomes. The addition of a set of large, specially designed full-color global maps greatly strengthened the text. The need for an abbreviated version of *Physical Geography* for use in one-semester courses was met in 1965 by introduction of a companion work, *Introduction to Physical Geography*. This two-tier system proved highly successful and continues today.

The Changing Focus The early 1970s witnessed a dramatic rise in public awareness of burgeoning problems of severe environmental impacts by human activity on natural systems in combination with an accelerating depletion of natural resources. In a new two-author collaboration, the Strahlers addressed these environmental/resource concerns in three avant-garde Wiley/ Hamilton textbooks: *Environmental Geoscience* (1973), *Introduction to Environmental Science* (1974), and *Geography and Man's Environment* (1977). Much of this exciting new material soon found its way into the new Strahler physical geography texts.

Expanding Concepts The late 1970s brought a thoroughly updated approach to physical geography in our contemporary two-tier textbook series: *Modern Physical Geography* and *Elements of Physical Geography*, the latter being the abbreviated version. Major conceptual advances incorporated in this series, and still in place in the current versions, included expanded treatment of the soil-water balance and an entirely new Strahler system of global climate classification based on the soil-water balance. Equally important has been the introduction of the unifying concept of open systems of energy and matter, using a series of unique flow diagrams throughout the text. At the same time, the new paradigm of plate tectonics, which had come into full acceptance in the geosciences, was incorporated in the Strahler texts, along with new emphasis on tectonic landforms. Superimposed on these new dimensions of global environmental systems has been our greatly increased attention to environmental impacts and natural resource problems. Severe environmental hazards—storms, floods, volcanic eruptions, and earthquakes—have been even more fully documented. Revised editions of these two Strahler texts, each serving its own academic level, have kept pace with new findings of scientific research and their applications in all of these vital areas.

ABOUT THE NEW FOURTH EDITION

Our Fourth Edition of *Modern Physical Geography* features welcome changes in four categories. First is the updating of text and figures, including the introduction of new topics of current interest. Second is a group of rearrangements of subjects and chapters. Third is the expansion of environmental and resource topics of vital concern to humans. Fourth is a new design that allows the use of full-color illustrations throughout the entire book.

Updating of Text and Figures As in past editions, this edition endeavors to present the most up-to-date information available. New topics of current importance have been inserted in various places throughout the revised work. As examples, you will find new coverage of warm-core and cold-core eddies in the North Atlantic, illustrated by color imagery; new maps of upper-air atmospheric circulation patterns in the Asiatic monsoon system; and a new section on the thunderstorm microburst, made visible at ground level by Doppler radar imagery. We have also updated our early chapter on the earth's radiation balance, where SI units of watts-per-square meter have replaced the langley in text and figures. Another example is the modernization of graphs and maps of the frequency and distribution of severe weather phenomena—hail storms and tornadoes. New information also appears throughout the special environmental/resource topics.

Rearrangements of Subjects and Chapters   The changes carried the strong approval of a majority of our manuscript reviewers and a panel of more than 30 active physical geography instructors who responded to a detailed authors' questionnaire. In general, these improvements tend to bring the new edition of *Modern Physical Geography* into the same sequential pattern as *Elements of Physical Geography*. The most important changes are:

- The sequence of topics in opening chapters has been tightened and focused more directly on the physical basis of atmospheric science. The revised first chapter (Our Rotating Planet) now focuses entirely on scientific principles of earth-sun relationships that are immediately applied in following chapters. Text and figures on the technical subjects of map projections and remote sensing that previously appeared in the opening chapters are now appendixes.

- Climatology is treated in four chapters (8–11), of which the first two cover classification principles and systems and the soil-water balance. The following two chapters describe global climate types in an integrated treatment with rich descriptive supplements on characteristic soils, vegetation, and plant resources. This arrangement, similar to that in the current edition of *Elements of Physical Geography*, adds a vital ecological content to the description of climates, offering a synthesis of each climate with its distinctive set of soils, landforms, and vegetation types.

- The former chapter on runoff and water resources has been transferred to a later position as Chapter 16, between the chapters on weathering and landforms of running water. It has been strengthened in coverage of ground water, floods, and lakes. This change, too, brings the topical arrangement into harmony with that of *Elements*.

- Deletion of the obsolete, half-century-old Marbut System of soil classification (USDA, 1938) was strongly endorsed by a majority of our reviewers and panelists. The chapter on World Soils now concentrates on the Comprehensive Soil Classification System and the Canadian System of Soil Classification.

Several brief topics, previously offered in the Second Edition (1983), and deleted from the Third Edition (1987), have been reinstated. Mostly, these are in meteorology, oceanography, and hydrology. The total effect is to improve the information content of the text as a whole, and particularly to provide essential background for the expanded environmental/resource units.

The list of suggested readings contains many recent articles in accessible journals, while the glossary includes many new terms.

Environmental/Resource Topics   Physical geography courses offered in our North American colleges and universities are admirably equipped to present today's students with the vital concepts of outstanding global environmental problems and issues. New courses intended for the same purposes are springing up in other departments, where they parade under such titles as "Earth Systems Science," "Planetary Ecosystems," "Global Environmental Science," and "Environmental Earth Science." To meet the challenge presented by these newcomers to our traditional and successful physical geography programs, we have expanded our coverage of salient global issues and problems by developing a full series of environmental/resource units. The following list illustrates the broad scope of these special features:

The Ozone Layer—A Shield to Earthly Life (Chapter 2)
The Greenhouse Effect and Global Climate Change (Chapter 4)
Wind as an Energy Resource (Chapter 5)
Cloud Cover, Precipitation, and Global Warming (Chapter 6)
Forecasting Two Hurricanes—Gilbert and Hugo (Chapter 7)
Air Pollution and Its Effects (Chapter 6)
Acid Deposition and Its Effects (Chapter 6)
Remote Sensing of Weather Phenomena (Chapter 7)
El Niño and the Southern Oscillation (Chapter 7)
Geothermal Energy Resources (Chapter 14)
Earthquakes as Environmental Hazards (Chapter 14)
Environmental Problems Related to Ground-Water Withdrawal (Chapter 16)
The Aral Sea—A Dying Saline Lake (Chapter 16)
Chemical Sources of Water Pollution (Chapter 16)
Flood Abatement Measures (Chapter 17)
Rising Sea Level and Coastal Inundation—A Threat from Global Warming (Chapter 20)
Deflation Induced by Human Activity—The Dust Bowl (Chapter 21)
Ice Sheets and Global Warming (Chapter 22)
The Green Revolution—Success or Failure? (Chapter 25)
The Great Yellowstone Fire—Disastrous or Beneficial? (Chapter 26)
Biomass Burning and Its Impacts on the Atmosphere (Chapter 26)
Exploitation of the Low-Latitude Rainforest Ecosystem (Chapter 27)
Forests and Global Warming (Chapter 27)
Drought and Land Degradation in the African Sahel (Chapter 27)

Improvements in Design   Turning finally to the conspicuous improvements in design and visual presentation, we are pleased that use of full-color printing throughout the book has eliminated the color plates, formerly inserted in rather awkward places and difficult to refer to in context. The benefits of full-color printing throughout the book are immediately evident from the appearance of the familiar world maps of temperatures, pressures and winds, precipitation, climates, soils, and vegetation at points in the chapters where they are first introduced. Color photographs and remote-sensing images in generous numbers also appear now in their appropriate places in the text. Many of these photos are new; others are taken from our shorter companion work, *Elements of Physical Geography*. In replacing many of the black/white photographs with color, we have made every effort to select scenes primarily for their science content. Many of the line drawings have been rendered more effective by use of multiple colors, and it has been possible to use many of the illustrations previously color-adapted for use in *Elements*. The new page design features a clearer, more open type.

As never before, the relevance of the updated new Fourth Edition of *Modern Physical Geography* to contemporary environmental and resource problems invites its use in a wide variety of college programs and courses. For briefer survey courses of one semester or one term, a viable instruction plan is to choose only what is needed from throughout this new revision. Select those chapters, sections, and paragraphs easily within range of the available study time and student background knowledge. A text exposition that is consistently lucid and fully illustrated by color graphics makes this flexible plan inviting and fully workable.

SUPPLEMENTARY MATERIALS

A revised *Study Guide* provides the student with chapter outlines and self-testing questions, along with sets of sample tests. The revised *Instructor's Manual* contains helpful teaching suggestions and background information, along with a full set of achievement tests. A bank of test questions in digital format will allow instructors using the new edition to prepare unique unit tests and final examinations. Our supporting manual, *Investigating Physical Geography*, provides a rich source of exercises and problems suitable for independent home study or supervised laboratory. Visual aids in the form of carefully selected overhead color transparencies and 35-mm slides are also available to instructors using the new edition.

ACKNOWLEDGMENTS

We are grateful for the collective advice of a large group of geography instructors who completed the authors' questionnaire on revision needs. Numerous comments and suggestions attached to the questionnaires by the respondents focused our attention on a wide range of possibilities for improving the book. The list of respondents includes the following geography professors who are currently using, or have recently used, *Modern Physical Geography* as a class textbook:

Susan W. Beatty, Nelson Caine, and David E. Greenland of the University of Colorado, Boulder; Robert M. Hordon and David A. Robinson, Rutgers University, New Brunswick, NJ; Robert K. Holz, University of Texas, Austin; Ronald W. Jenkins, Pennsylvania State University–York; Thomas S. Krabacher, California State University, Sacramento; Hsiang-Te Kung, Memphis State University, Memphis, TN; David R. Legates, University of Oklahoma, Norman; Michael E. Lewis, University of North Carolina, Greensboro; Robert B. McMaster, Syracuse University, NY.

We also thank the following reviewers of the full manuscript of the revised edition:

June M. Ryder, Karen Ewing, John Wolcott, Graham Thomas, and Margaret E.A. North of the University of British Columbia, Vancouver; Kenneth Hinkel, Nick Dunning, and Susanna Tong of the University of Cincinnati; Richard Hackett of Oklahoma State University; Donald E. Petzold of the University of Maryland; Barbara Borowiecki of the University of Wisconsin; Dr. Bruce Young of Santa Monica College; David Butler of the University of Georgia; Duane Nellis of Kansas State University; John Giardino of Texas A&M University; Stanley Nursworthy of California State University; and Michael L. Barnhardt of Memphis State University.

With such broad-based guidance and support we have been able to produce a revised edition that we hope will serve the needs of the large majority of physical geography instructors.

Alan H. Strahler
Arthur N. Strahler

CONTENTS

INTRODUCTION—
THE HUMAN ENVIRONMENT

Physical geography is an area of study that brings together and interrelates the important elements of the physical environment of humans. Physical geography draws on several natural sciences for its subject matter, among them sciences of the atmosphere (meteorology, climatology), oceans (oceanography), solid earth (geology), landforms (geomorphology), soils (soil science), and vegetation (plant ecology, biogeography). But physical geography is much more than a collection of topics drawn from other sciences; it weaves that information into patterns of interaction with humans in a way not expressed within each of the contributing sciences. As a branch of geography, physical geography also emphasizes spatial relationships—the systematic arrangements of environmental elements into regions over the earth's surface, and the causes for those patterns.

THE LIFE LAYER

The focus of physical geography is on the *life layer*, a shallow zone of the lands and oceans containing most of the world of organic life, or *biosphere*. The quality of the life layer is a major concern of physical geography; "quality" means the sum of the physical factors that make the life layer habitable for all forms of plants and animals, but most particularly for humans.

The quality of the physical environment of the lands is established by factors, forces, and inputs coming from both the atmosphere above and the solid earth below. The *atmosphere*, a gaseous envelope surrounding the solid earth, dictates climate, which governs the exchange of heat and water between atmosphere and ground. The atmosphere also supplies vital elements—carbon, hydrogen, oxygen, and nitrogen—needed to sustain all life of the lands.

The solid earth, or *lithosphere*, forms the stable platform for the life layer and is also shaped into landforms. These landscape features—mountains, hills, and plains—bring another dimension to the physical environment and provide varied habitats for plants. The solid earth is also the basic source of many nutrient ele-

ments, without which plants and animals cannot live. These elements pass from rock into the shallow soil layer, where they are held in forms available to organisms.

Water, another essential material of life, permeates the life layer, the overlying atmosphere, and the underlying solid earth. In all its forms, water on the earth constitutes the *hydrosphere*. Our study of physical geography can be described in the broadest of terms as a study of the atmosphere, hydrosphere, and lithosphere in relation to the biosphere.

NATURAL SYSTEMS IN PHYSICAL GEOGRAPHY

Understanding the activities and changes that go on everywhere within the life layer requires a physical geographer to think in terms of flow systems of matter and energy. Each flow system consists of connected pathways through which matter, or energy, or both move continuously. Most of the systems physical geography is concerned with are powered by solar energy and involve air, water, mineral matter, or living organisms. These concepts are explained early in the book and are followed in later chapters by examples of natural systems illustrated by a special type of schematic flow diagram.

AN ENVIRONMENTAL SCIENCE

Acting together, the inputs of energy and materials into the life layer from atmosphere and solid earth determine the quality of the environment and the richness or poverty of the organic life it can support. Thus an understanding of physical geography is vital when planning for survival of the earth's rapidly expanding human population. Survival will depend not only on how much fresh water and food is available; it will also depend on whether the environment is protected from pollution and destruction that reduce the capacity of the land to furnish those necessities.

Here we touch on another of the important goals of

physical geography, which is to evaluate the impact of humans on the natural environment. Many persons think of environmental science, the study of the interaction between humans and their environment, which is gaining wide recognition today, as a new discipline. Actually, geographers have been investigating environmental science for many decades. Physical geography has always been at the heart of environmental studies because it is strongly oriented toward the interaction between humans and their environment.

A PLAN FOR STUDY

We plan to begin our study of physical geography with the atmosphere and the ways in which it furnishes light, heat, and water to the life layer. An evaluation of climate follows; here we develop background information for a study of global environments. At this point we turn our attention to the hydrosphere, and the concept of a global water balance is introduced.

Place-to-place variations in the availability of water to plants are studied through the soil-water balance, an accounting system on which a rigorous and useful climate classification system can be built.

Next, we study the solid earth, or lithosphere, begin-

ning with the varieties of minerals and rocks that comprise the earth's crust. We then examine the global system of lithospheric plates and their interaction through the modern theory of plate tectonics. Plate interactions largely determine great lines of bending and breaking of crustal rock as well as long chains of active volcanoes. Tectonic and volcanic activities that elevate the earth's young mountain ranges are powered by internal sources of energy, supplied by the process of radioactivity in crustal rock. These internal forces produce their own distinctive landscape features.

Our next group of chapters investigates the solar powered processes that shape the earth's surface into landforms. These processes are weathering, mass wasting, and the activities of streams, waves and currents, glacial ice, and wind.

Chapters covering special phases of the biosphere follow. First, we study the nature of the soil layer and the varieties of soils found over the earth's land surfaces. Then, we investigate the relationships between organisms and their environments—a part of the science of ecology that is concerned with the flow of matter and energy through the biosphere. Biogeography, a branch of physical geography, adds an understanding of global patterns of natural vegetation types—such as forests and grasslands—as related to climate and soil.

OUR ROTATING PLANET

Our search for an understanding of the physical environment of humans starts with an inquiry into things astronomical. The most basic environmental controls relate to two facts: First, the earth is approximately spherical in form; second, the spherical earth is in motion, spinning on an axis and, at the same time, traveling in a nearly circular path around the sun. Let us look into these facts to derive significant interpretations relating to the quality of the earth's surface environment.

In this age of orbiting satellites, the spherical form of the earth is such an obvious fact that we may have difficulty imagining ourselves living in ancient times when the extent and configuration of the earth were entirely unknown. In that day, to sailors of the Mediterranean Sea, on their ships and out of sight of land, the sea surface looked perfectly flat and seemed to be terminated by a circular horizon. Based on this perception, the sailors might well have inferred that the earth has the form of a flat disk and that, if they traveled to its edge, their ships might fall off. Even so, these sailors could have sensed optical phenomena suggesting that the sea surface is not flat, but curved so as to be upwardly convex like a small part of the surface of a sphere. Even without optical instruments we can get some inkling that the earth's surface curves downward away from us through the observation that sunlight illuminates the tops of high clouds and high mountain summits after sunset and before sunrise.

PROOF OF THE EARTH'S SPHERICITY

Rigorous and convincing proof that the earth is spherical in form is not, however, as evident as it might seem if we limit ourselves to observations available to humans before the age of space vehicles. Consider the fact that a few survivors of Magellan's crew sailed completely around the globe and returned to their starting point in Spain. This discovery voyage, 1519–22, might seem at first thought to have proved the earth to be spherical; instead, it merely ascertained that the earth is a solid figure rather than a flat disk terminating in a sharp edge.

Circumnavigation could be performed on a cubical or cylindrical earth, or on an earth of any irregular solid form. So let us look for some possible proofs of earth sphericity available to earthbound humans. At least three basic experimental programs can achieve convincing results.

One proof of the earth's sphericity may be had from observations at sea. As a passing ship recedes farther and farther into the distance, it appears to sink slowly beneath the water level (Figure 1.1). Seen through binoculars or a telescope, the sea surface will appear to rise until the decks are awash, then gradually to submerge the masts. The explanation obviously is that the sea surface curves downward away from us. To prove that this curvature is spherical would require numerous observations in which measurements were made of the amount of apparent sinking of a vessel per unit of distance in many different directions away from the observing point.

A second proof is found by observing many lunar eclipses, at which times the earth's shadow falls on the

FIGURE 1.1 Because the sea surface is curved, a distant ship seen through a telescope appears to be partly submerged.

moon. The edge of the earth's shadow appears as an arc of a circle on the disk of the moon. It can be shown by geometrical proof that a sphere is the only solid body that will always cast a circular shadow. During successive eclipses, the earth is rarely oriented in just the same position. No matter what earth profile is cast on the moon, the circular shadows are all alike; therefore the earth must be spherical.

A third proof uses a simple principle of astronomy, known in ancient times and used effectively by the Arabs as early as the ninth century A.D. An observer at the north pole always sees the North Star, Polaris, in the zenith point in the sky because that star lies in line with the earth's axis of rotation (Figure 1.2). As the observer travels southward, Polaris appears to shift its position toward the horizon, so that halfway between the north pole and the equator (at latitude 45°N), the star lies halfway between the zenith and the horizon. Upon approaching the equator, the observer would find Polaris located close to the horizon. A series of measurements of the angle between the horizon and Polaris would show that the angle always decreases by 1° for each 111 km (69 mi) of southward travel. This observation would prove that the travel path has followed the arc of a circle. Repetition of these observations along many different north–south lines (meridians) would establish that the northern hemisphere is one-half of a true sphere. A similar set of observations could be carried out in the southern hemisphere, using a minor star in line with the south pole. In this way the earth could be proved to be spherical. This principle is actually used in navigation by the stars (celestial navigation). When we consider that, for more than two centuries, positions of vessels at sea have been accurately determined innumerable times by use of this basic method, the spherical form of the earth has obviously been demonstrated beyond question.

MEASURING THE EARTH'S CIRCUMFERENCE

Scholars among the ancient Greeks believed the earth to be spherical. Pythagoras (540 B.C.) and associates of Aristotle (384–322 B.C.) held this view. They also speculated on the length of the earth's circumference, but with highly erroneous guesses. It remained until about 200 B.C. for Eratosthenes, head of the library in Alexandria, Egypt, to perform a direct measurement of the earth's circumference based on a second principle of geometry. He observed that on a particular date of the year (close to the summer solstice, June 21) at Syene, a city located on the upper Nile River far to the south, the sun's rays at noon shone directly on the floor of a deep vertical well. In other words, the sun at noon was in the zenith point in the sky, and its rays were perpendicular to the earth's surface at that point on the globe (Figure 1.3). At Alexandria, on the same date, the rays of the noon sun made an angle with respect to the vertical. The magnitude of this angle was one-fiftieth of a complete circle, or $7\frac{1}{5}°$.

Eratosthenes needed only to know the north–south distance between Syene and Alexandria to calculate the earth's circumference: he would simply multiply the ground distance by 50 to obtain the circumference. In those days distances between cities were only crude estimates, based on travelers' reports. Eratosthenes took the distance to be 5000 stadia. It is not easy to rate his results in terms of accuracy because we are not sure of the length equivalent of the distance unit (the stadium) that he used. If it were the Attic stadium, equivalent to 185 m (607 ft), the circumference in modern units would come to about 43,000 km (26,700 mi). Considering that the true circumference is close to 40,000 km (25,000 mi), Eratosthenes' results seem offhand to be remarkably good.

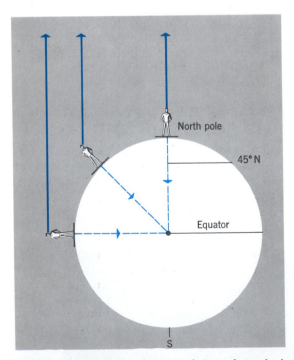

FIGURE 1.2 The height of the North Star above the horizon depends on one's position in the northern hemisphere. Because of the star's great distance from earth, its light rays are shown as parallel lines in the drawing.

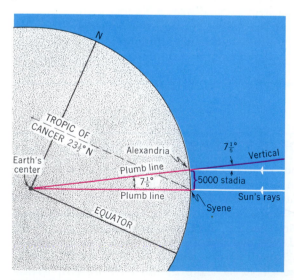

FIGURE 1.3 Eratosthenes' method of measuring the earth's circumference.

From Eratosthenes' classic experiment, it is an easy step to design an astronomical method of measuring the earth's figure using star positions instead of the sun. We need only to select a north–south line, whose length can be measured directly on level ground by surveying means. This line should be several tens of kilometers long. At the ends of the line, the angular position of any selected star can be measured at its highest point above the horizon or with respect to the vertical, using a level bubble or a plumb bob to establish a true horizontal or vertical reference. The difference in angular positions of the star will be equal to the arc of the earth's circumference lying between the ends of the measured line. This very procedure is believed to have been followed by ninth-century Arabs. Their measurements were probably much more accurate than those of Eratosthenes but, because the units of measure are not known in modern equivalents, their work cannot be checked.

In the many centuries following Eratosthenes' work, Western science lay in stagnation. Then, about 1615, Willebrord Snell, a professor of mathematics at the University of Leiden, developed methods of precise distance and angular measurement, which he applied to the problem of the earth's circumference. His work presaged the era of scientific geodesy ("geodesy" comes from the Greek word meaning "to divide the earth") and led to remarkably accurate measurements of the earth's figure about a century later.

GRAVITY IN THE ENVIRONMENT

From the standpoint of life on earth, what is significant about the fact that the earth's form closely approximates that of a true sphere? One answer is, of course, "gravity." *Gravity* is the force acting on a small unit of mass at the earth's surface, tending to draw that mass toward the earth's center. Gravity is a special case of the phenomenon of *gravitation*, the mutual attraction between any two masses—special in the sense that gravity refers to a unit of mass so small in comparison with the earth's mass that the attraction of the unit mass for the earth can be disregarded. Gravitational attraction varies inversely as the square of the distance separating the centers of two masses. So gravity depends on the distance of any particle of matter from the earth's center of mass. That center lies close to the geometrical center of the sphere.

Perhaps you recall a principle of geometry to the effect that a sphere is defined as a solid on which all surface points lie equidistant from a common point: the center of the sphere. Because of this principle, gravity turns out to be an almost constant value at all points at sea level over the entire globe. This is a fact of fundamental importance to all life on earth. Life has evolved through geologic time under the influence of a value of gravity uniform over the earth and probably little changed during the major evolutionary period of a billion years or so. Gravity thus represents the lowest common denominator of the planetary environment.

The force of gravity acts as an environmental factor in many ways. It separates substances of differing density into layers, the least dense being at the top and the most dense at the bottom. Air, liquid water, and rock are arranged in order of density because of their response to gravity. As a result, the life layer is defined as an interface between atmosphere and ocean and between atmosphere and the solid land surface.

Trees, animals, cliffs of rock, and human-made structures must have the strength to withstand the force of gravity, which tends to collapse and crush them. Under a weaker planetary gravity, such structures could rise higher or be constructed of weaker materials to attain a given height. Gravity supplies power for important physical systems of the life layer, particularly streams and glaciers that erode the land. To appreciate the importance of gravity as an environmental factor, we need only speculate as to what would happen if its influence were canceled out and a condition of weightlessness were to take its place. Environmental destruction would be total within a short time!

There are very small, systematic differences in the value of gravity from place to place over the earth. The value at the equator is slightly less than at the two poles; there is also a slight decrease in gravity as we rise to higher elevations above sea level. But, for practical purposes, gravity can be taken as a constant the world over.

The constancy of gravity over the earth's surface might be used in an experiment to prove the earth's sphericity. If we first assume that Newton's law of gravitation is valid, it follows that a given object should register the same weight at all places over the earth's surface. Using a spring balance as the scales, we might travel widely over the earth, repeatedly weighing a small mass of iron and recording the values. If they proved to be unvarying, we could conclude that we have taken our measurements at points all equidistant from the earth's center of mass and thus we are on a spherical surface. Actually, this same experiment, carried out with great precision and using highly refined instruments, has shown that the earth's true figure departs slightly from that of a true sphere.

THE EARTH AS AN OBLATE ELLIPSOID

In 1671 a French astronomer, Jean Richer, was sent by Louis XIV to the Island of Cayenne, French Guiana, to make certain astronomical observations. His clock had been so adjusted that its pendulum, 1 m (39.4 in.) long, beat the exact seconds in Paris (when a pendulum is made shorter, it beats faster; made longer, it beats slower). Upon arriving in Cayenne, which is near the equator, Richer found the clock to be losing about 2½ minutes per day. As soon as Newton's laws of gravitation and motion were published (1687), it became possible to attribute the slowing of the clock at Cayenne to a somewhat reduced value of gravity near the equator. It was soon realized that the smaller value of gravity could be accounted for by supposing that the equatorial region of the earth's surface lies farther from the earth's center than do more northerly regions.

Refined measurements have since revealed that the true form of the earth resembles a sphere that has been

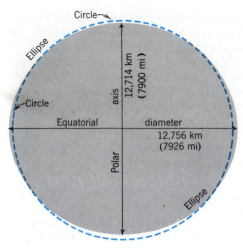

FIGURE 1.4 The earth's equatorial and polar diameters, according to the Geodetic Reference System.

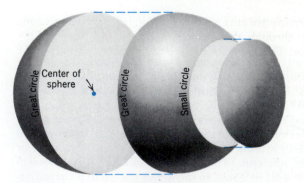

FIGURE 1.5 A great circle and a small circle.

compressed along the polar axis and made to bulge slightly around the equator (Figure 1.4). This form is known as an *oblate ellipsoid.* A cross section through the poles gives an *ellipse* rather than a circle. The equator remains a circle and is the largest possible circumference on the ellipsoid. The earth's oblateness is caused by the centrifugal force of the earth's rotation, which deforms the somewhat plastic earth into a form in equilibrium with respect to the forces of gravity and rotation.

Rounding off the earth's dimensions to the nearest whole kilometer, the equatorial diameter is 12,756 km (7926 mi), whereas the length of the polar axis is 12,714 km (7900 mi). The difference is just over 43 km (slightly less than 27 mi). The *oblateness* of the earth ellipsoid, or "flattening of the poles," is the ratio of this difference to the equatorial diameter, or roughly 43/12,756, which reduces to the approximate fraction 1/300. Thus we can say that the earth's polar axis is about 1/300 shorter than the equatorial diameter. Using these figures, the earth's equatorial circumference is about 40,075 km (24,900 mi). For rough calculations, 40,000 km (25,000 mi) is close enough.

GREAT AND SMALL CIRCLES

For many topics in physical geography, the earth can be treated as if it were a true sphere. For example, flattening of the poles can be disregarded to simplify an understanding of the important concept of the earth as an object turning under the sun's rays.

When a sphere is divided exactly in half by a plane passed through the center, the intersection of the plane with the sphere is the largest circle that can be drawn on the sphere and is known as a *great circle* (Figure 1.5). Circles produced by planes passing through a sphere anywhere except through the center are smaller than great circles and are designated *small circles.*

One important use of great circles is in navigation. Whenever ships must travel over vast expanses of open ocean between distant ports, or whenever planes must make long flights, it is desirable in the interests of saving fuel and time to follow the great-circle course between

the two points provided, of course, that there are no obstacles or other deterring factors preventing the use of the great circle. Navigators use special types of maps that have the property of always showing great circles as straight lines. To plot the shortest course between any two points the navigator simply draws a straight line between the two points on the chart.

MERIDIANS AND PARALLELS

The spinning of the earth on its axis provides two natural points—the poles—upon which to base the *geographic grid,* a network of intersecting lines inscribed on the globe for the purpose of fixing the locations of surface features. The grid consists of a set of north–south lines connecting the poles—the meridians—and a set of east–west lines running parallel with the equator—the parallels (Figure 1.6).

Meridians are halves of great circles, whose ends coincide with the earth's north and south poles. Though it is true that opposite meridians taken together comprise a complete great circle, note that a single meridian is only half of a great circle and contains 180° of arc. Additional characteristics of meridians are as follows:

1. All meridians run in a true north–south direction.
2. Meridians are spaced farthest apart at the equator and converge to a point at each pole.
3. An infinite number of meridians may be drawn on a globe. Thus a meridian exists for any point selected on the globe. For representation on maps and globes, however, selected meridians are spaced at equal distances apart.

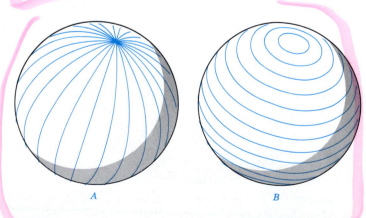

FIGURE 1.6 Meridians *(A)* and parallels *(B).*

Parallels are entire small circles, produced by passing planes through the earth parallel to the plane of the equator. They possess the following characteristics:

1. Parallels are always parallel to one another. Although they are circular lines, any two parallels always remain equal distances apart.
2. All parallels represent true east–west lines.
3. Parallels intersect meridians at right angles. This relationship holds true for any place on the globe, except the two poles, even though the parallels are strongly curved near the poles.
4. All parallels except the equator are small circles; the equator is unique in being a complete great circle.
5. An infinite number of parallels may be drawn on the globe. Therefore every point on the globe, except the north or south pole, lies on a parallel.

LONGITUDE

The location of points on the earth's surface follows a system in which lengths of arc are measured along meridians and parallels (Figure 1.7). Taking a selected meridian, or prime meridian, as a reference line, arcs are measured eastward or westward to the desired points. Taking the equator as the starting line, arcs are measured north or south to the desired points.

The *longitude* of a place is the arc, measured in degrees, of a parallel between that place and the *prime meridian* (Figure 1.7). The prime meridian is universally accepted as the meridian that passes through the old Royal Observatory at Greenwich, near London, England, and it is often referred to as the *Greenwich meridian*. The meridian has the value 0° longitude. The longitude of any given point on the globe is measured eastward or westward from this meridian, whichever is the shorter arc. Longitude may thus range from 0 to 180°, either east

or west. It is commonly written in the following form: long. 77° 03′ 41″W, which may be read "longitude 77 degrees, 3 minutes, 41 seconds west of Greenwich."

If only the longitude of a point is stated, we cannot tell its precise location because the same arc of measure applies to an entire meridian. For this reason, a meridian might be defined as a line representing all points having the same longitude. This definition explains why we often use the expression "a meridian of longitude." You may at first be confused by the statement that longitude is measured along a parallel of latitude, but this becomes clear when you realize that to measure the arc between a point and the prime meridian, it is necessary to follow a parallel eastward or westward (Figure 1.7).

The actual length, in kilometers or miles, of a degree of longitude will depend on where it is measured. At the equator the approximate length of one degree is computed by dividing the earth's circumference by 360°:

$$\frac{40,075 \text{ km}}{360°} = 111 \text{ km}$$

$$\frac{24,900 \text{ mi}}{360°} = 69 \text{ mi}$$

Because of the rapid convergence of the meridians northward or southward, this equivalent applies close to the equator only. It is also useful to know that the length of 1° of longitude is reduced to about one-half as much along the 60° parallels, or about 55½ km (34½ mi).

LATITUDE

The *latitude* of a place is the arc, measured in degrees, of a meridian between that place and the equator (Figure 1.7). Latitude thus ranges from 0° at the equator to 90° north or south at the poles. The latitude of a place, written as lat. 34° 10′ 31″N, may be read "latitude 34 degrees, 10 minutes, 31 seconds north." When both the latitude and longitude of a place are given, it is accurately and precisely located with respect to the geographic grid.

If the earth were a perfect sphere, the length of 1° of latitude (a one-degree arc of a meridian) would be a constant value everywhere on the earth. This length is almost the same as the length of a degree of longitude at the equator, so that the value of 111 km (69 mi) per degree may be used for ordinary purposes.

To be precise, and to take into account the oblateness of the earth, we must recognize that a degree of latitude changes slightly in length from equator to poles. The length of 1° of latitude at the equator is 110.6 km (68.7 mi); at the poles it is 111.7 km (69.4 mi), or 1.1 km (0.7 mi) longer. Thus, one degree of latitude at the poles is one percent longer than at the equator.

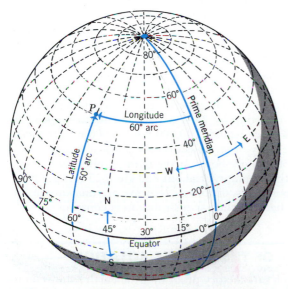

FIGURE 1.7 Longitude and latitude are measured along arcs of parallels and meridians, respectively. The point *P* has a longitude of 60°W, a latitude of 50°N.

THE NAUTICAL MILE

Both marine and air navigation use the nautical mile as the unit of length or distance. Meteorology (weather science) of the upper atmosphere has also used as the unit of wind speed the mariner's *knot*, which is a speed of one nautical mile per hour. It is therefore worthwhile for a geographer to understand the nautical mile.

The *nautical mile* can be quite simply defined as the length of one minute of arc of the earth's equator. Because measurements of the length of the equator have been refined many times during the past century, precise values of the nautical mile have also been revised. As adopted in 1954 by the U.S. Department of Defense, the equivalents in both kilometers and statute miles (English miles) are as follows:

1 nautical mi = 1.852 km = 1.1508 statute mi

For ordinary calculations, the nautical mile can be considered equal to 1.85 km (1.15 statute mi).

MAP PROJECTIONS

A *map projection* is an orderly system of parallels and meridians used as a basis for drawing a map on a flat surface. The basic problem of the mapmaker is to transfer the geographic grid from its actual spherical form to a flat surface, in such a way as to present the earth's surface or some part of it in the most advantageous way possible for the purposes desired.

We can avoid the map-projection problem only by using a globe. Unfortunately, a globe has shortcomings. First, we can see only one side of a globe at a time. Second, a globe is on too small a scale for many purposes. On globes ranging from a few centimeters to a meter in diameter, only the barest essentials of geography can be shown. A few large globes in existence, those 3 to 5 m in diameter, show considerable detail. But they serve also to accentuate a third shortcoming of globes—their lack of portability. Flat maps printed on paper can be folded compactly and put in a pocket, whereas even the smallest globe is a cumbersome and delicate object. Maps can be reproduced easily, whereas making a quality globe requires not only that a map be printed but also that the map be trimmed and carefully pasted in sections on a spherical shell.

The subject of map projections is treated in detail in Appendix I. Several useful varieties of grids are described and their advantages and special properties explained. You will find there the descriptions of the map projections used throughout this book for global maps showing various classes of information.

ROTATION OF THE EARTH

The varied environments of life on our planet depend to a large degree on the way the sun's rays strike a spherical earth. The angle of attack of the solar energy beam, varying greatly with latitude and with time of year, determines many commonplace phenomena—the daily path of the sun across the sky, the changing lengths of day and night, and the annual rhythm of the seasons. These daily and seasonal rhythms in turn act as fundamental controls of air temperatures, winds, ocean circulation,

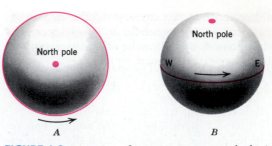

FIGURE 1.8 Direction of rotation is counterclockwise at the north pole *(A)* and eastward at the equator *(B)*.

precipitation, and storms—all of which, taken together, make up the earth's varied climates.

To understand earth–sun relationships requires thinking in three dimensions. You will need to visualize a spherical earth spinning like a top on its axis, but at the same time moving in a circular path around the sun. Superimposed on this simple system of motion is a tilt of the earth's axis with respect to the plane in which it travels. We can look at this earth–sun system from the purely imaginary viewpoint of an observer far out in space; but then the same motions must be recast into the real viewpoint of an observer on the earth's surface, turning with the earth and oriented by the earth's gravity.

The spinning of the earth on its polar axis is termed *earth rotation*. In the study of earth–sun relationships, we use a period of rotation, the *mean solar day*, consisting of 24 mean solar hours. This day is the average time required for the earth to make one complete turn with respect to the sun.

Direction of earth rotation can be determined by using one of the following rules: (1) Imagine yourself to be looking down on the north pole of the earth; the direction of turning is counterclockwise. (2) Place your finger on a point on a globe near the equator and push eastward. You will cause the globe to rotate in the correct direction (Figure 1.8); this explains the common expression "eastward rotation of the earth." (3) The direction of earth rotation is opposite that of the apparent motion of the sun, moon, and stars. Because these bodies appear to travel westward across the sky, the earth must be turning in an eastward direction.

The speed of travel of a point on the earth's surface in a circular path, because of rotation alone, is roughly computed by dividing the length of parallel at the latitude of the point by 24. Thus at the equator, where the circumference is about 40,000 km (25,000 mi), the eastward velocity of an object on the surface is about 1700 km (1050 mi) per hour. At the 60th parallel the velocity is half this amount, or about 850 km (525 mi) per hour. At the poles it is, of course, zero. We are unaware of this motion because the earth's rotation is constant.

PROOF OF THE EARTH'S ROTATION

Finding a satisfactory proof that the earth rotates on its axis was a frustrating problem of astronomy for several centuries. Those astronomers who believed that a fixed earth was the center of the universe—adherents to the Ptolemaic system—held a powerful grip on science until well into the fifteenth century, when the Polish astronomer Nicolaus Copernicus argued that the earth was one of several planets moving in orbits about the sun and experiencing axial rotation. The Copernican theory became generally accepted through the efforts of another leading astronomer, Johannes Kepler (1571–1630), and was put on a sound physical basis by use of Isaac Newton's laws of motion and gravitation (1687). Newton predicted that a rotating earth would be deformed into an oblate ellipsoid. By 1740 geodetic surveys had confirmed that the earth is, indeed, an oblate ellipsoid.

It remained for a French physicist, Jean Bernard Léon Foucault, to devise a public demonstration so forceful that even the layperson could be convinced of the reality of earth rotation. Known as the *Foucault pendulum*, the type of apparatus he used can be seen in action today in the United Nations Building in New York City and in many museums and universities (Figure 1.9). In 1851 Foucault suspended a cannonball from the dome of the Pantheon in Paris, using a slender wire about 60 m (200 feet) long. Once set in a swinging motion, the ball underwent a steady change in direction of swing. It was obvious that the earth's surface was slowly turning, but that this motion was not being transmitted to the cannonball. Free of mechanical coupling with the earth, the massive ball was maintaining its space path in conformance with Newton's first law of motion. (Every body remains in a

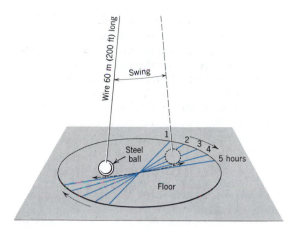

FIGURE 1.10 Principle of the Foucault pendulum.

state of rest or of uniform motion in a straight line unless compelled to change that state by an external force acting upon it.)

If a Foucault pendulum were set in operation at the north pole, the change of the pendulum swing would be clockwise, as shown in Figure 1.10, and would complete a 360° change of direction in 24 hours. There the change of direction is 15° per hour. At the equator, the swing path does not change direction at all. At intermediate latitudes, the rate of change in swing path ranges between 0° (equator) and 360° per 24 hours (pole). For example, at New York City, about lat. 40°N, the hourly change of direction is about 10°, and 37 hours are required for 360° of change. Table 1.1 gives values for 15° intervals of latitude.

ENVIRONMENTAL EFFECTS OF EARTH ROTATION

The physical and biological effects of earth rotation are truly profound in terms of the environmental processes of the life layer. First, and most obvious, is that rotation imposes a daily, or diurnal, rhythm on many phenomena to which plants and animals respond. These phenomena include light, heat, air humidity, and air motion. Plants respond to the daily rhythm by storing energy during

FIGURE 1.9 This handsome Foucault pendulum knocks over pins in succession to show that its direction of motion is changing. (Franklin Institute, Philadelphia, Pa.)

TABLE 1.1 The Foucault Pendulum at Various Latitudes

Latitude	Hourly Change in Pendulum Direction, Degrees*	Total Time for 360° Change in Direction, Hours
0°	None	None
15	3.9	93
30	7.5	48
45	10.6	34
60	13.0	28
75	14.5	25
90	15.0	24

*The following formula applies: Rate of direction change, degrees per hour = 15° × sine of latitude.

the day and releasing it at night. Animals adjust their activities to the daily rhythm, some preferring the day, others the night for food-gathering activities. The daily cycle of input of solar energy and a corresponding cycle of air temperature will be important topics for analysis in Chapters 3 and 4.

Second, as we shall find in the study of the earth's systems of winds and ocean currents, earth rotation causes the flow paths of both air and water to be consistently turned in compass direction: toward the right in the northern hemisphere and toward the left in the southern hemisphere. This phenomenon goes by the name of the Coriolis effect; we shall investigate it in Chapter 5.

A third physical effect of the earth's rotation is of environmental importance. Because the moon exerts its gravitational attraction on the earth, while at the same time the earth is turning with respect to the moon, tidal forces are generated. Tidal forces induce a rhythmic rise and fall of the ocean surface—the ocean tide. These motions in turn cause water currents of alternating direction to flow in shallow waters of the coastal zone. To a grain farmer in Kansas, the ocean tide may have no significance; but, for the clam digger and charter-boat captain on Cape Cod, the tidal cycle is a clock that regulates daily activities. For many kinds of plants and animals of saltwater estuaries, tidal currents are essential to maintain a suitable life environment. The tide and its currents are discussed in Chapter 20.

EARTH REVOLUTION

The motion of the earth in its travel path, or *orbit*, around the sun is termed *revolution*. The period of revolution, or *year*, is the time required for the earth to complete one circuit around the sun. The year is, however, defined by astronomers in several ways. For example, the time required for the earth to return to a given point in its orbit with reference to the fixed stars is called the sidereal year.

For earth–sun relationships we use the *tropical year*, which is the period of time from one vernal equinox to the next. The tropical year has a length of approximately 365¼ mean solar days. Every 4 years the extra one-fourth day difference between the tropical year and the calendar year of 365 days accumulates to nearly 1 whole day. By inserting a 29th day in February every leap year, we

are able to correct the calendar with respect to the tropical year. Further minor corrections are necessary to perfect this system.

In its orbit, the earth moves counterclockwise when viewed from a point in space above the north pole (Figure 1.11). This is the same direction of turning as the earth's rotation.

PERIHELION AND APHELION

The average distance between earth and sun is about 150 million km (93 million mi). Because the earth's orbit is an ellipse, rather than a circle, the distance may be 2½ million km (1½ million mi) greater or less than the average value (Figure 1.12). The distance is least, or about 147½ million km (91½ million mi) on about January 3, at which time the earth is said to be in *perihelion*. (This word comes from the Greek *peri*, around or near, and *helios*, the sun). On about July 4 the earth is at its farthest point from the sun, or in *aphelion* (Greek *ap*, away from, and *helios*), at a distance of 152½ million km (94½ million mi).

These differences in distance cause minor differences in the amount of solar energy received by the earth, but they are not the cause of summer and winter seasons. This is obvious because perihelion, when the earth receives more heat, falls at the coldest time of the year in the northern hemisphere. Moreover, opposite seasons exist simultaneously in the northern and southern hemispheres, proving that another cause exists. Instead, the seasons are the result of the tilt of the axis of rotation.

In theory, however, summers and winters should be slightly intensified in the southern hemisphere and slightly moderated in the northern hemisphere as a result of the coincidence of perihelion and aphelion with summer and winter seasons.

TILT OF THE EARTH'S AXIS

Imagine the earth's axis to be exactly perpendicular with the plane in which the earth revolves about the sun. As-

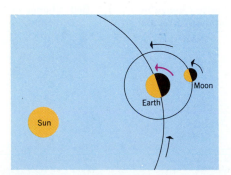

FIGURE 1.11 Direction of earth rotation and revolution.

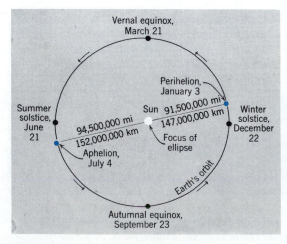

FIGURE 1.12 Earth's orbit and seasons.

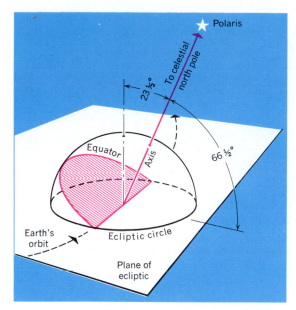

FIGURE 1.13 At all times the earth's axis maintains an angle of 66½° with the plane of the orbit (plane of the ecliptic).

tronomers call the plane containing the earth's orbit the *plane of the ecliptic*. Under these imagined conditions, the earth's equator would lie exactly in the plane of the ecliptic. The sun's rays, which furnish all energy for life processes on earth, would always strike the earth most directly at a point on the equator. In fact, at the equator, the rays would be exactly perpendicular to the earth's surface at noon. The rays would always just graze the north and south poles. Conditions on any given day would be exactly the same as conditions on every other day of the year (assuming the orbit to be circular). In other words, there would be no seasons.

In reality, the earth's axis is not perpendicular to the plane of the ecliptic; it is inclined by a substantial angle of tilt, measuring almost exactly 23½° from the perpendicular. Figure 1.13 shows this axial tilt in a three-dimensional perspective drawing. The angle between axis and ecliptic plane is then 66½° (90° − 23½° = 66½°).

To proceed we must couple the fact of axial tilt with a second fact: The earth's axis, while always holding the angle 66½° with the plane of the ecliptic, maintains a fixed orientation with respect to the stars. The north end of the earth's axis points constantly toward Polaris, the North Star. To visualize this movement, hold a globe so as to keep the axis tilted at 66½° with horizontal. Move the globe in a small horizontal circle, representing the orbit, at the same time keeping the axis pointed at the same point on the ceiling.

SOLSTICE AND EQUINOX

What is the consequence of the facts that (1) the earth's axis keeps a fixed angle with the plane of the ecliptic and (2) the axis always points to the same place among the stars? You will find that at one point in its orbit, the north end of the earth's axis is tilted toward the sun; at an opposite point in the orbit, it is tilted away from the sun. At the two intermediate points, the axis is not tilted with respect to the sun's rays (Figure 1.14). Next, consider the four critical positions in detail.

On June 21 or 22 the earth is so located in its orbit that the north polar end of its axis leans at the maximum angle 23½° toward the sun. The northern hemisphere is tipped toward the sun. This event is named the *summer solstice*. Six months later, on December 21 or 22, the earth is in an equivalent position on the opposite point in its orbit. At this time, known as the *winter solstice*, the

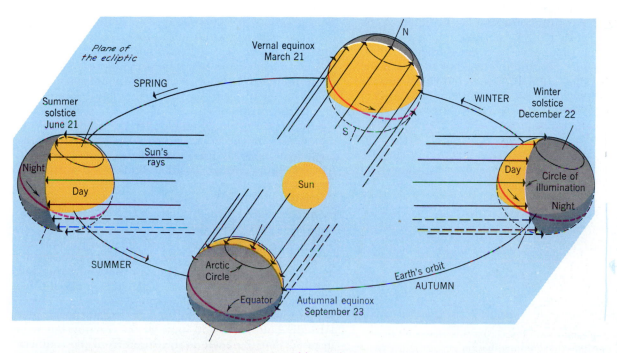

FIGURE 1.14 The seasons occur because the tilted earth's axis keeps a constant orientation in space as the earth revolves about the sun.

axis again is at a maximum inclination with respect to a line drawn to the sun, but now it is the southern hemisphere that is tipped toward the sun.

Midway between the dates of the solstices occur the *equinoxes*, at which time the earth's axis makes a 90° angle with a line drawn to the sun, and neither the north nor the south pole has any inclination toward the sun. The *vernal equinox* occurs on March 20 or 21; the *autumnal equinox* occurs on September 22 or 23. Conditions are identical on the two equinoxes as far as earth–sun relationships are concerned, whereas on the two solstices the conditions of one are the exact reverse of the other.

THE EQUINOXES

Consider first the conditions at the equinoxes because this is the simplest case. Figure 1.15 shows that the earth is at all times divided into two hemispheres with respect to the sun's rays. One hemisphere is lighted by the sun; the other lies in darkness. Separating the hemisphere is a circle, the *circle of illumination; it divides* day from night. Notice that at either equinox the circle of illumination cuts precisely through the north and south poles. Looking at the earth from the side, as in Figure 1.15, conditions at equinox are essentially as follows: The point at which the sun's noon rays are perpendicular to the earth, the *subsolar point*, is located exactly at the equator. Here the angle between the sun's rays and earth's surface is 90°. At both poles, the sun's rays graze the surface. As the earth turns, the equator receives the maximum intensity of solar energy; the poles receive none. At an intermediate latitude, such as lat. 40°N, the rays of the sun at noon make an acute angle (50°) with the earth's surface. The circle of illumination at the equinoxes passes through the poles and hence coincides with the meridians as the earth turns.

The path of the sun in the sky at the equinoxes is illustrated in Figure 1.16. To an earthbound observer, the earth's surface seems to be a flat, horizontal disk. The horizon lies at the circumference of the disk. The sun, moon, and stars seem to travel on the inner surface of a hemispherical dome. (This situation is simulated in

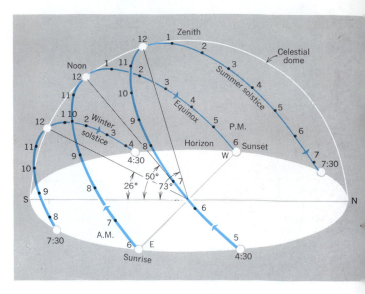

FIGURE 1.16 The sun's path in the sky at equinox and both solstices at lat. 40°N.

the design of a large planetarium, which uses a hemispheric ceiling.) At equinox, the sun rises at a point due east on the horizon and sets at a point due west on the horizon. This fact holds true at all latitudes except at the two poles.

Further details of equinox conditions are shown in Figure 1.17A. The parallels are divided into equal halves by the circle of illumination. Consequently, day and night are of exactly equal length, 12 hours each, at all latitudes. This fact explains the word *equinox*, from the Latin *aequus*, equal, and *nox*, night. (We are not taking into account twilight, which extends the period of daylight before sunrise and after sunset.) Conditions are the same for both the northern and southern hemispheres. Sunrise occurs at 6:00 A.M. (local solar time) and sunset occurs at 6:00 P.M. at all places on the globe, except at the poles, where special conditions prevail.

Solar noon occurs when the sun reaches its highest point in the sky. Noon occurs simultaneously at all points having the same longitude, that is, all points lying on a single meridian. The vertical angle of the sun above the horizon at noon, designated the *sun's noon altitude*, may be determined from Figure 1.17A by measuring the angle between a ray from the sun and a line tangent to the globe at a selected latitude. Sun's noon altitude in degrees is given for various latitudes. Because, within the limits of vision, the horizon appears to make a circle on a flat plane, a straight tangent line may be used for measuring the altitude angle on the diagram.

A few measurements of the sun's noon altitude should reveal that the altitude is always equal to the *colatitude*, or 90° minus the latitude. Thus, on the equinoxes, but not for any other time of year, the sun's noon altitude may be computed by a single simple subtraction, if only the latitude is given. The altitude is the same for similar latitudes both north and south of the equator, but keep in mind that the angle is measured from the southern

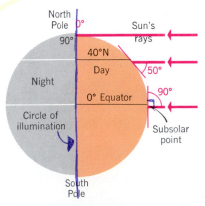

FIGURE 1.15 Equinox conditions. From this viewpoint the earth's axis appears to have no inclination.

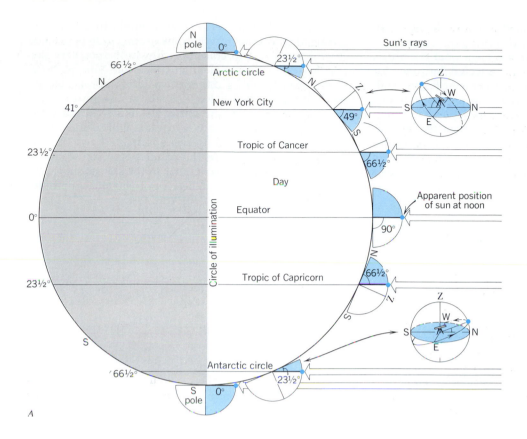

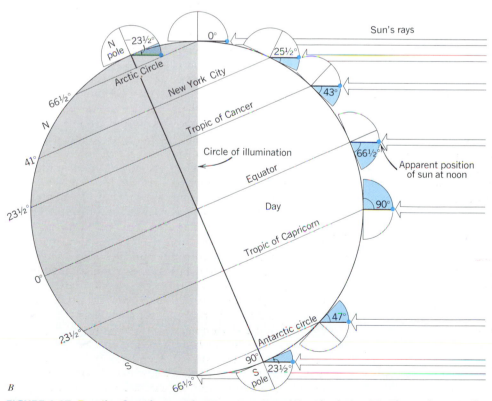

FIGURE 1.17 Details of earth-sun relations at equinox *(A)* and solstice *(B)*. The sun's noon altitude above the horizon is shown for various latitudes, the same as in Figure 1.18. (From A. N. Strahler, *The Earth Sciences*, 2nd ed., Harper & Row, Publishers. Copyright © by Arthur N. Strahler.)

horizon in the northern hemisphere, from the northern horizon in the southern hemisphere.

Figure 1.18 is a set of three-dimensional drawings, similar to Figure 1.16, covering selected latitudes for which the sun's noon altitude is shown in Figure 1.17. Imaginary solar paths are shown for the nighttime hours, when the sun is below the horizon. Whereas the conditions shown in Figure 1.16 for 40°N—about the latitude of New York City and Madrid—are most familiar to persons living in midlatitudes, the sun's path will seem strange as you travel far northward toward the pole or to places south of the equator.

A good plan will be to start in midlatitudes (Figure 1.16) and travel north. When the arctic circle is reached (*B*), the sun's path at equinox is low in the northern sky. Fort Yukon, Alaska, lies at this location. This path becomes even lower as the north pole is approached. Finally, at the north pole itself (*A*), the sun seems to follow the horizon through the entire 24 hours. Because the earth turns counterclockwise, the sun will seem to move clockwise (toward the observer's right).

As you travel toward the equator, the sun's path at equinox will seem to rise higher in the sky, as shown for the tropic of cancer (*C*). The city of Havana, Cuba, lies

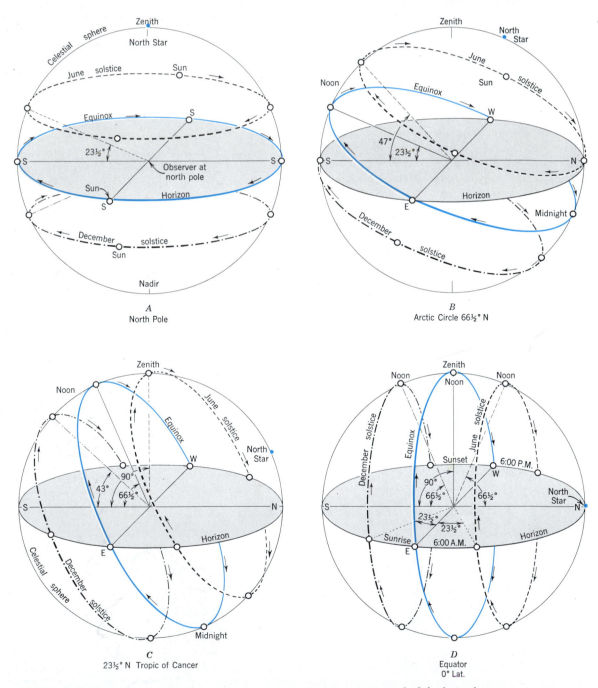

FIGURE 1.18 The sun's path in the sky at equinoxes and solstices at several of the latitudes shown in Figure 1.17.

close to this tropic. At the equator (*D*), the path is perpendicular to the horizon plane. Here the sun at noon reaches the zenith position. Then, crossing the equator into southern latitudes, the path inclines toward the northern horizon as shown for the tropic of capricorn (*E*). Rio de Janeiro, Brazil, lies close to this tropic.

Returning now to the equator—at Singapore, for example—we note that the sun has an altitude of 90° at noon. Because the path is in a plane perpendicular to the horizon plane, the sun changes altitude 15° per hour throughout the day. Furthermore, at the equator on these dates the shadow of a vertical rod will point due west from 6:00 A.M. until noon, will disappear precisely at noon, and will point due east from noon until 6:00 P.M.

THE SOLSTICES

Conditions at winter (December) solstice are shown in three dimensions in Figure 1.19. Because the maximum inclination of the axis is away from the sun, the entire area lying inside the *arctic circle*, lat. 66½°N, is on the dark side of the circle of illumination. Even though the earth rotates through a full circle during one day, this area poleward of the arctic circle remains in darkness. Conditions in the southern hemisphere during December solstice are shown in Figure 1.17*B*. All of the area lying south of the *antarctic circle*, lat. 66½°S, is under the sun's rays and enjoys 24 hours of day. The subsolar point has shifted to a position on the *tropic of capricorn*, lat. 23½°S.

At summer (June) solstice, conditions are exactly reversed from those of winter solstice. As Figure 1.20 shows, the subsolar point is now on the *tropic of cancer*, lat. 23½°N. Now the region poleward of the arctic circle

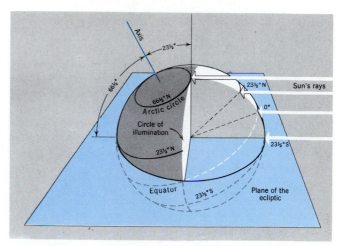

FIGURE 1.19 At winter solstice, the entire area between the arctic circle and the north pole is in darkness throughout the 24-hour period. (From A. N. Strahler, *The Earth Sciences*, 2nd ed., Harper & Row, Publishers. Copyright © by Arthur N. Strahler.)

has a 24-hour day (Figure 1.21). The region poleward of the antarctic circle has a 24-hour night. From one solstice to the next, the subsolar point has shifted over a latitude range of 47°. The progression of changes from equinox to solstice to equinox and back to solstice initiates the *astronomical seasons*: spring, summer, autumn, and winter. These are labeled on Figure 1.14.

Details of conditions at winter (December) solstice are shown in Figure 1.17*B*. Keep in mind that "winter" applies only to the northern hemisphere and that the southern hemisphere is experiencing its summer season. At this time, the circle of illumination divides all parallels of latitude that it crosses (except the equator) into unequal parts. The circle of illumination is now tangent to the arctic circle (lat. 66½°N) and the antarctic circle (lat. 66½°S), explaining why these two parallels are given a special designation. The circle of illumination bisects the equator in accordance with a law that any two intersecting great circles bisect each other.

At winter solstice, day and night are unequal in length over most of the globe. This inequality may be estimated from Figure 1.17*B* by noting what proportions of a given parallel lie to the left and to the right of the circle of illumination. The following facts are evident:

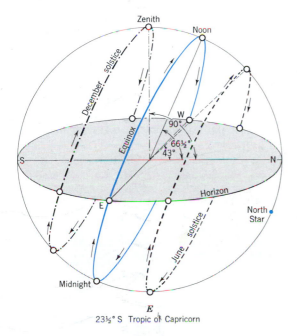

E
23½° S Tropic of Capricorn

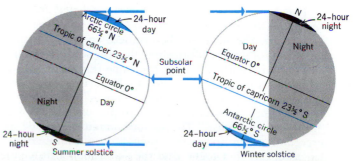

FIGURE 1.20 Solstice conditions.

FIGURE 1.21 The midnight sun seen in late July from Sunrise Point in Smith Sound, near Etah, Greenland, about lat. 78°N. Eight exposures were taken at intervals of 20 minutes, four before and four after midnight. (Courtesy Department Library Services, American Museum of Natural History, Neg. No. 230863.)

1. Night is longer than day in the northern hemisphere.
2. Day is longer than night in the southern hemisphere.
3. The inequality between day and night increases from the equator poleward.
4. At corresponding latitudes north and south of the equator, the relative lengths of day and night are in exact opposite relation.
5. Between the arctic circle, lat. 66½°N, and the north pole, night lasts the entire 24 hours. (This statement does not take into account twilight, which provides considerable light near the arctic circle.)
6. Between the antarctic circle, lat. 66½°S, and the south pole, day lasts the entire 24 hours.

At lat. 23½°S (the tropic of capricorn), the sun's rays at noon strike the earth at an angle of 90° above the horizon. Thus the noon sun is exactly in the zenith point. At the arctic circle, lat. 66½°N, the sun at noon is exactly on the horizon. At the south pole, the noon sun has an altitude of 23½° above the horizon and maintains this angle throughout the full 24 hours. The path of the sun in the sky at the winter solstice is illustrated for various latitudes in Figure 1.18, corresponding with Figure 1.17B.

In every way, conditions at summer solstice, June 21 or 22, are the exact reverse of conditions at winter solstice. Now, the northern hemisphere enjoys the same conditions of increased sunshine that the southern hemisphere had during the winter solstice (Figure 1.20). To see the relation of the sun's rays to the earth, turn Figure 1.17B upside down, exchanging "north" for "south," "arctic" for "antarctic," and "tropic of cancer" for "tropic of capricorn." The various statements made in preceding paragraphs concerning circle of illumination, length of day and night, and the sun's noon altitude at the winter solstice may be reread, with suitable wording

changes to fit the reversed conditions of the summer solstice.

The path of the sun in the sky on June 21 is shown in Figure 1.18. For all latitudes between the arctic and antarctic circles, the sun rises on the northeastern horizon and sets on the northwestern horizon.

THE SEASONAL CYCLE OF SUN'S DECLINATION

We have found that the subsolar point travels annually over a 47° latitude range, from lat. 23½°S at winter solstice to lat. 23½°N at summer solstice, crossing the equator twice yearly at the equinoxes. The latitude of the subsolar point at a given instant may be called the *sun's declination*. Figure 1.22 is a graph showing the sun's declination throughout the year. Notice that, at either solstice, the declination changes very slowly when it is undergoing a reversal in direction. For several days near solstice, the change of declination is too small to be detected without refined telescopic instruments. To the ancient Greeks and Romans, this was a time when the sun "stood still," which explains the derivation of the word *solstice*—from the Latin *sol*, "sun," and *stare*, "to stand."

In contrast, the rate of change of declination is rapid at the two periods of equinox. In each of the months before and after equinox, the declination change is nearly 12°. This fact explains the rapid shortening of days near the autumnal equinox and the rapid lengthening of days near the vernal equinox. Plants and animals respond strongly in various forms of activities during these two yearly periods of rapid declination change.

Some of the climate graphs in Chapter 10 reproduce the declination graph in simplified form, so you can use this annual cycle as a reference against which to judge the climatic seasons. A striking fact that will emerge is that the climatic seasons are not in phase with the astronomical seasons.

TIME

Global time relationships are important in this age of instantaneous communication and high-speed travel. Before the coming of the telegraph, problems of place-to-

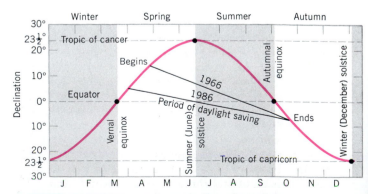

FIGURE 1.22 The sun's declination throughout the year.

place time differences were of little concern to people who lived most of their lives in one community. Even travelers were caused only the inconvenience of resetting their watches to the time used by local communities. The amount of time consumed in getting from one place to another was so much more than the difference in watch time between the two places that the time difference was of little consequence.

When it became possible to transmit messages instantaneously by telegraph, differences in local time resulting from differences in longitudinal position were immediately apparent. With the development of rapid travel, it became important to correct schedules for the gain or loss of time incurred by travel across the meridians. Jet aircraft can now fly fast enough to "keep pace with the sun" in midlatitudes. For example, a plane leaving New York at noon, eastern standard time, traveling about 1300 km per hour (800 mph), can arrive in San Francisco at noon, Pacific standard time.

LONGITUDE AND TIME

Our clocks run by the sun's schedule, but the sun sends us only one set of parallel light rays. Solar noon is defined by the subsolar point, at which the sun's rays are exactly perpendicular to the earth's surface. This means that the sun is in the observer's zenith, i.e., directly overhead. The meridian of longitude beneath the subsolar point can be called the *noon meridian*. Directly opposite the noon meridian, on the dark side of the globe, is a corresponding *midnight meridian*. As the earth rotates on its axis, the noon and midnight meridians sweep westward around the earth. The noon meridian separates the forenoon and afternoon of the same calendar day, while the midnight meridian is the dividing line between one calendar day and the next.

Because the noon meridian sweeps over 360° of longitude every 24 hours, it must cover 15° of longitude every hour, or 1° of longitude every 4 minutes. We therefore find it convenient to state that 1 hour of time is the equivalent of 15° of longitude. This equality forms the basis for all calculations concerning time belts of the globe. For example, if the noon meridian reaches one place on the globe 4 hours after it leaves another place, the two places are separated by 60° of longitude.

A working model of global time relations is illustrated in Figure 1.23. Two disks of different radii are attached at their centers in such a way that one disk can be turned while the other remains still. On the inner disk, radii are drawn to represent 15° meridians of a globe seen from a point above the north pole. On the outer disk, similar radii are marked in hours to represent the complete global set of time meridians. As a further refinement, the inner disk may be a hemispheric world map. As set in Figure 1.23, it is noon on the Greenwich meridian (long. 0°), while it is midnight on the 180° meridian in mid-Pacific.

Some persons become confused when trying to decide whether the watch time of places to the east (or west) of them is ahead of or behind their own watch time. The problem may be to calculate when a radio or television

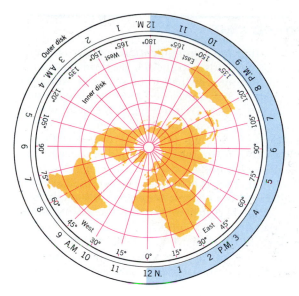

FIGURE 1.23 A working model of global time relationships.

program will be on the air in different parts of the country or to decide whether to set one's watch ahead or behind one hour when traveling from one standard time zone to another. To avoid confusion, visualize the time meridians moving westward around the globe. To see how this motion occurs, imagine that the inner disk of Figure 1.23 rotates eastward (counterclockwise) while the outer disk remains fixed on the page. Relative to the motion of the inner disk, the set of hours on the outer disk will seem to shift westward around the earth. Then consider, for example, that you are in New York City and the time there is noon. The noon meridian, which is at New York, left Greenwich, England, 5 hours earlier. Therefore, in England, 5 hours must have elapsed since noon and it must be 5:00 P.M. in that country. The rule is that places located to the east of you have a later hour. Again, consider that the noon meridian is at New York City. Because that meridian will require about 3 hours to travel westward to reach San Francisco, it must be 9:00 A.M. in that city. A counterpart of the rule just stated is that places located to the west of you have an earlier hour. (Both rules are subject to qualifications where the International Date Line lies between the places.)

LOCAL TIME

A century or more ago, the means of establishing a time system for a small community was to take the meridian of longitude passing through some central point in the town or city, for example, the courthouse or a cathedral. All clocks of the community were set to read 12:00 M. (noon) when the sun was directly over that meridian. The time system so derived is called *local time*, defined as mean solar time based on the local meridian. All places located on the same meridian, regardless of how far apart they may be, have the same local time; all places located on different meridians have unlike local times, differing by four minutes for every degree of longitude between them.

STANDARD TIME

As transportation and communication improved through the spread of railroads and the telegraph in the middle 1800s, local time systems had to be replaced. American railroads, about 1870, introduced a standardized system covering large belts of territory; but this system was developed by railroad companies for their own convenience. Consequently, where several railroads met or passed in a single town, the citizens had to contend with several different kinds of railroad time in addition to their own local time. It is said that before 1883 as many as five different time systems were used in a single town and that, altogether, the railroads of the United States followed 53 different systems.

The obvious solution to such problems is *standard time*, based on a *standard meridian*. In this system the local time of the standard meridian is arbitrarily given to wide strips of country on both sides. All clocks within the belt are set to a single time. By selecting standard meridians 15° apart, adjacent zones have standard times differing by exactly one hour. Furthermore, if these meridians represent longitudes that are multiples of 15 (e.g., 60°, 75°, 90°, or 105°), each successive standard time zone will differ from the standard time of Greenwich, England, by whole hour units.

STANDARD TIME IN THE UNITED STATES

The present system of standard time in the United States was put in operation on November 18, 1883, but it was not until March 19, 1918, that Congress passed legislation directing the Interstate Commerce Commission to determine time-zone boundaries. The standard meridians and boundaries are shown in Figure 1.24. The six standard time zones and their meridians are as follows:

Eastern	75°	Pacific	120°
Central	90°	Alaska–Hawaii	150°
Mountain	105°	Bering	165°

Had it been carried out precisely, the system would have resulted in belts extending exactly 7½° east and west of each standard meridian; but a glance at the map shows that great liberties have been taken in locating the boundaries. Wherever the time-zone boundary could conveniently be located along some already existing and widely recognized line, this was done. Natural boundaries have been used. For example, the eastern time–central time boundary line follows Lake Michigan down its center, and the mountain time–Pacific time boundary follows a ridgecrest line also used by the Idaho–Montana state boundary. Most frequently, the time-zone boundary follows state and county boundaries. For example, the eastern time–central time boundary follows the Alabama–Georgia state line, so as to place all of Georgia in the eastern standard zone.

The time belts are by no means equally distributed on both sides of the standard meridians, as a glance at Figure 1.24 will show. Such deviations of the boundaries permit entire states to operate under one kind of time.

DAYLIGHT SAVING TIME

Because many human activities, especially in urban areas, start well after sunrise but continue long after sun-

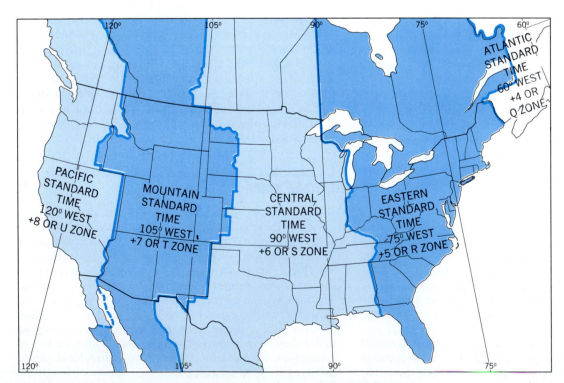

FIGURE 1.24 Time-zone map of the United States and southern Canada.

set, it is desirable to set forward the hours of daylight to utilize them to best advantage. A considerable saving in electric power can be made in summer when the early morning daylight period, wasted while schools, offices, and factories are closed, is transferred to the early evening when the majority of persons are awake and busy. The adjusted time system is known as *daylight saving time* and is obtained by setting ahead all timepieces by 1 hour. Thus, when the sun is over the standard meridian (i.e., noon by the sun), all clocks in that time zone read 1:00 P.M. Sunrise and sunset at the equinoxes or equator, instead of occurring at 6:00 A.M. and 6:00 P.M., would occur at 7:00 A.M. and 7:00 P.M., respectively.

Daylight saving time was adopted in the United States during World War I. After that, it was used locally throughout the United States where authorized by local legislation. During World War II, daylight saving time was used nationally throughout the entire period of February 1942 to October 1945 and was known as war time. During the same war period, England used a double daylight saving time in which clocks were running 2 hours ahead of Greenwich civil time. This practice was desirable because of the unusually long summer days that England enjoys as a result of her relatively far northerly latitude. Many European countries normally use daylight saving time during part or all of the year. Nations whose time is advanced by an hour throughout the entire year are Spain, France, the Netherlands, Belgium, and the Soviet Union.

In April 1966, the U.S. Congress passed the Uniform Time Act, which requires that daylight saving time be applied uniformly throughout each state, unless the legislature of that state votes to remain on standard time instead. In the latter case, standard time must be applied uniformly throughout the entire state.

Through 1986, daylight saving time went into effect at 2:00 A.M. of the last Sunday of April and continued until 2:00 A.M. of the last Sunday of October. Compare these dates with the dates of the vernal and autumnal equinoxes—March 21 and September 22, respectively (Figure 1.22). It is obvious that the period of daylight time was established by Congress to fit the warm season of the year, when outdoor recreation and vacation activity profit most from the added hour of evening daylight. At lat. 40°N (New York City), the sun sets at about 7:50 P.M. standard time on the last Sunday in April, whereas sunset occurs at about 5:05 P.M. standard time on the last Sunday in October. In 1986, Congress voted to begin daylight saving time on the first Sunday in April, a change that partially corrects the inequity in dates, as shown in Figure 1.22.

To conserve daylight uniformly according to times of sunrise and sunset, it would be necessary to select beginning and ending calendar dates having the same declination of sun. In that case, the line connecting beginning and ending dates on the graph, Figure 1.22, would be always horizontal. For example, use of the dates of vernal and autumnal equinox would satisfy this requirement and provide daylight saving time over one-half of the year. Some persons favor the use of daylight saving time throughout the year, as adopted in many European countries.

WORLD TIME ZONES

In 1884 an international congress was held in Washington, D.C., to consider the subject of world standard time. The result was that standard times of countries throughout the world are based on standard meridians, which are multiples of the unit 15° and thus differ from one another by whole hourly amounts. In all global time calculations, the prime meridian of Greenwich, England, is taken as the reference meridian. All time zones of the globe are described in terms of the number of hours between the standard meridian of that zone and the Greenwich meridian. To distinguish whether the time zones lie east or west of the Greenwich meridian, the time is designated *fast* (minus sign) for all places east of Greenwich (east longitude) and *slow* (plus sign) for all places west of Greenwich (west longitude). U.S. eastern standard time, for example, is said to be "five hours slow."

Figure 1.25 is a world map on which the 24 principal standard time zones of the world are shown. Fifteen-degree meridians are in black lines; 7½° meridians, which form large elements of the zone boundaries, are in color. Within each time zone is shown the number of hours between the zone time and Greenwich time. Some countries or islands lie about midway between 15° meridians. For these, a standard meridian is chosen halfway between the two and is thus a multiple of 7½°. The standard time of the country is therefore fast or slow by some multiple of a half hour. Iran (3½ hours fast) and India (5½ hours fast) illustrate this point. The country having the greatest east–west extent is the Soviet Union, with 11 standard time zones; but these are all advanced by 1 hour with respect to the standard time meridians in each zone, to give a perpetual daylight saving time. Canada occupies 6 time zones, one of which (Newfoundland) uses the time 3½ hours slow.

THE INTERNATIONAL DATE LINE

When we take a world map or globe with 15° meridians and count them in an eastward direction, starting with the Greenwich meridian as 0, we find that 180th meridian is number 12 and that the time of this meridian is, therefore, 12 hours fast. Counting in a similar manner westward from the Greenwich meridian, we find that the 180th meridian is again number 12, but that the time is 12 hours slow. Both results are, of course, correct; and the explanation becomes obvious when we note that the difference in time between 12 hours fast and 12 hours slow is 24 hours, or a full day. At the precise instant when the noon meridian coincides with the Greenwich meridian, the 180th meridian coincides with the midnight hour meridian. At this instant, and only at this instant, the same calendar day exists on both sides of the meridian. At all other times the calendar day on the west (Asiatic) side of the 180th meridian is one day ahead of that on the east (American) side. For example, if it is Monday on the Asiatic side of the 180th meridian, it is Sunday on the American side.

Because of these peculiar properties, the 180th meridian was designated the *International Date Line* by the International Meridian Conference held in Washington, D.C., in 1884. It is one of the fortuitous occurrences of modern civilization that, after the Greenwich meridian had come into widespread use in English-speaking countries as the international basis for the reckoning of longitude, the 180th meridian should have been found to fall in an almost ideal location: squarely in the middle of the world's largest expanse of ocean. Even so, the International Date Line must deviate both eastward and westward to permit certain land areas and groups of islands to have the same calendar day (see Figure 1.25).

OUR SPHERICAL HABITAT

Looking back over what we have covered in this chapter, certain concepts emerge that will be drawn upon in coming chapters.

Earth rotation generates the diurnal rhythms of the environment, because solar energy is applied from a single source and its parallel rays can be intercepted only by a hemisphere. Life processes show a strong response to the diurnal cycle of energy input. Earth rotation has the effect of turning the direction of motion of air and water. This Coriolis effect profoundly influences the surface environment. Earth rotation with respect to the moon generates the ocean tide and its ceaseless rhythmic water motions essential to the life of the shallow coastal waters.

The accident of a rotational axis inclined from the ecliptic plane generates a seasonal rhythm of changes in solar energy of the life layer and is a major factor in causing the environment to differ from one latitude zone to another.

In this study of the astronomical controls of environment we visualize the planet as a perfectly smooth, solid sphere of uniform surface quality, exposed to the near vacuum of outer space. In the next chapter we must take a step toward realism and recognize that our planet has an envelope of air, the atmosphere; an extensive water layer, the oceans; and large patches of exposed solid rock, the continents.

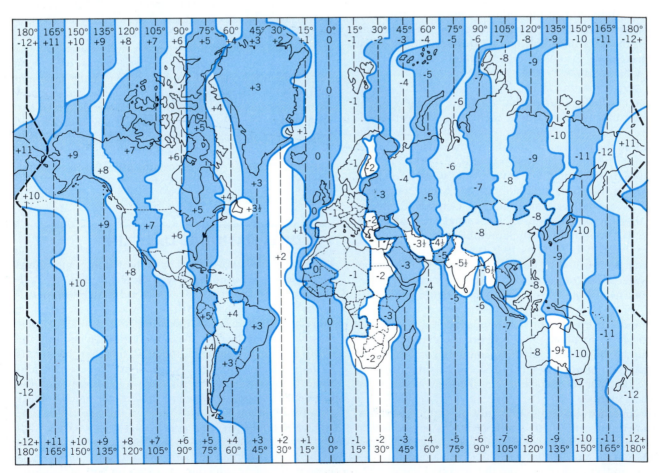

FIGURE 1.25 Time-zone map of the world.

THE METRIC SYSTEM

Metric to English (U.S.)

Rough equivalent: **Precise equivalent:**

Length, Distance

1 centimeter = 4/10 inch   1 cm = 0.394 in.
1 meter = 40 inches or 3¼ feet   1 m = 39.37 in. or 3.28 ft
1 kilometer = 6/10 mile   1 km = 0.621 mi

Area

1 square centimeter = 1.7 square inch   1 cm^2 = 0.155 in.^{2}
1 hectare = 2½ acres   1 ha = 2.47 a
1 square kilometer = 4/10 square mile   1 km^2 = 0.386 mi^2

Volume

1 cubic centimeter = 1/16 cubic inch   1 cc, 1 cm^3 = 0.061 in.^{3}
1 cubic meter = 35 cubic feet   1 m^3 = 35.3 ft^3
1 liter = 1 quart liquid (U.S.)   1 l = 1.057 qt

Weight

1 gram = 1/30 ounce (avoirdupois)   1 gm = 0.0353 oz
1 kilogram = 2 1/5 pounds (av.)   1 kg = 2.205 lb
1 tonne (metric ton) or 1000 kg = 1 1/10 short tons (U.S.)   1 t = 1.102 t

Temperature

1 Celsius degree = 9/5 Fahrenheit degrees   1 C° = 1.80 F°

English (U.S.) to Metric

Rough equivalent: **Precise equivalent:**

Length, Distance

1 inch = 2½ centimeters   1 in. = 2.540 cm
1 foot = 1/3 meter   1 ft = 0.305 m
1 mile = 1½ kilometers   1 mi = 1.609 km

Area

1 square inch = 6½ square centimeters   1 in.^{2} = 6.45 cm^2
1 acre = 4/10 hectare   1 a = 0.405 ha
1 square mile = 2½ square kilometers   1 mi^2 = 2.59 km^2

Volume

1 cubic inch = 16 cubic centimeters   1 in.^{3} = 16.39 cc, cm^3
1 cubic foot = 1/35 cubic meter   1 ft^3 = 0.0284 m^3
1 quart liquid (U.S.) = 1 liter   1 qt = 0.964 l

Weight

1 ounce (avoirdupois) = 28 grams   1 oz = 28.35 gm
1 pound (av.) = 450 grams, or ½ kilogram   1 lb = 453.6 gm, or 0.454 kg
1 short ton (U.S., 2000 pounds) = 9/10 tonnes (metric tons)   1 t = 0.907 t, or 907 kg

Temperature

1 Fahrenheit degree = 5/9 Celsius degree   1 F° = 0.556 C°

THE EARTH'S ATMOSPHERE AND OCEANS

The air, sea, and land constitute the major portions of four great material realms, or spheres, that comprise the total global environment (Figure 2.1). Three of these realms are inorganic: (1) *atmosphere;* (2) *hydrosphere;* and (3) *lithosphere.* The substances of which they consist are classified by chemists as inorganic, or nonliving, matter. The fourth realm, the *biosphere,* encompasses all living organisms of the earth. Because living organisms cannot exist except in a physical environment with which they interact, the biosphere includes those parts of the atmosphere, hydrosphere, and lithosphere comprising that physical environment.

Of the three inorganic spheres, the atmosphere is the gaseous realm. The hydrosphere is the water realm consisting of free water in gaseous, liquid, and solid states;

it includes fresh water of the atmosphere and the lands, as well as salt water of the oceans. The lithosphere is the solid realm composed of mineral matter. The three spheres of inorganic matter occupy layerlike shells over the globe because of differences in the density of the three types of substances. Each of the three spheres has a different and distinctive chemical makeup, inherited from its origin in the geologic past. The biosphere requires materials from all three of the inorganic spheres; these are elements used to form the molecules that make up organic matter.

Humans live at the bottom of an ocean of air. They are air-breathers dependent on favorable conditions of pressure, temperature, and chemical composition of the atmosphere that surrounds them. Humans also live on the solid outer surface of the earth, which they depend on for food, clothing, shelter, and means of movement from place to place. But the air and the land are not two entirely separate realms; they constitute an interface across which there is a continual flux of matter and energy. The surface environment, which we may call the life layer, is a shallow but highly complex zone in which atmospheric conditions exert control on the land surface, while at the same time the surface of the land exerts an influence on the properties of the adjacent atmosphere.

Essentially the same statements apply to the surface of the oceans and the atmospheric layer above it. Humans utilize the surface of the sea as a source of food and a means of transportation. There is a continual flow of energy and matter between the sea surface and the lower layer of the atmosphere. Here, again, we find an interface of vital concern to humans. The sea influences the atmosphere above it, whereas the atmosphere influences the sea beneath it.

Our objective in these early chapters is to examine the atmosphere and oceans with particular reference to these air–land and air–sea interfaces that are so vital to life on earth. To geographers, concerned as they are with spatial relationships on a global scale, the distributions of physical properties of the ocean and atmosphere are matters of special interest. Physical geographers describe and explain the ways in which the environmental ingredients of weather and climate change with latitude and season and with geographic position in relation to

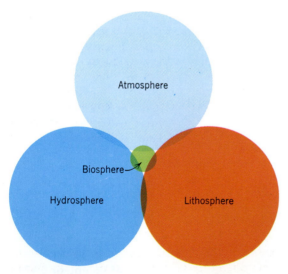

FIGURE 2.1 The earth realms shown as intersecting circles. The large outer circles represent the three great inorganic realms; each one overlaps the other two in a small area, suggesting that some of the substance of each realm is held within the other two. The biosphere, or organic realm, draws its substance from its inorganic environment, represented by the circle overlap into the three surrounding realms. The small diameter of the biospheric circle signifies that the total mass of matter held within the biosphere is only a small fraction of that in the other three realms.

oceans and continents. They seek out the broad patterns of similar regions and attempt to define their boundaries and organize them into systems of classes. More important, geographers try to evaluate the environmental qualities of each region, emphasizing the opportunities as well as the limitations of each for future development of natural resources such as food, water, energy, and minerals.

The principles we shall review in this chapter belong to two areas of natural science: *meteorology*, the science of the atmosphere, and *physical oceanography*, the physical-science aspect of the oceans.

COMPOSITION OF THE ATMOSPHERE

The earth's atmosphere consists of a mixture of various gases surrounding the earth to a height of many kilometers. Held to the earth by gravitational attraction, this envelope of air is densest at sea level and thins rapidly upward. Although almost all of the atmosphere (about 97%) lies within 30 km (18 mi) of the earth's surface, the upper limit of the atmosphere can be drawn approximately at a height of 10,000 km (6000 mi), a distance approaching the diameter of the earth itself. From the earth's surface upward to an altitude of about 80 km (50 mi), the chemical composition of the atmosphere is highly uniform throughout in terms of the proportions of its component gases.

Pure, dry air consists largely of nitrogen, about 78 percent by volume, and oxygen, about 21 percent (Figure 2.2). Nitrogen does not enter easily into chemical union with other substances and can be thought of as primarily a neutral substance. Very small amounts of nitrogen are extracted by soil bacteria and made available for use by plants. In contrast to nitrogen, oxygen is highly active chemically and combines readily with other elements in the process of oxidation. The combustion of fuels represents a rapid form of oxidation, whereas certain forms of rock decay (weathering) represent very slow forms of oxidation. Animals require oxygen to convert foods into energy.

The remaining 1 percent of the air is mostly argon, an inactive gas of little importance in natural processes.

Part of that 1 percent consists of a very small amount of carbon dioxide, about 0.033 percent. This gas is of great importance in atmospheric processes because of its ability to absorb radiant heat and so allow the lower atmosphere to be warmed by heat rays coming from the sun and from the earth's surface. Green plants, in the process of photosynthesis, utilize carbon dioxide from the atmosphere, converting it with water into solid carbohydrate.

Also present, but in extremely small amounts, are the following gases: neon, helium, krypton, xenon, hydrogen, methane, and nitrous oxide. All the component gases of the lower atmosphere are perfectly diffused among one another so as to give the pure, dry air a definite set of physical properties, just as if it were a single gas.

Besides these constituents of pure, dry air, the atmosphere holds varying amounts of water in the vapor (gaseous) state and as suspended liquid and solid particles (droplets and ice crystals). These important atmospheric ingredients are described in later chapters.

ATMOSPHERIC PRESSURE

Although we are not constantly aware of it, air is a tangible material substance, exerting *atmospheric pressure* on any solid or liquid surface exposed to it. This pressure is about 1 million dynes per square centimeter (about 1 kg/sq cm, or 15 lb/sq in.). Because atmospheric pressure is exactly counterbalanced by the pressure of air within liquids, hollow objects, or porous substances, its ever-present weight goes unnoticed. The pressure on 1 sq cm of surface can be thought of as the actual weight of a column of air 1 cm in cross section extending upward to the outer limits of the atmosphere (Figure 2.3A). Air is readily compressible. That which lies lowest is most greatly compressed and is, therefore, densest. In an upward direction, both density and pressure of air fall off rapidly. Figure 2.9 shows the upward decrease in pressure in terms of the fraction of the sea-level value.

Meteorologists have used another method of stating atmospheric pressure. It is based on a classic experiment of physics first performed by Evangelista Torricelli in 1643. A glass tube about 1 m (3 ft) long, sealed at one end, is completely filled with mercury. The open end is temporarily held closed. Then the tube is inverted and the end is immersed into a dish of mercury. When the opening is uncovered, the mercury in the tube falls a few centimeters, but then remains fixed at a level about 76 cm (30 in.) above the surface of the mercury in the dish (Figure 2.3B). Atmospheric pressure now balances the weight of the mercury column. When the air pressure increases or decreases, the mercury level rises or falls correspondingly. Here, then, is a device for measuring air pressure and its variations.

Any instrument that measures atmospheric pressure is a *barometer*. The type devised by Torricelli is known as the *mercurial barometer*. With various refinements over the original simple device, it has become the standard instrument. Pressure is read in centimeters or inches of mercury, the true measure of the height of the mercury column. Standard sea-level pressure is 76.0 cm (29.92 in.) on this scale.

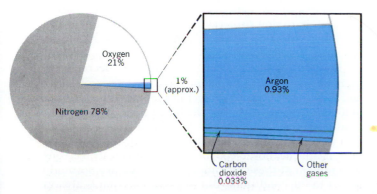

FIGURE 2.2 Component gases of the lower atmosphere. The figures tell approximate percentage by volume.

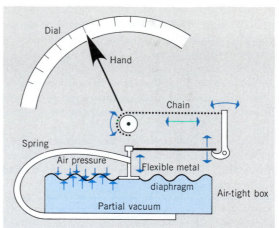

FIGURE 2.5   Principle of the aneroid barometer.

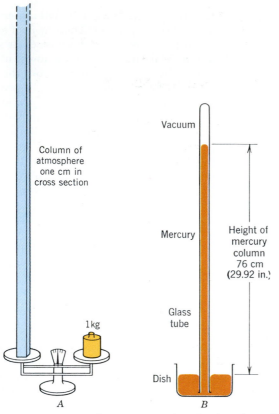

FIGURE 2.3 (*A*) Atmospheric pressure imagined as the weight of a column of air.
(*B*) Principle of the mercurial barometer.

A different unit of pressure is the working standard of meteorologists. This unit is the *millibar* (mb). One cm of mercury is equivalent to 13.3 mb (1 in. equals about 34 mb). Standard sea-level pressure is 1013.2 mb.

Another type of barometer is the *aneroid barometer* (Figure 2.4). It consists of a hollow metal chamber partly emptied of air and sealed (Figure 2.5). The upper wall of the chamber is a flexible diaphragm that moves up and down as the outside air pressure varies. These movements operate a hand, read against a calibrated circular dial. The aneroid is compact and sturdy. When calibrated in terms of altitude, the aneroid barometer becomes an altimeter, which is carried in all aircraft.

VERTICAL DISTRIBUTION OF ATMOSPHERIC PRESSURE

Figure 2.6 shows in detail the rate of decrease of atmospheric pressure (barometric pressure) with altitude up to a height of over 30 km (20 mi). For every 275 m (900 ft) of rise in altitude, pressure is diminished by 1/30. Steepening of the curve as it rises shows that the rate of decrease in pressure is rapid at first, but becomes less with increasing altitude.

As atmospheric pressure becomes less, the boiling point of water becomes lower. As the boiling point decreases, the time required to cook foods by boiling increases. As many experienced campers know, boiling potatoes is a slow process at altitudes above 2000 m (6600 ft). A pressure cooker is most useful at high altitudes.

The physiological effects of a pressure decrease on humans are well known from the experiences of flying and mountain climbing. Lowered pressure decreases the amount of oxygen entering the blood through the lungs. At altitudes of 3000 to 4500 m (10,000 to 15,000 ft) mountain sickness (altitude sickness), characterized by weakness, headache, nosebleed, or nausea, can occur. Persons who remain at these altitudes for a day or two normally adjust to the conditions, but physical exertion is always accompanied by shortness of breath.

FIGURE 2.4 This aneroid barometer is calibrated in both millibars (outer scale) and inches. (Taylor Instrument Co., Rochester, N.Y.)

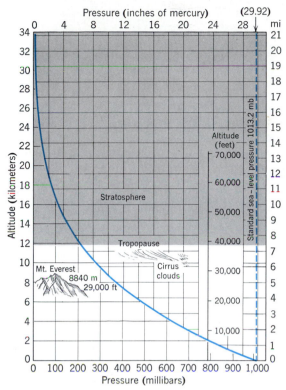

FIGURE 2.6 Decrease of atmospheric pressure with altitude.

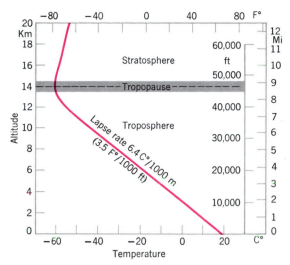

FIGURE 2.7 A typical environmental temperature lapse-rate curve for a summer day in midlatitudes.

Supplementary oxygen is needed for full physical activity above 5400 m (18,000 ft) in unpressurized aircraft. Lightweight oxygen containers enable climbers to reach the summit of Mount Everest (8800 m; 29,000 ft), a feat first accomplished in 1953. The cabins of commercial subsonic jet aircraft are pressurized to about 850 mb, equivalent to an altitude of about 1500 m (4500 ft), whereas the outside pressure at cruising altitude may be only 200 to 300 mb.

A record high altitude for jet aircraft in horizontal flight was set in 1976 by Captain Robert C. Helt, USAF, at 25.9 km (85,100 ft). For free manned balloons, a record altitude of 34.7 km (about 114,000 ft) was set in 1961 by Commander Malcolm D. Ross, USNR.

TEMPERATURE STRUCTURE OF THE LOWER ATMOSPHERE

The atmosphere has been subdivided into layers according to temperatures and zones of temperature change. Of greatest importance to humans and other life forms is the lowermost layer, the *troposphere*. If we sent up a sounding balloon carrying a recording thermometer and repeated this operation many times, we would obtain an average or representative profile of temperature. We would find that air temperature falls quite steadily with increasing altitude. The average rate of temperature decrease is about 6.4C°/1000 m of ascent (3½ F°/1000 ft). This rate is known as the *environmental temperature lapse rate*. When used repeatedly, this term is shortened to "lapse rate."

Figure 2.7 shows a typical air sounding into the troposphere at midlatitudes (lat. 45°N) on a summer day. Altitude is plotted on the vertical axis, temperature on the horizontal axis. The resulting curve is a sloping line. Temperature drops uniformly with altitude to a height of about 13 km (8 mi). Don't be surprised when the captain of your aircraft, flying at 12 km (40,000 ft) altitude, announces that the outside temperature is −50°C (−60°F).

The lapse rate changes abruptly at an altitude of about 14 km (9 mi). Instead of continuing to fall, temperature here remains constant with increasing altitude. This level of change is the *tropopause*; it marks a transition into the next higher temperature zone, known as the *stratosphere*.

The altitude of the tropopause is highest over the equator (17 km; 10 mi), and lowest over the poles (9–10 km; 5.5–6 mi). There are important seasonal changes in altitude of the tropopause in the middle and high latitudes, as shown in Figure 2.8. For example, at lat. 45° the average altitude in January is 12.5 km (8 mi), but rises to 15 km (9 mi) in July. Temperatures at the tropopause are much lower at the equator than at the poles, as shown in Figure 2.8. At first glance, this relationship may seem strange, accustomed as we are to considering the equatorial region as a warm zone and the polar regions as cold. On the other hand, because the lapse rate is quite uniform the world over, the thicker the troposphere, the colder the temperatures at the tropopause will be.

Above the tropopause, temperatures in the stratosphere gradually rise until a value of about 0°C (32°F) is reached at about 50 km (30 mi). Here at the *stratopause*, a temperature decrease sets in (Figure 2.9). Temperature decreases through the overlying *mesosphere*, a layer extending upward to about 80 km (50 mi), where a low point of −80°C (−120°F) is reached. This level of temperature minimum and reversal is called the *mesopause*. With further increase in altitude, a steep climb in tem-

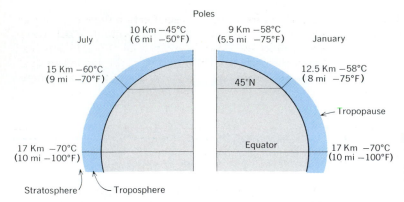

FIGURE 2.8 Average altitude and temperature of the tropopause for July (left) and January (right). (From A. N. Strahler, *The Earth Sciences*, 2nd ed., Harper & Row, Publishers. Copyright © by Arthur N. Strahler.)

perature takes place within the *thermosphere*. At this very high altitude, the air is extremely rarefied; the gas molecules are widely separated. This thin air holds very little heat, even though its temperature is high.

THE TROPOSPHERE AS AN ENVIRONMENTAL LAYER

The lowermost atmospheric layer, the troposphere, is of most direct importance to terrestrial life. Almost all phenomena of weather and climate that physically affect humans take place within the troposphere.

Besides the gases of pure dry air, the troposphere contains *water vapor*, a colorless, odorless, gaseous form of water that mixes perfectly with the other gases of the air. The quantity of water vapor present in the atmosphere is of primary importance in weather phenomena. Water vapor can condense into clouds and fog. When condensation is rapid, rain, snow, hail, or sleet—collectively termed precipitation—are produced and fall to earth.

Where water vapor is present only in small proportions, extremely dry deserts result. In addition, a most important function is performed by water vapor. Like carbon dioxide, it is a gas capable of absorbing heat in the radiant form coming from the sun and from the earth's surface. Water vapor gives to the troposphere the qualities of an insulating blanket, which inhibits the escape of heat from the earth's surface. This role of water vapor, along with several other gases, is well known as the *greenhouse effect*, explained in Chapter 3.

The troposphere contains myriad tiny dust particles, so small and light that the slightest movements of the air keep them aloft. They have been swept into the air from dry desert plains, lake beds, and beaches, or injected by explosive volcanoes. Strong winds blowing over the ocean lift droplets of spray into the air. These may dry out, leaving as residues extremely minute crystals of salt that are carried aloft. Forest and brush fires are another important source of atmospheric dust particles. Countless meteors, vaporizing from the heat of friction as they enter the upper layers of air, have con-

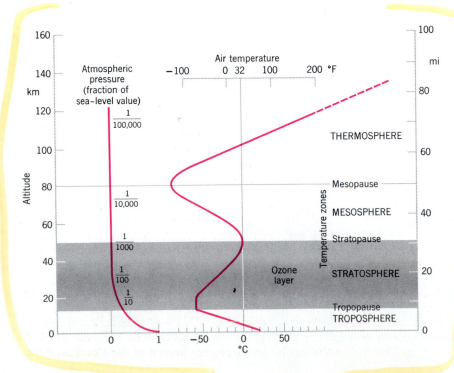

FIGURE 2.9 Temperature structure of the atmosphere.

tributed dust particles. Industrial processes involving combustion of fuels are also a major source of atmospheric dust.

Dust in the troposphere contributes to the occurrence of twilight and the red colors of sunrise and sunset, but the most important function of dust particles cannot be seen and is rarely appreciated. Certain types of dust particles serve as nuclei, or centers, around which water vapor condenses to produce cloud particles. In contrast, the stratosphere is largely free of water vapor and dust. Clouds are rare in the stratosphere, but there are high-speed winds in narrow zones within this layer.

THE OZONE LAYER—A SHIELD TO EARTHLY LIFE

Of vital concern to humans and all other life forms on earth is the presence of an *ozone layer* within the stratosphere. This layer sets in an altitude of about 15 km (9 mi) and extends upward to about 55 km (35 mi) (Figure 2.9). The ozone layer is a region of concentration of the form of an oxygen molecule known as *ozone* (O_3), in which three oxygen atoms are combined instead of the usual two atoms (O_2). Ozone is produced by the action of solar radiation on ordinary oxygen atoms.

The ozone layer serves as a shield, protecting the troposphere and earth's surface by absorbing most of the ultraviolet radiation found in the sun's rays. If these ultraviolet rays were to reach the earth's surface in full intensity, all exposed bacteria would be destroyed and animal tissues damaged severely. In this protective role, the presence of the ozone layer is an essential factor in the life environment.

A serious threat to the ozone layer is posed by the release into the atmosphere of Freons, synthetic compounds containing carbon, fluorine, and chlorine atoms. Compounds of this class are also called *halocarbons*. An alternative name is *chlorofluorocarbons*, or *CFCs*. Prior to a ban issued in 1976 in the United States, many aerosol spray cans used in the household were charged with halocarbons. They are also widely used as refrigerants, a practice that continues to contribute to the release of halocarbons.

Molecules of halocarbons drift upward through the troposphere and eventually reach the stratosphere. As these compounds absorb ultraviolet radiation (UVR) they are decomposed and chlorine is released. The chlorine in turn attacks molecules of ozone, converting them in large numbers by chain reaction into ordinary oxygen molecules. In this way the ozone concentration within the stratospheric ozone layer can be reduced, and the intensity of UVR reaching the earth's surface can be increased. A marked increase in the incidence of skin cancer in humans is one of the predicted effects. Other possible effects are reduction of crop yields of various plants and the killing of certain forms of aquatic life found in the surface layer of the oceans and in streams and lakes.

As studies of the stratospheric ozone layer based on satellite data accumulated during the 1980s, a substantial decline in the total global ozone began to be indicated. By 1987, following an 8 percent ozone decline in the preceding 8 years, scientists agreed that a serious and persistent global decrease was in progress at a rate far faster than what had been previously predicted.

Compounding the problem of global ozone decrease was the discovery in the mid-1970s of a curious "hole" in the ozone layer over the continent of Antarctica, where seasonal thinning of the layer occurs during the early spring of that hemisphere, reaching a minimum in the month of October. In 1986 a corresponding annual "hole" was first identified over the Arctic Ocean, occurring during February. In both places, the maximum thinning matches the time of lowest surface temperatures. The intensity of the Antarctic hole varies from one year to the next, but a strong general trend to greater seasonal thinning appears to be in progress. Although scientists now confidently attribute the seasonal thinning to presence of CFCs, the mechanisms by which they are made so effective at the poles have been strongly debated.

If the global ozone layer is indeed undergoing a substantial decrease in bulk, an increase in the rate of incoming ultraviolet solar radiation (UVR) would be expected and possibly subject to being detected by direct observations. Scientists have estimated that for each 1 percent decrease in global ozone an increase of UVR of 2 percent should result. By 1990, a slight increase in UVR—about 1 percent per year over the preceding 8 years—had been documented at an observing station in the Swiss Alps at an altitude of 3600 m (12,000 ft). In a contrary manner, records at eight stations in the United States over the period 1974–85 clearly showed an overall UVR decrease. Many other factors enter into year-to-year changes in recorded UVR, among them variations in average cloud cover, air pollution, and in the content of volcanic dust suspended in the upper atmosphere. For that reason, great caution is needed in linking any observed UVR increase to ozone depletion.

Scientists have warned that if the consumption of CFCs continues at its present rate, they will contribute to an additional atmospheric warming of 2C° (3.6F°) by the year 2050, over and above the warming predicted for projected increases in carbon dioxide alone (see Chapter 4). Responding to the global threat of ozone depletion and its anticipated environmental impacts on the biosphere, an international conference (the Vienna Conference), was convened in 1985 by the United Nations Environment Programme (UNEP). Representatives of 43 states brought forth the Vienna Convention for the Protection of the Ozone Layer. Their stated purpose was to promote the exchange of information, research, and monitoring data on human activities that might adversely affect human health and the environment. In 1987, 23 nations endorsed a UNEP plan for cutting global CFC consumption by 50 percent by the year 1999. In 1988, under an agreement known as the Montreal Protocol, the United States joined 31 other nations in approving a similar goal. In 1990, the international agreement was expanded and strengthened to a goal of 50 percent reduction of CFCs by 1995 and 85 percent by 1997. An important part of the program is the substitution for CFCs of the more benign hydrofluorocarbons (HCFCs). Implementing a worldwide phasing-out of the use of CFCs will prove difficult to achieve. Compliance will be particularly painful for the developing nations, because the new substitute chemicals are much more costly.

Another and quite different threat to the ozone layer has come from the air transport industry. In 1990, France and Great Britain proposed the building of a fleet of supersonic aircraft using a successor model to the present Concorde. The anticipated fleet of 200 such craft, operating in the stratosphere, is seen by environmentalists as not only contributing to the greenhouse effect by emissions of water vapor, but also destructive to the ozone layer by the emission of oxides of nitrogen.

THE EARTH'S MAGNETIC ATMOSPHERE

The earth can be thought of as a simple bar magnet, the axis of which approximately coincides with the earth's geographic axis (Figure 2.10). Magnetism is generated within the earth's metallic core, a central spherical body about half the earth's diameter. The earth's magnetic axis is inclined several degrees with respect to the geographic axis. Consequently, the north and south magnetic poles do not coincide with the geographic north and south poles, nor does the magnetic equator coincide with the earth's geographic equator.

Lines of force of the earth's magnetic field, shown in Figure 2.10, pass outward through the earth's surface and into surrounding space. A magnetic compass needle, which is nothing more than a delicately balanced bar magnet, orients itself in a position of rest parallel with the lines of force. Lines of force extending out into space comprise the earth's *external magnetic field*. This field can be thought of as a magnetic atmosphere. If we assume, for purposes of comparison, that the earth's gaseous atmosphere extends outward to a distance equal to twice its own radius, or 13,000 km (8000 mi), it is evident that the magnetic atmosphere extends far beyond the outermost limits of the gaseous atmosphere. All of the region within the limit of the magnetic field is described as the *magnetosphere;* its outer boundary has been named the *magnetopause* (Figure 2.11).

The simplest geometric model for the shape of the magnetosphere would be a doughnut-shaped ring surrounding the earth. The plane of the ring would lie in the plane of the magnetic equator, whereas the earth would occupy the opening in the center of the doughnut. Actually, this ideal shape does not exist because of the action of the *solar wind,* a more or less continual flow of electrons and protons emitted by the sun. Pressure of the solar wind acts to press the magnetopause close to the earth on the side nearest the sun (Figure 2.11). Lines of forces in this region are crowded together, and the magnetic field is intensified. On the opposite side of the earth, in a line pointing away from the sun, the magnetopause is drawn far out from the earth and the force lines are greatly attenuated. The entire shape of the magnetosphere has been described as resembling a comet.

An important environmental role of the magnetosphere is to shield the inner atmosphere and surface of the earth from an extremely powerful form of energy called *ionizing radiation.* This is the same form of dangerous radiation given off by such radioactive elements as uranium, used as a nuclear fuel. Ionizing radiation reaches the earth from the sun in the form of fast-moving electrons and protons, comprising the solar wind (Figure 2.11). Upon encountering the magnetosphere, these particles are trapped and retained within the force lines of the magnetic field. Here they tend to become greatly concentrated at times into a doughnut-shaped ring, called the *Van Allen radiation belt* (Figure 2.11). Trapped particles are also continually escaping from the tail of the magnetosphere. Without the protection of the magnetosphere, ionizing radiation would destroy all life exposed at the earth's surface.

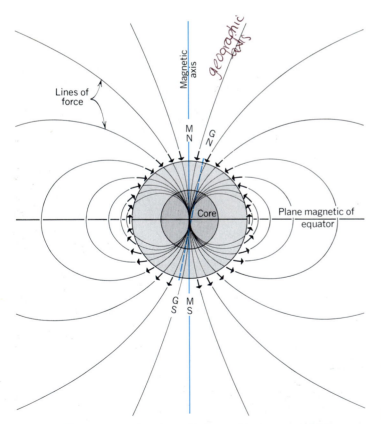

FIGURE 2.10 Lines of force of the earth's magnetic field are shown diagrammatically in this cross section drawn through the magnetic and geographic poles. The small arrows show the inclination of lines of force at surface points over the globe. (From A. N. Strahler, *The Earth Sciences,* 2nd ed., Harper & Row, Publishers. Copyright © by Arthur N. Strahler.)

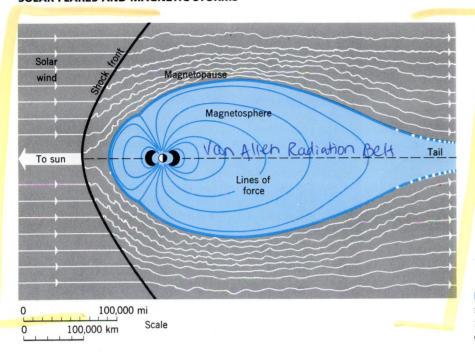

Solar wind

Shock front

Magnetopause

Magnetosphere

Van Allen Radiation Belt

To sun

Tail

Lines of force

0   100,000 mi

0   100,000 km   Scale

FIGURE 2.11 Magnetosphere and magnetopause. The Van Allen radiation belt is shown as a black area on either side of Earth.

SOLAR FLARES AND MAGNETIC STORMS

The environmental shield afforded by our magnetosphere is vulnerable to disturbances that severely impact the human environment through disruptions of our electronic systems. The villain is a *magnetic storm*, caused by the arrival of energetic particles from an intense solar flare. On the sun's surface are sunspots that have associated with them powerful magnetic fields, and these produce solar flares, which are emissions of ionized hydrogen gas from the vicinity of the sunspot. From such flares there emanate discharges of X rays, followed by streams of charged particles—electrons and protons comprising an ion gas cloud in a state described as a plasma. This energetic stream reaches our magnetosphere usually within 13 to 26 hours after the flare is first sighted. Upon arrival, the electrons are trapped by the earth's external magnetic field and are guided earthward along the lines of force, which they follow in general, but with strange corkscrew paths. These electrons temporarily intensify the lines of magnetic force and we have with us a magnetic storm. The magnetic field strength may thus be suddenly increased up to five times its normal intensity.

A magnetic storm simultaneously generates electrical currents in the *ionosphere*, a layer of ionized atmospheric gases that sets in at about 50 km (30 mi) and extends upward to about 1000 km (600 mi). These dynamo currents, which may measure in the thousands of amperes, flow in vast horizontal circular patterns. The terrestrial effect that we feel is from a system of related electrical currents induced in a shallow layer of the solid earth. These currents can be measured within transatlantic cables, and voltage fluctuations may reach 1000 to 2500 volts during a severe storm.

One of the most striking products of the magnetic storm is an optical phenomenon taking place in the ionosphere: the *aurora borealis* of the northern latitudes, and its southern hemisphere counterpart, the *aurora austra-*

lis. Seen at night in the northern sky, the aurora takes the form of light bands, rays, or draperies, continually shifting in pattern and intensity (Figure 2.12).

You should know that sunspot activity and solar flares follow an 11-year cycle of intensity. At peaks of

FIGURE 2.12 The aurora borealis. (Lee Snyder, Geophysical Institute, University of Alaska.)

this cycle, both sunspots and solar flares are much more numerous than at the minimum points. Peaks are not, however, equally intense, some being much greater than others. That of 1989–90 was a great one, rivaling the most intense peak previously observed, which was in 1957–59. As a result, the recent peak of solar activity brought some unpleasant and unexpected effects to an electronic technology that had emerged during cycles of relatively weak peaks.

Magnetic storms impact the modern electronic world in two ways. Currents within the ionosphere seriously disrupt radio communications which rely on the reflection of radio waves from the ionospheric layers. Satellites orbiting at much higher levels are affected by the intense fluxes of solar particles in the magnetosphere. Satellite television communication is thus disrupted, along with the functioning of navigation satellites on which modern aircraft operation relies. The degradation of solar power cells on satellites has been a serious problem. Even the gyroscopes that control the satellites can be adversely affected. Showers of energetic particles can at times penetrate deeply enough into the atmosphere to set off warning alarms aboard Concorde aircraft. The extent to which such showers of ionizing radiation can be a health hazard to aircraft passengers is a subject of investigation and debate.

At ground level, induced electric currents move in high-voltage power transmission lines, overheating transformers and causing shut-downs of entire power distribution systems. A great magnetic storm of March 1989 set off a major power failure in Quebec, when the power grid was operating under heavy load of electrical heating. Fortunately, loads elsewhere in North America were light at that time, or power outages might have spread widely over the country. It seems that electrical power grids as well as satellite power systems were designed during the preceding period of two weak sunspot cycles, when safety margins were set lower than would be required for times of maximum solar activity.

OCEANS AND THE HUMAN ENVIRONMENT

The importance of the oceans to humans is felt in a wide range of dimensions and scales. One environmental role played by the oceans is climatic. The huge water mass of the ocean stores a large quantity of heat. This heat is gained or lost very slowly. As we shall see, the oceans effectively moderate the seasonal extremes of temperature over much of the earth's surface. The oceans supply water vapor to the atmosphere and are the basic source of all rain that falls on the lands. This rainfall, the source of our vital freshwater supplies, originates from the ocean surface by a natural process of distillation of salt water.

The oceans sustain a vast and complex assemblage of marine life forms, both plant and animal. This organic production provides humans with a modest but important share of their food. Throughout history the oceans have served as trackless surfaces of transport of people and the commodities that sustain their civilizations. Winds, waves, currents, sea ice, and fog are environmental factors—sometimes favorable and sometimes hazardous—that the oceans impose on humans and their ships at sea.

The zone of contact between oceans and lands is a unique environment of the life layer. In Chapter 20 we shall investigate the processes by which ocean waves and currents shape coastal landscape features. We use and modify the coastal zone in various ways, ranging from ports to recreational facilities. This is a zone of great natural hazards from powerful storms that cause rapid coastal erosion and the flooding of low-lying coastal areas.

THE WORLD OCEAN

We shall use the term *world ocean* to refer to the combined ocean bodies and seas of the globe. Let us consider some statistics that emphasize the enormous extent and bulk of this great saltwater layer. The world ocean covers about 71 percent of the global surface (Figure 2.13); its average depth is about 3800 m (12,500 ft), when shallow seas are included with the deep main ocean basins. For major portions of the Atlantic, Pacific, and Indian oceans, the average depth is about 4000 m (13,000 ft).

The total volume of the world ocean is about 1.4 billion cu km (317 million cu mi), a quantity just over 97 percent of the world's free water. Of the small remaining volume, about 2 percent is locked up in the ice sheets of Antarctica and Greenland, and about 1 percent is represented by fresh water stored on the lands. These figures show the extent of the *hydrosphere*, a general term for the total free water of the earth, in all three of its states—gas, liquid, and solid. The hydrosphere is largely represented by the world ocean. To place the masses of the atmosphere and oceans in their proper planetary perspective, compare the following figures. (The unit of mass used here is 10^{21} kg.)

Entire earth	6000
World ocean	1.4
Atmosphere	0.005

FIGURE 2.13 Northern and southern hemispheres—a contrast in land–ocean distributions.

What are the basic differences between the world ocean and the atmosphere in terms of properties and behavior? How do atmosphere and ocean interact in the region of their interface? The answers to these questions are vital to an understanding of environmental processes because marine life depends on the exchanges of matter and energy across the atmosphere–ocean interface. It is also significant that the earliest life forms originated and developed in the shallow layer of water immediately beneath the interface.

The atmosphere, being composed of a gas that is easily compressed, has no distinct upper boundary; it becomes progressively denser toward its base under the load of the overlying gas. The world ocean has a sharply defined upper surface in contact with the densest layer of the overlying atmosphere.

Whereas the most active region of the atmosphere is the lowermost layer, the troposphere, the most active region of the ocean is its uppermost layer. At great ocean depths, water moves extremely slowly and maintains a uniformly low temperature. One reason for intense physical and biological activity in the uppermost ocean layer is that the input of energy from the overlying atmosphere drives water motions in the form of waves and currents. Another reason is that oxygen and carbon dioxide, gases vital to animal and plant growth, enter the ocean from the atmosphere. The atmosphere is also the source layer of heat and of condensed fresh water previously evaporated from the ocean. But the ocean surface also returns heat and water (in the vapor form) to the lower atmosphere, and this phenomenon is of primary importance in driving atmospheric motions. Interaction between atmosphere and ocean surface is a topic we shall explore in later chapters.

The oceans are formed into compartments by intervening continental masses. The land barriers inhibit the free global interchange of ocean waters, whereas the atmosphere is free to move globally. Another difference in the two bodies is that the atmosphere has little ability to resist unbalanced forces. The air therefore moves easily and rapidly, changing its velocity quickly from place to place. In contrast, ocean water can move only sluggishly and is very slow to respond to the changes in force applied by winds.

COMPOSITION OF SEAWATER

Seawater is a solution of salts—a brine—whose ingredients have maintained approximately fixed proportions over a considerable span of geologic time. Besides their importance in the chemical environment of marine life, these salts constitute a vast reservoir of mineral matter from which certain constituents may be extracted for industrial uses. One way to describe the composition of seawater is to state the principal ingredients that would be required to make an artificial brine approximately like seawater. These are listed in Table 2.1. Of the various elements combined in these salts, chlorine alone makes up 55 percent by weight of all the dissolved matter, and sodium makes up 31 percent. Magnesium, calcium, sul-

TABLE 2.1 Composition of Seawater

Name of Salt	Chemical Formula	Grams of Salt per 1000 g of Water
Sodium chloride	NaCl	23.0
Magnesium chloride	$MgCl_2$	5.0
Sodium sulfate	Na_2SO_4	4.0
Calcium chloride	$CaCl_2$	1.0
Potassium chloride	KCl	0.7
With other minor ingredients to total		34.5

fur, and potassium are the other four major elements in these salts. Important, but less abundant than elements of the five salts listed in Table 2.1 are bromine, carbon, strontium, boron, silicon, and fluorine. At least some trace of half of the known elements can be found in seawater. Seawater also holds in solution small amounts of all the gases of the atmosphere. Of these, oxygen and carbon dioxide are vital to organic processes in marine animals and plants.

LAYERED STRUCTURE OF THE OCEANS

As with the atmosphere, the ocean has a layered structure. Ocean layers are recognized in terms of both temperature and oxygen content. In the troposphere, air temperatures are generally highest at ground level and diminish upward. In the oceans, temperatures are generally highest at the sea surface and decline with depth. This trend is to be expected because the source of heat is from the sun's rays and from heat supplied by the overlying atmosphere.

With respect to temperature, the ocean persents a three-layered structure in cross section, as shown in the left-hand diagram of Figure 2.14. At low latitudes

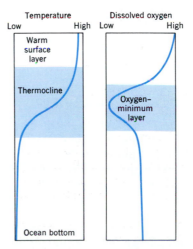

FIGURE 2.14 This schematic diagram shows the changes of temperature and oxygen content with depth typical of broad expanses of the oceans in lower latitudes.

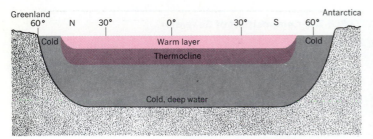

FIGURE 2.15 A schematic north-south cross section of the world ocean shows that the warm surface water layer disappears in arctic latitudes, where very cold water lies at the surface.

FIGURE 2.16 Planet Earth photographed in 1972 by astronauts of the Apollo 17 mission on their way to the moon. On this date the earth was near the December solstice, with its south polar region tilted toward the sun to receive maximum insolation. Africa is clearly outlined in the central and upper parts of the disk, its reddish-brown northern and southern desert belts conspicuously free of clouds. The intervening equatorial zone of the African continent reveals green vegetation between patches of clouds. Cyclonic storms form a white chain over the blue of the Southern Ocean. Antarctica, with its blanket of snow and clouds, is solid white. (NASA 72–HC–928.)

throughout the year and in midlatitudes in the summer, a warm surface layer develops. Here wave action mixes heated surface water with the water below it to give a warm layer that may be as thick as 500 m (1600 ft), with a temperature of 20° to 25°C (70° to 80°F) in oceans of the equatorial belt. Below the warm layer temperatures drop rapidly, making a second layer known as the *thermocline*. Below the thermocline is a third layer of very cold water extending to the deep ocean floor. Temperatures near the base of the deep layer range from 0° to 5°C (32° to 40°F). In arctic and antarctic regions, the three-layer system is replaced by a single layer of cold water, as Figure 2.15 shows. Temperature is a prime environmental factor affecting the abundance and variety of marine life, the bulk of which thrives in the shallow upper layer.

The content of free oxygen dissolved in seawater shows an oxygen-rich surface layer, accounted for by the availability of atmospheric oxygen and the activity of oxygen-releasing plant life in the sea. As shown in the right-hand diagram of Figure 2.14, oxygen content falls rapidly with depth; over large ocean areas, there is a distinct minimum zone of low oxygen content. Here oxygen has been consumed by biological activity. In deep water the oxygen content holds to a uniform and moderate value down to the ocean floor.

ATMOSPHERE AND OCEANS IN REVIEW

The broad overview of atmosphere and oceans given in this chapter shows us the major elements of physical structure and chemical composition of those two great global layers so vital to life on earth. Most of the information has been about static conditions we would encounter when probing upward into the atmosphere and downward into the oceans.

In the next several chapters we turn to the great systems of flow of matter and energy that continually involve the atmosphere and oceans, making them dynamic rather than static bodies. First, we shall trace the course of radiant energy from the sun as it passes through the atmosphere, reaches the earth, and is returned to outer space. The solar radiation system controls the thermal environment of the life layer, or biosphere. Solar radiation supplies the life layer with the energy needed for biological processes. We shall then follow with accounts of the vast systems of transport of the atmosphere and oceans. These great flows of air and water redistribute heat over the globe to provide more moderate and more favorable life conditions than would otherwise exist on our planet. Accompanying these circulation systems are intense disturbances of the air and sea—storms that are often severe and hazardous environmental stresses.

Looking over the other planets of the solar system, the uniqueness of Planet Earth as a life environment is most striking (Figure 2.16). Only Earth has both a great world ocean and a comparatively dense oxygen-rich atmosphere combined with a favorable temperature range. Mars, our nearest planet, has practically no free water in any form and only a very rarefied atmosphere with little oxygen. Venus, matching us closely in size, has a much denser atmosphere than Earth. But water in any form is almost totally lacking on Venus, and surface temperatures there are much higher than on Earth. Little Mercury, with no atmosphere and no water, roasts in the sun's rays. The greater outer planets—Jupiter, Saturn, Uranus, and Neptune—probably have large quantities of water, but it is frozen solid. Their atmospheres, composed in part of ammonia and methane, would be lethal to life such as ours even if the surface temperatures were not impossibly cold. The moon's surface has no free water or atmosphere to offer. So there is really no place for humans to live in large numbers other than Planet Earth.

FLOW SYSTEMS OF MATTER AND ENERGY

MATTER AND ENERGY

The earth realms with which physical geography is concerned consist of two components: matter and energy. Physics and chemistry are basic sciences that deal with the nature of matter and energy and the formulation of laws that govern their behavior. To define the terms *matter* and *energy* is not an easy task because they represent concepts that include everything in the real world. To start, we can substitute the world "substance" for the word "matter," but this only postpones the problem. Substance, in turn, has the attribute of occupying space. Matter is a tangible substance because it can be seen, felt, tasted, measured, weighed, or stored. Matter has the mysterious property of *gravitation*, the mutual attraction acting between any two aggregates (lumps, groups, or pieces) of matter.

Energy is often defined in terms of its effects. Perhaps the most common definition is that "energy is the ability to do work." Energy is somehow always involved in the motion of matter, but energy can be stored in matter that does not appear outwardly to have motion. A brick poised on a window ledge holds a store of energy even though the brick is at rest. That stored energy will rapidly be released in the form of motion if the brick is nudged slightly and falls to the sidewalk below. Whatever energy is, it can move or flow from place to place and it can also be held in storage in various ways. We frequently refer to energy as something expendable. For example, someone may say "I used up a lot of energy playing those two sets of tennis." Actually, energy cannot be destroyed by being "used"; it can only change from one form to another and move from one place to another. The same statement applies to matter, which cannot be destroyed, but only changed or moved from place to place. By "destroyed" we mean "removed from existence," or "removed from the universe."

STATES OF MATTER

The physical condition, or state, in which we find matter at a given place and time is a subject of great importance in physical geography. Certain *states of matter* are well known to everyone. Every day we drink water in its *liquid state*. Ice cubes represent water in the *solid state*. The *gaseous state* of water, called water vapor, can't be seen, but is easily sensed by the human skin in summer when the humidity of the air is relatively high.

The three common states of matter—solid, liquid, and gaseous—apply to both pure substances (elements and compounds) and to mixtures. Using only the simplest concepts of atoms and compounds we can describe the three states of matter in terms of observable behavior. For this purpose, the atoms or molecules that comprise matter can be visualized as uniform spheres, all physically alike.

A *gas* is a substance that expands easily and rapidly to fill any small empty container. Atoms or molecules of the gas, as the case may be, are in high-speed motion (Figure 2.17). Empty space between the atoms or molecules is vast in comparison with the dimensions of those particles. Particle motions take random directions; colli-

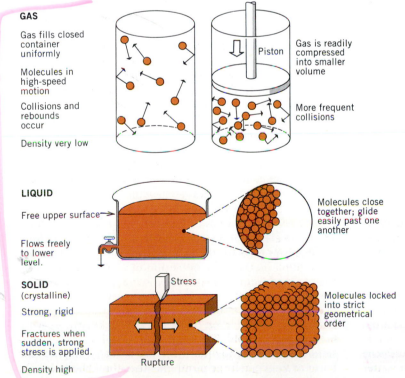

GAS

Gas fills closed container uniformly

Molecules in high-speed motion

Collisions and rebounds occur

Density very low

Piston

Gas is readily compressed into smaller volume

More frequent collisions

LIQUID

Free upper surface →

Flows freely to lower level.

Molecules close together; glide easily past one another

SOLID
(crystalline)

Strong, rigid

Fractures when sudden, strong stress is applied.

Density high

Stress

Rupture

Molecules locked into strict geometrical order

FIGURE 2.17 Some properties of gases, liquids, and solids. (From A. N. Strahler, *Physical Geology*. Harper & Row, Publishers. Copyright © by Arthur N. Strahler.)

sions between particles are frequent. The particles rebound like perfect spheres at each impact, changing direction abruptly. The particles also strike and rebound from the walls of the container. A gas is usually very much less dense than a liquid or solid consisting of the same chemical substance. For example, the gaseous water vapor in warm, moist air has a density only about 1/100,000 that of liquid water.

A *liquid* is a substance that flows freely in response to unbalanced forces but maintains a free upper surface so long as it does not fill the container or cavity in which it is held. The molecules of a liquid compound—water, for example—move more or less freely past one another as individuals or small groups. Under rather strong confining pressures (such as would exist at the bottom of the deep ocean) liquids are compressed only slightly into smaller volume. For many practical purposes, liquids can be considered to be incompressible (not capable of being compressed).

Both gases and liquids are classed as *fluids* because both substances freely undergo flowage. Put simply, both of these substances flow toward lower levels under the force of gravity wherever possible. As a result, fluids of different density tend to come to rest in layers with the fluid of greatest density at the bottom and that of least density at the top. This principle has several very important applications in sciences of the atmosphere and oceans.

A *solid* is a substance that resists changes of shape and volume. Solids are typically capable of withstanding large unbalanced forces (i.e., strong stresses) without yielding permanently, although they undergo a small amount of elastic bending. When yielding does occur, it is usually by sudden breakage (rupture). These principles have important applications in the study of glacier ice and rock, both of which are in the solid state.

Changes of state are accompanied by either an input or an output of energy into the substances undergoing change. We are all familiar with this principle in the preparation of food. To boil water and produce steam (change to vapor state) a great deal of heat must be applied. To freeze water, a great deal of heat must be removed from the water. Our next step will be to examine the nature and forms of energy that can be easily observed to cause changes in pure substances and mixtures behaving as gases, liquids, and solids.

FORMS OF ENERGY

We stated earlier that energy is commonly defined as the ability to do work. In strict terms of physics, energy is the product of force acting through distance. Thus, energy is the ability to move an object (exert a force) for a certain distance. Energy is stored and transported in a variety of ways. Some of the recognized forms of energy are: mechanical energy, heat energy, energy transmitted by radiation through space (electromagnetic energy), chemical energy, electrical energy, and nuclear energy.

Mechanical energy is energy associated with the motion of matter. There are two forms of mechanical energy: kinetic energy and potential energy. *Kinetic energy* is the ability of a mass in motion to do work. Thus, an

automobile traveling down a highway possesses kinetic energy because it is a mass in motion. Should this mass strike a telephone pole, its ability to do work upon its own body and upon the telephone pole will become quite obvious. The energy it will release in collision will increase with the weight (mass) of the car, and it will also increase with the square of the auto's speed. Kinetic energy, then, is proportional to the quantity of mass in motion multiplied by the square of its velocity. Kinetic energy is obvious in many kinds of natural processes acting at the earth's surface—a rolling boulder, a flowing stream, or a pounding surf.

Potential energy, or energy of position, is equal to the kinetic energy an object would attain if it were allowed to fall under the influence of gravity. Suppose a brick is balanced on the edge of a tabletop, then falls. The kinetic energy the brick possesses at the moment it hits the floor, as we have seen, is proportional to the mass of the brick multiplied by the square of its velocity. If the brick is again lifted to the tabletop, the work done in lifting gives the brick a quantity of potential energy. This energy will be released when the brick is again allowed to fall. It should be obvious at this point that the floor is merely a convenient stopping place for the brick; if a hole were opened in the floor, allowing the brick to fall farther, it would possess even more kinetic energy at its impact on the floor below. Therefore, potential energy must always be valued with respect to a given reference level, or base level. Looking around outdoors, we can spot many examples of the existence of potential energy in a landscape. A boulder poised at the top of a steep mountain face is a simple example. In fact, the entire mountain represents a large reservoir of potential energy judged in reference to the level of the floor of an adjacent valley.

Mechanical energy can be transmitted from one place to another in the form of *wave motion*, in which kinetic energy is carried through matter by an impulse passed along from one particle to the next. A sound wave is one example—a push on air molecules at one point will be transmitted outward in all directions. Another example is the phenomenon of earthquake waves, which carry large amounts of energy for great distances, not only over the ground surface, but also in paths deep within the solid earth. The familiar Richter scale of earthquake intensity measures the quantity of energy released by an earthquake. In all mechanical forms of wave motion matter is displaced (up-and-down, sideways, or forward-and-backward) in a rhythmic manner. Frictional resistance within the moving substance withdraws energy and the waves gradually die out as they travel farther from the source.

Sensible heat is another form of energy of paramount importance. Kinetic energy can readily be converted into sensible heat through the mechanism of friction. A familiar example is the braking action of a moving automobile. As the automobile slows to a stop (losing kinetic energy), the brake drums become intensely hot.

Sensible heat represents kinetic energy, but it is of an internal form, rather than the external form seen in moving masses. Thus, a cupful of water resting completely motionless on the table has internal energy because of constant motion of the water molecules on a scale too small to be visible. This internal motion is the sensible heat of the substance and its level of intensity is measured by the thermometer. For gases, the internal motion is in the form of high-speed travel of free molecules in space but with frequent collisions with other molecules. The energy level in the gaseous state of water (water vapor) is thus higher than within the liquid water. Ice, on the other hand, represents a lower energy level than liquid water, for here the molecules are locked into place in a fixed geometrical arrangement (see Figure 2.17). For these molecules in the solid state the motion is one of vibration without relative motion.

Sensible heat moves through gases, liquids, or solids by the process of conduction. The direction of heat flow by conduction is always in the direction from higher temperature to lower temperature. In the process of conduction the faster moving molecules of the warmer matter pass along part of their kinetic energy to the adjacent cooler matter, causing an increase in the speed of molecular motion along the path of energy travel. In this way heat moves through the matter. By conduction, heat can pass from a gas into a liquid or solid, from a liquid into a solid or gas, and from a solid into a liquid or gas. Sensible heat can also be transported within a layer of gas or liquid by convection, a process in which currents redistribute heat by mixing warmer and cooler parts of the fluid.

When ice melts, work must be done to overcome the crystalline bonds between molecules. This work requires an input of energy, but it does not raise the temperature of the substance. Instead, the energy seems mysteriously to disappear. Since energy cannot be lost, it is actually placed in storage in a form known as *latent heat*. Should the water freeze and again become ice, the latent heat will be released as sensible heat. A similar transformation from sensible to latent heat takes place when a liquid evaporates into a gas, because work must be done to overcome the bonds between molecules of the liquid. When water vapor returns to the liquid state, a process of condensation, latent heat is released as sensible heat. Both sensible heat and latent heat represent forms of stored energy (as does potential energy).

Sensible heat stored in matter may be lost directly to the surroundings through conduction, but even in a vacuum objects can lose heat. A fundamental law of physics states that all matter at temperatures above absolute zero gives off *electromagnetic energy*. The process of giving off energy is called *radiation*. We can think of this radiation as taking the form of waves traveling in straight lines through space. The waves come in a very wide range of lengths, but all travel at the same speed— 300,000 km (186,000 mi) per second—regardless of their length. Together, the total assemblage of waves of all lengths constitutes the *electromagnetic spectrum*. It includes visible light with all its rainbow colors, and also invisible shorter waves such as ultraviolet rays, X rays, and gamma rays. Besides these, the spectrum includes invisible long waves known as infrared rays (sometimes called heat rays), and still longer microwaves and radio waves.

Electromagnetic energy received from the sun powers a group of important natural processes constantly at work at the earth's surface. Electromagnetic energy, upon arriving at the earth's surface, is continually converted into mechanical energy and sensible heat, which in turn are transformed in such activities as winds in the atmosphere or the breakup of exposed rock, and in the transport of the resulting rock particles to new places of rest.

Yet another form of energy is *chemical energy*, which is absorbed or released by matter when chemical reactions take place. These reactions involve the coming together of atoms to form molecules, the recombining of molecules into new compounds and the reverse changes back to simpler forms of matter. Green plants use the sun's electromagnetic energy to produce chemical energy, which can be stored within the leaves and stems of a plant in the form of complex organic molecules.

Finally, we take note of *electrical energy* and *nuclear energy*. The lightning bolt is a spectacular display of a release of electrical energy. Nuclear energy is produced by the spontaneous disruption of atoms of certain elements said to be radioactive. Nuclear energy is an important natural process deep within the solid earth, where it continually generates heat. This accumulated heat is believed to be responsible for dislocations of the earth's outer crust and the slow motions of vast platelike rock masses many tens of kilometers thick.

FLOW SYSTEMS OF MATTER AND ENERGY

Understanding the varied processes that affect the atmosphere, hydrosphere, lithosphere, and biosphere requires a physical geographer to think in terms of *flow systems* of both matter and energy. A flow system is simply a series of connected pathways through which energy and/ or matter move more or less continuously. We will deal first with the energy flow system. An *energy flow system* traces the flow path of energy from a point of entry to a point of emergence. As energy flows through such a system, it may change form many times and may be detained temporarily in storage from place to place. In this process the energy flow makes use of matter as the medium of motion and of storage.

Energy flow systems set in motion and sustain *material flow systems* (flow systems of matter). The matter involved in such systems is not only transported from place to place along certain pathways, but it can also undergo changes of state and chemical changes. The matter traveling through the system can also be held temporarily in storage at certain points.

Flow systems of energy and matter can be visualized by means of simple schematic diagrams making use of a set of standard symbols. Three such diagrams, using familiar household examples, illustrate the three basic kinds of flow systems (Figure 2.18). First is the energy flow system, here illustrated in diagram A by a container of food being heated over an electric heating coil. The outer surface of the container represents the *system boundary*, shown by a rectangular outer box in the diagram. Input of energy into the system is shown as coming from an *energy source*, for which a circle is our symbol. An arrow shows the flow path of energy. Let us assume that the energy supplied by the electric heating coil is in the radiant form and consists of infrared rays, which are part of the electromagnetic energy spectrum. Thus the energy is transferred by radiation from the heater to the container, which absorbs the energy. Here an *energy transformation* occurs, in which the radiant energy is transformed into sensible heat, raising the temperature within the container. A rectangle is our symbol for an energy transformation. The sensible heat enters temporary *energy storage* in the container and its contents. A figure shaped like a storage bin indicates storage of

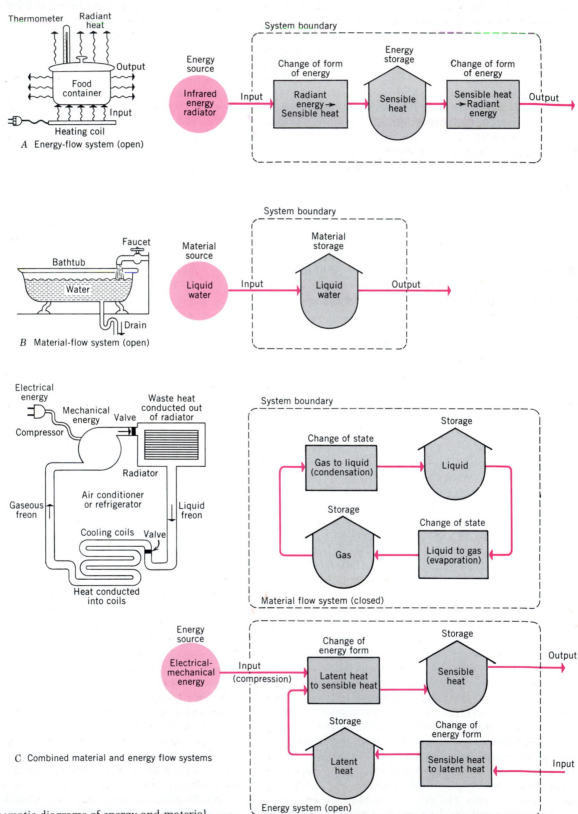

FIGURE 2.18 Schematic diagrams of energy and material flow systems using common household mechanical devices.

either energy or matter within a flow system. Energy continues its flow by again changing its form back to infrared radiation emitted by the surface of the container. This process represents the output of energy through the system boundary. (Other forms of energy flowing both into and out of the system may be operating simultaneously, but these have been omitted.)

With the rate of energy input held constant, a point in time is reached at which the amount of energy in storage becomes constant, as indicated by a constant temperature within the container. The rate of output of energy must then equal the rate of input. We say that the system is in a *steady state* of operation when the rates of input and output are equal and constant and the storage is also a constant quantity. If we were to increase the rate of energy input, the storage would increase (giving a higher temperature) and so would the rate of energy output. After a short time steady state would be reestablished at a new level of activity. When the energy input is turned off, the flow system simply collapses and ceases to exist. We have illustrated an *open system*, defined as a system that requires both an input and an output of energy (or matter) through the system boundary. An important principle is that all energy flow systems are open systems. No real energy system can operate as a closed system, entirely within the system boundary, because no perfect insulator exists to prevent the outflow of energy as heat through the enclosing boundary.

Diagram B of Figure 2.18 illustrates a simple material flow system, represented by a bathtub with its faucet and open drain. The system boundary is the outer surface of contact of the water with the walls of the tub and the free liquid surface. Input of water by means of the faucet can be regulated in such a way that the water level rises rapidly in the tub. As this happens, the rate at which water exits through the drain is steadily increased (because of the increasing hydraulic pressure of the deepening water). A point in time is reached at which the water level is constant and the quantity of matter in storage is thus a constant value. Rates of input and output are also constant and equal and the system is in steady state. This material flow system is an open system. When the input of water is shut off, the water drains completely out of the tub and the system simply disappears.

Diagram C of Figure 2.18 illustrates systems of matter flow and energy flow working simultaneously. The total system consists of the essential working parts of an air conditioner or a refrigerator using a circulating coolant fluid, such as Freon, to carry out the cooling process. A compressor, driven by an electric motor, receives coolant gas at low pressure and pumps it through a valve, placing the gas under high pressure. The compression process heats the gas to a temperature well above the air surrounding the radiator. In a radiator the gas cools and condenses to a liquid. As the liquid enters the cooling coils, it passes through another valve, which sprays it into the low pressure environment of the cooling coils. At low pressure, the coolant droplets quickly evaporate, becoming a very cold gas in the cooling coils and chilling the air near the coils. The gaseous coolant returns to the compressor to complete the cycle. The net effect has been

to extract heat from the environment near the cooling coils, store the heat in the gaseous Freon, then release the heat to the external surroundings during condensation in the radiator. In this way, heat is "pumped" from the cooling coils to the radiator.

The upper diagram shows the material flow system consisting of the flow of coolant and its changes of state. The changes of state are shown by the same rectangular boxes used as symbols of energy transformation in the energy system. Similarly, storages in the liquid and gaseous states are shown by the bin symbol. Beginning at the upper left corner, compression forces condensation of the gaseous coolant, a change from the gaseous to the liquid state. Liquid coolant then expands and evaporates, returning to the gaseous state. We have shown here a *closed system* because no matter enters or leaves the system.

The lower diagram illustrates the energy flow system that acts simultaneously with the material flow system. For each change in material state there is a change in form of energy. If the diagrams were printed on transparent acetate film, you could place one diagram over the other and see that the basic circuits of material flow and energy flow are perfectly superimposed. However, the energy system must have inputs and outputs of energy in addition to an internal closed flow circuit. Electrical energy is converted into mechanical energy of the compressor, but this transformation takes place outside the system. Mechanical energy is used to compress the gas, raising its internal temperature and adding to the store of sensible heat. At high pressure the gas condenses, releasing latent heat to add to the store of sensible heat. Part of the stored sensible heat is disposed of to the outside environment by conduction through the radiator and represents the energy output of the system. As the coolant evaporates, heat is drawn into the system from the outside and much of the sensible heat it holds is transformed into latent heat, which goes into temporary storage in the gaseous coolant. Condensation reverses the process and the latent heat is transformed back into sensible heat. Like the first example of an energy system, this one is an open system and must have both an energy input and an energy output.

The refrigerating machine is thus a combination of a closed material flow system and an open energy flow system. Most of the natural systems encountered in physical geography are also combinations of material and energy flow systems. No material system, whether open or closed, can operate without the expenditure of energy. The reverse of this statement is not true, however. It is possible to have an energy flow system without motion of matter being involved, because energy can flow through a body of matter by conduction or radiation, or a combination of the two, without any motion of the matter itself being required.

These concepts of flow systems of energy and matter will be put to good use at various points in our study of physical geography where the system concept is important. Later chapters include special flow diagrams to supplement the conventional description of several natural systems.

CHAPTER 3

THE EARTH'S RADIATION BALANCE

All life processes on Planet Earth are supported by radiant energy coming from the sun. The planetary circulation systems of atmosphere and oceans are driven by solar energy. Exchanges of water vapor and liquid water from place to place over the globe depend on this single energy source. It is true that some heat flows upward through the earth's crust to reach the surface, but the amount is trivial in comparison with the amount of energy that the earth intercepts from the sun's rays.

The flow of energy from sun to earth and then out into space is a complex system. It involves not only energy transmission by a form of radiation, but also energy storage and transport. Both storage and transport of heat occur in the gaseous, liquid, and solid matter of the atmosphere, hydrosphere, and lithosphere. We can simplify the study of this total system by first examining each of its parts. We shall start with the radiation process itself and develop the concept of a *radiation balance*, in which energy absorbed by our planet is matched by the planetary output of energy into outer space.

Solar energy is intercepted by our spherical planet, and the level of heat energy tends to be raised. At the same time, our planet radiates energy into outer space, a process that tends to diminish the level of heat energy. Incoming and outgoing radiation processes are simultaneously in action (Figure 3.1). In one place and time more

energy is being gained than lost; in another place and time more energy is being lost than gained.

The equatorial region receives much more energy through solar radiation than is lost directly to space. In contrast, polar regions lose much more energy by radiation into space than is received. So mechanisms of energy transfer must be included in the energy system. These mechanisms must be adequate to export energy from the region of surplus and to carry that energy into the regions of deficiency. On our planet, motions of the atmosphere and oceans act as heat-transfer mechanisms. A study of the earth's energy balance will not be complete until the global patterns of air and water circulation are described and explained in Chapter 5.

Thinking further, we realize that the movement of water through atmosphere and oceans and on the lands comprises a global system of transport of matter. This system is equal in environmental importance to the flow of energy. We also realize that the activities of these two systems are closely intermeshed. The concept of a water balance will be developed in Chapter 9 and will take its place beside the energy balance. Together, these two great flow systems of energy and matter form a single, grand planetary system and permit us to relate and explain many of the environmental phenomena of our earth within a unified framework.

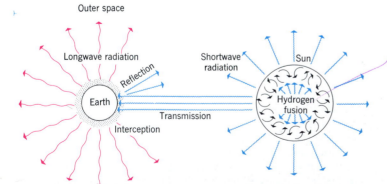

FIGURE 3.1 Simplified diagram of the sun–earth–space radiation system. (From A. N. Strahler, "The Life Layer," *Jour. Geography,* vol. 69, p. 72. © 1970 by *The Journal of Geography;* reproduced by permission.)

THE SUN'S ENERGY OUTPUT

A systematic approach to the earth's energy balance begins with an examination of the input, or source, of energy from solar radiation. We shall trace this radiation as it penetrates the earth's atmosphere and is absorbed or transformed. We then turn to the mechanism of output of energy by the earth as a secondary radiator.

Our sun, a star of about average size and temperature compared with the overall range of stars, has a surface temperature of about 6000°C (11,000°F). The highly heated, incandescent gas that comprises the sun's surface emits electromagnetic radiation moving in straight lines radially outward from the sun and requiring about 8⅓ minutes to travel the 150 million km (93 million mi)

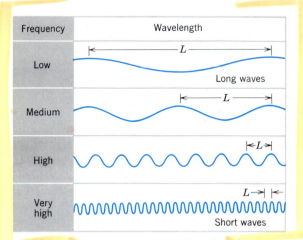

Frequency	Wavelength
Low	*L* — Long waves
Medium	*L*
High	*L*
Very high	*L* — Short waves

FIGURE 3.2 Relationship of wavelength to frequency. (From A. N. Strahler, *The Earth Sciences*, 2nd ed., Harper & Row, Publishers. Copyright © by Arthur N. Strahler.)

from sun to earth. The energy source for solar electromagnetic radiation is in the sun's interior where, under enormous confining pressure and high temperature, hydrogen is converted to helium. In this nuclear fusion process, a vast quantity of heat is generated and finds its way by convection and conduction to the sun's surface. Because the rate of internal production of energy is nearly constant, the energy output of the sun varies only slightly.

For many practical purposes, including the study of the earth's radiation balance, electromagnetic radiation can be treated as a wave form of energy transport. The waves can be visualized as resembling the bow waves generated by a ship moving through a perfectly calm sea. Waves of this type are known as sine waves.

Figure 3.2 illustrates how electromagnetic waves differ in dimensions throughout their entire range, or spec-

trum. Waves are described in terms of either wavelength or frequency. *Wavelength (L)* is the actual distance between two successive wave crests (or two successive wave troughs). Metric units of length, for example, centimeters or meters, are always used. Next, we know that all waves of the electromagnetic spectrum travel at the same speed. The number of waves passing a fixed point in a unit of time (1 second) is known as the *wave frequency*. Frequency depends on the wavelength. Long waves have low frequency; short waves have high frequency. Frequency is stated in terms of cycles per second; the unit of measure is the *hertz*. One hertz is a frequency of 1 cycle per second; 1 *megahertz* is a frequency of 1 million cycles per second. Knowing the speed of light to be 300,000 km per sec, it is easy to calculate that a frequency of 1 megahertz is associated with a wavelength of 300 m.

Equivalents between wavelength and frequency can be read directly on the scales of Figure 3.3, which shows the divisions of the electromagnetic spectrum—ranging from the shortest waves at the left to the longest waves at the right. The scale is a logarithmic scale (constant-ratio scale). Frequency in megahertz is given in powers of 10; each scale unit represents a tenfold increase from right to left. For example, 10^3 megahertz is "one thousand megahertz"; 10^9 is "one billion megahertz."

At the short end of the spectrum are gamma rays of extremely high frequency and short wavelength. It is customary to describe these short wavelengths in units of *angstroms*. One angstrom is equal to 0.000,000,01 cm (10^{-8} cm). Gamma rays are shorter than 0.03 angstroms. Next come the *X rays*. The shorter X rays, described as "hard," fall in the range 0.03 to about 0.6 angstroms. The longer X rays, described as "soft," range from about 0.6 to about 100 angstroms; they grade into *ultraviolet rays*, which extend to about 4000 angstroms.

For the visible light spectrum it is customary to

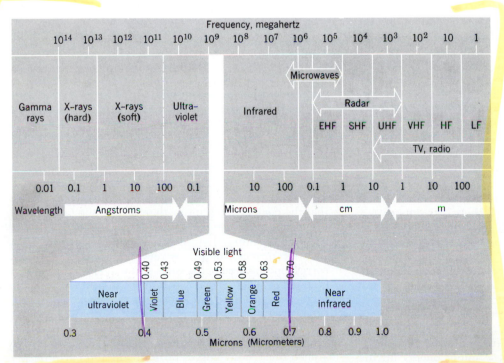

FIGURE 3.3 The electromagnetic radiation spectrum. (From A. N. Strahler, *The Earth Sciences*, 2nd ed., Harper & Row, Publishers. Copyright © by Arthur N. Strahler.)

switch to a longer unit of length, the *micron*. One micron equals 0.0001 cm (10^{-4} cm); 1 micron thus equals 10,000 angstroms. The *visible light* portion of the spectrum begins with violet at 0.4 microns. Colors then grade successively through blue, green, yellow, orange, and red, reaching the end of the visible spectrum at about 0.7 microns. In Figure 3.3, the visible light spectrum has been expanded to show the limits of the various colors.

Next in the spectrum come the *infrared rays*, consisting of wavelengths from about 0.7 to about 300 microns. These rays are not visible, but can sometimes be felt as "heat rays" from a hot object.

Longer wavelength regions of the spectrum, including microwaves, radar, and radio waves, are described in Appendix II under the subject of remote sensing. They are not important in the global energy balance.

Although the solar radiation travels through space without energy loss, the rays are diverging as they move away from the sun. Consequently, the intensity of radiation within a beam of given cross section (such as 1 sq m) decreases inversely as the square of the distance from the sun. The earth intercepts only about one two-billionth of the sun's total energy output.

MEASURING INCOMING SOLAR RADIATION

At the average distance from the sun, the intensity of solar radiation received on a unit area of surface held at right angles to the sun's rays is nearly constant. Known as the *solar constant*, this radiation intensity has an approximate value of 1400 watts per square meter (W/m^2).

[Note: The standard unit of energy (or work) is the *joule* (J). For most persons, the *calorie* is the more familiar energy unit, but it is no longer used to apply to the energy of electromagnetic radiation. Another unit, the *watt* (W), describes the rate at which energy flows in solar rays or through conducting metals. The watt is familiar to all as a measure of power of electrical devices, such as stereo amplifiers, toasters, and light bulbs. The watt is defined as 1 joule per second (1 W = 1 J/sec). Think of the watt as a measure of power. In stating the intensity of solar electromagnetic energy being received, we must specify the unit cross section of the beam, and this is assigned the area of 1 square meter (1 m^2). Thus the measure of intensity of received (or emitted) radiation is given as watts per square meter (W/m^2).]

In stating the value of the solar constant, we are assuming that incoming radiation is being measured beyond the limits of the earth's atmosphere. As the radiation passes through the atmosphere, its intensity is diminished, so that near sea level, it will have been reduced to a value commonly between 400 and 800 W/m^2.

Orbiting space satellites equipped with suitable instruments for measuring electromagnetic radiation intensity have provided precise data on the solar constant. Minor fluctuations in radiation intensity are observed, and these are attributed to variations in output of ultraviolet rays accompanying disturbances of the sun's surface. For the purpose of evaluating the earth's radiation balance, we can disregard these minor energy fluctuations.

ENERGY SPECTRA OF SUN AND EARTH

To understand the basic differences in the kind and intensity of electromagnetic energy emitted by the sun and the earth, you need to know the underlying physical laws governing the radiation of energy in relation to the temperature of a radiation-emitting object. Any substance whose temperature is above absolute zero ($-273°C$) emits electromagnetic radiation. The total energy emitted for each unit area of surface increases with increased temperature. Thus one square meter of the sun's surface at the extremely high temperature of several thousand degrees will emit a vastly greater quantity of energy each second than a square meter of the surface of the distant planet Pluto, which has a temperature close to absolute zero. Moreover, the radiation spectrum of emission of a cold surface differs greatly from that of a hot surface. For the cold surface, most of the energy emitted is in the long wavelengths; for the hot surface, most is concentrated in the short (visible and ultraviolet) wavelengths. Comparing the radiation properties of the sun with those of the earth shows how this principle applies.

Figure 3.4 is a graph comparing the intensity of emission of energy of the sun and earth. A logarithmic scale of wavelengths is used, as in Figure 3.3. The vertical scale, which is also logarithmic, shows intensity of energy emission (W/m^2) within each micron width of spectrum. Ideal energy curves are shown in Figure 3.4 for both sun (left) and earth (right) by dashed blue, smoothly arched lines. These represent the curves for ideal bodies that are perfect radiators of energy; they are called black bodies by physicists.

A *black body* will not only absorb all radiation falling on it; it will also radiate energy in a manner solely dependent on its surface temperature. The sun closely resembles a black body with a temperature of about 6000°K. (K stands for *absolute temperature* in *degrees Kelvin*. The degrees are of the same unit value as in the Celsius scale, but the zero point of the Kelvin scale is at $-273°C$, or absolute zero.) This ideal body has a peak intensity of energy output in the visible light region. The actual radiation intensity curve of the sun is shown by the solid blue line; it has minor irregularities and reduced intensity in the ultraviolet region.

Energy within the sun's spectrum is distributed about as follows: ultraviolet and shorter wavelengths, 9 percent; visible light, 41 percent; infrared, 50 percent.

For the earth, the ideal black-body radiation curve is for a planet with an average surface temperature of 300°K (27°C; 80°F). This curve lies entirely within the infrared region and peaks at about 10 microns. Notice that the peak is much lower than for the sun (about one-fifth as intense). The area under the earth curve is much smaller, meaning that the total energy emitted by a square centimeter of area is very much less than from a corresponding area of the sun's surface. This relationship is stated in strict terms by the *Stefan-Boltzmann law:* The total energy radiated by each unit of surface per unit of time varies as the fourth power of the absolute temperature (°K).

Referring again to Figure 3.4, we find beneath the arched black-body curves of both sun and earth deeply notched curves showing the intensity of energy measured

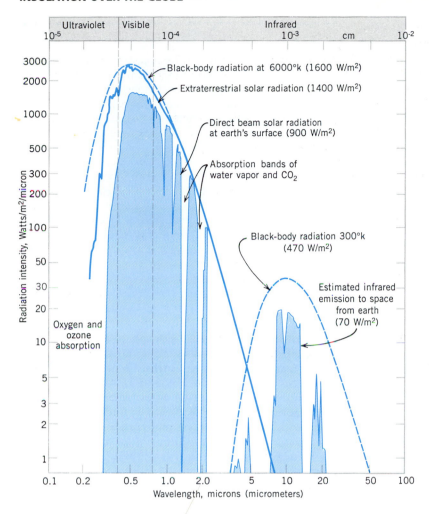

FIGURE 3.4 Spectra of solar and earth radiation. (Reprinted from *Physical Climatology*, p. 20, Figure 6, by W. D. Sellers by permission of the University of Chicago Press. © 1965 by The University of Chicago.)

after passage through the earth's atmosphere. The left-hand curve shows the solar spectrum as depleted by passage of the solar beam through the earth's atmosphere. Practically all ultraviolet rays have been removed, whereas certain bands of the infrared region have been cut off through absorption by atmospheric water vapor and carbon dioxide. The right-hand curve shows the earth's outgoing infrared radiation at the outer limits of the atmosphere. Much of the outgoing infrared energy has been absorbed by water vapor and carbon dioxide in the atmosphere. But in certain wavelength bands, known as *windows*, energy escapes to outer space. The major window, in terms of total energy escape, is between 8 and 14 microns. There are lesser windows in the range of 4 to 6 microns and 17 to 21 microns.

In atmospheric science, the entire solar spectrum is referred to as *shortwave radiation* because the peak of intensity lies in the visible region. In contrast, the earth's outgoing radiation spectrum is referred to as *longwave radiation*. We shall use these terms in describing the earth's energy budget.

INSOLATION OVER THE GLOBE

Referring back to Chapter 1 and Figure 1.15, showing equinox conditions, recall that at only the subsolar point does the earth's spherical surface present itself at right angles to the sun's rays. In all directions away from the subsolar point, the earth's curved surface becomes turned at a decreasing angle with respect to the rays until the circle of illumination is reached. Along that circle the rays are parallel with the surface.

Let us now assume that the earth is a perfectly uniform sphere with no atmosphere. Only at the subsolar point will solar energy be intercepted at the full value of the solar constant, 1400 W/m². We will now use the term *insolation* to mean the interception of solar shortwave energy by an exposed surface. At any particular place on the earth, insolation received in one day will depend on two factors: (1) the angle at which the sun's rays strike the earth, and (2) the length of time of exposure to the rays. These factors are varied by latitude and by the seasonal changes in the path of the sun in the sky.

Figure 3.5 shows that intensity of insolation is greatest where the sun's rays strike vertically, as they do at noon at some parallel between the tropic of cancer and the tropic of capricorn. With diminishing angle, the same amount of solar energy spreads over a greater area of ground surface. So, on the average, the polar regions receive the least insolation over a year's time.

Consider, now, a hypothetical situation in which the earth's axis is perpendicular to the plane of the ecliptic as the earth revolves around the sun. In other words, equinox conditions would prevail year-round. Under such conditions, the poles would not receive any inso-

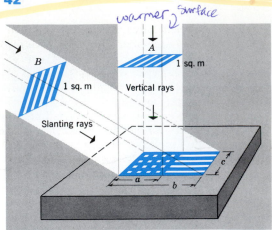

FIGURE 3.5 The angle of the sun's rays determines the intensity of the insolation on the ground. The energy of vertical rays *A* is concentrated in square *a*, but the same energy in the slanting rays *B* is spread over a rectangle, *b*.

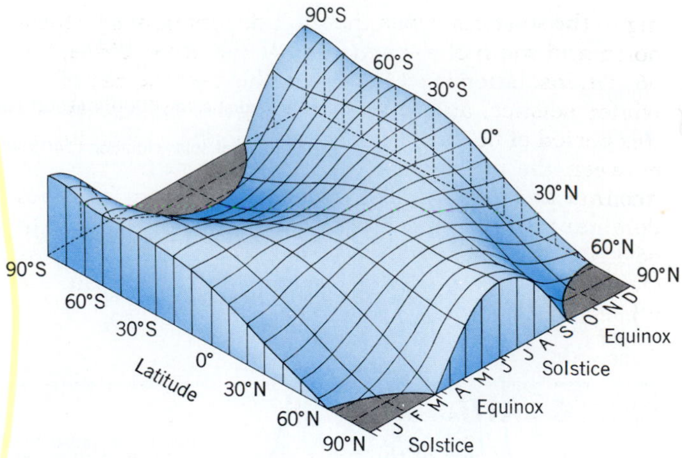

FIGURE 3.7 Variation of insolation with latitude and season. (After W. M. Davis.)

lation, regardless of time of year, whereas the equator would receive an unvarying maximum. But we know that the earth's axis is not perpendicular to the plane of the orbit.

As we found in Chapter 1, because of the tilt of the earth's axis, the subsolar point ranges yearly over a total latitude span of 47°, from the tropic of cancer (lat. 23½°N) at June solstice, to the tropic of capricorn (lat. 23½°S) at December solstice. This cycle does not make the yearly total insolation for the entire globe different from an imaginary situation in which the earth's axis is not inclined; but it does cause a great difference in the quantities received at various latitudes.

The intensity of insolation from equator to poles in W/m² is shown in Figure 3.6 by a solid line. A dashed line shows the insolation that would result if the earth's axis had no tilt. Notice how much insolation the polar regions actually receive—over 40 percent of the equatorial value. The added insolation at high latitudes is of major environmental importance because it brings heat in summer to allow forests to flourish and crops to be

grown over a much larger portion of the earth than would otherwise be the case.

A second effect of the axial tilt is to produce seasonal differences in insolation at any given latitude. These differences increase toward the poles, where the ultimate in opposites (6 months of day, 6 of night) is reached. Along with the variation in angle of the sun's rays another factor operates—the duration of daylight. At the season when the sun's path is highest in the sky, the length of time it is above the horizon is correspondingly greater. (This effect is clearly shown in Figure 1.18.) The two factors thus work hand in hand to intensify the contrast between amounts of insolation at opposite solstices.

A three-dimensional diagram (Figure 3.7) shows how insolation varies with latitude and season of year. Figure 3.8 shows graphs of insolation at selected latitudes from the equator to the north pole. (These diagrams show insolation at the outer limits of the atmosphere.) Notice that the equator receives two maximum periods (corresponding with the equinoxes, when the sun is overhead at the equator) and two minimum periods (correspond-

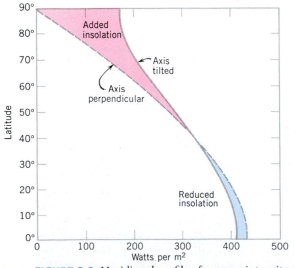

FIGURE 3.6 Meridional profile of average intensity of insolation (solid line).

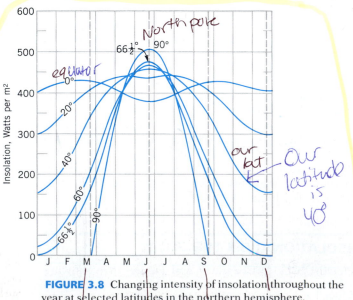

FIGURE 3.8 Changing intensity of insolation throughout the year at selected latitudes in the northern hemisphere.

ing to the solstices, when the sun's declination is farthest north and south of the equator). At the arctic circle, lat. 66½°N, insolation is reduced to nothing on the day of the winter solstice, and with increasing latitude poleward this period of no insolation becomes longer. All latitudes between the tropics of cancer and capricorn have two maxima and two minima, but one maximum becomes dominant as the tropic is approached. From lat. 23½° to 66½°N and S, there is a single continuous insolation cycle with maximum at one solstice, minimum at the other.

WORLD LATITUDE ZONES

The angle of attack of the sun's rays determines the flow of solar energy reaching a given unit of the earth's surface and so governs the thermal environment of the life layer. This concept provides a basis for dividing the globe into latitude zones (Figure 3.9). The specified zone limits should not be taken as absolute and binding. Instead, the system should be considered as a convenient terminology for identifying world geographic belts throughout the book.

The *equatorial zone* lies astride the equator and covers the latitude belt roughly 10°N to 10°S. Within this zone, the sun throughout the year provides intense insolation, while day and night are of roughly equal duration. Astride the tropics of cancer and capricorn are the *tropical zones*, spanning the latitude belts 10° to 25°N and S. In these zones, the sun takes a path close to the zenith at one solstice and is appreciably lower at the opposite solstice. Thus a marked seasonal cycle exists, but it is combined with a large total annual insolation.

Literary usage and that of many scientific works dif-

fer from what is described here; the terms "tropical zone" and "tropics" have been widely used to denote the entire belt of 47° of latitude between the tropics of cancer and capricorn. That is the definition of "tropics" you will find in most dictionaries. Even though correct, that definition is not well suited to a study of the physical environment, because it combines belts of very dissimilar climatic properties; it lumps the greatest of the world's deserts with the wettest of climates.

Immediately poleward of the tropical zones are transitional regions that have become widely accepted in geographers' usage as the *subtropical zones*. For convenience, we have assigned these zones the latitude belts 25° to 35°N and S, but it is understood that the adjective "subtropical" may be extended a few degrees farther poleward or equatorward of these parallels.

The *midlatitude zones*, lying between lat. 35° and 55°N and S, are belts in which the sun's angle of attack shifts through a relatively large range, so that seasonal contrasts in insolation are strong. Strong seasonal differences in lengths of day and night exist as compared with the tropical zones.

Bordering the midlatitude zones on the poleward side are the *subarctic* and *subantarctic* zones, lat. 55° to 60°N and S. Astride the arctic and antarctic circles, lat. 66½°N and S, lie the *arctic zone* and the *antarctic zone*. These zones have an extremely large yearly variation in lengths of day and night, yielding enormous contrasts in insolation from solstice to solstice. Notice that dictionaries define "arctic" as the entire area from arctic circle to north pole, and "antarctic" in a corresponding sense for the southern hemisphere.

The *polar zones*, north and south, are circular areas between about lat. 75° and the poles. Here the polar regime of a 6-month day and 6-month night is predominant. These zones experience the ultimate in seasonal contrasts of insolation.

INSOLATION LOSSES IN THE ATMOSPHERE

As the sun's radiation penetrates the earth's atmosphere, its energy is absorbed or diverted in various ways. At an altitude of 150 km (95 mi), the radiation spectrum possesses almost 100 percent of its original energy but, by the time rays have penetrated to an altitude of 88 km (55 mi), absorption of X rays is almost complete and some of the ultraviolet radiation has been absorbed as well.

As solar rays penetrate more deeply into the atmosphere, they reach the stratosphere. Here, as we explained in Chapter 2, ozone absorbs ultraviolet radiation within the ozone layer. The environmental shielding provided by this layer has already been stressed.

As radiation penetrates into deeper and denser atmospheric layers, gas molecules cause the visible light rays to be turned aside in all possible directions, a process known as *scattering*. Dust and cloud particles in the troposphere cause further scattering, described as *diffuse reflection*. Scattering and diffuse reflection send some energy into outer space and some down to the earth's surface.

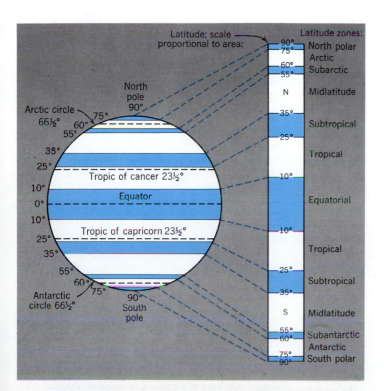

FIGURE 3.9 A geographic system of latitude zones.

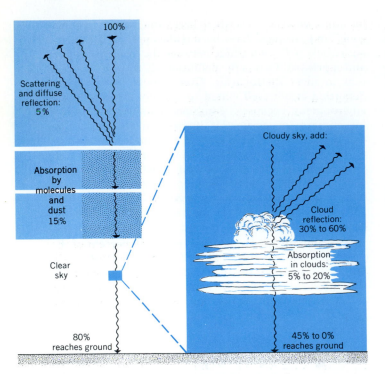

FIGURE 3.10 Losses of incoming solar energy by scattering, reflection, and absorption. (From A. N. Strahler, *The Earth Sciences*, 2nd ed., Harper & Row, Publishers. Copyright © by Arthur N. Strahler.)

As a result of all forms of shortwave scattering, about 5 percent of the total insolation is returned to space and forever lost, as shown in Figure 3.10. Scattered shortwave energy directed earthward is referred to as *downscatter*.

Another form of energy loss, *absorption*, takes place as the sun's rays penetrate the atmosphere. Both carbon dioxide and water vapor are capable of directly absorbing infrared radiation. Absorption results in a rise of temperature of the air. In this way some direct heating of the lower atmosphere takes place during incoming solar radiation. Although carbon dioxide is a nearly constant quantity in the air, the water vapor content varies greatly from place to place. Absorption correspondingly varies from one global environment to another.

All forms of direct energy absorption—by air molecules, including molecules of water vapor and carbon dioxide, and by dust—are estimated to total as little as 10 percent for conditions of clear, dry air to as high as 30 percent when a cloud cover exists. A global average of roughly 15 percent for absorption generally is shown in Figure 3.10. When skies are clear, reflection and absorption combined may total about 20 percent, leaving as much as 80 percent to reach the ground.

Yet another form of energy loss must be brought into the picture. The upper surfaces of clouds are extremely good reflectors of shortwave radiation. Air travelers are well aware of how painfully brilliant the sunlit upper surface of a cloud deck can be when seen from above. *Cloud reflection* can account for a direct turning back into space of from 30 to 60 percent of total incoming radiation (Figure 3.10). Clouds also absorb radiation. So we see that, under conditions of a heavy cloud layer, the com-

bined reflection and absorption from clouds alone can account for a loss of 35 to 80 percent of the incoming radiation and allow from 45 to 0 percent to reach the ground. A world average value for reflection from clouds to space is about 21 percent of the total insolation. Average absorption by clouds is much less—about 3 percent.

The surfaces of the land and ocean reflect some shortwave radiation directly back into the atmosphere. This small quantity, about 4 percent as a world average, may be combined with cloud reflection in evaluating total reflection losses. Table 3.1 lists the percentages given so far for the energy losses by reflection and absorption. Altogether the losses to space by reflection total 31 percent of the total insolation. Figure 3.11 (upper half) shows the same data in the form of arrows, whose widths are proportionate to the percentages. These percentages are estimates only; somewhat different sets of percentages have also been published.

The percentage of radiant energy reflected back by a surface is the *albedo*. Orbiting satellites, suitably equipped with instruments to measure the energy levels of shortwave and infrared radiation, both incoming from

TABLE 3.1 The Global Radiation Budget

Incoming Solar Radiation	Percentage	
Total at top of atmosphere		100
Diffuse reflection to space by scattering	6	
Reflection from clouds to space	21	
Direct reflection from earth's surface	4	
Total reflection loss to space by earth-atmosphere system (earth's albedo)		31
Absorbed by molecules, dust, water vapor, CO_2, clouds	21	
Absorbed by earth's surface	48	
Total absorbed by earth-atmosphere system		69
Sum of absorption and reflection		100

Outgoing Longwave Radiation	Percentage	
Radiation balance of earth's surface		
Loss directly to space	6	
Loss to atmosphere	107	
Total radiation from surface	113	
Gain by counterradiation from atmosphere	97	
Net outgoing radiation from surface	16	
Radiation balance of atmosphere		
Gain from surface radiation	107	
Loss by counterradiation	97	
Net gain from surface	10	
Gain from direct shortwave absorption	21	
Gain from latent heat transfer	22	
Gain from mechanical heat transfer	10	
Total net gain	63	
Radiated to space from atmosphere		63
Radiated directly to space from surface		6
Total radiation from planet to space		69

Data source: World Meteorological Organization, Frolich and London, 1986, as presented by R. G. Barry and R. J. Chorley, *Atmosphere, Weather, and Climate*, 5th ed., 1987.

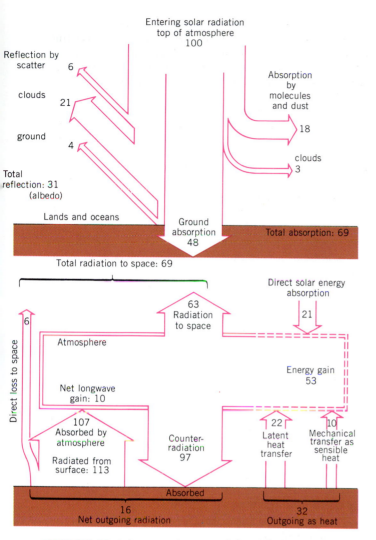

FIGURE 3.11 Schematic diagram of the global radiation balance. Figures correspond with those in Table 3.1.

the sun and outgoing from the atmosphere and earth's surface below, have provided data for estimating the earth's average albedo. Values between 29 and 34 percent have been obtained. The value of 31 percent given in Figure 3.11 lies between these limits.

The earth's albedo lies in a range intermediate between low values of the moon and inner planets (Mercury, 6%; Mars, 16%; moon, 7%) and high values of Venus (76%) and the great outer planets (73 to 94%). Here, again, we find another unique element of the environment of Planet Earth as compared with that of the other planets and the moon.

Figure 3.11 summarizes energy losses by absorption. A global figure of 21 percent is given for the combined losses through absorption by molecules, dust, water vapor, carbon dioxide, and clouds. About 48 percent of the original solar energy remains, and this amount is absorbed by the earth's land and water surfaces. We have now accounted for all the energy of insolation. Our next step is to deal with outgoing energy so as to balance the global energy budget.

LONGWAVE RADIATION

The earth's ground and ocean surfaces possess heat derived from absorption of insolation. This form of energy is referred to as sensible heat (heat that can be measured by a thermometer); it represents energy in temporary storage. This stored heat is being continually lost by longwave radiation into the overlying atmosphere, a process we can call *ground radiation*. (As used here, "ground" includes water surfaces.)

Because longwave radiation goes on throughout the night, it is possible to obtain *infrared imagery* of ground features, using special cameras mounted on aircraft. Figure 3.12 is an image obtained in the early morning hours

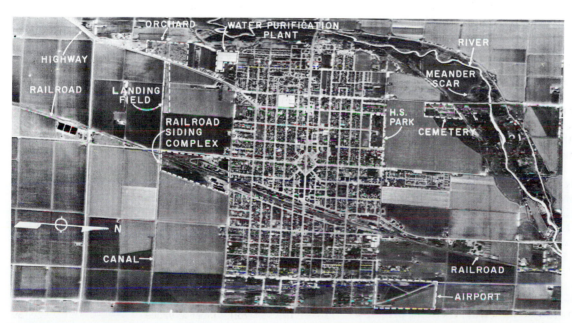

FIGURE 3.12 Infrared imagery of Brawley, California, a community in the Imperial Valley farming region of southernmost California. (Courtesy of Environmental Analysis Department HRB—Singer, Inc.)

in total darkness. Pavements and river appear bright because they are warm and radiate more intensely. In contrast, moist soil surfaces of fields are cooler and appear dark.

Infrared imagery is one of the kinds of information used in *remote sensing*, a modern research technology of major importance in physical geography. Remote sensing makes use of electromagnetic radiation in various bands over a wide range of the total spectrum (see Appendix II for details).

THE GLOBAL RADIATION BALANCE

The lower halves of Table 3.1 and Figure 3.11 show the components of outgoing (longwave) radiation for the earth's surface, the atmosphere, and the planet as a whole, using the same percentage units as for the incoming radiation. The total longwave radiation leaving the earth's land and ocean surface is equivalent to 113 percentage units. Of this, 6 units are lost to space, while 107 units are absorbed by the atmosphere. In turn, the atmosphere emits longwave radiation. For the box representing the atmosphere, the total of this emitted radiation is equivalent to 160 percentage units of the insolation (sum of 63 and 97). This figure may seem absurd until we note that this radiation is divided into two parts, one of which goes out into space (63 units) and the other of which is absorbed by the earth's surface as counterradiation (97 units).

The atmosphere also radiates longwave energy both downward to the ground and outward into space, where it is lost. Be sure to understand that longwave radiation is quite different from reflection in which the rays are turned back directly without being absorbed. Longwave radiation from both ground and atmosphere continues during the night, when no solar radiation is being received.

As Figure 3.4 shows, the intensity of longwave energy leaving our planet is only a small fraction of the intensity of incoming shortwave solar energy. Keep in mind that longwave radiation is constantly emitted from the entire spherical surface of the earth, whereas insolation falls on only one hemisphere. A single hemisphere presents the equivalent of only a cross section of the sphere to full insolation. Also, do not forget that about 31 percent of the incoming solar radiation is reflected directly back into space. Only the remaining 69 percent that is absorbed must be disposed of by longwave radiation. On the average, year in and year out, for the planet as a whole, the quantity of absorbed solar energy is balanced by an equal quantity of longwave emission to space.

Part of the ground radiation absorbed by the atmosphere is radiated back toward the earth's surface, a process called *counterradiation*. For this reason the lower atmosphere, with its water vapor and carbon dioxide, acts as a blanket that returns heat to the earth. This mechanism helps to keep surface temperatures from dropping excessively during the night or in winter at middle and high latitudes. Somewhat the same principle is used in greenhouses and in homes using direct solar heating; the glass windows permit entry of shortwave energy. Accumulated heat cannot escape by mixing with

cooler air outside. Meteorologists use the expression *greenhouse effect* to describe this atmospheric heating principle. Cloud layers are even more important in producing a blanketing effect to retain heat in the lower atmosphere because they are excellent absorbers and emitters of longwave radiation.

We must now introduce two new terms into the energy budget. Much of the heat energy passing from the ground to the atmosphere is carried upward by two mechanisms other than longwave radiation. First, sensible heat is conducted upward by turbulent motions of the air. As Figure 3.11 shows (lower right), this heat transport amounts to about 10 percentage units. Second, heat in latent form is conducted upward in water vapor that has evaporated from land and ocean surfaces. Latent heat transport accounts for about 22 percentage units. (Latent heat is explained in Chapter 6.)

Summarizing the information of the previous paragraph, energy leaving the earth's surface (both land and water surfaces) is divided up as follows:

	Percentage Units
Longwave radiation (net value)	16
Mechanical transport as sensible heat	10
Transport as latent heat	22
Total	48

Recall that 48 percent is the amount of insolation (shortwave energy) absorbed by the earth's surface. We have now balanced the annual energy budget so far as the surface is concerned.

The atmosphere in turn disposes of its heat by longwave radiation into outer space. Together with direct longwave emission from the ground, the total emission to space is 69 percent (Table 3.1). Add this figure to 31 percent loss by reflection (the earth's albedo), and the total is 100 percent.

THE GLOBAL RADIATION BALANCE AS AN OPEN ENERGY SYSTEM

The global radiation balance, shown in Figure 3.11 and Table 3.1, can be visualized as an open energy system. Figure 3.13 is a schematic diagram of this system, using symbols explained in Chapter 2. The outer limit of the atmosphere forms the system boundary. Part of the entering shortwave radiation is reflected directly to outer space by both atmosphere and ground. The portion that is absorbed directly by the atmosphere is converted into sensible heat and goes into atmospheric storage. Transformed into longwave energy, some of this energy leaves the system as direct output to space, while some is sent to the ground by counterradiation where it is absorbed, transformed into sensible heat, and stored.

Shortwave energy absorbed by the ground is transformed into sensible heat and stored. The ground gives up energy in three ways. (1) Energy is radiated to the overlying atmosphere as longwave radiation. (2) Energy is carried upward into the atmosphere as latent heat in water vapor. This process involves transformation from sensible to latent heat through evaporation, storage as

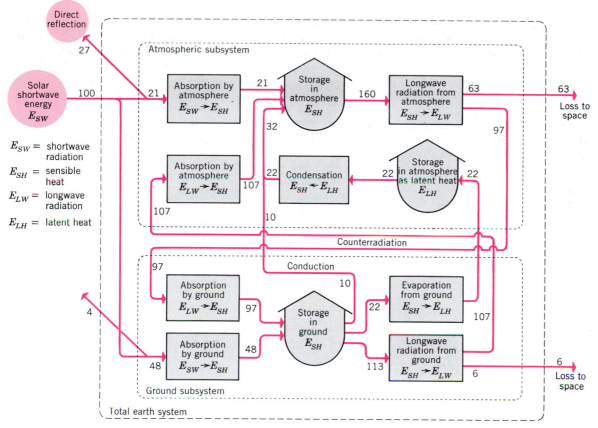

FIGURE 3.13 Schematic diagram of the global radiation balance as an open energy system.

latent heat, and reconversion to sensible heat during condensation of water vapor when clouds are formed. (3) Energy is transported upward as sensible heat by conduction and a mechanical process of mixing, involving no change of state. A small amount of energy exits the system directly from the ground surface as longwave radiation to space.

Note that there are two *energy subsystems,* shown by smaller rectangular boxes—atmospheric subsystem and ground subsystem. Each subsystem has its own input and output, but the two subsystems exchange energy by flow in both directions.

Although the flow paths and flow quantities are much the same in this diagram as in Figure 3.11, here we have added the necessary energy transformations and energy storages, without which the system concept would not be complete.

LATITUDE AND THE RADIATION BALANCE

Earlier in this chapter, in relating latitude to insolation, we showed that the tilt of the earth's axis causes a poleward redistribution of insolation, as compared with conditions that would apply if the axis were perpendicular to the orbital plane (Figure 3.6). Let us now look deeper into the wide range in rates of incoming and outgoing energy from low latitudes to the polar regions. We shall deal first with three components in the radiation balance: albedo (reflection), insolation, and outgoing longwave radiation.

Albedo

The percentage of shortwave radiation reflected by land and water surfaces is an important property because it determines the proportion of insolation that is absorbed at the surface and converted into sensible heat. Albedo

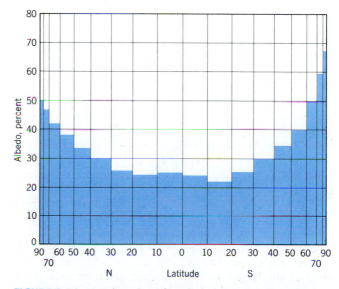

FIGURE 3.14 Meridional profile of earth's mean albedo based on satellite data. Latitude is scaled proportionally to earth's surface areas between successive 10-degree parallels of latitude. (Data from T. H. Vonder Haar and V. E. Suomi, *Science,* vol. 163, p. 667, Figure 1.)

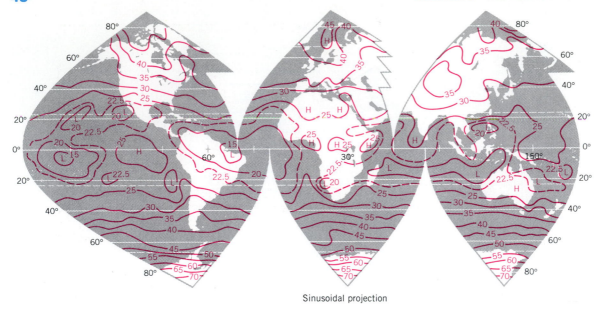

Sinusoidal projection

FIGURE 3.15 Mean planetary albedo compiled from satellite data. Values are in percent. Letters H and L designate centers of higher and lower values, respectively. (Data from T. H. Vonder Haar and V. E. Suomi, *Science*, vol. 163, p. 668, Figure 2*a*.)

of a water surface is very low (2%) for nearly vertical rays but high for low-angle rays. It is also extremely high for snow-covered land surfaces and ice-covered water surfaces (45 to 90%). Albedos of fields, forests, and bare ground are of intermediate values, ranging from as low as 5 percent to as high as 30 percent.

For the earth as a whole planet, albedo measured from an orbiting satellite depends on both surface reflection and cloud reflection because, as we noted before, clouds are excellent reflectors of shortwave radiation. Figure 3.14 is a meridional profile of the earth's mean albedo, averaged by latitude belts 10° wide; Figure 3.15, a world map of the planetary albedo. Notice that the

equatorial zone is a belt of low albedo, mostly in the 15 to 25 percent range. Low albedo here is explained by extensive ocean surfaces and forests and by the lack of persistent blanketlike cloud layers. Across midlatitudes, albedo increases steadily poleward. Maximum values are over the snow-covered surface of Antarctica.

Insolation

Shortwave radiation received at the earth's surface is shown in Figure 3.16. The average annual total is high in the equatorial belt, with values ranging from 160 to 200 W/m². Flanking this belt are tropical zones of excep-

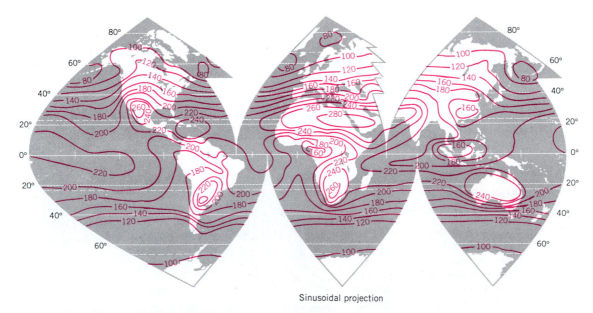

Sinusoidal projection

FIGURE 3.16 Highly generalized map of intensity of solar radiation received at the earth's surface. Units are watts per square meter. (Based on data from M. I. Budyko, *Atlas Teplovogo Balansa*, Moscow, USSR, Gidrometeorologicheskoe Izdatel'stvo.)

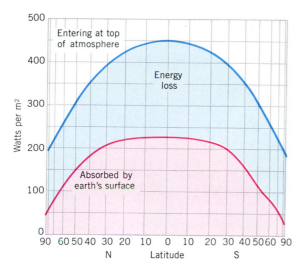

FIGURE 3.17 Meridional profiles of intensity of entering solar radiation at top of atmosphere and radiation absorbed by earth's surface. (Based on data of H. G. Houghton, M. I. Budyko, K. J. Hanson, and others in W. D. Sellers, *Physical Climatology*. From A. N. Strahler, *The Earth Sciences*, 2nd ed., Harper & Row, Publishers. Copyright © by Arthur N. Strahler.)

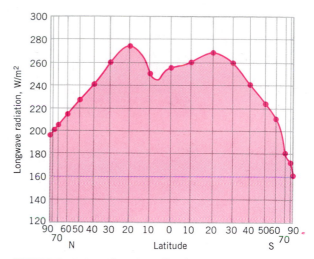

FIGURE 3.19 Meridional profile of mean intensity of longwave radiation from the earth, as measured by NOAA satellites. (NOAA data of A. Gruber and A. F. Krueger, 1984, *Bulletin of the American Meteorological Society*, vol. 65, no. 9.)

tionally large values over the continents—240 to 280 W/m²—representing the great tropical deserts, over which cloud cover is absent much of the time. Values decline rapidly poleward through midlatitudes and are low in the arctic and polar regions.

Figure 3.17, lower curve, is a meridional profile showing shortwave energy absorbed by the earth's surface. The data differ from Figure 3.16 in that shortwave energy reflected by the surface has been subtracted. The upper curve shows insolation at the top of the atmosphere; it is essentially the same information as in Figure 3.6. The zone between the upper and lower curves represents en-

ergy losses through absorption and reflection processes as the solar rays penetrate the atmosphere.

Longwave Radiation

Outgoing energy in the form of longwave radiation from the entire earth-atmosphere system is shown on the world map, Figure 3.18, based on data obtained by orbiting earth satellites. Radiation intensity is in units of watts per square meter (W/m²). The global radiation data are summarized in the meridional profile of Figure 3.19. Notice the maxima occurring in the tropical zones where the great deserts are situated. Here ground surface tem-

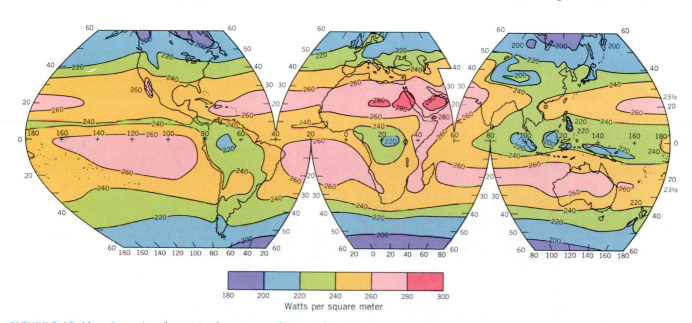

FIGURE 3.18 Mean intensity of outgoing longwave radiation. The units are watts per square meter. (NOAA data of A. Gruber and A. F. Krueger, 1984, *Bulletin of the American Meteorological Society*, vol. 65, no. 9.)

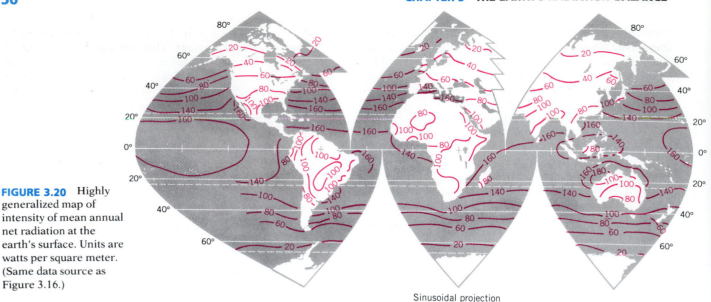

FIGURE 3.20 Highly generalized map of intensity of mean annual net radiation at the earth's surface. Units are watts per square meter. (Same data source as Figure 3.16.)

Sinusoidal projection

peratures are high and cloud cover is small, favoring intense outgoing longwave radiation. Over the equatorial belt, where cloud cover on the average is more frequent, radiation is somewhat reduced. Values decline poleward through the middle latitude zones to reach minimum quantities over the polar zones.

NET RADIATION

We now examine *net radiation*, which is the difference between all incoming energy and all outgoing energy carried by both shortwave and longwave radiation. Our analysis of the radiation balance has already shown that

for the entire globe as a unit, the net radiation is zero on an annual basis. In some places, however, energy is coming in faster than it is going out, and the energy balance is a positive quantity, or *energy surplus*. In other places energy is going out faster than it is coming in, and the energy balance is a negative quantity, or *energy deficit*.

Consider first the global distribution of net radiation at the earth's surface shown in Figure 3.20. In Figure 3.21 the uppermost meridional profile averages the same global data by latitude belts. There is a large surplus in low latitudes, with values well over 125 W/m^2 generally for the region between lat. 20°N and 20°S. Net radiation declines rapidly through midlatitudes, reaching zero in the vicinity of 70° in both hemispheres. (High-latitude data are missing from the world map, but show on the profile.) Negative values occur in two small polar zones, showing a radiation deficit.

For the atmosphere, net radiation shows a different global picture than that for the surface. The meridional profile of net radiation for the atmosphere is shown near the bottom of Figure 3.21. Notice that all values are negative and the deficit always exceeds −75 W/m^2. Next, we combine the data of the earth's surface with those of the

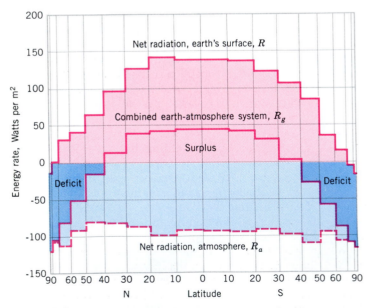

FIGURE 3.21 Meridional profiles of intensity of mean net radiation at earth's surface, from atmosphere, and from combined earth-atmosphere system. (Based on data of H. G. Houghton and M. I. Budyko in W. D. Sellers, *Physical Climatology.* From A. N. Strahler, *The Earth Sciences*, 2nd ed., Harper & Row, Publishers. Copyright © by Arthur N. Strahler.)

TABLE 3.2 Annual Meridional Heat Transport

Latitude (°N)	Heat Transport J/yr × 10²²
90	0.0
80	1.5
70	5.2
60	10.0
50	14.2
40	16.4
30	14.9
20	10.6
10	5.1
0	1.1

Data source: W. D. Sellers (1965), *Physical Climatology*, Univ. of Chicago Press, Chicago and London, Table 12.

atmosphere. In each latitude zone, the deficit quantity (negative value) is subtracted from the surplus quantity (positive value). This calculation produces net radiation values for the combined earth surface-atmosphere system, shown in the middle portion of Figure 3.21. There is a large region of surplus radiation from about lat. 40°N to 30°S and two high-latitude regions of deficit. On the diagram, the areas labeled "deficit" are together equal to the area labeled "surplus," as the radiation balance requires.

It is obvious from Figure 3.21 that the earth's energy balance can be maintained only if heat is transported from the low-latitude belt of surplus to the two high-latitude regions of deficit. This poleward movement of heat is described as *meridional transport*, moving north or south along the meridians of longitude. We should expect the rate of meridional heat transport to be greatest in midlatitudes, and this fact is shown by the figures in Table 3.2.

The meridional flow of heat is carried out by the circulation of the atmosphere and oceans. In the atmosphere, heat is transported both as sensible heat and as latent heat. These topics are explained in Chapters 5 and 6.

SOLAR ENERGY

Planet Earth intercepts solar energy at the rate of 1½ quadrillion megawatt-hours per year. This quantity of energy amounts to about 28,000 times as much as all the energy presently being consumed each year by humans. Thus, we realize that there is an enormous source of energy near at hand waiting to be used. That use would simply mean passing some of the flow of solar energy through human-made subsystems within the natural global energy system. Another remarkable virtue of solar energy is that its use by humans cannot increase the heat load on the atmosphere and hydrosphere. A major worry we express in Chapter 4 is that the combustion of hydrocarbon fuels will raise the global average temperature, both by emitting large amounts of heat and by raising the level of the carbon dioxide content of the atmosphere. Neither of these concerns applies to solar energy. Adding to these advantages the lack of environmental pollution—no emissions of sulfur dioxide or carbon particles—solar energy becomes a particularly attractive energy source.

Solar radiation provides useful energy in a variety of forms, both direct and indirect. Our concern in this chapter is with direct interception and conversion of solar energy in one of two ways: (1) direct absorption by a receiving surface, converting shortwave energy to sensible heat and raising the temperature of the receiving medium; (2) direct conversion of shortwave energy to electrical energy by the use of solar cells. Indirect, or secondary, sources of solar energy (described in later chapters) make use of converted solar energy stored in different forms. For example, air in motion as wind and water in motion as ocean waves or flowing rivers are forms of kinetic energy of matter in motion within flow systems powered by solar energy. Quite different in character is the solar energy that has been converted into chemical energy by plants and stored in plant tissues. These organic energy systems are described in Chapter 25. The fossil fuels—crude oil and coal—are derived from hydrocarbon compounds of organic origin that can be traced back to solar energy captured and stored in the distant geologic past.

Oldest and simplest of the forms of solar energy conversion is the direct interception of the warming rays of the sun by some kind of receiving surface or medium. Applications of this principle range from simple heating of buildings and domestic hot-water supplies, to the intense heating of boilers and furnaces by focusing the solar rays on a small target.

A great saving of expenditure of fossil fuels can be achieved through solar heating of buildings. Each home, school, or office building can have its own solar collecting system so that expensive energy transport, whether by pipeline, truck, or power line can be eliminated. In most cases, the goal of this application is to supplement, rather than to replace, the use of fossil fuels.

The simplest form of solar heating is the use of large glass panes to admit sunlight into a room—the greenhouse principle. Solar rays are admitted during the winter, when the sun's path is low in the sky, but excluded in summer by a suitable roof overhang. Of course, the same large glass panes will result in high heat losses at night and on cold, cloudy days by outgoing longwave radiation, unless thermal drapes or shutters are also used.

Practical solar heating of interior building space and hot-water systems makes use of *solar collectors*. The flat-plate collector consists of a network of metal tubes carrying circulating water. Aluminum or copper tubes, painted black, are efficient absorbers of solar energy. Water is pumped through the tubes to heat a large body of water in a storage tank. A cover of glass or clear plastic is used to obtain higher temperatures and to reduce heat loss to the atmosphere. Panels of solar collectors are usually placed on the roof of the building. If a house is designed around the solar heating system, the roof may be oriented and pitched at the optimum angle to intercept solar rays. Water can be heated to a temperature of 65°C (150°F) and can be used to transfer heat to a hot-water supply for laundering and bathing, as well as to a conventional space-heating system.

Where more intense heating of water or other conducting fluids is required, parabolic reflectors can be used. The reflector focuses the solar rays on a pipe, in which temperatures up to 550°C (950°F) can be attained. This temperature is sufficient to supply steam to a turbine that drives an electric generator. The waste heat can be used for space heating and hot-water heating. Parabolic reflector systems can be effectively used for industrial plants and shopping centers which not only need large amounts of energy during the day, but also

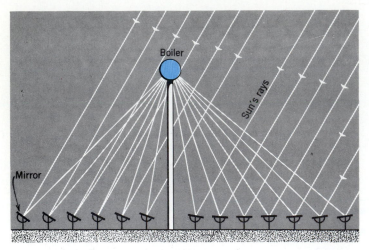

FIGURE 3.22 Design for a solar power plant using an array of movable mirrors (heliostats) to concentrate the sun's rays on a boiler. Each heliostat is controlled by a computer program so as to focus its rays on the tower as the sun travels its daily path across the sky.

have large, low buildings with flat roofs on which the reflectors can be installed.

Solar energy power plants have been designed to use reflecting mirror systems. A large number of movable mirrors, called heliostats, reflect solar rays to the top of a central tower where a boiler and electric generator are located (Figure 3.22). The extremely high temperatures and pressures produced in this way allow a number of kinds of gases and fluids to be used in the boiler. Hydrogen gas, which can be generated by this process, provides an ideal medium in which to store energy for later conversion to electricity when the solar input is cut off.

An estimate has been made of the feasibility of supplying all energy needs of the United States in the year 2000 through a system of heliostat power plants. The power plants would be located in the southwestern United States, where the annual number of hours of sunshine is large and the average solar radiation is high (Figure 3.23). In this region, a system with a 30-percent efficiency for converting solar energy into electrical en-

ergy would require a total land area equivalent to the area of a single square plot 280 km (175 mi) on a side. The plot would represent 0.86 percent of the area of the United States. Although the ecosystem of the entire area would be destroyed, there would be no output of pollutants, and there would be no significant water consumption. Scientists who have planned this system and computed its costs are of the opinion that it will become economically competitive in the near future with systems that depend on fossil fuels and nuclear energy. A heliostat power plant constructed in 1981 near Barstow, California, delivered 10 megawatts of electric power for 4 to 8 hours per day, depending on season (Figure 3.24). Power plants using parabolic reflectors have been constructed in the same area.

Direct heat absorption from solar rays can have an important energy-saving application in systems for distilling seawater to produce freshwater supplies. This prospect is particularly appealing because the tropical deserts in which the greatest intensity of solar energy is available are also the regions where pure, fresh water for urban and agricultural uses is in shortest supply. A solar energy system placed in the southwestern desert of the United States, for example, could supply enough electrical energy to desalinate seawater of the Pacific Ocean in a quantity sufficient for the needs of a population of 120 million people.

We turn now to the technology of generating electricity directly from the impact of the sun's rays. The photovoltaic effect is known to amateur photographers through the light meter. As you move the glass window of the meter across a scene, a small hand wavers over a calibrated dial to tell the varying intensities of the incoming light. The sensing device in this meter is a photovoltaic cell; it transforms light energy into electricity. The hand on the meter is actually showing the amount of electric current generated by the cell. Certain crystalline substances produce the photovoltaic effect. The problem is to construct from these substances solar cells that are high in efficiency and low in cost. Cells of crystalline silicon are the most widely used today, with an efficiency as great as 15 to 20 percent. Silicon cells have been used effectively in our space vehicles, where cost is not a sig-

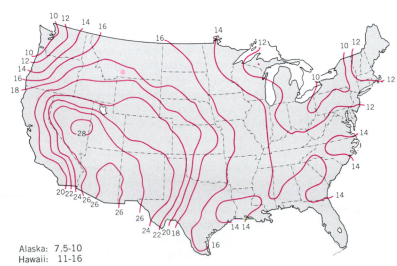

Alaska: 7.5–10
Hawaii: 11–16

FIGURE 3.23 Map of daily average direct solar radiation received at the ground in the United States. The unit used is 1000 joules per square meter. Data are from 235 National Weather Service stations over a 23-year period of observation. (Published by the Solar Energy Research Institute, Golden, Colorado, for the U.S. Department of Energy, 1980.)

FIGURE 3.24 Solar One, a pilot plant for generating electricity with assembled heliostats, began operation in 1982. Its 1818 heliostats present a total reflecting area of 70,000 sq m. (Courtesy of Southern California Edison Company.)

nificant factor. Solar cells can also be made using cadmium telluride or gallium arsenide with an efficiency as high as 30 percent. Because a single silicon cell generates only about one-half volt of electricity, a large number must be connected in an array to produce a high voltage output. Also, the current that is produced is direct current (DC) and requires transformation into alternating current (AC) for use in conventional power applications. A manufacturing process, in which silicon cells are produced from a continuous ribbon of molten silicon, has been developed and may eventually reduce costs greatly. The silicon used in solar cells must be of extremely high purity, a factor that makes mass production difficult.

As with direct heat-gathering solar panels, arrays of photovoltaic solar cells would occupy large areas of ground surface to produce important outputs of energy. Storage of the electricity is another problem, because the system does not work at night. Storage as hydrogen fuel is an attractive possibility for large-scale systems; batteries can be used to store power in small systems such as those of homes, farms, and ranches. One small-scale use is to turn electric water pumps needed for irrigation of fields in the sunny dry season of tropical countries, for example, in the Sahel region of Africa.

HUMAN'S IMPACT ON THE EARTH'S RADIATION BALANCE

Humans have profoundly altered the earth's land surfaces by deforestation, crop cultivation, and urbanization. By means of remote sensing imagery, the extent of these changes can be measured and evaluated on a scope never before possible. From our analysis of the earth's radiation balance, it should be obvious that the global radiation balance is a sensitive one, involving as it does a number of variable factors that determine how energy is transmitted and absorbed. We know that agriculture and urbanization have significantly changed the surface albedo and the capacity of the ground to absorb shortwave radiation and to emit longwave radiation. We know that combustion of fuels has already altered the content of the atmosphere through the release of carbon dioxide and dust particles. The nature of these changes and their possible impacts on urban climate and global climate are subjects we will investigate in the next chapter, which deals with heating and cooling of the earth's atmosphere, lands, and oceans.

HEAT AND COLD AT THE EARTH'S SURFACE

Temperature of the air and the soil layer is of major interest to the physical geographer. Temperature is a measure of heat energy available in the air and the soil. Organisms respond directly to heating and cooling of the environmental substance that surrounds them. We can refer to this influence as the *thermal environment.*

Our study of the radiation balance has shown that temperature changes result from the gain or loss of energy by the absorption or emission of radiant energy. When a substance absorbs radiant energy, the surface temperature of that substance is raised. This process represents a transformation of radiant energy into the energy of sensible heat, which is the physical property measured by the thermometer. Heat can also enter or leave a substance by conduction or can be lost as latent heat during evaporation.

Many of the biochemical processes taking place within organisms as well as many common inorganic chemical reactions are intensified by an increase in temperature of the solutions in which these reactions are occurring. Severe cold, which is simply the lack of heat energy within matter, may greatly reduce or even completely stop biochemical and inorganic reactions. This is why the vital environmental ingredient of heat—heat in the air, water, and soil—needs to be thoroughly understood. The thermal environment is surely a dominant part of our total physical environment.

We are all familiar with natural cycles of temperature change. There is a daily rhythm of rise and fall of air temperature as well as a seasonal rhythm. There are also systematic changes in average air temperatures from equatorial to polar latitudes and from oceanic to continental surfaces. These temperature changes require that the lower atmosphere and the surfaces of the lands and oceans receive and give up heat in daily and seasonal cycles. There must also be great differences in the quantities of heat received and given up annually in low latitudes as compared to high latitudes. Temperature cycles—daily and seasonal—and the influence of latitude on temperature will be dominant themes of this chapter.

MEASUREMENT OF AIR TEMPERATURES

Air temperature is one of the most familiar bits of weather information we hear and read about daily through news media. In the United States, information comes from National Weather Service observing stations and is taken following a carefully standardized procedure. Thermometers are mounted in an instrument shelter, shown in Figure 4.1.

The shelter shades the instruments from sunlight, but louvers allow air to circulate freely past the thermometers. The instruments are mounted from 1.2 to 1.8 m (4 to 6 ft) above ground level, at a height easy to read.

FIGURE 4.1 This instrument shelter houses thermometers and an instrument for measuring relative humidity. (Arthur N. Strahler.)

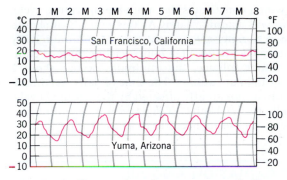

FIGURE 4.2 The recording thermometer made these continuous records of the rise and fall of air temperature over a period of one week.

DAILY CYCLES OF INSOLATION, NET RADIATION, AND AIR TEMPERATURE

Because the earth turns on its axis, insolation and net radiation undergo daily cycles of change. These cycles, in turn, produce the daily (diurnal) cycle of rising and falling air temperature with which we are all familiar. We now examine each of these three daily cycles to determine how they are linked together. The three graphs in Figure 4.4 show average curves of daily insolation, net radiation, and air temperature, greatly generalized for a typical observing station at lat. 40° to 45°N in the interior United States.

Graph A shows insolation; units are watts per square meter (W/m^2). At equinox, insolation begins about sunrise (6 A.M. by local time), rises to a peak value at noon, and declines to zero at sunset (6 P.M.). At June solstice, insolation begins about 2 hours earlier (4 A.M.) and ends 2 hours later (8 P.M.). The June peak is much greater than at equinox, and the total insolation for the day is also much greater. At December solstice, insolation begins about 2 hours later (8 A.M.) and ends 2 hours earlier (4 P.M.). Both the peak intensity and daily total insolation are greatly reduced in the winter solstice season.

Insolation, including both the direct solar beam and indirect sky shortwave radiation, can be measured continuously by an instrument known as a *pyranometer*. A sensing cell enclosed in a glass bulb receives radiation from the entire hemisphere of the sky. The pyranometer is the standard instrument used at radiation observing stations.

Graph B of Figure 4.4 shows net radiation in the same units as insolation. Net radiation shows a positive value—a surplus—shortly after sunrise and rises sharply to a peak at noon. The afternoon decline reaches zero shortly before sunset and becomes a negative quantity, or deficit. The deficit continues throughout the hours of darkness with a more-or-less constant value. At June solstice, a radiation surplus begins earlier and ceases later than the equinox, generating a much larger daily surplus. At December solstice, the surplus period is greatly shortened and the surplus is small. Now, because the nocturnal deficit period is long, the total deficit exceeds the surplus, so the net daily total is small negative quantity. (We shall investigate the annual cycle of net radiation in later paragraphs.)

Graph C of Figure 4.4 shows the typical, or average, daily temperature cycle. The minimum daily tempera-

At most stations, only the highest and lowest temperatures of the day are recorded. To save the observer's time, the *maximum–minimum thermometer* is used. This instrument uses two thermometers; one to show the highest temperature since it was last reset, the other to show the lowest. Another laborsaving device is the recording thermometer (thermograph), which draws a continuous record (a thermogram) on graph paper attached to a slowly moving drum. Figure 4.2 shows typical traces for several days of record.

When the maximum and minimum temperatures of a given day are added together and divided by two, we obtain the *mean daily temperature*. The mean daily temperatures of an entire month can be averaged to give the *mean monthly temperature*. Averaging the daily means (or monthly means) for a whole year gives the *mean annual temperature*. Usually such averages are compiled for records of many years' duration at a given observing station. These averages are used in describing the climate of the station and its surrounding area.

The *Celsius scale* (C) of temperature, in which 0°C is the freezing point and 100°C is the boiling point, is now used officially in all countries of the world except the United States; it is the scale used in physics, chemistry, and meteorology. The *Fahrenheit scale* (F) of temperature is the scale used by the general public in the United States and by the National Weather Service. The freezing temperature is 32°F; the boiling point is 212°F on this scale (Figure 4.3). We give Fahrenheit degrees in parentheses so you can relate temperatures to sensations of bodily comfort or discomfort.

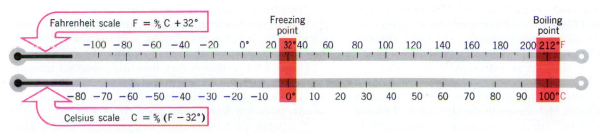

FIGURE 4.3 A comparison of the Celsius and Fahrenheit temperature scales.

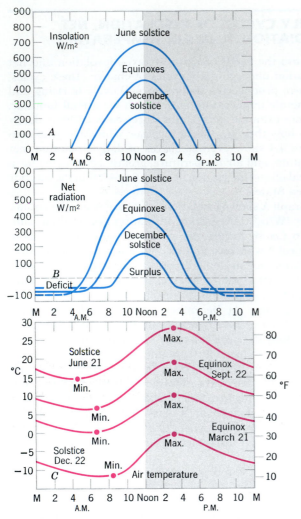

FIGURE 4.4 Idealized graphs of the daily cycles of insolation, net radiation, and air temperature for a midlatitude station in the interior United States.

ture usually occurs just after sunrise, corresponding with the onset of a radiation surplus. Heat now begins to flow upward from the ground surface to warm the lower air layer. Temperature rises sharply in the morning hours and continues to rise long after the noon peak of net radiation. We should expect the air temperature to rise as long as a radiation surplus is in effect and, in theory, this should produce a temperature maximum just before sunset. Another atmospheric process, however, begins to take effect in the early afternoon. Mixing of the lower air by eddies distributes sensible heat upward, offsetting the temperature rise and setting back the temperature peak to about 3 P.M. (The time of daily maximum varies usually between 2 and 4 P.M., according to local climatic conditions.) By sunset, air temperature is falling rapidly and continues to fall, but at a decreasing rate, throughout the entire night.

In midlatitudes and high latitudes, there is a strong seasonal change in the air-temperature curve. Upward displacement of the curve to warmer temperatures in summer and downward displacement to colder tempera-

tures in winter constitutes the annual temperature cycle, which we will examine later. At this point, you should note that at summer solstice, the time of minimum daily temperature occurs about 2 hours earlier than at equinox; at winter solstice, the minimum occurs about 2 hours later than at equinox. The time of daily maximum temperature is not greatly affected by the seasonal cycle. For simplicity, in Graph C of Figure 4.4 the maximum is shown as 3 P.M. throughout the year.

SOLAR RADIATION AND AIR TEMPERATURES AT HIGH ALTITUDES

The rapid decrease in density of the atmosphere with increasing altitude brings with it major changes in the environments of radiation and heat at the ground surface. Recall from Chapter 2 and Figure 2.6 that barometric pressure decreases by about one-thirtieth of itself for every 275 m (900 ft) of altitude increase. Thus, at 4600 m (15,000 ft), an altitude representative of mountain summits in the higher ranges of the western United States, pressure is only about 570 mb, or a little over half the sea-level value. Air density is correspondingly reduced, with the result that the overlying atmosphere reflects and absorbs a much smaller portion of the incoming solar radiation than at sea level.

Measurements of solar radiation at altitude 3600 to 4300 m (12,000 to 14,000 ft), taken near the summit of a mountain range in the desert region along the border between California and Nevada, showed noon peak values approaching 1400 W/m² under clear-sky conditions. Compare this value with a June maximum of about 500 W/m² typical for a sea-level station in midlatitudes in a humid climate. At the mountain location, the peak of net all-wave radiation reached almost 1100 W/m², a value about triple that at a sea-level station. Of course, the sea-level data include days with cloud cover as well as clear days, whereas the mountain measurements are for clear skies only. Nevertheless, the great daytime intensity of incoming and outgoing radiation at the high-altitude positions is truly remarkable.

Increasing intensity of incoming solar radiation with higher altitude has a profound influence on air and soil temperatures. Surfaces exposed to sunlight heat rapidly and intensely; shaded surfaces are quickly and severely cooled. This results in rapid air heating during the day and rapid cooling at night at high-mountain locations. This altitude effect is shown by the increasing spread between high and low daily air temperature readings from left to right in Figure 4.5.

The contrast between exposed and shaded surfaces is particularly noteworthy at high altitudes. It has been found that temperatures of objects in the sun and in the shade differ by as much as 22 to 28C° (40 to 50F°).

Figure 4.5 also shows how monthly mean air temperatures decrease with increasing altitude. The same 15-day period in July is used for all stations. For the total ascent, from sea level to an altitude of 4380 m (14,360 ft), the mean air temperature decreased by nearly 17C° (30F°).

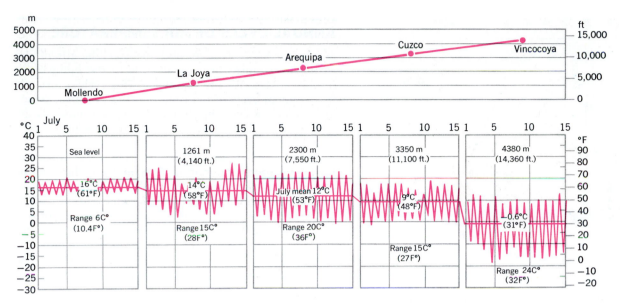

FIGURE 4.5 Daily maximum and minimum temperatures for mountain stations in Peru, lat. 15°S. The data cover the same 15-day observation period in July. (Data from Mark Jefferson.)

DAILY CYCLE OF SOIL TEMPERATURE

The daily cycle of air temperature can be strongly intensified close to the ground surface, as Figure 4.6 shows. Under the direct sun's rays, a bare soil surface or pavement heats to much higher temperatures than one reads in the thermometer shelter. At night, this surface layer cools to temperatures much lower than in the shelter. These effects are quite weak under a forest floor cover where the ground is shaded and moist. On the dry desert

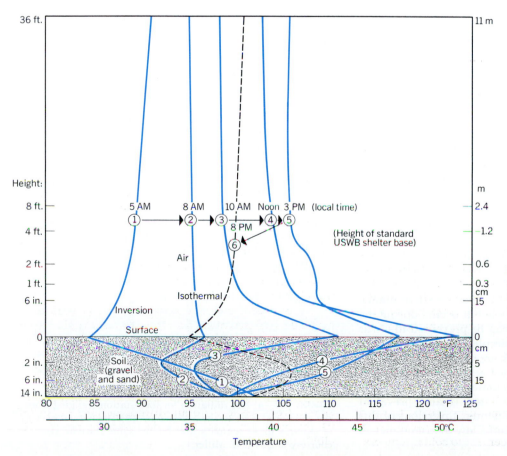

FIGURE 4.6 Air and soil temperatures at various levels above and below the ground surface throughout the day and night. Data are averages for July and August during a single summer. Height follows a square-root scale. (Data from Quartermaster Research and Development Branch, U.S. Army. From A. N. Strahler, *The Earth Sciences*, 2nd ed., Harper & Row, Publishers. Copyright © by Arthur N. Strahler.)

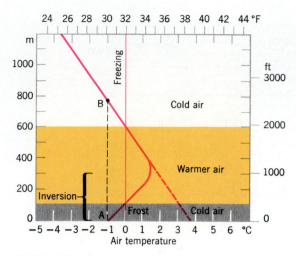

FIGURE 4.7 A low-level temperature inversion.

ANNUAL CYCLE OF AIR TEMPERATURE

As the earth revolves around the sun, the tilt of the earth's axis causes an annual cycle in the intensity of insolation, described in Chapter 3. This cycle is also felt in an annual cycle of net radiation that, in turn, generates an annual cycle in the mean daily and monthly air temperatures. In this way the climatic seasons are produced.

Figure 4.8*A* shows the yearly cycle of net radiation for four stations, ranging in latitude from the equator almost to the arctic circle. Figure 4.8*B* shows mean monthly air temperatures for these same stations. Starting with Manaus, a city on the Amazon River in Brazil, let us compare the net radiation graph with the air temperature graph. At Manaus, almost on the equator, the net radiation shows a large surplus in every month. The average surplus is about 100 W/m² but there are two

floor and on the pavements of city streets and parking lots, the surface temperature extremes are large.

Soil temperatures play a major role in the seasonal rhythms of plant physiology and in biological activity generally within the soil. Soil scientists regard soil temperature as a major factor in determining many soil properties. As Figure 4.6 suggests, the daily cycle of soil temperature will show its greatest range at the surface, whereas the daily cycle gradually dies out with increasing depth.

TEMPERATURE INVERSION AND FROST

During the night, when the sky is clear and the air is calm, the ground surface rapidly radiates longwave energy into the atmosphere above it. As we have just explained, the soil surface temperature drops rapidly and the overlying air layer becomes colder. When temperature is plotted against altitude, as in Figure 4.7, the straight, slanting line of the normal environmental lapse rate becomes bent to the left in a "J" hook. In the case shown, the air temperature at the surface, point A, has dropped to −1°C (30°F). This value is the same as at point B, some 750 m (2500 ft) aloft. As we move up from ground level, temperatures become warmer up to about 300 m (1000 ft). Here the curve reverses itself and the normal environmental lapse rate takes over.

The lower, reversed portion of the lapse rate curve is called a *low-level temperature inversion*. In the case shown, temperature of the lowermost air has fallen below the freezing point, 0°C (32°F). For sensitive plants, this condition is a *killing frost* when it occurs during the growing season. Killing frost can be prevented in citrus groves by setting up an air circulation that mixes the cold basal air with warmer air above. One method is to use oil-burning heaters; another is to operate powerful motor-driven propellors to circulate the air.

Low-level temperature inversion is also prevalent over snow-covered surfaces in winter. Inversions of this type are intense and often extend a few thousand meters into the air.

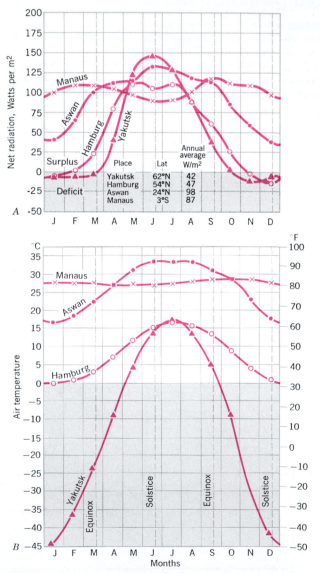

FIGURE 4.8 (*A*) Net radiation for four stations ranging from the equatorial zone to the arctic zone. (*B*) The annual cycles of monthly mean temperatures for the same stations. (Data courtesy of David H. Miller.)

minor maxima, approximately coinciding with the equinox, when the sun is nearly straight overhead. A look at the temperature graph of Manaus shows monotonously uniform air temperatures, averaging about 27°C (81°F) for the year. The *annual temperature range*, or difference between the highest mean monthly temperature and the lowest mean monthly temperature, is only 1.7C° (3F°). In other words, near the equator one month is much like the next, thermally. There are no temperature seasons.

We go next to Aswan, Arab Republic of Egypt, on the Nile river at lat. 24°N. Here we are in a very dry desert. The net radiation curve has a strong annual cycle, and the surplus is large in every month. The surplus rises to more than 125 W/m² in June and July, but falls to less than 50 W/m² in December and January. The temperature graph shows a corresponding annual cycle, with an annual range of about 17C° (30F°). June, July, and August are extremely hot, averaging over 32°C (90°F).

Moving farther north, we come to Hamburg, Germany, lat. 54°N. The net radiation cycle is strongly developed. The surplus lasts for 9 months, and there is a deficit for 3 winter months. The temperature cycle reflects the reduced total insolation at this latitude. Summer months reach a maximum of just over 16°C (60°F); winter months reach a minimum of just about freezing, 0°C (32°F). The annual range is 17C° (30F°).

Finally, we travel to Yakutsk, Siberia, lat. 62°N. During the long, dark winters, there is an energy deficit; it lasts about 6 months. During this time air temperatures drop to extremely low levels. For 3 of the winter months, monthly mean temperatures are between −35 and

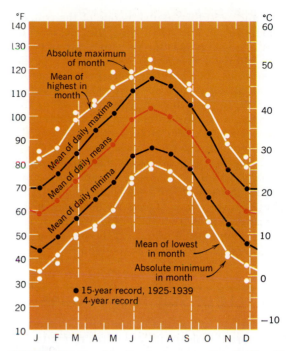

FIGURE 4.9 Monthly air temperature data for Bou-Bernous, Algeria, at lat. 27½°N in the heart of the Sahara Desert of North Africa.

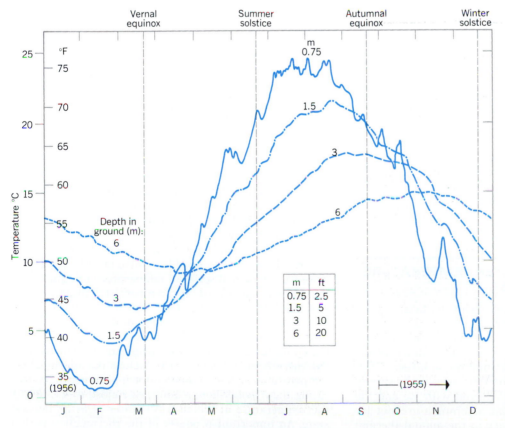

FIGURE 4.10 Soil temperatures recorded during one year at North Station, Brookhaven, Long Island, New York. (Data from I. A. Singer and R. M. Brown, 1956. From A. N. Strahler, *The Earth Sciences*, 2nd ed., Harper & Row, Publishers. Copyright © by Arthur N. Strahler.)

−45°C (−30 and −50°F). Yakutsk is one of the coldest places on earth. In summer, when daylight lasts most of the 24 hours, the energy surplus rises to a strong peak, reaching 150 W/m²—a value higher than for any of the other three stations. As a result, air temperatures show a phenomenal spring rise to summer-month values of over 13°C (55°F). In July the temperature is about the same as for Hamburg. The annual range at Yakutsk is enormous—over 61C° (110F°). No other region on earth, even the south pole, has so great an annual range.

The annual temperature range is also great in the inland deserts of the subtropical zone, astride the tropics of cancer and capricorn. Figure 4.9 shows monthly air temperatures at a station in Algeria, in the heart of the Sahara Desert. Also shown on the graph are the means of the daily maxima and minima and the means of the highest and lowest temperatures of the month.

ANNUAL CYCLE OF SOIL TEMPERATURE

Annual temperature range and mean annual temperature of the soil are important factors in the formation of various soil types. The annual cycle of heating and cooling of the soil is illustrated by data for an observing station on Long Island, New York, lat. 41°N, Figure 4.10. The soil here is sandy and porous. At the shallowest depth, 0.75 m (2.5 ft) the annual cycle is strong, with a range of about 25C° (45F°). The annual wave of warming decreases in amplitude with depth, while the peak of temperature comes progressively later with depth. At the 6 m (20 ft) depth, the warmest time is in November. The many minor irregularities in temperature at the shallowest depth reflect the effects of rainstorms, which cool the soil, and the short periods of warmer or cooler air temperature that are normal at all seasons for this climate.

You might anticipate that for a given climate, there is a depth below which the soil or rock temperature is unchanging the year around, and that this constant temperature will be close to the average air temperature near the ground surface. The facts bear out this supposition. For example, the air temperature in Endless Caverns, Virginia, which is constant at 13°C (56°F) throughout the year, is only a little higher than the annual average air temperature of 11°C (52°F). The exceptional constancy of subterranean temperatures makes an unusual thermal environment for animals that live in the total darkness of limestone caverns.

ANNUAL CYCLE OF WATER TEMPERATURE IN LAKES AND OCEANS

The environmental properties of freshwater lakes in middle and high latitudes pass through an annual thermal cycle in response to the strong annual cycle of net radiation. This thermal cycle in turn strongly influences the growth of aquatic plant and animal life in lakes.

Let us start with late winter or very early spring, when the lake surface is still covered with ice and the water beneath lies stagnant at a temperature not far above the freezing point. Stages in the annual thermal

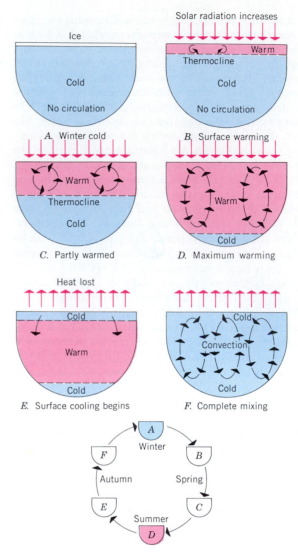

FIGURE 4.11 Schematic diagram of the annual cycle of heating, cooling, and mixing of lake water in the middle-latitude zone. (From A. N. Strahler, *The Earth Sciences*, 2nd ed., Harper & Row, Publishers. Copyright © by Arthur N. Strahler.)

cycle are shown in Figure 4.11 (Diagram A). As insolation rapidly increases, the ice is melted and warming of a surface layer begins (Diagram B). Once the layer has exceeded 4.4°C (39°F), it is less dense than the colder water below and remains on top. Winds blowing over the lake surface create small waves, and these cause a mixing that results in thickening of the warm layer (Diagram C). If we were to take a series of temperature readings from the surface downward, the temperature profile would look like that in Figure 4.12. The upper warm layer, known as the *epilimnion*, has a uniform temperature throughout; it is said to be an isothermal layer. Below the epilimnion, temperatures drop rapidly. This zone is the thermocline (Figure 2.14). Below the thermocline, temperatures again become uniform with depth, constituting the *hypolimnion*, a cold layer close to 4°C (39°F), the temperature at which fresh water is in its densest state. An important property of the thermocline is that

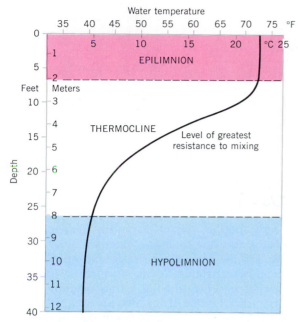

FIGURE 4.12 Summer temperature profile of a small lake in the middle-latitude zone. (From A. N. Strahler, *The Earth Sciences*, 2nd ed., Harper & Row, Publishers. Copyright © by Arthur N. Strahler.)

water does not easily rise or sink through this layer since water density changes with water temperature. A succession of horizontal fluid layers of rapidly changing density, whether they be of liquid or gas, tends to be stable and to resist mixing by vertical motions. (This principle is explained in Chapter 6 in connection with a related phenomenon known as an upper-level temperature inversion.)

As the summer progresses, the warm surface water layer, or epilimnion, becomes thicker and the thermocline is pushed down to greater depths (Diagram D). For a shallow lake, 10 to 20 m (30 to 60 ft) deep or less, the entire water body may be warmed, but for deeper lakes, a cold hypolimnion persists. Inflowing water of streams or springs may add to the mixing of warm and cold water.

As winter approaches and the net radiation declines to become negative in value (e.g., a radiation deficit), heat is lost from the surface layer (Diagram E). Being denser than the warmer water below, this cold surface water sinks to the bottom, setting up a general overturning, or convection, within the entire lake. The thermocline is destroyed and gradually the water becomes uniformly cold throughout. Mixing ceases when the point of maximum density 4°C (39°F) is reached. Further cooling produces a less dense surface water layer that remains on top, eventually freezing into a continuous ice layer.

In warm tropical and equatorial climates, where solar radiation is uniformly great throughout the year, lake water is comparatively warm down to the bottom. The coldest water that can sink to the lake floor can be no colder than the average surface water temperature at the coolest time of year. This seasonal cooling easily results in overturn and general mixing because the overall range of temperatures is small. For example, lakes at low elevation in Indonesia, situated near the equator, have a bottom temperature always close to 26°C (79°F), while the surface water temperature ranges annually from 26° to 29°C (79° to 84°F). When the surface water temperature is increasing, a weak thermocline develops, just as in lakes of middle latitudes, but it is destroyed when the seasonal cooling takes place. Because insolation at the top of the atmosphere is uniformly great throughout the year at low latitudes, the cooling of the surface water seasonally is brought about by the occurrence of a rainy season (a rainy monsoon) in which cloud cover reduces incoming radiation and copious rains have a cooling effect. In the dry season, insolation again becomes intense and the surface water is warmed.

The three-layer thermal structure of the oceans was described in Chapter 2 and pictured in Figures 2.14 and 2.15. Recall that a warm surface layer, a thermocline,

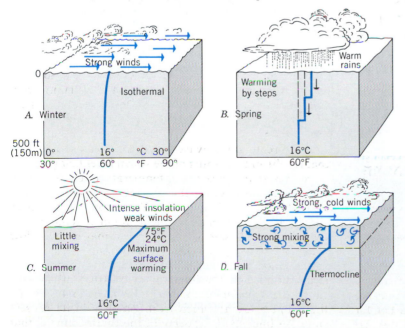

FIGURE 4.13 Seasonal water temperature changes and mixing typical of the ocean in the middle-latitude zone. (Based on data of E. C. LaFond, *Scientific Monthly*, August 1954. From A. N. Strahler, *The Earth Sciences*, 2nd ed., Harper & Row, Publishers. Copyright © by Arthur N. Strahler.)

and a cold deep layer are typical of middle-latitude oceans. The annual cycle of change in these latitudes is illustrated in Figure 4.13. During the winter (Diagram A) the upper layer is mixed by wave action under the force of strong winds. Temperatures are nearly isothermal in the upper 150 m (500 ft) although a gradual decrease with depth takes place until bottom temperatures close to 0°C (32°F) are reached. With the arrival of spring (Diagram B) increased insolation and the effects of warm rains cause warming by steps. Throughout the summer (Diagram C) intense insolation and lack of vigorous mixing develops a warm surface layer. In the fall (Diagram D), strong winds result in a mixing of the surface layer and the development of a pronounced thermocline. Radiation and conduction of heat from sea to atmosphere then cool the surface layer and the isothermal conditions of winter again set in.

LAND AND WATER TEMPERATURE CONTRASTS

The odd distribution of continents and ocean basins gives our planet its great variety of climates. Look back at the two global hemispheres—northern and southern—outlined in Figure 2.13. The northern hemisphere displays a polar sea surrounded by massive continents; the southern hemisphere shows the opposite—a pole-centered continent surrounded by a vast ocean. The Americas form a north-south barrier between two oceans—the Atlantic and the Pacific. The continents of Eurasia and Africa together form another great north-south barrier. Oceans and continents have quite different properties when it comes to absorbing and radiating energy.

Land surfaces behave quite differently from water surfaces. The important principle is this: The surface of any extensive deep body of water heats more slowly and cools more slowly than the surface of a large body of land when both are subject to the same intensity of insolation.

The slower rise of water-surface temperature can be attributed to four causes (Figure 4.14). (1) Solar radiation penetrates water, distributing the absorbed heat throughout a substantial water layer. (2) The specific heat of water is large (a gram of water heats up much

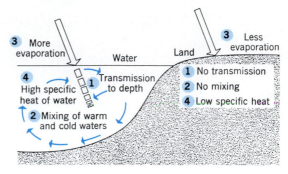

FIGURE 4.14

more slowly than a gram of rock). (3) Water is mixed through eddy motions, which carry the heat to lower depths. (4) Evaporation cools the water surface.

In contrast, the more rapid rise of land surface temperature can be attributed to these causes: (1) Soil or rock is opaque, concentrating the heat in a shallow layer; there is little transmission of heat downward. (2) Specific heat of mineral matter is much lower than water. (3) Soil, if it is dry, is a poor conductor of heat. (4) No mixing occurs in soil and rock.

The effect of land and water contrasts is seen in two sets of daily air temperature curves, shown in Figure 4.15. El Paso, Texas, exemplifies the temperature environment of an interior desert in midlatitudes. Soil moisture content is low, vegetation sparse, and cloud cover generally light. Responding to intense heating and cooling of the ground surface, air temperatures show an average daily range of 11 to 14C° (20 to 25F°). North Head, Washington, is a coastal station strongly influenced by air brought from the adjacent Pacific Ocean by prevailing westerly winds. Consequently, North Head exemplifies a maritime temperature environment. The average daily range at North Head is a mere 3C° (5F°) or less. Persistent fogs and cloud cover also contribute to the small daily range. Refer also to Figure 4.2, which shows the same environmental contrast when the record of Yuma is compared with that of San Francisco.

The principle of contrasts in the heating and cooling of water and land surfaces also explains the differences in the annual temperature cycles of coastal and interior stations. Figure 4.16 shows the annual cycle of mean

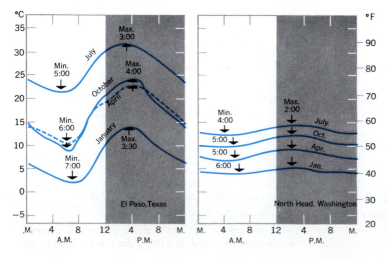

FIGURE 4.15 The average daily cycle of air temperature for four different months. Daily ranges and seasonal changes are great at El Paso, a station in the continental interior, but only weakly developed at North Head, close to the Pacific Ocean.

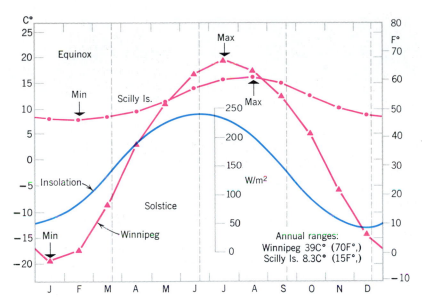

FIGURE 4.16 Annual cycles of monthly mean air temperature for two stations at lat. 50°N: Winnipeg, Manitoba, Canada, and Scilly Islands, England.

daily temperatures for two places at approximately the same latitude, where insolation is about the same for both. Winnipeg, Manitoba, Canada, is located in the heart of the North American continent. The Scilly Islands, England, are exposed to the Atlantic Ocean.

Although insolation reaches the maximum at summer solstice, the hottest part of the year for inland regions is about a month later because a net radiation surplus persists and heat energy continues to flow into the ground well into August. The air temperature maximum, closely corresponding with maximum ground output of longwave radiation, is correspondingly delayed. (Bear in mind that this cycle applies to middle and high latitudes, but not to the region between the tropics of cancer and capricorn.) Similarly, the coldest time of year for large land areas is January, about a month after winter solstice, because a net radiation deficit persists and the ground continues to lose heat even after insolation begins to rise.

Over the oceans there are two differences: (1) Maximum and minimum temperatures are reached about a month later than on land—in August and February, respectively—because water bodies heat or cool much more slowly than land areas; and (2) the yearly range is less than over land, following the law of temperature differences between land and water surfaces. Coastal regions are usually influenced by the oceans to the extent that maximum and minimum temperatures occur later than in the interior. This principle shows nicely for monthly temperatures of the Scilly Islands. February is slightly colder than January (Figure 4.16).

Figure 4.15 reinforces the evidence of Figure 4.16. The annual temperature range at El Paso is about 20C° (35F°), while that at North Head is only about 8C° (15F°).

AIR TEMPERATURE MAPS

The distribution of air temperatures over large areas can best be shown by a map composed of *isotherms*, lines drawn to connect all points having the same temperature. Figure 4.17 is a map on which the observed air temperatures have been recorded in the correct places. These may represent single readings taken at the same time everywhere, or they may represent the averages of many years of records for a particular day or month of a year, depending on the purposes of the map.

Usually, isotherms representing 5 or 10° differences are chosen, but isotherms can be drawn for any selected temperatures. The value of isothermal maps is that they make clearly visible the important features of the prevailing temperatures. Centers of high or low temperature are clearly outlined.

WORLD PATTERNS OF AIR TEMPERATURE

World maps of air temperature (Figure 4.18) allow us to compare thermal conditions in the two months of greatest extremes over large land areas—January and July. Patterns taken by the isotherms are largely explained by three controls: (1) latitude; (2) continent–ocean contrasts; and (3) altitude. (Ocean currents play an important, but secondary, role.) Important features to identify when interpreting these maps are as follows:

1. The trend of isotherms is east–west, with temperatures decreasing from equatorial zone to polar zones. This feature shows best in the southern hemisphere

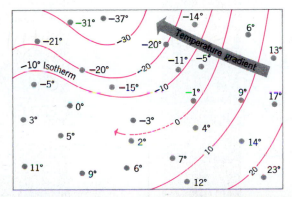

FIGURE 4.17 Isotherms are used to make temperature maps. Each line connects those points having the same temperature.

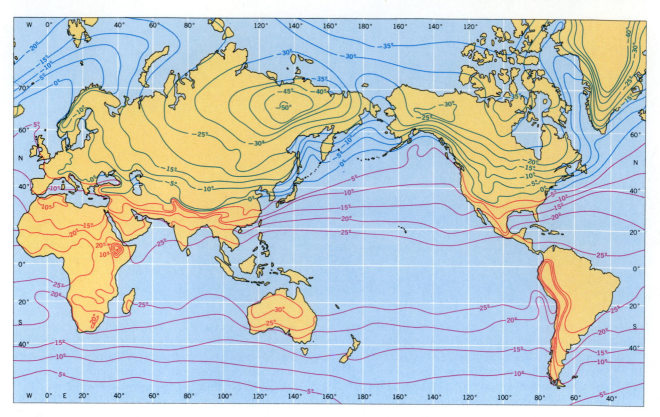

JANUARY

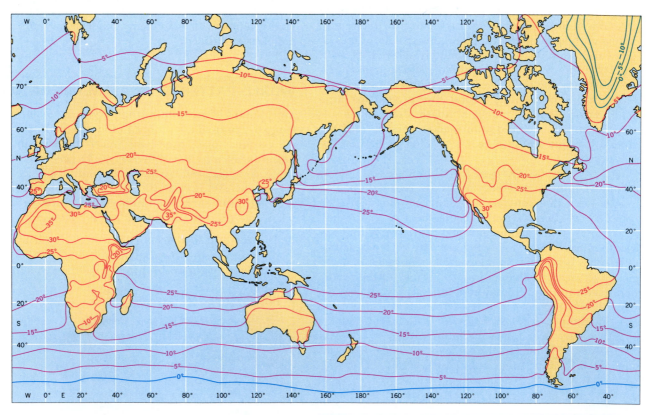

JULY

FIGURE 4.18 Mean monthly air temperatures (°C) for January and July, Mercator and stereographic projections. (Compiled by John E. Oliver from station data by World Climatology Branch, Meteorological Office, *Tables of Temperature*, 1958, Her Majesty's Stationery Office, London; U.S. Navy, 1955, *Marine Climate Atlas*, Washington, D.C.; and P. C. Dalrymple, 1966, American Geophysical Union.)

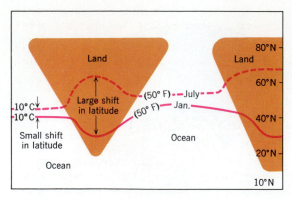

FIGURE 4.19 A schematic map of the seasonal shift of an isotherm.

because the great Southern Ocean girdles the globe as a uniform expanse of water, while the continent of Antarctica is squarely centered on the south pole. The parallel pattern of isotherms is explained, of course, by the general decrease in intensities of insolation and net radiation from equator to poles.

2. Large landmasses located in the subarctic and arctic zones develop centers of extremely low temperatures in winter. The two large landmasses we have in mind are North America and Eurasia. These centers are conspicuous on the January map.

3. Isotherms change very little in position from January to July over the equatorial zone, particularly over the oceans. This feature reflects the uniformity of insolation throughout the year near the equator.

4. Isotherms make a large north–south shift from January to July over continents in the midlatitude and subarctic zones. Figure 4.19 shows this principle. North America shows this effect well. In January the 15°C isotherm lies over central Florida; in July, this same isotherm has moved far north, cutting the southern shore of Hudson Bay and then looping far up into northwestern Canada. The 15°C isotherm on the Eurasian continent also shows the effect. We can explain the large north–south shift in position over these continents by the law of land and water contrasts.

5. Highlands are always colder than surrounding lowlands. An example is the Andes range, running along the western side of South America. The isotherms loop equatorward in long fingers over this lofty mountain chain.

6. Areas of perpetual ice and snow are always intensely cold. The Greenland and the Antarctic are the two great ice sheets. Notice how they stand out as cold centers in both January and July. Not only do these ice sheets have high surfaces, rising to over 3000 m (10,000 ft) in their centers, but the white snow surfaces have a high albedo and reflect away much of the insolation. The Arctic Ocean, bearing a cover of floating ice, also maintains its cold throughout the year; but the cold is much less intense in July than over the Greenland Ice Sheet.

THE ANNUAL RANGE OF AIR TEMPERATURES

Figure 4.20 is a world map showing the annual range of air temperatures. The lines resembling isotherms are *co-range lines;* they tell the difference between the January

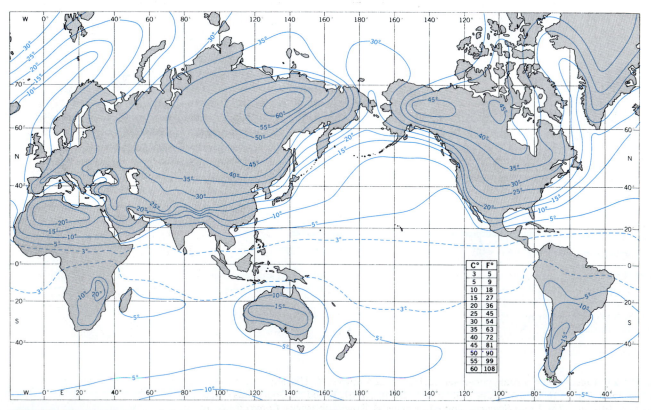

FIGURE 4.20 Annual range of air temperatures. Data shows differences between January and July means. (Same data sources as Figure 4.18.)

and July monthly means. This map serves as a summary of concepts we have previously emphasized. Note these features:

1. The annual range is extremely great in the subarctic and arctic zones of both Asia and North America. This effect is shown well in the annual temperature graph for Yakutsk (Figure 4.8).
2. The annual range is moderately large on land areas in the tropical zone, near the tropics of cancer and capricorn. North Africa, southern Africa, and Australia are examples. Here the annual ranges are substantially greater than over adjacent oceans.
3. The annual range is very small over oceans in the

equatorial zone. At all latitudes, the annual range over all oceans is generally less than over land areas.

This review of air temperatures over the globe will form the groundwork for an understanding of climates, developed in later chapters.

WORLD PATTERNS OF SEA SURFACE TEMPERATURES

Thermal environments of the ocean surface are summarized on a global basis by two maps, one showing surfacewater temperatures for August, the other for Febru-

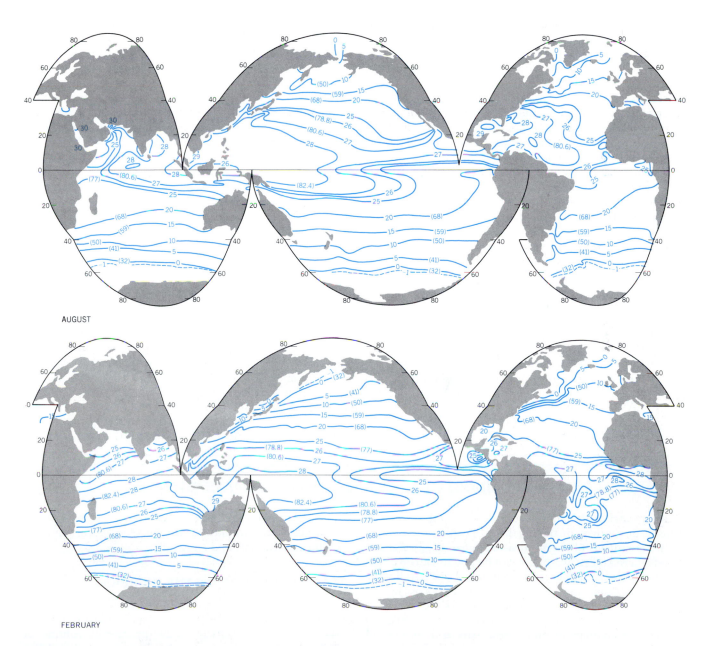

AUGUST

FEBRUARY

FIGURE 4.21 August and February mean sea-surface temperatures, °C. (Data of U.S. Navy Oceanographic Office, H. U. Sverdrup, 1942, and A. Defant, 1961. Based on Goode Base Map. From A. N. Strahler, *The Earth Sciences*, 2nd ed., Harper & Row, Publishers. Copyright © by Arthur N. Strahler.)

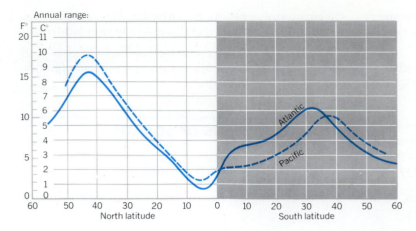

FIGURE 4.22 Mean annual range of sea-surface temperature shown in meridional profiles of the Atlantic and Pacific oceans. (Data of H. U. Sverdrup, M. W. Johnson, and R. H. Fleming, *The Oceans*. From A. N. Strahler, *The Earth Sciences*, 2nd ed., Harper & Row, Publishers. Copyright © by Arthur N. Strahler.)

ary (Figure 4.21). As you would expect, the isotherms trend east-to-west around the globe generally, with maximum values in a low-latitude belt and strong poleward decline both north and south. Large areas of equatorial oceans maintain a surface temperature higher than 26°C (80°F). In arctic and antarctic waters temperatures remain close to the freezing mark 0°C (32°F). But in middle latitudes there is a large annual range.

These relationships are shown in Figure 4.22 by two generalized profiles giving the annual range in sea-surface temperatures for the Atlantic and Pacific oceans. Notice the peak ranges between 40° and 50°N latitude, and between 30° and 40°S latitude. The range is greater and peaks at a higher latitude in the northern hemisphere because the North Atlantic and North Pacific oceans are hemmed in by landmasses, inhibiting mixing by ocean currents. In addition, the adjacent landmasses of North America and Eurasia accentuate the temperature contrasts between summer and winter. The South Atlantic and South Pacific, poleward of the 45th parallel of latitude, are joined with the Indian Ocean in a continuous ocean belt—the Southern Ocean—in which heat exchange by currents is free and no large land areas intervene. As a result, the sea-surface temperatures have a much smaller range there than in the northern hemisphere.

OCEAN THERMAL ENERGY

Ocean thermal energy conversion (OTEC) makes use of the temperature gradient from the surface downward in tropical waters. The temperature difference between the warm surface water layer and the cold deep water is on the order of 22C° (40F°). This does not seem like much of a difference to convert into energy, but the total solar heat input into the warm water layer is more than 10,000 times greater than needed for all human uses. A floating towerlike structure houses a heat exchanger using ammonia or propylene, which boil into gas at the temperature of the upper warm layer but condense into a liquid when chilled by deep water brought up from below. (A land-based plant close to the shoreline is also feasible.) The expanding gas would be used to drive a turbine and thus generate electricity. The cold water pipe would need to be 700 to 1000 m long, with a diameter of 30 m for a system capable of generating 250 megawatts of electric power. Although the scheme seems workable in theory, many problems will need to be solved to make it a major producer of energy. A demonstration plant producing 100 megawatts is planned for completion by the mid-1990s. The typical OTEC design also produces freshwater as a condensate of evaporated seawater, and can thus serve as a desalination plant.

THE GREENHOUSE EFFECT AND GLOBAL CLIMATE CHANGE

Human activity is capable of causing and sustaining long-term changes in the earth's atmosphere. Increases in the quantities of the natural component gases of the troposphere and stratosphere can be potent causes of climate change. Increase of the content of carbon dioxide is the most important of these. Other causes of climate change that we consider in other chapters are emissions of industrial gases and finely divided solid particles (dust), and changes in plant cover and soil surfaces of the lands. Our interest here is in human-induced changes in atmospheric temperature, which is one of the ingredients of global climate. Changes in other climate factors,

considered in later chapters, include the global patterns of atmospheric circulation (winds), cloud cover, and precipitation (rainfall and snowfall).

CARBON DIOXIDE ON THE INCREASE

Under preindustrial conditions of recent centuries, the atmospheric content of carbon dioxide (CO_2) was maintained at a level of about 0.0294 percent by volume, or 294 parts per million (ppm). An environmental problem arises because during the past century humans have greatly increased the rate of extraction and burning of

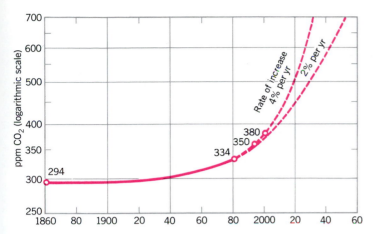

FIGURE 4.23 Increase in atmospheric carbon dioxide, observed to 1990 and predicted into the 21st century.

fossil fuels (coal, petroleum, natural gas), which had previously been locked in the earth's crust (see Chapter 12). Combustion of these fuels releases into the atmosphere both water and CO_2.

During the past 130 years (1860 to 1990), atmospheric CO_2 has increased about 22 percent by volume; it reached a level of about 350 ppm in 1990. As Figure 4.23 shows, the rate of increase, while slow at first, became much more rapid toward the end of the period. Projection of the present curve into the future suggests a CO_2 value of about 380 ppm by the year 2000. At that time, if the prediction is correct, the content of CO_2 will have increased by about 35 percent over the 1860 value. Doubling of the present level is predicted by about 2030, if the combustion of fossil fuels continues to increase at the present annual rate of 4 percent. Doubling time can be delayed, however, until about 2050 with a fuel combustion rate reduced to half the present value, and to well into the twenty-second century if the combustion rate remains fixed at the present level.

In making such predictions, researchers have also taken into account possible mechanisms whereby some proportion of the additional CO_2 might be absorbed and permanently held in reservoirs other than the atmosphere. Current estimates are that about 5 billion tons (gigatons, or Gt) of the element carbon (C) in CO_2 gas are released annually into the atmosphere by the combustion of fossil fuels. In addition, clearing and burning of

forests injects an estimated 1 to 3 Gt of carbon annually into the atmosphere, but at the same time growth of new trees removes a substantial part of that carbon from the atmosphere. It is considered possible that the land surfaces may actually withdraw more carbon than they give off. Using the figure of 1 Gt net contribution from burning, the total contribution is about 6 Gt.

Now, measurements of atmospheric CO_2 clearly show an annual increase of about 3 Gt of carbon during the 1980s. This is roughly half of the amount thought to be emitted by fossil fuel and forest combustion combined, leaving some 3 Gt to be accounted for. A very large amount of CO_2 is both absorbed and released by the oceans. Recent estimates place the net amount of carbon absorbed yearly by the oceans in excess of that released at no more than 1.6 Gt and possibly less than 1 Gt. This figure suggests that there must be terrestrial sinks that absorb the difference. In any case, there exists little doubt that the content of atmospheric CO_2 will continue to increase substantially.

THE ROLE OF TRACE GASES IN THE GREENHOUSE EFFECT

Besides carbon dioxide, several other atmospheric gases contribute a similar effect, which is to absorb outgoing longwave radiation throughout the troposphere (see Chapter 3). Known as *trace gases*, because of their relatively small concentration, they are methane (CH_4), nitrous oxide (N_2O), ozone (O_3), and two forms of chlorofluorocarbon (CFC). (Note that the ozone named here is in the troposphere and should not be confused with the stratospheric ozone described in Chapter 2.) Table 4.1 lists these gases along with CO_2, showing their relative concentration, rate of increase per year, and relative contribution to the greenhouse effect. Altogether, these four gases contribute 40 percent to that effect, and methane leads with 15 percent.

Methane is largely a product of natural biologic processes, but their output can be accelerated by human activities. Methane is emitted from the soil in which organic matter is in process of decay (hence its common name, "marsh gas"), and from the digestive tracts of animals. It is thought that much of the natural emission of methane comes from peat soils (histosols) of the arctic lands. The additional emission caused by human activities, designated as anthropogenic methane, has resulted

TABLE 4.1 Gases That Contribute to the Greenhouse Effect

Name of Gas	Concentration in ppm × 1000	Rate of Increase (% per year)	Relative Contribution (%)
Carbon dioxide (CO_2)	353,000	0.5	60
Methane (CH_4)	1700	1	15
Nitrous oxide (N_2O)	310	0.2	5
Ozone (O_3)	10–50	0.5	8
CFC-11	0.28	4	4
CFC-12	0.48	4	8

Data source: Henning Rodhe (1990) *Science*, vol. 248, p. 1218. Copyright © by the American Association for the Advancement of Science. Used by permission.

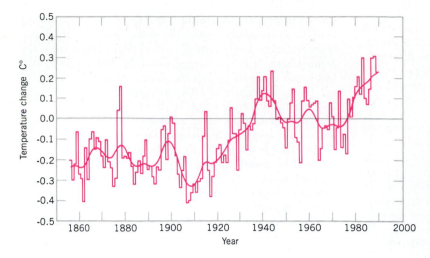

FIGURE 4.24 Mean annual surface temperature of the earth, 1856–1990. The vertical scale shows departures in C degrees from a zero line of reference representing the average from 1951 through 1980. The bar graph shows the mean for each year; the smooth curve is a ten-year running average. (From Philip D. Jones and Tom M. L. Wigley, 1990, University of East Anglia/UK Meteorology Office. Used by permission.)

from such processes as the expansion of rice agriculture, the increase in numbers of domesticated grazing animals (ruminants), disposal of organic wastes in landfills, various kinds of mining operations, and the burning of fossil fuels. The rate of increase of atmospheric methane has been extremely rapid since about 1900. Global warming, causing melting of frozen tundra soils, would tend initially to increase the rate of production of methane.

As Table 4.1 shows, the two forms of CFC have experienced a much faster rate of increase than the others listed, including CO_2, and their combined effect is 12 percent—second only to methane among the trace gases.

HAS A GLOBAL WARMING SET IN?

There seems to be no doubt that the observed increases in carbon dioxide and the trace gases can lead to a warming of the mean global atmospheric temperature—provided, of course, that the anticipated rise is not offset by some as yet unrecognized factor that may nullify or reverse that trend. Let us examine the available temperature data to see if a global temperature increase is already in progress, and to judge whether it represents

more than merely a cyclic phenomenon that may soon show a reversal.

Figure 4.24 is a graph prepared by climate research scientists of the University of East Anglia in Norwich, England, showing the record of mean global temperature from 1856 through 1990. The smooth line is the 10-year running average for the same data. An overall increase in temperature is unmistakable within the limits of the graph, but one can discern several wide swings from colder to warmer and back on a time scale of a decade or so.

In an attempt to get a better picture of climate changes over the past two centuries, scientists of the Lamont-Doherty Geological Observatory of Columbia University set up a tree-ring study project to measure the temperature response of old trees in North America, as revealed in their annual growth rings. The resulting graph, based on hundreds of trees from dozens of sites along the east-west northern tree line (subarctic limit of tree growth) is shown in Figure 4.25. From 1880 to 1970, the rising trend and its minor reversal in the 1960s are quite similar to that in Figure 4.24. Looking further back in time, however, another cycle of rise and decline comes

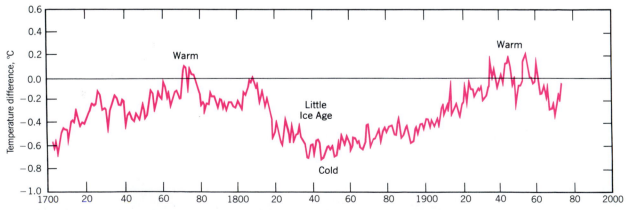

FIGURE 4.25 A reconstruction of the departures of northern hemisphere temperatures from the 1950–1965 mean, based on analyses of tree rings sampled along the northern tree limit of North America. (Courtesy of Gordon C. Jacoby of the Tree-Ring Laboratory of the Lamont-Doherty Geological Observatory of Columbia University.)

to view. The low point around 1840 correlates with a well-known cold event, the "Little Ice Age," which is documented by the advance of European alpine glaciers. Similar cycles of warming and cooling, each lasting on the order of 150 to 200 years, appear in Greenland ice-core records going back over 800 years, as shown in Figure 22.47.

The message of these long-term temperature records seems to be that the predicted warming because of increased industrial carbon dioxide, methane, ozone, and CFCs may not as yet have made a measurable impact. Opinions of leading climate scientists differ widely on this question of interpretation. In 1988, an international group of scientists published three scenarios of rates of global warming through the next century. The most conservative of these calls for warming at a rate of 0.06°C per decade, with no more than one-half degree of increase by the year 2100. A moderate scenario based on a rate of 0.3°C yields a 3-degree rise in the same period. An extreme scenario of 0.8°C calls for a catastrophic rise of five degrees by the middle of the next century, and this rate of change is some 10 to 100 times faster than rates seen in tree-ring and ice-core data records of the past few centuries. Many years of environmental monitoring may yet lie ahead before a consensus can be reached as to when and what degree the global environment will receive a full impact of human-induced warming that seems to many to be inevitable.

In coming chapters we augment this review of the predicted global warming with its possible impacts on other global phenomena, for example: changes in the global patterns of pressure systems and winds (Chapter 5); increased precipitation in some regions but expanded deserts in others (Chapter 6); melting of the arctic frozen ground (Chapter 15); an anticipated disastrous rise in sea level (Chapter 20); and changes in sea ice, glaciers, and ice-sheet margins (Chapter 22). Effects of biomass burning, forest cutting, and forest planting are reviewed in Chapters 26 and 27.

Data sources: Stephen H. Schneider (1990) *Bulletin of the American Meteorological Society*, vol. 71, no. 9, p. 1292–1304; Philip D. Jones and Tom M.L. Wigley (1990) *Scientific American*, vol. 263, no. 2, p. 84–91.

THE THERMAL ENVIRONMENT IN REVIEW

In this chapter we have covered a vital environmental factor in the life layer. Heat, as recorded by the thermometer, is an essential ingredient of climate near the ground. All organisms respond to changes in temperatures of the medium that surrounds them, whether air, soil, or water. Geographers study temperature records carefully, and they are interested in the mean values based on records of long periods of observation.

Cycles of air-temperature change are particularly important in the thermal environment. Both daily and seasonal changes in air temperature can be explained by daily and seasonal cycles of insolation and net radiation. Superimposed on these astronomical rhythms is the powerful effect of latitude. A zoning of thermal environments from the equatorial zone to the polar zones is one of the most striking features of the global climate. But equally important is the way large land areas, especially the continents of North America and Eurasia, subvert the simple latitude zones to cause great extremes of temperature. In contrast, the southern hemisphere is strongly dominated by the simple effects of latitude.

From the long-range standpoint, combustion of fuels by humans has serious implications, and these must be studied with care. Sustained research combined with intensified monitoring of the environment deserves high priority.

WINDS AND THE GLOBAL CIRCULATION

Our physical environment at the earth's surface depends for its quality as much on atmospheric motions as it does on the flow of heat energy by radiation. In the form of very strong winds—those in hurricanes and tornadoes—air in motion is a severe environmental hazard. Winds also transfer energy to the surface of the sea, as wind-driven waves. Wave energy, in turn, travels to the shores of continents, where it is transformed into vigorous surf and coastal currents capable of reshaping the coastline.

But air in motion has another, more basic, role to play in the planetary environment. Large-scale air circulation transports heat, both as sensible heat and as latent heat present in water vapor. Because of the global energy imbalance—a surplus in low latitudes and a deficit in high latitudes—atmospheric circulation must transport heat across the parallels of latitude from the region of surplus to the regions of deficit. Figure 5.1 shows this meridional transport in schematic form. Notice that circulation of ocean waters is also involved in the transport of sensible heat. But this oceanic circulation is a secondary mechanism, driven largely by surface winds.

Equally important to our environment are the rising and sinking motions of air. We shall find in Chapter 6

that precipitation, the source of all fresh water on the lands, requires massive lifting of large bodies of air heavily charged with water vapor. In reverse, large-scale sinking motions in the atmosphere lead to aridity and the occurrence of deserts. In this way the earth's surface becomes differentiated into regions of ample fresh water and regions of water scarcity.

With these broad concepts in mind, we examine the forces that set the atmosphere in motion and drive the ceaseless global circuits of air circulation.

WINDS AND THE PRESSURE GRADIENT FORCE

Wind is air motion with respect to the earth's surface, and it is dominantly horizontal. (Dominantly vertical air motions are referred to by other terms, such as updrafts or downdrafts.) To explain winds, we must first consider barometric pressure and its variations from place to place.

In Chapter 2 we found that barometric pressure falls with increasing altitude above the earth's surface. For an atmosphere at rest, the barometric pressure will be the same within a given horizontal surface at any chosen altitude above sea level. In that case, the surfaces of equal barometric pressure, called *isobaric surfaces*, are horizontal. In a cross section of a small portion of the atmosphere at rest, isobaric surfaces appear as horizontal lines, as shown in Figure 5.2A. (For convenience, we have shown surface pressure as 1000 mb.)

Suppose, now, that the rate of upward pressure decrease is more rapid in one place than another, as shown in Figure 5.2B. As we proceed from left to right across the diagram, the upward rate of pressure decrease is more rapid. The isobaric surfaces now slope down toward the right. At a selected altitude, say 1000 m (horizontal line), barometric pressure declines from left to right. Figure 5.2C is a type of map; it shows that the 1000-m horizontal surface cuts across successive pressure surfaces. The trace of each pressure surface is a line on the map; the line is known as an *isobar*. The isobar is thus a line showing the location on a map of all points having the same barometric pressure.

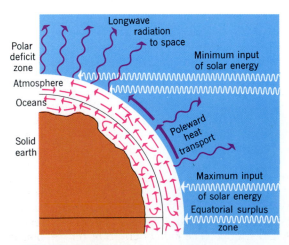

FIGURE 5.1 Circulation systems of the atmosphere and oceans are necessary to maintain the global heat balance.

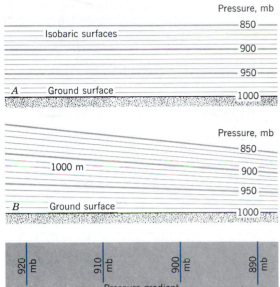

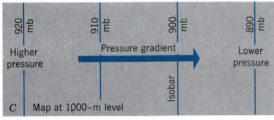

FIGURE 5.2 Isobaric surfaces and the pressure gradient. Diagrams A and B are vertical cross sections through the atmosphere. Diagram C is a map.

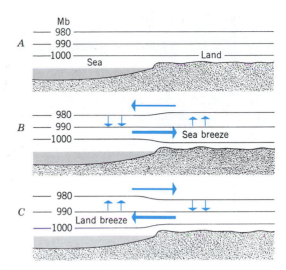

FIGURE 5.3 Sea breeze and land breeze. (From A. N. Strahler, *The Earth Sciences*, 2nd ed., Harper & Row, Publishers. Copyright © by Arthur N. Strahler.)

The change in barometric pressure across the horizontal surface of a map constitutes a *pressure gradient;* its direction is indicated by the broad arrow in Figure 5.2C. The gradient is in the direction from higher pressure (at the left) to lower pressure (at the right). You can think of the sloping pressure surface as a sloping hillside; the downward slope of the ground surface is analogous to the pressure gradient.

Where a pressure gradient exists, air molecules tend to drift in the same direction as that gradient. This tendency for mass movement of the air is referred to as the *pressure gradient force.* The magnitude of the force is directly proportional to the steepness of the gradient; that is, a steep gradient is associated with a strong force. Wind is the horizontal motion of air in response to the pressure gradient force.

SEA AND LAND BREEZES

Perhaps the simplest example of the relationship of wind to the pressure gradient force is a common phenomenon of coasts: the sea breeze and land breeze, illustrated in Figure 5.3. Diagram A shows the initial situation in which no pressure gradient exists. During the day, more rapid heating of the lower air layer over the land than over the ocean causes a pressure gradient from sea to land (Diagram B). Air moving landward in response to this gradient from higher to lower pressure constitutes the *sea breeze.* At higher levels, a reverse flow sets in. Together with weak rising and sinking air motions, a complete flow circuit is formed. During the night, when

radiational cooling of the land is rapid, the lower air becomes colder over the land than over the water. Higher pressure now develops over land and the barometric gradient is reversed. Air now moves from land to sea as a *land breeze* (Diagram C).

This illustration shows that a pressure gradient can be developed through unequal heating or cooling of a layer of the atmosphere. Air that is warmed also expands and becomes less dense. Air that is cooled contracts and becomes denser. The upward change in barometric pressure is then more rapid within the cooler air layer than within the warmer layer.

MEASUREMENT OF WINDS

A description of winds requires measurement of two quantities: direction and speed. Direction is easily determined by a wind vane, a common weather instrument (see Figure 5.5). Wind direction is stated in terms of the direction from which the wind is coming (Figure 5.4). Thus an east wind comes from the east, but the direction of air movement is toward the west. The direction of movement of low clouds is an excellent indicator of wind direction and can be observed without the aid of instruments.

Speed of wind is measured by an anemometer. There are several types. The commonest one at weather stations is the cup anemometer. It consists of hemispherical cups mounted as if at the ends of spokes of a horizontal wheel (Figure 5.5). The cups travel with a speed proportional to that of the wind. One type of anemometer turns a small electric generator, and the current it produces can be transmitted to a meter calibrated in units of wind speed. Units are meters per second or miles per hour.

For wind speeds at higher levels, a small hydrogen-filled balloon is released into the air and observed through a telescope. The rate of climb of the balloon is known in advance. Knowing the balloon's vertical position by measuring the elapsed time, an observer can cal-

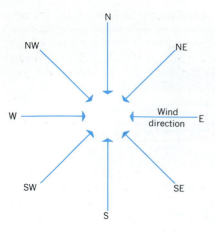

FIGURE 5.4 Winds are designated according to the compass point from which the wind comes. An east wind comes from the east, but the air is moving westward.

culate the horizontal drift of the balloon downwind. For upper-air measurements of wind speed and direction, the balloon carries a target that reflects radar waves and can be followed when the sky is overcast. Ground-based Doppler radar is now used to scan the region surrounding an airport to produce an image that shows horizontal wind directions and intensities (see Figure 6.21).

Measurement of wind directions and speeds throughout the troposphere by the National Weather Service has been greatly facilitated by the use of complex Doppler radar systems collectively dubbed "NEXt generation RADar" (NEXRAD). Installed on the ground, each radar system sends its beams upwards and outwards to sound air motions within a cylinder of the troposphere as much as 400 km (250 mi) in diameter and extending upward to a height of 3000 m (10,000 ft). The radar beams also

detect the presence and density of falling rain and hail. Although NEXRAD installations are expensive (as much as $2.5 million each), requiring powerful computers, the National Weather Service plans to cover the entire United States with a network of such stations.

THE CORIOLIS EFFECT AND THE WINDS

If the earth did not rotate on its axis, winds would follow the direction of pressure gradient. As mentioned in Chapter 1, earth rotation produces the *Coriolis effect*, which tends to turn the flow of air. The direction of action of this turning effect is stated in Ferrel's law: Any object or fluid moving horizontally in the northern hemisphere tends to be deflected to the right of its path of motion, regardless of the compass direction of the path. In the southern hemisphere, a similar deflection is toward the left of the path of motion. The Coriolis effect is absent at the equator but increases in strength toward the poles.

On Figure 5.6 the arrows within the circle show how an initial straight line of motion is modified by the Coriolis effect. Note especially that the compass direction is not of any consequence. If we face down the direction of motion, turning will always be toward our right in the northern hemisphere. Because the deflective effect is very weak, its action is conspicuous only in freely moving fluids such as air or water. Consequently, both winds and ocean current patterns are greatly affected by the Coriolis effect.

Our next step is to apply the Coriolis principle to winds close to the ground surface. Figure 5.7 shows isobars running east-west, forming a ridge of high pressure in each hemisphere. From each ridge, pressure decreases both to the north and south toward belts of low pressure. Broad arrows show the pressure gradient. The Coriolis effect turns the wind so that it crosses the isobars at an angle. For surface winds, the angle of turning is limited

FIGURE 5.5 A combination wind vane and cup anemometer. Wind speed and compass direction are displayed on the meter below. Also shown are a barometer and a maximum/minimum thermometer. (Courtesy of Taylor Instrument Company and Ward's Natural Science Establishment, Rochester, N.Y.)

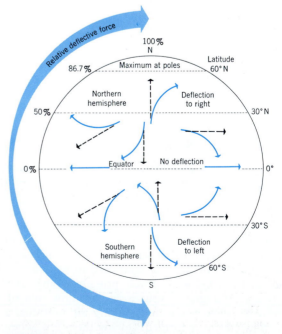

FIGURE 5.6 Deflective effect of the earth's rotation.

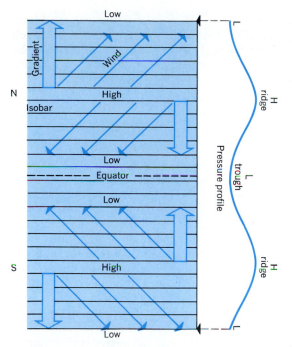

FIGURE 5.7 Surface winds cross isobars at an angle as the air moves from higher to lower pressure. Turn the figure sideways to view the pressure profile.

by the force of friction of the air with the ground. The diagram shows the wind making an angle of 45° with the isobars. In nature the angle is subject to some variation, depending on the character of the ground surface.

Looking first at the northern hemisphere case, the deflection is to the right. The northward pressure gradient gives a southwest wind. The southward gradient gives a northeast wind. In the southern hemisphere, winds are deflected to the left and the pattern is the mirror image of that in the northern hemisphere.

A general rule for the relationship of wind to pressure in the northern hemisphere is known as Ballot's law, which states: Stand with your back to the wind and the low pressure will be toward your left, the high toward your right.

Where isobars are curved, as shown on the simple weather map in Figure 5.8, the direction of pressure gra-

dient follows a curving trajectory, always cutting the isobars at a right angle. The map shows that where isobars are widely spaced, the gradient is weak; where isobars are closely spaced, the gradient is strong. Short arrows show wind directions. Wind speed is slow where the gradient is weak, fast where gradient is strong.

CYCLONES AND ANTICYCLONES

In the language of meteorology, a center of low pressure is called a *cyclone;* a center of high pressure is an *anticyclone.* Cyclones and anticyclones may be of a stationary type, or they may be rapidly moving pressure centers such as those that create the weather disturbances described in Chapter 7. Isobars form closed, circular patterns around both cyclones and anticyclones.

For surface winds, which move obliquely across the isobars, the systems for cyclones and anticyclones in both hemispheres are shown in Figure 5.9. Winds in a cyclone in the northern hemisphere show an anticlockwise inspiral. In an anticyclone, there is a clockwise outspiral. Note the reversal between the labels "anticlockwise" and "clockwise" in the southern hemisphere.

In both hemispheres the surface winds spiral inward on the center of the cyclone, so the air is converging on the center and must also rise to be disposed of at higher levels. For the anticyclone, by contrast, surface winds spiral out from the center. This motion represents a diverging of airflow and must be accompanied by a sinking (subsidence) of air in the center of the anticyclone to replace the out-moving air.

GLOBAL DISTRIBUTION OF SURFACE PRESSURE SYSTEMS

To understand the earth's surface wind systems, we must first study the global system of barometric pressure distribution. Once we grasp the patterns of isobars and

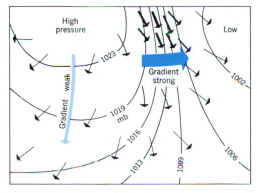

FIGURE 5.8 This small portion of a surface weather map shows curving isobars, with both weak and strong pressure gradients. The short arrows show the surface wind directions.

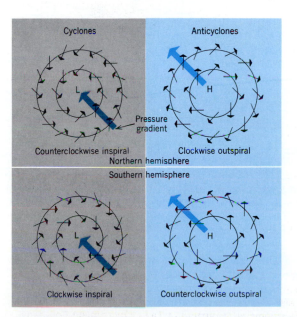

FIGURE 5.9 Surface winds in cyclones and anticyclones.

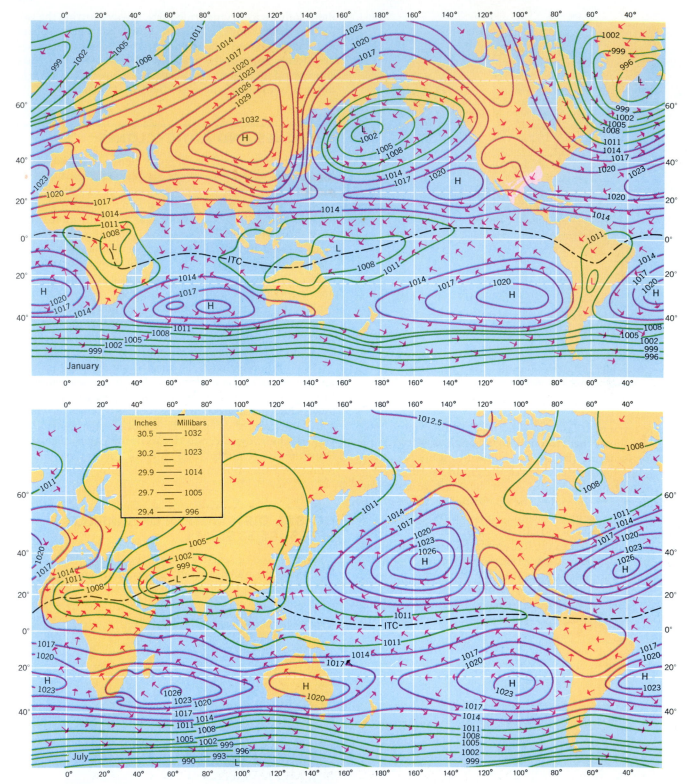

FIGURE 5.10 Mean monthly atmospheric pressure and prevailing surface winds for January and July. Pressure units are millibars reduced to sea level. Many of the wind arrows are inferred from isobars. (Compiled by John E. Oliver from published data by Y. Mintz, G. Dean, R. Geiger, and J. Blüthagen.)

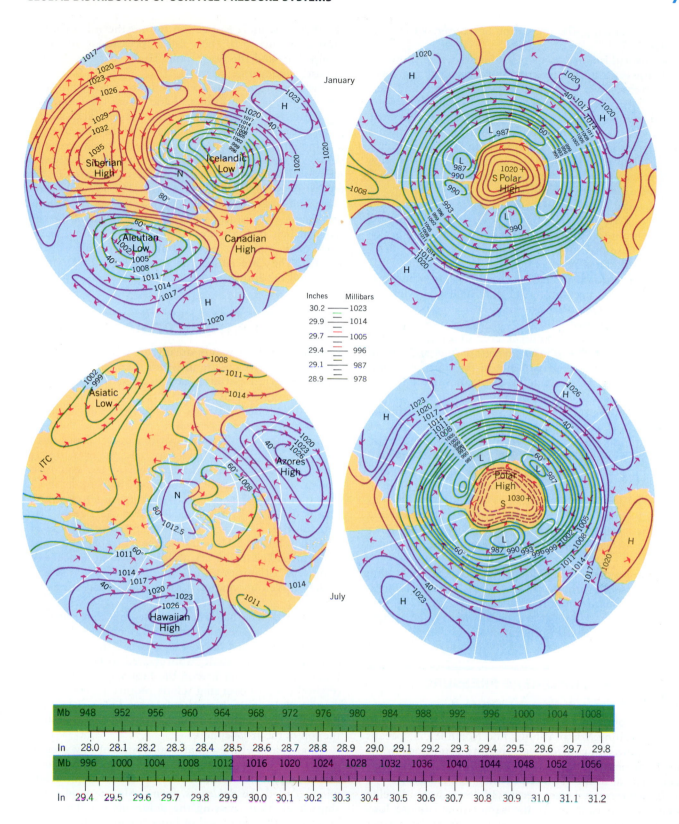

January

July

Inches		Millibars
30.2	——	1023
29.9	——	1014
29.7	——	1005
29.4	——	996
29.1	——	987
28.9	——	978

Mb	948	952	956	960	964	968	972	976	980	984	988	992	996	1000	1004	1008
In	28.0	28.1	28.2	28.3	28.4	28.5	28.6	28.7	28.8	28.9	29.0	29.1	29.2	29.3	29.4	29.5	29.6	29.7	29.8

Mb	996	1000	1004	1008	1012	1016	1020	1024	1028	1032	1036	1040	1044	1048	1052	1056
In	29.4	29.5	29.6	29.7	29.8	29.9	30.0	30.1	30.2	30.3	30.4	30.5	30.6	30.7	30.8	30.9	31.0	31.1	31.2

pressure gradients, the prevailing or average winds can be predicted.

Our world isobaric maps are constructed to show average pressures for the two months of seasonal temperature extremes over large landmasses—January and July (Figure 5.10). Because observing stations lie at various altitudes above sea level, their barometric readings must be reduced to sea-level equivalents, using the standard rate of pressure change with altitude (explained in Chapter 2). When this has been done and the daily readings are averaged over long periods of time, small but distinct pressure differences remain.

A reading of 1013 mb is taken as standard sea-level pressure. Readings higher than this will frequently be observed in midlatitudes, occasionally up to 1040 mb or higher. These pressures are designated as "high." Pressures ranging down to 982 mb or below are "low."

Over the equatorial zone is a belt of somewhat lower than normal pressure, 1011 and 1008 mb, which is known as the *equatorial trough*. Lower pressure is conspicuous by contrast with belts of higher pressure lying to the north and south and centered at about lat. 30°N and S. These are the *subtropical belts of high pressure*, in which pressures exceed 1020 mb. In the southern hemisphere this belt is clearly defined but contains centers of high pressure, called *pressure cells*.

In the southern hemisphere, south of the subtropical high-pressure belt, is a broad belt of low pressure, extending roughly from the midlatitude zone to the arctic zone. The axis of low pressure is centered at about lat. 65°S. This pressure trough is called the *subantarctic low-pressure belt*. Lying over the continuous expanse of Southern Ocean, this trough has average pressures as low as 984 mb. Over the continent of Antarctica is a permanent center of high pressure known as the *polar high*. It contrasts strongly with the encircling subantarctic low.

These pressure belts, along with isotherm belts, shift north and south with the seasons. These annual pressure-belt migrations are important in causing seasonal climate changes. We shall have several occasions to refer to these effects in analyzing world climates.

NORTHERN HEMISPHERE PRESSURE CENTERS

The vast continents of North America and Eurasia and the intervening North Atlantic and North Pacific oceans exert a powerful control over pressure conditions in the northern hemisphere. As a result, the belted arrangement typical of the southern hemisphere is absent.

In winter the large, very cold land areas develop high-pressure centers. At the same time, intense low-pressure centers form over the warmer oceans. Over north central Asia in winter we find the *Siberian high*, with pressure exceeding 1030 mb. Over central North America is a clearly defined, but much less intense, ridge of high pressure, called the *Canadian high*. Over the oceans are the *Aleutian low* and the *Icelandic low*, named after the localities over which they are centered. These two low-pressure areas have much cloudy, stormy

weather in winter. Notice, on the northern hemisphere map of Figure 5.10, how these pressure centers are grouped around the north pole. Highs and lows occupy opposite quadrants.

In summer, pressure conditions are exactly the opposite of winter conditions. Land areas develop low-pressure centers because at this season land–surface temperatures rise sharply above temperatures over the adjoining oceans. At the same time, the ocean areas develop strong centers of high pressure. This system of pressure opposites is striking on both the January and July isobaric maps of Figure 5.10. The low in Asia is intense: it is centered over Afghanistan. Over the Atlantic and Pacific oceans are two large, strong cells of the subtropical belt of high pressure. They are shifted northward of their winter position and are considerably expanded. They are called the *Azores high* (or Bermuda high) and the *Hawaiian high*.

THE GLOBAL PATTERN OF SURFACE WINDS

Prevailing surface winds during January and July are shown by arrows on the pressure maps of Figure 5.10. The basic wind patterns are stressed in Figure 5.11, which is a highly diagrammatic representation showing the earth as if no land areas existed to modify the belted arrangement of pressure zones.

Let us begin with winds of the tropical zones. From the two subtropical high-pressure belts the pressure gradient is equatorward, leading down to the equatorial trough of low pressure. Following the simple model shown in Figure 5.11, air moving from high to low pressure is deflected by the Coriolis effect. As a result, two belts of *trade winds*, or *trades*, are produced. These are labeled in Figure 5.11 as the northeast trades and southeast trades. The trades are highly persistent winds, deviating very little from a single compass direction. Sailing vessels traveling westward once made good use of the trades.

The pattern of the trades suggests that they must converge somewhere near the equator. Meeting of the trades takes place within a narrow zone called the *intertropical convergence zone*, usually abbreviated to *ITC*. The position of the ITC is marked on the January and July maps (Figure 5.10). Converging winds require a rise of air to dispose of the incoming volume of air. This rise takes the form of stalklike flow columns carrying the air toward the top of the troposphere.

Along parts of the equatorial trough of low pressure at certain times of year the trades do not come together in convergence. Instead, a belt of calms and variable winds, called the *doldrums*, forms. Mariners on sailing ships knew that crossing the doldrums was hazardous because of the likelihood of lying becalmed for long periods.

The trades, doldrums, and ITC all shift seasonally north and south, along with the shifting of pressure belts and isotherms. The ITC migrates north and south only a few degrees of latitude over the Pacific and Atlantic oceans, but covers as much as 20° to 30° of latitude over South America, Africa, and the large region of Southeast

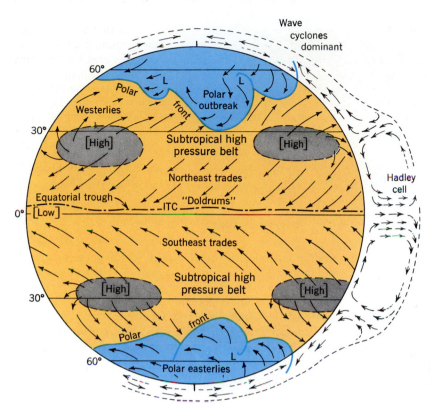

FIGURE 5.11 This schematic diagram of global surface winds disregards the disrupting effect of large continents in the northern hemisphere.

Asia and the Indian Ocean. Important seasonal changes in winds, cloudiness, and rainfall accompany these migrations of the ITC and the trades.

We now return to the subtropical high-pressure belt, which ranges between lat. 25° and 40°N and S. Here we encounter large, stagnant, high-pressure cells (anticyclones). In the centers of the cells, winds are weak and are distributed around a wide range of compass directions; calms prevail as much as a quarter of the time. Because of the high frequency of calms, mariners named this belt the *horse latitudes*. It is said that the name originated in colonial times from the experiences of New England traders, carrying cargoes of horses to the West Indies. When their ships were becalmed for long periods of time, the freshwater supplies ran low and the horses had to be thrown overboard.

Figure 5.12 is a schematic map of large anticyclonic cells centered over the oceans in two hemispheres. Winds make an outspiraling pattern that feeds into the converging trades. On the western sides of the cells, air flows poleward; on the eastern sides, the flow is equatorward. These flows have a strong influence on the climates of adjacent continental margins. Dryness of climate is a dominant general characteristic of the subtropical high-pressure belt and its cells, and we shall emphasize this feature again.

Between lat. 35° and 60°N and S is the belt of *prevailing westerly winds*, or *westerlies*. These surface winds are shown on Figure 5.11 as blowing from a southwesterly quarter in the northern hemisphere and from a northwesterly quarter in the southern hemisphere. This generalization is somewhat misleading, however, because winds from polar directions are frequent and strong. It

is more nearly accurate to say that within the westerlies, winds blow from all directions of the compass, but that the westerly components are definitely predominant. Rapidly moving cyclonic storms are common in this belt.

In the northern hemisphere, landmasses disrupt the westerlies but, in the southern hemisphere between lat. 40° and 60°S, there is an almost unbroken belt of ocean. Here the westerlies gather great strength and persistence. Sailors in the great clipper ships called these latitudes, "roaring forties," "furious fifties," and "screaming sixties." This belt was used extensively by sailing vessels traveling eastward from the South Atlantic Ocean to Australia, Tasmania, New Zealand, and the southern Pa-

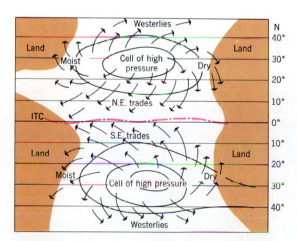

FIGURE 5.12 Over the oceans, surface winds spiral outward from the dominant cells of high pressure, feeding the trades and the westerlies.

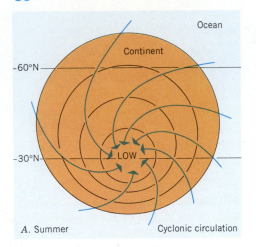
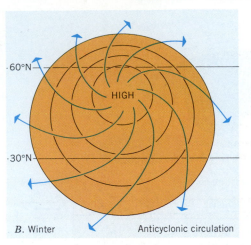

A. Summer   Cyclonic circulation

B. Winter   Anticyclonic circulation

FIGURE 5.13 Schematic diagram of yearly alternations in surface pressure and air flow over an imaginary middle-latitude continent in the Northern Hemisphere. (From S. Petterssen, *Introduction to Meteorology*, 3rd ed., p. 191, Figure 11.17. Copyright © 1969 by McGraw-Hill, Inc., New York. Reproduced by permission of the publisher.)

cific Islands. From these places it was then easier to continue eastward around the world to return to European ports. Rounding Cape Horn was relatively easy on an eastward voyage but, in the opposite direction, in the face of prevailing stormy westerly winds, it was a most dangerous operation.

A wind system called the *polar easterlies* has been said to be characteristic of the arctic and polar zones (Figure 5.11). The concept is at best greatly oversimplified and is certainly misleading when applied to the northern hemisphere. Winds in these high-latitude zones take a variety of directions, dictated by local weather disturbances. On the other hand, in the southern hemisphere, Antarctica is an ice-capped landmass resting squarely on the pole and surrounded by a vast oceanic expanse. Here the outward spiraling flow of polar easterlies seems to be a dominant feature of the circulation. The polar maps in Figure 5.10 show these easterly winds.

MONSOON WINDS OF SOUTHEAST ASIA

The powerful control exerted by the great landmass of Asia on air temperatures and pressures extends to the surface wind systems as well. Development of a summer low and a winter high over Asia in middle latitudes creates a seasonally alternating system of pressure gradients and with it a seasonally reversing set of surface winds which we call a *monsoon system*. The idealized wind system for an imaginary circular continent in the

middle latitudes of the Northern Hemisphere is shown in Figure 5.13. In summer (diagram A) the low-pressure center tends to develop in the southern part of the continent. This is a low associated with heating of the lower levels of the atmosphere; it does not extend high into the troposphere. Nevertheless pressure gradients at low levels are radially inward from sea to land, so that a cyclonic circulation forms. This is the *summer monsoon*, which in southern and eastern Asia is associated with the rainy season of the year.

In winter (diagram B), intense cooling produces the center of higher pressure in the northerly part of the landmass. Now an anticyclonic circulation is set up with surface winds blowing from continental interior toward the coast. In southern and eastern Asia this *winter monsoon* is associated with a period of dry, cool weather. Figure 5.14 shows pressures and surface winds in southern Asia during July and January, when the summer and winter monsoon seasons are strongly developed.

North America does not have the remarkable extremes of monsoon winds experienced by southeastern Asia but, even so, there is a distinct alternation of average temperature and pressure conditions between winter and summer. Wind records show that in summer there is a prevailing tendency for air originating in the Gulf of Mexico to move northward across the central and eastern parts of the United States, whereas in winter there is a prevailing tendency for air to move southward from high-pressure sources in Canada. Wind arrows in Figure 5.10 show this seasonal alternation in the airflow pattern.

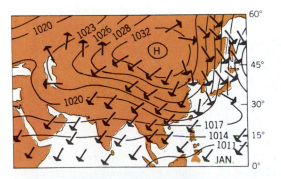
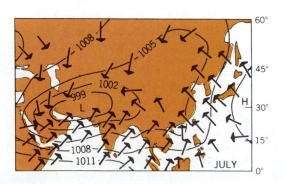

FIGURE 5.14 Surface maps of pressure and winds over Southeast Asia in January and July. Pressure in millibars.

LOCAL WINDS

In certain localities, *local winds* are generated by immediate influences of the surrounding terrain rather than by the large-scale pressure systems that produce global winds and large traveling storms. Local winds are of environmental importance in various ways. They can exert a powerful stress on animals and plants when the air is dry and extremely hot or cold. Local winds are also important in affecting the movement of atmospheric pollutants.

One class of local winds—sea breezes and land breezes—was explained earlier in this chapter. The cooling sea breeze (or lake breeze) of summer is an important environmental resource of coastal communities because it adds to the attraction of the shore zone as a residential or recreational site.

Mountain winds and *valley winds* are local winds following a daily alternation of direction in a manner similar to the land and sea breezes. During the day, air moves from the valleys, upward over rising mountain slopes, toward the summits. At this time hillslopes are intensely heated by the sun. At night the air then moves valleyward, down the hillslopes, which have been cooled at night by radiation of heat from ground to air. These winds are responding to local pressure gradients set up by heating or cooling of the lower air.

Still another group of local winds are known as *drainage winds*, or *katabatic winds*, in which cold air flows under the influence of gravity from higher to lower regions. Such cold, dense air may accumulate in winter over a high plateau or high interior valley. When general weather conditions are favorable, some of this cold air spills over low divides or through passes to flow out on adjacent lowlands as a strong, cold wind. Drainage winds occur in many mountainous regions of the world and go by various local names. The *mistral* of the Rhône valley in southern France is a well-known example; it is a cold, dry local wind. On the ice sheets of Greenland and Antarctica, powerful drainage winds move down the gradient of the ice surface and are funneled through coastal valleys to produce powerful blizzards lasting for days at a time.

Another type of local wind occurs when the outward flow of dry air from a strong high-pressure center (an anticyclone) is combined with the local effects of mountainous terrain. An example is the *Santa Ana*, a hot, dry easterly wind that, on occasion, blows from the interior desert region of southern California across coastal mountain ranges to reach the Pacific coast. Locally, this wind is funneled through narrow mountain gaps or canyon floors, where it gains great force. At times the Santa Ana wind carries large amounts of dust. It is greatly feared for its ability to fan intense brush fires out of control. The *bora* of the Adriatic coast of Yugoslavia is another wind of this class. It is a cold winter wind produced by the pressure gradient of a strong anticyclone located over the land. Descending from the coastal mountains to the sea, it can produce gusts up to 160 km (100 mi) per hour.

Still another type of local wind, bearing such names as *foehn* and *chinook*, results when strong regional winds passing over a mountain range are forced to descend on

the lee side with the result that the air is heated and dried. These winds are explained in Chapter 6.

WINDS ALOFT

The global surface wind systems we have examined represent only a shallow basal air layer a few hundred meters deep, whereas the troposphere is 10–15 kilometers deep. How does air move at these higher levels? Large, slowly moving high-pressure and low-pressure systems are found aloft, but these are generally simple in pattern, with smoothly curved isobars.

Winds high above the earth's surface are not affected by friction with the ground or water over which they move. The Coriolis effects turns the flow of air until it becomes parallel with the isobars, as Figure 5.15 shows. (The Coriolis effect is shown as if it were a force.) In this position the pressure gradient force and Coriolis force are exactly opposed and exactly balanced.

The ideal wind in this state of balance with respect to the two forces is termed the *geostrophic wind* for cases in which the isobars are straight. Where isobars are curved, centrifugal force must also be taken into account; but, in general, airflow at high altitudes parallels the isobars.

The upper part of Figure 5.15 is a simplified map of pressure and winds high in the troposphere. Notice how the wind arrows run parallel with the isobars, forming elliptic flow patterns around lows and highs. Our rules for winds in cyclones and anticyclones need to be slightly modified, as compared with surface winds. For an airline pilot flying in the northern hemisphere and wishing to

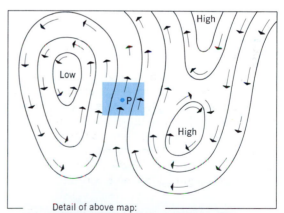

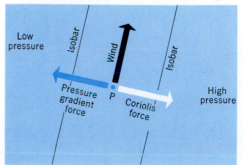

Detail of above map:

FIGURE 5.15 Wind follows isobars at high levels.

keep a tailwind at all times, the rule would be "keep the highs on your right and the lows on your left."

In describing pressure belts and winds at the earth's surface, we gave no explanation of their basic cause in terms of earth rotation and the Coriolis principle. To understand the surface winds, we need to investigate the workings of the entire troposphere, particularly at the higher levels, where surface friction is absent and the airflow closely follows the isobars.

WINDS ON A NONROTATING PLANET

Consider, first, an imaginary nonrotating planet that is heated uniformly around the equatorial belt (where a radiation surplus exists) but cooled severely at both polar regions (where radiation deficits exist). Figure 5.16 shows this ideal case. Air heated at the equator will expand and become less dense because all gases expand in volume when heated. Because the expanded lower air layer is now less dense, the atmospheric pressure at the earth's surface will be lower than average (L, Figure 5.16). The heated air will rise and upon reaching high levels in the atmosphere will spread horizontally, moving poleward. Chilled air over the poles will increase in density, creating high pressure at the earth's surface (H). The chilled air will sink and then spread horizontally, traveling toward the equatorial belt of low pressure. Once the airflow is established, a system of *meridional winds* results, as shown in Figure 5.16. The global winds will now be formed into two circulation cells, one in each hemisphere. As long as heat continues to be supplied to the equatorial belt, the wind system will remain in operation. We have here a *heat engine*, that is, a mechanical system driven by an input of heat energy.

Our model of a nonrotating earth serves to explain one very real feature of the earth's atmospheric circulation: the equatorial belt of low pressure, or equatorial

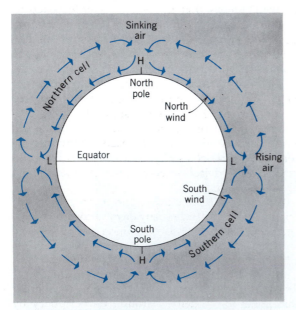

FIGURE 5.16 An imaginary atmospheric circulation system on a nonrotating planet.

trough, within which heated air rises to high levels. What actually happens to this rising air as it begins to move poleward in the upper atmosphere can be understood only by bringing into account the Coriolis effect.

THE HADLEY CELL CIRCULATON

When we introduce the Coriolis effect into the simple model of meridional circulation shown in Figure 5.16, the single hemispherical cell must break down into a much more complex air flow system. Warm air, rising at the equator and beginning to move northward, will be caused to veer toward the right—eastward, that is—and when it has reached a latitude of about 30° will be traveling due east as a westerly wind. (In the southern hemisphere, deflection toward the left would also result in a westerly wind at high levels.)

Because further poleward progress is prevented by the diversion to a westerly flow, the air in this zone tends to accumulate, or "pile up." A high-pressure belt is thus formed, and the air sinks to low levels, as shown by arrows in Figure 5.17. This continual subsidence forms what we have already identified as the prevailing subtropical high-pressure belt. Air reaching low levels is now required to move away from the down-sinking zone. Part of this air moves poleward, but much of it moves equatorward, as shown in the righthand part of Figure 5.17. Equatorward motion results in deflection to the west, setting up an easterly wind system. These winds are known as the *tropical easterlies;* near the surface, they are the trade winds. The easterlies form a broad, deep airstream over the entire equatorial belt, where they are referred to as the *equatorial easterlies*. In the lower troposphere, the westward-moving air is also slowly converging over the equator in the intertropical convergence zone (ITC). This converging air is rising slowly to complete the entire circuit.

Taking into account only the north-to-south, south-to-north, and vertical components of air movement—in other words, only the meridional flow—we find a cell of atmospheric circulation dominating the tropical and equatorial zones. This system has been named the *Hadley cell* in honor of George Hadley, who postulated its existence in 1735. Ideally there are two matching Hadley cells, one for each hemisphere, as shown in Figure 5.17.

THE UPPER-AIR WESTERLIES AND ROSSBY WAVES

Poleward of the Hadley cell in both hemispheres, flow in the troposphere is generally and persistently eastward, following the parallels of latitude and forming the global system of *upper-air westerlies*, as sketched in Figure 5.17.

Figure 5.18 shows in a simplified way the westerly winds in relation to the subtropical high-pressure belt and the tropical easterlies. Notice that a number of high-pressure cells comprise the high-pressure belt. The upper-air westerlies persist into the polar region, where they form a great vortex. Atmospheric pressure falls rap-

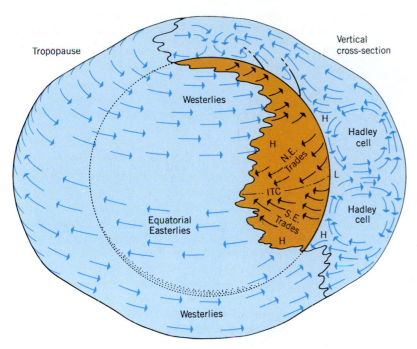

FIGURE 5.17 A schematic diagram of the Hadley cell circulation and its relationship to upper-air westerlies and equatorial easterlies.

idly toward the polar region, where the *polar low* is situated.

Westerly winds around the polar lows involve the entire depth of the troposphere. Recall from Chapter 2 (Figure 2.8) that the troposphere is much thinner at high latitudes than at low latitudes. The uniform flow of the upper-air westerlies is frequently disturbed by the formation of large undulations, called *Rossby waves* (Figure 5.19). These waves develop along a narrow zone of contact between the large body of cold polar air, forming the troposphere on the poleward side, and warm tropical air, which surrounds the globe on the equatorward side. This contact zone is called the *polar front;* it is an unstable zone, along which severe atmospheric disturbances are generated.

Diagram A of Figure 5.19 shows the polar front in a fairly smooth, stable condition. Minor undulations are, however, beginning to form along the front. These undulations deepen greatly (Diagrams B and C), becoming Rossby waves. Cold polar air pushes into lower latitudes in deep embayments, which form troughs of low pressure at high levels. Two such troughs are shown in Diagram C, their axes marked by dashed lines. Correspondingly, warm tropical air moves into higher latitudes in embayments between the troughs. Here, high-pressure ridges are formed in the troposphere. As we shall find in Chapter 7, this phase of wave development is associated with the outbreak of storms (cyclonic storms) near the surface.

Diagram D shows that the waves have deepened to the extent that the low-pressure troughs are constricting and becoming detached. When this happens, the detached mass of cold air becomes a *cut-off low,* which is now a cyclone in the upper atmosphere. A warm-air embayment may also become detached to form a *cut-off high,* which is an upper-air anticyclone. The importance of this cutoff process lies in the transport of cold air into lower latitudes and of warm air into high latitudes. Through this *advection process,* or horizontal mixing, the meridional transport of surplus heat into high latitudes is accomplished—a vital process in the maintenance of the global heat balance.

Rossby waves develop slowly. The cutoff process we have described may require many days to complete. The waves may weaken and dissolve without completing the cutoff cycle; they may remain stationary for long periods or may drift slowly in the east-west direction.

Figure 5.20 is an upper-air pressure map of the northern hemisphere for a day in April. It shows the great upper-air polar low indented around its periphery by several Rossby waves, some of which contain cut-off lows. Air flow is indicated by arrow points on the isobars.

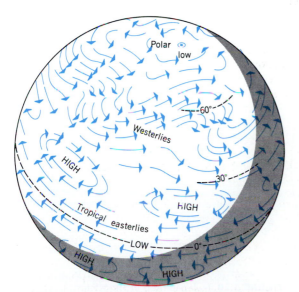

FIGURE 5.18 Schematic representation of circulation in the upper part of the troposphere, 6 to 12 km (20,000 to 40,000 ft). (From A. N. Strahler, *The Earth Sciences,* 2nd ed., Harper & Row, Publishers. Copyright © by Arthur N. Strahler.)

Rossby Waves and the Jet Stream

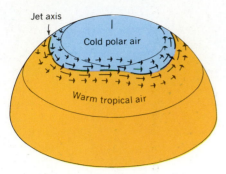

(A) The jet stream begins to undulate.

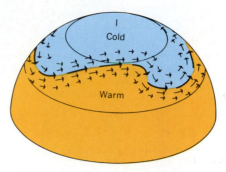

(B) Rossby waves begin to form.

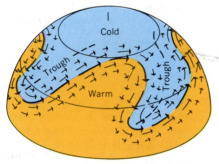

(C) Waves are strongly developed. The cold air occupies troughs of low pressure.

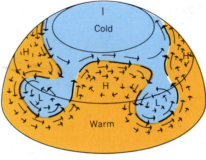

(D) When the waves are pinched off, they form cyclones of cold air.

FIGURE 5.19 Development of upper-air waves in the westerlies. (Data from J. Namias, NOAA, National Weather Service. From A. N. Strahler, *The Earth Sciences*, 2nd ed., Harper & Row, Publishers. Copyright © by Arthur N. Strahler.)

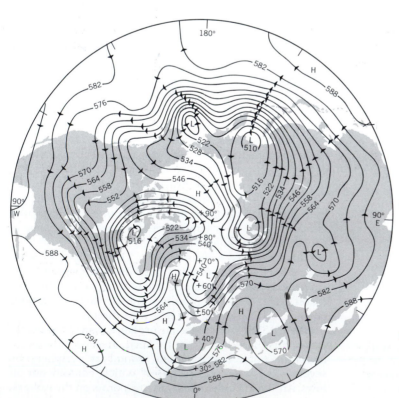

FIGURE 5.20 Isobaric map at the 5.5 km level (18,000 ft) for a day in April. Pressures are in millibars. Arrows show upper-air winds. (Based on satellite data from NOAA and NASA.)

THE POLAR FRONT JET STREAM

Associated with Rossby waves is a narrow zone of very high wind speed called the *jet stream*, formed along the line of contact of cold air and warm air. The position of the jet stream is shown in Figure 5.19 by a line of heavy arrows. The jet stream is a pulselike flow of air. It resembles a stream of water moving in a hose. Velocity is highest in the center line, or *core*, of the jet stream, which is surrounded by slower-moving zones. Figure 5.21 shows an actual jet stream in cross section, using lines of equal wind speed. Maximum speed in the core was about 300 km/hr (185 mph). Altitude of the core was about 11 km (36,000 ft). Figure 5.22 is a weather map showing that this jet was passing over the western United States, in a path curving southeast toward the Gulf of Mexico, then recurving to the northeast over Florida and outlining a Rossby wave. Position of the jet core is shown by a solid line. Lines of equal wind speed (isotachs) show that the pulse of highest speed was located over Wyoming.

The jet stream we have described forms at the level of the tropopause, along the polar front, the surface of contact between cold polar air and warm tropical air. For this reason, it is given the name *polar front jet stream*. The existence of the jet is fully explained by a very steep pressure gradient. Figure 5.23 shows the position of the jet core and the polar front at the tropopause in midlatitudes. The tropopause drops sharply in altitude at the polar front, being much lower over the cold troposphere layer than over the warm troposphere. Atmospheric pressure also changes abruptly at the polar front. Isobaric surfaces, shown by lines on the diagram, drop steeply at the polar front, the steepest drop being at the level of the jet-stream core. The steep pressure gradient at this point causes the high-speed flow of air to be local-

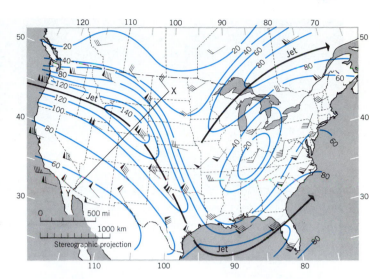

FIGURE 5.22 Isotachs (knots) and wind arrows at the 300-mb (9-km) level on a day in April. See Figure 7.10 for explanation of symbols. Solid arrows mark jet-stream axis. Cross section along the line X-Y is shown in Figure 5.21. (Same data source as in Figure 5.21.)

ized in a narrow zone. The polar front jet stream is extremely important through its control of surface weather patterns over midlatitudes. This is a subject we will develop in Chapter 7.

The polar front jet stream is a factor in the operation of jet aircraft in the range of their normal cruising altitudes in midlatitudes. In addition to strongly increasing or decreasing the ground speed of the aircraft the jet stream carries a form of air turbulence that at times reaches hazardous levels. This is clear air turbulence (CAT); it is avoided when known to be severe.

A second jet stream of major importance in the global circulation forms in the subtropical latitude zone. Called the *subtropical jet stream*, it occupies a position at the

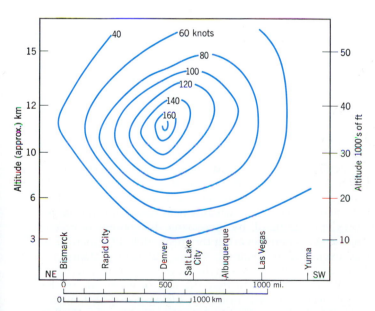

FIGURE 5.21 Cross-sectional diagram through a jet stream over the western United States. See Figure 5.22 for line of section and accompanying map data. (Based on data of National Weather Service in H. Riehl, 1962, *Jet Streams of the Atmosphere*, Colorado State Univ., Fort Collins.)

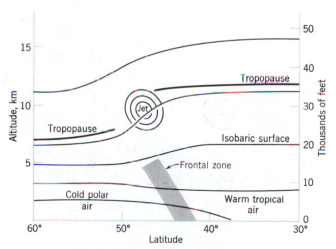

FIGURE 5.23 This schematic cross section through the polar front shows the position of the jet-stream core in relation to isobaric surfaces. (Based on data of National Weather Service in H. Riehl, 1962, *Jet Streams of the Atmosphere*, Colorado State Univ., Fort Collins.)

FIGURE 5.24 A strong subtropical jet stream is made visible in this photo through a narrow band of cirrus clouds generated by the turbulent air stream. This jet stream is moving from west to east at an altitude of about 12 km (40,000 ft). The cloud band lies at about lat. 25°N. In this view, astronauts aimed their camera toward the southeast, taking in the Nile River valley and the Red Sea. At the left we see the tip of the Sinai Peninsula. (NASA.)

tropopause just above the Hadley cell (Figure 5.24). Here westerly wind speeds reach maximum values of 345 to 385 km/hr (215 to 240 mph). Figure 5.25, a northern-hemisphere map, shows the average position of both the polar front jet stream and the subtropical jet stream in the winter season.

A third jet stream system is found at even lower latitudes. Known as the *tropical easterly jet stream*, it runs from east to west—opposite in direction to that of the polar-front and subtropical jet streams. The tropical easterly jet stream occurs only in the summer (high-sun) season and is limited to a northern-hemisphere location over Southeast Asia, India, and Africa. This jet stream is very high in altitude, about 15 km (50,000 ft), and has wind speeds over 180 km (115 mi) per hour. This tropical easterly jet stream is considered by some atmospheric scientists to play a major role in the rainy summer monsoon of Southeast Asia.

Figure 5.26 is a meridional cross section of the atmosphere showing in a schematic way the position of the three jet-stream systems. The diagram shows only the easterly and westerly components of the atmospheric circulation, disregarding the meridional components.

GREENHOUSE WARMING AND THE GLOBAL CIRCULATION

Will a substantial warming of global climate in coming decades affect the atmospheric circulation system? Some speculative answers can be suggested on the basis of principles of atmospheric science. We have seen in Chapters 3 and 4 that the global atmosphere is a sort of heat engine that responds to changes in the input of its fuel, which under natural conditions is incoming solar radiation. Adding heat from increased burning of fossil fuels and trees might seem to augment that fuel supply, but the actual quantity of sensible heat released is not large. On the other hand, by increasing the content of greenhouse gases and elevating the mean global temperature, as explained in Chapter 4, the radiation balance responds by shifting to a higher level of kinetic activity.

We can anticipate that the rise of warm air in the equatorial zone will intensify, lifting greater quantities of air to higher altitudes and strengthening the Hadley cells. These cells can be expected to enlarge in latitudinal extent, causing the subtropical zone of high pressure formed under the sinking air to widen and push into higher latitudes. Because a desert zone coincides with this high-pressure belt, it may be anticipated that the

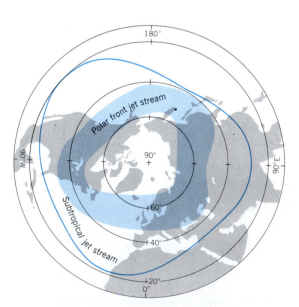

FIGURE 5.25 Mean winter position of the axis of the subtropical jet stream and the belt of principal winter activity of the polar front jet stream. (Based on data of H. Riehl, 1962, *Jet Streams of the Atmosphere*, Colorado State Univ., Fort Collins.)

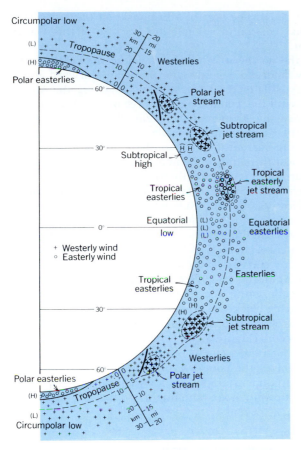

FIGURE 5.26 A schematic diagram of wind directions and jet streams along an average meridian from pole to pole. (From A. N. Strahler, *The Earth Sciences*, 2nd ed., Harper & Row, Publishers. Copyright © by Arthur N. Strahler.)

great subtropical deserts will shift poleward into what are now moister climate zones. We will come back to this possible effect in Chapter 6. Intensification of the tropical easterly winds, both at high levels and at the surface in the trade winds, is a related possibility. Perhaps, also, the seasonal Asiatic monsoon winds at low levels will be intensified.

Another effect of Hadley cell expansion may be to intensify the contrast between airmasses that are in contact along the polar front in both hemispheres. We might further predict that the polar jet stream will become more powerful and that weather disturbances originating along the polar front will become more frequent and more intense.

OCEAN CURRENTS

An *ocean current* is any persistent, dominantly horizontal flow of ocean water. Ocean currents are important regulators of thermal environments of the earth's surface. On a global scale, the vast current systems aid in exchange of heat between low and high latitudes and are essential in sustaining the global heat balance. On a local scale, warm water currents bring a moderating influence to coasts in arctic latitudes; cool currents greatly alleviate the heat of tropical deserts along narrow coastal belts.

Practically all the important surface currents of the oceans are set in motion by prevailing surface winds. Energy is transferred from wind to water by the frictional drag of the air blowing over the water surface. Because of the Coriolis effect, the water drift is impelled toward the right of its path of motion (northern hemisphere); therefore the current at the water surface is in a direction about 45° to the right of the wind direction. Under the influence of winds, currents may tend to bank up the water close to the coast of a continent, in which case the force of gravity, tending to equalize the water level, will cause other currents to be set up.

Density differences may also cause flow of ocean water. Such differences arise from greater heating by insolation, or greater cooling by radiation, in one place than in another. Thus surface water chilled in the arctic and polar seas will sink to the ocean floor, spreading equatorward and displacing upward the less dense, warmer water.

Still another controlling influence on water movements is the configuration of the ocean basins and coasts. Currents initially caused by winds impinge on a coast and are locally deflected.

GENERALIZED SCHEME OF OCEAN CURRENTS

To illustrate global surface water circulation, we can refer to an idealized ocean extending across the equator to latitudes of 60° or 70° on either side (Figure 5.27). Perhaps the most outstanding features are the circular movements, called *gyres*, around the subtropical highs,

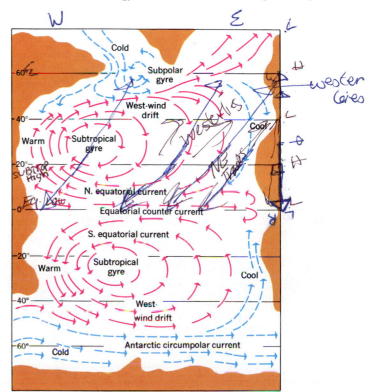

FIGURE 5.27 Schematic map of system of ocean currents. (From A. N. Strahler, *The Earth Sciences*, 2nd ed., Harper & Row, Publishers. Copyright © by Arthur N. Strahler.)

centered about 25° to 30°N and S. An *equatorial current* with westward flow marks the belt of the trades. Although the trades blow to the southwest and northwest, obliquely across the parallels of latitude, the water movement follows the parallels.

A slow, eastward movement of water over the zone of the westerly winds is named the *west-wind drift*. It covers a broad belt between 35° and 45° in the northern hemisphere, and between 30° or 35° and 70° in the southern, where open ocean exists in the higher latitudes. The equatorial currents are separated by an *equatorial countercurrent*. This is well developed in the Pacific, Atlantic, and Indian oceans as shown on the world map of ocean currents and drifts (Figure 5.28).

Along the west sides of the oceans in low latitudes the equatorial current turns poleward, forming a warm current paralleling the coast. Examples are the Gulf Stream (Florida or Caribbean stream), the Japan current (Kuroshio), and the Brazil current, which bring higher than average temperatures along these coasts.

The west-wind drift, upon approaching the east side of the ocean, is deflected both south and north along the coast. The equatorward flow is a cool current, produced by the upwelling of colder water from greater depths. It is well illustrated by the Humboldt current (Peru current) off the coast of Chile and Peru; by the Benguela current off the southwest African coast; by the California current off the west coast of the United States; and by the Canaries current off the Spanish and North African coast.

In the northeastern Atlantic Ocean, the west-wind drift is deflected poleward as a relatively warm current. This is the North Atlantic drift that spreads around the British Isles, into the North Sea, and along the Norwegian coast. The port of Murmansk, on the arctic circle, has year-round navigability because of this warm current.

In the northern hemisphere, where the polar sea is largely landlocked, cold water flows equatorward along the west side of the large straits connecting the Arctic Ocean with the Atlantic basin. Three principal cold currents are the Kamchatka current, flowing southward along the Kamchatka Peninsula and Kurile Islands; the Greenland current, flowing south along the east Greenland coast through the Denmark Strait; and the Labrador current, moving south from the Baffin Bay area through Davis Strait to reach the coasts of Newfoundland, Nova Scotia, and New England.

In both the north Atlantic and Pacific oceans, the Icelandic and Aleutian lows in a rough way coincide with two centers of counterclockwise circulation involving the cold arctic currents and the west-wind drifts.

The antarctic region has a relatively simple current scheme consisting of a single *antarctic circumpolar current* moving clockwise around the antarctic continent in latitudes 50° to 65°S, where a continuous expanse of open ocean occurs (Figure 5.29).

Oceanographers today recognize that oceanic circulation involves the complex motions of water masses of different temperature and salinity characteristics. Sinking and upwelling are both important motions in certain areas of the oceans.

OCEANIC STREAMS AND THEIR EDDIES

Where warm currents of the great middle-latitude gyres flow poleward in close proximity to cold currents that

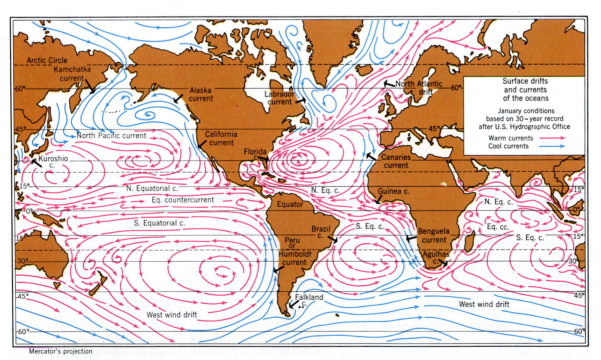

FIGURE 5.28 Surface drifts and currents of the oceans in January. (Based on data of U.S. Navy Oceanographic Office.)

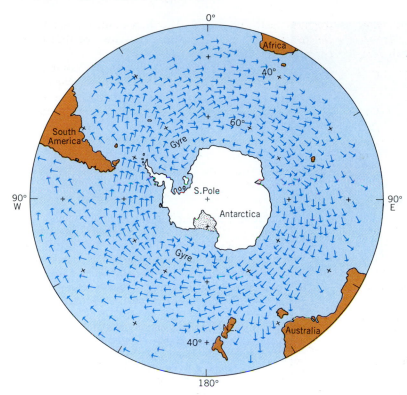

0°
40°
60°
Africa
Gyre
South America
S.Pole
90° W
90° E
Antarctica
Gyre
Australia
40°
180°

FIGURE 5.29 Surface water of the Southern Ocean flows continuously from west to east around Antarctica as a broad circumpolar current, but the pattern is complicated by the presence of faster jet-like streams and by two gyres close to the continent in which flow is reversed in direction. The map is greatly simplified from a computer-generated model. (Based on NOAA data as presented by D. Olbers and M. Wenzel in *EOS*, American Geophysical Union, vol. 71, no. 1, 1990.)

are moving equatorward from their arctic sources, a sharp line of demarcation can usually be drawn between the two. This line can be likened to the polar front of the troposphere, since both air and water are fluids and behave similarly in many respects. We can expect that the surface layer of the oceans will exhibit features resembling the Rossby waves of the atmosphere and their cut-off lows and highs, depicted in Figure 5.19.

A good example is the Gulf Stream, flowing in a northeasterly direction off the Atlantic coast of North America. Lying between this current and the coastline is

a southwesterly drifting mass of cold water—called Slope Water—that is fed by the cold Labrador Current. (Find this area on the world map, Figure 5.28.) Figure 5.30 is a set of schematic maps showing interaction between the warm Gulf Stream and the cold Slope Water. Notice that a large body of relatively stagnant warm water—the Sargasso Sea—lies adjacent to the Gulf Stream. In Stage 1 of Diagram A, the Gulf Stream is shown as resembling a meandering river; it develops a deep "meander bend" that pushes into the Slope Water. In Stage 2, the "meander loop" is closing on itself and

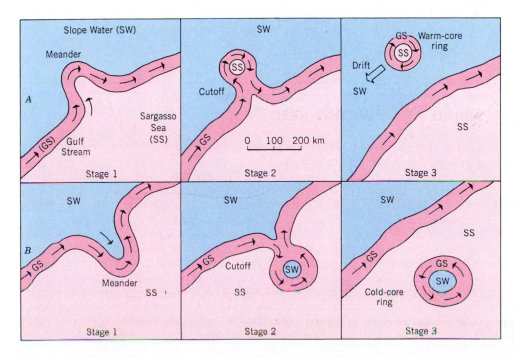

FIGURE 5.30 Schematic maps showing the formation of warm-core and cold-core rings from meanders of the Gulf Stream.

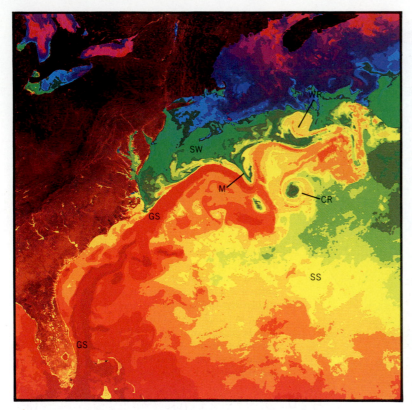

FIGURE 5.31 A satellite image in false colors showing the Gulf Stream (GS) and its interactions with cold water of the Continental Slope (SW), on the one side, and warm water of the Sargasso Sea (SS) on the other. Cold water appears in green and blue tones; warm water in red and yellow tones. Other features include a meander (M), a warm-core ring (WR), and a cold-core ring (CR). The image was made from data collected by NOAA-7 orbiting satellite. Data of 35 satellite passes during a week in April were processed so as to remove the effects of clouds. (Data of NOAA/NESDIS/NCDC/SDSD. Courtesy of Otis B. Brown, Robert Evans, and M. Carle, University of Miami, Rosenstiel School of Marine and Atmospheric Science, Florida.)

only a narrow neck connects it with the main stream. In Stage 3, a cutoff has occurred, leaving a detached eddy of warm Sargasso water surrounded by a closed ring of the Gulf Stream, which in turn is surrounded by cold Slope Water. This eddy is called a *warm-core ring*; it is rotating in a clockwise direction. In a similar manner, shown in Diagram B, a meander may develop in the opposite direction and be cut off (occluded) to generate a *cold-core ring* rotating in an anticlockwise direction, in which cold Slope Water occupies the center. In either case, a ring gradually moves away from the Gulf Stream and eventually dissolves.

Figure 5.31 is a false-color image generated from satellite data, showing the Gulf Stream and cores of the two kinds. One warm-core eddy and two cold-core eddies are easily identified in the area lying due south of the Bay of Fundy.

The interactions we have described here serve the same function as do the cut-off lows and highs of the Rossby waves, namely to mix warm and cold fluid masses and thus transport the heat surplus of low latitudes into the heat-deficient arctic latitudes. This form of mixing of the planetary fluids plays a major role in balancing the global energy budget.

WIND AS AN ENERGY RESOURCE

Wind power is an indirect form of solar energy that has been used for centuries. The windmill of the Low Countries of Europe played a major role in pumping water from the polders as they were reclaimed from tidal land. The windmill was also used to grind grain in low, flat areas where streams could not be adapted to waterpower. The design of new forms of windmills has intrigued inventors for many decades. The total supply of wind energy is enormous. The World Meteorological Organization has estimated that the combined electricity-generating wind power of favorable sites throughout the world comes to about 20 million megawatts, a figure about 100 times greater than the total electrical generating capacity of the United States. Many problems must

be solved, however, in developing wind power as a major resource.

Considerable interest has been expressed in using small, inexpensive windmills for irrigation pumping in India, where farm plots are small and the cost of operating diesel or electric pump motors is high.

Small wind turbines are a promising source of supplementary electric power for individual farms, ranches, and homes. The Darrieus rotor, with circular blades turning on a vertical axis, is well adapted to small generators—those with less than 50 kilowatts output. Wind turbines presently in operation and being developed are adjusted in scale according to the purposes they serve. Turbines capable of producing power in the range

FIGURE 5.32 A windfarm in San Gorgonio Pass, near Palm Springs, California. Wind speeds here average 27 km (17 mi) per hour. (A. N. Strahler.)

of 50 to 200 kilowatts are already in operation to serve small communities.

Wind turbines with a generating capacity in the range from 50 to 150 kilowatts have been assembled in large numbers at favorable locations to form "windfarms" (Figure 5.32). Arranged in rows along ridge crests, the turbines intercept local winds of exceptional frequency and strength. One such locality is the Tehachapi Pass in California, where winds moving southeastward in the San Joaquin Valley are funneled through a narrow mountain gap before entering the Mojave Desert. Another locality lies about 80 km (50 mi) east of San Francisco in the Altamont Pass area of Alameda and Contra Costa counties. Here daytime westerly winds of great persistence develop in response to the buildup of surface low pressure in the Great Valley to the east. A group of windfarms built here with a total of about 3000 turbines provides the Pacific Gas and Electric Company with a yield of over 500 million kilowatt hours of electricity per year.

By 1987, the total capacity of California windfarms stood at about 1400 megawatts, generated by about 17,000 wind turbines. Their total yearly output was the equivalent of 3.5 million barrels of crude oil.

Denmark has aggressively pursued the development of wind farms, and by 1990 had achieved the capability of producing ten percent of its country's electrical energy. The same year saw the completion of a group of wind turbines placed offshore in the Baltic Sea in water depths up to 5 meters. Holland, too, has an extensive system of wind turbines, and a similar development is in early stages in the United Kingdom.

Larger wind turbines, capable of generating from 2 to 4 megawatts, are under construction at a number of sites. In 1980, the Southern California Edison Company put in operation a 3-megawatt wind turbine in the San Gorgonio Pass, near Palm Springs, where wind speeds average 27 km (17 mi) per hour (Figure 5.33). This unit was designed to supply the electrical needs of about 1000 homes, saving annually the combustion of about 10,000 barrels of low-sulfur crude oil per year. The machine was dismantled after successful testing.

The world's largest wind turbine (as of 1990) using a blade with a horizontal axis was situated in Oahu, Hawaii; its 300-m (320-ft) propeller generates 3.2 megawatts. The largest Darrieus turbine, located near the south shore of the St. Lawrence estuary at Cap-Chat, Quebec, delivers 3.6 megawatts. It seems unlikely that

turbines of this size (2 to 4 megawatts) will become numerous.

Windpower machines adapted to operating from the ocean surface are based on special designs, using floating platforms or spar-buoy flotation to support propeller-driven turbines. Units capable of delivering several megawatts of power are envisioned as being grouped in offshore locations where winds are persistent and have much higher average speeds than at locations on land.

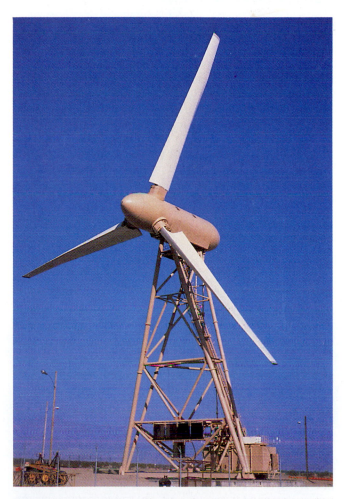

FIGURE 5.33 This giant wind turbine near Palm Springs, California, was designed to generate 3 megawatts of electric power in a 65-km (40-mi) wind. It has since been dismantled. (Southern California Edison.)

WAVE POWER

Wave energy is another indirect form of solar energy. Nearly all ocean waves are produced by the stress of wind blowing over the sea surface. Water waves are characterized by motion of water particles in vertical orbits. There is only a very slow downwind motion of the water, but a large flow of kinetic energy travels rapidly with the troughs and crests of the moving wave forms. Energy is extracted from wave motion by use of a floating object tethered to the seafloor. As the floating mass rises and falls, a mechanism is operated and drives a generator. Pneumatic systems use the principle of the bellows, operated by the changing pressure of the surrounding water as the water level rises and falls. A number of such devices have been devised and tested.

Another basic form of wave power uses a troughlike concrete structure with an upsloping floor to channel the powerful landward surge of a breaking wave into an elevated reservoir. The water then drains down through a turbine, after which it is returned to the ocean. Two such power stations were built and successfully operated in the latter half of the 1980s near Bergen, Norway. (One of these was destroyed by a storm in 1988.)

CURRENT POWER

Harnessing the vast power of a great current stream, such as the Gulf Stream or the Kuroshio, is a prospect that has not gone unnoticed. Called the Coriolis Program, after the effect that concentrates the Gulf Stream flow against the continental margin, one rather grandiose plan has called for a large number of current-driven turbines to be tethered to the ocean floor so as to operate below the ocean surface. Each turbine would be 170 m (550 ft) in diameter and capable of generating 83 megawatts of power. An array of 242 such turbines could deliver 10,000 megawatts—a large part of Florida's power requirement and equivalent to the use of about 130 million barrels of crude oil per year.

THE GLOBAL CIRCULATION AND THE HUMAN ENVIRONMENT

We have outlined the major mechanical circuits within which sensible heat is transported from low latitudes to high latitudes. In low latitudes, the Hadley cell operates like a simple heat engine to transport heat from the equatorial zone to the subtropical zone. Upper-air waves take up the transport and move warm air poleward in exchange for cold air. Ocean currents perform a similar function through the turning of the great gyres.

The global atmospheric circulation also transports heat in the latent form held by water vapor. This heat is released by condensation, a process we shall examine in the next chapter. The movement of water vapor also represents a mass transport of water and is part of the world water balance.

Wind systems of the lower troposphere have direct environmental significance. Arriving at a mountainous coastal zone after a long travel path over a great ocean, these winds carry a large amount of water vapor, which is deposited as precipitation on the coast. In this way the distribution of our water resources is partly determined by the atmospheric circulation patterns. Winds also transport atmospheric pollutants, carrying them tens and hundreds of kilometers from the source of pollution. These are environmental topics related to winds; we will investigate them in the next two chapters.

CHAPTER 6

ATMOSPHERIC MOISTURE AND PRECIPITATION

We have stressed that heat and water are vital ingredients of the environment of the biosphere, or life layer. Plant and animal life of the lands, on which humans depend for food, require fresh water. People use fresh water in many ways. The only basic source of fresh water is from the atmosphere through condensation of water vapor. In this chapter we are concerned mostly with water in the vapor state in the atmosphere and the processes by which it passes into the liquid or solid state and ultimately arrives at the surface of the ocean and the lands through the process of precipitation.

Water also leaves the land and ocean surfaces by evaporation and so returns to the atmosphere. Evidently, the global pathways of movement of water form a complex network. There is a global water balance, just as there is an energy balance; the water balance deals with flow of matter and so complements the energy balance. Let us first review basic terms and processes involved in an understanding of the water balance.

WATER STATES AND HEAT

As we explained in Chapter 2, water occurs in three states: (1) solid, frozen as crystalline ice; (2) liquid, as water; (3) gaseous, as water vapor (see Figure 2.17). From the gaseous state, molecules may pass into the liquid state by condensation; or, if temperatures are below the freezing point, they can pass by sublimation directly into the solid state to form ice crystals. By evaporation, molecules can leave a water surface to become gas molecules in water vapor. The analogous change from ice directly into water vapor is also designated sublimation. Then, of course, water may pass from liquid to solid state by freezing, and from solid state to liquid state by melting. These changes can be represented by a triangle in which the three states of water form the corners (Figure 6.1). Arrows show the six possible changes of state.

Of prime importance in meteorology are the exchanges of heat energy accompanying changes of state. For example, when water evaporates, sensible heat, which we can feel and measure by thermometer, passes into a hidden form held by the water vapor and known

as the *latent heat of vaporization*. This change results in a drop in temperature of the remaining liquid. The cooling effect produced by evaporation of perspiration from the skin is perhaps the most obvious example. For every gram of water that is evaporated, about 600 calories of sensible heat pass into the latent form. In the reverse process of condensation, an equal amount of energy is released to become sensible heat and the temperature rises correspondingly.

Similarly, the freezing process releases heat energy in the amount of about 80 calories per gram of water, whereas melting absorbs an equal quantity of heat. This heat is referred to as the *latent heat of fusion*. When sublimation occurs, the heat absorbed by vaporization, or released by crystallization, is approximately equal to the sum of the latent heats of vaporization and fusion.

HUMIDITY

The amount of water vapor that may be present in the air at a given time varies widely from place to place. It ranges from almost nothing in the cold, dry air of arctic regions in winter to as much as 4 or 5 percent of a given volume of the atmosphere in the humid equatorial zone.

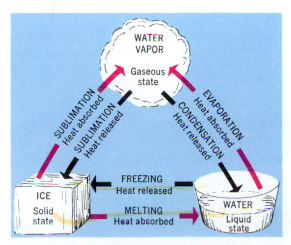

FIGURE 6.1 Three states of water. Changes of state involve absorption or release of heat.

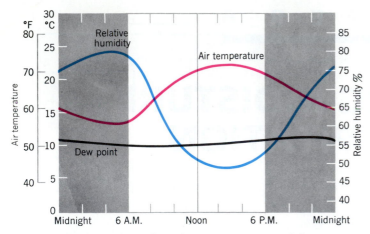

FIGURE 6.2 Relative humidity, temperature, and dew point for May in Washington, D.C. (Data from National Weather Service.)

The general term *humidity* refers to the amount of water vapor present in the air. At any specified temperature, the quantity of moisture that can be held by the air has a definite limit. This limit is the saturation quantity. The proportion of the amount of water vapor present at a given temperature relative to the maximum quantity that could be present is the *relative humidity;* it is expressed as a percentage. For *saturated air,* the relative humidity is 100 percent; when half the total possible quantity of vapor is present, relative humidity is 50 percent, and so on.

A change in relative humidity of the atmosphere can be caused in one of two ways. If an exposed water surface is present, the humidity can be increased by evaporation. This process is slow because it requires that the water vapor diffuse upward through the air. The second way is through a change of temperature. Even though no water vapor is added, a lowering of temperature results in a rise of relative humidity. This change is automatic because the capacity of the air to hold water vapor is lowered by cooling. After cooling, the existing amount of vapor represents a higher percentage to the total capacity of the air. Similarly, a rise of air temperature results in decreased relative humidity, even though no water vapor has been taken away.

The principle of relative-humidity change caused by temperature change is illustrated by a graph of these two properties throughout the day (Figure 6.2). As air temperature rises, relative humidity falls, and vice versa.

A simple example, shown in Figure 6.3, may help to illustrate these principles. At 10 A.M., the air temperature is 16°C (60°F) and the relative humidity is 50 percent. By 3 P.M., the air has become warmed by the sun to 32°C (90°F). The relative humidity has automatically dropped to 20 percent, which is very dry air. Next, the air becomes chilled during the night and, by 4 A.M., its temperature has fallen to 5°C (40°F). Now the relative humidity has automatically risen to 100 percent and the air is saturated. Any further cooling will cause condensation of the excess vapor into liquid or solid form. As the air temperature continues to fall, the humidity remains at 100 percent, but condensation continues and this may take the form of minute droplets of dew or fog. If the temperature falls below freezing, condensation occurs as frost on exposed surfaces.

Dew point is the critical temperature at which air becomes saturated during cooling. Below the dew point, condensation usually sets in, producing minute water droplets. An excellent illustration of condensation caused by cooling is seen in the summer, when beads of moisture form on the outside surface of a pitcher or glass filled with ice water. Air immediately adjacent to the cold glass or metal surface is chilled enough to fall below the dewpoint temperature, causing moisture to condense on the surface of the glass.

HOW RELATIVE HUMIDITY IS MEASURED

Humidity of the air can be measured in two ways. An instrument known as a *hygrometer* indicates relative humidity on a calibrated dial. One simple type uses a strand of human hair that lengthens and shortens according to the relative humidity and, in so doing, activates the dial (Figure 6.4). A continuous record of humid-

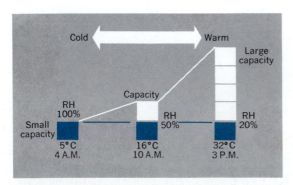

FIGURE 6.3 Relative humidity changes with temperature because capacity of warm air is greater than that of cold air.

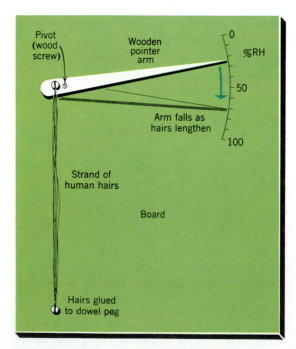

FIGURE 6.4 A simple, homemade hygrometer.

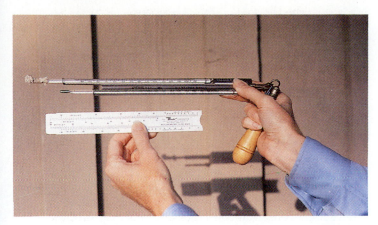

FIGURE 6.5 This standard National Weather Service psychrometer uses paired thermometers. The cloth-covered wet-bulb thermometer projects beyond the dry-bulb thermometer. The handle is used to swing the thermometers in the free air. Below it is a special slide rule that enables rapid calculation of relative humidity from dry bulb and wet bulb readings. (A. N. Strahler.)

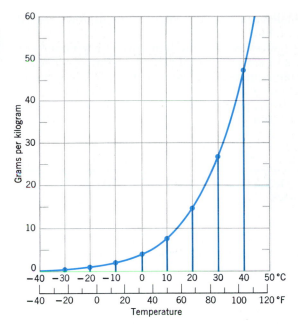

FIGURE 6.6 Maximum specific humidity of a mass of air increases sharply with rising temperature.

ity can be obtained by means of a *hygrograph.* Using the same basic mechanism as the hygrometer, a continuous, automatic record is drawn by a pen on a sheet of paper attached to a rotating drum.

A different principle is applied in the *sling psychrometer.* This instrument is simply a pair of thermometers mounted side by side (Figure 6.5). One is the ordinary type; the other has a piece of wet cloth around the bulb. When the air is fully saturated (relative humidity 100%), there is no evaporation from the wet cloth and both thermometers read the same. When the air is not fully saturated, evaporation occurs, cooling the cloth-covered thermometer below the temperature shown on the ordinary thermometer. Because the rate of evaporation depends on dryness of the air, the difference shown by the two thermometers will increase as relative humidity decreases.

Standard tables are available to show the relative humidity for a given combination of wet- and dry-bulb temperatures. To be sure that maximum possible evaporation is taking place, the two thermometers are attached by a swivel joint to a handle by which the thermometers can be swung around in a circle by hand. Other types of psychrometers use a fan to blow air past the wet thermometer bulb.

SPECIFIC HUMIDITY

Although relative humidity is an important indicator of the state of water vapor in the air, it is a statement only of the relative quantity present compared to a saturation quantity. The actual quantity of moisture present is denoted by *specific humidity,* defined as the mass of water vapor in grams contained in a kilogram of air. For any specified air temperature, there is a maximum mass of water vapor that a kilogram of air can hold (the saturation quantity). Figure 6.6 is a graph showing this maximum moisture content of air for a wide range of temperatures.

Specific humidity is often used to describe the moisture characteristics of a large mass of air. For example, extremely cold, dry air over arctic regions in winter may have a specific humidity of as low as 0.2 g/kg, whereas extremely warm moist air of equatorial regions often holds as much as 18 g/kg. The total natural range on a worldwide basis is such that the largest values of specific humidity are from 100 to 200 times as great as the least.

Figure 6.7 is a graph showing how relative humidity and specific humidity vary with latitude. Notice that the relative humidity curve has two saddles, one over each of the subtropical high-pressure belts where the world's tropical deserts are found. Humidity is high in equatorial and arctic zones. In contrast, the specific humidity curve

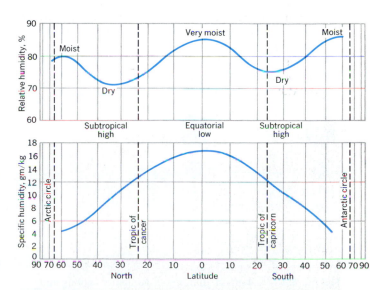

FIGURE 6.7 Pole-to-pole profiles of average relative humidity (above) and of specific humidity (below). (Data from Haurwitz and Austin.)

has a single peak, near the equator, and declines toward high latitudes.

In a real sense, specific humidity is a geographer's yardstick of a basic natural resource—water—to be applied from equatorial to polar regions. It is a measure of the quantity of water that can be extracted from the atmosphere as precipitation. Cold air can supply only a small quantity of rain or snow; warm air is capable of supplying very large quantities.

CONDENSATION AND THE ADIABATIC PROCESS

Falling rain, snow, sleet, and hail are referred to collectively as *precipitation*. Only where large masses of air are experiencing a steady drop in temperature below the dew point can precipitation occur in appreciable amounts. Precipitation cannot be brought about by the simple process of chilling of air through loss of heat by longwave radiation during the night. Precipitation requires that a large mass of air be rising to higher altitudes.

An important law of meteorology is that rising air experiences a drop in temperature, even though no heat energy is lost to the outside. The drop of temperature is a result of the decrease in air pressure at higher altitudes, permitting the rising air to expand. Because individual molecules of the gas are more widely diffused and do not move so fast, the sensible temperature of the expanding gas is lowered. This temperature drop is described as an *adiabatic process*, which simply means "occurring without any gain or loss of heat." In the adiabatic process, heat energy as well as matter remains within the system. The process is thus completely reversible. Expansion always results in cooling; compression always results in warming (Figure 6.8).

Within a rising body of air the rate of drop of temperature, termed the *dry adiabatic lapse rate*, is about 10C° per 1000 m of vertical rise. In English units the rate is 5½F° per 1000 ft. The dry rate applies only when no condensation is taking place. The dew point also declines gradually with rise of air: the rate is 2C° per 1000 m (1F° per 1000 ft).

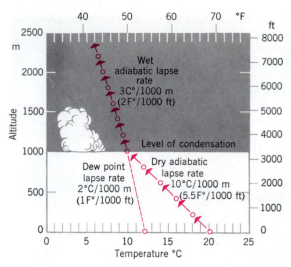

FIGURE 6.9 Adiabatic decrease of temperature in a rising air mass. (From A. N. Strahler, *The Earth Sciences*, 2nd ed., Harper & Row, Publishers. Copyright © by Arthur N. Strahler.)

Adiabatic cooling rate should not be confused with the environmental temperature lapse rate, explained in Chapter 2; that lapse rate applies only to nonrising air, whose temperature is measured at successively higher levels.

Lapse rates can be shown on a simple graph in which altitude is plotted on the vertical scale and temperature on the horizontal scale (Figure 6.9). The chain of circles connected by arrows represents air rising as if it were a bubble. Suppose that a body of air near the ground has a temperature of 20°C (68°F) and that its dew point temperature is 12°C (54°F); these are the conditions shown in Figure 6.9. If, now, a bubble of air undergoes a steady ascent, its air temperature will decrease much faster than its dew-point temperature. Consequently, the two lines on the graph are rapidly converging. At an altitude of 1000 m (3300 ft), the air temperature, now 10°C (50°F), will have met the dew-point temperature. The air bubble has now reached the saturation point. Further rise results in condensation of water vapor into minute liquid particles, and so a cloud is produced. The flat base of the cloud is a common visual indicator of the level of condensation.

As the bubble of saturated air continues to rise, further condensation takes place; but now a new principle comes into effect. When water vapor condenses, heat in the latent form is transformed into sensible heat, which is added to the existing heat content of the air. Recall that latent heat of vaporization amounts to about 600 calories for each gram of water. Originally, this stored energy was obtained during the process of evaporation at the time the water vapor entered the atmosphere.

As condensation continues within a rising mass of air, the latent heat liberated by that condensation partly offsets the temperature drop by adiabatic cooling. As a result, the adiabatic rate is substantially reduced. The reduced rate, which ranges between 3 and 6C° per 1000 m (2 to 3F° per 1000 ft), is termed the *wet adiabatic lapse*

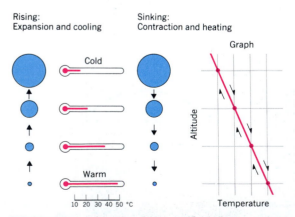

FIGURE 6.8 A schematic diagram of adiabatic cooling and heating accompanying the rising and sinking of a mass of air. (A. N. Strahler.)

rate. On the graph, this reduced rate is expressed by the more steeply inclined section of line above the level of condensation. The lower range of the lapse rate (close to 3C°/1000 m) applies when condensation is very rapid and is typical of comparatively warm rain clouds at low altitudes. As air ascends to high altitudes and becomes very cold, the rate of condensation decreases greatly and the wet adiabatic lapse rate increases to high values (up to 6C°/1000 m and greater), gradually approaching the dry adiabatic lapse rate of 10C°/1000 m.

CLOUD PARTICLES

A *cloud* is a dense mass of suspended water or ice particles in the diameter range of 20 to 50 microns. Each cloud particle has formed on a *nucleus* of solid matter, which is originally on the order of 0.1 to 1.0 micron in diameter. Nuclei of condensation must be present in large numbers; they must be *hygroscopic*, that is, of such a composition as to attract water vapor molecules. In Chapter 2 we referred to these minute suspended particles collectively as atmospheric dust and noted that one source is the surface of the sea. Droplets of spray from the crests of waves are carried rapidly upward in turbulent air. Evaporation of the water leaves a solid residue of crystalline salt, which is strongly hygroscopic. We are familiar with the way ordinary table salt becomes moist when exposed to warm humid air.

Although "clean air" is an environmental goal, the term is only relative because all air of the troposphere is charged with dust. As a result, there is no lack of suitable condensation nuclei. As we shall find in the discussion of air pollution, the heavy load of dust carried by polluted air over cities substantially aids in condensation and the formation of clouds and fog.

We are accustomed to finding that liquid water turns to ice when the surrounding temperature falls to the freezing point, 0°C (32°F), or below. Water in such minute particles as those comprising clouds remains in the liquid state at temperatures far below freezing, however. Such water is described as *supercooled*. Clouds consist entirely of water droplets at temperatures down to about −12°C (10°F). Between −12 and −30°C (10 and −20°F), the cloud is a mixture of water droplets and ice crystals. Below −30°C (−20°F), the cloud consists predominantly of ice crystals; below −40°C (−40°F), all the cloud particles are ice crystals. Very high, thin clouds, formed at altitudes of 6 to 12 km (20,000 to 40,000 ft) are composed of ice particles.

CLOUD FORMS

Clouds are classified according to altitude and form (Figure 6.10). On the basis of form there are two major classes of clouds: *stratiform*, or layered clouds, and *cumuliform*, or globular clouds.

The stratiform clouds are blanketlike and cover large areas. The important point about stratiform clouds is that they represent air layers being forced to rise gradually over stable underlying air layers of greater density. As forced rise continues, the rising layer of air is adiabatically cooled, and condensation is sustained over a large area. Stratiform clouds can yield substantial amounts of rain or snow. The cumuliform clouds are globular masses representing bubblelike bodies of warmer air spontaneously rising because they are less dense than the surrounding air. Precipitation yielded by a cumuliform cloud is concentrated within a small area.

Figure 6.10 shows that four *cloud families* are recognized. The first three families are defined according to altitude range—high, middle, low. The fourth family consists of the cumuliform types, which have a vertical

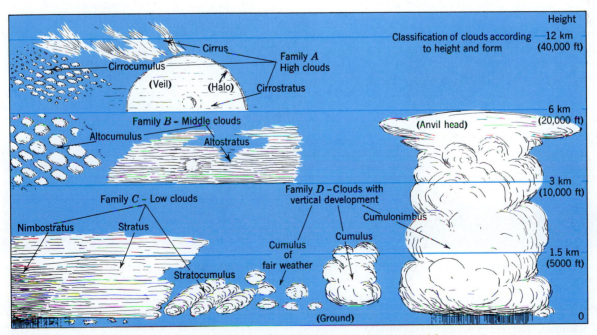

FIGURE 6.10 Cloud types are grouped into families according to altitude range and form. (Drawn by A. N. Strahler.)

(A) Fibrous cirrus, or mares' tails, above; cumulus of fair weather, below. (A. N. Strahler.)

(B) Cirrus in broad, parallel bands, indicating a jet stream aloft. (A. N. Strahler.)

(C) Cirrocumulus forming a mackerel sky. (Jerome Wyckoff.)

(D) Altocumulus. (Jerome Wyckoff.)

(E) An altostratus layer, feathering out in the distance. (A. N. Strahler.)

(F) A stratocumulus cloud deck seen from above. In the distance, a single active cumulus cloud is rising rapidly and, in time, may form a cumulonimbus cloud. (John S. Shelton.)

FIGURE 6.11 Common cloud forms.

development and may extend through a great altitude range. Figure 6.11 shows several of the cloud types.

The high cloud family includes cirrus and its related forms, cirrostratus and cirrocumulus, within the altitude range of 6 to 12 km (20,000 to 40,000 ft). They are composed of ice crystals. *Cirrus* is a delicate, wispy cloud, often forming streaks or stringers across the sky. It does not interfere with the passage of sunlight or moonlight and appears to the ground observer to be moving very slowly, if at all. Actually, cirrus clouds may be moving at high speed, as they often indicate the presence of the jet stream aloft. The observer on the ground can estimate direction of upper airflow by means of fibrous cirrus formations. *Cirrostratus* is a more complete layer of ice cloud, producing a halo about the sun or moon. Where the layer consists of closely packed globular pieces of cloud, arranged in groups or lines, the name *cirrocumulus* is given. This is the "mackerel sky" of popular description.

The middle cloud family, in the height range 2 to 6 km (6500 to 20,000 ft), includes altostratus and altocumulus. *Altostratus* is a blanket layer, often smoothly distributed over the entire sky. It is grayish in appearance, usually has a smooth underside, and will often show the sun as a bright spot in the cloud. *Altocumulus* is a layer of individual cloud masses, fitted closely together in geometric pattern. The masses appear white, or somewhat gray on the shaded sides, and blue sky can be seen between individual patches or rows. Altostratus is commonly associated with the approach of a period of precipitation, whereas altocumulus is characteristic of generally fair conditions.

In the low cloud family, from ground level to 2 km (6500 ft), are stratus, nimbostratus, and stratocumulus. *Stratus* is a dense, low-lying, dark-gray layer. When rain or snow is falling from this cloud, it is termed *nimbostratus;* the prefix *nimbo* is from the Latin word *nimbus,* for rainstorm. *Stratocumulus* is a low-lying cloud layer consisting of distinct grayish masses of cloud between which is open sky. The individual masses often take on the form of long rolls of cloud, oriented at right angles to the direction of wind and cloud motion. Stratocumulus is generally associated with fair or clearing weather, but sometimes rain or snow flurries issue from individual cloud masses.

Clouds of vertical development—the cumuliform clouds—tend to display a height as great as, or greater than, their horizontal dimensions. *Cumulus* is a white, woolpack cloud mass. Small cumulus clouds are associated with fair weather. Enlarged cumulus clouds (congested cumulus) show a flat base and a bumpy upper surface somewhat resembling a head of cauliflower. These clouds look pure white on the side illuminated by the sun, but may be gray or black on the shaded or underneath side.

Under different conditions, explained later in this chapter, individual cumulus masses grow into *cumulonimbus,* the thunderstorm cloud. This massive, towering cloud yields heavy rainfall, thunder and lightning, and gusty winds (see Figure 6.18). A large cumulonimbus cloud may extend from a height of 500 m (1600 ft) at the base up to 9 to 12 km (30,000 or 40,000 ft) at the top. When seen from a great distance, the top of the cumulo-

nimbus cloud is pure white; but to observers beneath, the sky may be darkened to almost a nighttime blackness. Processes occurring within the cumulonimbus cloud will be explained in connection with thunderstorms.

FOG

Fog is simply a cloud layer in contact with the land or sea surface or lying very close to the surface. Fog is a major environmental hazard to humans in an industrialized world. Dense fog on high-speed highways is a cause of terrifying chain-reaction accidents, sometimes involving dozens of vehicles and often taking a heavy toll in injuries and deaths. Landing delays and shutdowns at major airports caused by fog bring economic losses to airlines and inconvenience to thousands of travelers in a single day. For centuries fog at sea has been a navigational hazard. Today, with huge supertankers carrying oil, fog adds to the probability of ship collisions capable of creating enormous oil spills. Polluted fogs are a health hazard to urban dwellers and have at times taken a heavy toll in lives.

One type of fog, known as a *radiation fog,* is formed at night when temperature of the stagnant basal air falls below the dew point. This kind of fog is associated with a low-level temperature inversion (Figure 4.7). Another fog type, *advection fog,* can result from the movement of warm, moist air over a cold or snow-covered ground surface. Losing heat to the ground, the air layer undergoes a drop of temperature below the dew point and condensation sets in. A similar type of advection fog is formed over oceans where air from over a warm current blows across the cold surface of an adjacent cold current. Fogs of the Grand Banks off Newfoundland are largely of this origin because here the cold Labrador current comes in contact with warm water of Gulf Stream origin.

A special type of fog is prevalent along the California coast. It forms within a cool marine air layer in direct contact with cold water of the California current (Figure 6.12). Similar fog conditions prevail close to all continental west coasts in the tropical latitude zone, where cool equatorward currents parallel the shoreline.

FIGURE 6.12 Coastal fog bank enveloping the Bay Bridge, San Francisco. (Frank Lambrecht/Berg & Associates.)

PRECIPITATION FORMS

During the rapid ascent of a mass of air in the saturated state, cloud particles grow rapidly and attain a diameter of 50 to 100 microns. They then coalesce through collisions and grow quickly into droplets of about 500 microns diameter (about 1/50 in.). Droplets of this size reaching the ground constitute a drizzle, one of the recognized forms of precipitation. Further coalescence increases drop size and yields *rain*. Average raindrops have diameters of about 1000 to 2000 microns (1 to 2 mm; 1/25 to 1/10 in.), but they can reach a maximum diameter of about 7 mm (1/4 in.). Above this value they become unstable and break into smaller drops while falling. One kind of rain forms directly by liquid condensation and droplet coalescence in warm clouds of the equatorial and tropical zones. Rain of middle and high latitudes, however, is largely the product of the melting of snow as it makes its way to lower, warmer levels.

Snow is produced in clouds that are a mixture of ice crystals and supercooled water droplets. The falling crystals serve as nuclei to intercept water droplets. As these adhere, the water film freezes and is added to the crystalline structure (Figure 6.13). The crystals clot together readily to form larger snowflakes, and these fall more rapidly from the cloud. When the underlying air layer is below the freezing temperature, snow reaches the ground as a solid form of precipitation; otherwise, it will melt and arrive as rain. A reverse process, the fall of raindrops through a cold air layer, results in the freezing of rain and produces pellets or grains of ice. These are commonly referred to in North America as *sleet* but, among the British, "sleet" refers to a mixture of snow and rain.

Hail, another form of precipitation, consists of large pellets or spheres of ice. The formation of hail will be explained in our discussion of the thunderstorm.

FIGURE 6.14 Heavily coated wires and branches caused heavy damage in eastern New York State in January 1943 as a result of this ice storm. (National Weather Service and New York Power and Light Co., Albany, N.Y.)

When rain falls on a frozen ground surface that is covered by an air layer of below-freezing temperature, the water freezes into clear ice after striking the ground or other surfaces such as trees, houses, or wires (Figure 6.14). The coating of ice that results is called a *glaze*, and an *ice storm* is said to have occurred. Actually, no ice falls, so that ice glaze is not a form of precipitation. Ice storms cause great damage, especially to telephone and power wires and to tree limbs. Roads and sidewalks are made extremely hazardous.

HOW PRECIPITATION IS MEASURED

Precipitation is measured in units of depth of fall per unit of time: for example, centimeters or inches per hour or per day. One cm of rainfall is a quantity sufficient to cover the ground to a depth of 1 cm, provided that none is lost by runoff, evaporation, or sinking into the ground. A simple form of rain gauge can be operated merely by setting out a straight-sided, flat-bottomed pan and measuring the depth to which water accumulates during a particular period. Unless this period is short, however, evaporation seriously upsets the results.

A very small amount of rainfall, such as 2 mm (0.1 in.), makes too thin a layer to be accurately measured. To avoid this difficulty as well as to reduce evaporation loss, the *rain gauge* is made in the form of a cylinder whose base is a funnel leading into a narrow tube (Figure 6.15). A small amount of rainfall will fill the narrow pipe to a considerable height, thus making it easy to read accurately once a simple scale has been provided for the pipe. This gauge requires frequent emptying unless it is equipped with automatic devices for this purpose.

Snowfall is measured by melting a sample column of snow and reducing it to an equivalent in water. In this

FIGURE 6.13 These individual snow crystals, greatly magnified, were selected for their variety and beauty. They were photographed by W. A. Bentley, a Vermont farmer who devoted his life to snowflake photography. (National Weather Service.)

FIGURE 6.15 This 10-cm (4-in.) rain gauge is made of clear plastic, allowing the amount of rain caught in the inner cylinder to be read from the outside. Rain exceeding the capacity of the inner cylinder overflows into the outer cylinder and can be measured after the rain period has ended. (A. N. Strahler.)

way rainfall and snowfall records may be combined for purposes of comparison. Ordinarily, a 10-cm layer of snow is assumed to be equivalent to 1 cm of rainfall, but this ratio may range from 30 to 1 in very loose snow to 2 to 1 in old, partly melted snow.

HOW PRECIPITATION IS PRODUCED

Precipitation in substantial quantities is induced by two basic types of mechanisms. One is the spontaneous rise of moist air; the other is the forced rise of moist air.

Spontaneous rise of moist air is associated with *convection*, a form of atmospheric motion consisting of strong updrafts taking place with a *convection cell*. Air rises in the cell because it is less dense than the surrounding air. Perhaps a fair analogy is the updraft of heated air in a chimney but, unlike the steady airflow in a chimney, air rise in a convection cell takes place in pulses as bubblelike masses of air ascend in succession.

To illustrate the convection process, suppose that on a clear, warm summer morning the sun is shining on a landscape consisting of patches of open fields and woodlands. Certain of these types of surfaces, such as the bare

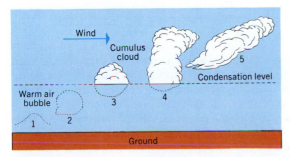

FIGURE 6.16 Rise of a bubble of heated air to form a cumulus cloud. (From A. N. Strahler, *The Earth Sciences*, 2nd ed., Harper & Row, Publishers. Copyright © by Arthur N. Strahler.)

ground, heat more rapidly and transmit heat by long-wave radiation to the overlying air. Air over a warmer patch becomes warmed more than adjacent air and begins to rise as a bubble, much as a hot-air balloon rises after being released. Vertical movements of this type are often called "thermals" by sailplane pilots, who use them to obtain lift.

As the air rises, it is cooled adiabatically so that eventually it is cooled below the dew point. Condensation begins at once, and the rising air column appears as a cumulus cloud. The flat base shows the critical level above which condensation is occurring (Figure 6.16). The bulging "cauliflower" top of the cloud represents the top of the rising warm air column, pushing into higher levels of the atmosphere. When the bubble of air is sufficiently cooled by the adiabatic process, it ceases to rise and condensation stops. Then the small cumulus cloud dissolves after drifting some distance downwind. Under a different set of atmospheric conditions, convection continues to develop unchecked and the cloud grows into a dense cumulonimbus mass, or thunderstorm, yielding heavy rain.

Why does spontaneous cloud growth sometimes take place and continue beyond the initial cumulus stage, long after the original input of heat energy is gone? Actually, the unequal heating of the lower air level served only as a trigger effect to release a spontaneous updraft, fed by latent heat energy liberated from the condensing water vapor. Recall that for every gram of water formed by condensation, 600 calories of heat are released. This heat acts like fuel in a bonfire.

Air capable of rising spontaneously during condensation is described as *unstable air*. In such air the updraft tends to increase in intensity as time goes on, much as a bonfire blazes with increasing ferocity as the updraft draws in greater supplies of oxygen. Of course, at very high altitudes, the bulk of the water vapor has condensed and fallen as precipitation, so the energy source is gone. When this happens the convection cell weakens and air rise finally ceases.

Unstable air, given to spontaneous convection in the form of heavy showers and thunderstorms, is most likely to be found in warm, humid areas, such as the equatorial and tropical zones throughout the year and the midlatitude regions during the summer season.

Figure 6.17 will help explain why spontaneous rise of air to produce intense convection can take place only when air properties are favorable. Diagram A is a plot of altitude against air temperature. The small circles represent a small parcel of air being forced to rise steadily higher, following the dry adiabatic rate of cooling shown. To the right of this line is a solid line showing the temperature of the undisturbed surrounding air; it decreases upward at the average environmental lapse rate.

Suppose that the air parcel is forcibly lifted from the ground, where its temperature is 30°C. After the air parcel has been carried up 500 m, its temperature has fallen about 5C° and is now 25°C, whereas the surrounding air (environment) is warmer by 2C° and has a temperature of 27°C. The air parcel would thus be cooler than the environment at 500 m and denser than the surrounding air. If no longer forcibly carried upward, it would tend to sink back toward the ground. These conditions represent

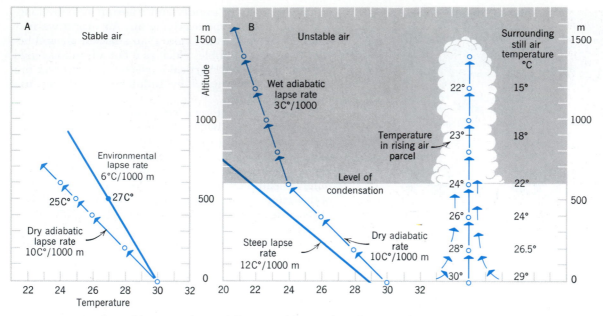

FIGURE 6.17 Under stable air conditions (A), air would resist forced rise. Under unstable conditions (B), rise is spontaneous.

stable air, not able to produce convection cells because the air would resist lifting.

When the lower air layer is excessively heated, the environmental lapse rate is increased and the air becomes unstable. Diagram B of Figure 6.17 shows a steep lapse rate of 12C°/1000 m. (As the graph is constructed, the steeper lapse rate is represented by a line of lower inclination.) An air parcel at ground level is shown to be 1C° warmer than the surrounding air. It begins to rise spontaneously because it is less dense than air over adjacent, less intensely heated ground areas. Although cooled adiabatically while rising, the air parcel at 600 m has a temperature of 24°C, but this is well above the temperature of the surrounding still air.

The air parcel, therefore, is always lighter than the surrounding air and continues to rise. At 600 m, the dew point is reached and condensation sets in. Now the rising air parcel is cooled at a reduced wet adiabatic rate, shown as 3C°/1000 m, because the latent heat liberated in condensation offsets the rate of drop due to expansion. At 1000 m, the rising air parcel is still several degrees warmer than the environment and therefore continues its spontaneous rise. Generally, steep lapse rates in air holding a large quantity of water vapor are associated with instability.

THUNDERSTORMS

Convection manifests itself in the *thunderstorm*, an intense local storm associated with a tall, dense cumulonimbus cloud in which there are very strong updrafts of air (Figure 6.18). Thunder and lightning normally accompany the storm, and rainfall is heavy, often of cloudburst intensity, for a short period. Violent surface winds may occur at the onset of the storm.

A single thunderstorm consists of individual cells. Air rises within each cell as a succession of bubblelike air bodies (Figure 6.19). As each bubble rises, air in its wake is brought in from the surrounding region, a process called *entrainment*. Rising air in the thunderstorm cell can reach vertical speeds up to 60 km (40 mi) per hour. Precipitation is in the form of rain in the lower levels, mixed water and snow at intermediate levels, and snow at high levels.

Upon reaching high levels, which may be 6 to 12 km (20,000 to 40,000 ft) or even higher, the rising rate diminishes and the cloud top is dragged downwind to form an *anvil top*. Ice particles falling from the cloud top act as nuclei for condensation at lower levels, a process called *cloud seeding*. The rapid fall of raindrops adjacent to the rising air bubbles exerts a frictional drag on the air and sets in motion a downdraft. Striking the ground where precipitation is heaviest, this downdraft forms a local squall wind, which is sometimes strong enough to fell trees and do severe structural damage to buildings.

MICROBURST! A THREAT TO AIRLINE PASSENGERS

Described as the greatest source of air carrier deaths in the United States, the *microburst* is a brief onset of intense winds close to the ground beneath a thunderstorm cell. Encounter with the sharply divergent pattern of these winds was the cause of airline crashes that took nearly 400 lives in the period between 1975 and 1987.

Figure 6.20A is a schematic cross-section of a microburst. A strong downdraft descends from the base of the cumulonimbus cloud, and may or may not contain rain. The outflow close to the ground produces a horizontal vortex in the form of a ring, as shown in the diagram.

(A) Active thunderstorm cells producing cumulonimbus clouds over the southern Rocky Mountains. (A. N. Strahler.)

(B) Torrential rain beneath the flat base of a cumulonimbus cloud, North Rim of Canyon, Arizona. (A. N. Strahler.)

(C) Rain shower falling from the flat base of a thunderstorm cloud, Colorado. (Daniel Zirinsky/Photo Researchers.)

(D) A succession of lightning strokes recorded by time exposure. (Deeks & Pribe/Science Source/Photo Researchers.)

FIGURE 6.18 Thunderstorms, cumulonimbus clouds, and lightning.

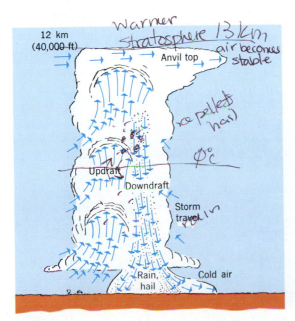

FIGURE 6.19 Schematic diagram of interior of a thunderstorm cell.

When no rain is present, the downdraft may be invisible, except for a ring of blowing dust produced by the turbulent winds.

Figure 6.21 is a Doppler radar image of the horizontal air flow as seen by the radar, looking toward the right from a position at the left of the frame. The radar can see only the wind component toward or away from it. The center of the downdraft lies about on the curved grid line. The brown and orange colors indicate wind from left to right; the green tones, wind from right to left. The difference in the two speeds is about 30 m/sec (70 mph).

Returning to Figure 6.20, the lower diagram (B) shows a takeoff flight profile within the wind structure immediately above it. The aircraft, climbing steeply, first encounters an increasing headwind, but this quickly decreases in intensity to reduce its performance. Upon encountering the outflow, the increasing tailwind further reduces lift, resulting in a crash. This is what happened to cause the July 1982 crash of Pan American Flight 759 immediately following takeoff from the New Orleans international airport, killing all 145 persons on board. For aircraft on a landing path, a somewhat similar set of

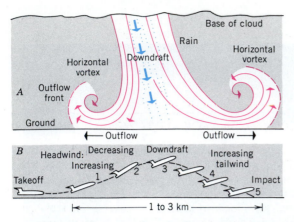

FIGURE 6.20 Anatomy of a microburst. (*A*) Schematic cross-section showing downdraft reaching ground and producing horizontal outflow with a sharply defined outflow front. (*B*) Schematic profile of aircraft takeoff leading to loss of control and crash. (Adapted from diagrams by Research Applications Program, National Center for Atmospheric Research, Boulder, Colorado.)

uncontrollable events occurs, forcing down the craft to impact the ground.

The best defense against this often invisible but lethal threat is to detect microbursts in the vicinity of an airport by means of a suitably positioned Doppler radar instrument, but this system is extremely expensive. Installation of a network of closely spaced anemometers also makes possible the detection of these small, short-lived microbursts as they pass across the airport.

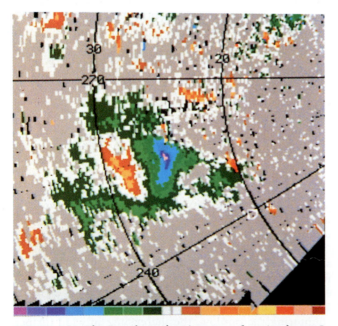

FIGURE 6.21 The Doppler radar signature of a microburst. In this computer-generated image, false colors are used to show horizontal wind speeds toward and away from the radar, as explained in the text. (Courtesy of the Research Applications Program, National Center for Atmospheric Research, Boulder, Colorado.)

CLOUD SEEDING

In efforts to modify weather phenomena in various ways, humans have made use of modern principles of meteorology. One goal has been to increase precipitation in areas experiencing drought; another has been to lessen the severity of storms. One method of inducing convectional precipitation is by artificial cloud seeding—the introduction of minute particles into dense cumulus clouds. The particles, which may be of silver iodide smoke, serve as nuclei that induce greater intensity of condensation, thereby increasing the size and height of cumulonimbus clouds, and perhaps resulting in increased rainfall. Carefully conducted tests over Florida in 1978 and 1981 failed to confirm that cloud seeding can increase rainfall. Among meteorologists, opinion remains divided as to its efficacy.

HAIL AND LIGHTNING—ENVIRONMENTAL HAZARDS

In addition to powerful wind gusts and torrential falls of rain, important environmental hazards connected with the thunderstorm are hail, lightning, and tornadoes. Hailstones are formed by the accumulation of ice layers on ice pellets suspended in the strong thunderstorm updrafts. The phenomenon is much like that of the icing of aircraft flying through a cloud of supercooled water droplets. After the hailstones have grown to diameters that often reach 3 to 5 cm (1 to 2 in.), they escape from the updraft and fall to the ground (Figure 6.22).

An important aspect of planned weather modification is that of reducing the severity of hailstorms. Annual losses from crop destruction by hailstorms amount to several hundred million dollars (Figure 6.23). Damage to wheat crops is particularly severe in a north–south belt of the High Plains, running through Nebraska, Kansas, and Oklahoma. Figure 6.24 is a map of frequency of severe hailstorms in the contiguous 48 United States. A much larger region of somewhat less hailstorm frequency extends eastward generally from the Rockies to

FIGURE 6.22 Hailstones produced by a summer thunderstorm. (Runk and Schoenberger/Grant Heilman.)

FIGURE 6.23 These stalks of full-grown corn have been severely damaged by a hailstorm. (Courtesy of the Illinois State Water Survey.)

the Ohio Valley. Corn is a major crop over much of this area. Scientists are studying the isolated cumulonimbus clouds that produce hail. Research has been directed to the development of cloud-seeding techniques by means of which the severity of hailstorms can be reduced.

Another effect of convection cell activity is to generate lightning, one of the environmental hazards that results annually in the death of many persons and livestock and in the setting of many forest and building fires. Lightning is a great electric arc—a gigantic spark—passing between cloud and ground, or between parts of a cloud mass (Figure 6.18D). During lightning discharge, a current of as much as 60,000 to 100,000 amperes may develop. Rapid heating and expansion of the air in the path of the lightning stroke sends out intense sound waves, which we recognize as a thunderclap. In the United States, lightning causes a yearly average of about 150 human deaths and property damage in hundreds of millions of dollars, including loss by structural fires and forest fires set by lightning.

OROGRAPHIC PRECIPITATION

Forced ascent of large masses of air occurs under two quite different sets of controlling conditions. When prevailing winds encounter a mountain range, the air layer as a whole is forced to rise to surmount the barrier. Precipitation produced in this way is described as *orographic precipitation*, meaning "related to mountains." A layer of cold air may act in much the same way as a mountain barrier. Warm air in motion often encounters a cold air layer. The cold air is denser than the warm and will remain close to the ground, acting as a barrier to the progress of the warm air. The warm air is then forced to rise over the barrier. In a related type of mechanism, the cold air layer is in motion and forces the warm air to rise over it. Precipitation resulting from these activities is explained in Chapter 7.

Figure 6.25 shows the steps associated with the production of orographic precipitation. Moist air arrives at the coast after passing over a large ocean surface. As the air rises on the windward side of the range, it is cooled at the adiabatic rate. When cooling reaches the dew point, precipitation sets in. After passing over the mountain summit, the air begins to descend the lee side of the range. Now it undergoes compressional warming through the same adiabatic process and, having no source from which to draw up moisture, becomes very dry. Upon reaching sea level, it is much warmer than at the start. A belt of dry climate, often called a *rainshadow desert*, exists on the lee side of the range. Several of the important deserts of the earth are of this type.

Dry, warm *chinook winds* often occur on the lee side of a mountain range in the western United States. These winds cause extremely rapid evaporation of snow or soil moisture. Chinook winds result from the turbulent mixing of lower and upper air in the lee of the range. The

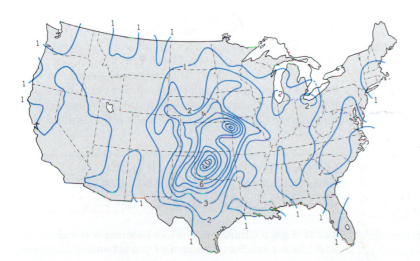

FIGURE 6.24 Map of frequency of severe hailstorms in the 48 contiguous United States. A severe hailstorm is defined as a local convective storm producing hailstones equal to or greater than 0.75 in. (1.9 cm) in diameter. The numbers are proportional to the density of severe hailstorms per 100 square kilometers over a 16-year period of record of the National Weather Service. (From Richard H. Skaggs, *Proc. Association of American Geographers*, vol. 6, Figure 2. Used by permission.)

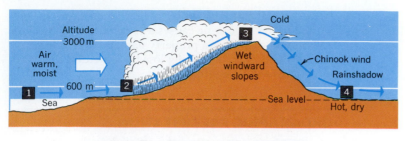

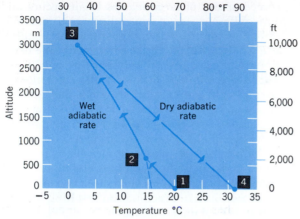

FIGURE 6.25 Forced ascent of oceanic air masses produces precipitation and a rainshadow desert. (From A. N. Strahler, *The Earth Sciences*, 2nd ed., Harper & Row, Publishers. Copyright © by Arthur N. Strahler.)

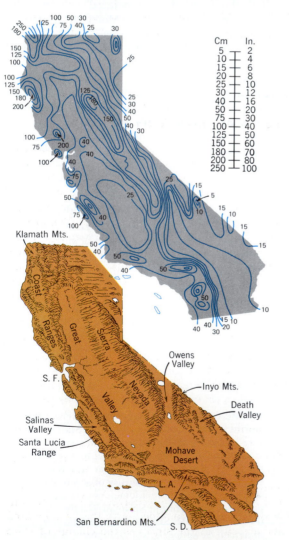

FIGURE 6.26 The effect of mountain topography on rainfall is shown well by the state of California. Isohyets in centimeters.

upper air, which has little moisture to begin with, is greatly dried and heated when swept down to low levels. A similar type of wind, the *foehn* (or *föhn*) occurs on the north side of the Austrian Alps.

An excellent illustration of orographic precipitation and rainshadow occurs in the far west of the United States. A map of California, Figure 6.26, shows mean annual precipitation using lines of equal precipitation, called *isohyets*. Prevailing westerly winds bring moist air from the Pacific Ocean over the Coast Ranges of central and northern California and the great Sierra Nevada, whose summits rise to 4000 m (14,000 ft) above sea level. Heavy rainfall on the windward slopes of these ranges nourishes rich forests. Passing down the steep eastern face of the Sierra Nevada, air must descend nearly to sea level, even below sea level in Death Valley. The resulting adiabatic heating lowers the humidity, producing part of America's great desert zone, covering a strip of eastern California and all of Nevada.

Much orographic rainfall at low latitudes is actually of the convectional type in that it takes the form of heavy showers and thunderstorms. The convectional storms are set off by the forced ascent of unstable air as it passes over the mountain barrier. The torrential monsoon rains of the Asiatic and East Indian mountain ranges are largely of this type. For example, Cherrapunji, a hill station facing the summer monsoon air drift in northeast India, averages 1082 cm (426 in.) of rainfall annually.

LATENT HEAT AND THE GLOBAL BALANCES OF ENERGY AND WATER

An understanding of the processes of condensation and precipitation allows us now to take another look at the global energy balance and to include the mechanism of latent heat transport. Figure 6.27 is a schematic diagram to show how evaporation and condensation are involved

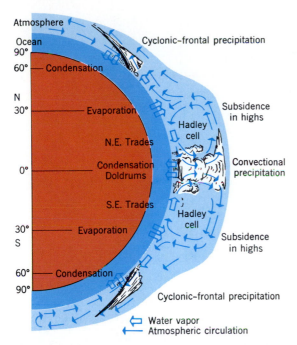

FIGURE 6.27 Schematic diagram of exchange of water vapor along a meridional profile. (From A. N. Strahler, "The Life Layer," *Journal of Geography*, vol. 69, no. 2, p. 74. Copyright © 1970 by *The Journal of Geography;* reproduced by permission.)

in energy exchange in each of the major latitude zones. We can now also summarize the pattern of transport of water vapor across the parallels of latitude; this is a transport of matter and is part of the global water balance. Figure 6.28 shows the average annual rate of vapor

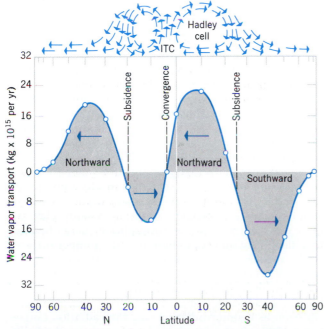

FIGURE 6.28 Graph of mean annual meridional transport of water vapor. (Data of M. I. Budyko and others in W. D. Sellers, *Physical Climatology*. From A. N. Strahler, *The Earth Sciences*, 2nd ed., Harper & Row, Publishers. Copyright © by Arthur N. Strahler.)

transport, in units of 10^{15} kg of water per year. Let us review the energy exchanges and water transports for each latitude zone.

The equatorial zone is characterized by a rise of moist, warm air in innumerable conventional cells reaching to the upper limits of the troposphere. As condensation and precipitation occur here, enormous amounts of latent heat energy are liberated. This zone has been called the "fire box" of the globe, in recognition of this intense production of sensible heat by condensation. Moving equatorward to replace the rising air are the tropical easterlies, or trades, within the Hadley cell circulation. Water evaporates from the ocean surface over the subtropical highs and is carried equatorward in vapor form. This transport shows in Figure 6.28 by peaks in the transport curve at about 10°N and 10°S lat. (Where the line of graph lies above the center line, the movement is northward; where below, movement is southward.)

Next we move into midlatitudes, where upper-air waves constantly form and dissolve in the westerlies. Here, by advection, cyclones and anticyclones exchange cold, polar air for warm, tropical air across the parallels of latitude. Water vapor is also transported poleward and reaches a peak rate of flow at about 40°N and 40°S lat., as Figure 6.28 shows. Condensation in cyclonic storms removes water vapor from the troposphere in increasing quantities into high latitudes, so that the movement of water vapor declines and reaches zero at the poles.

A CONVECTIVE STORM AS A FLOW SYSTEM OF ENERGY AND MATTER

A convective storm in which precipitation is occurring can be visualized as an open energy system, shown in Figure 6.29. The storm may consist of a single thunderstorm cell or a group of cells. The lateral boundary of the system is arbitrarily taken as an imaginary vertical surface surrounding the storm on the sides. The ground surface and the tropopause may be taken as lower and upper boundaries. Two forms of energy input can be recognized. First is direct incoming solar radiation, which warms the air, increasing the sensible heat in storage. Second is the importation of latent heat by evaporation from a moist ground surface or an ocean surface below. To simplify our analysis we have ignored the possible importation of both sensible and latent heat by horizontal air motion through the vertical side boundary. It can be assumed that an equal quantity of energy leaves the system in the same manner. Condensation within the rising air releases sensible heat, increasing the total air mass store of latent heat. From storage, energy leaves the system either by longwave radiation or by mass transport of sensible heat in falling raindrops or snowflakes. The small quantity of heat conducted directly to the ground from the overlying air is not shown.

An open material flow system, shown in the lower diagram of Figure 6.29, can be coordinated with the energy flow system. (Ice is included with the liquid state.) The major activity in the storm is the change of state of water from vapor to liquid, producing precipitation. Evaporation of the falling precipitation may take place,

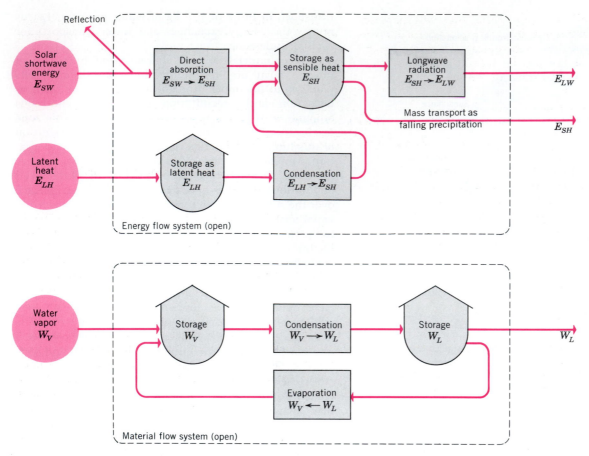

FIGURE 6.29   Schematic diagram of a convective storm as a flow system of energy and matter.

returning some of the liquid water to the vapor state, as shown by the return circuit in the diagram.

Enlarging our perspective, we can think of this system as representing the entire global precipitation process, averaged over long spans of time. The global energy system would have an input limited to solar shortwave energy and an output limited to longwave radiation to outer space. The global material flow system would show input by evaporation from liquid water storages on the continents and oceans and a return of liquid water to those storages. It would closely resemble the diagram of the hydrologic cycle (see Figure 9.3) except that transfer of runoff from continents to oceans would be excluded from the system.

CLOUD COVER, PRECIPITATION, AND GLOBAL WARMING

Your knowledge of water vapor, clouds, precipitation, and the global balances of energy and water can now be used to add to an understanding of the environmental problem of global warming, introduced in Chapter 4.

The observed rise in global surface temperatures described in Chapter 4 suggests a possible increase in sea surface temperature, as well. Satellite data seem to have detected a rise in temperature of the global ocean surface at the rate of about 0.1C° per year over the past decade or so. Any increase in sea surface temperature increases

the rate of evaporation. An increase in average atmospheric content of water vapor is thus one of the most important environmental factors connected with the impending global warming. Water vapor is correctly described as one of the "greenhouse gases" that absorbs and emits longwave radiation, and indeed it is the most important of them all. A global increase in this ingredient represents a feedback mechanism that will tend to intensify the temperature rise. It has been estimated that for a doubling of the carbon dioxide content, the increased water vapor would add an almost equal amount of temperature rise. Scientists realize that this relationship must be taken into account in computer modeling for prediction of global temperature rise.

An increase in average global water vapor content or the atmosphere can be expected to produce greater extent and density of the global cloud cover, as well as greater average global precipitation. Clouds play two quite different roles in the atmospheric energy balance. By reflection of solar radiation from their upper surfaces, clouds reduce the incoming energy and thus tend to cool the troposphere. On the other hand, cloud layers have a warming effect on the lower atmosphere by absorbing longwave radiation from the ground and returning that emission as counterradiation (see the radiation budget, Figure 3.11). Preliminary results of analysis of satellite data suggest that clouds in general may have a cooling effect greater than a warming effect.

An increase in global precipitation may mean that

year-round rainfall in the equatorial zone will increase, as will seasonal rainfall of the subtropical belts and the Asiatic monsoon region. Subtropical deserts and rain-shadow deserts of the middle latitudes, on the other hand, may intensify and spread poleward.

Much additional collection and model analysis of satellite data will be required in coming years to confirm or refute preliminary predictions such as we have suggested here and in earlier chapters.

AIR POLLUTION AND ITS EFFECTS

We can make good use of an understanding of the processes of condensation and precipitation when probing into changes made by humans in the atmospheric environment. These are inadvertent changes, largely the result of urbanization and the growth of industrial processes, which have enormously increased the rate of combustion of hydrocarbon (fossil) fuels in the past half-century or so. We have already considered the thermal effects of injecting increasing amounts of carbon dioxide into the atmosphere. Now we turn to consider the kinds of foreign matter, or *pollutants*, injected by human activities into the lower atmosphere, and the effects of that pollution on air quality and urban climates.

We recognize two classes of atmospheric pollutants. First, there are solid and liquid particles, which are designated collectively as *particulate matter*. Dusts found in the smoke of combustion, as well as droplets naturally occurring as cloud and fog, fall into the category of particles. Figure 6.30 gives the scale of sizes of various atmospheric particles. Second, there are components in the gaseous state, included under the general term *chemical pollutants* in that they are not normally present in measurable quantities in clean air remote from densely populated, industrialized regions. (Excess carbon dioxide produced by combustion is not usually classed as a pollutant.)

One group of chemical pollutants of industrial and urban areas is of primary origin, for example, produced directly from a source on the ground. Gases included in this group are carbon monoxide (CO), sulfur dioxide (SO_2), oxides of nitrogen (NO, NO_2, NO_3), and hydrocarbon compounds. These chemical pollutants cannot be treated separately from particulate matter, however, because they are often combined within a single suspended particle. We have already seen that certain dusts, particularly the sea salts, are hygroscopic and easily take on a covering water film. The water film in turn absorbs the chemical pollutants.

Much of the carbon monoxide, half the hydrocarbons, and about a third of the oxides of nitrogen come from exhausts of gasoline and diesel engines in vehicular traffic. Generation of electricity and various industrial processes contribute most of the sulfur oxides because the coal and lower-grade fuel oil used for these purposes are comparatively rich in sulfur. These same sources also supply most of the particulate matter. Fly ash consists of the coarser grades of soot particles emitted from smokestacks of generating plants. These particles settle out quite quickly within close range of the source. Combustion used for heating buildings is a comparatively minor contributor to pollution because the higher grades of fuel oil are low in sulfur and are usually burned efficiently. Finely divided carbon comprises much of the smoke of combustion and is capable of remaining in suspension almost indefinitely because of its colloidal size. Forest fires comprise a secondary contributor of particles. The burning of refuse is a minor contributor in all categories of pollutants.

In the smog of cities there are, in addition to the ingredients already mentioned, certain chemical elements contained in particles contributed by automobile and truck exhausts. Included are particles that contain lead, chlorine, and bromine.

When particles and chemical pollutants are present in considerable density over an urban area, the resultant mixture is known as *smog*. Almost everyone living in large cities is familiar with smog because of its irritating effects on the eyes and respiratory system and its ability to obscure distant objects. When concentrations of suspended matter are less dense, obscuring visibility of very

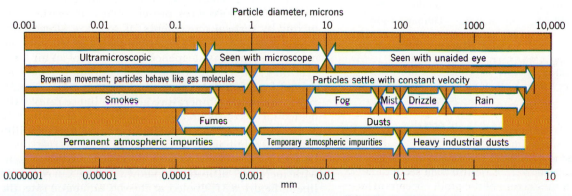

FIGURE 6.30 Sizes and physical properties of atmospheric particles. (Data from W. G. Frank, American Air Filter Company, Inc., Louisville, Ky.)

distant objects but not otherwise objectionable, the atmospheric condition is referred to as *haze*. Atmospheric haze builds up quite naturally in stagnant air masses as a result of the infusion of various surface materials. Haze is normally present whenever air reaches high relative humidity because water films grow on suspended hygroscopic nuclei. Nuclei of natural atmospheric haze particles consist of mineral dusts from the soil, crystals of salt blown from the sea surface, hydrocarbon compounds (pollens and terpenes) exuded by plant foliage, and smoke from forest and grass fires. Dusts from volcanoes may, on occasion, add to atmospheric haze.

It is evident at this point that what we are calling atmospheric pollutants are of both natural and human-made origin and that human activities can supplement the quantities of natural pollutants present. Table 6.1 illustrates this complexity by listing the primary pollutants according to sources.

Not all pollution generated by humans comes from the cities. Isolated industrial activities can produce pollutants far from urban areas. Particularly important are smelters and manufacturing plants in small towns and rural areas. Sulfide ores (metals in combination with sulfur compounds) are processed by heating in smelters close to the mine. There sulfur compounds are sent into the air in enormous concentrations from smokestacks. Fallout over the surrounding area is destructive to vegetation.

Mining and quarrying operations send mineral dusts into the air. For example, asbestos mines (together with asbestos processing and manufacturing plants) send into the air countless threadlike mineral particles, some of which are so small they can be seen only with the electron microscope. These particles travel widely and are inhaled by humans, lodging permanently in the lung tissue. Nuclear test explosions inject into the atmosphere a wide range of particles, including many radioactive substances capable of traveling thousands of kilometers in the atmospheric circulation.

Forest and grass fires caused by humans add greatly to smoke palls in certain seasons of the year. Plowing, grazing, and vehicular traffic raise large amounts of mineral dusts from dry soil surfaces. Bacteria and viruses, which we have not as yet mentioned, are borne aloft in air turbulence when winds blow over contaminated surfaces such as farmlands, grazing lands, city streets, and waste disposal sites.

Primary pollutants are conducted upward from the emission sources by rising air currents that are part of the normal convectional process. The larger particles settle under gravity and return to the surface as *fallout*. Particles too small to settle out are later swept down to earth by precipitation, a process called *washout*. By a combination of fallout and washout, the atmosphere tends to be cleaned of pollutants. In the long run a balance is achieved between input and output of pollutants, but there are large fluctuations in the quantities stored in the air at a given time. Pollutants are also eliminated from the air over their source areas by winds that disperse the particles into large volumes of cleaner air in the downwind direction. Strong winds can quickly sweep away most pollutants from an urban area but, during periods when a stagnant anticyclone is present, the concentrations rise to high values.

In polluted air certain chemical reactions take place among the components injected into the atmosphere, generating a secondary group of pollutants. For example, sulfur dioxide (SO_2) may combine with oxygen to produce sulfur trioxide (SO_3), which in turn reacts with water of suspended droplets to yield sulfuric acid (H_2SO_4). This acid is both irritating to organic tissues and corrosive to many inorganic materials. In other typical reactions, the action of sunlight on nitrogen oxides and organic compounds produces ozone (O_3), a toxic and destructive gas. Reactions brought about by the presence of sunlight are described as *photochemical reactions*. One toxic product of photochemical action is ethylene, produced from hydrocarbon compounds.

LOW-LEVEL INVERSIONS

The concentration of pollutants over a source area rises to its highest levels when the vertical mixing (convection) of the air is inhibited by a stable configuration of the vertical temperature profile of the air. The principles of stable and unstable air conditions were covered earlier in this chapter, and we shall now apply those principles to the problem of air pollution over cities.

When the normal environmental lapse rate of 6C°/1000 m (3.5F°/1000 ft) is present, there is resistance to mixing by vertical movements, as we have already discussed (Figure 6.31, left). Consider next that the environmental lapse rate is steepened by heating of an air layer near the ground because of excess heat radiated and conducted from hot pavements and rooftops (Figure 6.31, right). When the temperature gradient (lapse rate) of the heated air becomes greater than the dry adiabatic rate of 10C°/1000 m (5.5F°/1000 ft), a condition of instability exists and a bubble of warm air can begin to rise, like a helium-filled balloon. Assume that the lapse rate lessens with increased altitude, as shown by the curved solid line in Figure 6.31. Cooled at the dry adiabatic rate, the temperature of the rising bubble then falls faster than does the temperature of the surrounding air. When the

TABLE 6.1 Sources of Primary Atmospheric Pollutants

Pollutants from Natural Sources	Sources of Pollutants of Human Origin
Volcanic dusts	Fuel combustion (CO_2, SO_2, lead)
Sea salts from breaking waves	Chemical processes
Pollens and terpenes from plants	Nuclear fusion and fission
	Smelting and refining of ores
(Aggravated by human activities:)	
Smoke of forest and grass fires	Mining, quarrying
Blowing dust	Farming
Bacteria, viruses	

Data source: Association of American Geographers, *Air Pollution*, Commission on College Geography, Resource Paper No. 2.

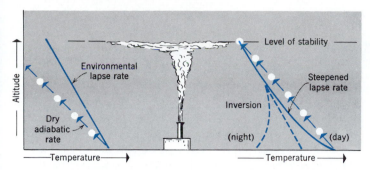

FIGURE 6.31 Relation of dry adiabatic lapse rate (circles) to various environmental lapse rates.

bubble has reached an altitude at which its temperature (and therefore also its density) matches that of the surrounding air, it can rise no further and convection ceases.

Now suppose that instead of the bubble of warm air we substitute the hot air from a smokestack (Figure 6.31). The rise of heated air follows essentially the same pattern, although initially faster and in the form of a vertical jet. Carrying up with it the pollutants of combustion, the rising hot air gradually cools and reaches a level of stability, where it spreads laterally. Cooling by longwave radiation and mixing with the surrounding air will reinforce the adiabatic cooling because a truly adiabatic system would not be realistic in nature.

Recall that at night, when the air is calm and the sky is clear, rapid cooling of the ground surface typically produces a low-level temperature inversion, illustrated in Figure 4.7. In cold air, the reversal of the temperature gradient may extend hundreds of meters into the air. A low-level temperature inversion represents an unusually stable air structure. When this type of inversion develops over an urban area, conditions are particularly favorable for the entrapment of pollutants to the degree that heavy smog or highly toxic fog can develop, as shown in Figure 6.32. The upper limit of the inversion layer coincides with the cap, or *inversion lid*, below which pollutants are held. The lid may be situated at a height of perhaps 150 to 300 m (500 to 1000 ft) above the ground.

Although situations dangerous to health during prolonged low-level inversion have occurred a number of times over European cities since the Industrial Revolution began, the first major tragedy of this kind in the United States occurred at Donora, Pennsylvania, in late October 1948. The city occupies a valley floor hemmed in by steeply rising valley walls which prevent the free mixing of the lower air layer with that of the surrounding region (Figure 6.32). Industrial smoke and gases from

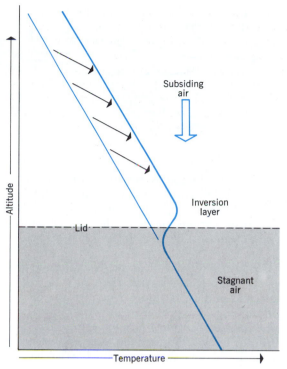

FIGURE 6.33 An upper-air inversion caused by subsidence. (From A. N. Strahler, *Planet Earth: Its Physical Systems Through Geologic Time*, Harper & Row, Publishers. Copyright © by Arthur N. Strahler.)

factories poured into the inversion layer for five days, increasing the pollution level. Because humidity was high, a poisonous fog formed and began to take its toll. In all, 20 persons died and several thousand persons were made ill before a change in weather pattern dispersed the smog layer.

UPPER-LEVEL INVERSIONS

Related to the low-level inversion, but caused in a somewhat different manner, is the *upper-level inversion*, illustrated in Figure 6.33. Recall that anticyclones are cells of subsiding air that diverges at low levels. Within the center of the cell, winds are calm or very gentle. As the air subsides, it is adiabatically warmed, so the normal temperature lapse rate is displaced to the right in the temperature-altitude graph, as shown in Figure 6.31 by the diagonal arrows. Below the level at which subsidence is occurring, the air layer remains stagnant. The temperature curve consequently develops a kink in which a part of the curve shows an inversion. The layer of inverted

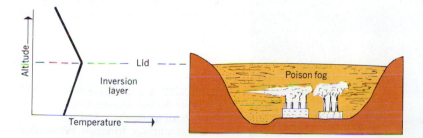

FIGURE 6.32 A low-level inversion was the predisposing condition for poison fog accumulation at Donora, Pennsylvania, in 1948. (From A. N. Strahler, *Planet Earth: Its Physical Systems Through Geologic Time*, Harper & Row, Publishers. Copyright © by Arthur N. Strahler.)

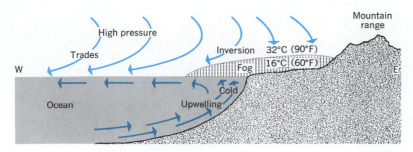

FIGURE 6.34 Subsiding air over a western continental coast produces a persistent upper-air temperature inversion, trapping cool air and fog in a surface layer near the shore. (From A. N. Strahler, *The Earth Sciences*, 2nd ed., Harper & Row, Publishers. Copyright © by Arthur N. Strahler.)

temperature structure strongly resists mixing and acts as a lid to prevent the continued upward movement and dispersal of pollutants. Upper-level inversions develop occasionally over various parts of the United States when an anticyclone stagnates for several days at a stretch.

For the Los Angeles Basin of southern California, and to a lesser degree the San Francisco Bay Area and over other west coasts in these latitudes generally, special climatic conditions produce prolonged upper-air inversions favorable to persistent smog accumulation. The Los Angeles Basin is a low, sloping plain, lying between the Pacific Ocean and a massive mountain barrier on the north side. Cool air is carried inland over the basin on weak winds from the south and southwest, but cannot move farther inland because of the mountain barrier. It is a characteristic of these latitudes that the air on the eastern sides of the subtropical high pressure cells is continually subsiding, creating a more or less permanent upper-air inversion that dominates the western coasts of the continents and extends far out to sea (Figure 6.34). The effect is particularly marked in the summer season, when the Azores and Bermuda highs are at their largest and strongest. The subsiding air over the Los Angeles Basin is warmed adiabatically as well as heated by direct absorption of solar radiation during the day, so it is markedly warmer than the cool stagnant air below the inversion lid, which lies at an altitude of about 600 m (2000 ft). Pollutants accumulate in the cool air layer, pro-

ducing the characteristic smog that first became noticeable for its irritating qualities in the early 1940s. Because water vapor content is generally low, Los Angeles smog is better described as a dense haze than as a fog, for it does not shut out the sun or cause a reduction in visibility to a degree that interferes with vehicular operation or the landing and takeoff of aircraft. Close to the coast, true fogs are frequent; but this condition is normally prevalent on dry west coasts adjacent to cold currents. The upper limit of the smog stands out sharply in contrast to the clear air above it, filling the basin like a lake and extending into valleys in the bordering mountains (Figure 6.35).

MODIFICATION OF URBAN CLIMATE

Applying principles of temperatures of the soil and lower air layer, we can anticipate the impact of humans as cities spread, replacing a richly vegetated countryside with blacktop and concrete. In the urban environment the absorption of solar radiation causes higher ground temperatures for two reasons. First, foliage of plants is absent, so the full quantity of solar energy falls on the bare ground. Absence of foliage also means absence of transpiration (evaporation from leaves), which elsewhere removes heat from the lower air layer. Second, roofs and pavements of concrete and asphalt hold no moisture, so evaporative cooling cannot occur as it would from a moist soil.

FIGURE 6.35 A dense layer of smog lying over the Los Angeles Basin. The view is from a point over the San Gabriel Mountains, looking southwest. (John S. Shelton.)

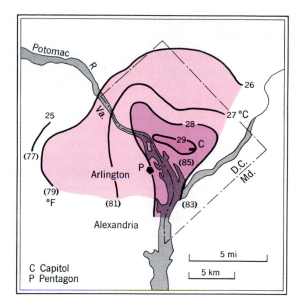

FIGURE 6.36 A heat island over Washington, D.C. Air temperatures were taken at 10:00 P.M. on a day in early August. (Data from H. E. Landsberg.)

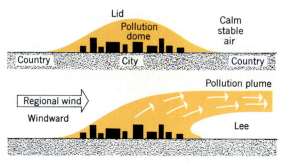

FIGURE 6.37 Pollution dome and pollution plume.

The thermal effect is that of converting the city into a hot desert. The summer temperature cycle close to the pavement of a city may be almost as extreme as that of the desert floor. This surface heat is conducted into the ground and stored there. The thermal effects within a city are actually more intense than on a sandy desert floor because the capacity of solid concrete, stone, or asphalt to conduct and hold heat is greater than that of loose, sandy soil. An additional thermal factor is that vertical masonry surfaces absorb insolation or reflect it to the ground and to other vertical surfaces. The absorbed heat is then radiated back into the air between buildings.

As a result of these changes in the energy balance, the central region of a city typically shows summer air temperatures several degrees higher than for the surrounding suburbs and countryside. Figure 6.36 is a map of the Washington, D.C., area, showing air temperatures

for a typical August afternoon. The lines of equal air temperature delineate a *heat island*. The heat island persists throughout the night because of the availability of a large quantity of heat stored in the ground during the daytime hours. In winter additional heat is radiated by walls and roofs of buildings, which conduct heat from the inside. Even in summer, we add to the city heat output through the use of air conditioners, which expend enormous amounts of energy at a time when the outside air is at its warmest.

Within the urban heat island pollutants are trapped beneath an inversion lid. The layer of polluted air takes the form of a broad *pollution dome* centered over the city when winds are very light or near calm (Figure 6.37). When there is general air movement in response to a pressure gradient, the pollutants are carried far downwind to form a *pollution plume*. Figure 6.38 has two maps showing plumes from the major cities of the Atlantic seaboard. The color bands show zones of fallout beneath the plumes and the color dot at the end of each band shows the distance traveled by the air in one day from the source. The left-hand map shows the effects of weak southerly winds on a day in June. In this situation pollution from one city affects another and generally the pollutants remain over the land, contaminating suburban and rural areas over a wide zone. The right-hand map, for a day in February, shows the effect of strong westerly winds, causing the pollution to be carried directly to sea.

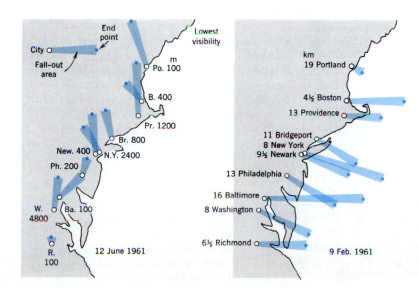

FIGURE 6.38 Pollution plumes from cities on the eastern seaboard under conditions of weak southerly winds (left) and strong westerly winds (right). The dot at the end of each plume represents the approximate distance traveled by pollutants at the end of a 24-hour period. (After H. E. Landsberg, 1962, in *Symposium—Air Over Cities*, Sanitary Engineering Center Technical Report A62-5, Cincinnati, Ohio.)

An important physical effect of urban air pollution is that it reduces visibility and illumination. A smog layer can cut illumination by 10 percent in summer and 20 percent in winter. Ultraviolet radiation is absorbed by smog, which at times completely prevents these wavelengths from reaching the ground. Reduced ultraviolet radiation may prove to be of importance in permitting increased bacterial activity at ground level. City smog cuts horizontal visibility to some one-fifth to one-tenth of the distance normal for clean air. Where atmospheric moisture is sufficient, the hygroscopic particles acquire water films and can cause formation of true fog with near-zero visibility. Over cities, winter fogs are much more frequent than over the surrounding countryside. Coastal airports, such as those in New York City, Newark, and Boston, suffer severely from a high incidence of fogs augmented by urban air pollution.

A related effect of the urban heat island is the general increase in cloudiness and precipitation over a city, as compared with the surrounding countryside. This increase results from intensified convection generated by heating of the lower air. For example, it has been found that thunderstorms over the city of London produce 30 percent more rainfall over the city than over the surrounding country. Increased precipitation over an urban area is estimated to average from 5 to 10 pecent over the normal for the region in which it lies.

SOME ENVIRONMENTAL EFFECTS OF AIR POLLUTION

A list of the harmful effects of atmospheric pollutants on plant and animal life and on inorganic substances would be lengthy if developed fully. We omit here the effects of air pollution on human health.

Ozone in urban smog has a most deleterious effect on plant tissues, and in some cases has caused the death or severe damage of ornamental trees and shrubs. Sulfur dioxide is injurious to certain plants and is a cause of loss of productivity in truck gardens and orchards in polluted air. Atmospheric sulfuric acid in cities has in some places largely wiped out lichen growth.

Although secondary in the sense that the loss is in dollars, rather than in lives and the health of animals and plants, the deterioration of various materials subjected to polluted air forms an important category of harmful effects. Building stones and masonry are susceptible to the corrosive action of sulfuric acid derived from the atmosphere. Metals, fabrics, leather, rubber, and paint deteriorate and discolor under the impact of exposure to urban air pollutants. In particular, natural rubber is vulnerable to ozone, which causes rubber to harden and crack. The sulfuric acid produced from sulfur dioxide corrodes exposed metals, particularly steel and copper.

ACID DEPOSITION AND ITS EFFECTS

Washout of sulfuric acid by precipitation results in rainwater with an abnormally high content of the sulfate ion, a condition known as *acid rain*. Nitric acid, formed by reactions involving pollutant nitrogen oxides, also contributes to the acidity of rainwater. Because fallout of solid pollutants in periods of dry weather also contributes these acids to the surfaces of plants, soil, streams, and lakes, the phenomenon is now described in broader terms as *acid deposition*. In earlier paragraphs on air pollution and its effects we covered the basic principles of acid deposition and made mention of its corrosive effects.

When tested for degree of acidity, measured in terms of the pH of the rainwater, acid rain shows pH values well below a pH of 5 to 6 typical of rainwater falling in unpolluted regions. (Rainwater is normally slightly acid because of the presence of carbon dioxide in solution, forming a weak concentration of carbonic acid.)

In the 1960s water chemists noted a significant lowering of the pH of rain in northwestern Europe. Values have been reduced to between pH 3 and 5. Because pH numbers are on a logarithmic scale, these values mean that rain in these areas is now often 100 to 1000 times more acid than previously.

The accompanying four maps show pH values for rainwater in northwestern Europe for the years 1956, 1959, 1961, and 1966 (Figure 6.39). This was the initial period in which rapid increase in acidity was first documented. The maps show both the dramatic lowering of pH levels and the widespread increase of areas receiving significantly acid precipitation.

American scientists studying the chemical quality of rainwater reported that since 1975 or thereabouts rainwater over a large area of the northeastern United States has had an average pH of about 4; they have observed pH values as low as 2.1 in rainwater of individual storms at certain localities. Other observations show that values less than pH 4 occur at times over many heavily industrialized United States cities, among them Boston, New York, Philadelphia, Birmingham, Chicago, Los Angeles, and San Francisco. Values between 4 and 5 have been observed near such urbanized areas as Tucson, Arizona; Helena, Montana; and Duluth, Minnesota—localities we do not usually associate with heavy air pollution.

Figure 6.40 is a map of the United States and Canada showing the average acidity of rainwater for the year 1982. Since about 1975 the level of acidity of rainwater seems to have leveled off in the heavily industrial northeastern United States and western Europe. At the same time, the contribution from nitrate has increased while that of sulfate has decreased, reflecting the change in proportion of these pollutants being injected into the atmosphere. Figure 6.41 shows the 1980 distribution of sulfur dioxide and nitrogen oxide in the United States. In recent years rainwater with pH below 5.0 has been identified locally in the southeastern and southwestern United States.

Some of the possible undesirable environmental effects of acid deposition upon natural systems are these: acidification of lakes and streams; excessive leaching of nutrients from plant foliage and the soil; various metabolic disturbances to organisms and upsetting of the balances of predators and prey in aquatic ecosystems.

One instance of the effect of acidification of stream water, observed in Norway, was the virtual elimination of salmon runs by inhibition of egg development. Now that the acidity of Norwegian streams has leveled off, salmon kills continue to occur, especially when rainstorms carry large amounts of sulfate into streams. In-

creased fish mortality observed in Canadian lakes was also attributed to acidification, and by 1980, Canada's Department of Environment reported that 140 Ontario lakes had no fish and that thousands of other lakes of the province were threatened with similar fate. Forests, too, appeared to have experienced damage from acid rain. In West Germany the impact was especially severe in the Harz Mountains and the Black Forest. It was reported in 1983 that about one-third of West German forests showed visible damage. In the eastern United States, pine and spruce trees have apparently experienced damage in recent years.

An important factor in the level of impact of acid deposition on the environment is the ability of the soil and surface water to absorb and neutralize acid. This

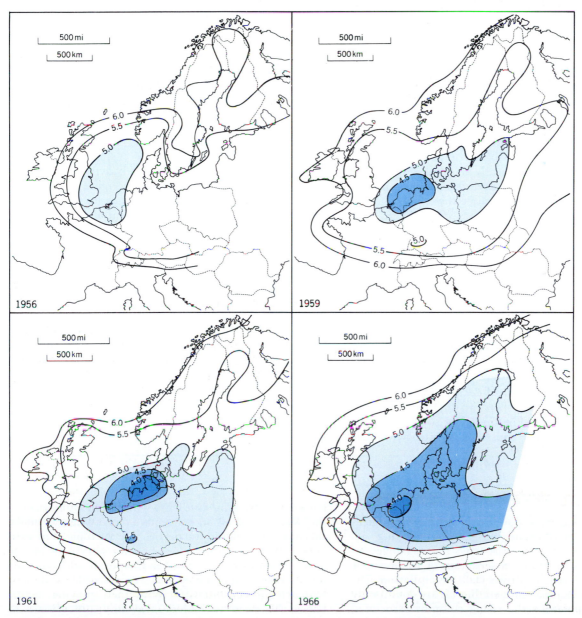

FIGURE 6.39 Rainwater acidity in northern Europe in the years 1956, 1959, 1961, and 1966. Figures give pH values. (Data of S. Odén, 1972; after G. E. Likens, et al., *Environment*, vol. 14, no. 2, p. 36, Figure 1.)

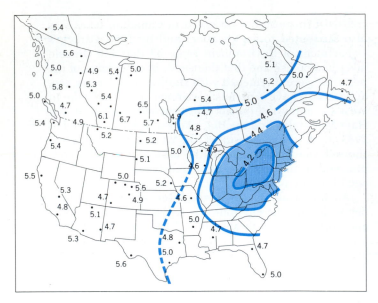

FIGURE 6.40 Distribution of acidity of rainwater over the United States and Canada, averaged for the year 1982. Figures give pH. Other recent years show essentially similar values. (From NOAA, Air Resources Laboratory.)

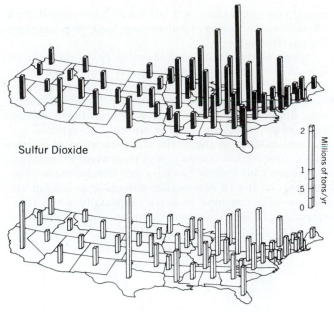

Sulfur Dioxide

FIGURE 6.41 Emissions of sulfur dioxide and nitrogen oxides by states for the year 1980. Essentially the same distribution continued through the 1980s. (Office of Technology Assessment, U.S. Congress, U.S. Government Printing Office, Washington, D.C.)

factor ranges widely over the United States. Areas of high alkalinity of surface waters are associated with aridity of climate and show low sensitivity to acid deposition. Areas where soil water is naturally acidic are highly sensitive to acid deposition. Such areas are associated with moist climates generally and include the eastern United States, high-mountain regions of the western states, and the Pacific northwest. Reasons for this distribution of sensitivity to acid deposition will emerge from the distributions of climates, soils, and vegetation classes, explained in later chapters.

In 1980, President Carter initiated the National Acid Precipitation Assessment Program (NAPAP) for the scientific study of the causes and effects of acid precipitation in the United States. A decade of investigation followed, and in 1990 NAPAP issued its report. On the sources and quantities of emissions, the report stated that in comparison with 1970 levels, emissions of sulfur dioxide and volatile organic compounds had declined 30 percent; of nitrogen oxides, by 8 to 16 percent. In 1990, electrical power plants continued to be the major source of sulfur oxides, which along with emissions of nitrogen oxides

and volatile organic compounds is heavily concentrated in the Northeast, about as shown in Figure 6.41. In 1990 an estimated 14 percent of Adirondack lakes were heavily acidic, along with 12 to 14 percent of streams in the Mid-Atlantic states. New controls on sulfur dioxide emissions called for by the 1990 Clean Air Act will speed the recovery of these aquatic environments.

Europe, in 1990, was continuing to experience heavy losses in timber, resulting from fallout of sulfur dioxide in combination with nitrogen oxide emissions from vehicles, industry, and farm wastes. Fallout of sulfur oxides was most heavily concentrated in Germany, Czechoslovakia, Poland, and Hungary, with the overall pattern of acidity generally similar to that shown on the 1966 map in Figure 6.39.

Today, scientists recognize that acid deposition will play a major role in the global environmental changes associated with the increase in emission of greenhouse gases.

THE ROLE OF ATMOSPHERIC MOISTURE

In this chapter we have covered basic meteorological principles of atmospheric moisture, including its changes of state and the energy flows those changes involve. We have seen that condensation of water vapor produces clouds, which in turn yield precipitation. The motions of air we have examined thus far are dominantly vertical displacements—the rising columns associated with dense cumulus clouds and thunderstorms, yielding heavy rain and hail.

Our next step will be to combine the principles of global atmospheric motions in wind systems driven by

barometric pressure gradients with the spontaneous rising of saturated air in convective cells. When we make this linkage, cyclonic storms become the topic of investigation. We replace more-or-less stationary cells of low and high pressure, described in Chapter 5, with comparatively rapidly moving cyclones on a smaller scale, but in much greater intensity. With these moving disturbances we need to include traveling anticyclones in which air is sinking and producing clear, dry weather. Our next chapter promises a fare heavy in atmospheric dynamics, which are often capable of imposing severe hazards on humans and their artifacts.

CHAPTER 7

AIR MASSES AND CYCLONIC STORMS

The atmosphere exerts stress, often severe, on humans and other life forms through weather disturbances involving extremes of wind speeds, cold, and precipitation. Under the category of storms, these phenomena generate environmental hazards directly by the impact of winds and precipitation and indirectly through their attendant phenomena—storm waves and storm surges on the seas and river floods, mudflows, and landslides on the lands. Weather disturbances of lesser magnitude are among the beneficial environmental phenomena because they bring precipitation to the land surfaces and so recharge the vital supplies of fresh water on which humans and all other terrestrial life forms depend.

An understanding of weather disturbances of all intensities enables us to predict their times and places of occurrence and therefore give warnings and allow protective measures to be taken. This function of the atmospheric scientist can be placed under the heading of environmental protection. It is also possible to a limited degree for us to modify atmospheric processes in a deliberate way to lessen the wind speeds of storms. These activities come under the heading of planned weather modification. Here, again, we find that the relationship of humans with the atmosphere is one of interaction.

TRAVELING CYCLONES

Much of the unsettled, cloudy weather experienced in middle and high latitudes is associated with traveling cyclones. The convergence of masses of air toward the cyclone centers is accompanied by lift of air and adiabatic cooling, which in turn produce cloudiness and precipitation. By contrast, much of the fair, sunny weather in these latitudes of westerly winds is associated with traveling anticyclones, in which the air subsides and spreads outward, causing adiabatic warming. This process produces stable air, unfavorable to the developments of clouds and precipitation.

Many cyclones are mild in intensity, passing with little more than a period of cloud cover and light rain or snow. On the other hand, when pressure gradients are strong, winds ranging in strength from moderate to gale

force accompany the cyclone. In such a case, the disturbance is called a *cyclonic storm*.

Traveling cyclones fall into three general classes: (1) The wave cyclone of midlatitude and arctic zones ranges in severity from a weak disturbance to a powerful storm. (2) The tropical cyclone of tropical and subtropical zones over ocean areas ranges from a mild disturbance to the highly destructive hurricane, or typhoon. (3) Although a very small storm, the tornado is an intense cyclonic vortex of enormously powerful winds; it is on a very much smaller scale of size than other types of cyclones and is related to severe convectional activity.

AIR MASSES

Cyclones of middle and high latitudes depend for their development on the coming together of large bodies of air of contrasting physical properties. A body of air in which the upward gradients of temperature and moisture are fairly uniform over a large area is known as an *air mass*. In horizontal extent, a single air mass may be as large as part of a continent; in vertical dimension, it may extend through the troposphere. A given air mass is characterized by a distinctive combination of temperature, environmental lapse rate, and specific humidity. Air masses differ widely in temperature from very warm to very cold and in moisture content from very dry to very moist.

A given air mass usually has a sharply defined boundary between itself and a neighboring air mass. This discontinuity is termed a *front*. We found an example of a front in the contact between polar and tropical air masses below the axis of the polar front jet stream in upper-air waves, as shown in Figure 5.19. This feature we called the polar front; it represents the highest degree of global generalization. Fronts may be nearly vertical, as in the case of air masses having little motion relative to one another. Fronts may be inclined at an angle not far from the horizontal in cases where one air mass is sliding over another. A front may be almost stationary with respect to the earth's surface, but, nevertheless, the adjacent air masses may be moving rapidly with respect to each other along the front.

The properties of an air mass are derived partly from the regions over which it passes. Because the entire troposphere is in more or less continuous motion, the particular air-mass properties at a given place may reflect the composite influence of a travel path covering thousands of kilometers and passing alternately over oceans and continents. This complexity of influences is particularly important in middle and high latitudes in the northern hemisphere, within the flow of the global westerlies.

Over vast tropical and equatorial areas, however, an air mass reflects quite simply the properties of an ocean or a land surface above which it moves slowly or tends to stagnate. Over a warm equatorial ocean surface, the lower levels of the overlying air mass develop a high water-vapor content. Over a large tropical desert, slowly subsiding air forms a warm air mass with low relative humidities. Over cold, snow-covered land surfaces in the arctic zone in winter, the lower layer of the air mass remains very cold with a very low water vapor content. Meteorologists have designated as *source regions* those land or ocean surfaces that strongly impress their temperature and moisture characteristics on overlying air masses.

Air masses move from one region to another, following the patterns of barometric pressure. During these migrations, lower levels of the air mass undergo gradual modification, taking up or losing heat to the surface beneath, and perhaps taking up or losing water vapor as well.

Air masses are classified according to two categories of source regions: (1) the latitudinal position on the globe, which primarily determines thermal properties; and (2) the underlying surface, whether continent or ocean, determining the moisture content. With respect to latitudinal position, five types of air masses are as follows:

Air Mass	Symbol	Source Region
Arctic	A	Arctic ocean and fringing lands
Antarctic	AA	Antarctica
Polar	P	Continents and oceans, lat. 50–60°N and S
Tropical	T	Continents and oceans, lat. 20–35°N and S
Equatorial	E	Oceans close to equator

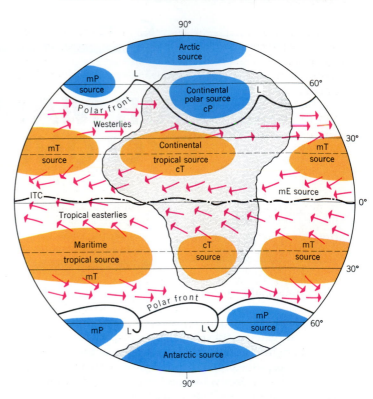

FIGURE 7.1 A schematic global diagram showing the source regions of air masses in relation to the polar front and the intertropical convergence zone.

With respect to the type of underlying surface, two further subdivisions are imposed on the preceding types:

Air Mass	Symbol	Source Region
Maritime	m	Oceans
Continental	c	Continents

By combining types based on latitudinal position with those based on underlying surface, a list of six important air masses results (Table 7.1). Figure 7.1 shows the global distribution of source regions of these air masses. Table 7.1 also gives some typical values of temperature and specific humidity at the surface, although a wider range in these properties may be expected, depending on season.

Note that the polar air masses (mP, cP) originate in the subarctic latitude zone, not in the polar latitude

TABLE 7.1 Properties of Typical Air Masses

Air Mass	Symbol	Properties	Temperature °C	(°F)	Specific Humidity g/kg
Continental arctic (and continental antarctic)	cA (cAA)	Very cold, very dry (winter)	−46°	(−50°)	0.1
Continental polar	cP	Cold, dry (winter)	−11°	(12°)	1.4
Maritime polar	mP	Cool, moist (winter)	4°	(39°)	4.4
Continental tropical	cT	Warm, dry	24°	(75°)	11.0
Maritime tropical	mT	Warm, moist	24°	(75°)	17.0
Maritime equatorial	mE	Warm, very moist	27°	(80°)	19.0

zone. The meteorological definition of the word "polar" for air masses has long been in use and has international acceptance; we cannot change this usage to conform with the geographic system of latitude zones defined in Chapter 3.

The maritime equatorial air mass (mE) holds about 200 times as much water vapor as the extremely cold and dry continental arctic and antarctic air masses, cA and cAA. The maritime tropical air mass (mT) and maritime equatorial air mass are quite similar in temperature and water vapor content. With very high values of specific humidity, both are capable of very heavy yields of precipitation. The continental tropical air mass (cT) has its source region over subtropical deserts of the continents. Although it may have a substantial water-vapor content, it tends to be stable and has low relative humidity when highly heated during the daytime. The maritime polar air mass (mP) originates over midlatitude oceans. Although the quantity of water vapor it holds is not large compared with the tropical air masses, the mP air mass can yield heavy precipitation. Much of this precipitation is of the orographic type, over mountain ranges on the western coasts of continents. The continental polar air mass (cP) originates over North America and Eurasia in the subarctic zone. It has low specific humidity and is very cold in winter.

NORTH AMERICAN AIR MASSES

The continental polar air mass of North America originates over north-central Canada (Figure 7.2). This air mass forms tongues of cold, dry air, which periodically extend south and east from the source region to produce anticyclones accompanied, in winter, by low temperatures and clear skies. Over the Arctic Ocean and its bordering lands of the arctic zone, the arctic air mass, which is extremely cold and stable, develops. When this air mass invades the United States, it produces a severe cold wave.

The maritime polar air mass originates over the North Pacific and Bering Strait, in the region of the persistent Aleutian low-pressure center. With ample opportunity to absorb moisture both over the source region and throughout its travel southeastward to the west coast of North America, this air mass is characteristically cool and moist, with a tendency in winter to become unstable, giving heavy precipitation over coastal ranges. Another maritime polar air mass of the North American region originates over the North Atlantic Ocean. It, too, is cool and moist.

Of the tropical air masses, the most common visitor to the central and eastern states is the maritime tropical air mass from the Gulf of Mexico. It moves northward, bringing warm, moist, unstable air over the eastern part of the country. In the summer, particularly, this air mass brings hot, sultry weather to the central and eastern United States. This air mass produces many thunderstorms. Closely related is a maritime tropical air mass from the Atlantic Ocean east of Florida, over the Bahamas.

A hot, dry continental tropical air mass originates over northern Mexico, western Texas, New Mexico, and Arizona during the summer. This air mass does not travel widely, but governs weather conditions over the source region.

Over the Pacific Ocean, in the cell of high pressure located to the southwest of Lower California, is a source region of another maritime tropical air mass. Occasionally, in summer, this moist, unstable air mass penetrates the southwestern desert region, bringing severe thunderstorms to southern California and southern Arizona.

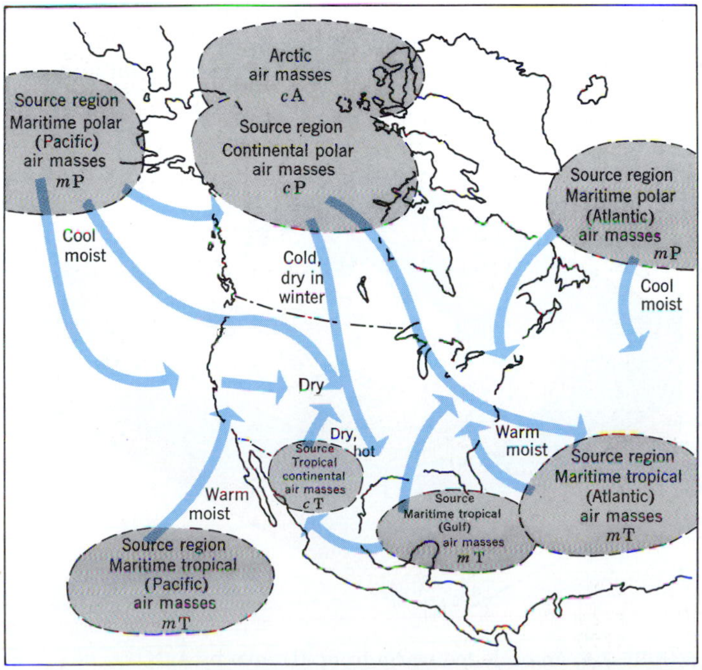

COLD AND WARM FRONTS

Figure 7.3 shows the structure of a front in which cold air is invading the warm-air zone. A front of this type is called a *cold front*. The colder air mass, being the denser,

FIGURE 7.2 North American air-mass source regions and trajectories. (Data from U.S. Department of Commerce.)

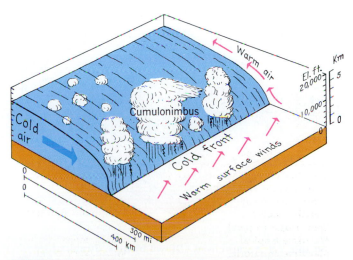

FIGURE 7.3 A cold front. (Drawn by A. N. Strahler.)

FIGURE 7.4 A line of cumulonimbus clouds marks the position of a cold front moving across the High Plains of eastern Colorado. The plane from which the photo was taken was flying in the cold air mass behind the front. (John S. Shelton.)

remains in contact with the ground and forces the warmer air mass to rise over it. The slope of the cold-front surface is greatly exaggerated in the figure, being actually of the order of slope of 1 in 40 (meaning that the slope rises 1 km vertically for every 40 km of horizontal distance). Cold fronts are associated with strong atmospheric disturbances. As the unstable warm air is lifted, it may break out in severe thunderstorms (Figure 7.4).

Figure 7.5 illustrates a *warm front*, in which warm air is moving into a region of colder air. Here, again, the cold air mass remains in contact with the ground and the warm air mass is forced to rise, as if it were ascending a long ramp. Warm fronts have lower slopes than cold fronts—on the order of 1 in 80, to as low as 1 in 200. Moreover, warm fronts commonly represent stable atmospheric conditions and lack the turbulent air motions of the cold front. Of course, if the warm air is unstable, it will develop convection cells and there will be heavy showers or thunderstorms.

Cold fronts normally move along the ground at a faster rate than warm fronts. So, when both types are in the same neighborhood, the cold front overtakes the warm front. An *occluded front* then results (Figure 7.6). The colder air of the fast-moving cold front remains next to the ground, forcing both the warm air and the less cold air to rise over it. The warm air mass is lifted completely free of the ground.

WAVE CYCLONES

The dominant type of weather disturbance of middle and high latitudes is a *wave cyclone*, a vortex that repeatedly forms, intensifies, and dissolves along the polar front between cold and warm air masses. At the time of World War I, the Norwegian meteorologist Jakob Bjerknes recognized the existence of atmospheric fronts and developed his *wave theory* of cyclones.

The term "front," used by Bjerknes, was particularly apt because of the resemblance of this feature to the fighting fronts in western Europe that were then active. Just as vast armies met along a sharply defined front that moved back and forth, so masses of cold polar air meet in conflict with warm, moist tropical air. Instead of mixing freely, these unlike air masses remain clearly defined; but they interact along the polar front in great spiraling whorls.

A situation favorable to the formation of a wave cyclone is shown by a surface weather map (Figure 7.7). A trough of low pressure lies between two large anticyclones (highs); one is made up of a cold, dry polar air mass, the other of a warm, moist maritime air mass. Airflow is converging from opposite directions on the two sides of the front, setting up an unstable situation.

A series of individual blocks (Figure 7.8) shows the sequence of stages in the life history of a wave cyclone. At the start of the cycle, the polar front is a smooth boundary along which air is moving in opposite directions. In Block A, the polar front shows a wave beginning to form. Cold air is turned in a southerly direction and

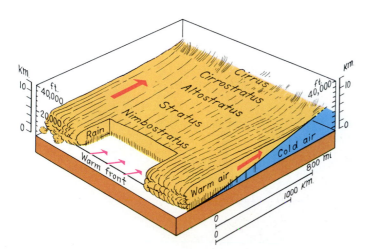

FIGURE 7.5 A warm front. (Drawn by A. N. Strahler.)

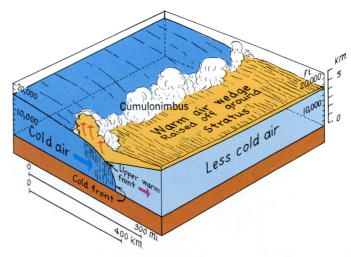

FIGURE 7.6 An occluded warm front. (Drawn by A. N. Strahler.)

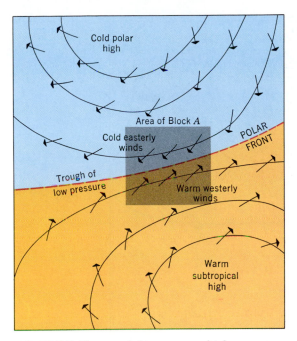

FIGURE 7.7 The trough between two high-pressure regions is a likely zone for the development of a wave cyclone.

warm air in a northerly direction, so each invades the domain of the other.

In Block B, the wave disturbance along the polar front has deepened and intensified. Cold air is now actively pushing southward along a cold front; warm air is actively moving northeastward along a warm front. The zone of precipitation is now extensive, but wider along the warm front than along the cold front. In Block C, the cold front has overtaken the warm front, reducing the zone of warm air to a narrow sector and producing an occluded front. Eventually, the warm air is forced off the ground (Block D), isolating it from the parent region of warm air to the south. The source of moisture and energy thus cut off, the cyclonic storm gradually dies out and the polar front is reestablished as originally (Block D).

CYCLONES ON THE DAILY WEATHER MAP

Further details of a wave cyclone are illustrated by surface weather maps for two successive days (Figure 7.9). Standard map symbols for wind directions and speeds are shown in Figure 7.10. Isobars are in millibars; the interval is 4 mb.

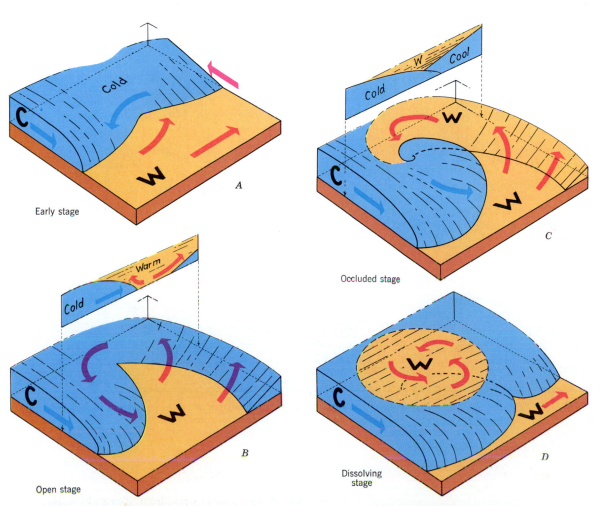

FIGURE 7.8 Stages in the development of a wave cyclone. (Drawn by A. N. Strahler.)

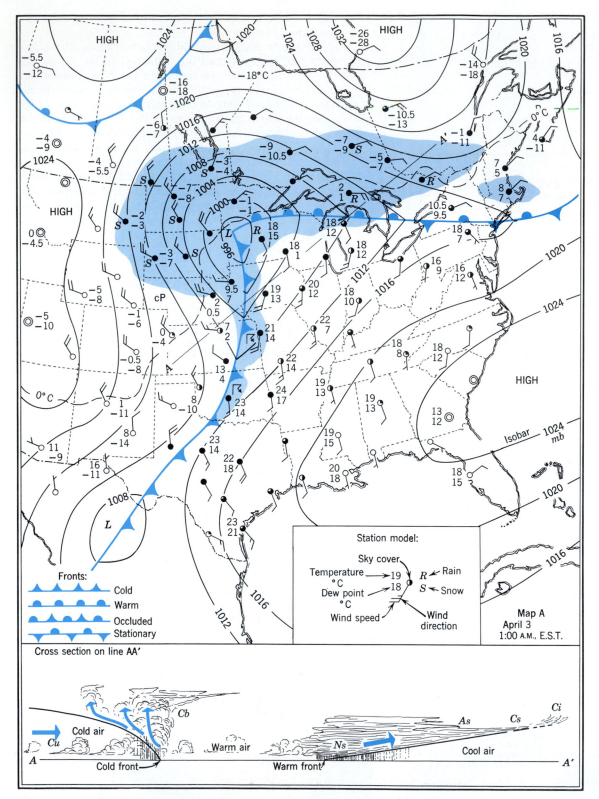

FIGURE 7.9 A wave cyclone shown on surface weather maps for successive days. Pressures are given in millibars; temperatures are given in Celsius degrees. Areas experiencing precipitation are shaded. See Figure 7.10 for explanation of wind symbols. (Modified and simplified from daily weather map of the National Weather Service.)

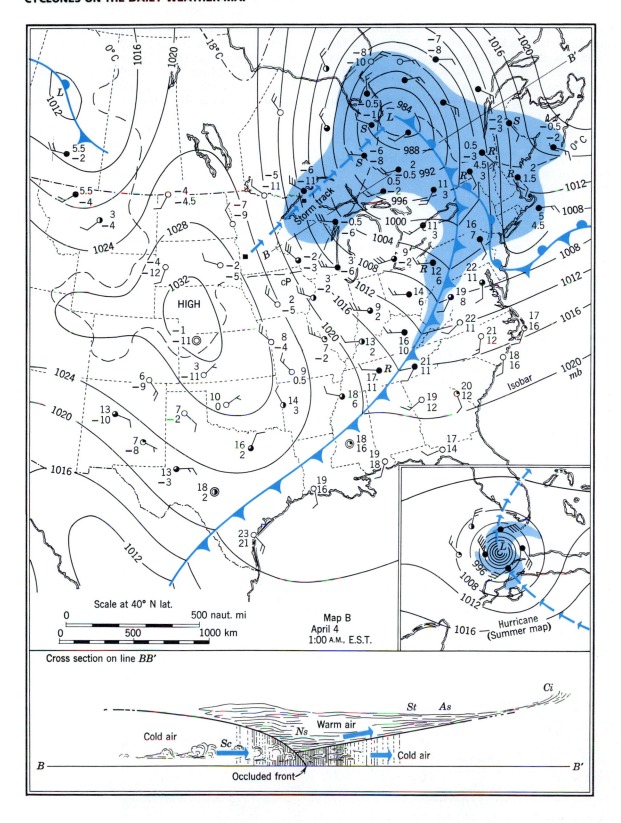

Scale at 40° N lat.

Map B
April 4
1:00 A.M., E.S.T.

Hurricane
(Summer map)

Cross section on line BB'

Cold air   Sc   Ns   Warm air   St   As   Ci

Cold air

Occluded front

B   B'

Map A of Figure 7.9 shows a cyclone in a stage approximately equivalent to Block B of Figure 7.8. The storm is centered over western Minnesota and is moving northeastward. Note the following features on Map A: (1) Isobars of the low are closed to form an oval-shaped pattern. (2) Isobars make a sharp V where crossing cold and warm fronts. (3) Wind directions, indicated by arrows, are at an angle to the trend of the isobars and form a pattern of counterclockwise inspiraling. (4) In the warm-air sector, warm, moist tropical air flows northward toward the warm front. (5) A sudden shift of wind direction accompanies the passage of the cold front. This wind shift is indicated by the widely different wind directions at stations close to the cold front, but on opposite sides. (6) A severe drop in temperature accompanies the passage of the cold front, as shown by differences in temperature readings at stations on opposite sides of the cold front. (7) Precipitation, shown by color shading, is occurring over a broad zone near the warm front and in the central area of the cyclone, but extends as a thin band down the length of the cold front. (8) Cloudiness, shown by the proportion of blackened area in station circles, extends over the entire cyclone. (9) The low is followed on the west by a high (anticyclone), in which low temperatures and clear skies prevail. (10) The 0°C (32°F) isotherm crosses the cyclone diagonally from northeast to southwest, showing that the southeastern part is warmer than the northwestern part.

A cross section through Map A of Figure 7.9 along the line AA' shows how the fronts and clouds are related. Along the warm front is a broad area of stratiform clouds. These take the form of a wedge with a thin leading edge of cirrus (Ci) and cirrostratus (Cs). Westward this thickens to altostratus (As), then to stratus (St), and finally to nimbostratus (Ns) with steady rain. Within the warm air mass sector, the sky may partially clear with scattered cumulus (Cu). Along the cold front are cumulonimbus (Cb) clouds with violent thunderstorms and heavy rains, but this activity is limited to a narrow belt and passes quickly.

The second weather map, Map B of Figure 7.9, shows conditions 24 hours later. The cyclone has moved rapidly

northeastward into Canada, its path shown by the line labeled "storm track." The center has moved about 1300 km (800 mi) in 24 hours, a speed of just over 65 km (40 mi) per hour. The cyclone has occluded. An occluded front replaces the separate warm and cold fronts in the central part of the disturbance. The high-pressure area, or tongue of cold polar air, has moved in to the west and south of the cyclone. The cold front is now passing over the eastern and Gulf Coast states. With closed isobars around the anticyclone, the skies are clear and the winds are weak. In another day the entire storm will have passed out to sea, leaving the eastern region in the grip of cold but clear weather. A cross section below the map shows conditions along the line BB', cutting through the occluded part of the storm. Note that the warm air mass is being lifted higher off the ground and is giving heavy precipitation.

WAVE-CYCLONE TRACKS

Observations, made over many decades, of the movements of cyclones and anticyclones have shown that certain tracks are most commonly followed. Figure 7.11 is a map of the United States and southern Canada showing these common paths. Notice that some cyclonic storms travel across the entire continent from places of origin in the North Pacific, such as the Aleutian low. Other cyclones originate in the Rocky Mountain region, the central states, or the Gulf Coast. Most of the tracks converge toward the northeastern states and the St. Lawrence lowland. These cyclones pass out into the North Atlantic, where they tend to concentrate in the region of the Icelandic low.

General world distribution of paths of wave cyclones is shown in Figure 7.12. Notice the heavy concentration in the neighborhood of the Aleutian and Icelandic lows.

Wave cyclones commonly form in succession to travel in a chain across the North Atlantic and North Pacific oceans. A chain of this kind is called a *cyclone family*. Figure 7.13 is a schematic diagram of cyclone families showing the frontal boundaries between cold and warm air masses. A polar weather map, Figure 7.14, shows four families of wave cyclones surrounding the great polar low. Rossby waves (R) are identified. Figure 7.15, a world

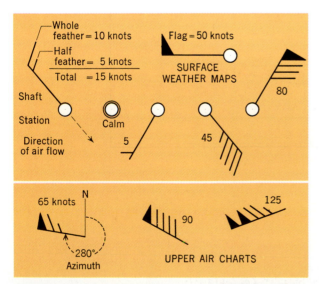

FIGURE 7.10 Standard weather-map symbols for winds.

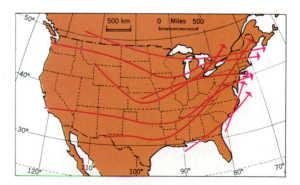

FIGURE 7.11 Common tracks taken by wave cyclones passing across United States and southern Canada. (Data from National Weather Service.)

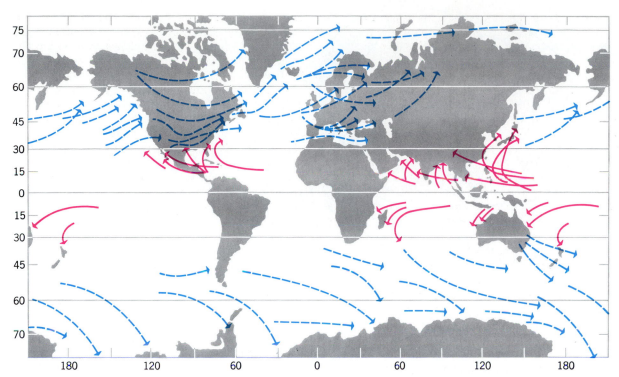

FIGURE 7.12 Typical paths of tropical cyclones (solid lines) and wave cyclones of middle latitudes (dashed lines). (Based on data of S. Petterssen, B. Haurwitz, and N. M. Austin, J. Namias, M. J. Rubin, and J-H. Chang.)

weather map, shows three families. As each cyclone moves northeastward it deepens and occludes, becoming an intense upper-air vortex. For this reason, cyclones arriving at the western coasts of North America and Europe are typically occluded.

In the southern hemisphere, storm tracks are more nearly along a single lane, following the parallels of latitude. This simplicity of pattern is the result of uniform ocean surface throughout midlatitudes. Only the southern tip of South America breaks the monotonous oceanic expanse. The polar-centered ice sheet of Antarctica provides a centralized source of polar air.

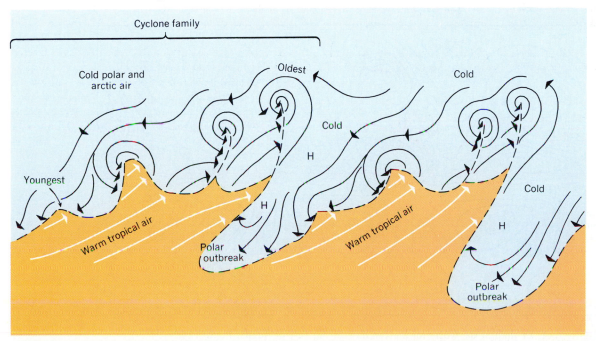

FIGURE 7.13 Two families of wave cyclones in the Northern Hemisphere, as seen on a schematic weather map. After Bjerknes and Solberg; Petterssen. (From A. N. Strahler, *The Earth Sciences*, 2nd ed., Harper & Row, Publishers. Copyright © by Arthur N. Strahler.)

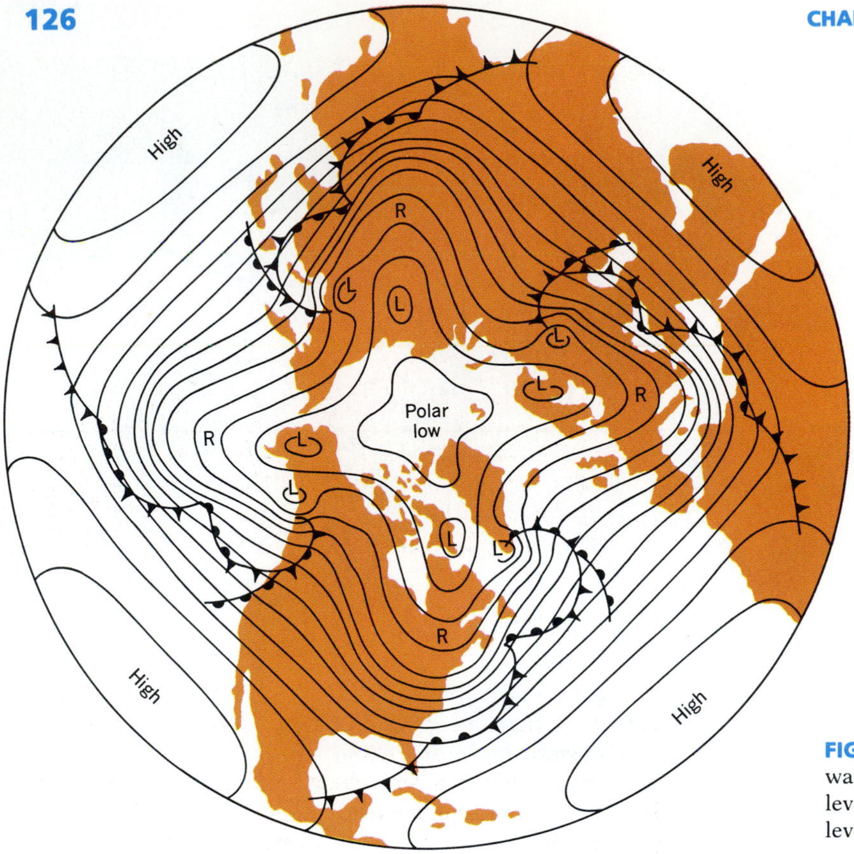

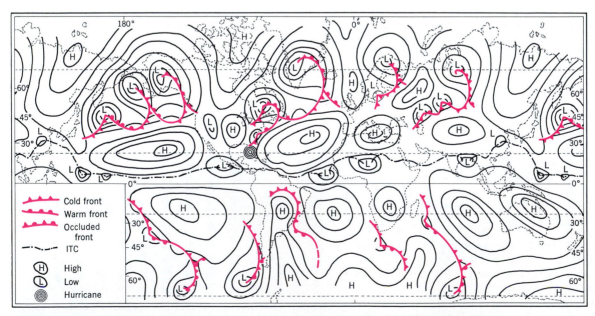

FIGURE 7.14 Polar map showing four families of wave cyclones. Cyclones are shown by fronts at sea level, whereas upper-air waves are shown by high-level isobars. (Data of Palmen and S. Petterssen.)

WAVE CYCLONES AND UPPER-AIR WAVES

How are wave cyclones and their accompanying anticyclones related physically to the Rossby waves described in Chapter 5? The wave cyclone with its fronts is a low-level phenomenon in the troposphere and gives way at high levels to the smooth westerly flow of air within the Rossby waves. Figure 7.16 shows the two systems superimposed. The upper diagram is a map; the lower diagram is a cross section along the line x–y of the map.

The key to a linkage of upper and lower systems lies in the streamlines of upper-air flow. Entering the west side (left) of the Rossby wave, the flow lines are coming together horizontally in a convergence pattern. The converging air is forced to descend to lower levels along the jet-stream axis, or core. As the air descends, it develops

FIGURE 7.15 A daily weather map of the world for a given day during July or August might look like this map, which is a composite of typical weather conditions. (Data from M. A. Garbell.)

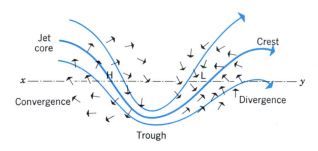

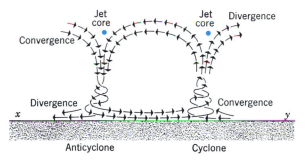

FIGURE 7.16 Schematic diagram of the coupling of upper-air flow paths with surface cyclones and anticyclones. (From A. N. Strahler, *The Earth Sciences*, 2nd ed., Harper & Row, Publishers. Copyright © by Arthur N. Strahler.)

FIGURE 7.17 A tornado funnel cloud. (Howie Bluestein/ Science Source/Photo Researchers.)

an anticyclonic outspiral and produces a low-level anticyclone (high), from which the air diverges horizontally. As the upper-air flow leaves the Rossby wave toward the east (right), the flow lines separate in a divergence pattern. Here, air must rise along the jet core. Air lifted from lower levels develops a cyclonic inspiral and is replaced by air moving into the cyclone near the surface. As soon as cyclonic flow begins, fronts are formed and a typical wave cyclone comes into existence. The cyclone is now "dragged along" beneath the jet core and moves northeastward. Thus we find that wave cyclones are both initiated and steered by the upper-level flow. Knowledge of upper-air conditions is of great value in forecasting surface weather.

Recall from Chapter 5 that, in an advanced stage of development, the Rossby wave can form a cut-off upperair low with a cold mass of air circulating within it. At ground level, the cut-off low is represented by a strongly occluded cyclone, so the low-pressure center extends through the entire troposphere. Deep lows of this kind may persist for many days at a stretch.

TORNADOES

The smallest but most intense of all known storms is the *tornado*. It seems to be a typically American storm because it is most frequent and violent in the United States. Tornadoes also occur in Australia in substantial numbers and are reported occasionally in other places in midlatitudes.

The tornado is a small, intense cyclone in which the air is spiraling at tremendous speed. It appears as a dark *funnel cloud* (Figure 7.17) hanging from a cumulonimbus cloud. At its lower end, the funnel may be 100 to 500 m

(300 to 1500 ft) in diameter. The funnel appears dark because of the density of condensing moisture, dust, and debris swept up by the wind.

Wind speeds in a tornado exceed anything known in other storms. Estimates of wind speed run as high as 400 km (250 mi) per hour. As the tornado moves across the country, the funnel writhes and twists. The end of the funnel cloud may alternately sweep the ground, causing complete destruction of anything in its path, and rise in the air to leave the ground below unharmed. Tornado destruction occurs from the great wind stress. A popular theory of years past is that tornado destruction occurs in part from the sudden reduction of air pressure in the vortex of the cyclonic spiral, and that closed houses literally explode as a result. The importance of this mechanism has been greatly downgraded as a result of recent studies.

Tornadoes occur as parts of dense cumulonimbus clouds traveling in advance of a cold front. They seem to

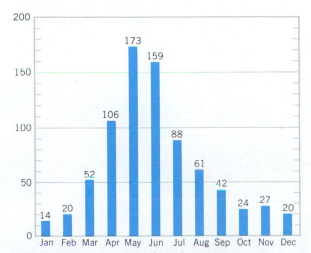

FIGURE 7.18 Average number of tornadoes reported in each month in the United States for the period 1960–89. (Data of NOAA National Weather Service, National Severe Storms Forecast Center.)

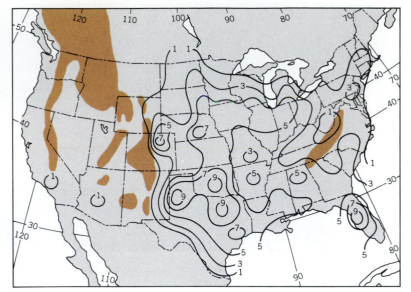

FIGURE 7.19 Frequency of occurrence of observed tornadoes in the 48 contiguous United States. Numbers on the isopleths tell average annual number of tornadoes per 10,000 square miles (equivalent to a square 100 × 100 mi; 160 × 160 km). The data span a 30-year record, 1960 through 1989. (Courtesy of Edward W. Ferguson, National Severe Storms Forecast Center, National Weather Service.)

originate where turbulence is greatest. They are most common in the spring and summer, but can occur in any month (Figure 7.18). Where maritime polar air lifts warm, moist tropical air on a cold front, conditions may become favorable for tornadoes. Current theory suggests that a strong horizontal, cylindrical vortex, high in the frontal zone, becomes turned on its side to become a vertical vortex, and this is turn descends to ground level, becoming greatly reduced in diameter and intensity, forming the tornado funnel cloud.

Tornadoes occur in greatest numbers in the central and southeastern states and are rare over mountainous and forested regions. They rarely occur from the Rocky Mountains westward and there are relatively few on the eastern seaboard (Figure 7.19).

Devastation from a tornado is often complete within the narrow limits of its path (Figure 7.20). Only the strongest buildings, constructed of concrete and steel, can resist major structural damage. A tornado can often be seen or heard approaching, but those that come during the night may give no warning. The National Weather Service maintains a tornado forecasting and warning system. Whenever weather conditions conspire to favor tornado development, the danger area is alerted

FIGURE 7.20 Swath of destruction left by a tornado that swept through Louisville, Kentucky. (George Hall/Woodfin Camp.)

and systems for observing and reporting a tornado are set in readiness.

Waterspouts are similar in structure to tornadoes but form at sea under cumulonimbus clouds. They are smaller and less powerful than tornadoes. Seawater may be lifted 3 m (10 ft) above the sea surface. Rapid condensation of water vapor makes a visible column reaching to the overlying cloud base. Waterspouts are commonly found in subtropical waters of the Gulf of Mexico and off the southeastern coast of the United States. They result from air turbulence along a cold front moving into lower latitudes.

TROPICAL AND EQUATORIAL WEATHER DISTURBANCES

Weather systems of the tropical and equatorial zones show some basic differences from those of midlatitudes. The Coriolis effect is weak close to the equator, and there is a lack of strong contrast between air masses. Consequently, clearly defined fronts and large, intense wave cyclones are missing. On the other hand, there is intense atmospheric activity in the form of convection cells because of the high moisture content of the maritime air masses in these latitudes. In other words, there is an enormous reservoir of energy in the latent form. This same energy source powers the most formidable of all storms—the tropical cyclone.

The world weather map (Figure 7.15) shows a typical set of weather conditions in the subtropical and equatorial zones. Fundamentally, the arrangement consists of two rows of high-pressure cells, one or two cells to each land or ocean body. The northern row lies approximately along the tropic of cancer; the southern row lies along the tropic of capricorn. Between the subtropical highs lies the equatorial trough of low pressure. Toward this trough the northeast and southeast trades converge along the intertropical convergence zone (ITC). At higher levels in the troposphere, the airflow is almost directly from east to west in the form of persistent tropical easterlies, described in Chapter 5.

One of the simplest forms of weather disturbance is an *easterly wave*, a slowly moving trough of low pressure within the belt of tropical easterlies (trades). These waves occur in lat. 5° to 30°N and S over oceans, but not over the equator itself. Figure 7.21 is a simplified weather map of an easterly wave showing isobars, winds, and the zone of rain. The wave is simply a series of indentations in the isobars to form a shallow pressure trough. The wave travels westward at a rate of 325 to 500 km (200 to 300 mi) per day. Air near the surface converges on the eastern, or rear, side of the wave axis. Moist air is lifted and produces scattered showers and thunderstorms. The rainy period may last for a day or two.

Another related disturbance is the *weak equatorial low*, which forms near the center of the equatorial trough. Moist equatorial air masses converge on the center of the low, causing rainfall from many individual convectional storms. Several such weak lows are shown on the world weather map (Figure 7.15), lying along the

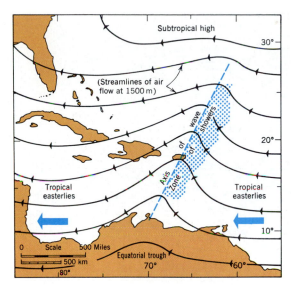

FIGURE 7.21 An easterly wave passing over the West Indies. (Data from H. Riehl, *Tropical Meteorology*, McGraw-Hill Book Co., New York.)

ITC. Because the map is for a day in July or August, the ITC is shifted well north of the equator. At this season, the rainy monsoon is in progress in Southeast Asia.

Another distinctive feature of tropical-zone weather is the occasional penetration of powerful tongues of cold polar air from the midlatitudes into very low latitudes. These tongues are known as *polar outbreaks;* they bring unusually cool, clear weather with strong, steady winds moving behind a cold front with squalls. The polar outbreak is best developed in the Americas. Outbreaks that move southward from the United States over the Caribbean Sea and Central America are called "northers" or "nortes"; those that move north from Patagonia into South America are called "pamperos." One such outbreak is shown over South America on the world weather map (Figure 7.15).

THE ASIATIC MONSOON

In Chapter 5 we described the Asiatic monsoon system of surface winds, shown in Figure 5.14, associating this seasonal reversal in direction with the alternation of the north-central Asiatic high in winter and the southern Asian low in summer. To understand how the high-sun (summer) monsoon rains of southern and eastern Asia are generated requires that we investigate seasonal changes in air flow throughout the overlying troposphere.

Two maps show the main features of the Asiatic tropospheric circulation (Figure 7.22). In January, when pressure and wind belts have been shifted farthest southward, the ITC lies south of the equator, beyond the limits of the map (see Figure 5.10, January). The trades at low level lie over the East Indies. In a zone between 5° and 15°N, the upper-air easterlies follow closely the parallels of latitude. Farther north, over peninsular India and

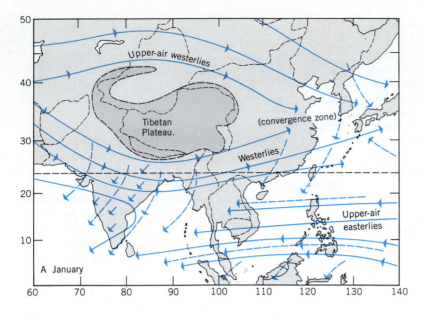

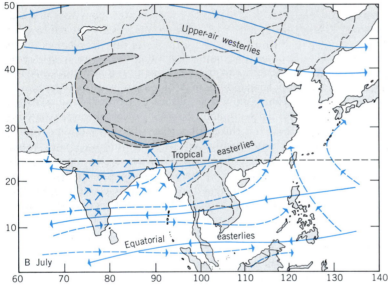

FIGURE 7.22 Generalized maps showing atmospheric circulation over southern and southeastern Asia at times of low-sun (Map A) and high-sun (Map B). Long solid arrows show air flow at about 6 km (20,000 ft); dashed lines show flow at about 0.6 km (2000 ft). Short arrows show surface winds. (Redrawn from R. G. Barry and R. J. Chorley, 1987, *Atmosphere, Weather and Climate,* 5th ed., Methuen & Co., London, Figures 6.18 and 6.22. Used by permission.)

Burma, the low-level northeasterly winds are overlain by upper-air westerlies. These prevailing westerlies contain jet streams that flow around both north and south sides of the lofty Tibetan plateau; they converge on the eastern side of that plateau, over northern China. The entire system favors the southerly and southeasterly outflow of cold, dry continental polar (cP) airmasses from sources in central Asia, causing a dry winter season over the continental margins.

As winter gives way to spring, the pressure and wind belts shift northward. By July (Map B of Figure 7.22), the ITC has reached a position in the northern part of the Indian peninsula and lying across Burma, Thailand, Vietnam, and the Philippines. Over the ITC, upper-air flow is dominated by tropical easterlies, while the upper-air westerlies have retreated into northern Asia. Southward, the tropical easterlies merge into the equatorial easterlies. Above them flows the easterly tropical jet stream. What may seem strange to those of us accustomed to weather patterns of the middle latitudes, is that

low-level air flow at about 600 m (2000 ft), shown by dashed arrows, is west-to-east, i.e., traveling directly opposite to the easterlies and their jet stream. Surface winds, because of the effect of drag, are generally southwesterly over the Indian peninsula.

During this summer season maritime tropical (mT) airmasses are carried from over the Indian Ocean and China Sea into southern and eastern Asia. Rainfall in the tropical zone occurs in weak lows that tend to drift westward, steered by the upper-air easterlies. Convectional precipitation within the lows is typically heavy, but because of their movement, rainfree periods are experienced between their passages.

The onset of the rainy monsoon season advances over southern and eastern Asia in a regular time progression. It arrives in late May at the island of Ceylon and southern tip of the Indian peninsula, and by mid-June has reached the 20th parallel of latitude. For China, the onset is early in May along its southern border, reaching north China by mid-July.

TROPICAL CYCLONES

One of the most powerful and destructive types of cyclonic storms is the *tropical cyclone*, otherwise known as the *hurricane* or *typhoon*. The storm develops over oceans in latitudes 8° to 15°N and S, but not close to the equator, where the Coriolis effect is extremely weak. In many cases an easterly wave simply deepens and intensifies, growing into a deep, circular low. High sea-surface temperatures that are over 27°C (80°F) in these latitudes are important in the environment of storm origin because warming of air at low level creates instability. Once formed, the storm moves westward through the trade-wind belt. It may then curve poleward and penetrate well into the belt of westerly winds.

The tropical cyclone is an almost circular storm center of extremely low pressure into which winds are spiraling at high speed, accompanied by very heavy rainfall (Figures 7.23 and 7.24). Storm diameter may be 150 to 500 km (100 to 300 mi). Wind speeds range from 120 to 200 km (75 to 125 mi) per hour, sometimes much higher. Barometric pressure in the storm center commonly falls to 950 mb or lower (Figure 7.25).

A characteristic feature of the tropical cyclone is its *central eye*, in which calm prevails (Figures 7.26 and 7.31). The eye is a cloud-free vortex produced by the intense spiraling of the storm. In the eye, air descends from high altitude and is adiabatically warmed. Passage of the central eye may take about half an hour, after which the storm strikes with renewed ferocity, but with winds in the opposite direction.

World distribution of tropical cyclones is limited to six regions, all of them over tropical and subtropical oceans (Figure 7.27): (1) West Indies, Gulf of Mexico, and Caribbean Sea; (2) western North Pacific, including the Philippine Islands, China Sea, and Japanese Islands; (3) Arabian Sea and Bay of Bengal; (4) eastern Pacific coastal region off Mexico and Central America; (5) south Indian Ocean, off Madagascar; and (6) western South Pacific, in the region of Samoa and the Fiji Islands and the east coast of Australia (see also Figure 7.12). Curiously, these storms are unknown in the South Atlantic. Tropical cyclones never originate over land, although they often penetrate well into the margins of continents.

Tracks of tropical cyclones of the North Atlantic are shown in Figure 7.28. Most of the storms originate at lat. 10° to 20°, travel westward and northwestward through the trades, then turn northeast at about lat. 30° to 35° into the zone of the westerlies. Here the intensity lessens and the storms change into typical midlatitude wave cyclones. In the trade-wind belt, the cyclones travel 10 to 20 km (6 to 12 mi) per hour.

The occurrence of tropical cyclones is restricted to certain seasons of the year, and these vary according to the global location of the storm region. For hurricanes of the North Atlantic, the season runs from May through November, with maximum frequency in late summer or early autumn. The general rule is that tropical cyclones of the northern hemisphere occur in the season during which the ITC has moved north; those of the southern hemisphere occur when it has moved south.

The environmental importance of tropical cyclones lies in their tremendously destructive effect on inhabited islands and coasts (Figure 7.29). Wholesale destruction of cities and their inhabitants has been reported on several occasions. A terrible hurricane that struck Barbados in the West Indies in 1780 is reported to have torn stone buildings from their foundations, destroyed forts, and carried cannons more than 30 m (100 ft) from their loca-

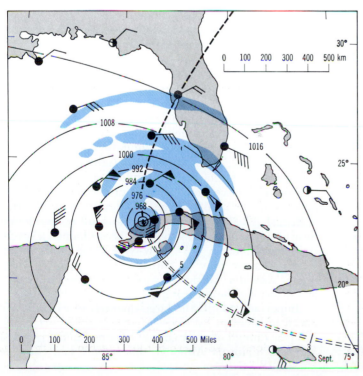

FIGURE 7.23 Surface weather map of a typical hurricane passing over the western tip of Cuba. See Figure 7.10 for interpretation of wind arrows.

FIGURE 7.24 Hurricane Gladys, photographed from the Apollo 7 spacecraft on October 17, 1968, at an altitude of about 180 km (110 mi). Bands of cumulonimbus clouds spiral around the central eye. The large, smooth cloud patches are high cirrostratus layers derived from the tops of the rising convection cells. (NASA AS7-7-1877)

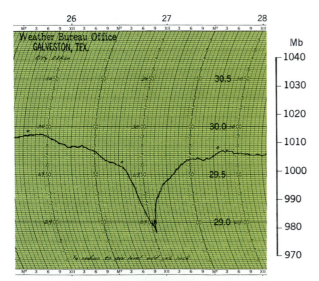

FIGURE 7.25 Trace sheet from a barograph at Galveston, Texas, during the hurricane of July 27, 1943. (National Weather Service.)

tions. Trees were torn up and stripped of their bark, and more than 6000 persons perished.

Coastal destruction by storm waves and greatly raised sea level is perhaps the most serious effect of tropical cyclones. Where water level is raised by strong wind pressure, great storm surf attacks ground ordinarily far inland of the limits of wave action. A sudden rise of water level, known as *storm surge*, may take place as the hurricane moves over a coastline. Ships are lifted and carried inland to become stranded. If high tide accompanies the storm, the limits reached by inundation are even higher. The terrible hurricane disaster at Galveston, Texas, in 1900 was wrought largely by a sudden storm surge, inundating the low coastal city and drowning about 6000 persons. At the mouth of the Hooghly River on the Bay of Bengal, 300,000 persons died as a result of inundation by a 12-m (40-ft) storm surge that accompanied a severe tropical cyclone in 1737. Low-lying coral atolls of the western Pacific may be entirely swept over by wind-driven seawater, washing away palm trees and houses and drowning the inhabitants.

Important, too, is the large quantity of rainfall produced by tropical cyclones. A considerable part of the summer rainfall of some coastal regions can be traced to a few such storms. Although this rainfall is a valuable

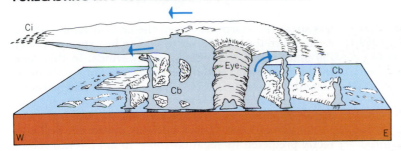

FIGURE 7.26 Schematic diagram of a hurricane. Cumulonimbus clouds in concentric rings rise through dense stratiform clouds. Width of the diagram represents about 1000 km (600 mi). (National Oceanic and Atmospheric Administration, National Weather Service, R. C. Gentry, *Weatherwise*, vol. 17, p. 182.)

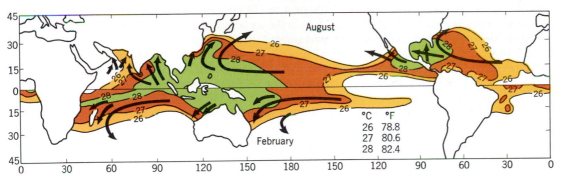

FIGURE 7.27 Typical paths of tropical cyclones in relation to sea surface temperatures (°C) in summer of the respective hemisphere. (Data from Palmén.)

water resource, it may prove a menace as a producer of unwanted river floods and, in areas of steep mountain slopes, gives rise to disastrous mudslides and landslides.

Attempts to reduce the severity of hurricanes by cloud seeding were made during the 1960s by scientists of the National Oceanic and Atmospheric Administration (NOAA) under an experimental program known as Project Stormfury. Seeding of four hurricanes on eight different days may have produced an observed reduction in wind speeds of between 10 and 30 percent on four of the days. Subsequent research led to the conclusion that hurricane clouds contain too much natural ice and too little supercooled water to respond satisfactorily to seeding. The project was abandoned in 1983, but much was learned about the physics of condensation processes.

FORECASTING TWO HURRICANES— GILBERT AND HUGO

At the National Hurricane Center (NHC) in Coral Gables, Florida, meteorologists of the U.S. National Weather Service scrutinize the progress of each tropical storm and hurricane as it develops in the North Atlantic and heads toward the North American mainland.

The NHC uses the Saffir-Simpson scale of hurricane intensities, which emphasizes the potential damage they can cause. Intensities range from that of a minimal-sized hurricane (category 1), to the most powerful (category 5), which has a central pressure less than 920 mb. Prior to 1988, only two hurricanes of category 5 had come

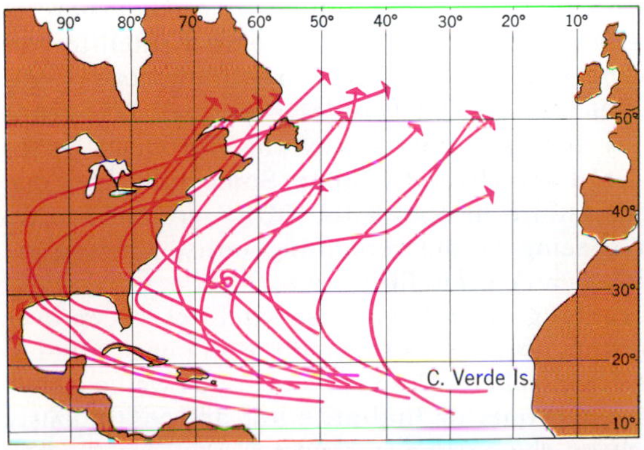

FIGURE 7.28 Tracks of some typical hurricanes occurring during August. (After U.S. Navy Oceanographic Office.)

FIGURE 7.29 Only the foundations remain from a building leveled by winds and storm swash during the hurricane of September 1960 at Plantation Key, Florida. (Leslie F. Conover/Photo Researchers.)

ashore along the Gulf Coast: an unnamed hurricane that devastated Key West in 1935; and Camille, that in 1969 swept over the shoreline of Mississippi and Louisiana.

Hurricane Gilbert, category 5, swept over Jamaica on September 12, 1988, destroying one-fifth of its 500,000 homes. Two days later, Gilbert devastated the Yucatan Peninsula, then continued across open waters of the Gulf to threaten the Gulf Coast with winds up to 260 km (160 mi) per hour. Where would he come ashore? Potential targets ranged from Mexico, on the south, to Brownsville, Corpus Christi, and Galveston, Texas, farther to the northeast. Six hours before the anticipated landfall, the NHC chose a coastal point in Mexico 640 km southwest of Galveston. At the same time, a private weather forecaster had selected for the landfall a point between Corpus Christi and Galveston. Uncertain which forecast to accept, many residents of Galveston fled inland. The NHC forecast proved correct. Controversy followed over the question of whether private forecasting companies should be allowed to issue warnings that differ widely from those of the NHC.

Hurricane Hugo, in September 1989, was a category 5 storm as it approached the Caribbean Sea, but weakened to a rating of 3 as it passed over Puerto Rico and headed northwesterly in the general direction of the Georgia-Carolinas coastline. Up to that point, NHC forecasts had been quite accurate, but now began to prove

less so, as the storm traversed the critical region where recurvature toward the north was to be expected under the influence of upper-air westerly winds. In this zone the storm's trajectory underwent some minor changes. Meantime, Hugo was intensifying and soon reached category 4 with wind speeds up to 220 km (135 mi) per hour. Accuracy of the forecasts made 24 and 48 hours previous to the actual landfall had by then improved, and correctly predicted its location some 50 km (30 mi) southwest of Charleston. That city experienced a storm surge about 4 m (13 ft) high, while farther along the coast to the northeast it peaked at over 6 m (20 ft). High winds and coastal flooding did damage at an estimated total of $7 billion, but with deaths limited to 21 persons, thanks to accurate forecasting and widely broadcast warnings. Continuing its forecasting, the NHC correctly predicted that the center of Hugo would pass over West Virginia, 1000 km (600 mi) distant.

Forecasting of hurricane paths is made difficult by rapid changes in direction and strength of the upper-air steering winds. Improvements in hurricane forecasting have come about through the use of a computer model known as ETA (from the Greek letter *eta*), which was used by the NHC in tracking Hugo. This model is better able to incorporate information on the steering winds and other significant regional meteorological data.

REMOTE SENSING OF WEATHER PHENOMENA

Remote sensing methods, explained in Appendix II, have been intensively developed for weather observation through the use of specialized orbiting satellites. The first of these, Nimbus-1, was launched into a sun-synchronous polar orbit in 1964. It transmitted TV photographs as well as high-resolution infrared images, the latter taken in darkness. Later Nimbus satellites and the subsequent NOAA TIROS series, all polar orbiters, have been continuously improved in their data-collecting ability in many categories. Besides transmitting images, these weather satellites record the vertical profile of temperatures in the atmosphere. They also provide data of ozone, water vapor, albedo, cloud and snow cover, precipitation, and sea ice. Data are received, processed, and distributed at a central facility in Suitland, Maryland. The satellites collect data from several hundred stations, including not only fixed and floating platforms, but balloons as well. Both the U.S. National Weather Service and the World Weather Program use the TIROS data for forecasting purposes. The satellite images are particularly useful in tracking hurricanes and typhoons, a service that has helped to save countless lives and forestall a great amount of property damage.

Figure 7.30 shows two satellite images of an occluded wave cyclone over the eastern Pacific Ocean. These images were prepared in such a way as to form a stereoscopic pair. When viewed simultaneously, one frame with each eye, a strong three-dimensional effect results.

A simple lens stereoscope (used in viewing air photos) can be placed over the page to achieve this 3-D effect.

Another class of weather satellites uses a *geostationary orbit*, in which the satellite holds a fixed position over a selected point on the earth's equator. These earth-synchronous satellites are in an equatorial orbit at a height of 36,200 km (22,300 mi); there the satellite speed exactly matches the speed of eastward rotation of the globe.

The first of the geostationary satellites, called the Synchronous Meteorological Satellite (SMS), was launched in 1974 and was followed by several more of the same type. Five geostationary weather satellites, all launched by NASA in 1979, were in simultaneous operation: GOES-East and GOES-West, operated by NOAA; INSAT, by India; Meteostat, by the European Space Agency; GMS Sunflower, by Japan. Some difficulties due to systems failure in one of the GOES group have required respacing of the remaining units, pending replacement by a new satellite.

The scanning systems of these satellites sweep over the globe from north to south, yielding an image equal to about one hemisphere. Tropical storms, as well as wave cyclones and fronts of midlatitudes, are continuously monitored by the earth-synchronous satellites, supplementing imagery returned by the polar-orbiting satellites (Figure 7.31). Currently, these satellites are being used in the World Climate Research Program (WCRP).

FIGURE 7.30 An occluded cyclone over the eastern Pacific Ocean shows on this satellite image as a tight cloud spiral. The cold front makes a dense, narrow cloud band sweeping to the south and southwest of the cyclone center. View these frames with a stereoscope for strong three-dimensional effect. (NOAA-2 satellite image, courtesy of National Environmental Satellite Service.)

In 1985, satellite data collection using these satellites and a polar-orbiting satellite was begun under the International Satellite Cloud Climatology Project, an activity of the WCRP. Meteorological data from NOAA's satellites is under the direction of its National Environmental Satellite, Data, and Information Service (NESDIS) in Washington, D.C.

Figure 7.32 shows three full-earth GOES photographs taken simultaneously at sunrise. The upper photo (A) utilizes the visible wavelengths and shows the position of the terminator (line of contact of day and night) as well as many cloud patterns. The second photo (B) is the thermal-infrared image, on which the terminator does not appear. In this image, the light-colored clouds are warm compared to the surrounding cold atmosphere. The third image (C) is processed from infrared images to show regions of the upper troposphere with abundant

water vapor (light areas). The intertropical convergence zone, in which rising air carries moisture to high altitudes, is clearly shown as a light-colored band near the equator.

One of the instruments on Nimbus satellites, the Total Ozone Mapping Spectrometer, has played an essential role in providing information on the arctic and antarctic ozone holes and the extent to which global ozone depletion has been taking place (see Chapter 2). Nimbus-7 has recently contributed to oceanography and meteorology through precise measurements of sea-surface temperatures and the speeds of ocean-surface winds. These data are obtained by the Scanning Multichannel Microwave Radiometer (SMMR), which can "see" through a cloud cover. Sea-surface temperature data will play a major role in prediction and detection of global climate change.

FIGURE 7.31 Hurricane Anita, over the western Gulf of Mexico, as observed by GOES-2 on September 1, 1977. The eye of the storm was about 175 km (100 mi) southeast of Brownsville, Texas. The hurricane was moving slowly in a southwesterly direction and later crossed the coast 230 km (140 mi) south of Brownsville. (Satellite Services Division, Environmental Data and Information Service, NOAA.)

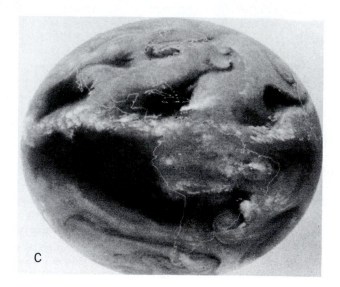

FIGURE 7.32 Simultaneous full-earth images from GOES at sunrise. (A) Image from the visible region of the spectrum, showing half of the earth in darkness and the remainder in light. (B) Thermal infrared image, showing thermal emission from both dark and light halves of the globe. (C) Infrared image processed to show tropospheric water vapor "cloud" patterns. (Courtesy of NASA.)

EL NIÑO AND THE SOUTHERN OSCILLATION

At intervals of some 3 to 8 years there occurs a remarkable disturbance of ocean and atmosphere that begins in the eastern Pacific Ocean and spreads its effects widely over the globe for more than a year's time, bringing with it unseasonal weather patterns with abnormalities in the form of droughts, heavy rainfalls, severe spells of heat and cold, or a high incidence of cyclonic storms. This phenomenon is called *El Niño*. Back in the 1890s it was reported that Peruvian fishermen were using the expression *"Corriente del Niño"* ("Current of the Christ Child") to describe an invasion of warm surface water that occurred once every few years around Christmas and greatly depleted their catch of fish.

The occurrences of El Niño have been at irregular intervals and with varying degrees of intensity, the most notable ones being in 1891, 1925, 1940–41, 1965, 1972–73, and 1982–83. As we mentioned in Chapter 5, the cool Humboldt (Peru) current flows northward off the South American coast, then, about at the equator, it turns westward across the Pacific as the South Equatorial current. The Humboldt current is characterized by upwelling of cold, deep water, which carries nutrient ions that serve as food for plankton. Fish feed on these high concentrations of plankton. The anchoveta, a small fish used commercially to produce fish meal for animal feed, thrives in great numbers and is harvested by the Peruvian fishermen. With the onset of El Niño, upwelling ceases, the cool water is replaced by warm water from the west, and the plankton and their anchoveta predators disappear. Vast numbers of birds that feed on the anchoveta die of starvation. Only since the event of 1982–83 has the full story of El Niño been uncovered, although certain synoptic meteorological phenomena that accompany El Niño have long been linked with it.

To understand what sets El Niño in action, review the January average conditions of pressure and winds in the equatorial and subtropical zones of the Pacific, using the map in Figure 7.33, which is basically the same as in Figure 5.10. Note the position of the intertropical convergence zone (ITC) along the axis of the equatorial trough of low pressure. Note the pattern of easterly winds, or trades. Refer to Figure 5.28 to see how the South Equatorial current is related to the pressure and wind system. Note especially that low pressure along the

ITC decreases steadily from east to west and that starting at about 160°W, the low-pressure trough broadens greatly in latitude and deepens over a large area that includes Indonesia, New Guinea, and northern Australia. This area in January (high-sun season in the southern hemisphere) normally experiences heavy rainfall. For northern Australia this is the rainy or monsoon season of the tropical wet-dry climate (see Chapter 10). As for the Peruvian coast, a desert climate prevails throughout the year and January is usually almost rainless.

Gilbert Walker, a British mathematician of broad interests, who served in a government scientific post in India starting in 1904, is credited with observing an occasional strange reversal in the barometric pressure system we have just described. In some years, the Indonesian/Australian low pressure region was replaced by high pressure while low pressure deepened over the eastern Pacific portion of the equatorial trough. Walker called this phenomenon the *Southern Oscillation* (SO). He observed that the onset of high pressure brought drought to the western regions while the eastern Pacific region experienced abnormally high rainfall.

Since Walker's time, his observations have been confirmed as precursory to the onset of El Niño. Two observing stations have been singled out as end points on this meteorological seesaw: Darwin, on the northern coast of Australia at about long. 130°E; and the island of Tahiti, at about long. 150°W. Figure 7.34 is a special map showing the level of correlation in barometric pressure readings over the oceanic region. At any given point over the entire map area, pressure during a SO event is compared with the pressure at Darwin to obtain a pressure anomaly. Lines with large numbers (6 to 8, positive or negative) indicate a high level of correlation between Darwin and Tahiti, the former having high pressure and the latter having correspondingly low pressure. Today the SO is combined with El Niño, "EN," under the acronym ENSO.

The onset of the SO pressure anomaly in May 1982 was soon followed by two important changes in the surface environment. First, the trades ceased to blow; second, sea-surface temperatures rose sharply just south of the equator in the eastern Pacific. The temperature rise started along the South American coast in October 1982,

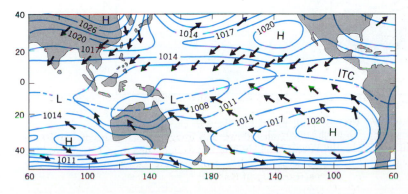

FIGURE 7.33 January average surface barometric pressure and winds of the Pacific equatorial zone.

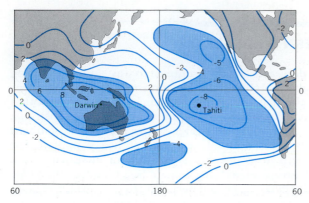

FIGURE 7.34 Correlation of annual mean sea-level pressures with the pressure at Darwin, Australia. Isopleths give correlation coefficients. (From Eugene M. Rasmusson, 1985, *American Scientist*, vol. 73, p. 169, Figure 2. Used by permission.)

when upwelling ceased. The warm zone then extended progressively farther west between the equator and lat. 5°S, reaching almost to long. 180° by December 1982 and persisting through February 1983. The sea surface temperature had risen to more than 2C° (3.6F°) above the normal value in what is referred to as the "mature phase" of the ENSO, lasting from December 1982 through February 1983.

Cessation of the trades, beginning in the previous June (1982) was followed by a curious event, the onset of low-level equatorial westerly winds about in mid-Pacific in the longitude range from 160°E to 150°W. At the same time, the precipitation zone of the southwestern Pacific shifted far eastward into the area of equatorial westerlies, reaching its maximum intensity in January 1983 between long. 140° and 160°W. Blowing over the sea surface, the westerlies set up a succession of eastward-moving Kelvin waves, which drove warm water in surges to the Pacific shores of South America in what can be thought of as a kind of sloshing effect, raising sea level along those coasts. A consequence of the warm-water zone off the South American coast was the occurrence of torrential rainfalls, causing devastating floods in the Andean highlands of Peru, Bolivia, and Colombia. The warm surface layer spread northward along

the Central American and North American coasts, reaching latitudes as far north as Oregon. One effect of the rather sudden rise in water temperature and water level was the rapid heat death in December 1982 of reef corals along more than 3000 km of reef-bound coasts. This event is now regarded by some ecologists as a nonreversible ecological disaster leading to ultimate reef destruction.

Associated with the mature phase of the ENSO was an important change in the pattern of the northern hemisphere polar jet, as shown in Figure 7.35. Notice that the ENSO jet swept across Mexico and the Caribbean and continued strong across the Atlantic to enter North Africa and continue in strength across the Arabian Peninsula and northern India. One effect of the far-southward position of the polar jet over North America was to bring frequent and intense invasions of cyclonic storms. These dumped heavy rain on coastal areas and heavy snow over the western mountain ranges, as well as heavy rains over the Gulf Coast and Florida. Precipitation in January and February 1983 exceeded 200 percent of normal in these locations.

Another probable effect of ENSO was the unusual occurrence, between December 1982 and April 1983, of several tropical cyclones in the south-central Pacific, where few of these storms normally occur.

Early in 1983 the dry high-pressure area of the western Pacific moved slowly northward to cover the Philippines and spread westward to include southern India and Sri Lanka. Meantime, a severe drought was occurring in southeast Africa.

Other kinds of disturbances of the atmosphere—collectively called weather anomalies—occurring in higher latitudes during ENSO are beyond the scope of our account. The inferred chains of worldwide effects belong to a class of meteorological events called "teleconnections." The interactions involved in teleconnections are somewhat obscure and are the object of continued research. ENSO may have been different in important ways from previous El Niño events.

In the years following the great 1982–83 El Niño, research scientists looked for signs of a recurrence of the phenomenon, as well as into records of the past for more information on the intervals of recurrence. A weakly de-

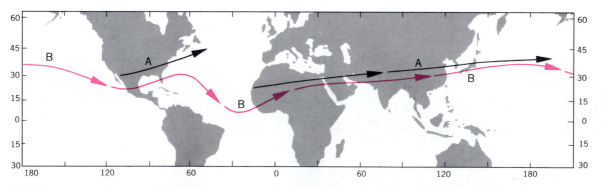

FIGURE 7.35 (A) The typical prevailing jet stream pattern for December through February. (B) Pattern during ENSO December 1982 through February 1983. (Data of Eugene M. Rasmusson, 1984, *Oceanus*, vol. 27, no. 2, p. 12, Figure 9. Copyright © 1984 by the Woods Hole Oceanographic Institution. Used by permission.)

veloped El Niño occurred in 1987 and served as a test of computer models devised for prediction.

Perhaps the most significant development has been the recognition of a distinctive set of contrasting, or complementary, conditions that prevails between El Niño events. Known as *La Niña* (the girl), it is a period during which sea-surface temperatures in the central and eastern Pacific Ocean fall to lower levels; i.e., to below 25°C (77°F). The South Pacific subtropical high becomes very strongly developed during the high-sun season of that hemisphere and causes abnormally strong southeast trade winds. The force of these winds drags warm surface water westward, exposing cooler water at the surface. La Niña conditions were recognized during 1988 and were suspected of having been causally related to a severe drought experienced over North America in the summer of 1988. In some manner not yet well understood, the conditions of a La Niña give rise to the onset of the following El Niño, which in turn gives rise to the next La Niña, perpetuating an unending succession of these events.

Scientists have also recognized in the Atlantic Ocean a phenomenon similar in some respects to the Pacific El Niño. In 1984, warm water spread southward along the western coast of southern Africa, replacing the cool Benguela Current and bringing rainfall to what is normally a desert coastal zone similar to that along the western coastal zone of Peru.

Significant changes from year to year in sea surface temperatures over large low-latitude areas of the Pacific ocean and in other oceans will, of course, be reflected in the mean global surface temperature of the given year (see Figure 4.24). Accordingly, those annual values require correction, if a significant global warming trend is to be inferred.

Data sources: George Philander (1989) EL Niño and La Niña, *American Scientist*, vol. 77, p. 451–59; Ants Leetmaa (1989) The Interplay of El Niño and La Niña, *Oceanus*, vol. 32, no. 2, p. 30–34.

ATMOSPHERIC SCIENCE IN REVIEW

At the close of this group of chapters on atmospheric science we can look back to analyze the broad pattern of concepts covered. As any coin with two sides, environmental science has two opposite but complementary phases. One is the impact or influence of the environment upon humans; the other is the human influence or impact upon the environment.

The influences of the atmospheric environment upon humans are to a large extent favorable, even downright essential to survival. The atmosphere furnishes a favorable environment in which all life forms derive energy and materials needed to sustain growth. But, at various times and places, extremes of atmospheric activity are harmful, impacting humans and other life forms unfavorably. Severe weather phenomena are in this category. Perhaps we have placed more emphasis upon the harmful impacts of atmospheric phenomena than upon the day-to-day beneficial phenomena that sustain life; if so, we need to reorganize our perspective.

Turning over the environmental coin, we have given major emphasis to our impact upon the atmospheric environment. That impact largely concerns climate change. Two forms of climate change have been analyzed: inadvertent changes in global climate and deliberately induced changes. Certain inadvertent (unintentional, or unplanned) changes have global scope; these changes are the most difficult to recognize and predict. Are we inducing a global warming trend? Are observed changes natural, or are they caused by industrial activity? Inadvertent changes in urban climates, on the other hand, are clearly evident and their causes well understood. The deliberate forms of change humans try to impose upon the atmosphere deal entirely with specific weather phenomena; for example, they are attempts to produce more rain, to produce less hail, to disperse fog, or to lessen storm wind intensities.

Our next four chapters continue to deal with the atmosphere, but focus on global patterns of monthly and yearly averages of air temperature and precipitation, i.e., on climate and climatology. Fresh water of the lands will assume a more dominant role, however, and we shall be taking into account not only precipitation, but also the role of water as runoff, soil water, and ground water. Evaporation of this surface water will play a major role in climate, and we can establish an open system of movement of water in vapor and liquid states as a guiding model.

C H A P T E R   8

GLOBAL CLIMATE SYSTEMS

A goal of physical geography is to recognize and describe environmental regions of significance to humans and to all life forms generally. Which are the vital components of the environment of the life layer with respect to humans in particular and to the biosphere in general? Keep in mind that the basic food supply of the biosphere is organic matter synthesized by plants. Plants are the primary producers of organic matter, which sustains them as well as other forms of life.

In view of our dependence on the primary producers, it seems logical to emphasize those components of the environment that are vital to plant growth. Plants occupy both marine and terrestrial environments. Because plant life of the lands is most directly exposed to the atmosphere for exchanges of energy and matter, and because most of our food is derived from terrestrial plants, we will want to concentrate on the land–atmosphere interface.

Two vital, atmospherically derived components vary greatly from place to place and from season to season—energy and water. Both must be in forms available to plants. Plants use the energy of solar radiation to carry out photosynthesis. (Photosynthesis is the process by which carbohydrate molecules are synthesized.) Plants also require carbon dioxide and water to synthesize carbohydrate compounds. Carbon dioxide, however, is largely uniform in its atmospheric concentration over the globe at all seasons, so we can omit it as a variable factor in the environment.

Plants require the presence of sensible heat as measured by air and soil temperatures within specified limits. Plants also require water in the root zone of the soil. Although land animals are consumers they, too, require fresh water and a tolerable range of surrounding temperatures. Plants also release energy and water to the atmosphere through the process of respiration, in which the carbohydrate molecules are broken down into carbon dioxide and water. (Photosynthesis and respiration are discussed in Chapter 25.)

Besides energy, carbon dioxide, and water, plants need nutrients, which they derive from the soil. The nutrients are released by the plants when the plant tissue dies. In this way the nutrients are returned, or recycled, to the soil. (Nutrient cycling is discussed in Chapter 25.)

To summarize, land plants require light energy, carbon dioxide, water, and nutrients. The input of these ingredients comes from two basic sources: (1) the adjacent atmosphere and (2) the soil. Input of light energy, heat energy, and water from the atmosphere is encompassed by the concept of climate. In short, plants depend on climate and soil. Of these two sources of energy and matter, the soil is the more strongly affected by the plants it serves through the recycling of matter. Climate, when broadly defined as a source of energy and water, is an independent agent of control. Climate is determined by latitude and by large-scale air motions and air mass interactions within the troposphere. If we are to establish a pyramid of priorities and interactions, climate occupies the apex as the independent control (Figure 8.1). Below it and forming the base of the pyramid are (1) the organic process of plants and (2) the soil process. Plants and soil interact with one another at the basal level.

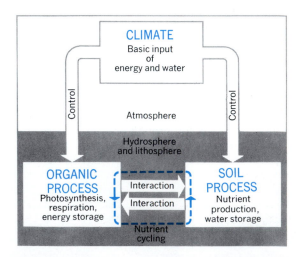

FIGURE 8.1 The role of climate in environmental processes of the life layer.

CLIMATE AND CLIMATE CLASSIFICATION

Climate has always been a keystone in physical geography and has formed the basis for defining physical regions of the globe. Let us first examine the content of traditional *climatology*, the science of climate. In the broadest sense, *climate* is the characteristic condition of the atmosphere near the earth's surface at a given place or over a given region. Components that enter into the description of climate are mostly the same as weather components used to describe the state of the atmosphere at a given instant. If weather information deals with the specific event, climate represents a generalization of weather. A statement of the climate of a given observing station, or of a designated region, is described through the medium of weather observations accumulated over many years' time. Not only are mean, or average, values examined, but the departures from those means and the probabilities that such departures will occur are also taken into account.

Earlier chapters contained much information falling within the definition of climate. For example, world maps showing average January and July barometric pressures, winds, and air temperatures are expressions of climate. The physical components of climate are many; they include measurable quantities such as radiation, sensible heat, barometric pressure, winds, relative and specific humidity, dew point, cloud cover and type, fog, precipitation type and intensity, evaporation and transpiration, incidence of cyclones and anticyclones, and frequency of frontal passages. Which of these components are really significant in our analysis of climate? Approaching climatology as geographers, we shall take only what we need.

Our problem in devising a meaningful system of climate classes (i.e., a classification of climates) is to select categories of available information that correlate closely with the needs of life on the lands. We then use those categories to define and delineate regional classes of climates. If we have done our work well, the regional units within the climate system will reflect strongly the control of the atmosphere on terrestrial life. This same system will give some indication of the opportunities and constraints that the atmospheric environment imposes on humans as they seek to increase their food supplies and water resources at the same time that they extend the areas of urban and industrial land use. The climate classification we seek must have utility in guiding land-use planning and population growth as well as describing the natural environment. This role is particularly crucial in guiding the progress of the developing nations as they seek to augment inadequate food resources.

AIR TEMPERATURE AS A BASIS FOR CLIMATE CLASSIFICATION

All species of organisms, whether plants or animals, are subject to limiting temperatures of the surrounding air, water, or soil; survival is not possible above or below these limits. Few growing plants can survive temperatures over 50°C (120°F) for more than a few minutes. Many species of plants native to the tropical and equatorial latitude zones die if they experience even a short period of below-freezing temperature (0°C; 32°F). Freezing of water in the plant tissues causes physical disruption in plants not adapted to such conditions. Alternate freezing and thawing of the soil can have a disruptive effect on plant roots and is an important factor in limiting plant growth in the arctic and alpine zones.

Apart from extreme limits of temperature tolerance, plants react to an increase in air and water temperature through an increase in the intensity of physical and chemical activity. The optimum rates at which both photosynthesis and respiration take place increase with rising temperature in the range from near-freezing to 20° to 25°C (70° to 80°F), after which the rate of photosynthesis begins to decline. Respiration, however, continues to increase in rate to considerably higher temperatures. Knowing that plants are strongly influenced by the factor of sensible air and soil temperatures that surround them, we include air temperature as one of the essential climate factors.

Besides being an important environmental factor in plant physiology and reproduction, air temperature enters into many activities of animal life (e.g., hibernation and migration). For humans, air temperature is an important physiological factor and relates directly to the quantity of energy expended in space heating and air conditioning of buildings. Nevertheless, air temperature alone does not define meaningful climate classes because the ingredient of water availability is missing.

Temperature of the lower air layer, as measured in the standard thermometer shelter (Chapter 4), has long provided one essential variable quantity in leading systems of climate classification. Monthly data based on daily readings of the maximum–minimum thermometer have been accumulated for many decades at thousands of observing stations the world over. Consequently, availability has been an important factor in favoring the use of air temperature in climate classification.

THERMAL REGIMES

In Chapter 4 we studied the annual air temperature cycle and its relationship to the annual energy balance cycle. Figure 4.8 showed annual temperature cycles for four stations, ranging from the equatorial zone to the subarctic zone. Let us now carry this investigation of annual temperature cycles a step further to explore the global range of annual cycles.

A comparison of annual air temperature cycles for many observing stations over the globe allows us to recognize a number of distinctive types, which can be called *thermal regimes*. A few of these are shown in Figure 8.2. Each regime has been labeled according to the latitude zone in which it lies; for example, equatorial, tropical, midlatitude, and subarctic. Some labels also include words descriptive of the location in terms of position with respect to landmass. "Continental" refers to a continental interior location; "marine west coast" refers to a location close to the ocean and on the western side of the continent.

The equatorial regime is uniformly very warm; temperatures are close to 27°C (80°F) year-round, and there are no temperature seasons. The tropical continental regime ranges from very hot when the sun is high, near one solstice, to mild at the opposite solstice. Close to the ocean, however, we find the tropical west coast regime with only a weak annual cycle and no extreme heat. The same weak annual cycle can be traced into the midlatitude west coast regime, and it persists even much farther poleward. In the continental interiors, the strong annual cycle prevails through the midlatitude continental regime into the subarctic continental regime, where the annual range is enormous. The ice sheet regime of Greenland is in a class by itself, with severe cold all year. Other regimes can be identified and, because they grade into one another, the list could be expanded indefinitely.

Two important concepts are illustrated in Figure 8.2. One concept is that of *continentality*, the tendency of a large landmass to impose a large annual range on the annual cycle. Continentality becomes stronger with higher latitude because the insolation cycle shows stronger seasonal extremes with increasing latitude. An-

other concept is that of the marine influence, which tends to weaken the annual cycle and holds temperatures in a moderate range. This effect is, of course, due to the capacity of the oceans to hold a large amount of heat in storage and to take up and give off that heat very slowly, as compared with land areas.

PRECIPITATION AS A BASIS FOR CLIMATE CLASSIFICATION

Precipitation data, obtained by the simple rain gauge (Chapter 6), are abundantly available for long periods of record for thousands of observing stations widely distributed over the globe. Small wonder, then, that monthly and annual precipitation data form the cornerstone of most of the widely used climate classifications.

Average annual precipitation is shown on a world map, Figure 8.3. This map uses isohyets, lines drawn through all points having the same annual quantity. For regions where all or most of the precipitation is rain, we use the word "rainfall"; for regions where snow is an important part of the annual total, we use the word "precipitation."

Global patterns of precipitation are related to air-mass source regions and prevailing movements of air

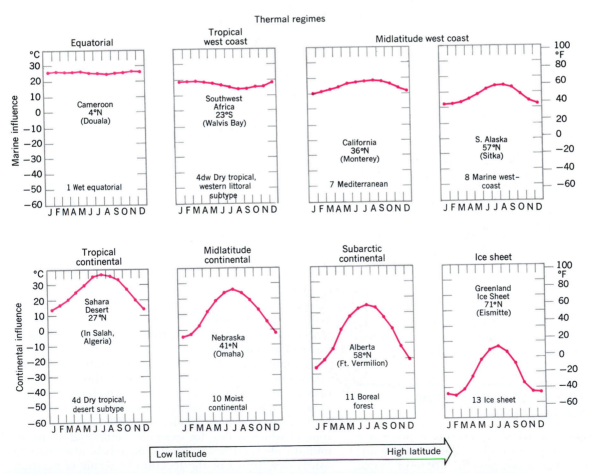

FIGURE 8.2 (above and right) Some important thermal regimes, represented by annual cycles of air temperature. (Based on Goode Base Map.)

TABLE 8.1 **World Precipitation Regions**

Name	Latitude Range	Continental Location	Prevailing Air Mass	Annual Precipitation Centimeters	Annual Precipitation Inches
1. Wet equatorial belt	10°N to 10°S	Interiors, coasts	mE	Over 200	Over 80
2. Trade-wind coasts (windward tropical coasts)	5–30°N and S	Narrow coastal zones	mT	Over 150	Over 60
3. Tropical deserts	10–35°N and S	Interiors, west coasts	cT	Under 25	Under 10
4. Midlatitude deserts and steppes	30–50°N and S	Interiors	cT, cP	10–50	4–20
5. Moist subtropical regions	25–45°N and S	Interiors, coasts	mT (summer)	100–150	40–60
6. Midlatitude west coasts	35–65°N and S	West coasts	mP	Over 100	Over 40
7. Arctic and polar deserts	60–90°N and S	Interiors, coasts	cP, cA	Under 30	Under 12

masses. Seven precipitation regions can be recognized in terms of annual total in combination with location. Details are given in Table 8.1. Figure 8.4 is a schematic diagram of the global distribution of precipitation, simplifying the picture as much as possible in terms of an imaginary continent. Notice that the adjectives "wet," "humid," "subhumid," "semiarid," and "arid" have

been applied to each of five levels of annual precipitation, which range from 200 cm (80 in.) to zero.

1. The wet equatorial belt of heavy rainfall, over 200 cm (80 in.) annually, straddles the equator and includes the Amazon River basin in South America, the Congo River basin of equatorial Africa, much of the African

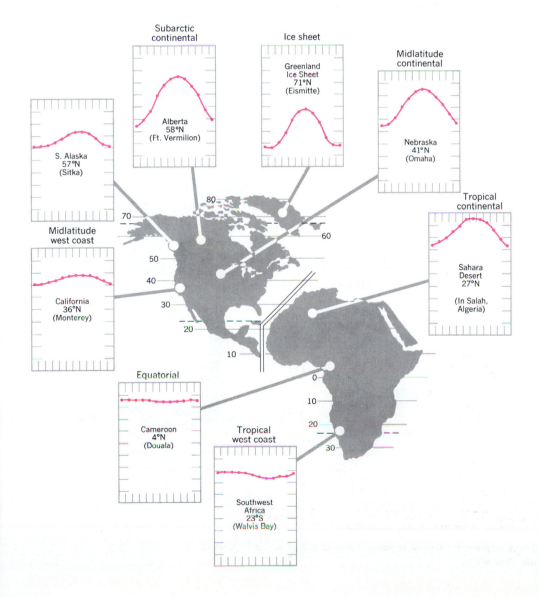

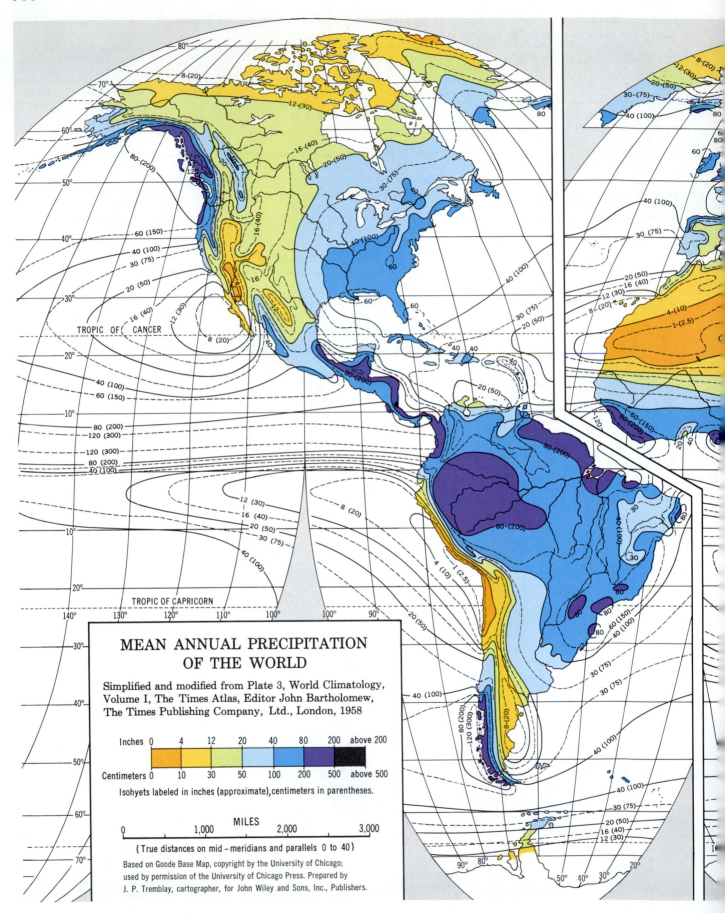

MEAN ANNUAL PRECIPITATION OF THE WORLD

Simplified and modified from Plate 3, World Climatology, Volume I, The Times Atlas, Editor John Bartholomew, The Times Publishing Company, Ltd., London, 1958

Inches 0   4   12   20   40   80   200   above 200

Centimeters 0   10   30   50   100   200   500   above 500

Isohyets labeled in inches (approximate), centimeters in parentheses.

MILES
0   1,000   2,000   3,000

(True distances on mid – meridians and parallels 0 to 40)

Based on Goode Base Map, copyright by the University of Chicago; used by permission of the University of Chicago Press. Prepared by J. P. Tremblay, cartographer, for John Wiley and Sons, Inc., Publishers.

FIGURE 8.3 World precipitation.

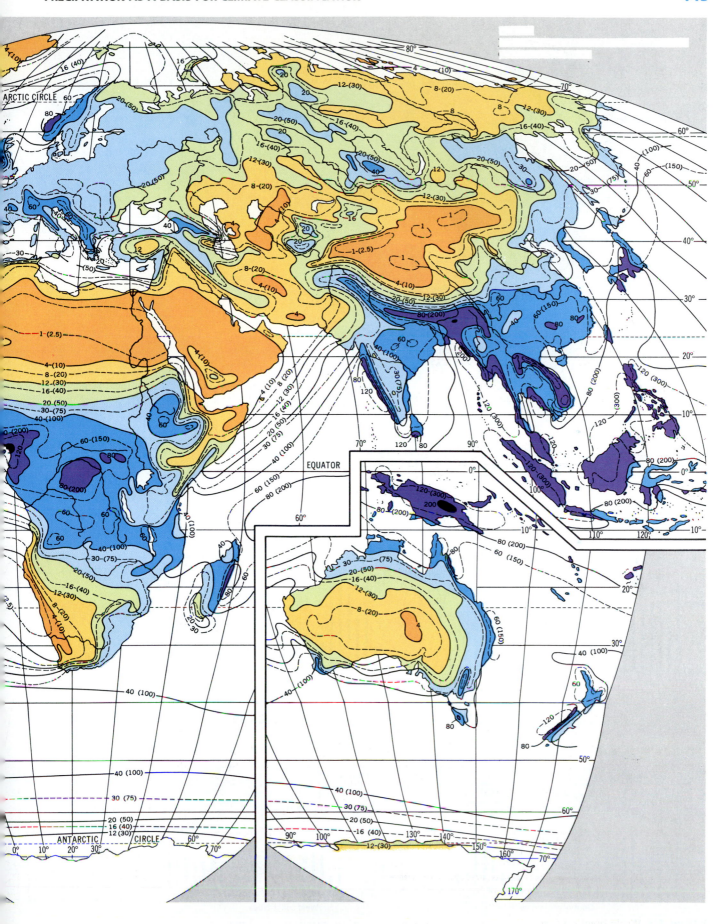

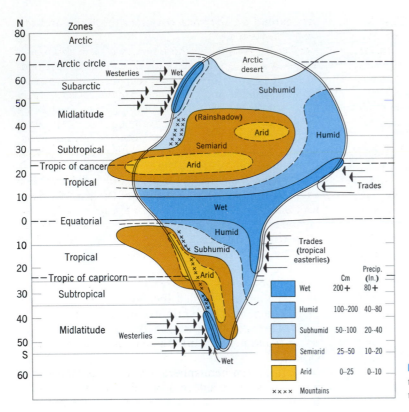

FIGURE 8.4 A schematic diagram of the distribution of annual precipitation over an idealized continent and adjoining oceans.

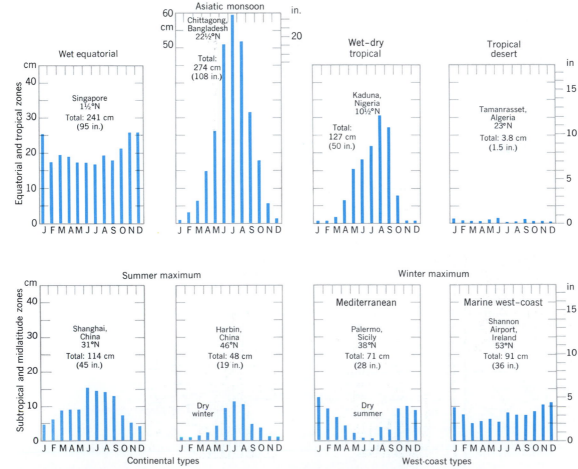

FIGURE 8.5 (above and right) Eight precipitation types selected to show various seasonal patterns. (Based on Goode Base Map.)

coast from Nigeria west to Guinea, and the East Indies. Here the prevailingly warm temperatures and high moisture content of the maritime equatorial (mE) air mass favor abundant convectional rainfall. Thunderstorms are frequent year-round.

2. Narrow coastal belts of high rainfall, 150 to 200 cm (60 to 80 in.), and locally even more, extend from near the equator to latitudes of about 25° to 30°N and S on the eastern sides of every continent or large island. For example, see the eastern coasts of Brazil, Central America, Madagascar, and northeastern Australia. These are the trade-wind coasts, where the moist maritime tropical (mT) air mass from warm oceans is brought over the land by the trades (tropical easterlies). Encountering coastal hills and mountains, this air mass produces heavy orographic rainfall.

3. In striking contrast to the wet equatorial belt astride the equator are the two zones of vast tropical deserts lying approximately on the tropics of cancer and capricorn. These are hot, barren deserts, with less than 25 cm (10 in.) of rainfall annually and in many places with less than 5 cm (2 in.). They are located under and are caused by the subtropical cells of high pressure, where the subsiding continental tropical (cT) air mass is adiabatically warmed and dried. Such little rain as these areas experience is largely convectional and extremely unreliable. Note that the tropical des-

erts extend off the west coasts of the lands and out over the oceans.

4. Farther northward, in the interiors of Asia and North America between lat. 30° and 50°, are great continental midlatitude deserts and expanses of semiarid grasslands known as steppes. Annual precipitation ranges from less than 10 cm (4 in.) in the driest areas to 50 cm (20 in.) in the moister steppes. Dryness here results from remoteness from ocean sources of moisture. Located in a region of prevailing westerly winds, these arid lands occupy the position of rainshadows in the lee of coastal mountains and highlands. For example, the Cordilleran Ranges of Oregon, Washington, British Columbia, and Alaska shield the interior of North America from the moist maritime polar (mP) air mass originating in the Pacific. While descending into the intermontane basins and interior plains, the mP air mass is warmed and dried.

Similarly, mountains of Europe and the Scandinavian peninsula obstruct the flow of moist mP air masses from the North Atlantic into western Asia. The great southern Asiatic ranges likewise prevent the entry of moist mT and mE air masses from the Indian Ocean.

The southern hemisphere has too little land in the midlatitudes to produce a broad continental desert, but the dry north–south strip of Patagonia, lying on

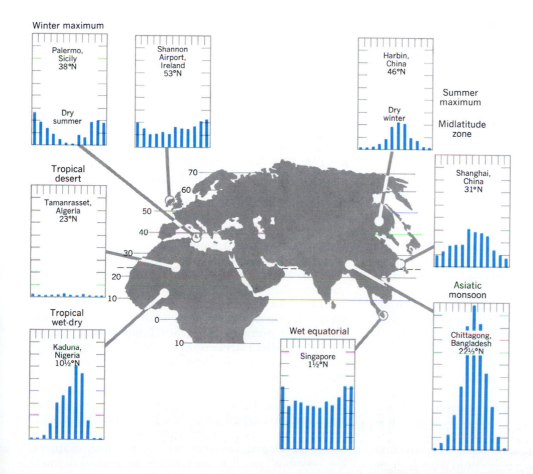

the lee side of the Andean chain, is roughly the counterpart of the North American deserts and steppes of Oregon and northern Nevada.

5. On the southeastern sides of the continents of North America and Asia, in lat. 25° to 45°N, are the moist subtropical regions, with 100 to 150 cm (40 to 60 in.) of precipitation annually. Smaller areas of the same kind are found in the southern hemisphere in Uruguay, Argentina, and southeastern Australia. These regions lie on the moist western sides of the subtropical high-pressure centers in such a position that the humid mT air mass from the tropical ocean is carried poleward over the adjoining land. Commonly, too, these areas receive very heavy rains from tropical cyclones.

6. Another distinctive wet location is on midlatitude west coasts of all continents and large islands lying between about 35° and 65° in the region of prevailing westerly winds. These zones were mentioned in Chapter 7 as good examples of coasts on which abundant orographic precipitation falls as a result of forced lift of mP air masses. Where the coasts are mountainous, as in Alaska and British Columbia, southern Chile, Scotland, Norway, and South Island of New Zealand, the annual precipitation is over 200 cm (80 in.). These rugged coasts formerly supported great valley glaciers that carved deep bays (fiords).

7. A seventh precipitation region is formed by the arctic and polar deserts. Northward of the 60th parallel, annual precipitation is largely under 30 cm (12 in.), except for the west-coast belts. Cold continental polar (cP) and continental arctic (cA) air masses cannot hold much moisture; consequently, they do not yield large amounts of precipitation. At the same time, however, evaporation rates are low.

As you would expect, zones of gradation lie between these precipitation regions. Moreover, this list does not recognize the fact that seasonal patterns of precipitation differ from region to region.

SEASONAL PATTERNS OF PRECIPITATION

Although the annual total precipitation is a useful quantity in establishing the characteristics of a climate type, it can be a misleading statistic because there may be strong seasonal cycles of precipitation. It makes a great deal of difference in terms of native plants and agricultural crops if there are alternately dry and wet seasons instead of a uniform distribution of precipitation throughout the year. It also makes a great deal of difference whether the wet season coincides with a season of higher temperatures or with a season of lower temperatures because plants need both water and heat.

The seasonal aspects of precipitation can be largely covered by three major patterns: (1) uniformly distributed precipitation; (2) a precipitation maximum during the summer (or season of high sun), when insolation is at its peak; and (3) a precipitation maximum during the winter or cooler season, when insolation is least. The first pattern can include a range of possibilities from little or no precipitation in any month to abundant precipitation in all months.

A study of monthly means of precipitation throughout the year at many observing stations over the globe shows a number of outstanding precipitation types. Some of these are illustrated in Figure 8.5. For each, the average monthly precipitation is shown by the height of the bar in the small graphs.

Dominant in very low latitudes is an equatorial rainy type, illustrated by Singapore, near the equator. Rain is abundant in all months, but some months have considerably more than others. The tropical desert type may have so little rain in any month that it cannot be shown on the graph. The wet-dry tropical type has a very wet season at the time of high sun (summer solstice) and a very dry season at the time of low sun (winter solstice). This seasonal alternation is carried to its greatest extreme in the Asiatic monsoon type, with an extremely wet high-sun monsoon season and a very dry period at low sun.

The summer precipitation maximum is carried into higher latitudes on the eastern sides of continents in a subtropical moist type. This same feature persists into the midlatitude continental type, which has a long, dry winter but a marked summer rain period.

A cycle with a winter precipitation maximum is typical of the west sides of continents in midlatitudes. A Mediterranean type, named for its prevalence in the Mediterranean region, has a very dry summer but a moist winter. This same cycle is carried into higher midlatitudes along narrow strips of west coasts. There is precipitation throughout the year in the marine west-coast type, but with a distinct maximum in winter and a distinct minimum in summer.

Each of these types will appear in more detail in later descriptions of individual climates. The reasons for annual precipitation cycles will be developed further in terms of shifting belts of pressure and winds and the seasonal changes in domination by different air masses.

A CLIMATE SYSTEM BASED ON AIR MASSES AND FRONTAL ZONES

Recall from Chapter 7 that air masses are classified according to the general latitude of their source regions and the surface qualities of the source region, whether land or ocean. This air-mass classification has built into it a recognition of the factors of temperature and precipitation because (1) air-mass temperature decreases poleward and (2) precipitation capability of an air mass is related to sources of moisture. Source regions of air masses change seasonally along with shifting belts of air temperature, pressure, and winds. Frontal zones also migrate seasonally. Thus seasonal cycles of both air temperature and precipitation can be described in terms of changing air-mass activity.

The climate system we shall develop uses the concepts of thermal regimes and precipitation types, but explains them in terms of air-mass activity. This explanatory system makes use of all the global climatic information developed systematically in previous chapters. The system allows that information to be used to inter-

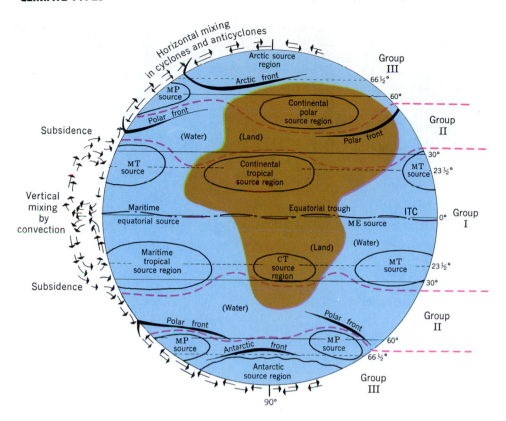

FIGURE 8.6 The three major climate groups are related to air-mass source regions and frontal zones.

pret a given climate and to provide the reasons for its occurrence in a particular location.

Our climate system is based on location of air-mass source regions and the nature and movement of air masses, fronts, and cyclonic storms. A schematic global diagram (Figure 8.6) shows the principles of the classification. Notice that this figure is basically the same as Figure 7.1.

Group I: Low-Latitude Climates

Group I includes the tropical air-mass source regions and the intertropical convergence zone (ITC) that lies between. Climates of Group I are controlled by the subtropical high-pressure cells, or anticyclones, which are regions of air subsidence and are basically dry, and by the great equatorial trough of convergence that lies between them. Although air of polar origin occasionally invades the tropical and equatorial zones, the climates of Group I are almost wholly dominated by the tropical and equatorial air masses. Tropical cyclones are important in this climate group.

Group II: Midlatitude Climates

Climates of Group II are in a zone of intense interaction between unlike air masses: the *polar front zone*. Tropical air masses moving poleward and polar air masses moving equatorward are in conflict in this zone, which contains a procession of eastward-moving wave cyclones. Locally and seasonally, either tropical or polar air masses may dominate in these regions, but neither has exclusive control.

Group III: High-Latitude Climates

Climates of Group III are dominated by polar and arctic (including antarctic) air masses. The two polar continental air-mass source regions of northern Canada and Siberia fall into this group, but there is no southern hemisphere counterpart to these continental centers. In the arctic belt of the 60th to 70th parallels, air masses of arctic origin meet polar continental air masses along an *arctic front zone*, creating a series of eastward-moving wave cyclones.

CLIMATE TYPES

Within each climate group are a number of *climate types* (or simply, *climates*): four in Group I, six in Group II, and three in Group III, for a total of 13 climate types. For ease in identification, the climate types are numbered; but the numbers have no significance as a code. Ordering of climate within each group is in many cases quite arbitrary, particularly for the climates in Group III. You need use only the climate names because these are descriptive of climate quality and global location.

In presenting the 13 climate types, we shall make use of a graphic device called a *climograph* (Figure 8.7). It shows simultaneously the cycles of monthly mean air temperature and monthly mean precipitation, combining the temperature and precipitation graphs used in Figures 8.2 and 8.5. The climograph allows you to integrate at a glance the important basic features of the climate.

Another useful graphic device is the generalized diagram of the climate groups, climate types, and climate

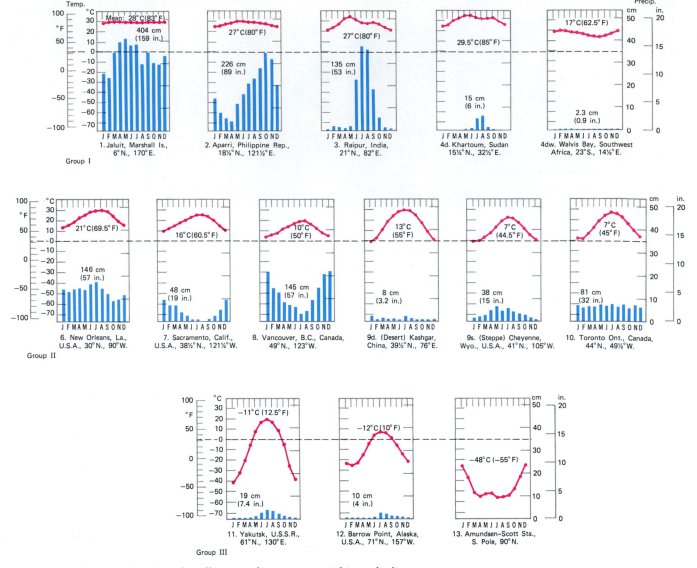

FIGURE 8.7 Climographs selected to illustrate climate types within each climate group.

subtypes shown in Figure 8.8. Boundaries are drawn in smooth, simple lines on an imaginary supercontinent, as we did for precipitation (Figure 8.4). Coastal highlands are shown, because they have strong local effects in producing orographic precipitation belts and rainshadows. The diagram shows typical positioning of each climate type and the common boundaries between climates. Boundaries between groups are labeled.

A world map of climates, Figure 8.9, shows the actual distribution of climate types and subtypes on the continents. This map is based on available data of a large number of observing stations. It is, however, a simplified map; the climate boundaries are uncertain in many areas where observing stations are thinly distributed. Figure 8.10 shows the same climate types and boundaries for the contiguous 48 United States and southernmost Canada. Precise definitions of the climates and their boundaries are presented in Chapter 9. In the remainder of this chapter, we will analyze the 13 climates in purely descriptive terms, using actual figures of temperature

and precipitation very broadly and only for purposes of example.

DRY AND MOIST CLIMATES

All but two of the 13 climate types are classified as either dry climates or moist climates. *Dry climates* are those in which evaporative water losses from the soil and from plant cover exceed the incoming precipitation by a wide margin. Generally speaking, the dry climates do not support permanent streams and the land surface is clothed with sparse plant cover—either grasses or shrubs—or lacks a plant cover. *Moist climates* are those with suffi-

FIGURE 8.8 (right) A schematic diagram of the placement of climates on an idealized continent. Compare with the world map, Figure 8.9.

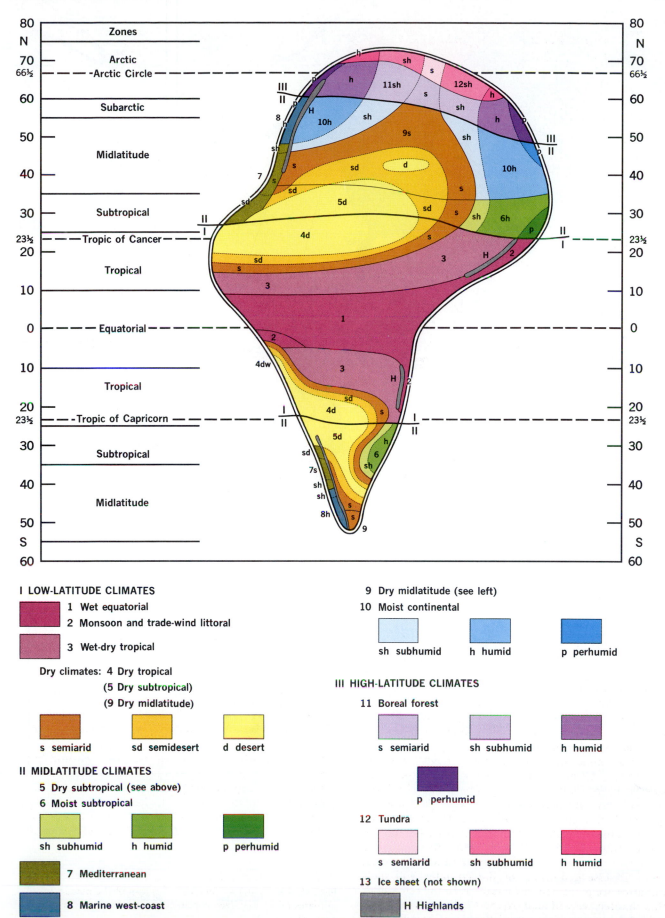

Zones

80 N	
70	Arctic
66½	Arctic Circle
60	Subarctic
50	
40	Midlatitude
30	Subtropical
23½	Tropic of Cancer
20	
10	Tropical
0	Equatorial
10	Tropical
20	
23½	Tropic of Capricorn
30	Subtropical
40	Midlatitude
50 S	
60	

I LOW-LATITUDE CLIMATES

1 Wet equatorial
2 Monsoon and trade-wind littoral

3 Wet-dry tropical

Dry climates: 4 Dry tropical
(5 Dry subtropical)
(9 Dry midlatitude)

s semiarid   sd semidesert   d desert

II MIDLATITUDE CLIMATES

5 Dry subtropical (see above)
6 Moist subtropical

sh subhumid   h humid   p perhumid

7 Mediterranean

8 Marine west-coast

9 Dry midlatitude (see left)
10 Moist continental

sh subhumid   h humid   p perhumid

III HIGH-LATITUDE CLIMATES

11 Boreal forest

s semiarid   sh subhumid   h humid

p perhumid

12 Tundra

s semiarid   sh subhumid   h humid

13 Ice sheet (not shown)

H Highlands

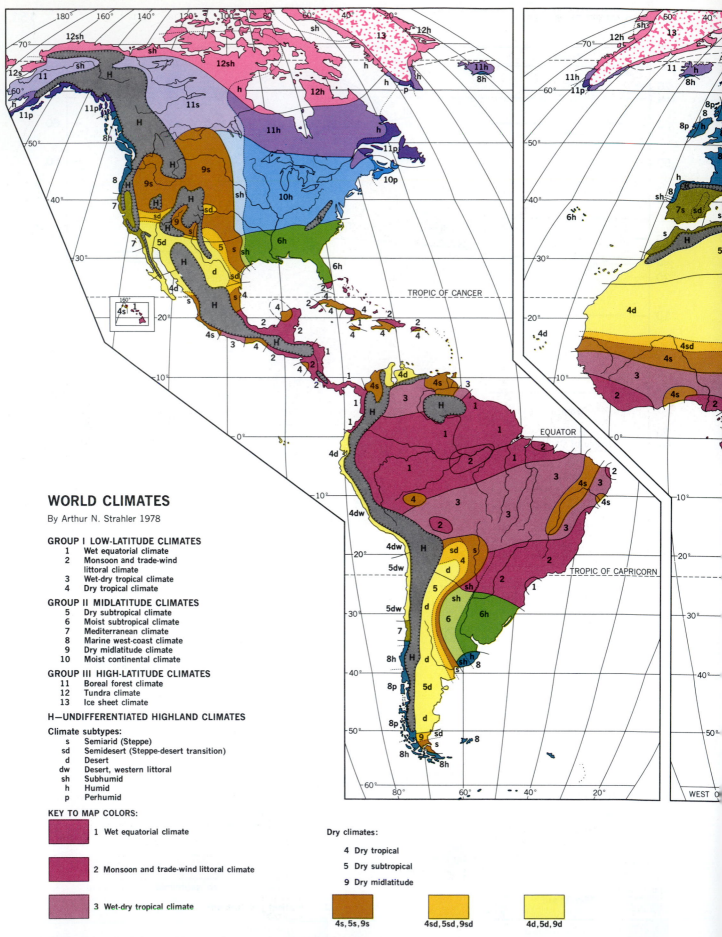

WORLD CLIMATES

By Arthur N. Strahler 1978

GROUP I LOW-LATITUDE CLIMATES
1 Wet equatorial climate
2 Monsoon and trade-wind
   littoral climate
3 Wet-dry tropical climate
4 Dry tropical climate

GROUP II MIDLATITUDE CLIMATES
5 Dry subtropical climate
6 Moist subtropical climate
7 Mediterranean climate
8 Marine west-coast climate
9 Dry midlatitude climate
10 Moist continental climate

GROUP III HIGH-LATITUDE CLIMATES
11 Boreal forest climate
12 Tundra climate
13 Ice sheet climate

H—UNDIFFERENTIATED HIGHLAND CLIMATES

Climate subtypes:
s   Semiarid (Steppe)
sd Semidesert (Steppe-desert transition)
d   Desert
dw Desert, western littoral
sh Subhumid
h   Humid
p   Perhumid

KEY TO MAP COLORS:

1 Wet equatorial climate

2 Monsoon and trade-wind littoral climate

3 Wet-dry tropical climate

Dry climates:

4 Dry tropical

5 Dry subtropical

9 Dry midlatitude

4s, 5s, 9s

4sd, 5sd, 9sd

4d, 5d, 9d

FIGURE 8.9 Climates of the world. (Based on Goode Base Map.)

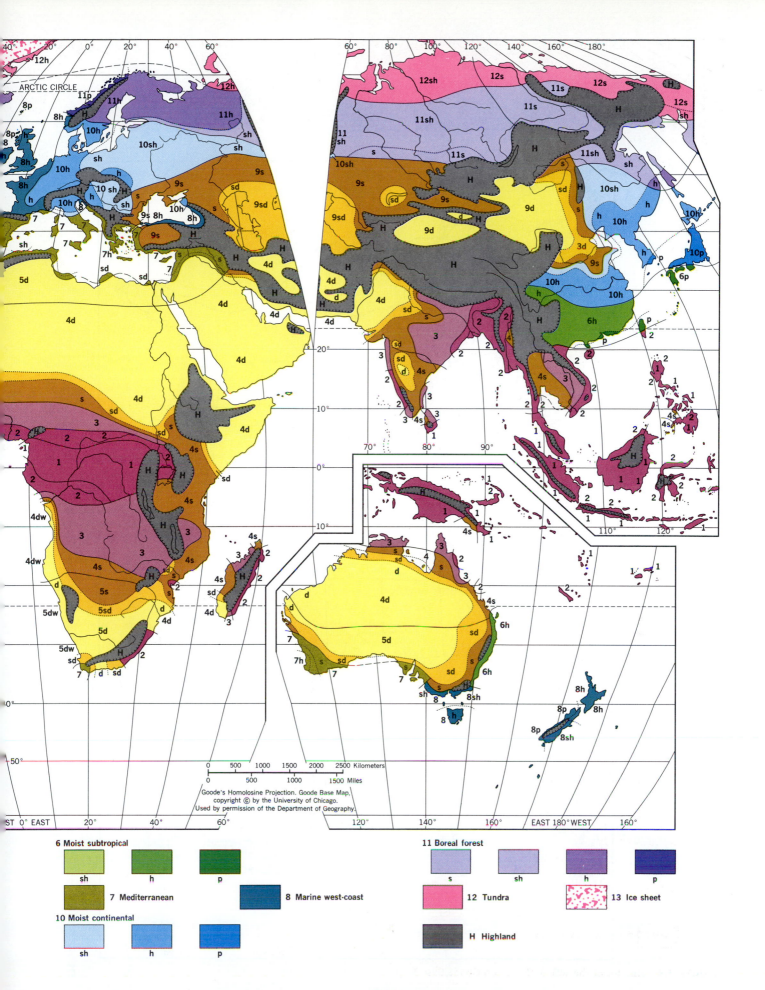

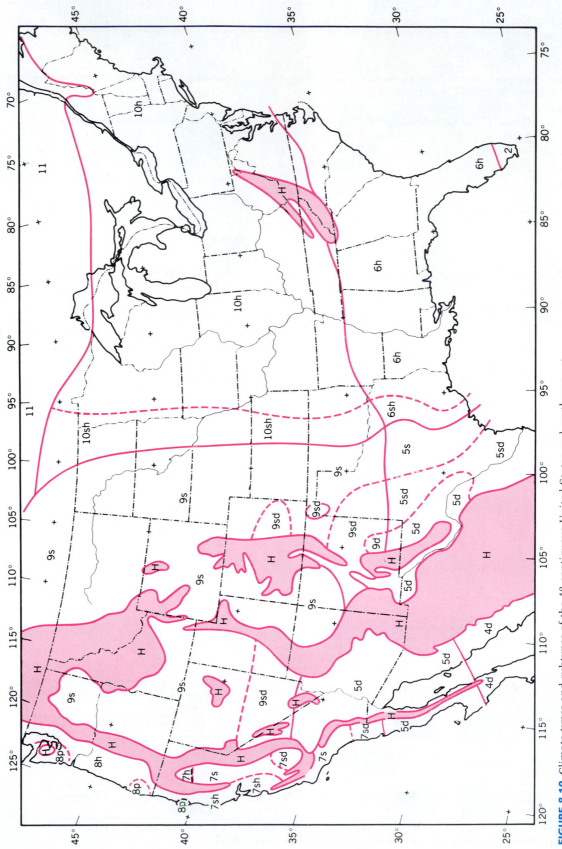

FIGURE 8.10 Climate types and subtypes of the 48 contiguous United States and southernmost Canada. For legend see Figure 8.9. (A. N. Strahler.)

cient rainfall to maintain the soil in a moist condition through much of the year and to sustain the year-round flow of the larger streams. Moist climates support forests or prairies of tall grasses. A precise distinction between dry climates and moist climates is given in connection with the Köppen climate system (discussed later in this chapter) and the newer climate system based on the soil-water balance (Chapter 9).

Within the dry climates is a wide range of degree of aridity, ranging from very dry deserts through transitional levels of aridity to arrive at adjacent moist climates. We will refer to three dry climate subtypes: semiarid (or steppe), semidesert, and desert. The *semiarid subtype* (steppe subtype), designated by the letter *s*, has enough precipitation to support sparse grasslands of a variety referred to by geographers as *steppes*, or *steppe grasslands*. The steppe climate subtype will be found ad-

jacent to moist climates. The *desert subtype (d)* is extremely dry and has so little precipitation that only scattered hardy plants can grow. The *semidesert subtype (sd)* is transitional between semiarid and desert subtypes. (Precise definitions of these subtypes and their boundaries are stated in Chapter 9.)

Within the moist climates is a wide range of degree of wetness. On the one hand is a *subhumid subtype*, designated by the letters *sh*, in which the evaporative losses from soil and plant cover approximately balance the precipitation on an annual average basis. Where precipitation is great enough to produce stream flow through most of the year and to support forests, the *humid subtype* prevails, designated by the letter *h*. Where precipitation is very heavy, with copious stream flow, the *perhumid subtype* prevails *(p)*. (Precise definitions and boundaries of these subtypes are given in Chapter 9.)

THE KÖPPEN CLIMATE SYSTEM

Some professional geographers use a climate classification system devised originally by Dr. Wladimir Köppen in 1918 and subsequently revised by his students, R. Geiger and W. Pohl. The version we present here was published in 1953 and is sometimes called the Köppen–Geiger–Pohl system. Köppen was both a climatologist and a plant geographer, so his main interest lay in finding climate boundaries that coincided approximately with boundaries between major vegetation types. Although he was not entirely successful in achieving this goal, his climate system has appealed to geographers because it is strictly empirical and allows no room for subjective decisions.

Under the Köppen system, each climate is defined according to assigned values of temperature and precipitation, computed in terms of annual or monthly values. Any given station can be assigned to its particular climate group and subgroup solely on the basis of the rec-

ords of temperature and precipitation at that place provided, of course, that the period of record is long enough to yield meaningful averages. To use the Köppen system, it is not necessary to understand climate.

The Köppen system features a shorthand code of letters designating major climate groups, subgroups within the major groups, and further subdivisions to distinguish particular seasonal characteristics of temperature and precipitation.

The five major climate groups are designated by capital letters as follows:

A *Tropical rainy climates.* Average temperature of every month is above 18°C (64.4°F). These climates have no winter season. Annual rainfall is large and exceeds annual evaporation.

B *Dry climates.* Evaporation exceeds precipitation on the average throughout the year. There is no water

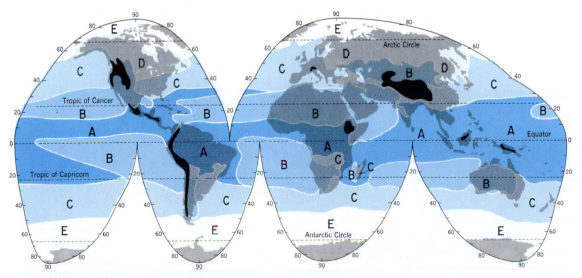

FIGURE 8.11 Highly generalized world map of major climatic regions according to the Köppen classification. Highland areas are in black. (Based on Goode Base Map.)

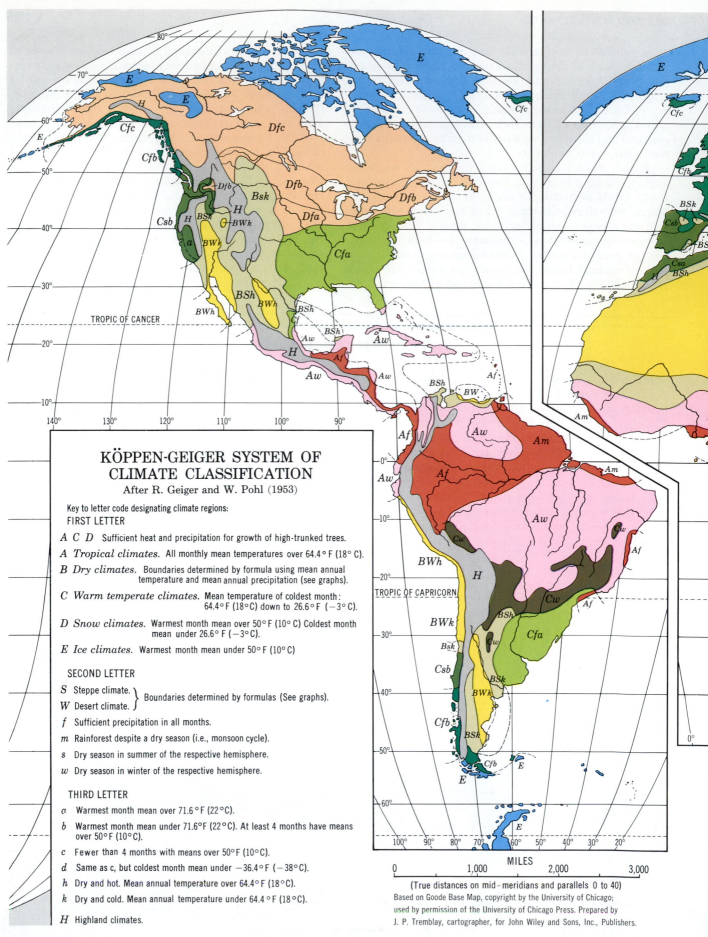

KÖPPEN-GEIGER SYSTEM OF
CLIMATE CLASSIFICATION

After R. Geiger and W. Pohl (1953)

Key to letter code designating climate regions:

FIRST LETTER

A C D Sufficient heat and precipitation for growth of high-trunked trees.

A *Tropical climates.* All monthly mean temperatures over 64.4° F (18° C).

B *Dry climates.* Boundaries determined by formula using mean annual temperature and mean annual precipitation (see graphs).

C *Warm temperate climates.* Mean temperature of coldest month: 64.4° F (18° C) down to 26.6° F (−3° C).

D *Snow climates.* Warmest month mean over 50° F (10° C) Coldest month mean under 26.6° F (−3° C).

E *Ice climates.* Warmest month mean under 50° F (10° C)

SECOND LETTER

S Steppe climate. ⎫
 ⎬ Boundaries determined by formulas (See graphs).
W Desert climate. ⎭

f Sufficient precipitation in all months.

m Rainforest despite a dry season (i.e., monsoon cycle).

s Dry season in summer of the respective hemisphere.

w Dry season in winter of the respective hemisphere.

THIRD LETTER

a Warmest month mean over 71.6° F (22° C).

b Warmest month mean under 71.6°F (22° C). At least 4 months have means over 50° F (10° C).

c Fewer than 4 months with means over 50° F (10° C).

d Same as c, but coldest month mean under −36.4° F (−38° C).

h Dry and hot. Mean annual temperature over 64.4° F (18° C).

k Dry and cold. Mean annual temperature under 64.4° F (18° C).

H Highland climates.

MILES

0   1,000   2,000   3,000

(True distances on mid-meridians and parallels 0 to 40)

Based on Goode Base Map, copyright by the University of Chicago; used by permission of the University of Chicago Press. Prepared by J. P. Tremblay, cartographer, for John Wiley and Sons, Inc., Publishers.

FIGURE 8.12 World map of climates according to the Köppen–Geiger–Kohl system.

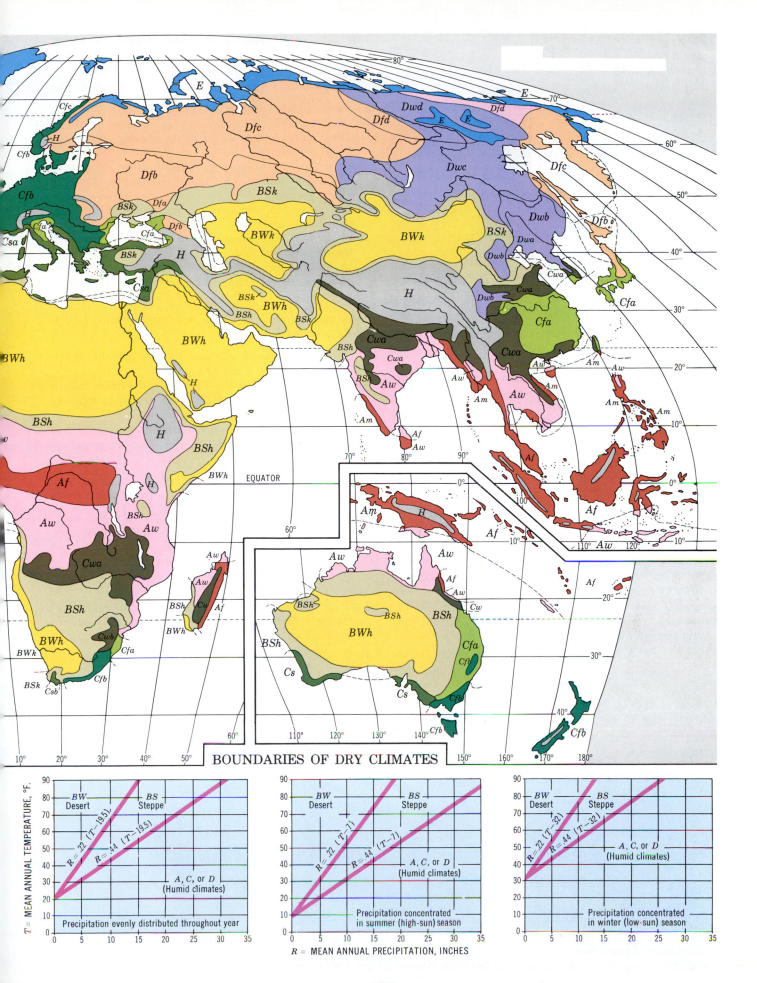

BOUNDARIES OF DRY CLIMATES

Chart 1: Precipitation evenly distributed throughout year
- BW Desert / BS Steppe
- R = 22 (T−19.5)
- R = 44 (T−19.5)
- A, C, or D (Humid climates)

Chart 2: Precipitation concentrated in summer (high-sun) season
- BW Desert / BS Steppe
- R = 22 (T−7)
- R = 44 (T−7)
- A, C, or D (Humid climates)

Chart 3: Precipitation concentrated in winter (low-sun) season
- BW Desert / BS Steppe
- R = 22 (T−32)
- R = 44 (T−32)
- A, C, or D (Humid climates)

T = MEAN ANNUAL TEMPERATURE, °F.

R = MEAN ANNUAL PRECIPITATION, INCHES

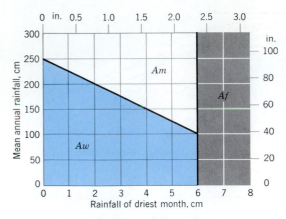

FIGURE 8.13 Boundaries of the A climates.

surplus; hence no permanent streams originate in B climate zones.

C *Mild, humid (mesothermal) climates.* Coldest month has an average temperature under 18°C (64.4°F), but above −3°C (26.6°F); at least one month has an average temperature above 10°C (50°F). The C climates thus have both a summer and a winter season.

D *Snowy-forest (microthermal) climates.* Coldest month average temperature is under −3°C (26.6°F). Average temperature of the warmest month is above 10°C (50°F), that isotherm coinciding approximately with the poleward limit of forest growth.

E *Polar climates.* Average temperature of the warmest month is below 10°C (50°F). These climates have no true summer.

Note that four of these five groups (A, C, D, and E) are defined by temperature averages, whereas one (B) is defined by precipitation-to-evaporation ratios. Groups A C, and D have sufficient heat and precipitation for the growth of forest and woodland vegetation. Figure 8.11 shows the boundaries of the five major climate groups. Figure 8.12 is a world map of Köppen climates.

Subgroups within the five major groups are designated by a second letter, according to the following code:

S Semiarid (steppe).

W Arid (desert).
    (The capital letters S and W are applied only to the dry B climates.)

f Moist. There is adequate precipitation in all months and no dry season. This modifier is applied to A, C, and D groups.

w Dry season in winter of the respective hemisphere (low-sun season).

s Dry season in summer of the respective hemisphere (high-sun season).

m Rainforest climate despite short, dry season in monsoon type of precipitation cycle. This applies only to A climates.

From combinations of the two letter groups, 12 distinct climates emerge:

Af *Tropical rainforest climate.* Rainfall of the driest month is 6 cm (2.4 in.) or more.

Am *Monsoon variety of Af.* Rainfall of the driest month is less than 6 cm (2.4 in.). The dry season is strongly developed.

Aw *Tropical savanna climate.* At least one month has rainfall less than 6 cm (2.4 in.). The dry season is strongly developed.

Figure 8.13 shows the boundaries between Af, Am, and Aw climates as determined by both annual rainfall and rainfall of the driest month.

BS *Steppe climate.* A semiarid climate characterized by grasslands. It occupies an intermediate position between the desert climate (BW) and the more humid climates of the A, C, and D groups. Boundaries are determined by formulas given in Figure 8.14.

BW *Desert climate.* An arid climate with annual precipitation usually less than 40 cm (15 in.). The boundary with the adjacent steppe climate (BS) is determined by formulas given in Figure 8.14.

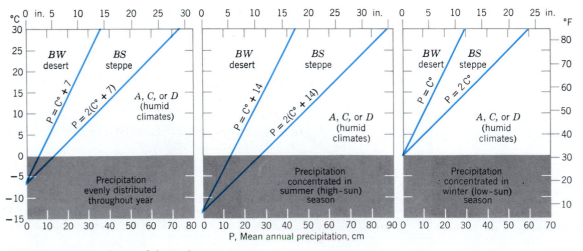

FIGURE 8.14 Boundaries of the B climates.

TABLE 8.2 Köppen Equivalents to Climates Classified by Air Masses and Front Zones

Group 1: Low-Latitude Climates

1. Wet equatorial climate	Af:	Tropical rainforest climate
2. Monsoon and trade-wind littoral climate	Am:	Tropical rainforest climate, monsoon type; and some areas of
	Af:	Tropical rainforest climate
3. Wet-dry tropical climate	Aw:	Tropical savanna climate; and
	Cw:	Mild climate with dry winter
4. Dry tropical climate	BWh:	Desert climate, hot; and
	BSh:	Steppe climate, hot

Group II: Midlatitude Climates

5. Dry subtropical climate	BWh:	Desert climate, hot; and
	BSh:	Steppe, climate hot
6. Moist subtropical climate	Cfa:	Temperate rainy climate, hot summer
7. Mediterranean climate	Csa:	Temperate rainy climate, dry hot summer; and
	Csb:	same, with dry warm summer
8. Marine west-coast climate	Cfb:	Temperate rainy climate with warm summers; also some parts of
	Csb:	Temperate rainy climate, dry warm summer
9. Dry midlatitude climate	BWk:	Desert climate, cold, and
	BSk:	Steppe climate, cold
10. Moist continental climate	Dfa:	Cold snowy-forest climate, hot summers; and
	Dfb:	same, with warm summers

Group III: High-Latitude Climates

11. Boreal forest climate	Dfc:	Cold snowy-forest climate, cool summers; and
	Dw:	Cold snowy-forest climate, dry winters; and
	Cfc:	Temperate rainy climate, cool short summers
12. Tundra climate	ET:	Polar tundra climate
13. Ice-sheet climate	EF:	Polar climate, perpetual frost

Cf *Mild humid climate with no dry season.* Precipitation of the driest month averages more than 3 cm (1.2 in.).

Cw *Mild humid climate with a dry winter.* The wettest month of summer has at least 10 times the precipitation of the driest month of winter. (Alternate definition: Seventy percent or more of the mean annual precipitation falls in the warmer six months.)

Cs *Mild humid climate with a dry summer.* Precipitation of the driest month of summer is less than 3 cm (1.2 in.). Precipitation of the wettest month of winter is at least three times as much as the driest month of summer. (Alternate definition: Seventy percent or more of the mean annual precipitation falls in the six months of winter.)

Df *Snowy-forest climate with a moist winter.* No dry season.

Dw *Snowy-forest climate with a dry winter.*

ET *Tundra climate.* Mean temperature of the warmest month is above 0°C (32°F) but below 10°C (50°F).

EF *Perpetual frost climate.* Ice-sheet climate. Mean monthly temperatures of all months are below 0°C (32°F).

To denote further variations in climate, Köppen added a third letter to the code group. Meanings are as follows:

a With hot summer; warmest month is over 22°C (71.6°F); C and D climates.

b With warm summer; warmest month is below 22°C (71.6°F); C and D climates.

c With cool, short summer; less than four months are over 10°C (50°F); C and D climates.

d With very cold winter; coldest month is below −38°C (−36.4°F); D climates only.

h Dry-hot; mean annual temperature is over 18°C (64.4°F); B climates only.

k Dry-cold; mean annual temperature is under 18°C (64.4°F); B climates only.

As an example of a complete Köppen climate code, BWk refers to a "cool desert climate," and Dfc refers to a "cold, snowy forest climate with cool, short summer."

Table 8.2 gives approximate Köppen equivalents for the 13 climate types based on air masses and frontal zones.

CLIMATE, VEGETATION, AND SOILS

Plant geographers are keenly aware that plants are responsive to differences in climate. A particular combination of climatic factors most favorable to its growth is associated with each plant species, as are certain extremes of heat, cold, or drought beyond which it cannot survive. Plants tend to adapt their physical forms to meet the stresses of climate. We find a wide range of forms taken by assemblages of the dominant plant species. These overall form-patterns, or habits, closely reflect climate controls.

Since the late nineteenth century, soil scientists have recognized that the several fundamental classes of mature soils, seen on a worldwide scope, are strongly controlled by climatic elements. But plants also contribute to determining the properties of the very soils on which they depend.

To develop the concept of an interplay among climate controls, vegetation, and soils, we must make a special study of water within the soil layer. Although precipitation is an essential climate ingredient, what really counts in the growth of plants is the amount of water available in the soil zone where the plant roots are situated. Evaporation from soil and plant surfaces can return a large proportion of the incoming precipitation to the atmosphere. Within the soil is a balance between incoming and outgoing water. As in the case of the energy balance, there can be seasonal water surpluses and water deficits. In the next chapter we shall investigate the soil-water balance as an important step in increasing our understanding of the varied environments of the life layer.

THE SOIL-WATER BALANCE AND CLIMATE

A basic concept of physical geography is that the availability of water to plants is a more important factor in the environment than precipitation itself. Much of the water received as precipitation is lost in a variety of ways and is not usable for plants. Just as in a fiscal budget, when the monthly or yearly loss of moisture exceeds the precipitation, a budgetary deficit results; when precipitation exceeds the losses, a budgetary surplus results. An analysis of the water budget of a given area of the earth's surface is actually arrived at in much the same way as a fiscal budget. The accounting requires only additions and subtractions of amounts within fixed periods of time, such as the calendar month or year.

To understand why such moisture gains and losses occur, we need to study the physical processes affecting water in its vapor, liquid, and solid states—not only in the atmosphere but also in the soil and rock and in the exposed water of streams, lakes, and glaciers. The science of *hydrology* treats such relationships of water as a complex but unified system on the earth. The primary concern of this chapter is the hydrology of the soil zone.

SURFACE AND SUBSURFACE WATER

We classify water on the land according to whether it is *surface water,* flowing exposed or ponded on the land, or *subsurface water,* occupying openings in the soil or rock. Water held in the soil within a meter or two of the surface is the *soil water;* it is the particular concern of the plant ecologist, soil scientist, and agricultural engineer. Water held in the openings of the bedrock is referred to as ground water and is studied by the geologist. Surface water and ground water are the subjects of Chapter 16.

THE HYROLOGIC CYCLE AND THE GLOBAL WATER BALANCE

Water of oceans, atmosphere, and lands move in a great series of continuous interchanges of both geographic position and physical state, known as the *hydrologic cycle* (Figure 9.1). A particular molecule of water might, if we

could trace it continuously, travel through any one of a number of possible circuits involving alternately the water vapor state and the liquid or solid state.

The pictorial diagram of the hydrologic cycle given in Figure 9.1 can be quantified for the earth as a whole. Figure 9.2 is a mass-flow diagram relating the principal pathways of the water circuit. We can start with the oceans, which are the basic reservoir of free water. Evaporation from the ocean surfaces totals about 419,000 cu km per year (English equivalents are given in Figure 9.2). At the same time, evaporation from soil, plants, and water surfaces of the continents totals about 69,000 cu km. Thus the total evaporation is 488,000 cu km; it represents the quantity of water that must be returned annually to the liquid or solid state.

Precipitation is unevenly divided between continents and oceans; 106,000 cu km is received by the land surfaces and 382,000 cu km by the ocean surfaces. Notice that the continents receive about 37,000 cu km more water as precipitation than they lose by evaporation.

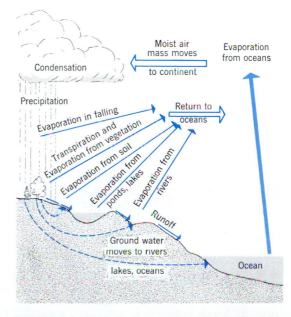

FIGURE 9.1 The hydrologic cycle. (After Holtzman.)

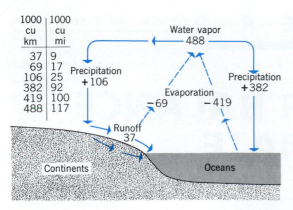

FIGURE 9.2 The global water balance. Figures give average annual water flux in and out of world land areas and world oceans. (Data of John R. Mather, 1974. From A. N. Strahler, *The Earth Sciences*, 2nd ed., Harper & Row, Publishers. Copyright © by Arthur N. Strahler.)

This excess quantity flows over or under the ground surface to reach the sea; it is collectively termed *runoff*.

We can state the *global water balance* as follows:

$$P = E + G + R$$

where P = precipitation
      E = evaporation
      G = net gain or loss of water in the system, a storage term
      R = runoff (positive when out of the continents, negative when into the oceans)

All terms have the dimensions of volume per unit time (e.g., cubic kilometers per year). When applied over the span of a year, and averaged over many years, the storage term G can be considered as zero because the global system is essentially closed so far as matter is concerned. The quantities of water in storage in the atmosphere, on the lands, and in the oceans will remain about constant from year to year.

The equation then simplifies to

$$P = E + R$$

Using the figures given for the continents,

$$106,000 = 69,000 + 37,000$$

and for the oceans,

$$382,000 = 419,000 - 37,000$$

For the globe as a whole, combining continents and ocean basins, the runoff terms cancel out:

$$106,000 + 382,000 = 69,000 + 419,000$$
$$488,000 = 488,000$$

THE HYDROLOGIC CYCLE AS A CLOSED MATERIAL FLOW SYSTEM

The global water balance can be treated as a material flow system, using the type of schematic diagram described in Chapter 2. Figure 9.3 shows a closed system representing the total global hydrosphere. Three subsystems are present: atmosphere, continents, and oceans. For the sake of simplicity, glacier ice is included with liquid water as a single form of water. In the atmosphere subsystem the input is in the form of water vapor derived by evaporation from the ocean surface and by evaporation and evapotranspiration from the continental surface. Stored water vapor undergoes condensation to become precipitation in the liquid form as rain (or solid state as snow), leaving the atmosphere subsystem and entering either the continent subsystem or the ocean subsystem. By runoff, water in storage on the continents is transferred to the oceans without change of state.

THE GLOBAL WATER BALANCE BY LATITUDE ZONES

From a global water balance in terms of water volume per unit of time (cu km/yr; cu mi/yr), we now turn to water balance data in terms of water depth per unit of time (cm/yr; in./yr). Whereas the first set of data tells us the total quantity of water transported into or out of a specified area in one year, the second set will tell us the intensity of water flow, independent of total surface area and total water quantity.

The water balance can be estimated for latitude belts of 10° width. This treatment reveals a response to the latitudinal changes in radiation and heat balances from the equatorial zone to the polar zones. Figure 9.4 shows

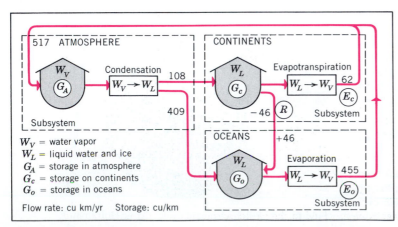

FIGURE 9.3 Schematic diagram of the hydrologic cycle as a closed material system.

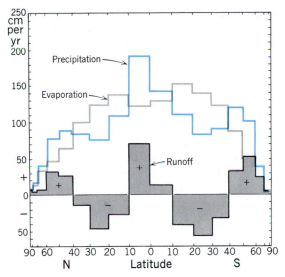

FIGURE 9.4 The water balance for 10° latitude zones. (Data from several sources compiled by W. D. Sellers. From A. N. Strahler, *The Earth Sciences*, 2nd ed., Harper & Row, Publishers. Copyright © by Arthur N. Strahler.)

the average annual values of precipitation, evaporation, and runoff. The runoff term R in this case includes import or export of water in or out of the belt by ocean currents as well as by stream flow. Notice the equatorial zone of water surplus with positive values of R, in contrast to the subtropical belts, with an excess of evaporation. The runoff term R, which has a negative sign in the subtropical belts, represents the importation of water by ocean currents to furnish the quantity needed for evaporation. Water surpluses occur poleward of the 40th parallel; but the values of all three terms decline rapidly at high latitudes, going nearly to zero at the poles. This graph should be compared with Figure 6.28, which shows water-vapor transport across parallels of latitude. Notice that the precipitation surpluses are sustained by importation of water vapor and that evaporation surpluses are sustained by export of water vapor.

GLOBAL WATER IN STORAGE

The total global water resource is stored in the gaseous, liquid, and solid states. Water of the oceans constitutes

over 97 percent of the total, as we would expect (Figure 9.5). Next comes water in storage in glaciers, a little over 2 percent. Of the remaining quantity, almost all is ground water, so surface water in lakes and streams is a very small quantity indeed. But it is surface water, together with the very small quantity of soil water, that sustains all life of the lands. Some environmentalists consider that fresh surface water will prove to be the limiting factor in the capacity of our planet to support the rapidly expanding human population. The quantity of water vapor held in the atmosphere is also very small—only about 10 times greater than that held in streams; but this atmospheric moisture is the source of all fresh water of the lands.

GLOBAL WATER BALANCE AND CLIMATE CHANGE

Long-term changes in water-balance quantities are associated with atmospheric environmental changes. For example, atmospheric cooling on a global scale would bring a reduction of atmospheric water-vapor storage. This change would, in turn, reduce precipitation and runoff generally. But at the same time, a greater proportion of that precipitation would be in the form of snow, so the storage in ice accumulations would rise and the storage in ocean waters would fall. These changes describe the changing water balance associated with onset of an ice age, or glaciation—a major environmental change already experienced by our planet at least four times in the past two million years (see Chapter 22).

INFILTRATION AND RUNOFF

Most soil surfaces in their undisturbed, natural states are capable of absorbing the water from light or moderate rains. This absorption process is known as *infiltration*. Most soils have natural passageways between poorly fitting soil particles, as well as larger openings, such as earth cracks resulting from soil drying, borings of worms and animals, cavities left from decay of plant roots, or openings made by growth and melting of frost crystals. A mat of decaying leaves and stems breaks the force of falling drops and helps keep these openings clear. If rain falls too rapidly to be passed downward through these soil openings, the excess amount flows as a surface-water

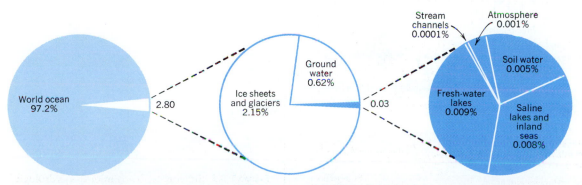

FIGURE 9.5 The total volume of global water in storage is largely held in the world ocean.

layer down the direction of ground slope. This surface runoff, called overland flow, is described in Chapter 16.

As stated in Chapter 6, rainfall is measured in units of centimeters or inches per hour. This is the depth to which water will accumulate in each hour if rain is caught in a flat-bottomed, straight-sided container, assuming none to be lost by evaporation or splashing out (see Figure 16.2). Similarly, infiltration is stated in centimeters or inches per hour and can be thought of as the rate at which the water level in the same container might drop if the water were leaking through a porous base. Runoff, also stated in cm (in.) per hour, may be thought of as the amount of overflow of the container per hour when rain falls too fast to be disposed of by leaking through the base.

EVAPORATION AND TRANSPIRATION

Between periods of rain, water held in the soil is gradually given up by a twofold drying process. First, direct evaporation into the open air occurs at the soil surface and progresses downward. Air also enters the soil freely and may actually be forced alternately in and out of the soil by atmospheric pressure changes. Even if the soil did not "breathe" in this way, there would be a slow diffusion of water vapor surfaceward through the open soil pores. Ordinarily only the upper 30 cm (1 ft) of soil is dried by evaporation in a single dry season. In the prolonged drought of deserts, a dry condition extends to depths of many meters.

Second, plants draw the soil water into their systems through vast networks of tiny rootlets. This water, after being carried upward through the trunk and branches into the leaves, is discharged through leaf pores into the atmosphere in the form of water vapor. The process is termed *transpiration*.

In studies of climatology and hydrology, it is convenient to use the term *evapotranspiration* to cover the combined water loss from direct evaporation and the transpiration of plants. The rate of evapotranspiration slows down as soil-water supply becomes depleted during a dry summer period because plants use various devices to reduce transpiration. In general, the less water remaining, the slower the loss through evapotranspiration.

Figure 9.6 illustrates the various terms explained up to this point and serves to give a more detailed picture of that part of the hydrologic cycle involving the soil. The soil layer from which plants can draw moisture is the *soil-water belt*. This belt gains water through precipitation and infiltration. As the minus signs show, the soil loses water through transpiration, evaporation, and overland flow. Excess water also leaves the soil by downward *gravity percolation* to the ground-water zone below. Between the soil-water belt and the ground-water zone is an *intermediate belt*. Here water is held at a depth too great to be returned to the surface by evapotranspiration because it is below the level of plant roots.

GROUND WATER AND STREAM FLOW

Gravity percolation carries excess water down to the ground-water zone, in which all pore spaces are fully saturated (Figure 9.6). Within this zone water moves slowly in deep paths, eventually emerging by seepage into streams, ponds, and lakes.

Excess water leaves the area by stream flow. Streams are fed directly by overland flow in periods of heavy, prolonged rain or rapid snowmelt. Streams that flow throughout the year—perennial streams—derive much of their water from ground water seepage. Streams fed only by overland flow run intermittently, and their channels are dry much of the time between rain periods. We shall investigate ground-water flow and stream flow in detail in Chapter 16.

WATER IN THE SOIL

When infiltration occurs, the water is drawn downward through the soil pores, wetting successively lower layers. This activity is called *soil-water recharge*. Eventually the soil layer holds the maximum possible quantity of water, although the larger pores remain filled with air. Water movement then continues downward through the underlying intermediate belt.

Suppose now that the rain stops and several days of dry weather follow. Excess soil water continues to drain downward under gravity, but some water clings to the

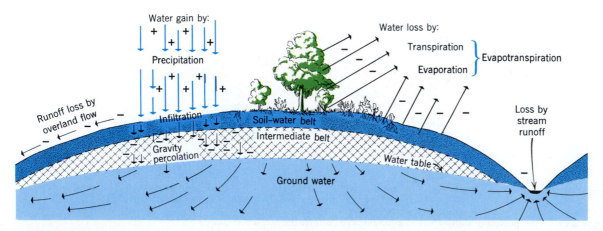

FIGURE 9.6 The soil-water belt occupies an important position in the hydrologic cycle.

soil particles and effectively resists the pull of gravity because of the force of capillary tension. We are all familiar with the way a water droplet seems to be enclosed in a "skin" of surface molecules, drawing the droplet together into a spherical shape so that it clings to the side of a glass indefinitely without flowing down. Similarly, tiny films of water adhere to the soil grains, particularly at the points of grain contacts, and will stay until disposed of by evaporation or absorption into plant rootlets.

When a soil has first been saturated by water, then allowed to drain under gravity until no more water moves downward, the soil is said to be holding its *storage capacity* of water. (Among soil scientists, storage capacity is referred to as *field capacity*.) Drainage takes no more than two or three days for most soils. Most excess water is drained out within one day. Storage capacity is measured in units of depth, usually centimeters or inches, just as is precipitation. For example, a storage capacity of 2 cm in the uppermost 10 cm of the soil means that for a given cube of soil, 10 cm on a side (1 cu decimeter), all the extractable water would form a layer of water 2 cm deep in a 10 × 10 cm container with flat bottom and vertical sides. This depth would be equivalent to complete absorption of a 2-cm rainfall by a completely dry 10-cm layer of soil.

Storage capacity of a given soil depends largely on its texture. Sandy soil has a very small storage capacity; clay soil has a large capacity.

THE SOIL-WATER CYCLE

We can turn next to consider the annual water budget of the soil. Figure 9.7 shows the annual cycle of soil water for a single year at an agricultural experiment station in Ohio. This example can be considered generally representative of conditions in moist, midlatitude climates where there is a strong temperature contrast between winter and summer. Let us start with the early spring

(March). At this time the evaporation rate is low because of low air temperatures. The abundance of melting snows and rains has restored the soil water to a surplus quantity. For two months the quantity of water precolating through the soil and entering the ground water keeps the soil pores nearly filled with water. This is the time of year when one encounters soft, muddy ground conditions, whether driving on dirt roads or walking across country. This, too, is the season when runoff is copious and major floods may be expected on larger streams and rivers. In terms of the soil-water budget, a *water surplus* exists.

By May the rising air temperatures, increasing evaporation, and full growth of plant foliage bring on intense evapotranspiration. Now the soil water falls below the storage capacity, although it may be restored temporarily by unusually heavy rains in some years. By midsummer a large *water deficit* exists in the water budget. Even the occasional heavy thunderstorm rains of summer cannot restore the water lost by heavy evapotranspiration. Small springs and streams dry up, and the soil becomes firm and dry. By November, however, the soil water again begins to increase. This is because the plants go into a dormant state, sharply reducing transpiration losses. At the same time, falling air temperatures reduce evaporation. By late winter, usually in February at this location, the storage capacity of the soil is again fully restored.

THE SOIL-WATER BALANCE

From the example of soil-water change in a single year, we move forward to a more generalized concept. The gain, loss, and storage of soil water are accounted for in the *soil-water balance*. Figure 9.8 is a flow diagram that illustrates the components of the balance. *Soil-water storage* (S), the actual quantity of water held in the soil-water zone, is increased by recharge during *precipitation* (P) but decreased by use through *evapotranspiration* (E). Any

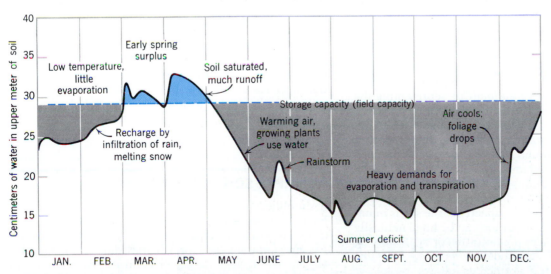

FIGURE 9.7 A typical annual cycle of soil-water change in the Midwest shows a short period of surplus in the spring and a long period of deficit in the summer and fall. (Data of Thornthwaite and Mather.)

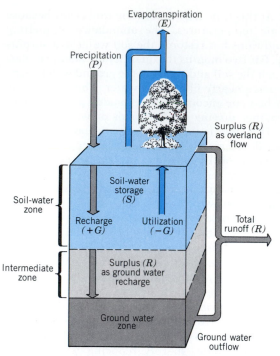

FIGURE 9.8 Schematic diagram of the soil-water balance in a soil column. (From A. N. Strahler, *The Earth Sciences*, 2nd ed., Harper & Row, Publishers. Copyright © by Arthur N. Strahler.)

water surplus (R) is disposed of by downward percolation to the ground-water zone or by overland flow.

The soil column is assumed to have a unit cross-sectional area, for example, 1 sq cm. We can therefore use flow units of depth per unit time (e.g., cm/month or cm/year).

The water-balance equation is as follows:

$$P = E + G + R$$

where P = precipitation
E = evaporation
G = change in soil-water storage
R = water surplus

To proceed, we must recognize two ways to define evapotranspiration. First is *actual evapotranspiration (Ea)*, which is the true or real rate of water-vapor return to the atmosphere from the ground and its plant cover. Second is *potential evapotranspiration (Ep)*, representing the water-vapor flux under an ideal set of conditions. One condition is that there be present a complete (or closed) cover of uniform vegetation consisting of fresh green leaves, and no bare ground exposed through that cover. The leaf cover is assumed to have a uniform height above ground—whether the plants be trees, shrubs, or grasses. A second condition is that there be an adequate water supply, such that the storage capacity of the soil is maintained at all times. This condition can be filled naturally by abundant and frequent precipitation or artificially by irrigation. To simplify the ponderous terms we have just defined, they may be transformed as follows:

"actual evapotranspiration" is *water use (Ea)*
"potential evapotranspiration" is *water need (Ep)*

The word "need" signifies the quantity of soil water needed if plant growth is to be maximized for the given conditions of solar radiation and air temperature and the available supply of nutrients.

The difference between water use and water need is the *soil-water shortage (D)*. This is the quantity of water that must be furnished by irrigation to achieve maximum crop growth within an agricultural system.

All the terms in the soil-water budget are stated in centimeters of water depth, the same as for precipitation. (We will not give English equivalents on the following pages.) A particular value of storage capacity of the soil is established in advance. In the soil-water budgets we present here, it has been assumed that the soil layer has a storage capacity of 30 cm (about 12 in.).

Several rigorous methods have been devised for estimating the monthly water need (potential evapotranspiration) at any given location on the globe. The method used in this chapter was developed by C. Warren Thornthwaite, a distinguished climatologist and geographer. Thornthwaite's method is based on air temperature, as well as latitude and date. Latitude and season determine intensity and duration of solar radiation received at the ground. Put another way, water need is a measure of the maximum capability of a land surface to return energy from the surface to the atmosphere by the mechanism of latent heat flow under the defined conditions of plant cover and water supply.

THE SOIL-WATER BALANCE AS AN OPEN MATERIAL FLOW SYSTEM

The soil-water balance can be treated as an open material flow system, as shown in Figure 9.9. System input is by precipitation. Rain or snow falling on plant foliage and stems is intercepted and goes into temporary storage; it may be lost by evaporation (change of state), or it may drip or flow down to the ground surface. Some water is temporarily stored as *surface detention* in depressions on the soil surface, from which it may be lost by evaporation, or may leave the system as overland flow. By infiltration, water enters storage in the soil-water zone. Some of this stored water may evaporate from the soil surface, leaving the system as water vapor. Uptake of water through the roots of plants is another way in which soil-water storage is depleted. Transpiration disposes of this water in the vapor state. Surplus soil water may infiltrate downward through the intermediate zone, entering the ground-water zone and leaving the system as ground-water flow.

A SIMPLE SOIL-WATER BUDGET

A simplified *soil-water budget* for the year is shown in Figure 9.10, a graph on which monthly means are plotted as points and connected by smooth lines. In this example, precipitation (P) is much the same in all months, with no strong yearly cycle. In contrast, water need (Ep) shows a strong seasonal cycle: low values in winter and a high summer peak. For midlatitudes in a moist climate, this model is about right.

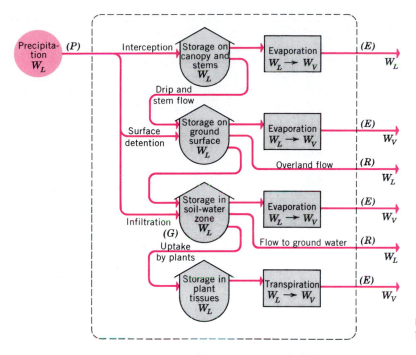

FIGURE 9.9 Schematic diagram of the soil-water balance as an open material flow system.

At the start of the year a large water surplus (R) exists, disposed of by runoff. By May conditions have switched over to a water deficit. In this month plants begin to withdraw soil water from storage. *Storage withdrawal* (−G) is represented by the difference between the water-use curve and the precipitation curve (−G is calculated as Ea − P, when Ea exceeds P). As storage withdrawal continues, the plants reduce their water use to less than the optimum quantity, so without irrigation the water-use curve departs from the water-need curve. Storage withdrawal continues throughout the summer. The deficit period lasts through September. The area labeled "soil-water shortage" represents the total quantity of water needed by irrigation to ensure maximum growth throughout the deficit period. Soil-water shortage (D), is calculated by subtracting Ea from Ep (D = Ep − Ea).

In October precipitation (P) again exceeds water need (Ep), but the soil must first absorb an amount equal to the summer storage withdrawal (−G). So we next have a period of *storage recharge* (+G); it lasts through November. In December the soil has reached its full storage capacity (30 cm). Now a water surplus (R) again sets in, lasting throughout the winter.

Thornthwaite was concerned with practical problems of crop irrigation. He developed the calculation of the soil-water budget to place crop irrigation on a precise, accurate basis. The Thornthwaite method was later adopted by soil scientists of the U.S. Department of Agriculture to identify classes of soils associated with various soil-water budgets. Water need (potential evapotranspiration) is a difficult quantity to measure, and several other methods have been developed to estimate its true value.

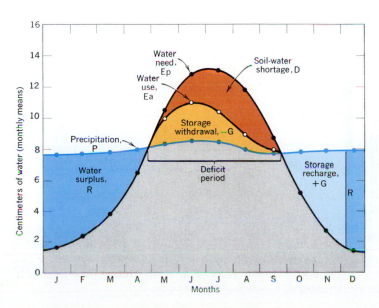

FIGURE 9.10 A simplified soil-water budget typical of a humid midlatitude climate.

TABLE 9.1 Simplified Example of a Soil-Water Budget

Equation	P = Ea	+G	+R	Ep	(Ep − Ea) = D
January	11.0 = 1.0		+ 10.0	1.0	0.0
February	9.0 = 2.0		+ 7.0	2.0	0.0
March	6.0 = 3.5		+ 2.5	3.5	0.0
April	3.0 = 6.0	− 3.0		6.0	0.0
May	2.5 = 7.0	− 4.5		8.5	1.5
June	2.0 = 6.0	− 4.0		9.5	3.5
July	2.5 = 5.0	− 2.5		9.0	4.0
August	4.0 = 4.5	− 0.5		7.0	2.5
September	7.0 = 4.5	+ 2.5		4.5	0.0
October	9.0 = 3.0	+ 6.0		3.0	0.0
November	10.5 = 1.5	+ 6.0	+ 3.0	1.5	0.0
December	12.0 = 1.5		+ 10.5	1.5	0.0
Totals	78.5 = 45.5	− 14.5 + 14.5	+ 33.0	57.0	11.5
	78.5 = 78.5				

In a world beset by severe and prolonged food shortages, the Thornthwaite concepts and calculations are of great value in assessing the benefits to be gained by increased irrigation. Only in a few parts of the tropical and midlatitude zones is precipitation ample to fulfill the water need during the growing season. In contrast, the equatorial zone generally has a large water surplus throughout the year.

CALCULATING A SIMPLE SOIL-WATER BUDGET

Table 9.1 gives simplified data for a hypothetical example of a soil-water budget. Columns are arranged so that the terms appear in the same sequence as in the soil-water balance equation. Values are rounded to multiples of 0.5 cm to simplify addition and subtraction. Figure 9.11 shows the same data plotted in graphic form.

By means of simple arithmetic, we have calculated the soil-water budget for each month singly and for the year as a whole. Tallied separately at the right are the monthly differences between Ep and Ea, giving a total soil-water shortage, D, of 11.5 cm for the year. This is the quantity of water that would have to be supplied by irrigation in an average year to sustain the full value of Ep. The importance of the water budget calculation in estimating the need for irrigation of crops should be obvious. We have also been able to calculate the annual surplus (R = 33 cm). This information can be used to estimate the recharge of ground water and runoff into streams. In this way an assessment of the water-resource

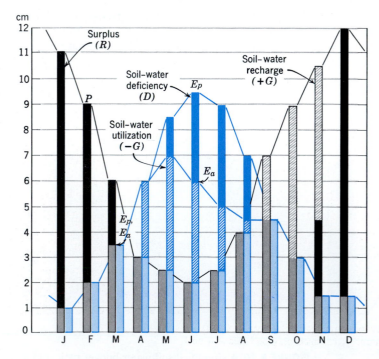

FIGURE 9.11 A model soil-water budget. Bars are correctly scaled to agree with the figures in Table 9.1.

potential of a region can be made—a vital consideration in planning regional economic development and resource management.

THE GLOBAL RANGE OF WATER NEED

Figure 9.12 is a world map showing mean annual water need (Ep, potential evapotranspiration) calculated by the Thornthwaite method. Isopleths of water need in centimeters per year are greatly generalized and are deleted from major highland regions. What counts is the general pattern of poleward decrease of Ep from high values in the equatorial and tropical zones to very low values in the arctic zone. A second important feature is the effect of deserts, which produce abnormally high values compared with areas of moist climate in which temperatures are moderated by precipitation and persistent cloud cover.

Figure 9.13 illustrates the effect of latitude on the annual cycle of water need. Both air-temperature cycles and insolation cycles are reflected in these cycles. Each chart has been labeled as to climate, according to the system given in Chapter 8.

In the wet equatorial climate (1), water need is high all year and the total is over 150 cm (60 in.). For the monsoon and trade-wind littoral climate (2) in the tropical latitude zone, there is a pronounced annual cycle, but

a large annual total water need: almost 150 cm (60 in.). In the tropical desert climate (4d), the annual cycle is very strongly developed and the total water need remains high: nearly 127 cm (50 in.). In midlatitudes on the west coast, the Mediterranean climate (7) shows a well-developed annual cycle; but monthly values of water need remain high throughout the mild winter. In the moist continental climate (10) winter months have no water need, but the summer peak is high; the total is 75 cm (30 in.). In the marine west-coast climate (8), close to the Pacific Ocean, however, water need remains substantial throughout the winter because of the mild temperatures. Farther poleward in the boreal forest climate (11), the number of months of zero water need increases to six and the annual total water need diminishes to 40 cm (16 in.). North of the arctic circle, the tundra climate station (12) shows nine consecutive winter months of zero water need and a very narrow summer peak. The total water need here is least of all: 20 cm (8 in.).

The details of these examples are not important. What counts is the trends they show. Total annual water need diminishes from a maximum in the equatorial zone to a minimum in the arctic zone. At the same time, the annual cycle becomes stronger and stronger in the poleward direction. As the cold of winter makes its effects more pronounced, the number of months of zero water need increases. Of course, plant growth is essentially zero when soil water is frozen, so the growing season

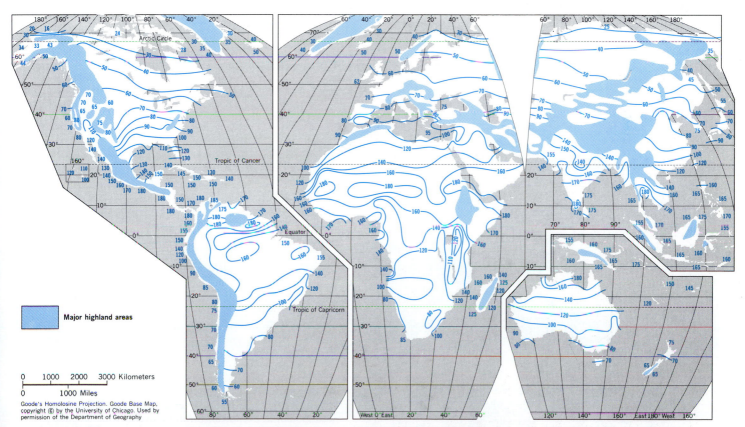

FIGURE 9.12 World map of annual water need (potential evapotranspiration). (Data of C. W. Thornthwaite Associates, Laboratory of Climatology, Centerton, N.J. Based on Goode Base Map.)

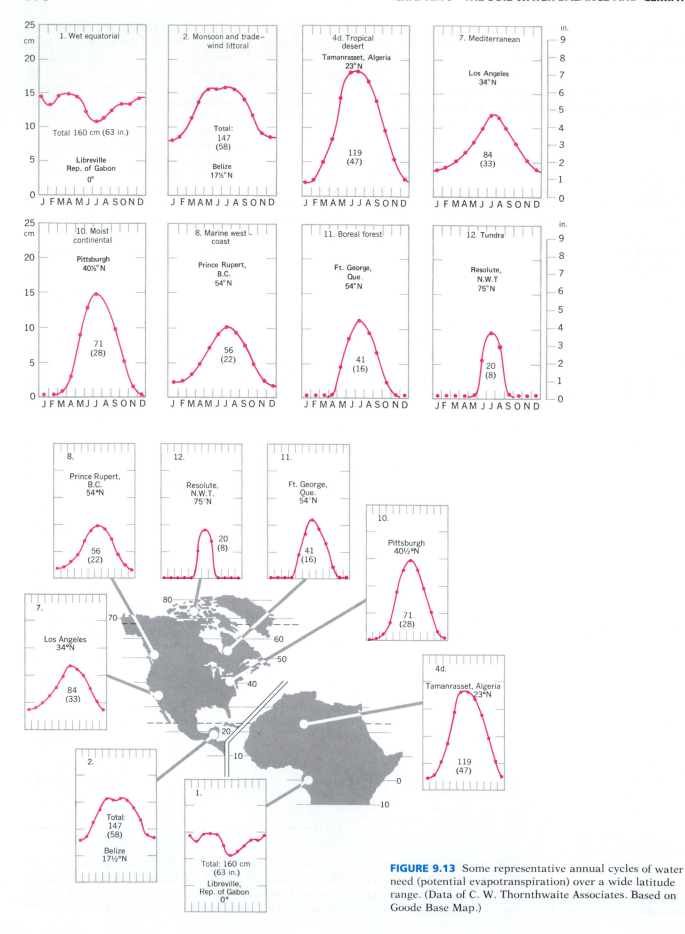

FIGURE 9.13 Some representative annual cycles of water need (potential evapotranspiration) over a wide latitude range. (Data of C. W. Thornthwaite Associates. Based on Goode Base Map.)

is nicely reflected in graphs of water need. Water need persists throughout the winter far poleward along narrow western coastal zones (littorals), in contrast to the interior regions.

THE SOIL-WATER BALANCE AS A BASIS OF CLIMATE CLASSIFICATION

C. W. Thornthwaite's soil-water balance is precise and quantitative in its use of station data of monthly mean precipitation and monthly mean air temperature (used to derive water need). The soil-water balance offers us the opportunity to set up a quantitative climate classification. The Köppen climate system, as we noted, has been held in esteem for many decades by geographers because it uses precise, quantitative definitions of climate types based on monthly mean precipitation and temperature data. But a climate system based on the soil-water balance has certain advantages over the Köppen system. First, the Thornthwaite method gives information of direct value in assessing the conditions favorable or unfavorable to the growth of plants—both native plants and cultivated crops. Second, the soil-water balance has been applied by soil scientists to a modern system of soil classification, which we will present in Chapters 23 and 24. Thus a quantitative water-balance climatology is a necessity. The Köppen system does not fill this need and is not used in the U.S. comprehensive soil classification system.

We offer here a climate system based on recognition of distinctive soil-water regimes. All that is needed to establish this climate classification is a set of definitions. The classification retains the 13 climate classes recognized in Chapter 8 on the basis of air-mass source regions and frontal zones. The names of the 13 classes remain valid—only the frame of reference is changed. The new quantitative frame of reference simply reinforces the explanatory–descriptive framework already established.

SOURCE OF DATA

A climate system based on the soil-water balance requires a large base of station data derived in a consistent manner from long records of monthly mean precipitation and monthly mean air temperature. We make use of published water-balance data of the continents prepared by the Laboratory of Climatology in Centerton, New Jersey, which was originally under the direction of Dr. Thornthwaite. Over a 13-year period, the staff of the Laboratory of Climatology computed the water balances of more than 13,000 stations. Representative station data have been published in a series of 8 volumes, covering the world's inhabited land areas. These data were used in drawing our world map of climates, Figure 8.9. The Thornthwaite data have also been used by the Soil Conservation Service of the U.S. Department of Agriculture as the basis for recognizing soil-water regimes associated with the major classes of soils (see Chapters 23 and 24).

WATER NEED AS AN INDICATOR OF AVAILABLE HEAT

Our three major climate groups—I. Low-Latitude, II. Midlatitude, and III. High-Latitude—can be put on a firm quantitative basis using the total annual water need, Ep. In a general way, Ep grades from maximum values in low latitudes to minimum values in the arctic zone. This gradation reflects the poleward decrease in both annual net radiation and mean annual air temperature.

The Low-Latitude Climates, Group I, are those climates with total annual Ep greater than 130 cm. This criterion applies rigorously to the dry tropical climate (4) and the wet-dry tropical climate (3); but Ep may be somewhat less (down to 110 cm) in some parts of the wet equatorial climate (1) and in the monsoon and tradewind littoral climate (2). On our world climate map, Figure 8.9, the poleward boundaries of climates 3 and 4 are based on the 130-cm isopleth of Ep. This boundary is particularly important in separating the dry tropical climate (4) from the dry subtropical climate (5) in northern Africa, North America, Australia, South America, and southern Africa.

The Midlatitude Climates, Group II, are those climates with total annual Ep ranging from 130 cm down to 52.5 cm. Thus the poleward boundary of Group II climates follows the 52.5-cm isopleth of Ep. This boundary is found in both North America and Eurasia, separating the moist continental climate (10) from the boreal forest climate (11). It is a line approximately coinciding with the southern limit of the great boreal forests. To the south lie deciduous forests and grasslands of warmer climates.

The High-Latitude Climates, Group III, have a total annual water need, Ep, of less than 52.5 cm. This group includes the boreal forest climate (11), the tundra climate (12), and the ice-sheet climate (13). In the tundra climate (12), Ep is less than 35 cm. Thus the boundary between boreal forest climate (11) and tundra climate (12) is the 35-cm isopleth of Ep. It is a line marking the northern limit of the boreal forests in North America and Eurasia.

Figure 9.14 is a schematic diagram showing the boundaries of the three climate groups in terms of total annual water need, Ep. The horizontal layering in the diagram expresses the availability of heat to plants in the life layer. The second climate ingredient, availability of soil water, is expressed by the vertical columns, defined in terms of soil-water shortage (D), soil-water storage (S), and water surplus (R). We turn next to the definitions of climate types in terms of availability of soil water.

DEFINING DRY AND MOIST CLIMATES

An essential step in designing a climate system based on the soil-water balance is to distinguish precisely a dry climate from a moist climate. A *dry climate* is one in which the total annual soil-water shortage, D, is 15 cm

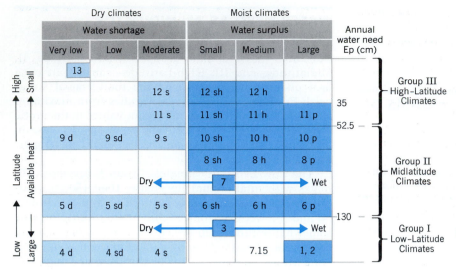

FIGURE 9.14 Schematic diagram of world climates in terms of water need and degree of dryness and wetness.

(5.9 in.) or larger, while at the same time there is no water surplus, R. A *moist climate* is one in which the soil-water shortage, D, is less than 15 cm. (There is no requirement that a moist climate must show a water surplus.) These definitions enable us to draw a map boundary between dry and moist climates. We will find that this boundary, the 15-cm isopleth of D, coincides in midlatitudes quite well with the boundary between steppe grasslands (short-grass prairie) and tall-grass prairie.

SUBDIVISIONS OF THE DRY CLIMATES

Within the dry climates, three degrees of dryness (aridity) are defined precisely in terms of monthly values of soil-water storage, S. Recall that the Thornthwaite system of calculating the soil-water budget assumes a soil storage capacity (field capacity) of 30 cm (about 12 in.). Values of soil-water storage, S, can thus range from 0 to 30 cm. Values of S are zero, or very close to zero,

throughout the year in dry desert locations. In very moist climates S runs at, or close to, 30 cm in all months.

Table 9.2 defines three levels of aridity within the dry climates; these form three climate subtypes described in Chapter 8.

Boundaries between each of the three subtypes are drawn on the world climate map, Figure 8.9, in accordance with these definitions. Areas falling within each subtype are assigned a different color in the range from light brown to pale yellow. Three climates fall in the dry category; these are distinguished on the map by their numbers (4, 5, and 9).

SUBDIVISIONS OF THE MOIST CLIMATES

The degree of wetness of the moist climates spans a wide range—from climates with no water surplus, R, to those with a very large water surplus. Our system uses three levels of wetness based on total annual water surplus, R, defining precisely the three climate subtypes already

TABLE 9.2 Definition of Climate Subtypes

Symbol	Name	Definition
s	Semiarid subtype (Steppe subtype)	At least two months in which soil-water storage, S, is equal to or exceeds 6 cm.
sd	Semidesert subtype	Fewer than two months with S greater than 6 cm, but at least one month with S greater than 2 cm.
d	Desert subtype	No month with S greater than 2 cm.
sh	Subhumid subtype	Soil-water shortage, D, is greater than 0 but less than 15 cm, when there is no water surplus, R. Otherwise, D is greater than R, where R is not 0.
h	Humid subtype	R is 1 mm or greater but less than 60 cm. (R is always greater than D.)
p	Perhumid subtype	R is 60 cm or greater.

described in Chapter 8. Definitions of subtypes are given in Table 9.2.

Boundaries separating each of the three moist subtypes are shown on the climate map, Figure 8.9. Each subtype is assigned a different color intensity, from light to dark. Note that these subtypes are recognized only for moist climates in midlatitude and high-latitude groups.

The schematic diagram, Figure 9.14, shows the subtypes of dry and moist climates, arranged from the greatest water shortage (left) to the greatest water surplus (right).

CLIMATES WITH STRONG MOISTURE SEASONS

Two of the 13 climate types cannot be described as either dry or moist by the definitions we have given. Instead, these two types have a very wet season alternating with a very dry season, with the result that they show both a substantial soil-water shortage, D, and a substantial soil-water surplus, R. In the low-latitude group, the wet-dry tropical climate (3) has D greater than 20 cm and R greater than 10 cm. In the humid subtype of the Mediterranean climate (7), both D and R exceed 15 cm. In the schematic diagram, Figure 9.14, these two wet-dry climates (3 and 7) are shown as alternating between dry and moist extremes.

SOIL-WATER CLIMATOLOGY AND HUMAN FOOD RESOURCES

Concepts of the soil-water balance must be applied to two great questions facing the human race: (1) Can the developing nations increase their food production fast enough to stave off starvation? (Answers given by well-informed specialists span the range from extreme pessimism to extreme optimism.) (2) Will there be enough fresh water to supply the rapidly increasing demands of the energy-consuming industrial nations? An understanding of current problems of agriculture and fresh-water supplies over the lands of the globe can be greatly illuminated by use of water-balance climatology.

Perhaps the most important practical lesson of this chapter and the next two is that few soil-water budgets provide the full water need of food plants during the growing season, with neither too great a shortage nor too great a surplus. Severe soil-water shortages require irrigation to achieve crop production. On the other hand, large year-round surpluses of water lead to removal of plant nutrients, so costly applications of fertilizers are needed. Irrigation in a dry climate has serious side effects, such as the accumulation of salts in the soil and the rise of ground water to saturate the soil. The soil-water budget sets stringent limitations on the expansion of agricultural resources. Intelligent global planning for environmental management depends heavily on an understanding of all phases of the soil-water balance.

LOW-LATITUDE CLIMATES

Climates of Group I, the low-latitude climates, lie for the most part between the tropics of cancer and capricorn, but with poleward extensions of their domain in North Africa and southwestern Asia reaching to latitudes as far north as about the 35th parallel. Similar but narrow poleward extensions of latitude occur in the southern hemisphere along east coasts windward to the trades. In terms of world latitude zones, defined in Figure 3.9, the low-latitude climates occupy all the equatorial zone (10° N to 10°S), most of the tropical zone (10–15° N and S), and part of the subtropical zone.

In terms of prevailing pressure and wind systems, the low-latitude climates occupy the equatorial belt of doldrums and intertropical convergence zone (ITC), the belt of tropical easterlies (northeast and southeast trades), and large portions of the oceanic subtropical highs.

As a group, the four low-latitude climates include types ranging from extremely dry to extremely moist, which is to say that the group includes climates with extremely large annual water deficits as well as those with extremely large water surpluses. Low-latitude climates span the range from a climate of extreme seasonality of precipitation—the wet-dry tropical climate—to one with heavy precipitation throughout the year—the wet equatorial climate. Thermal regimes are likewise quite varied, from the monotonous uniformity (equability) found in the wet equatorial climate to a strong annual range in the interior tropical deserts. Cycles of water need (potential evapotranspiration) are correspondingly varied in range. Perhaps the only climatic criterion shared by all the low-latitude climates is that the total annual water need is always greater than 130 cm.

Europeans who colonized the regions of low-latitude climates found themselves in very strange surroundings, difficult to cope with in terms of conventional social and agricultural practices with which they were familiar. Soils were unlike any they cultivated effectively in Europe and, in many areas, proved to be either infertile or difficult to till. Abundant rainfall that might have seemed at first to be a great asset to crop cultivation proved detrimental to that end. Extremes of heat proved difficult to bear and many strange, lethal diseases were endemic. Even today, native inhabitants of the Third World nations of the low-latitude climates have not yet learned to adapt their borrowed agricultural practices and economic systems to the unrelenting limitations of climate, soils, and native vegetation—limitations their ancestors understood and respected.

WET EQUATORIAL CLIMATE (1)

This climate of the intertropical convergence zone (ITC) is dominated by warm, moist maritime equatorial (mE) and maritime tropical (mT) air masses yielding heavy convectional rainfall. Rainfall is copious in all months, annual total often being over 250 cm (100 in.). Strong seasonal differences in monthly rainfall are typical and can be attributed to changes in position of the ITC and to local orographic effects. Remarkably uniform temper-

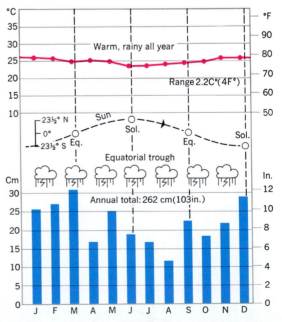

FIGURE 10.1 Iquitos, Peru, is a rainforest station near the equator. Temperatures differ very little from month to month, and there is copious rainfall throughout the year.

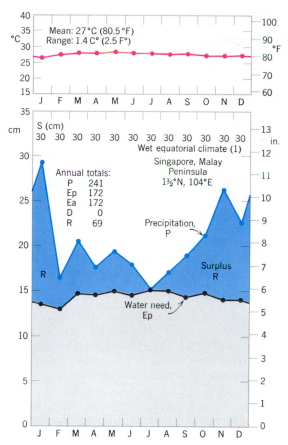

FIGURE 10.2 Soil-water budget for Singapore, Malay Peninsula, lat. 1½°N. (Data from C. W. Thornthwaite Associates, Laboratory of Climatology, Centerton, N.J.)

atures prevail throughout the year. Both monthly mean and mean annual temperatures are typically close to 27°C (80°F).

Latitude Range: 10°N to 10°S.

Major Regions of Occurrence: Amazon lowland of South America; Congo Basin of equatorial Africa; and East Indies, from Sumatra to New Guinea.

Example: Figure 10.1 is a climograph for Iquitos, Peru, a typical wet equatorial station located close to the equator in the broad, low basin of the upper Amazon River. Note that the annual range in temperature is only 2.2C°

(4F°) and that the annual rainfall total is just over 250 cm (100 in.). In all but one month, the mean monthly rainfall is more than 15 cm (6 in.).

Soil-Water Budget: Precipitation, P, is heavy throughout the entire year. In nearly all months, P exceeds water need, Ep. As a result, there is a substantial water surplus, R, in all or nearly all months. Soil-water storage, S, is large in all months and is usually over 25 cm in 10 or more months. For many stations, S is 30 cm in all 12 months.

Example: Singapore, Malay Peninsula, lat. 1½°N (Figure 10.2). Precipitation, P, at Singapore is large in every month, the least being 15 cm, the greatest 29 cm. The annual total precipitation is very great: 240 cm. Ep is highly uniform in monthly values, which range from 13 cm to 15 cm. The yearly total Ep is thus very great: 172 cm. P is much greater than Ep in all months but one (July), so 11 months show a water surplus. Total annual R is nearly 70 cm. There is no soil-water shortage, D, in any month. Moreover, soil-water storage, S, stands at the maximum value of 30 cm in every month. It is obvious that optimum conditions for plant growth exist year-round at Singapore.

Daily Range in Air Temperature

A remarkable feature of the wet equatorial climate is the extreme uniformity of monthly mean air temperature, which is typically between 26° and 29°C (79° and 84°F) for stations at low elevation in the equatorial zone. The equatorial thermal regime is illustrated by graphs of minimum and maximum daily temperatures for two months in the city of Panama, lat. 9° N (Figure 10.3). The selected months—July and February—represent the extremes of the yearly temperature cycle, yet the means differ by only 0.16C° (0.3F°). The daily temperature range, in contrast, is usually from 8 to 11C° (15 to 30F°). Thus the daily range greatly exceeds the annual range. Geographers have a saying about this: "Night is the winter of the tropics."

MONSOON AND TRADE-WIND LITTORAL CLIMATE (2)

Trade winds (tropical easterlies) bring maritime tropical (mT) air masses from moist western sides of oceanic high-pressure cells to give heavy orographic rainfall

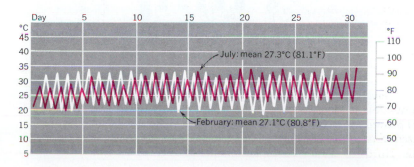

FIGURE 10.3 July and February temperatures at Panama City, lat. 9°N. The sawtooth graph shows daily maximum and minimum readings for each day of the month. (Data from Mark Jefferson.)

along narrow eastern littorals. (*Littoral* means a "narrow coastal zone.") Shower activity is intensified by arrivals of easterly waves. Rainfall shows a strong annual cycle, peaking in the season of high sun, when the ITC is close by. A marked short season of reduced rainfall occurs at time of low sun. In Southeast Asia, the rainy period represents the summer monsoon season. Temperatures are very warm throughout the year, but with a marked annual cycle. Minimum temperatures occur at time of low sun.

Latitude Range: 5° to 25°N and S.

Major Regions of Occurrence: Trade-wind littorals occur along the east sides of Central and South America, the Caribbean Islands, Madagascar (Malagasy), Indochina, the Philippines, and northeast Australia. Monsoon west coasts are found in western India (Malabar Coast) and Burma. Large inland extensions of the monsoon climate occur in Bangladesh and Assam, central and western Africa, and southern Brazil.

Example: Figure 10.4 is a climograph for Belize City, Belize, on the Central American coast at lat. 17°N, exposed to the tropical easterlies. Rainfall is copious from June through November, when the ITC is in this latitude zone. Easterly waves are common in this season, bringing torrential rainfall. Following the winter solstice (end of December), rainfall is greatly reduced, with minimum values in March and April. At this time, the ITC lies farthest away and the climate is dominated by the subtropical high-pressure cell. Air temperatures show an annual range of 5C° (9F°), with maximum in the high-sun months.

Asiatic Monsoon Variety

A special monsoon variety or subtype of the monsoon and trade-wind littoral climate is associated with the Asiatic monsoon system. It features an extreme peak of rainfall during the high-sun period and a well-developed dry season with two or three months of only small rainfall amounts.

Example: Figure 10.5 is a climograph for Cochin, India, lat. 10°N. Located on the west coast of lower peninsular India, Cochin occupies a windward location with respect to the southwest winds of the rainy (summer) monsoon season. In the rainy season, monthly rainfall is extremely high—over 64 cm (25 in.) per month in June and July. A strongly pronounced season of low rainfall occurs at time of low sun—December through March. Air temperatures show only a very weak annual cycle, cooling a bit during the rains; but the annual range is small at this low latitude.

Soil-Water Budget: A small soil-water shortage, D, develops in the short dry season. Water need, Ep, exceeds 4 cm in every month or the yearly total Ep is more than 130 cm, or both. A large water surplus, R, is generated in the rainy months. Soil-water storage, S, is greater than 20 cm through 6 to 9 consecutive months.

Example: Aparri, Luzon Island, Philippine Islands, lat. 18°N (Figure 10.6). This trade-wind littoral station shows

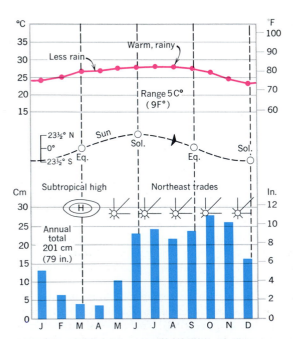

FIGURE 10.4 This climograph for Belize City, a Central American east-coast city at lat. 17°N, shows a marked season of low rainfall following the period of low sun.

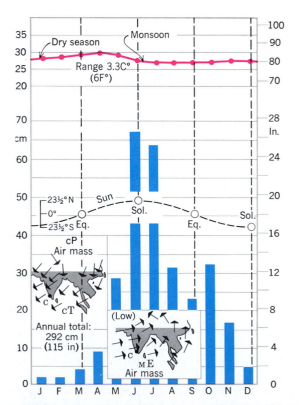

FIGURE 10.5 Cochin, India, on a windward coast at lat. 10°N, shows an extreme peak of rainfall during the rainy monsoon, contrasting with a short dry season at time of low sun.

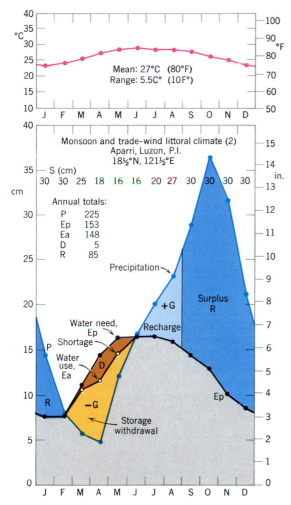

FIGURE 10.6 Soil-water budget for Aparri, Luzon Island, Philippine Islands, lat. 18°N. (Data from C. W. Thornthwaite Associates, Laboratory of Climatology, Centerton, N.J.)

FIGURE 10.7 Unsurpassed as an agricultural wonder is this rice terrace system at Banaue in northern Luzon, Philippine Islands. The bright green patches are young rice plants, ready to be uprooted and transplanted in rows in the flooded rice paddies. These remarkable terraces have been maintained for centuries. (S. Vidler/Leo de Wys, Inc.)

a strong annual precipitation cycle, peaking strongly following the high–sun period in a wet season, with P greater than 25 cm in 3 consecutive months. Following the period of low sun, when the ITC is in the southern hemisphere, a period of greatly reduced rainfall develops, with P falling below 10 cm in 3 consecutive months. The annual cycle of soil-water need, Ep, also has a strong annual cycle, but the maximum values occur from May through August, which is well in advance of the rainfall peak. As a result, a very small soil-water shortage, D, is developed through May, June, and July. When P rises to exceed Ep, recharge begins. By September a large surplus is developed. The total water surplus, R, is very large: 85 cm. Soil-water storage, S, remains at 30 cm for 6 consecutive months (September through February). The lowest monthly values of S, about 16 cm, occur in May and June, which is the brief water-shortage period. Thus, despite a short dry season, soil water remains ample for plant growth. Rainforest thrives in this environment. Stream flow, which reaches flood proportions in the wet season, diminishes greatly during the brief dry season. The climate here is ideal for rice, which needs very wet conditions for growth, as well as a dry harvest season (Figure 10.7).

The Rainforest Environment

Our first two climates—wet equatorial (1) and monsoon and trade-wind littoral (2)—have in common certain environmental properties of great interest and importance to geographers. First is the extreme uniformity of monthly mean air temperature. Second is the combination of a very large annual total rainfall, a great annual water surplus, and high soil-water storage throughout the year. Both of these factors create a special environment for the development of a unique combination of soil type and native vegetation. We can call this the *low-latitude rainforest environment.*

Streams flow copiously throughout most of the year and river channels are lined along the banks with dense forest vegetation (Figure 10.8). Natives of the rainforest traveled the rivers in dugout canoes, and today they enjoy easier travel by attaching an outboard motor. Over a century ago, larger shallow-draft river craft turned the

FIGURE 10.8 Rainforest of the western Amazon lowland, northwest of Iquitos, Peru. A petroleum exploration well is being drilled from a barge, floated 3700 km (2300 mi) up the Amazon River and its tributary, seen here, the Cuinico River. (Phillips Petroleum Company.)

major waterways into the main arteries of trade; towns and a few cities were built on the river banks. Aircraft have added a new dimension in mobility, crisscrossing the almost trackless green sea of forest to find landings in clearings or on broad reaches of the larger rivers.

In the low-latitude rainforest environment, the large water surplus and prevailing warm soil temperatures promote the decay and decomposition of rock to great depths, so that a thick soil layer is usually present. The soil is typically rich in oxides of iron and has a deep red color; it belongs to a soil class called the Oxisols (Chapter 24). This kind of soil has largely lost its ability to hold nutrient substances (ions) needed by plants such as grasses and grain crops important in modern agriculture practiced in higher latitudes. There are, however, many kinds of plants capable of surviving on Oxisols. Most conspicuous of these are the great forest trees with broad leaves that do not undergo a seasonal shedding; these comprise the *broadleaf evergreen rainforest* (Figure 10.8). The key to the success of this kind of forest lies in the ability of the trees to quickly reuse (recycle) the essential plant nutrients that are released by the decay of fallen leaves and branches.

A major difference between the low-latitude rainforest and forests of higher latitudes is the great diversity of species that it possesses. In the low-latitude rainforest one can find as many as 3000 different tree species in an area of only a few square kilometers—midlatitude forests would possess perhaps only one-tenth that number.

The animal assemblage, or fauna, of the rainforest is also very rich. A 16-km² (6-mi²) area in the Canal Zone, for example, contains about 20,000 species of insects, whereas there are only a few hundred in all of France. This very large number of species occurs because the rainforest environment is so uniform and free of physical stress.

Animal life of the rainforest is most abundant in the upper layers of the rainforest. Above the canopy, birds and bats are important carnivores and feed on insects above and within the topmost leaf canopy. Below this level live a wide variety of birds, mammals, reptiles, and invertebrates. These animals feed on the leaves, fruit,

and nectar abundantly available in the main part of the canopy. Ranging between the canopy and the ground are small climbing animals—monkeys, for example—that forage in both layers. At the surface are the large ground animals, including herbivores that graze the low leaves and fallen fruits, and carnivores that prey on the abundant animals.

Plant Products and Food Resources of the Rainforest

Several forest products are of economic value. Rainforest lumber, such as mahogany, ebony, or balsawood, is an important export. Quinine, cocaine, and other drugs come from the bark and leaves of tropical plants; cocoa comes from the seed kernel of the cacao plant. Natural rubber is made from the sap of the rubber tree. The tree comes from South America, where it was first exploited. Rubber trees also are widely distributed through the rainforest of Africa. Today, the principal production is from plantations in Indonesia, Malaysia, Thailand, Vietnam, and Sri Lanka (Ceylon).

An important class of food plants native to the wet low-latitude environment are starchy staples; some are root structures and others are fruits. Manioc, also known as cassava, is one of these staples. The plant has a tuberous root—something like a sweet potato—that reaches a length over 0.3 m (1 ft) and may weigh several kilograms (Figure 10.9). It is cultivated in small plantations placed in forest clearings. When properly prepared to remove a poisonous cyanide compound, the manioc roots yield a starchy food with very little (1% or less) protein, a factor that contributes to malnutrition. Manioc is used as a food in the Amazon basin of Brazil; it was taken to Africa in the sixteenth century by the Portuguese. In the present century, its use in Africa has increased widely. Manioc is now also important in Indonesia.

Yams are another starchy staple of the wet low-latitude regions. Like the manioc, the yam is a large underground tuber. Yams are a major source of food in West Africa. The plant was introduced into the Caribbean region during the era of slavery and is an important

FIGURE 10.9 Tuberous manioc roots are soaked in a river after being peeled. The roots may be prepared for eating by roasting or boiling, or they may be used to make a coarse meal. This scene is in the Amazon basin of Brazil, where manioc is an important starchy staple. (Francois Gohier/Photo Researchers.)

food in the region today. The yam has a higher protein content than the manioc.

Taro is another starchy staple of wet low-latitude climates. The taro plant has large leaves, which are edible, but the food value (largely starch) lies mostly in an enlarged underground portion of the plant, called a "corm." Visitors to Hawaii know the taro plant through its transformation into poi, a fermented paste. Few mainland tourists eat poi a second time, but it has long been a favored food of native Hawaiians. Taro was imported into Africa from Southeast Asia and eventually reached the Caribbean region.

The banana and plantain are starchy staples familiar to everyone. The banana plant lacks woody tissue and is, in fact, a perennial herb; the fruit is classed as a berry by botanists. The banana was first cultivated for food in Southeast Asia, then spread to Africa. It was imported into the Americas in the sixteenth century and quickly became well established. The plantain is a coarse variety of the banana that is starchy, has little sugar, and requires cooking.

Perhaps the plant most important to humans in the low latitudes has been the coconut palm. Besides being a staple food, it provides a multitude of useful products in the form of fiber and structural materials. The coconut palm flourishes on islands and coastal fringes of the wet equatorial and tropical climates (Figure 10.10). Copra, the dried meat of the coconut, is a valuable source of vegetable oil. Copra and coconut oil are major products of Indonesia, the Philippines, and New Guinea. Palm oil and palm kernels of other palm species are important products of the equatorial zone of West Africa and the Congo River basin.

In Chapter 27, we present a special section on exploitation of the low-latitude rainforests, contrasting the primitive practice of shifting cultivation of small forest plots with the wholesale destruction of the rainforests in progress today over large areas in Central and South America, Africa, and the East Indies.

WET-DRY TROPICAL CLIMATE (3)

Seasonal alternation occurs between dominance of moist mT or mE air masses along the ITC and dry continental tropical (cT) air masses of the subtropical high-pressure belt. As a result, there is a very wet season at times of high sun and a very dry season at times of low sun. Cooler temperatures accompany the dry season, but give way to a very hot period before the rains begin.

FIGURE 10.10 A coastal grove of coconut palms, City of Refuge, Kona Coast, Island of Hawaii. (A. N. Strahler.)

Latitude Range: 5° to 20°N and S (Asia: 10° to 30°N).

Major Regions of Occurrence: India; Indochina; West Africa; southern Africa; South America, both north and south of the Amazon lowland; and north coast of Australia.

Example: Figure 10.11 is a climograph for Timbo, Guinea, at lat. 10½°N in West Africa. The rainy season begins just after vernal equinox, reaching a peak in July and August, or about 2 months following summer solstice. At this time, the ITC has migrated to its most northerly position and large infusions of moist mE air masses occur from the ocean lying to the south. Rainfall then declines as the low-sun season arrives. Three months—December through February—are practically rainless. At this season the subtropical high-pressure cell dominates the climate, with a stable subsiding cT air mass over the area. The temperature cycle is linked closely with precipitation. In February and March air temperature rises sharply, bringing on a brief hot season. As soon as the rains set in, the effect of a cloud cover and evaporation of rain causes the temperature to decline. By July temperatures have resumed an even level at about 22°C (72°F).

Example: Figure 10.12 is a climograph for Allahabad, India, showing the wet-dry tropical climate with a strong Asiatic monsoon influence. Notice that this city lies at a much higher latitude (25°N) than Timbo, Guinea (10½°N), or about 1450 km (900 mi) farther north. The onset of the rainy season at Allahabad is much more sudden than at Timbo. The annual range of temperature at Allahabad, about 18C° (32F°), is much greater than at

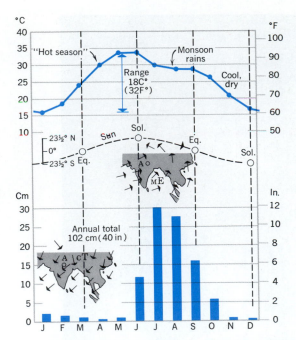

FIGURE 10.12 Allahabad, India, lat. 25°N, lies in the rich agricultural lowland of the Ganges River. Here the monsoon rains are usually ample, but there is a long dry season. Note the very high temperatures in May and June, just before the onset of rains.

Timbo. The hot season at Allahabad reaches extreme values—over 32°C (90°F)—and there is a marked cool season at time of low sun, when monthly mean temperature falls to 18°C (65°F).

Soil-Water Budget: Water need, Ep, exceeds 4 cm in every month or the yearly total Ep is more than 130 cm, or both. A substantial water surplus, R, occurs in the wet season; a substantial soil-water shortage, D, occurs in the dry season. R is 10 cm or greater; D is 20 cm or greater. Soil-water storage, S, exceeds 20 cm in 5 months or fewer.

Example: Raipur, Madhya Pradesh, India, lat 21°N (Figure 10.13). Dominated by the Asiatic monsoon system, Raipur shows a very wet high-sun season with P over 20 cm in 4 consecutive months (June through September), which together account for 117 cm, or 87 percent, of the total annual rainfall. During the long, cool dry season, 7 consecutive months have P less than 3 cm. The annual cycle of water need, Ep, is strongly developed and the annual total Ep, 152 cm, is very large. But because Ep rises to high values while P is very low in March, April, and May, a large soil-water shortage, D, is accumulated: 43 cm. This shortage period is also the hot season. Storage recharge, +G, takes place in June and July, so by August a substantial water surplus, R, is developed. The annual surplus, R, however, is only 25 cm, an amount about half as much as the shortage, D. This ratio of D to R (about 2 to 1) is typical of the climate. Soil-water storage, S, is greater than 20 cm in 5 consecutive months (July through November) and attains 30 cm

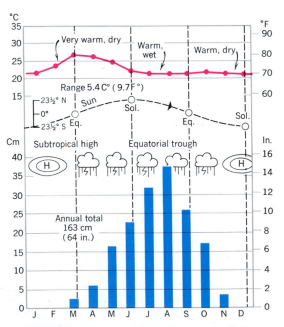

FIGURE 10.11 Timbo, Guinea, at lat. 10½°N, is in the wet-dry tropical climate of West Africa. A long, wet season at time of high sun alternates with an almost rainless dry season at time of low sun.

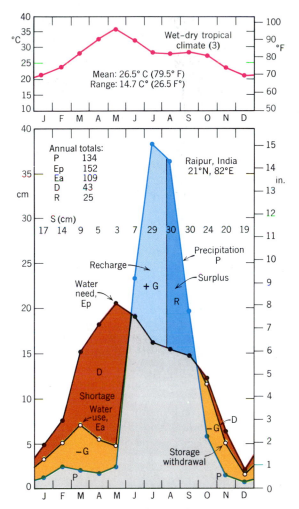

FIGURE 10.13 Soil-water budget for Raipur, Madhya Pradesh, India, lat. 21°N. (Data from C. W. Thornthwaite Associates, Laboratory of Climatology, Centerton, N.J.)

FIGURE 10.14 Savanna woodland of the Serengeti Plain, Tanzania, East Africa. Acacia trees with flattened crowns remain green, although the coarse grasses have turned to straw in the dry season. (H. Barad/Photo Researchers, Inc.)

(storage capacity) in August and September. The period of high values of S is the season of rapid plant growth. The long drought period requires that native plants also be adapted to very low soil-water levels. Grasses, deciduous trees, and certain shrubs can survive this cycle of extreme conditions.

The Savanna Environment

Because of the great sweep of the annual alternations of wet and dry seasons in this climate, the native vegetation of the wet-dry tropical climate is characterized by plants capable of surviving through several consecutive months of drought, then bursting into leaf and bloom to grow rapidly in a rainy season that comes with the high-sun period of the year. For that reason the native plant cover can be described as *rain-green vegetation*, a name that covers a wide range in terms of the kinds of plants present and the ways in which they are distributed over the land surface.

Rain-green vegetation consists of two basic types. First is *savanna woodland*. By "woodland" we mean an open forest in which trees are widely spaced apart. In the savanna woodland, coarse grasses occupy the open space between the trees, which are typically coarse-barked and may bear large thorns (Figure 10.14). Large expanses of open grassland may also be present. The grasses turn to straw in the dry season, and many of the tree species shed their leaves to cope with the drought. Second, in the more arid parts of the climate region, small thorny trees and large shrubs form dense patches; this is referred to as thorntree–tall-grass savanna. (Further details are given in Chapter 27.) Because of the prevalence of the savanna vegetation in the tropical wet-dry climate, it can be identified as the *savanna environment* when we wish to refer to the unique package of features of climate, soils, and vegetation that it presents.

In the savanna environment, most river channels are nearly or completely dry in the low-sun dry season. (An exception would be rivers fed from moist mountain regions.) In the rainy season, these river channels become filled to their banks with swiftly flowing, turbid water. The rains are not reliable and agriculture without irrigation is hazardous at best. When the rains fail, a devastating famine can ensue.

Soils of the savanna environment are similar in their physical characteristics and fertility to those of the rain-forest environment, but this statement would apply only when we compare soils that occupy well-drained upland surfaces. Oxisols can be found in this environment along with a related soil group that shares the low fertility and red color of the Oxisols. In contrast, substantial areas of the savanna environment have fertile soils developed and sustained by the slow infall of windblown dusts from adjacent deserts. Equally important are highly fertile alluvial soils of major rivers that flow through the regions of tropical wet-dry climate. Annual flooding of these rivers leaves deposits of fertile silt carried down from distant mountain ranges.

Animal Life of the African Savanna

Closely adapted to the vegetation and climate is the natural animal life of the savanna grasslands and woodlands. These are the regions of the carnivorous game animals and a vast multitude of grazing animals on which they feed (Figure 10.15). The savanna of Africa is the natural home of herbivores, such as wildebeest, gazelle, deer, antelope, buffalo, rhinoceros, zebra, giraffe, and elephant. Their predators are the lion, leopard, hyena, wild dog, and jackal. Some of the herbivores depend on fleetness of foot to escape the predators. Others, such as the rhinoceros, buffalo, and elephant, defend themselves by their size, strength, or armor-thick hide. The giraffe is a peculiar adaptation to savanna woodlands; its long neck permits browsing on the higher foliage of scattered trees.

The dry season brings a severe struggle for existence to animals of the African savanna. As streams and hollows dry up, the few muddy waterholes must supply all drinking water. Danger of attack by carnivores is greatly increased.

The savanna ecosystem in Africa faces the prospect of widespread destruction. Parks set aside to preserve the ecosystem confine the grazing animals to a narrow range and prevent their seasonal migrations in search of food. In some instances the growth in animal populations has been phenomenal because they have been protected from hunting. In confined preserves they rapidly consume all available vegetation in futile attempts to survive. Rapidly growing human populations are bringing increasing pressure on management agencies to allow encroachment on game preserves in order to expand cattle grazing and agriculture. Poaching on a large scale is now threatening the extinction of the elephant and rhinoceros in many areas.

Agricultural Resources of the Savanna Environment

In terms of agricultural resources and practices, no one sweeping statement can apply meaningfully to the entire savanna environment wherever it occurs. Instead, we achieve a truer picture by considering Southeast Asia separately from Africa and other world areas of wet-dry tropical climate. The Asiatic monsoon system imparts a special quality to agricultural patterns of Southeast Asia. There are, moreover, vastly larger human populations in the Asiatic monsoon nations than elsewhere in low latitudes, and these peoples have evolved their own unique culture patterns.

Of the staple foods grown widely in Southeast Asia, rice is dominant. About one-third of the human race subsists on rice, and most of this vast population is crowded into the arable lands of Southeast Asia. Intensive rice cultivation of this part of the world actually spans three climate types: monsoon and trade-wind littoral climate (2), wet-dry tropical climate (3), and moist subtropical climate (6). Rice requires flooding of the ground at the time the seedling plants are cultivated, and this activity has been traditionally timed to coincide with the peak of the rainy monsoon season (Figure 10.7). The crop matures and is harvested in the dry season. Sugarcane is another important crop that grows rapidly during the rainy season and is harvested in the dry season.

Sorghum, also called kaffir corn or guinea corn, is an important food crop of the savanna environment in Africa and India. It is a grain capable of survival under conditions of a short wet season and a long, hot dry season. The peanut is another major food crop of the savanna environment in India, and it has been introduced into a corresponding climate zone of West Africa.

Food and Woodland Resources of the African Savanna

Because the savanna environment in Africa is a transition zone between rainforest and desert environments, there is a corresponding gradation of plant resources and the ways in which they are used to support human life. In the moister zones, those of the savanna woodland with a long wet season, agriculture follows a pattern known as *bush-fallow farming*. This practice is related in some

FIGURE 10.15 Vast herds of wildebeest and zebra graze the lush vegetation of the Serengeti Plain in Tanzania, still green from the rains. These animals migrate long distances in search of food and water. (M. P. Kahl/Photo Researchers.)

FIGURE 10.16 Masai herdsmen in southwestern Kenya, East Africa. Cattle represent a form of wealth and prestige; they also supply food in the form of milk and blood. Quality of these animals is secondary in importance to their numbers. (Marvin E. Newman/Woodfin Camp.)

ways to the slash-and-burn agriculture of the rainforest. Trees are cut from a small area, piled up, and burned. The ash provides fertilizer for cultivated plots. After a few years, the land is allowed to revert to shrub and tree growth. Some of the same crops that are grown in the rainforest agriculture are grown in wetter areas of the bordering savanna. Crops include grains, yams, soybeans, and sugarcane. Tobacco and cotton are also grown.

In the drier savanna grassland and thorntree-savanna, where the wet season is short, the main subsistence crops of uplands are sorghum, millet (a kind of grain), peanuts, and corn. Cotton, along with peanuts, is an important cash crop for export. Besides the agricultural system of the permanent farmers, there exists a shifting cattle culture, in which large numbers of cattle are maintained as a display of wealth (Figure 10.16). The cattle provide food in the form of milk, butter, and blood. Following a nomadic pattern, the cattle are moved into the semidesert zone in the rainy season to graze on grasses, then returned to the savanna grassland zone in the dry season, where they graze on fallow croplands and rely on water holes for survival.

The savanna woodland belt of Africa provides a number of other plant resources besides cultivated food crops. Trees are cut for firewood, which is the only fuel for cooking available to most inhabitants. Trees also provide construction poles for dwellings. Among the savanna woodland trees that furnish important export products are the cashew-nut tree and the kapok tree. Gum arabic is taken from acacia trees. Most native trees of the savanna woodland have little commercial value as export lumber (an exception is black teak), but plantations of exotic tree species have been introduced in some areas. Both pine and eucalyptus have been successfully introduced as sources of pulpwood. These trees grow very rapidly, as compared with pulpwood trees in high latitudes.

DRY TROPICAL CLIMATE (4)

The dry tropical climate occupies source regions of the cT air mass in high-pressure cells centered over the tropics of cancer and capricorn. This subsiding air mass is stable and dry, becoming highly heated at the surface under intense insolation. A strong annual temperature cycle follows the changing declination of the sun. The high-sun period brings extreme heat; the low-sun season, a comparatively cool season. Extremely dry areas, recognized as desert subtype (4d), are over the tropics of cancer and capricorn. This dry zone grades on the equatorward side through a narrow zone of semidesert subtype (4sd), into a semiarid, or steppe climate subtype (4s). In this transition zone, a short rainy season occurs, grading into the long rainy season of the wet-dry tropical climate (3). A narrow western littoral zone (4dw) has a cool, uniform thermal regime because of the presence of a cool marine air layer.

Latitude Range: 15° to 25°N and S.

Major Regions of Occurrence: Sahara–Arabia–Iran–Thar desert belt of North Africa and southern Asia; a large part of Australia; small areas in Central America; South America; and South Africa. Important areas of the steppe subtype (4s) are found in India and Thailand, with many small, scattered dry areas to the lee of highlands in the trade-wind belt.

Examples Figure 10.17 is a climograph for a tropical desert (4d) station in the heart of the North African desert. Wadi Halfa, Sudan, lies at lat. 22°N, almost on the tropic of cancer. The temperature record shows a strong annual cycle with a very hot period at the time of high sun, when 3 consecutive months average 32°C (90°F). Daytime maximum air temperatures are frequently between 43 and 48°C (110 to 120°F) in the warmer months. There is a comparatively cool season at time of low sun, but the coolest month averages a mild 16°C (60°F) and freezing temperatures are very rarely recorded. Precipi-

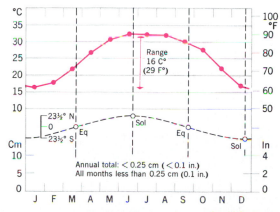

FIGURE 10.17 Wadi Halfa is a city on the Nile River in Sudan at lat. 22°N, close to the Egyptian border. Too little rain falls to be shown on the graph. Air temperatures are very high during the high-sun months.

tation averages less than 0.25 cm (0.1 in.) in all months; over a 39-year period the maximum rainfall recorded in a 24-hour period was only 0.75 cm (0.3 in.).

Another example of a station in the tropical desert (4d) of North Africa is BouBernous, Algeria, at lat. 27½° N, referred to in Chapter 4. Figure 4.9 shows details of the monthly temperature cycle and includes the extreme values of the record. Notice that the annual range of the mean of daily means is substantially greater than that for Wadi Halfa, a result of the high-sun months being hotter than at Wadi Halfa. On rare occasions in January, temperatures fall close to the freezing mark.

Soil-Water Budget: Annual total water need, Ep, exceeds 130 cm or all monthly values of Ep exceed 4 cm, or both. Large areas of North Africa between lat. 10 and 20°N have total Ep between 160 and 180 cm, with all monthly values of Ep over 10 cm.

Example: Khartoum, Sudan, lat. 15½°N (Figure 10.18). Khartoum lies in the heart of an enormous area of tropical desert climate (4d) that spans North Africa and the Arabian Peninsula. The total annual soil-water need, Ep, is extremely large: 183 cm. The annual cycle of Ep is strongly developed, while all monthly values of Ep exceed 7 cm. Monthly values in excess of 15 cm occur in 7 successive months, during which time very high air temperatures prevail. Precipitation, P, shows a sharp peak in July and August, with more than 6 cm in each of those months. Total annual P is, however, only 18 cm. Note that water use, Ea, is identical with P in every month. Every month shows a substantial soil-water shortage, D. The annual total of D is very large: 165 cm.

Soil-water storage, S, is zero in all months. Few plants can survive in this extremely dry, hot environment; those that can are adapted to profit from rains in July and August, which briefly moisten the uppermost soil layer.

Western Littoral Subtype

Along the western coasts of South America and southern Africa, in a narrow coastal strip, we find the western littoral subtype (4dw) well developed. Here, the cool Humboldt and Benguela currents flow close to the shore-line (Figure 5.28). Upwelling of cold, deep water is occurring within these currents, chilling the lower air layer. This condition explains both coolness and uniformity of air temperatures. Coastal fog is a persistent feature of this littoral climate. Figure 10.19 shows a stretch of this coastal desert in northern Chile (the Atacama Desert).

Example: Figure 10.20 shows the desert climate typical of narrow western littorals (4dw). The station is Walvis Bay, a port city on the west coast of Namibia, at lat. 23°S. (The yearly cycle begins with July because this is a southern hemisphere station.) For a location nearly on the tropic of capricorn (23½°S), the monthly temperatures are remarkably cool; the warmest month mean is only 19°C (67°F). The coolest month is 14°C (57°F), making an annual range of only 5C° (10F°).

FIGURE 10.19 The desert coast of northern Chile. Known as the Atacama Desert, this barren coastal strip between the Andes Mountains and the Pacific Ocean comes close to being a truly rainless zone. (George Gerster/Comstock.)

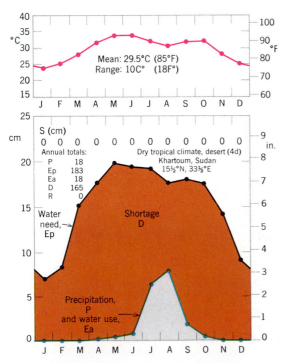

FIGURE 10.18 Soil-water budget for Khartoum, Sudan, lat. 15½°N. (Data from C. W. Thornthwaite Associates, Laboratory of Climatology, Centerton, N.J.)

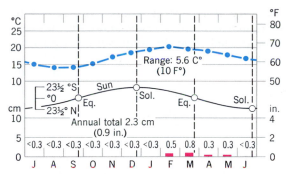

FIGURE 10.20 Walvis Bay, Namibia, is a desert station on the west coast of southwest Africa at lat. 23°S. Air temperatures are cool and remarkably uniform throughout the year.

Semiarid Subtype (Tropical Steppe)

The semiarid, or steppe subtype of the dry subtropical climate (4s) forms part of the transition zone between the dry desert and adjoining areas of the wet-dry tropical climate (3). A short wet season, starting at the time of solstice (June in northern hemisphere), brings a moderate amount of rainfall. The long dry season is nearly rainless for four or more consecutive months. Extremely high temperatures characterize the hot season that precedes the rains.

Example: Figure 10.21 is a climograph for Kayes, Mali, located at lat. 14½°N in West Africa. This station lies in the tropical semiarid or steppe subtype (4s) and is in the transition zone between the dry Sahara Desert to the north and the wet-dry tropical climate (3) immediately to the south (see climograph of Timbo, Figure 10.11). The

short wet season brings a total annual rainfall of 75 cm (29 in.).

Soil-Water Budget: The soil-water budget of the semiarid, or steppe subtype differs from that of the desert subtype in one particular feature. Precipitation (P) briefly exceeds water need (Ep) during the short rainy season. This results in a small surplus (+G) that goes to recharge the soil water, but it is not sufficient to generate the surplus that is typical of the adjacent wet-dry tropical climate (3).

Example: Fort Lamy, Chad, lat. 12°N (Figure 10.22). Fort Lamy lies in a narrow belt of tropical steppe (4s) climate subtype crossing North Africa from west to east; it lies immediately north of the belt of wet-dry tropical climate (3). Total annual precipitation, P, is 62 cm; it occurs mostly in a short rainy season that peaks sharply in August, with a monthly mean of about 25 cm. The total water need, Ep, is very high (175 cm) and shows an annual cycle similar to that of Khartoum, but with a sharp dip in August, coinciding with the brief period of rain. Because P exceeds Ep in August by a substantial margin, soil-water recharge, +G, occurs in August in the amount of about 11 cm. The remainder of the year sees small monthly withdrawals from soil-water storage. Soil-water shortage, D, is heavy throughout most months, with a large annual total of 113 cm. No water surplus, R, is developed. Soil-water storage, S, exceeds 6 cm in 2 consecutive months (August and September). This criterion establishes the climate subtype as semiarid, or steppe (4s).

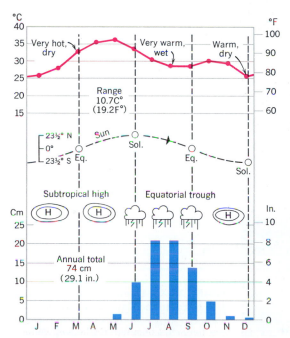

FIGURE 10.21 Kayes, Mali, at lat. 14½°N, lies in the Sahel of West Africa. There is a short wet season in normal years. The dry season is long, with a succession of rainless months.

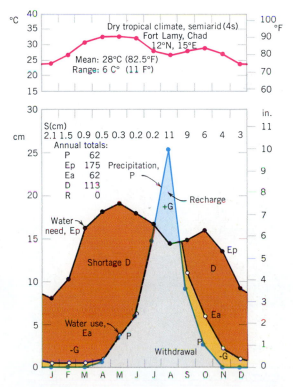

FIGURE 10.22 Soil-water budget for Fort Lamy, Chad, lat. 12°N. (Data from C. W. Thornthwaite Associates, Laboratory of Climatology, Centerton, N.J.)

The Tropical Desert Environment

The tropical deserts and their bordering semiarid zones comprise a global environmental region sustained by subsiding air masses of the continental high-pressure cells.

Because desert rainfall is unreliable in its occurrence, even at those seasons when it is most likely to occur, river channels and the beds of smaller streams are dry most of the time. However, a sudden and intense desert downpour can cause local flooding of brief duration that transports large amounts of silt, sand, gravel, and boulders. These events can be called "flash floods." Major river channels often end in flat-floored basins having no outlet. Here clay and silt are deposited and accumulate, along with layers of soluble salts. Shallow salt lakes occupy some of these basins. The action of desert streams and the deposition of their salts by evaporation are covered in some detail in Chapter 18.

Large expanses of the very dry low-latitude desert appear to be entirely devoid of plant life except, perhaps, along the banks of some of the larger dry stream channels. Large barren areas bear a cover of loosely fitted rock particles (sometimes called "desert pavement"), whereas loose drifting sand completely mantles other large areas. These kinds of surfaces are described in Chapter 21.

Plants that are capable of survival in the true desert (4d) are able to do so because they are quick to take advantage of a rare rainfall that may arrive only once in several years. Many of these plants are annuals. Some kinds of plants survive because they are hard-leaved or spiny shrubs equipped to resist water loss by transpiration. Other kinds are succulent plants—the cacti, for example—that store water in their spongy tissues. The tamarisk, a shrub or small tree of the Sahara Desert, thrives along the dry beds of major watercourses by sending its roots deep into the sand and gravel beneath to reach stored water that accumulated during a river flood.

Native vegetation of the tropical deserts can be divided into two general classes. First is the *dry desert*, described in the foregoing paragraphs and associated with the dry desert climate subtype (4d). Second is *semidesert*, associated with the semiarid climate subtype (4sd). A particularly important occurrence is the *thorntree semidesert* of Africa, characterized by thorny trees and shrubs that shed their leaves for the long dry season. Figure 10.23 shows this thorntree vegetation in the Kalahari Desert of southern Africa. Few, if any, native Bushmen survive in this natural setting today.

RAINFALL VARIABILITY OF THE LOW-LATITUDE CLIMATES

One important criterion by means of which we can set apart the low-latitude climates into three basic classes is the degree of year-to-year reliability of rainfall, or *rainfall variability*. Our three classes would be as follows:

Low variability (high reliability), as in the wet equatorial climate (1).
Moderate variability (moderate reliability), as in monsoon and wet-dry tropical climates (2 and 3).
High variability (low reliability), as in the dry tropical climate (4).

The concept of variability of precipitation is one of great basic importance in physical geography because it enables us to understand more clearly the hazards humans face in raising the food crops on which their lives depend.

FIGURE 10.23 Thorntree semidesert of Botswana, in the Kalahari region of southern Africa. In this dry-season scene, the larger tree has shed its leaves. The native inhabitants are Bushmen, whose temporary shelters are seen here. (S. Trevor/Bruce Coleman.)

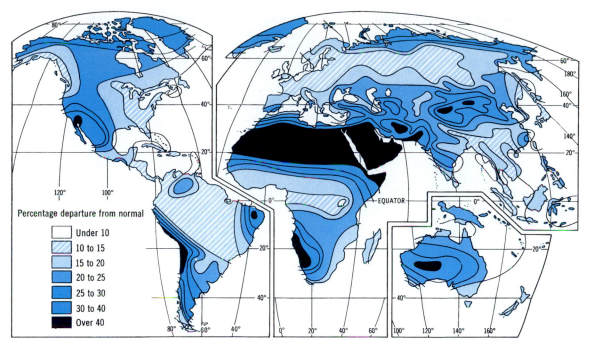

FIGURE 10.24 Precipitation variability map of the world. (After William Van Royen, *Atlas of the World's Agricultural Resources*, Prentice-Hall, Englewood Cliffs, N.J. Based on Goode Base Map.)

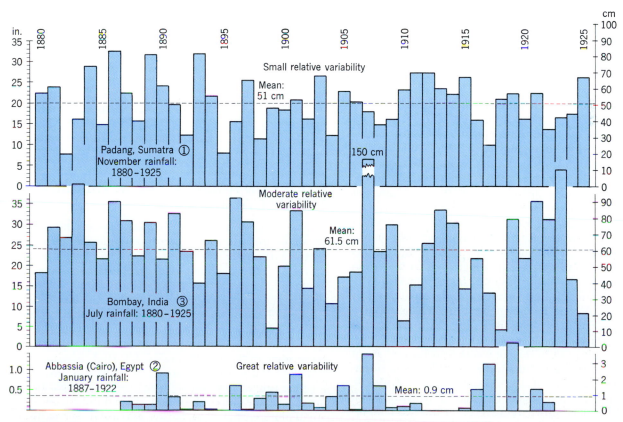

FIGURE 10.25 These graphs for three locations illustrate the concept for variability of rainfall by showing the actual amount of rain received each year during the month that, on the average, is the rainiest month at each location. (Data from H. H. Clayton, Smithsonian Institution.)

The concept is one of mathematical probability: What is the probability that the coming year will bring a devastating drought, rather than a plentiful harvest? Geographers express precipitation variability in terms of the long-term percentage of departure from the norm, or average value. The isopleths on our world map, Figure 10.24, show these percentages. This map is highly generalized and is little better than a schematic diagram. The tropical deserts show the highest variability. Very low values are associated with the moist climates, both in the equatorial and midlatitude zones. The lowest values shown are for coastal belts, for example, the marine west-coast climate of western Europe. Figure 10.25 compares the variability of the tropical desert with that of the other low-latitude climates. Rainfall of the wettest month of the year is shown. The lowermost graph shows precipitation received at Abbassia, Egypt (near Cairo), through a 36-year period. Five of the years had more than twice the average, two years had four or more times the average, and 10 years had only a trace or none at all. This record is one of extreme variability. The middle graph shows a 46-year record for Bombay, India, in the wet-dry tropical climate. A moderately large variability is shown. Only one year had more than twice the average, and there was some rain in every year. The uppermost graph shows the corresponding record for Padang, Sumatra, a wet equatorial station. The variability is obviously small compared with the other records.

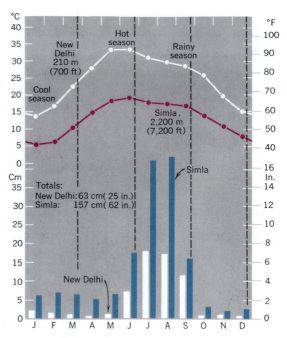

FIGURE 10.26 Climographs for New Delhi and Simla, both in northern India. Simla is a welcome refuge from the intense heat of the Gangetic Plain in May and June.

HIGHLAND CLIMATES OF LOW LATITUDES

Highland climates, shown by a distinctive pattern on the world climate map, are cool to cold, usually moist climates occupying mountains and high plateaus. Many occur as narrow belts, with strong climatic gradients, becoming steadily colder with increasing altitude. The climate of a given highland area is usually closely related to the climate of the surrounding lowland in seasonal characteristics, particularly the form of the annual temperature cycle and the times of occurrence of wet and dry seasons. In lowlands of arid climate, narrow mountain ranges tend to be islands of moist climate. Highland climates are not usually included in the broad schemes of climate classification. Designation of highland areas on most climate maps is somewhat arbitrary. Many small highland areas are simply not shown on a world map.

The upward decrease in monthly mean temperatures was explained in Chapter 4 and illustrated with a graph for several stations in the Andes of South America (Figure 4.5). In Figure 10.26, an example of the altitude effect in the tropical zone is shown by climographs for two stations in close geographic proximity. New Delhi, the capital city of India, lies in the Ganges lowland. Simla, a mountain refuge from the hot weather, is located about 2000 m (6500 ft) in altitude, in the foothills of the Himalayas. When hot-season averages are over 32°C (90°F), Simla is enjoying a pleasant 18°C (65°F), which is a full 14C° (25F°) cooler. Notice, however, that the two temperature cycles are quite similar in phase, with the minimum month being January for both.

The general effect of increased altitude is to bring an increase in precipitation, at least for the first few kilometers of altitude increase. This change is due to the production of orographic rainfall, generated by the forced ascent of air masses (see Chapter 6). The altitude effect shows nicely in the monthly rainfalls of New Delhi and Simla (Figure 10.26). New Delhi shows the typical rainfall pattern of the tropical climate of Southeast Asia, with monsoon rains peaking in July and August. Simla has the same pattern, but the amounts are larger in every month and the monsoon peak is very strong. Simla's annual total is well over twice that of New Delhi. Note that Simla also receives rainfall in the low-sun season.

MOUNTAIN AGRICULTURE IN THE HIGH EQUATORIAL ANDES

In Chapter 4, we studied the effect of increasing altitude on air temperature for a series of mountain stations in Peru at latitude 15° N (Figure 4.5). As daily mean temperatures steadily decline to reach the freezing mark at about 8400 m (14,000 ft), the daily range increases greatly in comparison with the sea-level value.

Agriculture at high altitudes in the equatorial Andes of Peru and Bolivia has been practiced for centuries despite a hostile environment that includes cold and intense solar radiation (Figure 10.27). High intermontane basins of the Bolivian Plateau, or Altiplano, are in the altitude range 3200 to 4300 m (10,500 to 14,000 ft) and are above the upper limits of the rainforest. Here, in small plots, corn, wheat, barley, and potatoes can be cultivated on a limited scale, and nearby mountain pastures provide grazing for domesticated animals (Figure 10.28).

FIGURE 10.27 Snowclad peaks of the Bolivian Andes rise to altitudes over 6000 m (20,000 ft). An arctic-type climate above the snowline nourishes living glaciers, even though the location is only a few degrees of latitude removed from the equator. The Indian village in the foreground lies in an altitude zone between 3600 and 4200 m (12,000 and 14,000 ft) in which wheat, barley, and potatoes can be grown. Here, the air temperature averages about 10°C (50°F) from one month to the next throughout the entire year. (Courtesy American University Field Staff.)

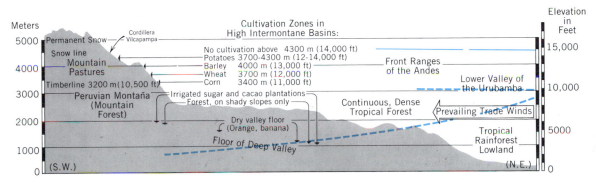

FIGURE 10.28 Altitude zoning of vegetation in the equatorial Andes of southern Peru, lat. 10° to 15°S. (After Isaiah Bowman, *The Andes of Peru*.)

THE LOW-LATITUDE CLIMATES IN REVIEW

No generalization is meaningful for the four low-latitude climates considered together, because they include great extremes of the climatic spectrum. Anyone who makes a sweeping statement about the "tropics" is very poorly informed. How can the extreme aridity of the tropical deserts be equated with the extreme wetness of the equatorial zone in reference to the thermal environment, the soil-water balance, or the food resource? If diversity is the outstanding characteristic of the low-latitude climates, shall we then look to the midlatitudes for some sweeping simplicity of character? The midlatitude zone has for centuries been called the "temperate zone." How "temperate" is the midlatitude zone? Keep this question in mind as we turn next to the climates of the midlatitude zone.

MIDLATITUDE AND HIGH-LATITUDE CLIMATES

Climates of Group II, the midlatitude climates, almost fully occupy land areas of the midlatitude zone and a large proportion of the subtropical latitude zone. Along the western fringe of Europe, they extend into the subarctic latitude zone as well, reaching to the 60th parallel. Unlike the low-latitude climates, which are about equally distributed in northern and southern hemispheres, the great preponderance of the midlatitude climate area is in the northern hemisphere. In the southern hemisphere, land area poleward of the 40th parallel south is so small that the climates are under dominance of the great Southern Ocean and lack the continentality of their northern hemisphere counterparts.

In terms of air masses and frontal zones, the midlatitude climates of the northern hemisphere lie in a broad zone of intense interaction between two groups of very unlike air masses. From the subtropical zone, tongues of maritime tropical air masses enter the midlatitude zone, where they meet in conflict with tongues of maritime polar and continental polar air masses along the discontinuous and shifting polar front zone.

In terms of prevailing pressure and wind systems, the midlatitude climates include the poleward halves of the great oceanic subtropical high pressure systems, and much of the belt of prevailing westerly winds. As a result, weather systems, such as traveling cyclones and their fronts, characteristically move from west to east. This dominant circumglobal eastward airflow gives a strong asymmetry to the distribution of climates from west to east across the great North American and Eurasian continental masses. In this respect, then, the midlatitude climates are the direct opposites of the low-latitude climates, which are dominated by the tropical easterlies (trades), maintaining a westward circumglobal airflow.

The six midlatitude climate types range from two that are very dry to three that are extremely moist, so that the group includes climates with large annual water deficits and those with large water surpluses. The midlatitude climates span the range from a climate of extreme seasonality of precipitation—the Mediterranean climate—to those of more-or-less uniformly distributed precipitation. Thermal regimes are likewise quite varied, the low annual ranges seen along the windward west coasts contrasting with the great annual ranges in the continental interiors. Annual cycles of water need (potential evapotranspiration) are correspondingly varied, but not to the great degree seen in the low-latitude climates. Total annual water need lies between 130 cm in the warmer subtropical zone to about 50 cm in the colder northern fringe, and this is a rather strong gradient of change poleward across the entire midlatitude climate zone.

Clearly, the midlatitude climates exhibit a diversity so great that no single common characteristic stands out to unify them. Europeans who migrated to North America quickly discovered this remarkable diversity of climate and its effects on the natural plant cover they encountered and the soils they began to cultivate.

DRY SUBTROPICAL CLIMATE (5)

Caused by the same air-mass patterns, the dry subtropical climate is simply a poleward extension of the dry tropical climate. The annual temperature range is greater than that for the dry tropical climate. There is a distinct cool season in the lower latitude portions and a cold season in the higher latitude portions. The cold season, occurring at time of low sun, is due in part to invasions of continental polar (cP) air masses from higher latitudes, and in this way the climate shows the influence of polar air masses. Precipitation that occurs in the low-sun season is produced by midlatitude cyclones that make incursions into the subtropical zone. As in the case of the tropical climate, subtypes include steppe subtype (5s), semidesert subtype (5sd), and desert subtype (5d).

Latitude Range: 25° to 35°N and S.

Major Regions of Occurrence: North Africa; Near East (Jordan, Syria, Iraq); southwestern United States and northern Mexico; southern part of Australia; Argentina (Pampa and Patagonia); and South Africa.

Example: Figure 11.1 is a climograph for Yuma, Arizona, a city close to the Mexican border at lat. 33°N. The annual temperatures show a strong seasonal cycle with

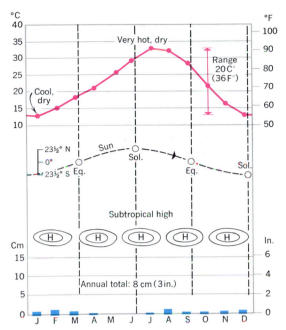

FIGURE 11.1 Yuma, Arizona, lat. 33°N, is a North American example of the subtropical desert climate.

hot summers. A cold season brings monthly means as low as 13°C (55°F). The annual range is 20C° (36F°). Precipitation, which totals about 8 cm (3 in.), is small in all months, but with two maxima. The August maximum is caused by invasions of mT air masses, which result in

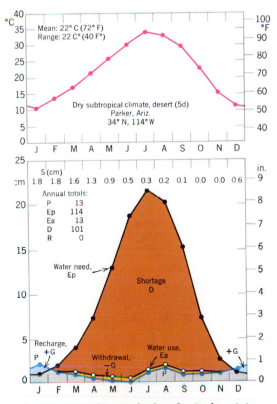

FIGURE 11.2 Soil-water budget for Parker, Arizona, lat. 34°N. (Data from C. W. Thornthwaite Associates, Laboratory of Climatology, Centerton, N.J.)

thunderstorm activity. Higher rainfalls from December through March are produced by midlatitude cyclones following a southerly path. Two months, May and June, are nearly rainless.

Soil-Water Budget: Water need, Ep, of the coolest month is less than 4 cm, but greater than 0.7 cm. Annual total Ep is less than 130 cm. Semiarid (5s), semidesert (5sd), and desert (5d) subtypes are recognized.

Example: Parker, Arizona, lat. 34°N (Figure 11.2). Located on the Colorado River, Parker lies in the desert subtype (5d) of the dry subtropical climate. Certain distinctive differences exist between this climate and the dry tropical desert subtype (4d), illustrated by the graph for Khartoum (Figure 10.18). Total annual water need, Ep, is 114 cm, much less than at Khartoum and well under the boundary value of 130 cm separating the two climates. The annual cycle of Ep is very strongly developed, with a sharp summer peak. Monthly Ep in the cold season falls to low values (1 cm) in December and January—much lower than in the tropical dry climates. Precipitation, P, which totals 13 cm, shows two distinctive maxima: a cold-season maximum when midlatitudes cyclones pass across the region; and a summer maximum, when maritime tropical air masses that penetrate from the Gulf of Mexico or the Gulf of California set off thunderstorms. P exceeds Ea by a small margin in December and January, allowing a small amount of storage recharge (+G). This water is utilized in very small monthly amounts during the remainder of the year. The total soil-water shortage, D, is very large, and there is no water surplus, R. As required for the desert subtype, soil-water storage, S, does not exceed 2 cm in any month.

The Subtropical Desert Environment

Much of what we said about the tropical desert environment applies to the adjacent subtropical desert. The boundary between these two climate types is arbitrarily drawn and would be invisible to an observer traveling from one to the other. But if we were to travel northward in the subtropical climate zone of North America, arriving at about 34° N in the interior Mojave Desert of southeastern California, we would encounter environmental features significantly different from those of the low-latitude deserts of tropical Africa, Arabia, and northern Australia. Differences would be found in soils, native vegetation, and animal life.

Although the great summer heat of the low-elevation basin floors of the Mojave Desert is comparable with that experienced in the Sahara Desert, the low sun brings a clearly recognizable winter season not found in the tropical deserts. In the Mojave Desert cyclonic precipitation can occur in most months, including the cool low-sun months.

In the Mojave Desert (and adjacent Sonoran Desert), plants are often large and numerous, in some places giving the appearance of an open woodland. One example is the occurrence of forestlike stands of the tall, cylindrical saguaro cactus (Figure 27.33); another is the woodland of

FIGURE 11.3 The many-branched ocotillo (*Fouquiera splendens*) is shown here with its thin, soft leaves, which appear only after rain has moistened the ground. Later, they will fall off. It will display flame-red flower clusters in March and April. The creosote bush is seen to left and right. Anza Borrega State Park, California. (A. N. Strahler.)

Joshua trees found in higher parts of the Mojave Desert. Other large shrubs or small trees include the prickly pear cactus (Figure 26.1), the ocotillo plant, the creosote bush, and the smoke tree (Figure 11.3).

Animals of the Mojave and Sonoran deserts show many interesting adaptations to the dry environment. Many of the invertebrates follow much the same life pattern as the ephemeral annual plants, which is to remain dormant in dry periods but to emerge when rain falls and take advantage of the event. For example, the tiny brine shrimp of the American desert may wait many years in dormancy until normally dry lake beds fill with water, an event that occurs perhaps three to four times per century. The shrimp then emerge and complete their life cycles before the lake evaporates.

The mammals are by nature poorly adapted to the desert environment, yet many survive there by employing a variety of mechanisms to avoid water loss. Many desert mammals do not sweat through skin glands; they rely instead on other methods of cooling. For example, the huge ears of the jackrabbit serve as efficient radiators of heat to the clear sky. Many desert mammals conserve water by excreting highly concentrated urine and relatively dry feces. The desert mammals also conserve water by limiting their physical activity to the night. In this respect, they are joined by most of the rest of the desert fauna in spending their days in cool burrows in the soil and their nights foraging for food.

Humans adapt to the American subtropical desert environment by importing the environment to which they are accustomed. Irrigation projects allow water to be imported in abundance and lavishly applied to croplands, where most of it is lost by intense evapotranspiration. Pipelines and highways facilitate the importation of building materials, machinery, home appliances (the air conditioner), and fuels; transmission lines bring electricity from distant dams and coal-fired power plants. These human technological adaptations have run into

serious environmental difficulties, one of which we will present in depth in Chapter 16. It is the deterioration of croplands by deposition of salts left by evaporation of irrigation waters and the accompanying waterlogging of the soil.

MOIST SUBTROPICAL CLIMATE (6)

Subtropical eastern continental margins are dominated by the mT air mass, flowing out from the moist western sides of the oceanic high-pressure cells. This air mass brings copious summer rainfall, much of it convectional. An occasional tropical storm brings heavy rainfall. Winter precipitation is also copious, produced in midlatitude cyclones. Invasions of the cP air mass are frequent in winter, bringing spells of subfreezing weather. Summers are warm, with persistent high humidity. No winter month has a mean temperature lower than 0°C (32°F). Subhumid, humid, and perhumid (very wet) subtypes are identified. This climate is characterized in Southeast Asia by a strong monsoon effect, with increased summer rainfall.

Latitude Range: 20° to 35°N and S.

Major Regions of Occurrence: Southeastern United States, southern China; Formosa, southernmost Japan, Uruguay and adjoining parts of Brazil and Argentina, and eastern coast of Australia.

Example: Figure 11.4 is a climograph of Charleston, South Carolina, located on the eastern seaboard at lat. 33°N. In this region, a marked summer maximum of precipitation is typical. Total annual rainfall is copious (120

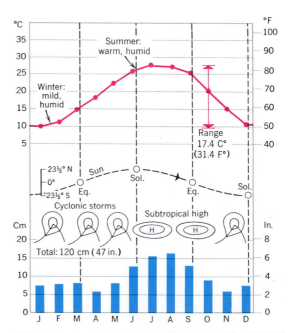

FIGURE 11.4 Charleston, South Carolina, lat. 33°N, has a mild winter and a warm summer. There is ample precipitation in all months, but a definite summer maximum.

cm, 47 in.); at least 5 cm (2 in.) of precipitation falls in every month. The annual temperature cycle is strongly developed, with an annual range of about 17C° (31F°). Winters are mild, with the January mean temperature a full 10C° (18F°) above the freezing mark.

Soil-Water Budget: This moist climate is characterized by a moderate to large water surplus, R, and a small seasonal soil-water shortage, D. The annual cycle of Ep is strongly seasonal, with fairly low values in the winter months. Ep is less than 4 cm in at least 1 month. In all months, Ep is at least 0.8 cm. Thus the coldest month temperature mean is not below 0°C. In the subhumid subtype (6sh), D is greater than zero but less than 15 cm when R is zero; otherwise, D is greater than R when R is not zero. In the humid subtype (6h) R is greater than zero but not over 60 cm, while R is greater than D. In the perhumid subtype (6p), R is greater than 60 cm.

Example: Baton Rouge, Louisiana, lat. 30°N (Figure 11.5). Precipitation, P, is copious in all months, but with a pronounced dip in late summer and autumn. There is a small summer precipitation peak. Otherwise monthly values of P are uniformly about 12 cm and the annual total is large: 143 cm. There is a strong cycle of soil-water need, Ep, with summer-month values exceeding 16 cm and winter values falling to about 2 cm. Because winter-month mean temperatures are well above freezing, however, some plant growth can continue through the mild winter with active transpiration. Storage withdrawal

(−G) occurs in 4 summer months, but the shortage, D, is small in every month and totals only 2 cm for the year. In contrast, the water surplus, R, is large (39 cm). Thus the climate qualifies as the humid subtype (6h).

East-Asian Monsoon Variety

In Chapter 7, we noted that the summer monsoon advances northward across China and southern Japan during May and June. This effect is clearly shown in the strong summer rainfall peak of stations in southern China and Japan. Nagasaki, located on the island of Kyushu, at the extreme southern end of Japan, lat. 33°N, is a good example (Figure 11.6). This climate falls in the perhumid subtype, with a total annual precipitation over 190 cm (75 in.). Rainfall peaks strongly in summer. Both June and July receive over 25 cm (10 in.), reflecting the Asiatic monsoon peak. In contrast, precipitation is low in the winter months, when a dry air mass from interior Asia dominates the weather. The air temperature cycle at Nagasaki shows a greater range than at Charleston, although summer temperatures are almost the same in both places. The winter months in Nagasaki are colder than at Charleston; even so, the coldest month, January, has an average well above the freezing point.

The Moist Subtropical Forest Environment

With a very small summer water shortage and a large winter water surplus, soil water remains adequate for plant growth without irrigation in most years. Rivers and streams flow copiously through much of the year. Flooding can be severe from tropical cyclones (hurricanes in the Gulf Coast region; typhoons in southeastern

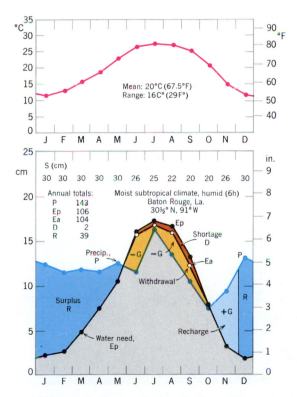

FIGURE 11.5 Soil-water budget for Baton Rouge, Louisiana, lat. 30½°N. (Data from C. W. Thornthwaite Associates, Laboratory of Climatology, Centerton, N.J.)

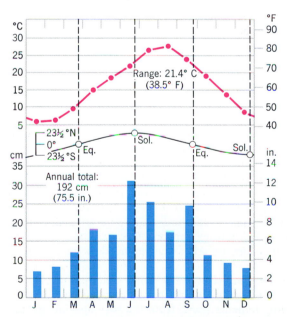

FIGURE 11.6 Nagasaki, in southern Japan, lat. 33°N, shows the effect of the monsoon by the strong rainfall peak in summer.

FIGURE 11.7 Broadleaf evergreen forest of the Gulf Coast region is represented here by Evangeline oaks bearing Spanish "moss," an epiphyte that forms long beardlike streamers. The ground beneath is maintained as a lawn. Evangeline State Park, Bayou Teche, Louisiana. (A. N. Strahler.)

China and southern Japan) that come inland and produce torrential rains in the high-sun months.

In such a moist regime, much of the natural vegetation present when explorers and colonists first visited this area in the New World consisted of broadleaf deciduous forest. In a narrow Gulf coastal zone of the United States and in a large part of southern China and the south island of Japan, the native broadleaf forest was of the evergreen type, in which a leaf canopy remains green throughout the year. This is the broadleaf evergreen forest. In Louisiana the evergreen Evangeline oak and the magnolia flourished as representative trees of the evergreen broadleaf forest (Figure 11.7). Farther inland in the Gulf states and in Florida, the forest vegetation is of an entirely different kind—southern pine forest—adapted to sandy soils (see Figure 27.23). Near its colder northern limits, vegetation of the moist subtropical climate grades into broadleaf deciduous forest. (In a deciduous forest, leaf shedding occurs annually and the trees are bare throughout the winter season.) Today, large areas of these former forests have been replaced by agricultural croplands.

The comparatively warm climate and high rainfall of the moist subtropical environment favor the leaching out of nutrient elements (ions) from the soil layer. As in the soils of the moist low-latitude climates, iron oxides accumulate in the soil, giving it colors ranging from yellow to red. In terms of agricultural crops, especially the grains or cereals, the soils of the moist subtropical envi-

ronment rate as low in fertility and require large applications of fertilizers. Another unfavorable factor is the susceptibility of these soils to severe erosion and gullying when exposed by forest removal and intensive cultivation.

Agricultural Resources of the Moist Subtropical Forest Environment

No simple statement can cover agricultural adaptation to the moist subtropical forest environment in both North America and Southeast Asia. Differences in these two widely separated areas are partly historical and cultural, but partly reflect the stronger monsoon effect in Asia, which causes a stronger concentration of precipitation in the summer. The human population is vastly denser in Southeast Asia than in the New World and subsists largely on rice, the dominant staple food crop. Two and even three rice crops are harvested annually in southern China. The rice crop is often followed by planting of wheat, winter legumes, peas, or green fertilizer crops. Except in the Mississippi delta region, little rice is produced in the southern United States. Both American and Asiatic regions produce sugarcane, peanuts, tobacco, and cotton, although not on an equal intensity in both regions. One striking difference is that tea is widely cultivated in Southeast Asia, but not at all in the southern United States (Figure 11.8). Corn is an important crop in the southern United States but is not important in southern China.

The potential of the moist subtropical climate to produce more food rests in more intensive land use instead of expansion of the area now under cultivation. The best land is already in use and, in Southeast Asia, elaborate terrace systems have been used for centuries to allow

FIGURE 11.8 Tea leaves are carefully picked from new growth of a cultivated evergreen shrub (Camellia) on this terraced tea plantation on Honshu Island, Japan. (Thomas Hopker/Woodfin Camp.)

FIGURE 11.9 Japanese farmers planting rice seedlings in a flooded rice paddy. By the time this rice crop was harvested, a single acre may have consumed a thousand person-hours of hand labor. Today, mechanical paddy transplanters can perform this operation. (W. H. Hodge/Peter Arnold.)

farming of steep hill slopes. New genetic strains of rice and corn offer promise of increased yields when the necessary fertilizers are used. Japan and Taiwan apply very high levels of fertilizers and achieve high rice yields (Figure 11.9). The People's Republic of China is only now in the process of sharply increasing its production and use of fertilizers from comparatively low levels of the recent past. China is also developing independently some new high-yielding strains of rice and wheat.

In the southern United States, cattle production is another source of increased food production and makes use of soils too sandy for field crops. With soil water frequently replenished through the long, warm summer, pasture and range land can be continuously productive. Tree farming is also an important use of sandy soils. Pines are well adapted to rapid growth on sandy soils and thrive where nutrient bases are in short supply (Figure 27.23).

The large water surplus of the moist subtropical climate has important implications in terms of economic development. The large flows of rivers can furnish abundant freshwater resources for urbanization and industry without competition from irrigation demands. Evaporative losses from reservoirs are much less important than in arid lands. The maintenance of copious stream flows tends to reduce the dangers of severe water pollution and its adverse effects on ecosystems of streams and estuaries.

MEDITERRANEAN CLIMATE (7)

This wet-winter, dry-summer climate results from a seasonal alternation of the same conditions that cause the dry subtropical climate (5), which lies at lower latitudes, and the moist marine west-coast climate (8), which lies on the poleward side. The moist maritime polar (mP) air mass invades in winter with cyclonic storms, generating ample rainfall. In summer, subsiding cT and mT air

masses are dominant, with extreme drought of several months' duration. In terms of total annual rainfall, the Mediterranean climate spans a wide range from arid to humid, depending on location. Temperature range is moderate, with warm to hot summers and mild winters. Coastal zones between 30° and 35° show a smaller annual range, with very mild winters.

Latitude Range: 30° to 45°N and S.

Major Regions of Occurrence: Central and southern California; coastal zones bordering the Mediterranean Sea; Western Australia and South Australia; Chilean coast; and Cape Town region of South Africa.

Example: Figure 11.10 is a climograph for Monterey, California, a coastal city at lat. 36½°N. The annual temperature cycle is very weak, with a range of only about 7C° (12F°), reflecting the strong control of the cold California current and its cool marine air layer. Coolness of the summer, with monthly means not exceeding 17°C (62°F) is typical of the narrow western littorals. The minimum-month winter temperature of 10.5°C (51°F) is very high compared with inland locations at this latitude. Rainfall drops to nearly zero for four consecutive summer months, but rises to more than 7.5 cm (3 in.) per month in the rainy winter season. The total annual rainfall is only about 42 cm (17 in.), and the climate as a whole can be classed as semiarid.

Example: Figure 11.11 is a climograph for Naples, Italy, located at lat. 40½°N on the west coast of the Italian peninsula. Compared with Monterey, the annual temperature range is much greater—16C° (28.5C°)—and summer temperatures are very much warmer. No cold coastal ocean current is present to subdue the temperature cycle. Total annual rainfall (86 cm, 34 in.) is much

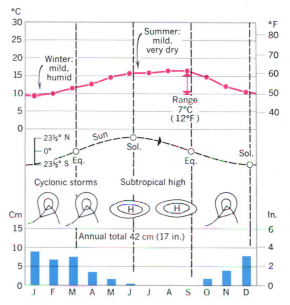

FIGURE 11.10 Monterey, California, lat. 36½°N, has a very weak annual temperature cycle because of its closeness to the Pacific Ocean. The summer is very dry.

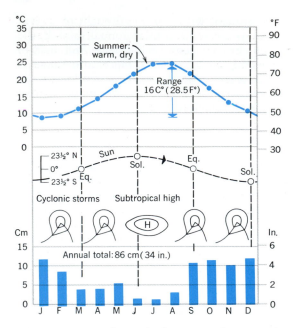

FIGURE 11.11 Naples, Italy, lat. 40½°N, has a much warmer summer than Monterey, but winter temperatures are much the same.

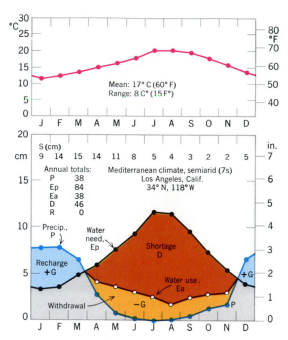

FIGURE 11.12 Soil-water budget for Los Angeles, California, lat. 34°N. (Data from C. W. Thornthwaite Associates, Laboratory of Climatology, Centerton, N.J.)

greater than at Monterey, and there is a small amount of rain, on the average, in every summer month.

Soil-Water Budget: Water need, Ep, is 0.8 cm or greater in every month, separating the Mediterranean climate from colder climates (9 and 10) bordering on the interior side. The dry aspect of the Mediterranean climate is expressed in a soil-water shortage, D, always 15 cm or larger. Water surplus, R, may be zero (in the semiarid and semidesert subtypes). A diagnostic feature of the climate is the large amplitude of the annual cycle of soil-water storage, S. For all subtypes, the storage index* is 75 percent or greater. The following subtypes are recognized:

7sd	semidesert	Soil-water storage, S, exceeds 6 cm in fewer than 2 months. S exceeds 2 cm in at least 1 month.
7s	semiarid (steppe)	S is 6 cm or greater in 2 or more months.
7sh	subhumid	Water surplus, R, ranges from 0 to 15 cm, so that D is larger than R.
7h	humid	R exceeds 15 cm.

Both semidesert (7sd) and semiarid (7s) subtypes are true dry climates. No desert subtype is included. Adjacent desert climates showing the winter rainfall maximum fall into the dry subtropical climate (5d) or the dry midlatitude climate (9d). The Mediterranean precipitation cycle also extends into the adjacent marine west-coast climate (8), in which D is always less than 15 cm.

Example: Los Angeles, California, lat. 34°N (Figure 11.12). Los Angeles is typical of the semiarid subtype (7s) of the Mediterranean climate found along the entire California coast from Monterey to Oceanside. The Medi-

terranean precipitation cycle, with its long dry summer, is a striking feature that contrasts with the annual cycle of water need, Ep, in opposite phase. Thus the summer peak of Ep coincides with the season of almost zero rainfall, greatly accentuating the soil-water shortage, D. Storage withdrawal, −G, sets in early in the year (April) and continues until December. The large accumulated shortage, D, totals 46 cm. Storage recharge, +G, begins in December and continues throughout the winter, but no water surplus, R, is generated. Native plants—mostly grasses and hard-leaved shrubs or trees—are adapted to the long dry summer. Massive amounts of summer irrigation are required to maintain green lawns and garden crops.

Example: Perth, Western Australia, lat. 32°S (Figure 11.13). This coastal city represents the humid subtype (7h) of the Mediterranean climate. (Note that the graph begins with July at the left.) Water surplus, R, is 28 cm, a large quantity. Soil-water shortage, D, is also large: 27

---

*The storage index, stated as a percentage, is calculated from maximum and minimum monthly values of S, as follows:

$$\frac{S_{max} - S_{min}}{S_{max}} \times 100$$

An alternative method of identifying the Mediterranean climate uses monthly values of P and Ea. For July (northern hemisphere) or January (southern hemisphere), the ratio of P to Ea is calculated in percent, as follows:

$$\frac{P}{Ea} \times 100$$

Stations with ratios lower than 40 percent are assigned to the Mediterranean climate. For nearly all stations, there is agreement between the two methods.

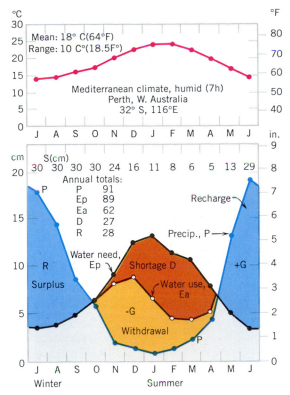

FIGURE 11.13 Soil-water budget for Perth, Western Australia, lat. 32°S. (Data from C. W. Thornthwaite Associates, Laboratory of Climatology, Centerton, N.J.)

cm. The result is a true wet-dry climate—a subtropical equivalent to the wet-dry tropical climate (3), from which it is far separated in latitude. Compared with Los Angeles, which is of the semiarid subtype (7s), Perth has much greater winter rainfall and a total annual P over twice as large as at Los Angeles. Perth has small monthly amounts of rainfall throughout the summer, whereas rainfall is nearly zero at Los Angeles in the equivalent period. Perth has a large water surplus, R, whereas Los Angeles has none. The humid climate of the Perth region supports an evergreen forest that is composed largely of species of eucalyptus trees.

The Mediterranean Climate Environment

Like the wet-dry tropical climate of the low latitudes, the Mediterranean climate offers a "feast-or-famine" environment for plants. Along with that incongruous linkage of a very dry climate with a wet one there arise a number of special environmental problems for opportunistic humans who have been strongly attracted to it. The attraction lies in the benign thermal cycle, especially in the narrow coastal zones, or littorals. There, the mild winters with considerable sunshine (despite periods of substantial rainfall) are a most welcome refuge from the severe winters of the midlatitude continental interiors of Eurasia and North America. The catch lies in the scarcity of local freshwater supplies to support the heavy load of humanity that insists upon the same lavish use of water that is easily afforded in the moist soil-water regimes.

Soil fertility in valley and lowland areas of the Medi-

terranean climate is naturally high, as it usually is in semiarid climates of the midlatitudes. Aridity has permitted the retention in the soil layer of nutrients essential to forage grasses and grains, fruit trees, vegetables, and many other varieties of plants. Despite winter rains, these nutrients have not been leached away. (Soils native to this environment belong to a special category of subclasses, explained in Chapter 24.)

The native vegetation of the Mediterranean climate environment is adapted to survival through the long summer drought. Shrubs and trees that can survive such drought are characteristically equipped with small, hard or thick leaves that resist water loss through transpiration. These plants are called *sclerophylls*; the prefix *scler*, from the Greek for "hard," is combined with *phyllo*, which is Greek for "leaf." (Compare with "atherosclerosis," the disease of hardening of the arteries.) Sclerophylls of the Mediterranean environment are typically evergreen and retain green leaves through the entire yearly cycle. Examples are the evergreen oaks, of which there are several common species in California, and the cork oak of the Mediterranean lands (Figure 11.14). Another is the olive tree, native to the Mediterranean lands (Figure 11.15). In Australia, the thick-leaved eucalyptus tree is the dominant sclerophyll. Oak woodland of California bears a ground cover of grasses that turn to straw in the summer (Figure 11.16). Another form of native vegetation in this environment is a cover of drought-resistant shrubs, including sclerophylls and spiny-leaved species. In the Mediterranean lands, this scrub vegetation goes under the name of maquis or garrigue. In California, where it is called *chaparral*, it clothes steep hill and mountain slopes too dry to support oak woodland or oak forest (see Figure 27.24).

Wildfire is an integral part of the Mediterranean environment of California. Chaparral is extremely flammable during the long fire season of summer. Brush fires rage

FIGURE 11.14 This bark, stripped from the cork oak (*Quercus suber*), will be ground up and cemented into cork board and other structural products. Thick bark of choice quality is used for wine corks. Algeria, North Africa. (A. N. Strahler.)

FIGURE 11.15 Olive trees on a steep, barren mountain slope. This scene is in the Atlas Mountains of Algeria. (A. N. Strahler.)

FIGURE 11.16 Evergreen–oak woodland with grassland, Santa Ynez Valley, Santa Barbara County, California. (above) At the end of the long, dry summer the grasses are dormant, but the oak trees are green. (below) At the end of the cool, wet winter, grasses are a lush green whereas a few deciduous oaks in the foreground are nearly leafless. (A. N. Strahler.)

through chaparral and oak forests and leave the soil surface bare and unprotected. When torrential rains occur in winter, large quantities of coarse mineral debris are swept downslope by overland flow and carried long distances by streams in flood. Mudflows and debris floods (usually called "mudslides" in the news media) are particularly destructive to human habitations on canyon floors and on piedmont fan surfaces that are often heavily urbanized.

Throughout the Mediterranean lands of Europe, North Africa, and the Near East, devastating soil erosion, induced by human activity over the past 2000 years

or longer, has left its scars on the landscape. Many hillsides have been denuded of their soils and present a barren rocky aspect. Sediment, representing the displaced soil, has formed thick layers of sand and silt in adjacent valley floors.

Lands bordering the Mediterranean Sea produce cereals—wheat, oats, and barley—where arable soils are

FIGURE 11.17 Groves of lemon, orange, and avocado trees surround the homes and stables of the wealthy in Montecito, California. Chaparral covers the steep slopes of the Santa Ynez Mountains in the background. (A. N. Strahler.)

extensive enough to be cultivated. However, we usually think of that region as an important source of citrus fruits, grapes, and olives for European markets. Cork from the bark of the cork oak is also a product of economic value (Figure 11.14). In central and southern California, citrus, grapes, avocados, nuts (almond, walnut), and deciduous fruits are grown extensively (Figure 11.17). Irrigated alluvial soils are also highly productive of vegetable crops, such as carrots, lettuce, cauliflower, broccoli, artichokes, and strawberries, as well as sugar beets and forage crops (alfalfa). Cattle ranching and sheep grazing are of major importance on grassy hill slopes unsuited to field crops and orchards (see Figure 11.16) and on irrigated lowland pastures.

Because the Mediterranean environment is limited in global extent to comparatively small land areas, it offers little prospect for the expansion of croplands to provide major additions to the world's food supply. Irrigation is essential for high productivity, but there are hazards associated with heavy irrigation of lowland soils: salt accumulation and waterlogging. Urbanization and industrial development also face major problems of obtaining additional water supplies through importation over long distances by aqueduct. Nevertheless, the mild, sunny climate of southern California has proved a powerful population magnet, and water importation has already been developed on a mammoth scale. Expansion of suburban housing has, however, begun to take over rich, flat croplands, reducing the agricultural potential in a number of areas.

MARINE WEST-COAST CLIMATE (8)

Midlatitude west coasts, windward to the prevailing westerlies, receive frequent cyclonic storms involving the cool, moist mP air mass. In this humid climate, precipitation is copious in all months, but with a distinct winter maximum. Where the coast is mountainous, the orographic effect causes a very large annual precipitation, or perhumid subtype. The annual temperature range is comparatively small for midlatitudes. Winter temperatures are very mild compared with inland locations at equivalent latitudes.

Latitude Range: 35° to 60°N and S.

Major Regions of Occurrence: Western coast of North America, spanning Oregon, Washington, and British Columbia; western Europe and the British Isles; Victoria and Tasmania; New Zealand; and Chile, south of 35°S.

Example: Figure 11.18 is a climograph for Vancouver, British Columbia, a west-coast port city at 49°20′N, just north of the United States–Canada border. The total annual precipitation—146 cm (57 in.)—establishes this as a perhumid subtype of a moist climate. Most of the precipitation falls in the winter season and there is a strong summer minimum. The temperature cycle shows a remarkably small annual range—15C° (27.5F°)—for this latitude. Even the winter months have temperature averages above the freezing mark.

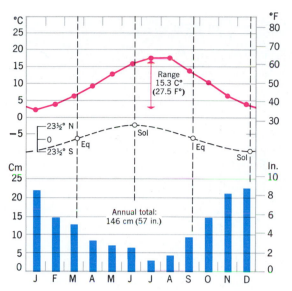

FIGURE 11.18 Vancouver, British Columbia, lat. 49°N, has a large annual total precipitation, but with greatly reduced amounts in the summer. The annual temperature range is small and winters are very mild for this latitude.

Soil-Water Budget: This cool, moist climate has a soil-water surplus, R, ranging from small to very large. Soil-water shortage, D, ranges from moderate to very small and may be zero. Annual total water need, Ep, is less than 80 cm, largely because summers are exceptionally cool for the higher midlatitudes. Monthly Ep is 0.8 cm or greater in every month, so no month has a mean temperature as low as 0°C.

Example: Cork, Ireland, lat. 52°N (Figure 11.19). Located on the south coast of Ireland and exposed to west-

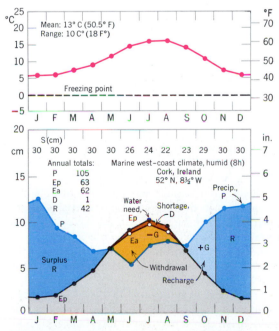

FIGURE 11.19 Soil-water budget for Cork, Ireland, lat. 52°N. (Data from C. W. Thornthwaite Associates, Laboratory of Climatology, Centerton, N.J.)

FIGURE 11.20 Needleleaf forest of Douglas fir and hemlock in Snoqualmie National Forest, Washington. The barren mountainside illustrates the practice of block cutting, in which all trees are removed. Severe soil erosion can follow, causing stream channels to be choked with debris. (Jay Lurie/Black Star.)

erly winds of the North Atlantic, Cork illustrates the humid subtype (8h) of the marine west-coast climate. Precipitation, P, shows a strong annual cycle, with a winter maximum and a summer minimum, reflecting a poleward persistence of the Mediterranean cycle. P is copious in all months, and the annual total is large: 105 cm. Water need, Ep, shows a strong annual cycle, but in all winter months Ep is 1.8 cm or greater. Thus the lowest mean monthly temperature is only about 6°C; growth of many perennial plants can continue slowly throughout the winter months. In the summer months, Ep is not much greater than P so the water shortage, D, is very small, totaling only 1 cm. The water surplus, R, is large: 42 cm. Even small streams can flow continuously throughout the year. The humid climate supports a native vegetation of forest.

The Marine West-Coast Environment

Because of the copious precipitation, large soil-water surplus, and small water shortage, lowland soils of the marine west-coast climate regions show the effect of leaching out of nutrients, but in Europe have retained moderate fertility. Applications of fertilizers and lime are needed for bountiful crop production, and in Europe these soils have been successfully cultivated for centuries. Much of the land surface within the marine west-coast environments in northern Europe, British Columbia, southern Chile, and the South Island of New Zealand is on mountainous slopes that have been heavily scoured by the ice sheets and mountain glaciers of the recent Ice Age. Soils of these glaciated areas are extremely young and are poorly developed.

Forest is the native vegetation of this environment. In the perhumid mountainous areas of the northern Pacific coast there flourish dense needleleaf forests of redwood, fir, cedar, hemlock, and spruce (Figure 11.20). Under the lower precipitation regime of Ireland, southern England,

France, and the Low Countries a broadleaf deciduous forest was the native vegetation, but much of it disappeared many centuries ago under cultivation, so that only scattered forest plots or groves remain (Figure 11.21). Sometimes called "summergreen" deciduous forest, it is dominated by tall broadleaf trees that provide a continuous and dense canopy in summer but shed their leaves completely in winter. Dominant tree species of this forest type in western Europe are oak and ash, with beech in the cooler and moister areas. The marine west-coast environment of western Europe and the British Isles has been intensively developed for centuries for such diverse uses as crop farming, dairying, orchards, and forest (Figure 11.21). It is an environment in most

FIGURE 11.21 A rural scene in northern Scotland, showing diversified farming around the small village of Rhymie. The barren uplands are moors bearing heath, a cover of small plants. (Mark A. Melton.)

respects similar in agricultural character and productivity to the moist continental climate environment with which it merges on the east.

In North America the mountainous terrain of the coastal belt offers only limited valley floors for agriculture, which is generally diversified farming. Forests are the primary plant resource, and here they constitute perhaps the greatest structural and pulpwood timber resource on earth. Douglas fir, western cedar, and western hemlock are the principal lumber trees of the Pacific Northwest (Figure 11.20). The same mountainous terrain that limits agriculture in the Pacific Northwest is a producer of enormous water surpluses that run to the sea in rivers. Including now the ranges of the northern Rocky Mountains with the Pacific coastal ranges, the potential for long-distance transfer of this excess water to dry regions of the western United States has not passed unnoticed.

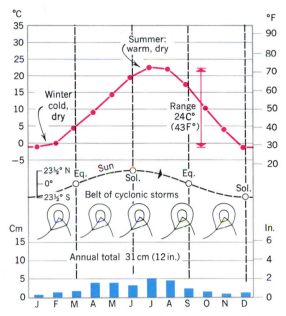

FIGURE 11.22 Pueblo, Colorado, lat. 38°N, lies in the semiarid (steppe) continental climate, just east of the Rockies. Rainfall peaks strongly in the summer months.

DRY MIDLATITUDE CLIMATE (9)

Limited almost exclusively to North America and Eurasia, this continental dry climate occupies a rainshadow position with respect to mountain ranges on the west or south. Maritime air masses are blocked effectively much of the time, so the continental polar cP air mass dominates the climate in winter. In summer, a dry continental air mass of local origin is dominant. Summer rainfall is caused by sporadic invasions of maritime air masses. Steppe and semidesert subtypes (9s, 9sd) are extensive; true desert (9d) occurs only in basins of interior Asia. The annual temperature cycle is strongly developed, with a large annual range. Summers are warm to hot, but winters are very cold.

Latitude Range: 35° to 55°N.

Major Regions of Occurrence: Western North America (Great Basin, Columbia Plateau, Great Plains); Eurasian interior, from steppes of eastern Europe to Gobi Desert and North China; and a small area in southern Patagonia.

Example: Figure 11.22 is a climograph for Pueblo, Colorado, located at 38°N, just east of the Rocky Mountains. The climate is of the semiarid (steppe) subtype (9s), with a total annual precipitation of 31 cm (12 in.). Most of this precipitation is in the form of summer rainfall, when the moist mT air mass invades from the Gulf of Mexico and causes thunderstorms. In winter, because the mP air mass is effectively dried by passage over mountain ranges, snowfall is light and yields only small monthly precipitation averages. The temperature cycle has a large annual range—24C° (43F°)—with warm summers and cold winters. January, the coldest winter month, has a mean below 0°C (32°F).

Example: Figure 11.23 is a climograph for Kazalinsk, Russia, a city at lat. 46°N, located just east of the Aral Sea in Kazakhstan. The climate is semidesert (9sd), with

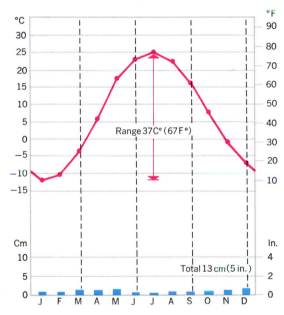

FIGURE 11.23 At Kazalinsk, Russia, lat. 46°N, winters are bitterly cold and summers are warm. Little precipitation falls in any month in this interior desert.

an annual precipitation total of only 13 cm (5 in.), or less than half that of Pueblo. The annual temperature range is extremely great—37C° (67F°)—and winters are bitterly cold. Winter-month temperatures average below freezing (0°C, 32°F) for five consecutive months.

Soil-Water Budget: Monthly Ep is 0.7 cm or less in at least 1 month. Thus winters are cold; usually at least one month has a mean temperature of 0°C or lower. Near the northern limit, Ep is zero for as many as five consecutive

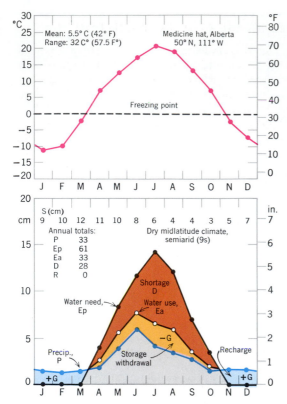

FIGURE 11.24 Soil-water budget for Medicine Hill, Alberta, Canada, lat. 50°N. (Data from C. W. Thornthwaite Associates, Laboratory of Climatology, Centerton, N.J.)

FIGURE 11.25 Harvesting wheat on the rolling Palouse Hills of eastern Washington. The rich dark soil is formed on a thick layer of silt carried to this area by wind near the close of the Ice Age. As a natural prairie grassland, this area is ideally suited to wheat farming. (Grant Heilman.)

months. The lower limiting value of annual total Ep is 52.5 cm, occurring along the boundary with the boreal forest climate (11). Semiarid (9s), semidesert (9sd), and desert (9d) subtypes are recognized.

Example: Medicine Hat, Alberta, Canada, lat. 50°N (Figure 11.24). Located near the northern limit of the Great Plains, Medicine Hat illustrates the semiarid (steppe) subtype (9s) of the dry midlatitude climate. The annual cycle of water need, Ep, peaks strongly in summer, following five consecutive months in which Ep is zero because of severe winter cold. Precipitation shows a distinct annual cycle; the summer months have about double the precipitation of the winter months. A substantial soil-water shortage, D, develops in summer and totals 28 cm. But soil-water storage, S, remains above 3 cm, even in the late summer and early autumn months. Storage recharge, +G, is not sufficient to bring S to levels approaching storage capacity, so no surplus, R, is generated. Recharge accumulates in the frozen state throughout the winter, to be released rapidly in the spring thaw. Spring wheat, grown here, uses the accumulated soil water of early spring and matures in summer.

The Dry Midlatitude Environment

Low annual precipitation combined with a large soil-water shortage under a strongly continental thermal regime has produced soils of high natural fertility that re-

tain large supplies of the nutrient elements (positively charged ions), known to soil scientists as "bases." The principal nutrient bases are calcium, magnesium, potassium, and sodium. These soils are moderately to strongly alkaline, in contrast to soils of the humid and perhumid midlatitude climates that are acid in chemical balance (see Chapter 24).

Grasses thrive on large supplies of these soil nutrients and a mildly alkaline condition of the soil. Thus the native vegetation of the semiarid subtype (9s) consists principally of hardy perennial short grasses capable of enduring severe summer drought. We refer to this vegetation type as *short-grass prairie*. Among geographers, the Asiatic plains landscapes of these short grasses are known as *steppes*. (The singular form, *steppe*, is used as a general term for the total environment.)

Soils of the short-grass prairie are dominantly of a major class known as *Mollisols*. The prefix of this recently coined word uses the Greek root *mollis*, meaning "soft." (Compare with the word "mollify," meaning "to soften.") This refers to the rather loose soil texture, consisting of small soil particles and giving the soil the property of being easily tilled. Soil color is brown to pale brown, and these soils are known variously as chestnut soils and brown soils.

Wheat is perhaps the most important single crop produced in unirrigated areas of short-grass prairie bordering the subhumid zone. One such important wheat-producing region lies in southern Alberta and Saskatchewan and in the northern border region of Montana. Another is the Palouse Hills region of southeastern Washington state and western Idaho (Figure 11.25). Here the crop is spring wheat, which is planted in the spring of the year. Using the soil water that has been recharged in early spring and the precipitation that falls in late spring and early summer, the crop is able to reach maturity for midsummer harvesting.

In Russia, the rich wheat region of the Ukraine continues in a narrow zone far eastward across the steppes of Kazakhstan. In northern China wheat is grown within a steppe region bordering the moist continental climate.

Wheat production of the midlatitude steppes is very much at the mercy of variations in seasonal rainfall. Good years and poor years follow cyclic variations. Soil water, not soil fertility, is the key to wheat production over these vast steppe lands lying beyond the practical limits of upland irrigation systems fed by major rivers.

Semiarid steppes form the great sheep and cattle ranges of the world. The steppes of central Asia have for centuries supported a nomadic population whose sheep and goats find subsistence on the scanty grassland (Figure 11.26). On the vast expanses of the High Plains, the American bison lived in great numbers until being almost exterminated by hunters. The short-grass veldt of South Africa also supported much game at one time.

Steppe grasses do not form a complete sod cover; loose, bare soil is exposed between grass clumps. For this reason, overgrazing during a series of dry years can easily reduce the hold of grasses enough to permit destructive deflation (wind erosion), followed by water erosion and gullying.

MOIST CONTINENTAL CLIMATE (10)

Located in central and eastern parts of North America and Eurasia in the midlatitude zone, this climate is in the polar-front zone—the battleground of polar and tropical air masses. Seasonal temperature contrasts are strong, while day-to-day weather is highly variable. Ample precipitation throughout the year is increased in summer by the invading mT air mass. Eastern maritime locations are perhumid. Cold winters are dominated by cP and continental arctic (cA) air masses from subarctic source regions. In eastern Asia (China, Korea, Japan), the monsoon effect is strongly evident in a summer rainfall maximum and a relatively dry winter. In Europe, the moist continental climate lies in a higher latitude belt (45° to 60°N) and receives precipitation from the mP air mass coming from the North Atlantic.

Latitude Range: 30° to 55°N (Europe: 45° to 60°).

Major Regions of Occurrence: Eastern parts of the United States and southern Canada, northern China, Korea, Japan, and central and eastern Europe.

Example: Figure 11.27 is a climograph for Madison, Wisconsin, lat. 43°N, in the American Midwest. The annual temperature range is very large (31C°, 56F°). Summers are warm but winters are cold, with 3 consecutive monthly means well below freezing (0°C, 32°F). Precipitation is ample in all months and totals 81 cm (32 in.), giving a humid climate. There is a well-developed summer maximum of precipitation when the mT air mass

FIGURE 11.26 These nomads of northern Afghanistan are encamped on the floor of an arid valley, close to a snow-fed stream. Camels will carry the tents and other possessions to a new location when necessary. (Victor Englebert/Photo Researchers.)

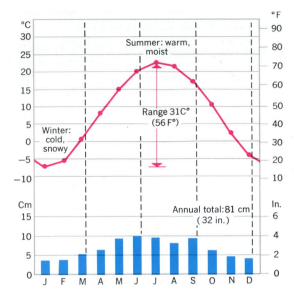

FIGURE 11.27 Madison, Wisconsin, lat. 43°N, has cold winters and warm summers, making the annual temperature range very large.

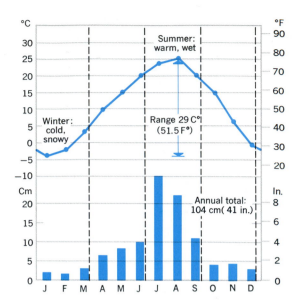

FIGURE 11.29 Inchon, Korea, lat. 37½°N, has two very rainy months during the summer monsoon, but very little precipitation in winter.

invades and thunderstorms are formed along moving cold fronts and squall lines. Much of the winter precipitation is in the form of snow, which remains on the ground for long periods.

Example: Figure 11.28 is a climograph for Moscow, Russia, at lat. 56°N. This location is over 600 km (1000 mi) farther north than Madison, so Moscow summers are not as warm and winters are colder. Annual total precipitation at Moscow is less (54 cm, 21 in.) than at Madison, with smaller monthly amounts throughout the year. Winter precipitation, largely snow, remains on the ground for many months.

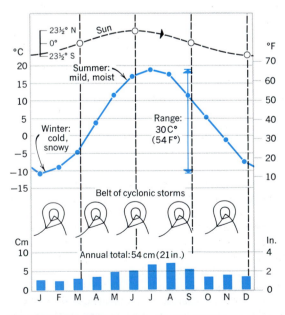

FIGURE 11.28 Moscow, Russia, lat. 56°N, has an annual range about the same as Madison; but summers in Moscow are not as warm.

Example: Figure 11.29 is a climograph for Inchon, Korea, at lat. 37½°N. Compared with Madison, Inchon has somewhat warmer summers; but winters are not as cold. The major difference in the two stations is in the precipitation cycle. Inchon shows the Asiatic monsoon effect with two very rainy summer months and greatly diminished precipitation in winter. Because of the summer monsoon, with influx of a maritime air mass from the western Pacific Ocean, Inchon has a larger total annual precipitation (104 cm, 41 in.) than either Madison or Moscow.

Soil-Water Budget: Water surplus, R, ranges from small to large, except in a narrow subhumid subtype (10sh) bordering the dry continental climate (9). Total annual water need, Ep, is more than 52.5 cm, while R is greater than D. In the perhumid subtype Ep is typically zero in 1 to 5 winter months. A summer maximum of precipitation, P, is typical, but is not present everywhere. In the subhumid subtype (10sh), D is greater than zero, but less than 15 cm when R is zero; otherwise D is greater than R when R is not zero. In the humid subtype (10h), R is greater than zero but not over 60 cm, while R is greater than D. In the perhumid subtype (10p), R is greater than 60 cm.

Example: Pittsburgh, Pennsylvania, lat. 40°N (Figure 11.30). Located in the interior eastern region, Pittsburgh illustrates the humid subtype (10h) of the moist continental climate. Precipitation, P, runs rather uniformly throughout the year, though with a slight summer maximum. Total annual P is substantial: 92 cm. The annual cycle of soil-water need, Ep, peaks strongly in summer, after three consecutive winter months of zero value, when soil water is frozen and plants are dormant. A small soil-water shortage, D, develops in summer but is small in every month and totals only 4 cm. A substantial water surplus, R, occurs in winter and early spring; the

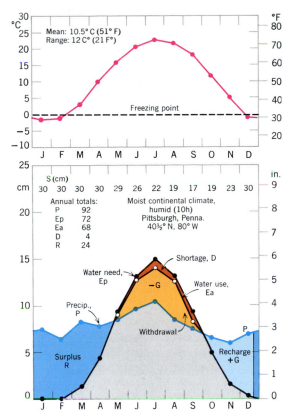

FIGURE 11.30 Soil-water budget for Pittsburgh, Pennsylvania, lat. 40°N. (Data from C. W. Thornthwaite Associates, Laboratory of Climatology, Centerton, N.J.)

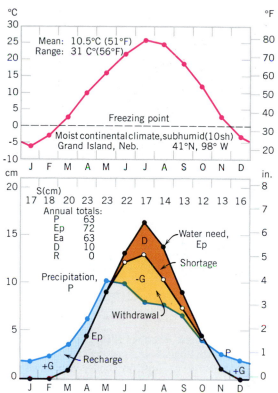

FIGURE 11.31 Soil-water budget for Grand Island, Nebraska, lat. 41°N. (Data from C. W. Thornthwaite Associates, Laboratory of Climatology, Centerton, N.J.)

yearly total is 24 cm. Some of this water is held in the frozen state in winter, to be released rapidly in the early spring when thawing occurs. Spring floods are highly probable. Larger streams maintain their flow throughout the summer. Forest is the natural plant cover.

Example: Grand Island, Nebraska, lat. 41°N (Figure 11.31). Located almost exactly midway between the Atlantic and Pacific oceans, Grand Island illustrates the subhumid subtype (10sh) of the moist continental climate. Water need, Ep, peaks very sharply in summer, but falls to zero in three winter months. Precipitation shows a strong summer maximum in the warmer months, but the peak months occur in early summer—May and June. A soil-water shortage, D, sets in by June and continues throughout October, with a total value of 10 cm. Storage recharge, +G, in the fall, winter, and early spring is not enough to bring about a surplus. Soil-water storage, S, is high in late spring, then falls off throughout the summer. This area lies close to the boundary of dry and moist climates.

The Moist Continental Forest and Prairie Environment

With ample precipitation throughout the year and only a small summer water shortage, the humid and perhumid subtypes of the moist continental climate support forests as the native vegetation. Soils beneath these forests show

effects of the moist environment through the leaching out of soil bases and other soil components and a strong tendency to soil acidity.

These effects are most severe in the colder, more northerly parts of the climate zone, where strongly acidic soils are found on sandy surface layers. Here the evergreen needleleaf forest dominates. For example, pine forest was found in the Great Lakes region. Throughout much of the northeastern United States and southeastern Canada mixed coniferous and deciduous forest was the native type. It graded southward into broadleaf deciduous forest, which was found in a large area of the eastern United States. Here, moderately leached forest soils are found, and these have retained a high level of natural fertility suitable for crop cultivation. The deciduous forests were also found in this climate in central and eastern Europe, and in a narrow belt penetrating far eastward into Siberia; they are also found in north-central China and in Korea.

An important environmental factor has been the activity of the great ice sheets that recently covered the northern parts of this climate region. The ice had a profound effect on the landforms of the region as well as on parent materials of the soils, which are of recent origin and poorly developed in many parts of the area. These effects are explained in Chapter 22.

A most important feature of the moist continental climate is that, when traced into the continental interior, it grades into a progressively less humid climate. In North America the gradation to greater aridity is seen in a narrow north–south zone of a subhumid subtype (10sh)

FIGURE 11.32 Farmlands, marked off in a square township grid, extend for tens of kilometers across a former glacial lake bed near Aberdeen, South Dakota. The agricultural soils here are Borolls, a cold-climate suborder of the Mollisols. A river floodplain with meanders occupies the foreground. (John S. Shelton.)

adjacent to the semiarid (steppe) climate (9s) (Figure 11.32). The effect of this westward gradation is profound on both soils and native vegetation. There is a large region of the Middle West, starting about in Illinois and continuing west through Iowa and into Nebraska, where a variety of the Mollisols is the dominant soil type. Also called prairie soil, this soil is rich in nutrient bases and formerly supported a natural cover of tall, dense grasses; it was the *tall-grass prairie* (see Figure 27.30). This kind of prairie once extended from about the United States–Canada border southward to the Gulf Coast, but today only a few small remnants can be found.

Throughout Europe, large areas of the moist conti-nental forest environment have been under field crops, pastures, and vineyards for centuries, while at the same time forests have been carefully cultivated over large areas (Figure 11.33). In this environment the potentials for both crop agriculture and forest culture (sylviculture) have reached a near-optimum adjustment in terms of the soils and terrain.

Because of the availability of soil water through a warm summer growing season, the moist continental environment has an enormous potential for food production. The cooler, more northerly sections in North America and Europe support dairy farming on a large scale. A combination of acid soils, known as Spodosols,

FIGURE 11.33 The Mosel River in its winding, entrenched meandering gorge through the Rhineland–Pfalz province of western Germany. On the steep, undercut valley wall at the right are vineyards and forested land. (Porterfield/Chickering Photo Researchers.)

and unfavorable glacial terrain in the form of bogs and lakes, rocky hills, and stony soils has deterred crop farming in many parts.

Farther south, plains formed on the former lake floors and on undulating uplands are ideally suited to crop farming. Here, soils are of high fertility. Cereals grown extensively in North America and Europe include corn, wheat, rye (especially in Europe), oats, and barley. Corn is also an important crop in Hungary and Rumania. Beet sugar is an important product of this environmental region in Europe, but not in North America. On the other hand, soybeans are intensively cultivated in the midwestern United States and in northern China and Manchuria, but very little in Europe. Rice is a dominant crop in both South Korea and Japan, much farther poleward than elsewhere in Asia. The rice seedlings can be planted in paddies flooded during the brief but copious rains of midsummer, then harvested in the dry autumn. (Among geographers, this northern rice area is often included in the region called Monsoon Asia.) Agricultural productivity of the tall-grass prairie lands in the United States is now legendary under the name of the "corn belt." Corn production is concentrated most heavily in the prairie plains of Illinois, Iowa, and eastern Nebraska. Wheat is also a major crop near the western limits of the tall-grass prairie in Kansas and Oklahoma.

THE MIDLATITUDE CLIMATES IN REVIEW

The midlatitude climates run the gamut from very wet to very dry and from mild marine coastal climates to strongly seasonal continental climates. Soils and natural vegetation cover an equally great range of types befitting the spectrum of climate. No useful generalization is possible for such a diverse group of environments.

Production of food resources in the midlatitude regions spans as wide a range of intensities as the climates themselves. Here we have the richest food-producing regions of the world, fully developed by the Western nations through massive inputs of fertilizers, pesticides, and fuels and guided by the most advanced technology. But there are also unproductive deserts at the same latitudes. These midlatitude environments, taken as a whole, are neither temperate in climate nor uniform in plant resources. Perhaps this heterogeneous quality of the midlatitudes is just what we should expect of a global zone where polar and tropical air masses wage war incessantly over vast land areas that cut across the latitude zones and their prevailing westerly airflow.

HIGH-LATITUDE CLIMATES

Climates in Group III, the high-latitude climates, exclusive of the ice sheet climate, lie almost entirely in northern hemisphere lands of North America and Eurasia. They occupy the northern subarctic and arctic latitude zones, but extend southward into the midlatitude zone as far south as about the 47th parallel in eastern North America and Asia. In terms of air masses and frontal zones, the high-latitude climates of the northern hemisphere lie in a zone of intense interaction between unlike air masses. Maritime polar (mP) air masses interact violently with continental polar (cP) and arctic (A) air masses in a discontinuous and constantly fluctuating arctic front zone. In summer, tongues of maritime tropical air masses (mT) reach the subarctic latitudes to interact with polar air masses and yield important precipitation.

In terms of prevailing pressure and wind systems, the high-latitude climates coincide closely with the belt of prevailing westerly winds that form the periphery of the circumpolar flow of the great upper-air polar vortex. Local reversals of surface airflow to east winds accompany traveling cyclones and extend upward to high levels in cut-off lows that are part of the air-mass exchange system capable of transporting water vapor into high latitudes.

The high-latitude climates have low annual total evapotranspiration, always less than about 50 cm, reflecting the prevailing low air and soil temperatures, and declining sharply poleward to extremely low values in the tundra climate and effectively to zero in the ice sheet climate. The frozen condition of the soil in several consecutive winter months causes plant growth to virtually cease, cutting off evapotranspiration. Snow that falls in this period is retained in surface storage until the spring thaw releases it for infiltration and runoff. Needless to say, the growing season for crops is short in the subarctic zone, but low air and soil temperatures are partly compensated for by the great increase in day length.

Europeans who came to settle the high latitudes of North America found familiar counterparts to their native climates in northern Europe. Experience in the boreal forest lands and tundra of Scandinavia, Finland, and northern Russia served them well in exploiting the natural environmental resources and they encountered few surprises.

BOREAL FOREST CLIMATE (11)

This is a continental climate with long, bitterly cold winters and short, cool summers. This climate occupies the source region of the cP air mass, which is cold, dry, and stable in the winter. Invasions of the very cold cA air mass are common. The annual range of temperature is greater than that for any other climate, attaining a value of 60C° (110F°) in Siberia. Precipitation is substantially increased in summer, when maritime air masses penetrate the continent with traveling cyclones; but the total annual precipitation is small. Although much of the boreal forest climate is classed as humid, with precipitation of 50 to 100 cm (20 to 40 in.), large areas in western Canada and Siberia have annual precipitation totals of less than 40 cm (16 in.) and are classed as subhumid (11sh) or semiarid (11s) subtypes.

Latitude Range: 50° to 70°N.

Major Regions of Occurrence: Central and western Alaska; Canada, from Yukon Territory to Labrador;

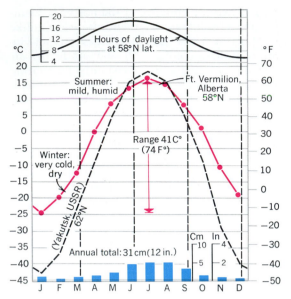

FIGURE 11.34 Extreme winter cold and a very great annual range in temperature characterize the boreal forest climate, illustrated by these climographs for Fort Vermilion in Alberta, Canada, and Yakutsk, Russia.

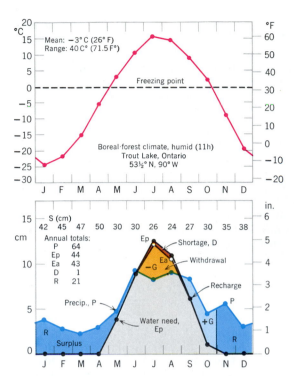

FIGURE 11.35 Soil-water budget for Trout Lake, Ontario, Canada, lat. 53½°N. (Data from C. W. Thornthwaite Associates, Laboratory of Climatology, Centerton, N.J.)

southernmost Greenland; Iceland; and Eurasia, from northern Europe across all of Siberia to the Pacific Coast.

Example: Figure 11.34 is a climograph for Fort Vermilion in Alberta, Canada, at lat. 58°N. The very great annual temperature range (41C°, 74F°) is typical for North America. Monthly mean air temperatures are below freezing (0°C, 32°F) for 7 consecutive months. The summers are short and cool. Precipitation shows a marked annual cycle with a strong summer maximum; but the total annual precipitation is only 31 cm (12 in.), and the climate can be characterized as subhumid (11sh). Although precipitation in winter is small, a snow cover remains over solidly frozen ground throughout the entire winter. On the same climograph, temperature data are shown for Yakutsk, U.S.S.R., a Siberian city at lat. 62°N. The enormous annual range is evident, also the extremely low winter-month means. January reaches a mean of about −45°C (−50°F), making this region the coldest on earth except for the ice-sheet interiors of Antarctica and Greenland. Precipitation is not shown for Yakutsk, but the annual total is only about 18 cm (7 in.)—much less than for Fort Vermilion—and indicates a semiarid climate subtype (11s).

Soil-Water Budget: The total annual water need, Ep, ranges from 35 to 52.5 cm and is effectively zero for 5, 6, or 7 consecutive winter months. During the short summer, Ep peaks sharply. Across Canada, subtypes are arranged in order from west to east from subhumid (11sh) in the Yukon and Northwest Territories, through humid (11h), to perhumid (10p) in Labrador and Newfoundland. (See climate 10 for a definition of these subtypes.) Across Eurasia, this order is reversed in direction. An area in eastern Siberia qualifies as a dry subtype (11s), with D greater than 15 cm.

Example: Trout Lake, Ontario, Canada, lat. 53½°N (Figure 11.35). Trout Lake lies in central Ontario, in the heart of the Canadian Shield not far south of Hudson Bay. This station illustrates the humid subtype (11h) of the boreal forest climate. The cycle of water need, Ep, rises to a sharp peak following 6 consecutive winter months of zero evapotranspiration. Precipitation, P, also rises sharply during the summer, but there is a brief period of soil-water use, − G. Water shortage, D, is extremely small: 1 cm. Recharge is completed in October, after which time soil water becomes solidly frozen. Throughout the remainder of the winter, precipitation accumulates as snow. Soil-water storage, S, is shown on the graph to exceed 30 cm from November through April. The excess represents accumulated snow; it is released as runoff in May during the spring thaw. The soil retains a high level of water storage throughout the summer. Streams flow copiously during the summer. The growing season for crops is very short.

The Boreal Forest Environment

Land surface features of much of the region of boreal forest climate were shaped beneath the great Pleistocene ice sheets, which had their centers over the Hudson Bay–Labrador region, the northern Cordilleran Ranges, the Baltic region, and highland centers in Siberia. Severe ice erosion exposed hard bedrock over vast areas and created numerous shallow rock basins. Bouldery rock debris, called glacial till, mantles the rock surface in many places. Many of the shallower rock basins have been filled by organic bog materials forming a matlike layer

FIGURE 11.36 A peat bog in boreal forest near the border between Norway and Sweden. Blocks of peat are dried out on a crude rack of poles and will be used as household fuel. (John S. Shelton.)

FIGURE 11.38 Lichen woodland near Lac Cambrien, lat. 57°N, in northern Quebec. The trees are black spruce. Between the trees is a carpet of lichen. (R. N. Drummond.)

called muskeg. Peat, a black substance consisting of partly decomposed plant matter, has accumulated in numerous bogs. These have provided a low-grade fuel in northern Europe (Figure 11.36). Peat and those mineral soils poorly developed in the recent glacial materials are low in readily available plant nutrients and are acid in chemical balance.

The dominant upland vegetation of this climate region is boreal forest, consisting of needleleaf trees. In North America and Europe, these are evergreen needleleaf trees, mostly pine, spruce, and fir (Figures 11.37 and 27.20). In central and eastern Siberia, the boreal forest is dominated by the larch, which sheds its needles in winter and is thus a deciduous tree (Figure 27.21). Associated with the needleleaf trees are stands of aspen, balsam poplar, willow, and birch.

Along the northern fringe of boreal forest lies a zone of woodland in which low trees, such as black spruce, are spaced widely apart. The open areas are covered by

a surface layer of lichens and mosses (Figure 11.38). This cold woodland is referred to by geographers as the *taiga*.

Crop farming in the continental subarctic environment is largely limited to lands surrounding the Baltic Sea in bordering Finland and Sweden. Cereals grown in this area include barley, oats, rye, and wheat. Along with dairying, these crops primarily provide food for subsistence. The principal nonmineral economic product throughout the subarctic lands and eastern Canada is pulpwood from the needleleaf forests. Logs are carried down the principal rivers to pulp mills and lumber mills. Forests of pine and fir in Sweden, Finland, and European Russia are the primary plant resource (Figure 11.37). The wood products are exported in the form of paper, pulp, cellulose, and construction lumber.

TUNDRA CLIMATE (12)

The tundra climate occupies the arctic coastal fringes, dominated by cP, mP, and cA air masses, and with frequent cyclonic storms. Winters are long and severe. There is a very short mild season, which many climatologists do not recognize as a true summer. A moderating influence of the nearby ocean water prevents winter temperatures from falling to the extreme lows found in the continental interior. The tundra climate ranges from a humid subtype (12h) bordering the Atlantic Ocean to subhumid (12sh) and semiarid (12s) subtypes bordering the Arctic Ocean.

FIGURE 11.37 A narrow road passes through dense boreal forest of spruce and fir in central Sweden. (John S. Shelton.)

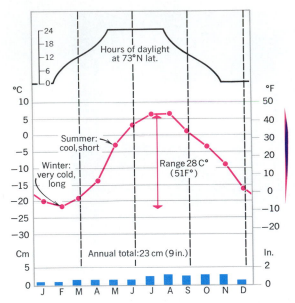

FIGURE 11.39 Upernivik, Greenland, lat. 73°N, is a typical arctic tundra station.

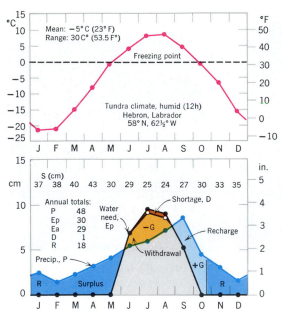

FIGURE 11.40 Soil-water budget for Hebron, Labrador, Canada, lat. 58°N. (Data from C. W. Thornthwaite Associates, Laboratory of Climatology, Centerton, N.J.)

Latitude Range: 60° to 75°N and S.

Major Regions of Occurrence: Arctic slope of North America, Hudson Bay region and Baffin Island, Greenland coast, northern Siberia bordering the Arctic Ocean, and Antarctic Peninsula.

Example: Figure 11.39 is a climograph for Upernivik, located on the west coast of Greenland at lat. 73°N. A short milder season, equivalent to a summer season in lower latitudes, has monthly means barely over 5°C (41°F). The long, very cold winters bring monthly means as low as −20°C (−9°F). The annual temperature range (28C°, 51°F) is not as large as for the boreal forest climate to the south. Total annual precipitation is small—23 cm (9 in.)—representing the subhumid subtype (12 sh). Increased precipitation beginning in July is explained by the melting of the sea-ice cover and a warming of ocean water temperatures, increasing the moisture content of the local air mass.

Soil-Water Budget: Total annual water need, Ep, is less than 35 cm. During 8 or more consecutive months, soil water is frozen and Ep is effectively zero. Some parts of the tundra climate are humid (12h), with a substantial water surplus, R; other areas are subhumid (12sh). (See climate 10 for a definition of these subtypes.) In eastern Siberia, the tundra belt qualifies as a dry climate of the semiarid subtype (12s).

Example: Hebron, Labrador, Canada, lat. 58°N (Figure 11.40). Located on the Atlantic coast, Hebron illustrates the humid subtype (12h) of the tundra climate. Because of the great length and severity of the winter, water need, Ep, is effectively zero for 8 consecutive months. Total annual Ep is only 30 cm. Ep peaks sharply in the short warm season, when the sun is in the sky for much of the 24-hour day. Total annual precipitation is 48 cm, a

substantial quantity. Precipitation shows a strong annual cycle with a summer maximum, peaking in September, just prior to the onset of winter. Storage withdrawal, −G, occurs during June, July, and August, and the soil-water shortage, D, is extremely small (actually less than 1 cm). Soil-water storage, S, remains very high throughout the summer period. Storage capacity (30 cm) is reached before the end of October. Snow that falls throughout the winter is held for release in May, when the spring thaw occurs.

The Arctic Tundra Environment

The term *tundra* describes both an environmental region and a major class of vegetation. Figure 11.41 is a polar map showing the extent of the arctic tundra. (An equivalent climatic environment—called alpine tundra—prevails in many global locations in high mountains above the timberline.) Soils of the arctic tundra are poorly developed and consist of freshly broken mineral particles and varying amounts of humus (finely divided, partially decomposed plant matter). Peat bogs are numerous. Because soil water is solidly and permanently frozen not far below the surface, the summer thaw brings a condition of water saturation to the soil.

Trees exist in the tundra only as small, shrublike features because of the seasonal damage to roots by freeze and thaw of the soil layer and to branches exposed to the abrading action of wind-driven snow. Vegetation of the treeless tundra consists of grasses, sedges, and lichens, along with shrubs of willow. Traced southward, the vegetation changes into birch–lichen woodland, then into the needleleaf boreal forest.

In some places a distinct tree line separates the forest and tundra. It coincides approximately with the 10°C (50°F) isotherm of the warmest month and has been used

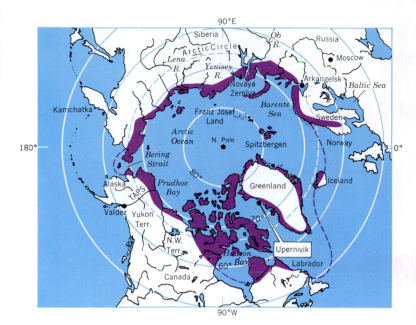

FIGURE 11.41 The tundra of the northern hemisphere.

by geographers as a boundary between boreal forest and tundra.

Vegetation is scarce on dry, exposed slopes and summits—the rocky pavement of these areas gives them the name of "fell-field," the Danish term meaning "rock desert."

The number of species in the tundra ecosystem is small, but the abundance of individuals is high. Among the animals, vast herds of caribou in North America or reindeer (their Eurasian relatives) roam the tundra, lightly grazing the lichens and plants and moving con-

stantly (Figure 11.42). A smaller number of musk-oxen are also consumers of the tundra vegetation. Wolves and wolverines, arctic foxes, and polar bears are predators. Among the smaller mammals, snowshoe rabbits and lemmings are important herbivores. Invertebrates are scarce in the tundra, except for a small number of insect species. Black flies, deerflies, mosquitoes, and "no-see-ums" (tiny biting midges) are all abundant and can make July on the tundra most uncomfortable for humans and animals. Reptiles and amphibians are also rare. The boggy tundra, however, offers an ideal summer environ-

FIGURE 11.42 Caribou migration across the arctic tundra of northern Alaska. (Warren Garst/Tom Stack and Assoc.)

FIGURE 11.43 Reindeer moss, a variety of lichen, seen here on rocky tundra of Alaska. (Steve McCutcheon.)

ment for many migratory birds such as waterfowl, sandpipers, and plovers.

The food chain of the tundra ecosystem is simple and direct. The important producer is "reindeer moss," a lichen (Figure 11.43). In addition to the caribou and reindeer, lemmings, ptarmigan (arctic grouse), and snowshoe rabbits are important lichen grazers. The important predators are the fox, wolf, lynx, and bear, although all these animals may feed directly on plants as well. During the summer, the abundant insects help support the migratory waterfowl populations. The directness of the tundra food chain makes it particularly vulnerable to fluctuations in the populations of a few species.

ICE SHEET CLIMATE (13)

Source regions of arctic (A) and antarctic (AA) air masses are situated on vast, high ice sheets, and over polar sea ice of the Arctic Ocean. Mean annual temperature is much lower than that of any other climate, with no above-freezing monthly mean. Strong temperature inversions develop over the ice sheets. The strong net radiation deficit in winter at high surface altitudes intensifies the cold. Strong cyclones with blizzard winds are frequent. Precipitation, almost all occurring as snow, is very small, but accumulates because of the continuous cold.

Latitude Range:　65° to 90°N and S.

Example:　Figure 11.44 shows temperature graphs for five representative ice-sheet stations. The graph for Eismitte, Greenland, shows the northern-hemisphere temperature cycle, whereas the other four examples are all from Antarctica. Temperatures in the interior of Antarctica have proved to be far lower than at any other place on earth. The Russian meteorological station Vostok, located about 1300 km (800 mi) from the south pole at an altitude of about 3500 m (11,400 ft) may be the world's coldest spot. Here a record low of −88.3°C (−127°F) was

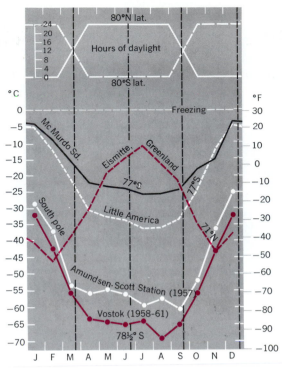

FIGURE 11.44 Temperature graphs for five ice-sheet stations.

observed. At the pole itself (Amundsen-Scott Station), July, August, and September of 1957 had averages of about −60°C (76°F). Temperatures run roughly 22C° (40F°) higher, month for month, at Little America because it is located close to the Ross Sea and is at low altitude.

The Ice-Sheet Environment

Because of low monthly mean temperatures throughout the year over the ice sheets, this environment is devoid of vegetation and soils. The few species of animals found on the ice margins are associated with a marine habitat. In terms of habitation by humans, the ice-sheet environment is extremely hostile because of extreme cold, high winds, and a total lack of food and fuel resources. Enormous expenditures of energy are required to import these necessities of life and to provide shelter. These efforts are justified because of the need for scientific research, but in the foreseeable future there is little prospect that this icy environment will provide useful supplies of energy or minerals.

GLOBAL CLIMATES, LANDFORMS, SOILS, AND NATURAL VEGETATION

We have now surveyed the principal climates of the globe. Each climate zone, together with its characteristic landforms, soils, and native vegetation, comprises a unique natural environmental region. In the order named, we will make an in-depth study of each of these other ingredients, or factors, that together make up physical geography as a natural science.

Landforms—those diverse relief features of the continental surfaces, such as hills, mountains, plains, and valleys—strongly influence soil development and will be our next target. First, however, materials of the earth's rock crust and geologic processes that shape and move that crust require study, because landforms are sculpted from these mineral substances and crustal structures.

Since late in the nineteenth century, soil scientists have recognized that the several major classes of mature soils, seen on a worldwide scope, are strongly controlled by climate as well as by landforms and the mineral content of the rock that underlies the land surface.

Plant geographers are keenly aware that plants are responsive to differences in climate, landforms, and soils. With each plant species is associated a particular combination of climatic factors most favorable to its growth. Plants tend to adapt their forms to meet the stresses of climate.

As we move through the coming chapters, many interactions among these ingredients of physical geography will become evident. These interactions form a web of interesting pathways through which flow various forms of energy and matter.

C H A P T E R 12

MATERIALS OF THE EARTH'S CRUST

With this chapter we turn to the lithosphere, or solid mineral realm of our planet. Which materials and structures of the lithosphere are most important to humans and to the processes of the life layer? The inorganic mineral matter of the lithosphere is a vitally important nutrient reservoir for life processes, as we will see in later chapters on soils and nutrient cycling in the biosphere. Rocks exposed at the earth's surface are the primary source of these nutrient materials.

The lithosphere is also a platform for life on the lands. The platform consists basically of the continents. Continents and ocean basins are global features of first order of magnitude. What materials make up the continents? Why do the continental outlines show such great diversity? Why is the global distribution of continents seemingly unrelated to the latitude zones? These are important questions in the background of physical geography. The surfaces of the continents are shaped into a remarkable variety of surface configurations, called *landforms*. Landforms strongly influence the distribution of ecosystems; landforms exert strong controls over human occupation of the lands.

Our plan in this series of 11 chapters (Chapters 12–22) is first to examine the lithosphere on a large scale to determine its composition and structure and to interpret the major features of the continents. This section deals first with *geology*, the science of the solid earth. Then we investigate landforms in terms of their origin and environmental qualities. In following this plan, we will start with sources of energy and matter arising from deep within the planet, but we will end up with surface features shaped by solar energy through activity of the atmosphere and hydrosphere. In this way you will be prepared to investigate the soil layer and its cover of vegetation.

COMPOSITION OF THE EARTH'S CRUST

From the standpoint of the environment of humans, the most significant zone of the solid earth is the thin outermost layer—the earth's crust. This mineral skin, averaging about 17 km (10 mi) thickness for the globe as a whole, contains the continents and ocean basins and is the source of soil and sediment, salts of the sea, gases of the atmosphere, and all free water of the oceans, atmosphere, and lands.

Figure 12.1 displays in order the eight most abundant elements of the earth's crust in terms of percentage by weight. Oxygen (O), the predominant element, accounts for about half the total weight. It occurs in combination with silicon (Si), the second most abundant element. Oxygen is a major element in organic substances. Silicon is used in minor amounts by plants.

Aluminum (Al) and iron (Fe) are third and fourth on the list. Both are nutrients required by plants, although in smaller quantities. Both metals are of primary importance in our industrial civilization, and it is most fortunate that they are comparatively abundant elements. Four metallic elements follow: calcium (Ca), sodium (Na), potassium (K), and magnesium (Mg). All four are on the same order of abundance—2 to 4 percent. Calcium, potassium, and magnesium are major plant nutrients; their presence is essential if a soil is to have a high level of fertility.

If we were to extend the list, the ninth-place element would be titanium, followed in order by hydrogen, phos-

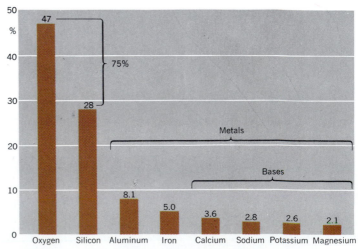

FIGURE 12.1 The average composition of the earth's crust is given here in terms of the percentage by weight of the eight most abundant elements.

214

phorus, barium, and strontium. Both hydrogen (H) and phosphorus (P) are essential nutrient elements in plant growth. Hydrogen, combined with oxygen in the form of water (H_2O), is used by plants to form organic molecules.

ROCKS AND MINERALS

The elements of the earth's crust are organized into compounds that we recognize as minerals. A *mineral* is a naturally occurring, inorganic substance, usually possessing a definite chemical composition and a characteristic atomic structure. Vast numbers of mineral varieties exist, together with a great number of their combinations into rocks.

Rock is broadly defined as any aggregate of minerals in the solid state. Rock comes in a wide range of compositions, physical characteristics, and ages. A given rock is usually composed of two or more minerals, and usually many minerals are present; however, a few rock varieties

consist almost entirely of one mineral. Most rock of the earth's crust is extremely old in terms of human standards, the time of formation ranging back many millions of years. But rock is also being formed at this very hour as a volcano emits lava that solidifies on contact with the atmosphere.

Rocks of the earth's crust fall into three major classes: (1) *Igneous rocks* are solidified from mineral matter in a high-temperature molten state, that is, from a *magma*. (2) *Sedimentary rocks* are layered accumulations of mineral particles derived in various ways from preexisting rocks. (3) *Metamorphic rocks* are igneous or sedimentary rocks that have been physically and chemically changed by application of heat and high pressures during mountain-making events. Although we have listed these classes in a conventional sequence, it will become obvious in this chapter that no one class has first place in terms of origin. Instead, they form a continuous circuit through which the crustal minerals have been recycled during many millions of years of geologic time.

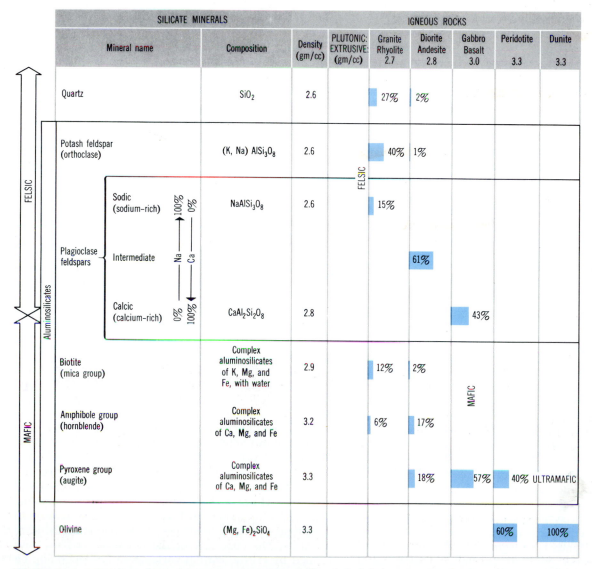

FIGURE 12.2 Simplified chart of common silicate minerals and abundant igneous rocks. (From A. N. Strahler, *Planet Earth: Its Physical Systems Through Geologic Time*, Harper & Row, Publishers. Copyright © by Arthur N. Strahler.)

THE SILICATE MINERALS

Igneous rocks make up the vast bulk of the earth's crust. Practically all igneous rock consists of *silicate minerals*, which are all compounds containing silicon atoms in combination with oxygen atoms in a close linkage. In the crystal structure of silicate minerals, one atom of silicon is linked with four atoms of oxygen as the unit building block of the compound. Most of the silicate minerals also contain one, two, or more of the metallic elements listed in Figure 12.1.

Figure 12.2 gives the names and chemical compositions of seven major silicate minerals, or groups of silicate minerals. The large bulk of all igneous rocks consists of two or more of these silicate minerals in varying proportions.

Among the commonest minerals of most rock types is *quartz;* its composition is silicon dioxide (SiO_2). There follow five mineral groups, collectively forming the *aluminosilicates* because all contain aluminum. Two groups of *feldspars* are set apart: the *potash feldspars* contain potassium (K) as the dominant metallic ion, but sodium (Na) is commonly present in various proportions. The *plagioclase feldspars* form a continuous series, beginning with the sodic, or sodium-rich varieties, and grading through with increasing proportions of calcium to the calcic, or calcium-rich varieties.

Belonging to the *mica group,* which is familiar because of its property of splitting into very thin, flexible layers, is biotite, a dark-colored mica with a complex chemical formula. Potassium, magnesium, and iron are present in biotite, along with some water. The *amphibole group,* of which the common black mineral hornblende is a common representative, is a complex aluminosilicate containing calcium, magnesium, iron, and water. Similar in outward appearance and having essentially the same component elements is the *pyroxene group,* with the mineral augite as a representative. Last on the list is *olivine,* a dense greenish mineral that is a silicate of magnesium and iron, but without aluminum.

Density is an important property of a given mineral. *Density* is defined as the mass of substance per unit of volume and is stated in grams per cubic centimeter (gm/cc) (Figure 12.3). Density of pure water is 1.0 gm/cc.

Looking down the list of densities of the silicate mineral in Figure 12.2, you will notice that there is a progressive increase from the least dense (quartz, 2.6 gm/cc) to the most dense (olivine, 3.3 gm/cc). This change reflects the decreasing proportion of aluminum and sodium, elements of low atomic weights, and the increasing proportion of calcium and iron, elements of considerably greater atomic weights.

The list as a whole is conveniently divided into two major groups of silicate minerals: *felsic minerals,* consisting of quartz and the feldspars, and *mafic minerals,* consisting of those silicates rich in magnesium and iron. The coined word *felsic* is easily recognized as a combination of "fel," for feldspar, and "si," for silica. The syllable "ma" in *mafic* stands for magnesium, the letter "f" for iron (Fe). The felsic minerals are light in color and of comparatively low density; the mafic minerals are dark in color and of comparatively high density.

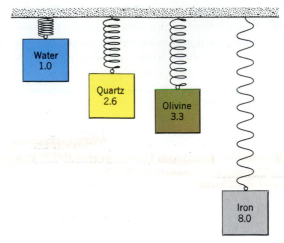

FIGURE 12.3 The concept of density is illustrated by several cubes of the same size, but of different substances, hung from a coil spring. The extent to which the spring is stretched is directly proportional to the weight of the cube and to its density in grams per cubic centimeter. Quartz and olivine are common minerals. (After A. N. Strahler, *Principles of Earth Science,* Harper & Row, New York. Copyright © by Arthur N. Strahler.)

Two important mafic minerals that are not silicates occur in many igneous rocks. These are magnetite, an oxide of iron (Fe_3O_4) and ilmenite, an oxide of iron and titanium ($FeTiO_3$). Both of these minerals are black and have high densities—4½ to 5½ gm/cc.

SILICATE MAGMAS

From the geologic standpoint, the silicate minerals can be viewed as the fundamental materials out of which other rock groups—sedimentary and metamorphic—are created. About 99 percent of the bulk of the igneous rocks of the earth's crust consists of the seven silicate minerals or mineral groups listed in Figure 12.2. The remainder consists of minerals of secondary importance in bulk, although their number is very large. The eight silicate minerals or groups combine to form a dozen or so igneous rock varieties having widespread occurrence. We shall simplify the list to five representative rock types.

Igneous rocks are derived from *silicate magma* formed at depths of many kilometers in the earth under conditions of very high temperatures and pressures. Here, silicate magmas probably have temperatures in the range from 500 to 1200°C (900 to 2200°F) and are under pressures 6000 to 12,000 times as great as atmospheric pressure at sea level.

As magma cools at or near the surface, crystallization occurs over a certain critical range of temperatures and pressures. Through a complex series of interactions, the eight elements named in Figure 12.1 are gathered into compounds as individual crystals of the various silicate minerals. The character of the igneous rocks that are formed by crystallization varies greatly, depending on both the initial composition of the magma and its cooling history.

VOLATILES IN MAGMAS

Of paramount importance in terms of the origin of the atmosphere and hydrosphere is the presence in magmas of substances besides the elements that comprise solidified silicate rock. These substances are known as *volatiles* because they remain in a gaseous or liquid state at much lower temperatures than the silicate compounds. Consequently, the volatiles are separated from the magma as it cools and solidifies. The volatiles escape into the atmosphere from volcanoes and from gas and stream vents in geothermal localities.

From analyses of gas samples collected from volcanic emissions, we know that water is the major constituent of the volatile group. Table 12.1 lists the volatiles found in gases emanating from magma of active Hawaiian volcanoes. For comparison, the table lists the proportion of these volatiles in the atmosphere and hydrosphere. Notice that the abundances of the various constituents are of the same general order of magnitude in both lists.

The emanation of volatiles from the earth's crust is called *outgassing* and is the source of free water of the earth's hydrosphere as well as the atmospheric gases carbon dioxide, nitrogen, argon, and hydrogen. Chlorine and sulfur compounds of seawater have derived their chlorine and sulfur from outgassing. We can conclude, then, that silicate magmas, with their enclosed volatiles, have through geologic time supplied almost all the essential components of the atmosphere, hydrosphere, and lithosphere.

TEXTURES OF IGNEOUS ROCKS

Igneous rocks are classified not only on the basis of mineral composition, but also on the basis of grades of sizes of the component crystals. The term *rock texture* applies both to crystal size and to the arrangement of crystals of mixed sizes.

Very gradual cooling of a magma enclosed in solid crustal rock results in growth of large mineral crystals and forms *intrusive igneous rock* with coarse-grained texture (Figure 12.4). Magma that reaches the earth's surface, pouring out through gaping cracks to form stream-

FIGURE 12.4 A fresh exposure of massive gray granite high in the Sierra Nevada, California. The white bands are igneous dikes of felsic composition. The vertical boreholes were drilled to blast out the rock with explosives. (A. N. Strahler.)

like masses, is called *lava* (Figure 12.5). Lava is classed as *extrusive igneous rock*. Rapid cooling of lava results in very small mineral crystals, not usually distinguishable with the unaided eye, and gives the rock a fine-grained texture. Very sudden cooling of outpouring magma yields a natural volcanic glass; the black variety is called *obsidian* (Figure 12.6, right). Frothing of magma as its enclosed gases expand gives a porous, spongelike rock known as *scoria* or *pumice* (Figure 12.6, left).

Active volcanoes emit great volumes of solid particles formed by rapid cooling of emerging magma and impelled by the explosive release of gases under great pressure. These particles fly through the air and land on the ground at varying distances from the vent, depending on

TABLE 12.1 Volatiles in Gases of Magmas

	Lava Gases from Mauna Loa and Kilauea Volcanoes (Percent by Weight)	Volatiles Free in Earth's Atmosphere and Hydrosphere (Percent by Weight)
Water, H_2O	60	93
Carbon, as CO_2 gas	24	5.1
Sulfur, S_2	13	0.13
Nitrogen, N_2	5.7	0.24
Argon, A	0.3	Trace
Chlorine, Cl_2	0.1	1.7
Hydrogen, H_2	0.04	0.07
Fluorine, F_2	—	Trace

Data source: W. W. Rubey, *Geol. Soc. Amer. Bull.*, vol. 62, p. 1137. Figures have been rounded off.

FIGURE 12.5 This lava flow of basalt inundated a road near Kilauea Volcano in Hawaii National Park. The rough, blocky surface resulted from the continual solidification and fracturing of the lava surface as the flow moved over the ground. (John S. Shelton.)

FIGURE 12.6 A frothy, gaseous lava solidifies into a light, porous scoria (left). Rapidly cooled lava may form a dark volcanic glass (right). (A. N. Strahler.)

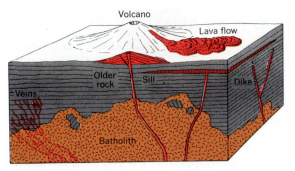

FIGURE 12.7 Forms of igneous rock bodies. (Drawn by A. N. Strahler.)

their weight and the strength of prevailing winds. The finest particles, which are in the form of minute glass shards, travel long distances and accumulate in layers as *volcanic ash*. Later we will include volcanic ash as one of the classes of sediment (pyroclastic sediment). Particles the size of gravel or pebbles fall close to the vent. *Tephra* is the collective term for all sizes of solid particles blown from a volcanic vent.

CLASSIFICATION OF IGNEOUS ROCKS

Using the simplest possible classification of igneous rocks, we recognize five coarse-grained types. These are named on the top line above the columns in Figure 12.2. For the first three rocks listed, important equivalent extrusive types (lavas) are recognized. Bars of varying width, with attached percentages, show the typical mineral compositions of these igneous rocks.

Granite and its extrusive equivalent, *rhyolite*, are rich in quartz and potash feldspar, with lesser amounts of sodic plagioclase, biotite, and amphibole. *Diorite* and its extrusive equivalent, *andesite*, are almost totally lacking in quartz and potash feldspar, but consist dominantly of intermediate plagioclase and lesser amounts of the mafic minerals.

Going progressively in the direction of domination by mafic minerals, we come to *gabbro* and its lava equivalent, *basalt*. In these, plagioclase feldspar is of the calcic type, making up nearly half the rock, while pyroxene makes up the remainder. In a common variety of basalt, olivine is present at the expense of part of the feldspar.

The next rock, *peridotite*, is not abundant in the crust, but probably makes up the bulk of the next lower layer, or mantle. It is composed mostly of pyroxene and olivine. Finally, we list *dunite*, a rare rock composed almost entirely of olivine, as an example of the extreme mafic end of the mineral series.

Granite and diorite, rocks rich in felsic minerals, are collectively described as *felsic rocks*, whereas gabbro and basalt are *mafic rocks;* the extreme mafic types are *ultramafic rocks*.

Densities of the igneous rocks are proportional to the densities of the component minerals. Thus granite has a density of about 2.7 gm/cc; gabbro and basalt, about 3.0; and peridotite and dunite, 3.3.

FORMS OF IGNEOUS ROCK BODIES

Intrusive igneous rock bodies are called *plutons*. The largest of these is the *batholith*. Figure 12.7 shows the relationship of a batholith to the overlying rock. While forcing its way upward, the magma makes room for its bulk by dissolving and incorporating the older rock above it. Batholiths are several kilometers in depth and may extend beneath an area of several thousand square kilometers.

Figure 12.7 shows two other common forms of plutons. One is a *sill*, a platelike layer formed when magma forced its way between two horizontal rock layers, lifting the overlying rock to make room. A second is the *dike*, a near-vertical wall-like body formed by the spreading apart of a vertical rock fracture (Figure 12.8). The dike rock is fine textured because of rapid cooling. Dikes are commonly the conduits through which magma reaches the surface. Magma entering small fractures in the overlying rock solidifies in a branching network of thin veins.

Volcanoes are conical or dome-shaped structures built by the emission of lava and its contained gases from a restricted vent in the earth's surface (Figure 12.7). The magma rises in a narrow, pipelike conduit from a

FIGURE 12.8 A dike of mafic igneous rock with nearly vertical parallel sides, intruded into flat-lying sedimentary rock layers. Arrows mark the contact between igneous rock and sedimentary rock. Spanish Peaks region, Colorado. (A. N. Strahler.)

magma reservoir far below. Reaching the surface, igneous material may pour out in tonguelike *lava flows* or may be ejected as tephra under pressure of confined gases.

DISINTEGRATION AND DECAY OF IGNEOUS ROCKS

Poets and advertising copywriters assure us that the highly polished granite slab is an everlasting monument in a changing world. But in truth, the surface environment is poorly suited to the preservation of an igneous rock formed under conditions of high pressure and high temperature. Most silicate minerals formed in magmas do not last long, geologically speaking, in the low temperatures and pressures of atmospheric exposure, particularly because free oxygen, carbon dioxide, and water are abundantly available.

Rock surfaces are also acted on by physical forces of distintegration, tending to break up the igneous rock into small fragments and to separate the component minerals, grain from grain. The fragmentation process is explained in Chapter 15 under the topic of physical weathering. Fragmentation is essential for the chemical reactions that follow because it results in a great increase in mineral surface area exposed to attack by chemically active solutions. The products of rock disintegration and decay tend to accumulate in a soft surface layer, the *regolith*, that overlies the solid, unaltered *bedrock* beneath it (see Figure 15.1).

SIZE GRADES OF MINERAL PARTICLES

Physical weathering, as well as the grinding and crushing processes to which mineral particles are subjected when transported by streams, waves and currents, wind, and ice, continually reduce those particles to smaller and smaller diameters. The mineral particles are collectively called *sediment*, which may eventually become sedimentary rock. Large particles, such as pebbles, are reduced in size by *abrasion*, the grinding away of the surface.

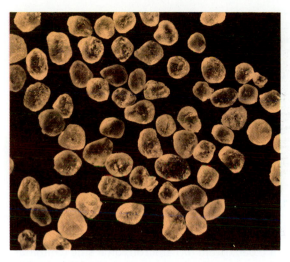

FIGURE 12.9 Rounded quartz grains from an ancient sandstone. The grains average about 1 mm (0.04 in.) in diameter. (Andrew McIntyre, Columbia University.)

Abrasion thus produces countless very small particles. When a particle is reduced by abrasion to the size of a sand grain, it is quickly pulverized between much larger fragments. To describe this material, a system of particle grade sizes is needed.

The U.S. Department of Agriculture has established size limits and names for various grades of mineral particles in soils. Table 12.2 shows these grades. The three basic categories are *sand* (2.0 to 0.05 mm), *silt* (0.05 to 0.002 mm), and *clay* (finer than 0.002 mm). Sand is subdivided into five subgrades; size limits are given in the table. Although larger particles are commonly present in a sediment sample, they are removed by screening when the sample is analyzed. The larger size grades of sediment particles are pebbles (2 to 64 mm), cobbles (64 to 256 mm), and boulders (larger than 256 mm).

Clay particles range downward in diameter from 0.002 mm, or 2 microns, to smaller than 0.001 micron. Clay particles finer than about 0.01 micron are classed as *colloids*. A property of mineral colloids is their ability to remain in suspension indefinitely in water, once the particles have become dispersed (separated one from another). A colloidal suspension appears clouded or murky.

Grains of sand, as well as pebbles and cobbles, are often well rounded in shape as a result of slow mechanical abrasion during transport by water or wind. Spherical quartz grains of coarse sand size, shown in Figure 12.9, were shaped by wind action in moving dunes. Silt grains and the coarser grades of clay are usually highly angular and may appear under the microscope as if they were particles of crushed glass. Fine clay particles, those of colloidal dimension, are typically in the form of thin scales and plates (Figure 12.10).

Decreasing particle size is accompanied by a great increase in surface area of particles contained within a given volume; when the colloidal size is reached, the surface area is enormous. Table 12.3 illustrates this point by assuming that we start with a single cube, 1 cm on a side (pebble size); it has a volume of 1 cc and a surface area of 6 sq cm. Consider next that we slice the cube into cubes 0.1 cm on a side (size of coarse sand); this move yields 1000 cubes with a total surface area of 60 sq cm.

TABLE 12.2 Grade Sizes of Sediment Particles, U.S. Department of Agriculture System

Grade Name	Diameter Limits
	2.0 mm
Very coarse sand	
	1.0 mm
Coarse sand	
	0.5 mm
Medium sand	
	0.25 mm
Fine sand	
	0.10 mm
Very fine sand	
	0.05 mm
Silt	
	0.002 mm (2 microns)
Clay noncolloidal	2 to 0.01 microns
colloidal	Below 0.01 microns

TABLE 12.3 Surface Area of Cubes Obtained by Subdividing a One-Centimeter Cube

Cube Dimensions Length of Side (cm)	Grade Equivalent	Number of Particles (per cc)	Total Surface Area (sq cm)
1	Pebble	1	6
0.1	Coarse sand	10^3	60
0.0001	Fine clay	10^{12}	60,000
0.000,01 (0.1 micron)	Colloidal clay	10^{15}	600,000 (8 m × 8 m)

Subdividing the cube into 10^{12} cubes, each 0.0001 cm on a side (fine clay size) gives a total surface area of 60,000 sq cm. Finally, subdivision of the original cube into 10^{15} cubes, each 0.000,01 cm (0.1 micron) on a side (colloidal clay size), yields a total surface area of 600,000 sq cm; if spread out into a continuous horizontal surface, this would be an area about 8 m by 8 m. This relationship between surface area and particle size is important because nutrients and water are held on soil particle surfaces. We shall investigate this phenomenon in Chapter 23.

CHEMICAL ALTERATION OF SILICATE MINERALS

Mineral alteration consists of a number of chemical reactions; all these reactions change the original silicate minerals of igneous rock, the *primary minerals*, into new compounds, the *secondary minerals*, that are stable in the surface environment. Chemical alteration also affects various types of sedimentary and metamorphic rocks. The total alteration process is also referred to as "rock decay."

Chemical alteration includes several types of reactions, all of which take place more or less simultaneously. Surface water—whether it be raindrops, water of streams and lakes, or water held in the soil—contains gases of the atmosphere. The presence of atmospheric gases dissolved in natural water is a matter of great environmental importance because of the inorganic reactions with mineral matter and because the presence of two gases in particular—oxygen and carbon dioxide—is vital to life processes of plants and animals living in water.

The presence of dissolved oxygen in water in contact with mineral surfaces leads to *oxidation*, which is the chemical union of oxygen atoms with atoms of those metallic elements (calcium, potassium, magnesium, iron) abundant in the silicate minerals. At the same time, carbon dioxide (CO_2) in solution forms a weak acid, *carbonic acid*, capable of reaction with several minerals. In addition, where decaying vegetation is present, soil water containing complex organic acids is capable of interaction with mineral compounds. Some common minerals, such as rock salt (sodium chloride, NaCl), dissolve directly in water; but direct solution is not particularly effective for the silicate minerals.

Water itself combines with certain mineral compounds in a reaction known as *hydrolysis*. This process is not merely a soaking or wetting of the mineral, but a true chemical change producing a different compound and a different mineral. The reaction is not readily reversible under atmospheric conditions, so the products of hydrolysis are stable and long-lasting—as are the products of oxidation. In other words, these changes represent a permanent adjustment of mineral matter to a new environment of pressures and temperatures.

Because water is required for mineral alteration, you might think that the rate of rock decay would be directly proportional to the amount of free water available in the rock and soil and that the dry deserts would be environments of very limited rock decay. To some extent this is a valid conclusion. Polished stone surfaces of the ancient Egyptian monuments have been almost perfectly preserved throughout the centuries in a dry, hot desert climate; these same monuments, taken to midlatitude cities of humid climate and exposed to the atmosphere, undergo rapid disintegration. (Frost action and the action of sulfuric acid derived from polluted air are also factors in rapid disintegration.) Although mineral alteration is perhaps much slower in dry deserts than in humid lands, there is nevertheless enough water present as water vapor and as dew to allow alteration to proceed; we observe the decay products in abundance in igneous rock surfaces in most deserts.

The effect of cold is quite another matter. When soil water is frozen, chemical reactions are greatly slowed. Minerals in arctic regions of perennially frozen soil and rock (permafrost) show little chemical decay because seasonal thaw affects only a thin surface layer. Most chemical reactions take place more rapidly at high temperatures than at low. Consequently, chemical alteration of minerals is most rapid in warm (and moist) climates of low latitudes.

MINERAL PRODUCTS OF HYDROLYSIS AND OXIDATION

We now investigate some important alteration products of the common silicate minerals and mineral groups. Certain of these products are clay materials. A *clay mineral* is one that has plastic properties when moist because it consists of minute thin flakes lubricated by layers of water molecules. Potash feldspar undergoes hydrolysis to become *kaolinite*, a soft, white clay mineral with the composition $Al_2Si_2O_5(OH)_4$. Crystals of kaolinite, greatly magnified, are shown in Figure 12.10. Kaolinite becomes plastic when moistened. It is an important ceramic mineral used to make chinaware, porcelain, and tile. Kaolinite can also be derived from the plagioclase feldspars.

Bauxite is an important alteration product of feldspars, occurring typically in warm climates of tropical and equatorial zones where rainfall is abundant year-round or in a rainy season. Bauxite is actually a mixture of minerals, the dominant constituent being diaspore, with the formula $Al_2O_3 \cdot H_2O$. The combination of two atoms of aluminum with three atoms of oxygen is known as *sesquioxide of aluminum*. Full oxidation has taken place, yielding an unusually stable compound. Unlike kaolinite, which is a true clay with plastic properties,

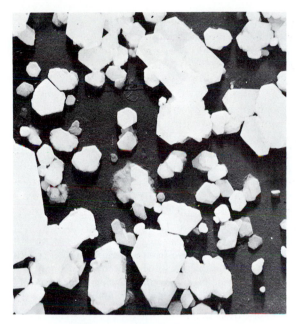

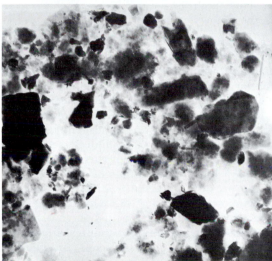

FIGURE 12.10 (upper) Electron microscope photo of kaolinite crystals magnified about 20,000 times. (Paul F. Kerr.) (lower) Enlarged here about 20,000 times are tiny flakes of the clay minerals illite (sharp outlines) and montmorillonite (fuzzy outlines). These particles have settled out from suspension in San Francisco Bay. (Harry Gold: Courtesy of R. B. Krone, San Francisco District Corps of Engineers, U.S. Army.)

bauxite forms massive rocklike lumps and layers below the soil surface.

A second clay mineral is *illite*, formed as an alteration product of feldspars and muscovite mica. Illite is a hydrous aluminosilicate of potassium. It occurs as minute thin flakes of colloidal dimensions and is carried long distances in streams (Figure 12.10). *Montmorillonite*, a common clay mineral (more correctly a group of minerals), is derived from the alteration of feldspar, or certain of the mafic minerals, or volcanic ash. Fragments of montmorillonite are seen together with illite in Figure 12.10.

Another group of clay minerals important in the soil is *vermiculite*, which is similar in composition to mont-

morillonite. Whereas montmorillonite expands greatly when it absorbs water, vermiculite does not. Vermiculite is a hydrous aluminosilicate rich in magnesium and iron. It is formed by hydrolysis of mafic silicate minerals, such as biotite mica and hornblende, and is a common product of the chemical weathering of mafic volcanic rocks.

Two important alteration products of the mafic minerals are *hematite*, sesquioxide of iron (Fe_2O_3), and *limonite*, a hydrous iron compound with the formula $2Fe_2O_3 \cdot H_2O$. Iron sesquioxide in hematite and limonite is a stable form of iron and is found widely distributed in rocks and soils. It is closely associated with bauxite. Hematite supplies the typical reddish to chocolate-brown colors of soils and rocks.

Silicate minerals differ in their susceptibility to alteration. Olivine is most susceptible to alteration, followed by the pyroxenes, amphiboles, biotite, and sodic plagioclase feldspar. Potash feldspars are somewhat less susceptible. Muscovite mica is comparatively resistant to alteration, while quartz is immune to hydrolysis and to further oxidation.

As hydrolysis takes place, *silica* (silicon dioxide, SiO_2) is released. It has the same composition as the mineral quartz found in the felsic igneous rocks. (We can refer to quartz of igneous origin as primary quartz.) Grains of primary quartz released from decomposing rock are also subject to being dissolved. Dissolved silica, whether primary or derived from hydrolysis, is commonly redeposited in finely crystalline forms adhering to other minerals. Several kinds of reformed silica are found in the regolith and soil and are recognized as distinct minerals. For example, *chalcedony* is a form of silica with crystals too small to be individually separated. The banded varieties of chalcedony are familiar to us as agate, an ornamental or gem stone.

The minerals we have described here are of great importance in the composition of the soil; we shall refer to them again in Chapters 23 and 24.

CLASTIC SEDIMENT

Sediment deposited in layers after transport by streams, waves and currents, wind, or ice, is an important class of parent material of the soil. Sediment deposits are environmentally important in forming the physical base, or substrate, for many forms of plant and animal life. Water-saturated sands, soils, silts, and clays from the life environments of thousands of species of aquatic organisms under the shallow seas, in tidal estuaries, in lake bottoms and stream beds, and in swamps and marshes.

A principle of paramount importance is that organisms modify the sediment in which they live and feed; they also create sediment through life processes, for example, as shells and skeletons. Consequently, a major class of sediments is organically derived, in contrast to the chemically derived products of rock alteration and the physically derived particles of rock disintegration.

Clastic and nonclastic are the two major divisions of sediments (Figure 12.11). *Clastic sediments* are those derived directly as particles broken from a parent rock source, in contrast to the *nonclastic sediments*, which are

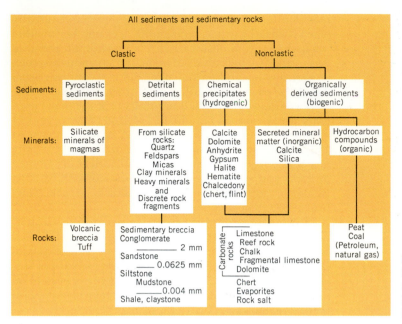

FIGURE 12.11 Composition and classification of sediments and sedimentary rocks. (Redrawn from A. N. Strahler, *Principles of Physical Geology*, Harper & Row, Publishers. Copyright © by Arthur N. Strahler.)

of newly created mineral matter precipitated from chemical solutions or from organic activity. The clastic sediments are, in turn, divided into two kinds. The *pyroclastic sediments* we have already covered under the term *tephra*, which consists of particles of igneous minerals and rocks. In contrast, *detrital sediments* are fragments derived by the decomposition and/or disintegration of preexisting rocks of any type. Detrital fragments may consist of individual or broken mineral crystals or groups of mineral particles as rock fragments. Detrital sediment includes transported products of chemical alteration, such as the clay minerals, the oxides, and the hydroxides.

Because the detrital sediments are derived from any one of the rock groups—igneous, sedimentary, and metamorphic—there is a wide range of parent minerals. One sediment source is from the primary silicate minerals and the secondary alteration products of those minerals. The highly susceptible primary minerals—mostly mafic—are often altered prior to transportation, whereas quartz is immune to such alteration. Consequently, the most important single component of coarse detrital sediment is quartz (see Figure 12.9). Second in abundance are fragments of unaltered fine-grained parent rocks. Feldspar and mica are also commonly present. Clay minerals, particularly kaolinite, illite, and montmorillonite, are major constituents of very fine sediment.

The grade size of particles in a particular body of detrital sediment determines the ease and distance of travel of the particles in transport by water currents. Obviously, the finer the particles, the more easily they are held in suspension in the fluid; the coarser particles tend to settle to the bottom of the fluid layer. In this way a separation of grades, called *sorting*, occurs and determines the texture of the sediment deposit and of the sedimentary rock derived from that sediment. Colloidal clays do not settle out unless they are made to clot together into larger groups. This process, called *floccula-*

tion, takes place when fresh water carrying the sediment mixes with salt water of the ocean.

The concept of sorting of size grades in detrital sediment is extremely important when applied to the parent mineral matter of the soil. On the one hand, when sorting is carried almost to perfection, all the particles in a sample fall into a narrow range of diameters. For example, we may find a sediment consisting almost entirely of fine sand and medium sand, with only small percentages of very fine sand and coarse sand—not even a trace of silt or clay and no pebbles. This sample of extremely good sorting would prove to be taken from a sand dune, in which the winnowing action of the wind has been extremely effective in isolating grains in a narrow range of diameters. In contrast, we may find that a sediment has substantial proportions of clay, silt, sand, and pebbles,

FIGURE 12.12 A boulder of quartzitic conglomerate, cut through and polished, reveals rounded pebbles of quartz and chert. The specimen is about 12 cm across; it comes from the Appalachian ridges of southeastern New York State and is of Silurian age. (A. N. Strahler.)

FIGURE 12.13 These flat-lying strata consist of banded shale layers that are easily eroded by running water. San Rafael Group, Upper Jurassic age, Paria River, southern Utah, west of Glen Canyon City. (A. N. Strahler.)

all well mixed together. In this sediment, sorting is lacking because the process of transportation did not involve any physical mechanism for separating size grades. An example of unsorted sediment would be the material dragged along beneath a moving glacier. The degree of sorting varies greatly according to the process by which the particles are transported and deposited.

CLASTIC SEDIMENTARY ROCKS

Streams carry sediment to lower levels and to locations where permanent accumulation is possible. (Wind and glacial ice also transport sediment, but not necessarily to lower elevations or to places suitable for accumulation.) Usually these sites of accumulation are shallow seas bordering the continent, but they may also be inland seas and lakes. Here the sediment is reworked and redistributed by wave and current action. Over long spans of time, thick masses of sediments undergo physical or chemical changes, or both, to become compacted and hardened, producing sedimentary rock. The process of compaction and hardening is referred to as *lithification*.

Of the kinds of rocks listed in Figure 12.11, those composed of pyroclastic sediments include *volcanic breccia*—a crude mixture of large and small pyroclastic fragments that have fallen close to a volcanic vent—and *tuff*, formed of volcanic ash that has been transported by wind or water and deposited in layers. Of the detrital sedimentary rocks, the coarsest are the *sedimentary breccia* (angular rock fragments in a matrix of fine particles) and *conglomerate*, a common rock consisting of pebbles, cobbles, or boulders (usually well rounded) in a fine-grained matrix of sand or silt (Figure 12.12).

Sandstone is formed of the sand grades of sediment, cemented into a solid rock by silica (SiO_2) or calcium carbonate ($CaCO_3$). The sand grains are commonly of quartz, as shown in Figure 12.9, and sometimes of feldspar; but in other cases they are sand-sized fragments of fine-grained rock containing several minerals. (*Siltstone* is similar to sandstone, but composed of silt-sized grains.)

A mixture of water with particles of silt and clay sizes, along with some sand grains, is called *mud*. The sedimentary rock hardened from such a mixture is called *mudstone*. Compacted and consolidated clay layers become *claystone*.

Sedimentary rocks of mud composition are commonly laminated in such a way that they break apart into small flakes and plates; they are described as being *fissile*. Rock with this structure is called *shale* (Figure 12.13).

Shale is the most abundant of the sedimentary rocks. It is formed largely of the clay minerals kaolinite, illite, and montmorillonite. The compaction of the original clay and mud involves a considerable loss of volume as water is driven out.

A characteristic feature of sedimentary rocks is their layered arrangement, the layers being called *strata*. Typically, layers of different textural composition are alternated or interlayered. The planes of separation between layers are known as *bedding planes*. Bedding planes of strata deposited on the ocean floor (marine strata) are usually horizontal in attitude, or nearly so (Figure 12.14).

FIGURE 12.14 Cliffs and ledges of upper Paleozoic sedimentary rock formations in the North Rim of Grand Canyon, Arizona. The great sheer cliff of sandstone (center) is the Coconino Formation, an ancient dune deposit. Below it are deep red shales and thin sandstones of the Hermit and Supai formations. Above lie marine limestones of the Kaibab Formation, forming the canyon rim. (A. N. Strahler.)

NONCLASTIC SEDIMENTS AND SEDIMENTARY ROCKS

The nonclastic sediments are of particular importance in the environment because they represent enormous storages of carbon, obtained from atmospheric carbon dioxide and changed into carbonate and hydrocarbon compounds by both inorganic and organic processes.

We recognize two major divisions of the nonclastic sediments: (1) *Chemical precipitates* are compounds precipitated directly from water in which the ions are transported; these sediments are described as *hydrogenic*. (2) *Organically derived sediments* are created by the life processes of plants and animals; these are described in *biogenic* (see Figure 12.11).

Chemical precipitates are of major importance in sediments of the seafloor; but a second important environment of deposition is in inland salt lakes of desert regions, where evaporation greatly exceeds precipitation (see Chapter 16).

The most important hydrogenic and biogenic minerals (hydrocarbon compounds are not included) are listed in Table 12.4, together with their compositions. The first three—calcite, aragonite, and dolomite—are classed as *carbonates*. *Calcite* is the dominant carbonate mineral and occurs in many forms. *Aragonite* is of the same composition as calcite, but of different crystal structure; it is secreted in the shells of certain invertebrate animals. *Dolomite* contains magnesium as well as calcium. All three carbonate minerals are soft substances, as compared with the silicate minerals.

The carbonate minerals and rocks are highly susceptible to a form of chemical weathering known as *carbonation*. Atmospheric carbon dioxide (CO_2), dissolved in surface water, forms a weak acid—carbonic acid (H_2CO_3). This acid reacts with calcium carbonate ($CaCO_3$) to produce a soluble product, calcium bicarbonate, which is removed with surplus water and conducted out of the region by runoff. This process is reversible, so calcium-bearing solutions, when undergoing evaporation, deposit calcium carbonate in crystalline form in

FIGURE 12.15 Evaporites in a salt pan on the floor of Death Valley, California. Water partially fills the small depressions between sharp ridges. (Copyright © 1990 by Mark A. Melton.)

the soil, regolith, or alluvium. Carbonate deposition is a dominant geologic process in regions of dry climates; it is an important soil-forming process as well.

The second group of minerals listed in Table 12.4 are hydrogenic. They are *evaporites*, typically formed where seawater is evaporated in shallow bays and gulfs. *Anhydrite* and *gypsum* are composed of calcium sulfate, the latter in combination with water. The third evaporite, *halite*, is commonly known as rock salt, with the composition sodium chloride ($NaCl$). In refined form, it is the table salt we use in cooking and flavoring. Evaporites formed on the floors of shallow, temporary lakes in a desert climate include many varieties of salts of sodium, calcium, magnesium, and potassium (Figure 12.15).

Hematite, a sesquioxide of iron (Fe_2O_3) is a common hydrogenic mineral in sedimentary rocks; it is a major ore of iron. No less important than the carbonates and evaporities is chalcedony, which we have described as commonly formed during mineral alteration in regolith and soil.

The minerals listed in Table 12.4 can be chemically precipitated from seawater or the water of saline lakes, or they can be secreted by organisms to produce the nonclastic sedimentary rocks. Most common of the carbonate rocks is *limestone*, of which there are many varieties. One source is in reefs built by corals and algae (*reef rock*). Another form is *chalk*, deposited in shallow water; it is made up of the skeletons of a marine form of the algae (Figure 12.16). Other limestones are formed of shell fragments or other broken carbonate matter. Some limestones are densely crystalline. *Dolomite*, a rock of the same name as the mineral that composes it, may have been derived through the alteration of limestone, as magnesium ions of seawater gradually replaced calcium ions.

Chert, composed of chalcedony, is an important siliceous sedimentary rock. It occurs as nodules in limestone and, in some cases, as solid rock layers referred to as *bedded cherts*. Gypsum, anhydrite, and halite are layered

TABLE 12.4 Common Hydrogenic and Biogenic Minerals

	Mineral Name	Composition
EVAPORITES	Calcite	Calcium carbonate $CaCO_3$
	Aragonite	Calcium carbonate $CaCO_3$
	Dolomite	Calcium-magnesium carbonate $CaMg(CO_3)_2$
	Anhydrite	Calcium sulfate $CaSO_4$
CARBONATES	Gypsum	Hydrous calcium sulfate $CaSO_4 \cdot 2\,H_2O$
	Halite	Sodium chloride $NaCl$
	Hematite	Sesquioxide of iron Fe_2O_3
	Chalcedony (chert, flint)	Silica SiO_2

FIGURE 12.16 This great wave-cut cliff of chalk forms a promontory on the English Channel (La Manche) coast of Normandy, France, near Etretat. Waves have carved an arch through the promontory, revealing a distant stack of the same rock. The chalk layers are composed of the calcareous skeletons of microscopic plankton that accumulated on the seafloor in Cretaceous time. The Dover-to-Calais tunnel beneath the Channel was drilled within this formation throughout its entire length. (Snowden Hoyer/Woodfin Camp.)

rocks of their respective mineral compositions and occur in association with clastic sedimentary rocks.

HYDROCARBON COMPOUNDS IN SEDIMENTARY ROCKS

Hydrocarbon compounds form a second group of biogenic sediments. These organic substances occur both as solids (peat and coal) and as liquids and gases (petroleum and natural gas), but only coal qualifies physically for designation as a rock.

Peat, a soft, fibrous substance of brown to black color, accumulates in a bog environment where the continual presence of water inhibits decay and oxidation of plant remains. One form of peat is of freshwater origin and represents the filling of shallow lakes (see Figure 26.2). Thousands of such peat bogs are found in North America and Europe; they occur in depressions remaining after recession of the great ice sheets of the Pleistocene Epoch (Chapter 22). This peat has been used for centuries as a low-grade fuel (Figure 11.35). Peat of a different sort is formed in the saltwater environment of tidal marshes (Chapter 20).

Coal, petroleum, and natural gas are collectively called *fossil fuels* because they originated from organic matter produced by plants or animals that lived in the geological past. *Coal* was formed by large-scale accumulation of plant remains, accompanied by subsidence of the area and burial of the compacted organic matter under thick layers of inorganic sediments. In this way, *coal seams* interbedded with shale, sandstone, and limestone strata came into existence (see Figure 19.17). Groups of strata containing coal seams are referred to as *coal mea-*

sures. Individual seams range in thickness from a fraction of an inch to as great as 12 m (50 ft) in the exceptional case.

Coal is classified into three types, representing a developmental sequence. *Lignite*, or brown coal, is soft and has a woody texture. It represents an intermediate stage between peat and true coal. Further compaction accompanying deep burial resulted in the transformation of lignite into *bituminous coal*, often called "soft coal." In areas where the crust was compressed and folded by mountain-making forces, bituminous coal was further changed, becoming *anthracite*. Whereas bituminous coal typically breaks into blocklike fragments, anthracite exhibits a glassy type of fracture.

The coals consist largely of the elements carbon, hydrogen, and oxygen, with small amounts of sulfur also present. Inorganic impurities form ash, the noncombustible residue remaining after coal is completely burned.

The general term *petroleum* spans the range from liquid crude oil to natural gas in the one direction, and to asphalt and related semisolid hydrocarbon substances in the other. *Crude oil* in the natural state is a mixture of a large number of hydrocarbon compounds. More than 200 compounds have been isolated and analyzed in crude oil.

Crude oils differ in terms of the relative abundances of various hydrocarbon groups. Generally speaking, the paraffin compounds are the most abundant of hydrocarbons in both liquid petroleum and natural gas. Crude oil is described as paraffin-base when paraffins are dominant; it is of low density and typically yields good lubricants and a large proportion of kerosene. An example is the paraffin-base crude oil of the Pennsylvania fields. Asphalt-base crude oil has a high density and is referred to as *heavy oil*; its primary yield is fuel oils.

Natural gas, found in close association with accumulations of crude oil, is a mixture of gases. The principal gas is methane (marsh gas, CH_4) and there are minor amounts of ethane, propane, and butane, all of which are hydrocarbon compounds. Small amounts of carbon dioxide, nitrogen, and oxygen are also present, and sometimes helium.

The amount of sulfur present in crude oil (and in coal) is a matter of great environmental importance, since during combustion of the fuel the sulfur becomes oxidized into sulfur dioxide gas (SO_2). As we learned in Chapter 6, sulfur dioxide is a major air pollutant. The sulfur content of crude oil varies greatly; the highest percentage is 55 times greater than the lowest. The sulfur content of natural gas is generally much lower than for crude oil. For this reason natural gas is preferred as a fuel in urban areas and is described as being "cleaner" than crude oil.

Geologists are in general agreement that petroleum originates from organic matter buried within thick marine sediments that accumulate along continental margins. Clays and muds deposited in a chemically reducing environment where oxygen is lacking are favorable for petroleum accumulation. As the organic matter is buried, it is converted into hydrocarbon compounds of types related to those found in petroleum. Heating at great depth plays an important role in this chemical conversion. A second phase involves the movement, or migration, of petroleum from the source rock to a *reservoir rock*. Petroleum moves both upward and laterally out of the compacting sediment layers and eventually becomes concentrated in a porous rock mass. Most reservoir rocks are sedimentary types, with sand and sandstone often being excellent reservoirs. It is essential that the reservoir rock be capped or surrounded by a dense rock through which the oil cannot pass. Thus a *reservoir trap* is formed and prevents the petroleum from escaping to the surface. Shale is a common kind of cap rock. Figure 19.36 illustrates one of the simplest geologic traps, consisting of an uparching of strata (an anticline or dome structure).

Everyone interested in energy resources has heard of *oil shale* and of the tremendous reserve of hydrocarbon fuel it holds. The fact is that this sedimentary rock in the Rocky Mountain region is not really shale at all, and the hydrocarbon it holds is not really petroleum. Strata of the Rocky Mountains called "oil shales" are composed of calcium carbonate and magnesium carbonate. The strata were formed as lake deposits of lime mud in a Cenozoic lake. These soft, laminated deposits belong to the Green River Formation. The oil shale beds occur largely in northwestern Utah, northwestern Colorado, and southwestern Wyoming.

The hydrocarbon matter of the Green River Formation is a waxy substance, called *kerogen*, which adheres to the tiny grains of carbonate material. When the shale is crushed and heated to a temperature of 480°C (900°F) the kerogen is altered to petroleum and driven off as a liquid. The rock may be mined and processed in surface plants, or burned in underground mines, from which the oil is pumped to the surface.

Yet another form of occurrence of hydrocarbon fuels is *bitumen*, a variety of petroleum that behaves much as a solid, although it is actually a highly viscous liquid. Bitumen goes by other common names, such as tar, asphalt, or pitch. In some localities bitumen occupies pore spaces in layers of sand or porous sandstone. It remains immobile in the enclosing sand and will flow only when heated. Outcrops of *bituminous sand* (*oil sand*) exposed to the sun will show bleeding of the bitumen. Perhaps the best known of the great bituminous sand deposits are those occurring in Alberta, Canada. Where exposed along the banks of the Athabasca River, the oil sand is being extracted from surface mines. Extraction of oil from wells will require that the sand be heated by steam or other heat sources.

METAMORPHIC ROCKS

Any of the types of igneous or sedimentary rocks may be altered by the tremendous pressures and high temperatures that accompany mountain-building processes of the earth's crust. The result is a rock so changed in appearance and structure as to be classified as a metamorphic rock. Typically, metamorphism results in the development of new textures and structures within the rock. Mineral components of the parent rock are, in many cases, reconstituted into different mineral varieties. Recrystallization of the original minerals can also occur.

Shale, after being subjected to the unequal application of stress, is altered to *slate*, a fine-textured rock that splits neatly into thin cleavage plates so familiar as roofing shingles and as flagstones of patios and walks.

With application of increased heat and unequal stress, slate changes into *schist*, representing a more advanced stage of metamorphism. Schist and slate have a structure called *foliation*. In schist, foliation consists of thin and irregularly curved planes of parting in the rock (Figure 12.17). Schist is set apart from slate by the coarse texture of the mineral grains, the abundance of mica, and occasionally the presence of scattered large crystals of newly formed minerals, such as garnet.

At a still more advanced level of metamorphism, schist may be altered into *gneiss*, a rock characterized by alternate light and dark streaks or bands (Figure 12.18). The light bands are composed largely of felsic minerals (quartz and feldspar); the dark bands, of mafic minerals (biotite, amphibole, and pyroxene). Schist and gneiss can also form from igneous rock such as basalt or granite. In gneiss formed directly from granite (granite gneiss), mineral grains may be simply drawn out into

FIGURE 12.17 This freshly exposed schist shows shiny surfaces (slickensides) along which shearing has occurred. Known as "greenschist," it represents the end product of deep deformation of black mud once deposited on the floor of an ancient oceanic trench. Franciscan Formation, Gold Beach, Oregon. (A. N. Strahler.)

FIGURE 12.18 Intensely deformed structures in gneiss of Precambrian age, revealed by weathering of this glacially polished rock surface near Breckenridge, Colorado, in the Rocky Mountains. (Copyright © 1988 by Mark A. Melton.)

long, pencil-like shapes, aligned in the direction in which the rock mass was elongated.

Two common metamorphic rocks are recognized on the basis of being composed largely of a single mineral. The metamorphic equivalent of pure quartz sandstone or chert is *quartzite*, formed by recrystallization of the quartz as a result of heat and pressure. *Marble* is formed by the recrystallization or growth in crystal size of the mineral calcite in limestone during the application of heat and pressure. Marble has a coarse texture on freshly broken surfaces. Bedding planes are obscured, and masses of mineral impurities may be drawn out into darker streaks or bands.

THE CYCLE OF ROCK TRANSFORMATION

We can now bring together the formative processes of rocks into a single unified concept of recycling of matter through geologic time. A schematic diagram, Figure 12.19, distinguishes between a *surface environment* of low pressures and temperatures and a *deep environment* of high pressures and temperatures. The deep environment is the realm of the intrusive igneous and metamorphic rocks. The surface environment is one of mineral alteration and sediment deposition.

Seen in its complete form, the total circuit of rock changes in response to environmental stress constitutes the *cycle of rock transformation*. Figure 12.19 emphasizes that mineral matter is continually recycled through the three major rock classes. Igneous rocks are by no means the "original" rocks of the earth's crust. Actually, there is no known record of the rocks that first formed the earth's crust; they were consumed and recycled long ago.

This brief introduction to common minerals and rocks that make up the earth's crust prepares us to look at a dynamic geologic system—plate tectonics—which provides a new perspective on the cycle of rock transformation.

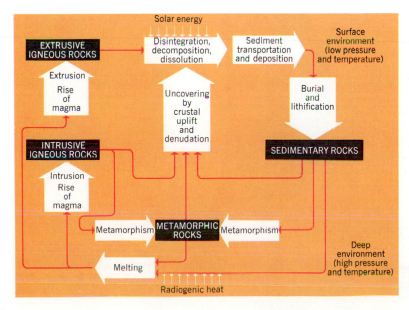

FIGURE 12.19 Schematic diagram of the cycle of rock transformation. (From A. N. Strahler, *The Earth Sciences*, 2nd ed., Harper & Row, Publishers. Copyright © by Arthur N. Strahler.)

CHAPTER 13

THE LITHOSPHERE AND PLATE TECTONICS

Environmental regions of the globe depend for their distribution on configurations of the earth's crust dictated by geologic processes. These processes are powered by energy sources deep within the earth. The internal earth forces have shaped the continents and ocean basins without conforming in the least to the orderly latitude zones of climate. We can think of latitude zones of temperature, winds, and precipitation as concentric color bands painted on a circular dinner plate as the potter's wheel spins. Suppose that the dinner plate falls to the floor and is shattered and that the fragments are put back into place and cemented together. The fracture patterns cut across the circular color zones in a discordant and random pattern.

This same unique combination of disorder on order characterizes the earth's solid surface. We find volcanoes erupting today in the cold desert of Antarctica as well as near the equator in Africa. An alpine mountain range has been pushed up in the cold subarctic zone of Alaska, where it trends east–west; but another lies astride the equator in South America and runs north–south. Both ranges lie in belts of intense crustal activity, where numerous strong earthquakes are generated. Although the geologic processes that create high mountains are quite insensitive to latitude and climate, climates do respond to mountain ranges through the orographic effect. In this way, the chance configurations of the earth's relief features bring diversity to the global climate.

In this chapter we will survey the major geologic features of our planet, starting with its deep interior as a layered structure. We then examine the outermost layer, or crust, comparing crustal forms of the continents with those of the ocean basins. Only within the past two decades have geologists provided a unified theory to explain the differences between continents and ocean basins and to interpret the major forms of crustal unrest in a meaningful way. Fortunately, we are reviewing geologic processes just at the moment in history when a new revolution in geology has come to completion. The revolutionary findings now furnish us with a complete scenario of earth history on a grand scale of both time and spatial dimensions.

THE EARTH'S INTERIOR

Figure 13.1 is a cutaway diagram of the earth to show its major parts. The earth is an almost spherical body approximately 6400 km (4000 mi) in radius. The center is occupied by the *core*, a spherical zone about 3500 km (2200 mi) in radius. Because of the sudden change in behavior of earthquake waves upon reaching this zone, it has been concluded that the outer core has the properties of a liquid, in abrupt contrast to the solid state of the rock shell that surrounds it. The innermost part of the core, with a radius of 1255 km (780 mi), is probably in the solid state.

Astronomical calculations show that the earth has an average density of about 5½ gm/cc, whereas the surface rocks average 3 gm/cc or less (Chapter 12). This observation must mean that density increases greatly toward

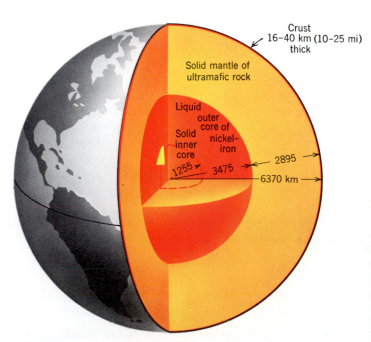

FIGURE 13.1 Concentric zones of the earth's interior.

228

the earth's center, where it may be about 10–15 gm/cc. Iron, with a small proportion of nickel, is considered to be the substance comprising the entire core. This conclusion is supported by the fact that many meteorites, representing disrupted fragments of our solar system, are of iron–nickel composition. Temperatures in the earth's core may lie between 2200° and 2750°C (4000° and 5000°F); pressures are as high as three to four million times the pressure of the atmosphere at sea level.

Outside the core lies the *mantle*, a layer about 2895 km (1800 mi) thick, composed of mineral matter in a solid state. Judging from the behavior of earthquake waves, the mantle is probably composed largely of the mineral olivine (magnesium iron silicate). A comparatively rare surface rock of this composition is dunite (see Figure 12.2). The upper zone of the mantle is probably of the composition of peridotite, another ultramafic rock described in Chapter 12.

THE EARTH'S CRUST

Outermost and thinnest of the earth zones is the *crust,* a layer 5 to 40 km (3 to 25 mi) thick, formed largely of igneous and metamorphic rocks. The base of the crust, where it contacts the mantle, is sharply defined. This contact is established from the way in which earthquake waves change velocity abruptly at that level (Figure 13.2). The surface of separation between crust and mantle is called the *Moho,* a simplification of the name of the seismologist who discovered it.

From a study of earthquake waves it is concluded that the crust beneath the continents consists of two rock layers: a lower, continuous rock layer of mafic composition; and an upper layer of felsic rock. Because the felsic layer has a chemical composition about like that of granite, it is commonly described as being *granitic rock.* There is no sharply defined surface of separation between these layers. The crust of the ocean basins consists of basalt; the felsic continental layer is absent. Figure 13.3 shows schematically a small part of the crust near the margin of a continent.

The crust is much thicker beneath the continents than beneath the ocean floors, as Figure 13.3 shows. Whereas 40 km (25 mi) is a good average figure for crustal thickness beneath the continents, 5 km (3 mi) is an average figure for thickness of the basaltic crust beneath the deep ocean floors. The reasons for differences in both thickness

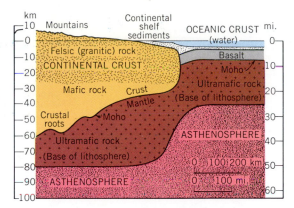

FIGURE 13.3 A schematic diagram showing the composition and thickness of the crust and mantle.

and rock composition between continental and oceanic crust are explained later in terms of processes that have created the crust.

LITHOSPHERE AND ASTHENOSPHERE

A new concept must now be added to what has been said thus far about the earth's structure. Evidence derived from the study of earthquake waves shows that within the upper mantle is a *soft layer* in which the mantle rock is at a temperature close to the melting point. A good analogy would be a bar of cast iron heated at one end until it is white hot, but not quite hot enough to melt. The cold part of the bar is brittle and will snap with a sharp fracture if it is struck. The hot part of the bar is quite soft and can be shaped under pressure of a vise or by hammer blows. Beneath the continental crust, the soft layer of the mantle sets in at an average depth of about 80 km (50 mi), which is well below the base of the continental crust (see Figure 13.3). Beneath the oceanic crust, the soft layer sets in at a depth of about 40 km (25 mi).

In modern geologic language, the soft layer of the upper mantle is the *asthenosphere;* the rigid layer above it is the *lithosphere* (Figure 13.4). Geologists have thus restricted the meaning of lithosphere to an outer rock layer; but this does not conflict with our use of the word to mean the solid earth realm. The asthenosphere, or soft layer, extends to a depth of about 300 km (185 mi). The upper and lower limits are gradational. The weakest portion of the asthenosphere is at a depth of roughly 200 km (125 mi).

The important concept we derive from these facts is that the rigid lithosphere has the capability of moving bodily over the asthenosphere. The asthenosphere yields by slow plastic flowage distributed through a thickness of many tens of kilometers.

If the lithosphere formed a single continuous shell over the entire earth, that shell would—in theory at least—be capable of rotating bodily over the deeper mantle and core. Instead, the lithosphere is broken into large units called *lithospheric plates.* A single plate is of continental dimensions and is capable of moving inde-

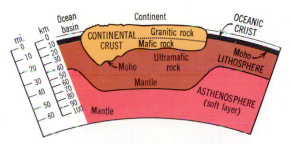

FIGURE 13.2 The earth's crust is much thicker under continents than beneath the ocean basins.

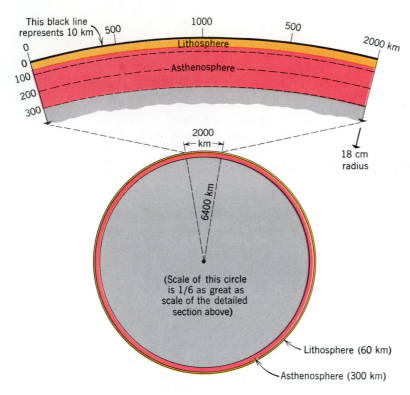

FIGURE 13.4 The earth's lithosphere and asthenosphere drawn to true scale. The curvature of the upper diagram fits a circle 18 cm in radius. The black line at the top is scaled to represent a thickness of 10 km (6 mi); it will accommodate about 98 percent of the earth's surface features, from ocean floors to high mountains and plateaus. Only a few lofty mountains would project above the black line and only a few deep ocean trenches would project below the line. The complete circle below is drawn on a scale ⅙ as large as the upper diagram. Here we gain an appreciation of the extreme thinness of the mobile lithospheric plates that move over the asthenosphere.

pendently of the plates that surround it. Plate motions and boundaries are a topic we shall elaborate on later in this chapter.

DISTRIBUTION OF CONTINENTS AND OCEAN BASINS

The first-order relief features of the earth are the continents and ocean basins. Using a globe, we can compute that about 29 percent of the globe is land, 71 percent is oceans. If the seas were to drain away, however, it would become obvious that broad areas lying close to the continental shores are actually covered by shallow water, less than 180 m (600 ft) deep. From these relatively shal-

low continental shelves the ocean floor drops rapidly to depths of thousands of meters. In a way, then, the ocean basins are brimful of water. The oceans have even spread over the margins of ground that would otherwise be assigned to the continents. If the ocean level were to drop by 180 m (600 ft), the surface area of continents would increase to 35 percent; the ocean basins would decrease to 65 percent. We can use these figures as representative of the true relative proportions.

Figure 13.5 shows graphically the percentage distribution of the earth's surface area with respect to elevation both above and below sea level. Note that most of the land surface of the continents is less than 1 km (3300 ft) above sea level. There is a rapid drop-off from about −1 to −3 km (−3000 to −10,000 ft) until the ocean floor

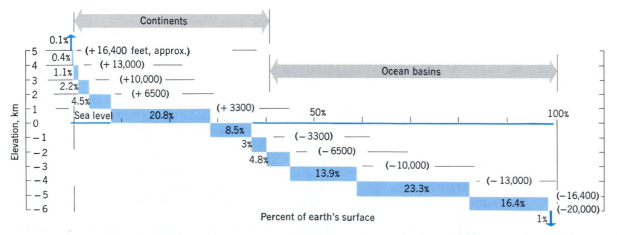

FIGURE 13.5 Distribution of the earth's solid surface in successively lower elevation zones.

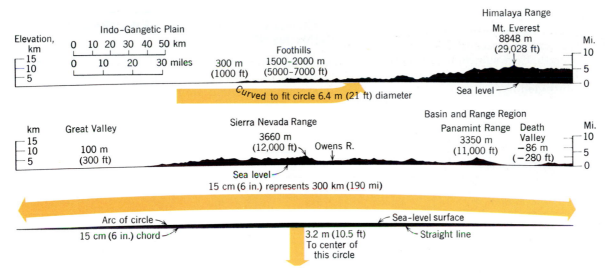

FIGURE 13.6 These profiles show the earth's great relief features in true scale, with sea-level curvature fitted to a globe 6.4 m (21 ft) in diameter.

is reached. A predominant part of the ocean floor lies between 3 and 6 km (10,000 to 20,000 ft) below sea level. Disregarding the earth's curvature, the continents can be visualized as platformlike masses, the oceans as broad, flat-floored basins.

SCALE OF THE EARTH'S RELIEF FEATURES

Before turning to a description of the major subdivisions of the continents and ocean basins, we need to grasp the true scale of the earth's relief features in comparison with the earth as a sphere (Figure 13.6). Most relief globes and relief maps are greatly exaggerated in vertical scale.

For a true-scale profile around the earth, we might draw a chalk-line circle 6.4 m (21 ft) in diameter, representing the earth's circumference on a scale of 1: 2,000,000. A chalk line 1 cm (³/₈ in.) wide would include within its limits not only the highest point on the earth, Mount Everest (8840 m; 29,000 ft), but also the deepest known ocean trenches, somewhat deeper than 11,000 m (36,000 ft).

Figure 13.6 shows profiles correctly curved and scaled to fit a globe whose diameter is 6.4 m (21 ft). The surface profile is drawn to natural scale, without vertical exaggeration. Although the most imposing landscape features of Asia and North America are shown, they are only trivial irregularities on the great global circle.

THE GEOLOGIC TIME SCALE

To place crustal rocks and structures in their positions in time, we need to refer to some major units in the scale of geologic time. Table 13.1 is a greatly abbreviated list of the major time divisions and their ages. All time older

than 570 million years is *Precambrian time*. Three *eras* of time follow: *Paleozoic*, *Mesozoic*, and *Cenozoic*. These eras saw the evolution of life forms in the oceans and on the lands.

Geologic eras are subdivided into *periods*. Name, duration, and age of each period are given in Table 13.1. Individual periods will be mentioned on later pages in connection with the movements of lithospheric plates and the breakup of the ancient continents. Throughout the entire course of geologic time, brief but intense episodes of crustal deformation occurred in which strata were crumpled and broken by tectonic activity. These events of geologic history are called *orogenies*. The names of several important orogenies are shown in Table 13.1. Precambrian orogenies named in the table are those of the Canadian Shield region of North America. Names of orogenies throughout the three younger eras also apply to North America; equivalent European names are given in parentheses.

The Cenozoic Era is particularly important in terms of continental landscapes because nearly all landforms seen today were produced in the 66 million years (m.y.) since that era began. Because the Cenozoic Era is comparatively short in duration—scarcely more than the average duration of a single period within older eras—it is subdivided directly into *epochs*.

In recent years, geologists have found it convenient to refer to the first three Cenozoic epochs as the *Paleogene Period*, while lumping the Miocene and Pliocene epochs under the name of the *Neogene Period*.

Notice that on Table 13.1, the Cenozoic Era is also shown as subdivided into a *Tertiary Period* (Paleocene through Pliocene) and a *Quaternary Period* that lumps together the Pleistocene and Holocene epochs. Although archaic, this grouping is widely used today by many geologists.

Ages given in Table 13.1 have been determined through chemical analysis of radioactive elements enclosed in rocks of all ages. Although the method of age determination is subject to a small range of error, ages are well established and are accepted by geologists. Events of Precambrian time are subject to greater uncertainties because these ancient rocks have been repeatedly deformed and intruded by magmas.

SECOND-ORDER RELIEF FEATURES OF THE CONTINENTS

When the continents and ocean basins are considered as first-order relief features, we can then recognize subdivisions within each that are relief features of a second or-

TABLE 13.1 The Geologic Time Scale

Era	Period	Epoch	Duration m.y.	Age m.y.	Orogenies
CENOZOIC	(Quaternary)	Holocene	(10,000 yr)		
		Pleistocene	2		
				2	
	(Tertiary) / Neogene	Pliocene	3		
				5	
		Miocene	19		
				24	
	Paleogene	Oligocene	13		
				37	
		Eocene	21		
				58	
		Paleocene	8		
				66	
MESOZOIC	Cretaceous		78		Laramide
				144	
	Jurassic		64		Nevadan
				208	
	Triassic		37		
				245	
PALEOZOIC	Permian		41		Allegheny (Hercynian)
				286	
	Carboniferous	Pennsylvanian	34		
				320	
		Mississippian	40		
				360	Acadian
	Devonian		48		
				408	(Caledonian)
	Silurian		30		
				438	Taconic
	Ordovician		67		
				505	
	Cambrian		65		
				570	

Era	Period		Duration b.y.	Age b.y.	Orogenies
PRECAMBRIAN TIME	Late Precambrian		0.3–0.4		
				0.9–1.0	Grenville
	Middle Precambrian		0.6–0.8		
				1.6–1.7	Hudsonian
			0.7–0.9		
				2.4–2.5	Kenoran
	Early Precambrian		0.9–1.0		
		Oldest dated rocks		3.6–3.8	
		Earth accretion completed		4.6–4.7	
		Age of universe		17–18	

Data source: 1983 Geologic Time Scale, Decade of North American Geology, *Geology,* vol. 11, p. 504, 1983.

der of magnitude. We shall deal first with the relief features of the continents, familiar to most persons from direct experiences of travel and from the medium of photography.

Broadly viewed, the continental masses consist of two basic subdivisions: (1) active belts of mountain making and (2) inactive regions of old rocks. The growth of mountain ranges occurs through one or two very different geologic processes. First is *volcanism*, the formation of massive accumulations of volcanic rock by extrusion of magma (Chapter 12). Many lofty mountain ranges consist of chains of volcanoes built of lava and tephra. Second of the mountain-building processes is *tectonic activity*, the breaking and bending of the earth's crust under internal earth forces. Crustal masses that are raised by tectonic activity form mountains and plateaus; masses that are lowered form crustal depressions. In many instances, volcanism and tectonic activity have combined to produce a mountain range.

ALPINE CHAINS

Active mountain-making belts are narrow zones; most lie along continental margins. These belts are sometimes referred to as *alpine chains* because they are characterized by high, rugged mountains, such as the Alps of central Europe. These mountain belts are formed in the Cenozoic Era by volcanic or tectonic activity, or a combination of both, and this activity has continued to the present day in many places. The alpine chains are characterized by broadly curved patterns on the world map (Figure 13.7). Each curved section of an alpine chain is referred to as a *mountain arc;* the arcs are linked in sequence to form the two principal mountain belts. One is the *circum-Pacific belt;* it rings the Pacific Ocean basin. In North and South America, this belt is largely on the continents and includes the Andes and Cordilleran ranges. In the western part of the Pacific basin, the mountain arcs lie well offshore from the continents and take the form of *island arcs*, running through the Aleutians, the Kuriles, Japan, the Philippines, and many lesser islands. Between the large islands, these arcs are represented by volcanoes rising above the sea as small, isolated islands.

The second chain of major mountain arcs forms the *Eurasian-Indonesian belt*, starting in the west at the Atlas Mountains of North Africa and running through the Near East and Iran to join the Himalayas. The belt then continues through Southeast Asia into Indonesia, where it meets the circum-Pacific belt. We shall return later to these active belts of mountain-making, explaining them in terms of lithospheric plate motions.

A world map of structural regions, Figure 13.8, recognizes the alpine chains as belonging to an *Alpine system*. Because the Alpine system also includes some adjacent inactive regions produced by orogenies of the Mesozoic Era, it appears on the world map in broad belts rather than as the narrow, linear arcs of late Cenozoic activity suggested in Figure 13.7.

CONTINENTAL SHIELDS AND MOUNTAIN ROOTS

Belts of recent and active mountain-making account for only a small portion of the continental crust. The remainder consists of inactive regions of much older rock. Within these inactive regions, we recognize two structural types of crust: shields and mountain roots. *Continental shields* are low-lying continental surfaces beneath which lie igneous and metamorphic rocks in a complex arrangement (Figure 13.8). The rocks are very old, mostly of Precambrian age, and have had a very involved geologic history. For the most part, the shields are regions of low hills and low plateaus, although there are some exceptions where large crustal blocks have been uplifted. Many thousands of meters of rock have been eroded from the shields during their exposure throughout a half-billion years.

Large areas of the continental shields are under a cover of younger sedimentary layers, ranging in age from the Paleozoic through the Cenozoic eras. These strata accumulated at times when the shields subsided and were inundated by shallow seas. Marine sediments were laid down on the ancient shield rocks in thicknesses ranging from hundreds to thousands of meters. These shield areas were then arched and again became land surfaces. Fluvial denudation has since removed large sections of the sedimentary cover, but it remains intact over vast areas. We refer to such areas as *covered shields* to distinguish them from the *exposed shields*, in which the Precambrian rocks lie bare. Figure 13.8 shows the covered and exposed shields in relation to the mountain belts. An example of an exposed shield is the Canadian Shield of North America. Exposed shields are also extensive in Scandinavia, South America, Africa, peninsular India, and Australia.

Crustal movements of the shields in later geologic time have been of a type known as *epeirogenic movements*, that is, rising or sinking of the crust over broad areas without appreciably breaking or bending the rocks. Epeirogenic movements reflect crustal stability generally, in contrast to tectonic activity affecting mountain arcs.

Remains of older mountain belts lie within the shields in many places. These *mountain roots* are formed mostly of Paleozoic and Mesozoic sedimentary rocks that have been intensely deformed and locally changed into metamorphic rocks.

One important system of mountain roots was formed in the Caledonian orogeny, occurring about the end of the Silurian Period (Table 13.1). Often referred to as Caledonides, these roots are found in a belt extending from northern Ireland, through Scotland, into Scandinavia (Figure 13.8). A second important root system was formed in the Allegheny orogeny that closed the Paleozoic Era. In North America this system is represented by the newer Appalachian Mountains. In Europe, a roughly equivalent event called the Hercynian orogeny formed a mountain belt (the Hercynides) extending across the southern British Isles and northern Europe. Our world

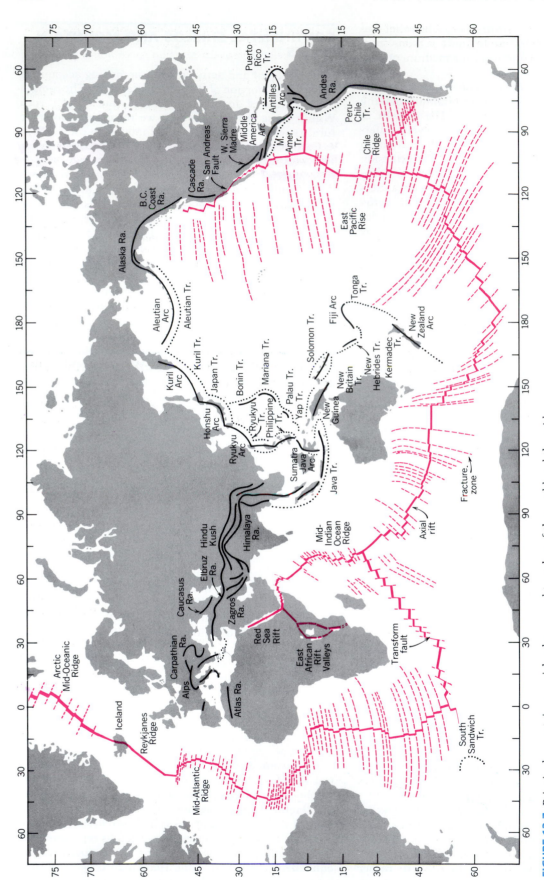

FIGURE 13.7 Principal mountain arcs, island arcs, and trenches of the world and the mid-oceanic ridge. (Mid-oceanic ridge map from A. N. Strahler, *Physical Geology*, Harper & Row, Publishers. Copyright © by Arthur N. Strahler.)

FIGURE 13.8 World structural regions. (From R. E. Murphy, *Annals*, *Association of American Geographers*, Map Supplement No. 9. Based on Goode Base Map.)

A. Alpine system. World-girdling system of mountain chains formed since late Mesozoic time (Cretaceous Period, or younger). Faulted areas, plateaus, basins, and coastal plains enclosed by such ranges are included in the system.

C. Caledonian and Hercynian mountain roots, the remains of mountain chains and ranges formed during the Paleozoic and Mesozoic eras, prior to the Cretaceous Period. Faulted areas, plateaus, basins, and coastal plains enclosed by these mountain remnants are included.

G. Gondwana shields. Areas of stable, massive blocks of continental crust, lying south of the great east–west portion of the Alpine system, where Precambrian rocks form either the entire surface rock or an encircling enclosure with no gap of more than 320 km (200 mi) between outcroppings of younger rock layers.

L. Laurasian shields. Areas of stable, massive blocks of continental crust lying north of the great east–west portion of the Alpine system. (Remainder of definition same as in G, above.)

R. Rifted shield areas. Block-faulted areas of shields forming grabens together with associated horsts and volcanic features. Rifting is a result of extension and thinning of continental lithospheric plates.

V. Isolated volcanic areas. Areas of volcanoes, active or extinct, with associated volcanic features, lying outside the Alpine or older mountain systems and outside the rifted shield areas. Volcanism is an expression of hot spots above mantle plumes.

S. Sedimentary covers. Areas of sedimentary layers that have not been subjected to orogeny and that lie outside either the crystalline rock enclosures of the shields or the enclosing mountains and hills of the Alpine or older orogenic systems. These areas of sedimentary rock form continuous covers over underlying structures.

(Illustration on pages 236/237.)

map of structural regions, Figure 13.8, recognizes the Paleozoic mountain roots of both Caledonian and Allegheny (Hercynian) orogenies as a distinct region. Thousands of meters of overlying rocks have been removed from these old tectonic belts, so only basal structures remain. Roots appear as chains of long, narrow ridges, rarely rising over a thousand meters above sea level. Landforms of mountain roots are described in Chapter 19.

A somewhat younger generation of mountain belts was produced by orogenies of the Mesozoic Era. Named in Table 13.1 are the Nevadan and Laramide orogenies that affected strata of Mesozoic age in the western Cordilleran ranges, including the Rocky Mountains. Although these ranges have experienced a great amount of erosion throughout the Cenozoic Era, they remain lofty mountains because of a series of later uplifts. They cannot be classified as ancient mountain roots and are best described as being intermediate in evolution between the ancient mountain roots of Paleozoic age and the alpine belts of late Cenozoic age. On the world map of structural regions (Figure 13.8), the Mesozoic mountain belts are included within the Alpine system.

SECOND-ORDER RELIEF FEATURES OF THE OCEAN BASINS

Crustal rock beneath the ocean floors consists almost entirely of basalt, covered over large areas by a comparatively thin accumulation of sediments. Age determinations of the basalt and its sediment cover show that the oceanic crust is quite young, geologically speaking. Much of that crust was formed during the Cenozoic Era and is less than 60 million years old. Over large areas, the rock age is Mesozoic and falls into the Cretaceous Period (−65 to −136 m.y.). Small areas belong to the Jurassic Period (−136 to −190 m.y.). These data mean that the ocean basins began to come into existence early in the Mesozoic Era. When we consider that the bulk of the continental crust is of Precambrian age—mostly over 1 billion years old—the young age of the oceanic crust is all the more remarkable. We will need to fit this fact into the general theory of global tectonic activity.

In overall plan, the ocean basins are characterized by a central ridge structure that divides the basin about in half. This feature is shown by schematic diagram in Figure 13.9. The *mid-oceanic ridge* consists of submarine

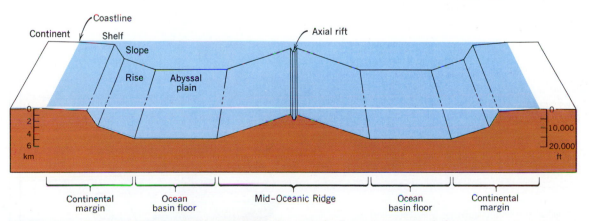

FIGURE 13.9 This schematic block diagram shows the ocean basins as symmetrical elements on a central axis. The model applies particularly well to the North and South Atlantic oceans.

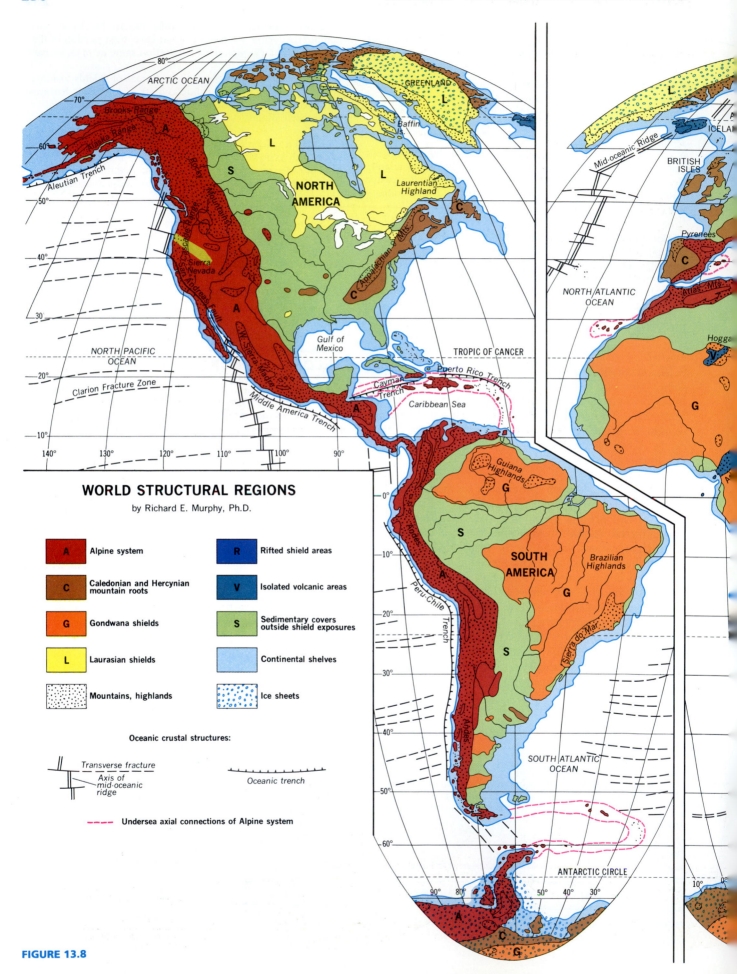

WORLD STRUCTURAL REGIONS

by Richard E. Murphy, Ph.D.

A	Alpine system	
C	Caledonian and Hercynian mountain roots	
G	Gondwana shields	
L	Laurasian shields	
	Mountains, highlands	
R	Rifted shield areas	
V	Isolated volcanic areas	
S	Sedimentary covers outside shield exposures	
	Continental shelves	
	Ice sheets	

Oceanic crustal structures:

Transverse fracture

Axis of mid-oceanic ridge

Oceanic trench

– – – Undersea axial connections of Alpine system

FIGURE 13.8

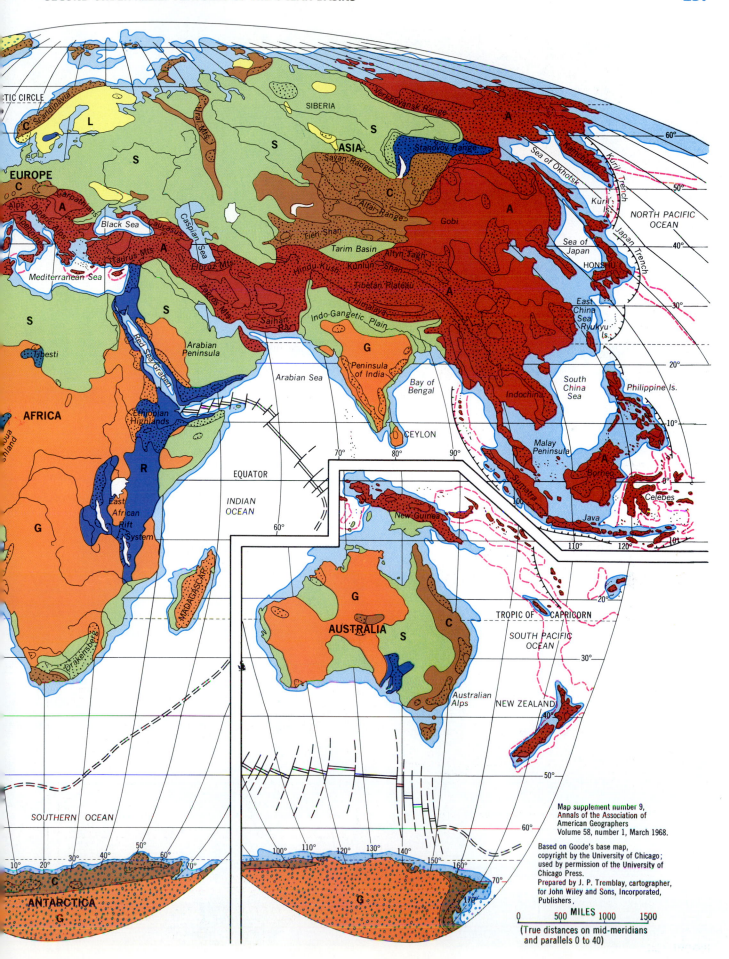

Map supplement number 9,
Annals of the Association of
American Geographers
Volume 58, number 1, March 1968.

Based on Goode's base map,
copyright by the University of
Chicago; used by permission of the University of
Chicago Press.
Prepared by J. P. Tremblay, cartographer,
for John Wiley and Sons, Incorporated,
Publishers,

MILES
0   500   1000   1500
(True distances on mid-meridians
and parallels 0 to 40)

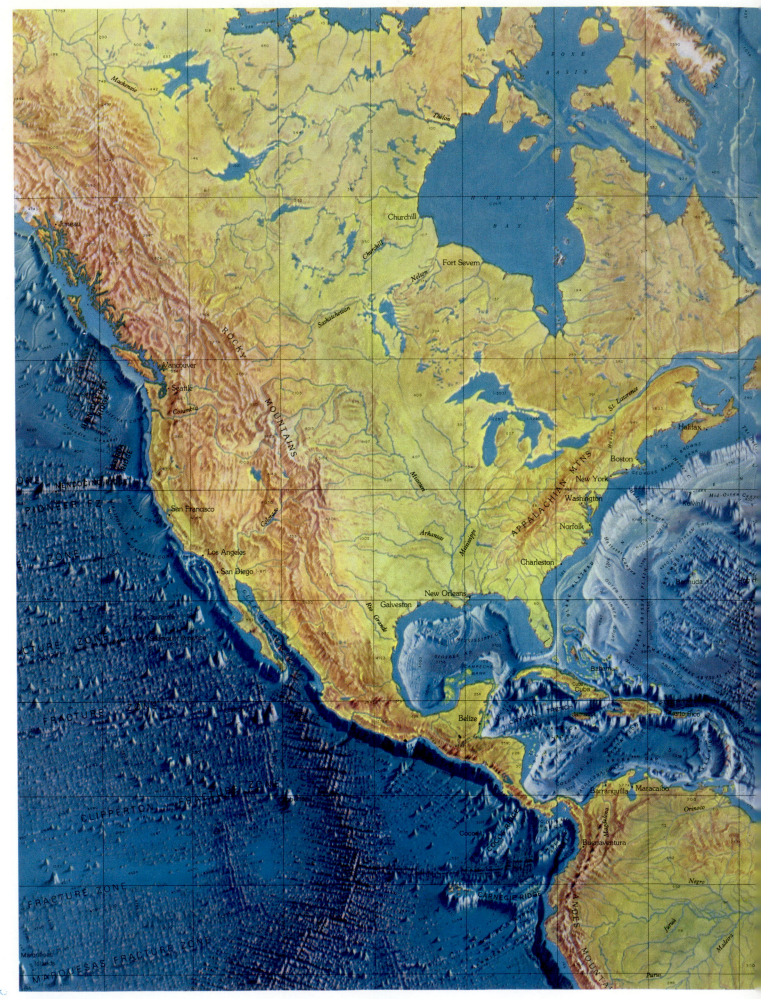

FIGURE 13.10 238

WORLD OCEAN FLOOR
by B. C. Heezen and Marie Tharp

copyright Marie Tharp 1977

FIGURE 13.10 A portion of Map of the World Ocean Floor by Bruce C. Heezen and Marie Tharp. Based on Mercator map projection. (Copyright © 1977 by Marie Tharp. Used by permission.)
(Illustration on pages 238/239.)

hills rising gradually to a mountainous central zone (Figure 13.10). Precisely in the center of the ridge, at its highest point, is an *axial rift*, which is a trenchlike feature. The form of this rift suggests that the crust is being pulled apart along the line of the rift axis. The midoceanic ridge and its axial rift can be traced through the ocean basins for a total distance of about 64,000 km (40,000 mi). Figure 13.7 shows the extent of the ridge. From the South Atlantic, the ridge turns east and enters the Indian Ocean. There, one branch penetrates Africa while the other continues east between Australia and Antarctica, then swings across the South Pacific. Nearing South America, it turns north and penetrates North America at the head of the Gulf of California, as seen in Figure 13.10.

The axial rift is broken in many places along its length by crustal fractures. Motion on these fracture lines (faults) has caused the rift to be sharply offset. Traces of the offsetting fractures extend far out on either side of the midoceanic ridge. As we shall explain later, the axial rift and its offsetting fractures represent the boundary between adjacent lithospheric plates that are undergoing separation.

On either side of the mid-oceanic ridge are broad, deep plains and hill belts belonging to the *ocean basin floor* (Figure 13.9). Their average depth is about 5 km

(17,000 ft). The flat surfaces are called *abyssal plains;* they are extremely smooth because they have been built up of fine sediment.

THE PASSIVE CONTINENTAL MARGINS

Nearing the continents the ocean floor begins to slope gradually upward, forming the *continental rise*. The floor then steepens greatly in the *continental slope*. At the top of this slope we arrive at the brink of the *continental shelf*, a gently sloping platform some 120 to 160 km (75 to 100 mi) wide along the eastern margin of North America. Water depth is about 180 m (600 ft) at the outer edge of the shelf.

The *continental margin*, shown in Figure 13.9 as the third element of the typical ocean basin, can be defined as the narrow zone in which oceanic lithosphere is in contact with continental lithosphere (see Figure 13.3). Thus, the continental margin is a feature shared by continent and ocean basin.

The symmetrical model illustrated in Figure 13.9 is fully exemplified in the North Atlantic and South Atlantic ocean basins (Figure 13.11). It also applies rather well to the Indian Ocean and Arctic Ocean basins. The margins of these symmetrical basins are described as *passive continental margins*, meaning that they have not been subjected to Cenozoic tectonic and volcanic activity.

The passive continent margins are underlain by great thicknesses of sedimentary strata derived from the continents. The strata range in age from the late Mesozoic (Jurassic, Cretaceous) through the Cenozoic. The shelf strata form a wedge-shaped deposit, thinning landward and feathering out over the continental shield (Figure

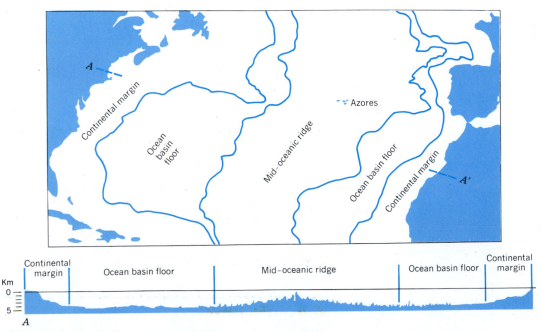

FIGURE 13.11 Major divisions of the North Atlantic Ocean basin (above), and a representative profile from New England to the coast of Africa (below). Profile exaggeration is about 40 times. (Data of B. C. Heezen, M. Tharp, and M. Ewing, *The Floors of the Oceans*, Geological Society of America Special Paper, 65.

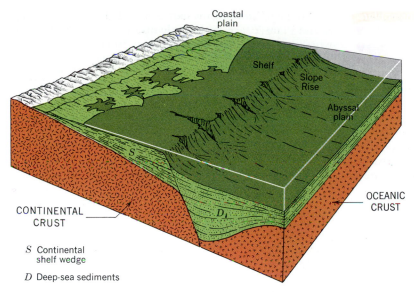

S Continental
   shelf wedge

D Deep-sea sediments

FIGURE 13.12 This block diagram shows a passive continental margin with an inner wedge of sediments beneath the continental shelf and an outer wedge of deep-sea sediments beneath the continental rise and abyssal plain. (Drawn by A. N. Strahler.)

13.12). The sediments have been brought from the land by rivers and spread over the shallow seafloor by currents. A great deal of attention is now being paid to the continental-shelf wedge as a potential source of rich petroleum accumulations, reached only from offshore drilling platforms. Below the continental rise and its adjacent abyssal plain is another thick sediment deposit; it is formed of deep-sea sediments carried down the continental slope by swift muddy currents, called *turbidity currents*.

Major rivers of the continents bring large amounts of sediment to the inner continental shelves, where large deltas are formed (Chapter 20). Seaward of the delta there is often present a narrow troughlike feature called a *submarine canyon*, representing the seaward extension of the river channel across the outer shelf. (Both the Hudson River and the Congo River have major submarine canyons of this kind.) Sediment carried down this canyon by turbidity currents descends the continental slope and accumulates on the continental rise in the form of a *submarine fan* (or *cone*). The fan of a large river may extend out over the deep ocean floor for a distance of many hundreds of kilometers (Figure 13.13). Two of the

greatest known fans are those of the Indus and Ganges-Brahmaputra rivers. A similar, but smaller, fan lies in deep water of the Gulf of Mexico, off the Mississippi delta (Figure 13.10).

THE ACTIVE CONTINENTAL MARGINS

The Pacific Ocean basin, while having a mid-oceanic ridge with ocean basin floors on either side, has continental margins quite different from those of the Atlantic and Indian oceans; they are characterized by mountain arcs or island arcs with offshore *oceanic trenches*. Geologists refer to these ocean basin limits as *active continental margins*.

The locations of the major trenches are shown in Figure 13.7. Trench floors reach depths of 7 km (23,000 ft) and even more (see Figure 13.10). Many lines of scientific evidence show that the oceanic crust is sharply downbent to form these trenches and that they mark the boundary between two lithospheric plates that are being brought together.

In the western Pacific Ocean basin are several subdivisions of the deep ocean floor known as *back-arc basins*. A typical back-arc basin is bounded on the continental side by either a deep trench or a passive continental margin and on the oceanward side by an island arc and its adjacent trench. One example is the Bering Abyssal Plain lying between the Alaskan–Siberian continental margin and the Aleutian volcanic island arc with its bordering Aleutian Trench. A second example is the deep ocean basin that lies between the Ryukyu–Philippine Trench and the Bonin–Mariana–Yap–Palau island arc system.

PLATE TECTONICS

Both crustal spreading along the axial rift of the mid-oceanic ridge and crustal downbending beneath oceanic trenches involve the entire thickness of lithospheric plates. The general theory of lithospheric plates with

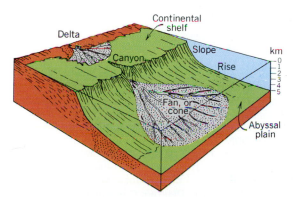

FIGURE 13.13 Block diagram of a deep-sea fan (cone) and its relationship to the continental shelf. (From A. N. Strahler, *Physical Geology*, Harper & Row, Publishers. Copyright © by Arthur N. Strahler.)

their relative motions and boundary interactions is *plate tectonics. Tectonics* is a noun meaning "the study of tectonic activity." Tectonic activity, in turn, refers to all forms of breaking and bending of the entire lithosphere, including the crust.

Figure 13.14 shows the major features of plate interactions. The vertical dimensions of the block diagram (A) is greatly exaggerated, as are the landforms. A true-scale cross section (B) shows the correct relationships between crust and lithosphere, but surface relief features can scarcely be shown. A global diagram (C), with greatly exaggerated relief, gives the correct impression of plates

as being curved to fit the earth-sphere. Two plates, Plate X and Plate Y, are both made up of *oceanic lithosphere*, which is comparatively thin (about 50 km; 30 mi). Plate Z, made up of *continental lithosphere*, is much thicker (150 km; 95 mi). Because the continental lithosphere bears a thick crust, much of which is felsic (granitic) rock, it is comparatively buoyant. Oceanic lithosphere, on the other hand, is made up only of mafic and ultramafic rock; it is comparatively dense and has a low-lying upper surface.

Plates X and Y are pulling apart along their common boundary, which lies along the axis of a mid-oceanic

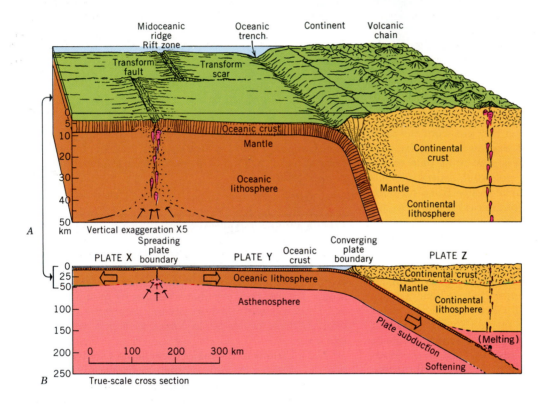

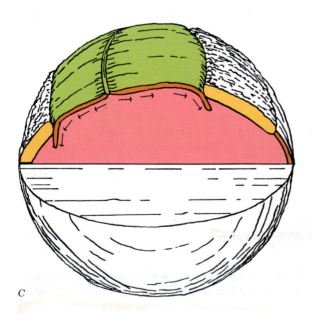

FIGURE 13.14 Schematic cross sections showing some of the important elements of plate tectonics. Diagram A is greatly exaggerated in vertical scale so as to emphasize surface and crustal features. Only the uppermost 30 km (20 mi) is shown. Diagram B is drawn to true scale and shows conditions to a depth of 250 km (155 mi). Here the actual relationships between lithospheric plates can be examined, but surface features can scarcely be shown. Diagram C is a pictorial rendition of plates on a spherical earth and is not to scale. (From A. N. Strahler, *Physical Geology*, Harper & Row, Publishers. Copyright © by Arthur N. Strahler.)

ridge. This plate activity tends to create a gaping crack in the crust, but magma continually rises from the mantle beneath. The magma appears as basaltic lava in the floor of the rift and quickly congeals. At greater depth under the rift, magma solidifies into gabbro, an intrusive rock of the same composition as basalt. Together, the basalt and gabbro continually form new oceanic crust.

The oceanic lithosphere of Plate Y is moving toward the thick mass of continental lithosphere that comprises Plate Z. Because the oceanic plate is comparatively thin and dense, in contrast to the thick, buoyant continental plate, the oceanic lithosphere bends down and plunges into the soft asthenosphere. The process of downplunging of one plate beneath another is called *subduction*.

The leading edge of the descending plate is cooler than the surrounding asthenosphere—enough cooler, in fact, that this descending slab of brittle rock is denser than the surrounding asthenosphere. Consequently, once subduction has begun, the slab "sinks under its own weight," so to speak. Gradually, however, the slab is heated by the surrounding asthenosphere and thus it softens at greater depth. The underportion, which is mantle rock in composition, simply reverts to asthenosphere as it softens. The thin upper crust, formed of less dense mineral matter, actually melts and becomes magma. This magma tends to rise because it is less dense than the surrounding material. Diagram B of Figure 13.14 shows some magma pockets formed from the upper edge of the slab. They are pictured as rising like hot-air balloons through the overlying continental lithosphere. Reaching the earth's surface, quantities of this magma build volcanoes, which tend to form a volcano chain lying about parallel with the deep oceanic trench that marks the line of descent of the oceanic plate.

Viewing Plate Y as a unit in Figure 13.14, it appears that a single lithospheric plate is simultaneously undergoing *accretion* (growth by addition) and *consumption* (by softening and melting), so that the plate might conceivably maintain its size, without necessarily having to expand or diminish. Actually, plate tectonics includes the possibility that a plate of oceanic lithosphere can either grow or diminish in extent. There are also tectonic models that allow for the creation of new plates of oceanic lithosphere where none existed and for plates to disappear completely. In this respect the theory is quite flexible.

PLATE BOUNDARIES

We have yet to consider a third type of lithospheric plate boundary. Two lithospheric plates may be in contact along a common boundary on which one plate merely slides past the other with no motion that would cause the plates either to separate or to converge (Figure 13.15). The plane along which motion occurs is a nearly vertical fracture, a *fault*, extending down through the entire lithosphere; it is called a *transform fault*. A *fault* is a plane of rock fracture along which there is motion of the rock mass on one side with respect to that on the other. The transform fault belongs to a general class of

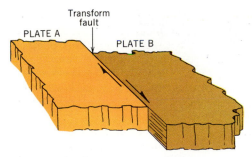

FIGURE 13.15 A transform fault involves the horizontal motion of two adjacent lithospheric plates, one sliding past the other. (From A. N. Strahler, *Physical Geology*, Harper & Row, Publishers. Copyright © by Arthur N. Strahler.)

faults called *transcurrent faults*, on which all motion is in the horizontal direction (see Chapter 14).

In summary, there are three major kinds of active plate boundaries:

Spreading boundaries. New lithosphere is being formed by accretion.
Converging boundaries. Subduction is in progress; lithosphere is being consumed.
Transform boundaries. Plates are gliding past one another on a transform fault.

Let us put these three boundaries into a pattern to include an entire lithospheric plate. As shown in Figure 13.16, we have visualized a moving rectangular plate set in the middle of a surrounding stationary plate, that is, the plate resembles a window. The moving plate is bounded by transform faults on two parallel sides. Spreading and converging boundaries form the other two parallel sides. Several familiar mechanical devices come to mind in visualizing this model. One is the sunroof in the top of an automobile; it is a window that opens by sliding backward along parallel side tracks to disappear under the fixed roof area at the rear. Another familiar device is the old-fashioned rolltop desk. Boundaries can be curved as well as straight, while individual plates can pivot as they move. There are many geometric variations in the shapes and motions of individual plates.

THE GLOBAL SYSTEM OF LITHOSPHERIC PLATES

The first global map of lithospheric plates appeared in published form in 1968, and there have since been many minor changes and revisions in the map. Boundaries have been relocated and new plates identified; differences have appeared in the naming of plates, as well. Today, however, a fairly good consensus exists in the geologic community as to the numbers and names of the major plates, the nature of their boundaries, and their relative motions. Differences of interpretation persist in many boundary details. Also, a few sections of certain plate boundaries are of uncertain classification or location.

For a particular lithospheric plate to be identified and named, its boundaries should all be active. In other

TABLE 13.2 The Lithospheric Plates

Great plates:	Lesser plates:
Pacific	Nazca
American (North, South)	Cocos
	Philippine
Eurasian	Caribbean
Persian subplate	Arabian
African	Juan de Fuca
Somalian subplate	Caroline
Austral-Indian	Bismark
Antarctic	Scotia

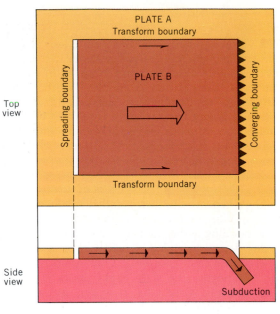

FIGURE 13.16 A schematic diagram of a single rectangular lithospheric plate with two transform boundaries. (From A. N. Strahler, *Physical Geology*, Harper & Row, Publishers. Copyright © by Arthur N. Strahler.)

words, there must be good evidence of present or recent relative motion between the plate and all its contiguous (adjoining) plates. The global system of lithospheric plates consists of six great plates. These are listed in Table 13.2 and shown on a world map in Figure 13.17. Several lesser plates are also recognized, ranging from intermediate in size to comparatively small. Several subplates are also recognized within the great plates. Two shown in Figure 13.17 are the Persian subplate and the Somalian subplate. Plate boundaries are shown by standard symbols, explained in the key accompanying the map. Keep in mind that the Mercator grid distorts the areas of the plates, making them appear greatly expanded in high latitudes.

Figure 13.18 is a schematic circular cross section of the lithosphere along a great circle in low latitudes. It shows several of the great plates and their boundaries.

The great Pacific plate occupies much of the Pacific Ocean Basin and consists almost entirely of oceanic lithosphere. Its relative motion is northwesterly, so that it has a converging (subduction) boundary along most of its western and northern edge. The eastern and southern

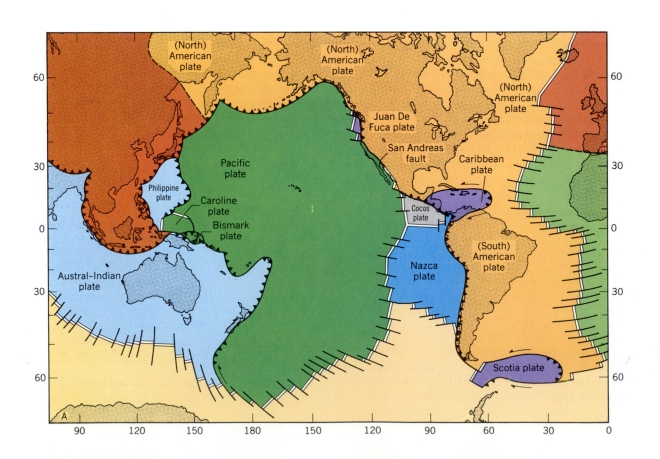

edge is mostly spreading boundary. A sliver of continental lithosphere is included, making up the coastal portion of California and all of Baja California. The California portion of the plate boundary is the San Andreas Fault, an active transform fault.

The American plate includes most of the continental lithosphere of North and South America, as well as the entire oceanic lithosphere lying west of the mid-oceanic ridge (Mid-Atlantic Ridge) that divides the Atlantic

Ocean basin down the middle. For the most part, the western edge of the American plate is a converging boundary with active subduction extending from Alaska through Central America to southernmost South America. Many geologists recognize a North American plate and a South American plate, with the boundary running east–west at about lat. 15°N. That boundary does not, however, appear to be tectonically active and is simply an arbitrary line.

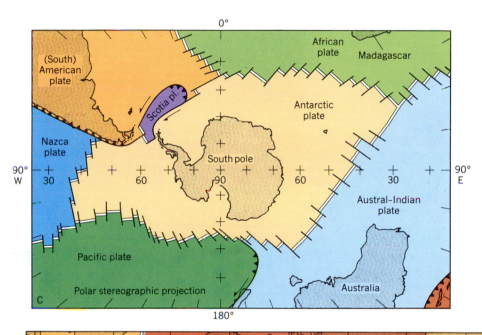

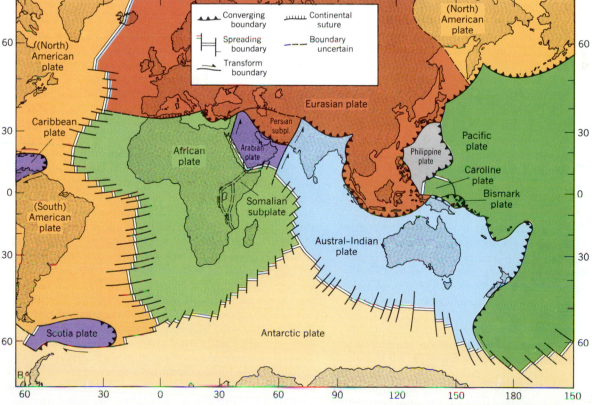

FIGURE 13.17 Generalized map of the major lithospheric plates and their boundaries. (A. N. Strahler.)

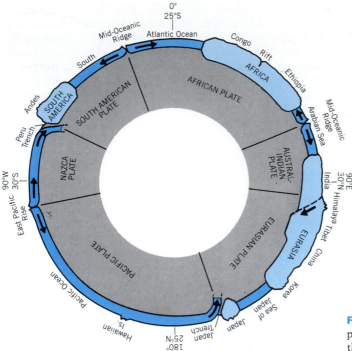

FIGURE 13.18 Schematic circular cross section of the major plates on a great circle tilted about 30 degrees with respect to the equator. (A. N. Strahler.)

The Eurasian plate is largely continental lithosphere, but is fringed on the east and north by a belt of oceanic lithosphere. The African plate can be visualized as having a central core of continental lithosphere nearly surrounded by oceanic lithosphere. The Austral–Indian plate (also called the Australian plate) takes the form of an elongate rectangle. It is mostly oceanic lithosphere, but contains two cores of continental lithosphere—Australia and peninsular India. The Antarctic plate has an elliptical outline and is almost completely enclosed by a spreading boundary. The continent of Antarctica forms a central core of continental lithosphere completely surrounded by oceanic lithosphere.

Of the remaining six plates, the Nazca and Cocos plates of the eastern Pacific are rather simple fragments of oceanic lithosphere bounded by the Pacific mid-oceanic spreading boundary on the west and by a subduction boundary on the east. The Philippine plate is noteworthy as having subduction boundaries on both east and west edges. Two small but distinct lesser plates—Caroline and Bismark—lie to the southeast of the Philippine plate, but these can be included within the Pacific plate. The Arabian plate resembles the "sunroof" model shown in Figure 13.16; it has two transform fault boundaries and its relative motion is northeasterly. Both the Caribbean and Scotia plates have important transform boundaries, as well as an active subduction boundary. The tiny Juan de Fuca plate is steadily diminishing in size and will eventually disappear by subduction beneath the American plate.

Geologists recognize one or more subplates within certain of the major plates. A *subplate* is a plate of secondary importance set apart from the main plate by a boundary that is uncertain or questionable, either as to its true nature or the level of its activity. An example is the Somalian subplate of the African plate. It is bounded by the East African Rift Valley system and there is good reason to think that this portion of the African plate is beginning to split apart and will become an independent plate.

SUBDUCTION TECTONICS AND VOLCANIC ARCS

Converging plate boundaries, with subduction in progress, are zones of intense tectonic and volcanic activity. The narrow zone of a continent that lies above a plate undergoing subduction is therefore an active continental margin. Figure 13.19 shows some details of the geologic processes that are associated with plate subduction. Two diagrams are used: (A) exaggerated to show crustal and surface details, (B) drawn to true-scale to show the lithospheric plates.

The trench axis represents the line of meeting of sediment coming from two sources. Carried along on the moving oceanic plate is deep-ocean sediment—fine clay and ooze—that has settled to the ocean floor. From the continent comes terrestrial sediment in the form of sand and mud brought by streams to the shore and then swept into deep water by currents. In the bottom of the trench both types of sediment are intensely deformed and are carried downward with the moving plate. The deformed sediment is then shaped into wedges that ride, one over the other, on steep fault planes. The wedges accumulate into an *accretionary prism* in which metamorphism takes place. In this way, the continental margin is built outward and new continental crust of metamorphic rock is formed. The accretionary prism is of relatively low density and tends to rise, forming a *tectonic crest*. The tectonic crest is shown to be submerged, but in some cases it forms an island chain paralleling the coast; that is, a

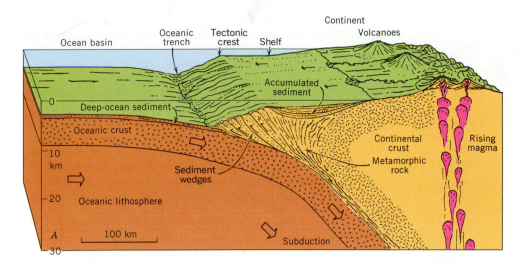

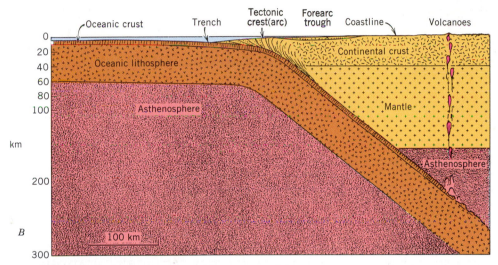

FIGURE 13.19 Some typical features of an active subduction zone. Diagram A uses a great vertical exaggeration to show surface and crustal details. Sediments scraped off the moving plate form tilted wedges that accumulate in a rising tectonic mass. Between the tectonic crest and the mainland is a shallow trough in which sediment brought from the land is accumulating. Metamorphic rock is forming above the descending plate. Magma rising from the top of the descending plate reaches the surface to build a chain of volcanoes. Diagram B is a true-scale cross section showing the entire thickness of the lithospheric plates. (From A. N. Strahler, *Physical Geology*, Harper & Row, Publishers. Copyright © by Arthur N. Strahler.)

tectonic arc. Between the tectonic crest and the mainland is a shallow trough, the *forearc trough*. This trough traps a great deal of terrestrial sediment, which accumulates in a basinlike structure. The bottom of the forearc trough continually subsides under the load of the added sediment. In some cases, the seafloor of the trough is flat and shallow, forming a type of continental shelf. Sediment carried across the shelf moves down the steep outer slope of the accretionary prism in tonguelike flows of turbidity currents.

The lower diagram of Figure 13.19 shows the descending lithospheric plate entering the asthenosphere. Intense heating of the upper surface of the plate melts the oceanic crust, forming basaltic magma. As the magma rises, it is changed in chemical composition at the base of the crust and becomes andesite magma. The rising andesite magma reaches the surface to form volcanoes of andesite lava, such as those we see in the Andes of South America.

OROGENY

Generally speaking, high-standing mountain masses (other than volcanic mountains) are elevated by one of two basic tectonic processes: compression and extension. Compressional tectonic activity—"squeezing together" or "crushing"—acts at convergent plate boundaries; extensional tectonic activity—"pulling apart"—occurs where oceanic plates are separating or where a continental plate is undergoing breakup into fragments. We turn first to compressional tectonic processes.

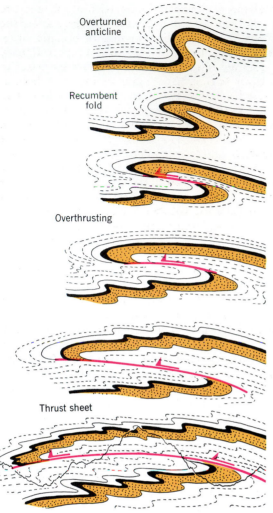

FIGURE 13.20 In alpine structure, the tightly folded strata are broken by overthrust faults; one slice is heaped on the next. (From A. N. Strahler, *The Earth Sciences*, 2nd ed., Harper & Row, Publishers. Copyright © by Arthur N. Strahler.)

The alpine mountain chains typically consist of intensely deformed strata of marine origin. The strata are tightly compressed into wavelike structures, called *folds*. Typically, the folds become overturned and take on a recumbent attitude, as shown in Figure 13.20. Accompanying the folding is a form of faulting in which slices of rock move over the underlying rock on fault surfaces of low inclination; these are *overthrust faults*. Individual

rock slices, called *thrust sheets*, are carried many tens of kilometers over the underlying rock. In the European Alps, thrust sheets of this kind were named *nappes* (from the French word meaning "cover sheet" or "tablecloth"). Nappes may be thrust one over the other to form a great pile (Figure 13.21). The entire deformed rock mass produced by such compressional mountain-making is called an *orogen*; the event that produced it is an *orogeny*.

Thrust sheets are known to have moved nearly horizontally for distances of several tens of kilometers. As one theory of thrust-sheet development, geologists postulate that at the time the sheets were in motion the thrust planes actually sloped downward in the direction of motion of the sheet. You might visualize the thrust sheets as sliding "downhill" under the force of gravity. The phenomenon is known as *gravity gliding*. The presence of ground water under great pressure beneath the thrust sheet is thought to have reduced the friction so much that the enormous rock layer could glide easily for a long distance.

OROGENS AND COLLISIONS

As you can easily visualize in a situation where two lithospheric plates are converging along a subduction boundary, any high-standing mass of continental lithosphere projecting above the level of the oceanic lithosphere will come closer and closer to the continent beneath which that plate is descending. Ultimately, the two masses of continental lithosphere must collide, because the impacting mass is too thick and too buoyant to pass down beneath the continental plate it is impacting. The result is orogeny in which various kinds of crustal rocks are crumpled into folds and sliced into nappes. Quite aptly, this process has been called "telescoping."

Two types of orogens are recognized. One type results from the impact of a relatively small mass of high-standing continental lithosphere against a full-sized mass of continental lithosphere. Following this type of collision, the small mass is firmly welded to the continent and will become a permanent part of the continental shield. The second type of orogen results from the collision of two very large bodies of continental lithosphere (full-sized plates), permanently uniting them and terminating further tectonic activity along that collision zone.

To prepare for the first type of orogeny, a simple mechanical model will be useful. Imagine we are in a custom bakery that makes cakes on special order. As each

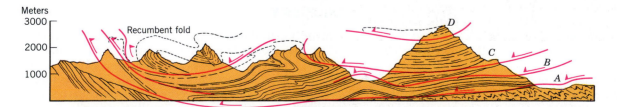

FIGURE 13.21 A classic cross section through a portion of the Helvetian Alps, Switzerland, shows thrust slices. Horizontal and vertical scales are the same. (Simplified from A. Helm, *Geologie der Schweiz*, vol. II-1, Tauschnitz, Leipzig.)

cake is completed, it is placed on a continually moving conveyer belt that transports it to the packaging department. There, the cake slides off onto a short table surface at the level of the belt. The table will accommodate only a few cakes; if the attendant leaves the station for too long, disaster soon sets in. As another cake arrives, it slams into those already on the table, crushing them severely and compacting them into a single mass.

The conveyer belt in our analog represents oceanic lithospheric disappearing into a subduction zone. Each cake represents one of several possible crustal features that can project above the otherwise uniform abyssal surface. Small projecting objects, such as seamounts, usually cause no problem; they may be knocked off at the base by impact with the overlying plate to become incorporated into the accretionary prism. The problem is with much larger crustal protrusions, too massive and too firmly rooted to be subducted.

What kinds of traveling crustal masses do we have in mind as capable of colliding with a large continental mass and adhering to it? One is the volcanic island arc; for example, the Kuril arc or the Lesser Antilles arc. An entirely different class of objects consists of small fragments or bits of continental crust called *microcontinents*. Some of the older island arcs with a long history of accretion qualify as microcontinents—Honshu, the Philippines, or Hispanola, for example. In other cases, a fragment of continent has been pulled away from the mainland on a widening rift that develops into a back-arc basin. These islands of continental crust are thus surrounded by oceanic crust.

The objects described above can arrive at the active margin of a large body of continental lithosphere as an oceanic plate moves toward a subduction boundary adjacent to continental lithosphere. Over millions of years, the impacting object can travel thousands of kilometers.

The impact of an island arc is called an *arc-continent collision*; it is illustrated by a set of diagrams in Figure 13.22. The setting, shown as Stage A, is one of a passive continental margin (right) closing with a volcanic arc (left). In this case, the subduction is occurring along the volcanic arc, narrowing the expanse of oceanic lithosphere separating the two features. An accretionary prism is growing larger as the volcanic arc increases in size and depth. In Stage B, the intervening ocean is completely closed and the accretionary prism is forced to ride up the continental margin, gliding on a basal thrust fault. As shown in Stage C, impact of the volcanic arc causes the overthrusting to "telescope" the strata of the passive margin, and to extend the low-angle thrust slice far over the continental shield. This kind of thrust plane is called a *décollement*, as it is located at the interface of the strata and the ancient crystalline basement rocks beneath. Rocks of the volcanic arc are also severely deformed and converted into metamorphic rock.

Notice also that in Stage C, a new subduction boundary has come into existence on the oceanward side of the former volcanic arc. Magmas rising from the downplunging plate penetrate the metamorphic zone of the orogen. These give rise to bodies of felsic magma that rise farther and become emplaced as granitic batholiths. Magma also reaches the surface to produce outpourings of andesitic and rhyolitic lava. The process of intrusion by magma has greatly thickened the crust beneath the orogen, making it a permanent addition to the continent.

Examples of orogens resulting from arc-continent collisions can be found on both eastern and western sides of North America. The relationships shown in Figure 13.22 are patterned in a general way after Paleozoic arc-continent collisions that affected the Appalachian and Ouachita mountain belts. Another example is the Cordilleran Orogeny that (as the name indicates) built the Cordilleran ranges of western North America. The inner limit of this orogenic belt is seen today in a belt of low-angle overthrusting shown on the map in Figure 13.24. In the northern Rocky Mountains of Montana, Alberta, and British Columbia the overthrust zone is remarkably displayed in high, glaciated mountains that

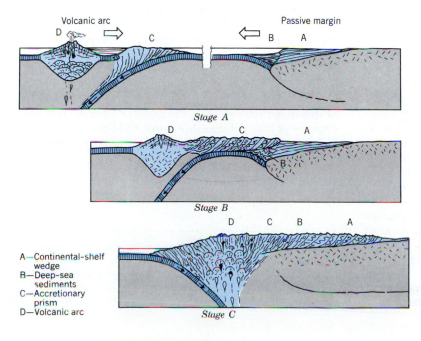

A—Continental-shelf wedge
B—Deep-sea sediments
C—Accretionary prism
D—Volcanic arc

Stage A

Stage B

Stage C

FIGURE 13.22 Schematic cross section of an arc-continent collision. Not shown to true scale. (Drawn by A. N. Strahler.)

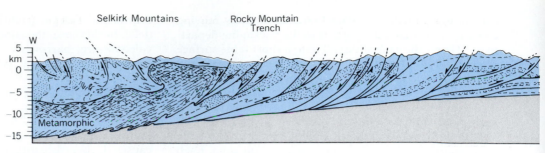

FIGURE 13.23 A structural cross section through the Canadian Rockies between the Bow and Athabasca rivers, Alberta. (Data of Geological Survey of Canada, presented by R. A. Price and E. Montjoy, Geological Association of Canada; Special Paper No. 6, Figure 2–1, 1970. (From A. N. Strahler, *Physical Geology*, Harper & Row, Publishers. Copyright © by Arthur N. Strahler.)

hold Waterton-Glacier International Peace Park, Banff Park, and Jasper Park. Figure 13.23 is a structural cross section of this overthrust zone in which strata of Cretaceous age are intensely deformed by steep, upwardly inclined thrust faults of a type known as imbricate faults. Another feature of the Cordilleran Orogeny was the intrusion of numerous batholiths in late Cretaceous time.

ACCRETED TERRANES OF WESTERN NORTH AMERICA

In recent years, geologists who have been involved in mapping the bedrock features of western North America from Mexico to Alaska have come to recognize that it consists of a mosaic of crustal patches called *terranes*. A particular terrane is quite distinct geologically from those that surround it, in that it has its own special rock type or suite of component rock types. If we think of this assemblage of terranes as a mosaic of tiles, we see that whereas most of the tiles are unique, others are duplicated in widely separated places. Figure 13.24 is a map of these terranes. There are at least 50 of them and each is separated from its neighbors by a fault contact. In many cases, the fault is of the transcurrent type and trends more or less parallel with the continental margin.

Each terrane has a name. For example, the terrane called Wrangellia, shown on the map by solid color, occurs in four separate patches. It is a terrane consisting of basaltic island–arc volcanics and sedimentary rocks of types that include both cherts of deep-sea origin and algal limestones of shallow-depth origin. Using paleomagnetic methods, geologists have been able to establish the original global location of certain of the terranes. Much to their initial surprise, Wrangellia was found to have come from a location at lat. 10° and as far distant as a spot equivalent to where New Guinea is now located—traveling since then a distance of nearly 10,000 km (6000 mi)! It occupied that location during Triassic time, when some 3 km (1000 ft) thickness of basaltic island–arc lavas was deposited.

Each of the many terranes is now regarded as a microcontinent—using that term in a general sense. Obviously, a terrane can only travel by being carried along with the lithospheric plate in which it is embedded.

These bits of continental lithosphere were embedded in oceanic lithosphere, which moved toward North America, eventually bringing each microcontinent to a location close to a subduction boundary at the active western margin of North America. Here a collision took place and the microcontinent was welded to the continent. Many such collisions took place, eventually forming the mosaic of terranes.

Once the microcontinents were added to the continent, they were subjected to being sheared into two or more small fragments by transcurrent faults trending about parallel with the continental margin. Faults of this type are induced to form in continental crust when the subducting lithospheric plate is approaching at an oblique angle to the subduction boundary. (The San Andreas Fault is an example of this type of transcurrent fault.) Dragged along within the transcurrent-fault slices, terrane fragments become separated and distributed up and down the entire continental marginal zone. This is what seems to have happened in the case of Wrangellia. The accretion of microcontinents to form a mosaic pattern may have also been one of the modes by which the Precambrian shield was constructed.

Whether the arrival and impact of a microcontinent was capable of generating a full-blown orogeny is questionable. For a small microcontinent, a few tens of kilometers across, the tectonic effect would perhaps have been localized. In the case of an impacting volcanic island arc thousands of kilometers long a major orogeny would have ensued, strongly affecting thousands of kilometers length of the continental margin. A phenomenon of this magnitude seems quite distinct from the repeated impacts of many microcontinents. Evidently, such collisions come in a wide range of magnitudes and produce a wide range of effects.

FIGURE 13.24 Map of microcontinent terranes of western North America. Fragments of Wrangellia are shown in solid color. Names of several of the terranes are given. (Based on data of U.S. Geological Survey; M. Beck, A. Cox, and D. L. Jones, *Geology*, vol. 8, p. 455; and others.)

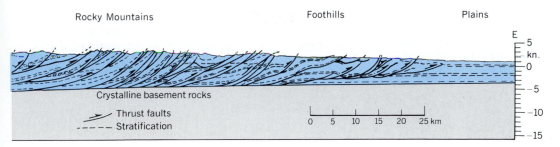

Rocky Mountains   Foothills   Plains

Crystalline basement rocks

➤ Thrust faults
--- Stratification

0 5 10 15 20 25 km

E
5
kn.
0
−5
−10
−15

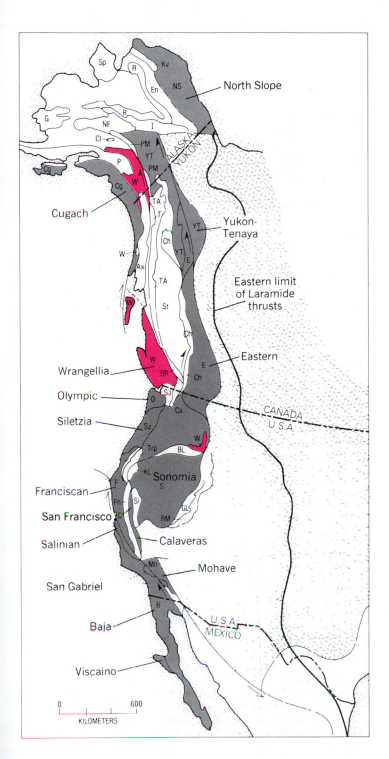

North Slope
Cugach
Yukon–Tenaya
Eastern limit of Laramide thrusts
Eastern
Wrangellia
Olympic
Siletzia
Franciscan
San Francisco
Salinian
Calaveras
Mohave
San Gabriel
Baja
Viscaino

ALASKA / YUKON
CANADA U.S.A.
U.S.A. MEXICO

0   600
KILOMETERS

CONTINENT–CONTINENT COLLISIONS

A second type of orogen results from the collision of two full-sized masses of continental lithosphere, both being of the dimensions of a great plate or a subplate. Named the Eurasian-type, this orogen is described as being the result of a *continent–continent collision*. Collisions of this type have occurred in the Cenozoic Era along a great tectonic line that marks the southern boundary of the Eurasian plate (Figure 13.25). The line begins with the Atlas Mountains of North Africa, runs through the European Alps, and extends across the Aegean Sea region into western Turkey. Beyond a major gap in Turkey, the line takes up again in the Zagros Mountains of Iran. Jumping another gap in southeastern Iran and Pakistan, the collision line sets in again in the great Himalayan Range.

Thus, we are dealing with three collision segments: European, Persian, and Himalayan. Each segment represents the collision of a different north-moving plate against the single and relatively immobile Eurasian plate. The European segment was formed by collision of the African plate with the Eurasian plate in the Mediterranean region. The Persian segment resulted from the collision of the Arabian plate with the Persian subplate of the Eurasian plate. The Himalayan segment represents the collision of the Indian continental portion of the Austral–Indian plate with the Eurasian plate. The tectonic structures of the three collision segments differ markedly.

Figure 13.26 is a series of cross sections in which the tectonic events of a typical continent–continent collision are reconstructed. Diagram A shows a passive margin at the left, an active subduction margin at the right. As the ocean between the converging continents is eliminated, a succession of overthrust faults cuts through the oceanic crust (diagram B). The imbricate thrust slices ride up, one over the other, telescoping the oceanic crust and the sediments above it. As the slices become more and more tightly squeezed, they are forced upward. The upper part of each thrust sheet assumes a horizontal attitude to form a nappe, which then glides forward under gravity on a low downgrade. The final nappes in the series consist in part of the remains of slices of oceanic crust. A mass of metamorphic rock is formed between the joined continental plates, welding them together. This new rock mass is called a *continental suture;* it is a distinctive type of orogen.

Sutures of the three collision segments are shown by

a special line pattern in Figure 13.25. The Himalayan collision segment is still highly active, with the Indian plate being underthrust deeply beneath the mountain range. Immediately to the north lies the Tibetan Plateau, with an extremely thick continental crust, perhaps re-

sulting from compressional shortening and thickening. To the north, compression has produced several major transcurrent faults along which great segments of the Asiatic crust have moved eastward, forming the China subplate.

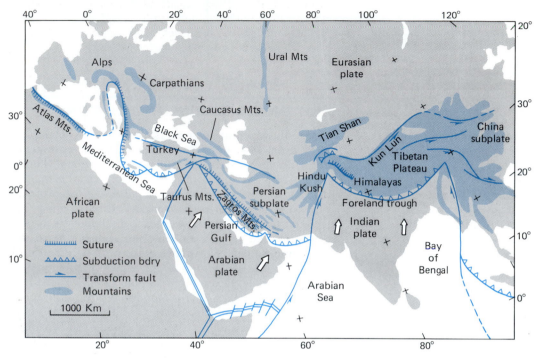

FIGURE 13.25 Sketch map of Eurasian collision segments. (From A. N. Strahler, *Physical Geology*, Harper & Row, Publishers. Copyright © by Arthur N. Strahler.)

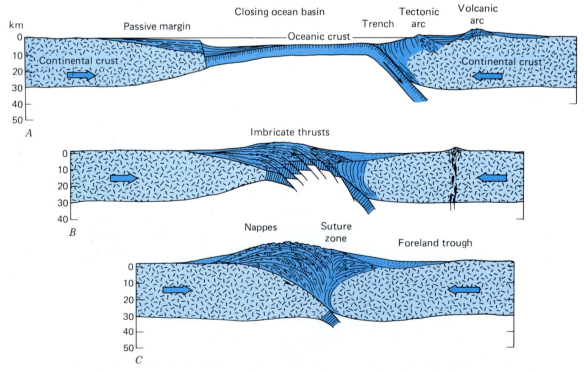

FIGURE 13.26 Schematic cross sections showing continent–continent collision and the formation of a suture zone with nappes. (From A. N. Strahler, *Physical Geology*, Harper & Row, Publishers. Copyright © by Arthur N. Strahler.)

Continent–continent collisions have occurred since the late Precambrian. Many ancient sutures have been identified in the continental shields. The Ural Mountains is one such suture, trending north–south at the arbitrary dividing line between Europe and Asia. It was formed near the close of the Paleozoic Era.

CONTINENTAL RUPTURE AND NEW OCEAN BASINS

We have already noted that the continental margins bordering the Atlantic Ocean basin on both its eastern and western sides are very different from the active margin of a subduction zone. The Atlantic margins have no tectonic activity at present; they are passive continental margins. Even so, they represent the contact between continental lithosphere and oceanic lithosphere, with continental crust meeting oceanic crust, as shown in Figure 13.3. To understand how passive margins are formed, we must go back into tectonic history that involved the rifting apart of a single continental lithospheric plate.

This process is called *continental rupture*. Figure 13.27 shows by three schematic block diagrams how continental rupture takes place and leads to the development of passive continental margins. At first the crust is both lifted and stretched apart as the lithospheric plate is arched upward. In this stage, *fault-block mountains* are formed. They are the result of extensional tectonics. Next, a long narrow valley, called a *rift valley*, appears (Block A). The widening crack in its center is continually filled in with magma rising from the mantle below. The magma solidifies to form new crust in the floor of the rift valley. Crustal blocks slip down along a succession of steep faults, maintaining a mountainous landscape. As separation continues, a narrow ocean appears; down its center runs a spreading plate boundary (Block B). Plate accretion takes place to produce new oceanic crust and lithosphere. We find in the Red Sea today an example of a narrow ocean formed by continental rupture (Figure 13.28). Its straight, steep coasts are features we would expect of such a history. The widening of the ocean basin can continue until a large ocean has formed and the continents are widely separated (Block C).

During the process of opening of an ocean basin, the spreading boundary develops a series of offsets, one of

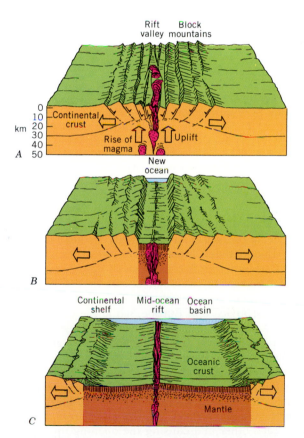

FIGURE 13.27 Schematic block diagrams showing stages in continental rupture and the opening up of a new ocean basin. The vertical scale is greatly exaggerated to emphasize surface features. (A) The crust is uplifted and stretched apart, causing it to break into blocks that become tilted on faults. (B) A narrow ocean is formed, floored by new oceanic crust. (C) The ocean basin widens, while the passive continental margins subside and receive sediments from the continents. (From A. N. Strahler, *Physical Geology*, Harper & Row, Publishers. Copyright © by Arthur N. Strahler.)

FIGURE 13.28 Astronauts aboard the Gemini XII space vehicle took this south-looking photo of the Red Sea, which separates the Arabian Peninsula (left) from Africa (right). The narrow sea is about 200 km (125 mi) wide. Its straight, parallel coastlines suggest its origin—as a widening belt of new ocean between separating lithospheric plates. At the lower left we see the triangular Sinai Peninsula. bounded by two narrower fault depressions—the Gulf of Suez (bottom of photo) and the Gulf of Aqaba (left). (NASA)

which is shown in the upper left-hand part of Diagram A of Figure 13.14. The offset ends of the axial rift are connected by an active transform fault. As spreading continues, a scarlike feature is formed on the ocean floor as an extension of the transform fault. These *transform scars* take the form of narrow ridges or scarps (clifflike features) and may extend for hundreds of kilometers across the ocean floors. Before the true nature of these scarps was understood, they were named *fracture zones*. That name still persists and can be seen on maps of the ocean floor (Figure 13.10). The scars are not, for the most part, associated with active faults, although some undergo minor vertical fault movements at a later time.

Notice in Block C of Figure 13.27 that the overall appearance of the ocean basin and its continental margins resembles the schematic diagram in Figure 13.9. The passive margins are accumulating terrestrial sediment in the form of a continental shelf that rests on the continental crust, while the oceanic crust is accumulating a wedge of deep-sea sediments. The continental margin gradually subsides as these sediment wedges thicken, until the total sediment thickness reaches several kilometers (Figure 13.12). A wide, shallow continental shelf is typical of the passive continental margins. Large deltas built by rivers contribute a great deal of the shelf sediment. Turbidity currents carry the sediment down the steep continental slope and spread it out on the continental rise, producing deep-sea fans (Figure 13.13).

THE BREAKUP OF PANGAEA

Although modern plate tectonics became an acceptable scientific theory within only the past two decades, the concept of breakup of an early supercontinent into fragments that drifted apart is many decades old. Almost as soon as good navigational charts became available to show the continental outlines, persons of learning became intrigued with the close correspondence in outline between the eastern coastline of South America and the western coastline of Africa. In 1668, a Frenchman interpreted the matching coastlines as proof that the two continents became separated during the biblical flood. In 1858 Antonio Snider-Pelligrini produced a map to show the American continents nested closely against Africa and Europe. He went beyond the purely geometrical fitting to suggest that the reconstructed single continent explains the close similarity of fossil plant types in coal-bearing rocks in both Europe and North America.

Moving ahead to the early twentieth century, we come to the ideas of two Americans, Frank B. Taylor and Howard B. Baker, whose published articles presented evidence favoring the hypothesis that the New World and Old World continents had drifted apart. Nevertheless, credit for a full-scale hypothesis of breakup of a single supercontinent and the drifting apart of individual continents belongs to a German scientist, Alfred Wegener, a meteorologist and geophysicist who became interested in the various lines of geologic evidence that the continents had once been united. He first presented his ideas in 1912 and his major work on the subject appeared in

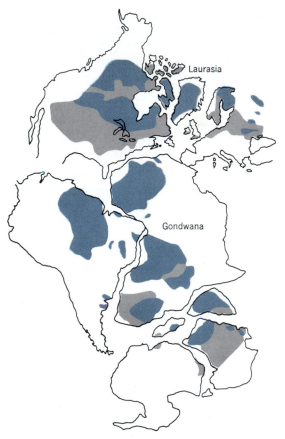

FIGURE 13.29 Reassembled continents, prior to start of continental drift. Dark pattern: areas of oldest shield rocks (older than −1.7 b.y.). Light pattern: rocks in the age range −0.8 to −1.7 b.y. (Redrawn from P. M. Hurley and J. R. Rand, *Science*, vol. 164, p. 1237. Copyright © 1969 by the American Association for the Advancement of Science.)

1922. A storm of controversy followed, and many American geologists denounced the hypothesis.

Wegener had reconstructed a supercontinent named *Pangaea*, which existed intact about 300 million years ago in the Carboniferous Period. Figure 13.29 is a modern version of Pangaea; Figure 13.30 shows the breakup of Pangaea. Wegener visualized the Americas as fitted closely against Africa and Europe, while the continents of Antarctica and Australia, together with the subcontinents of peninsular India and Madagascar, were grouped closely around the southern tip of Africa. Starting about 200 million years ago, continental rifting began as the Americas pulled away from the rest of Pangaea, leaving a great rift that became the Atlantic Ocean. Later, the other fragments rifted apart, pulling away from Africa and from each other and causing the opening up of the ancestral Indian Ocean, as shown in Figure 13.30.

Several lines of hard geologic evidence favored the existence of Pangaea. The evidence for a single supercontinent, based on matching tectonic structures and fossil assemblages, seemed quite convincing to many geologists during the 1920s and 1930s, but the separation of the continents—a process then known as *continental drift*—was strongly opposed on physical grounds. Weg-

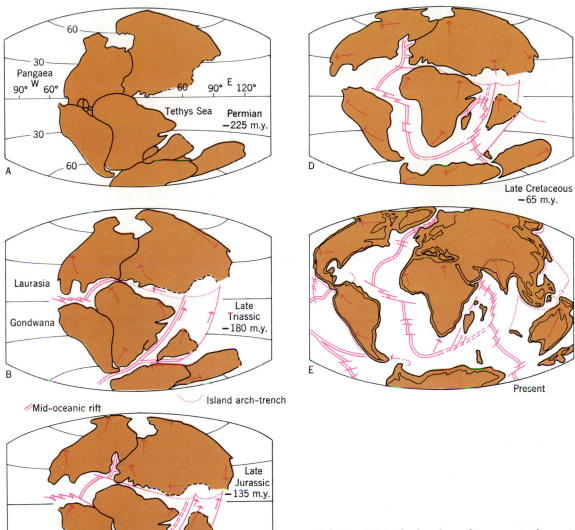

FIGURE 13.30 The breakup of Pangaea is shown in five stages. Inferred motion of lithospheric plates is indicated by arrows. (Redrawn and simplified from maps by R. S. Dietz and J. C. Holden, *Journal of Geophysical Research*, vol. 75, pp. 4943–51, Figures 2 to 6, copyrighted by the American Geophysical Union.)

ener had proposed that the continental layer of less dense rock, called "sial," had moved like a great floating raft through a "sea" of denser oceanic crustal rock, called "sima." (The syllables si, al, and ma refer to silica, alumina, and mafic, respectively.) Geologists could show by use of laws of physics that this mechanism was physically impossible, because rigid crustal rock could not behave in such a fashion.

Wegener's scenario of continental drift took on new meaning in the 1960s and 1970s, when plate tectonics emerged as a leading theory. The modern interpretation is, of course, that continental drift involves entire lithospheric plates, much thicker than merely the outer crust of either the continents or the ocean basins. Plate motions over a soft, plastic asthenosphere have allowed the continents to be carried along according to the general timetable postulated by Wegener. Some changes have been made in Wegener's timetable of events. There

have also been numerous improvements in the fitting together of the original pieces of the supercontinent Pangaea.

MODELING THE EARTH'S INTERNAL ENERGY SYSTEM

Tectonic and volcanic activity represent expenditures of internal energy stored within the nuclei of atoms of radioactive elements, such as uranium and thorium. As Figure 13.31 shows, this energy source lies within the system boundary. It is an inheritance from the time of accretion of the planet some 4.6 billion years ago. Spontaneous decay of these atoms transforms the stored atomic energy into sensible heat stored in the surrounding rock, which may be in the solid state as crystalline rock or in the liquid state as magma. Most of this

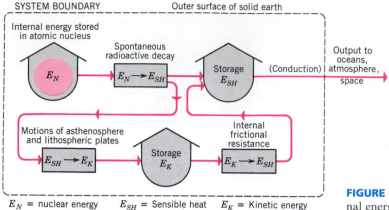

Internal energy stored
in atomic nucleus

Spontaneous
radioactive decay

E_N

$E_N \rightarrow E_{SH}$

Storage
E_{SH}

Output to
oceans,
(Conduction) | atmosphere,
space

Motions of asthenosphere
and lithospheric plates

Internal
frictional
resistance

$E_{SH} \rightarrow E_K$

Storage
E_K

$E_K \rightarrow E_{SH}$

E_N = nuclear energy E_{SH} = Sensible heat E_K = Kinetic energy

FIGURE 13.31 Flow diagram of the earth's internal energy system.

sensible heat is slowly conducted to the earth's surface, where it is lost to the oceans and atmosphere, and ultimately to outer space.

Some of the sensible heat in the mantle is, however, used to drive slow currents within the plastic asthenosphere and move the brittle lithospheric plates. Thus some sensible heat is transformed into kinetic energy of matter in motion. Kinetic energy is converted back into sensible heat through internal friction as the asthenosphere is dragged against the underlying strong mantle rock and the overlying rigid plates. This sensible heat also enters storage and follows the conduction pathway to the surface.

Because there is no external energy input in this system, the total system energy decreases with time; it is called an *exponential decay system*. We have not taken into account the possibility that some energy enters the system through tidal flexing of the earth. There is also the possibility that new energy inputs have been made from time to time by infall and impact of asteroids and large meteoroids.

Because the earth's initial store of radioactive elements is slowly but steadily diminishing through spontaneous decay, the total energy available to operate the moving plate system must be decreasing. From this conclusion we can speculate that plate motions and tectonic

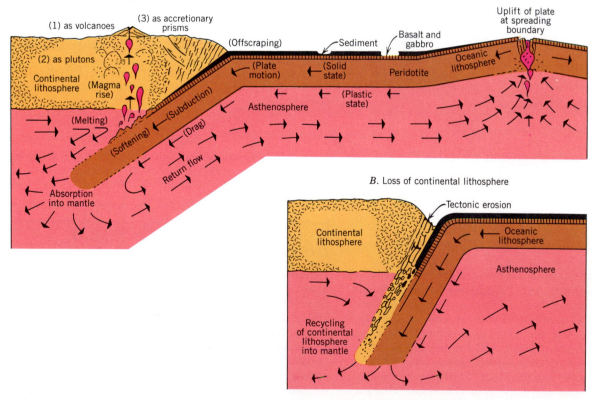

FIGURE 13.32 Schematic cross sections to show how the plate tectonic system results in gain or loss of crustal material to the continental lithosphere. The diagram is not drawn to scale.

activity will become less vigorous as eons of time pass. Correspondingly, igneous activity will decline in intensity. There will be fewer orogenies and new mountains will not rise as high above sea level. Erosion of the continents will ultimately dominate the scene and the continents will be maintained at a lower average elevation and will show only subdued relief features.

THE TECTONIC SYSTEM IN REVIEW

The system of lithospheric plates in motion represents an enormous material-flow system powered by an internal energy-flow system. The scheme of the cycling of mineral matter is fairly well understood in a general way, although many details remain speculative.

Figure 13.32 is a schematic diagram (not to scale) showing some of the major features of the material-flow system. Diagram A shows how a plate of oceanic lithosphere undergoing subduction transfers matter to the margin of the continental lithosphere by volcanic and tectonic processes. Magma formed by melting of the upper surface of the plate penetrates the continental lithosphere and is added to the continental crust in the form of igneous plutons and extrusive masses (volcanoes). Off-scraping of the upper surface of the subducting plate contributes to the growth of accretionary prisms, which become permanent additions to the continental crust as metamorphic rock.

Most of the downgoing plate is softened by heating and is reabsorbed into the asthenosphere. Slow currents deep within the asthenosphere, moving in a direction generally opposite to the plate motion, return the enriched mantle rock to the spreading plate boundaries.

Diagram B of Figure 13.32 shows that under certain conditions, the dragging action of the downgoing plate tears loose blocks and slabs of the adjacent continental lithosphere. This material is carried down into the asthenosphere. Thus by *tectonic erosion* much felsic rock of the continental crust can enter the mantle and be recycled.

The energy system that causes plate motions is generally agreed to have its source in the phenomenon of radioactivity. Radioactive elements in the crust and upper mantle constantly give off heat. This is a process of transformation of matter into energy. As the temperature of mantle rock rises, the rock expands. As in the case of the atmosphere, upward motion of less dense material takes place by convection. It is thought that mantle rock is rising steadily beneath the spreading plate boundaries. How this rise of heated rock causes plates to move is not well understood, but one hypothesis states that as the lithospheric plate is lifted to a higher elevation above the rising mantle it tends to move horizontally away from the spreading axis under the influence of gravity. At the opposite edge of the plate, subduction goes on because the oceanic plate is denser than the asthenosphere through which it is sinking. Motion of the plate exerts a drag on the underlying asthenosphere, setting in motion flow currents in the upper mantle. Thus slow convection currents probably exist in the asthenosphere beneath the moving plates, but their pathways and depths of operation are not well understood.

VOLCANIC AND TECTONIC LANDFORMS

In this chapter we continue to investigate volcanic and tectonic processes, but with emphasis on the detailed crustal features created by those processes, in contrast to the global perspective of enormous lithospheric plates. Volcanic and tectonic processes create a wide variety of both crustal rock masses and landforms.

Landforms are the surface configurations of the land, for example, mountain peaks, cliffs, canyons, and plains. *Geomorphology* is the scientific study of landforms, including their history and processes of origin. Starting with this chapter, we examine landforms produced directly by volcanic and tectonic processes.

In chapters to follow, landforms shaped by processes acting through the medium of the atmosphere and hydrosphere will be the objects of study. These activities of land sculpture can be described collectively as a process of *denudation*, the lowering of the continental surfaces by removal and transportation of mineral matter through the action of running water, waves and currents, glacial ice, and wind.

ing that they follow in sequence after the initial landforms are created and a crustal mass—a *landmass*—has been raised to an elevated position. As shown in Figure 14.1, a single uplifted crustal block (an initial landform) is attacked by agents of denudation and carved up into a large number of sequential landforms.

Any landscape is really nothing more than the existing stage in a great contest. As lithospheric plates collide or pull apart, the internal earth forces spasmodically elevate parts of the crust to create initial landforms. The external agents persistently keep wearing these masses down and carving them into vast numbers of smaller sequential landforms.

All stages of this struggle can be seen in various parts of the world. High alpine mountains and volcanic chains exist where the internal earth forces have recently dominated. Rolling low plains of the continental interiors reflect the ultimate victory of agents of denudation. All intermediate stages can be found. Because the internal earth forces act repeatedly, new initial landforms keep coming into existence as old ones are subdued.

INITIAL AND SEQUENTIAL LANDFORMS

The configuration of continental surfaces reflects the balance of power, so to speak, between internal earth forces, acting through volcanic and tectonic processes, and external forces, acting through the agents of denudation. Seen in this perspective, landforms in general fall into two basic categories.

Landforms produced directly by volcanic and tectonic activity are *initial landforms* (Figure 14.1). Initial landforms include volcanoes and lava flows, down-dropped rift valleys, and alpine ranges elevated in zones of recent crustal deformation. The energy for lifting molten rock and rigid crustal masses to produce the initial landforms has an internal heat source. This heat is believed to be produced largely by natural radioactivity in rock of the earth's crust and mantle; it is the fundamental energy source for the motions of lithospheric plates.

Landforms shaped by processes and agents of denudation belong to the class of *sequential landforms*, mean-

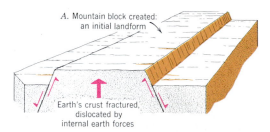

A. Mountain block created: an initial landform

Earth's crust fractured, dislocated by internal earth forces

B. Mountain block carved into sequential landforms
(a) Erosional (canyon)
(divide)
(b) Depositional (fan)
Earth forces dormant

FIGURE 14.1 Initial and sequential landforms. (Drawn by A. N. Strahler.)

FIGURE 14.2 Mount Mayon, in southeastern Luzon, the Philippines, is often considered the world's most nearly perfect composite volcanic cone. An active volcano, its summit rises to an altitude of nearly 2400 m (8000 ft). Volcanic ash, which forms a fresh layer with each eruption, is rapidly furrowed by water erosion and later acquires soil and a forest cover. (Consular General of the Philippines.)

VOLCANIC ACTIVITY

We have identified volcanism as one of the forms of mountain-building. The extrusion of magma builds landforms, and these collectively can accumulate both as volcanoes and as thick lava flows to make imposing mountain ranges. Most of these volcanic chains are within the circum-Pacific belt. Here, subduction of the Pacific, Nazca, Cocos, and Juan de Fuca plates is active. The Cascade Mountains of northern California, Oregon, and Washington represent one such chain. The Aleutian Range of Alaska is another. Important segments of the Andes Mountains in South America and the island of Java in Indonesia consist of volcanoes.

The form and dimensions of a volcano are quite varied, depending on the type of lava and the presence or absence of tephra. The nature of volcanic eruption, whether explosive or quiet, depends on the type of magma.

An important point is that the felsic lavas (rhyolite and andesite) have a high degree of viscosity (property of tackiness, resisting flowage) and hold large amounts of gas under pressure. As a result, these lavas produce explosive eruptions. In contrast, mafic lava (basalt) is highly fluid (low viscosity) and holds little gas, with the result that the eruptions are quiet and the lava can travel long distances to spread out in thin layers.

COMPOSITE VOLCANOES

Tall, steep-sided volcanic cones are produced by felsic lavas (Figure 14.2). These cones usually steepen toward the summit, where a depression, the *crater*, is located. In these volcanic eruptions tephra falls on the area surrounding the crater and contributes to the structure of the cone (Figure 14.3). The interlayering of ash layers and lava streams produces a *composite volcano*. Included are volcanic bombs. These solidified masses of lava range up to the size of large boulders and fall close to the crater. Very fine volcanic dust rises high into the tropo-

sphere and stratosphere, where it remains suspended for years (Figure 14.4).

Another important form of emission from the explosive types of volcanoes is a cloud of incandescent gases and fine ash. Known as a *nuée ardente* (French for "glowing cloud"), this intensely hot cloud travels rapidly down the flank of the volcanic cone, searing everything in its path. On the island of Martinique, in 1902, a glowing cloud issued without warning from Mount Pelée; it swept down on St. Pierre, destroying the city and killing all but two of its 30,000 inhabitants.

Most lofty conical volcanoes, well known for their scenic beauty, are of the composite type. Examples are Mount Shasta in the Cascade Range (see Figure 19.47), Fujiyama in Japan, Mount Mayon in the Philippines (Figure 14.2), and Mount Shishaldin in the Aleutian Islands.

Many of the world's active composite volcanoes lie above subduction zones. In Chapter 13 we explained the rise of andesitic magmas beneath volcanic arcs of active continental margins and island arcs. One good example is the volcanic arc of Sumatra and Java, lying over the subduction zone between the Australian plate and the Eurasian plate; another is the Aleutian volcanic arc, lo-

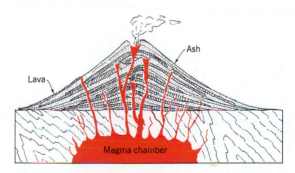

FIGURE 14.3 Idealized cross section of a composite volcanic cone with feeders from magma chamber beneath. (From A. N. Strahler, *Planet Earth: Its Physical Systems Through Geologic Time*, Harper & Row, Publishers. Copyright © by Arthur N. Strahler.)

FIGURE 14.4 Mount St. Helens, a composite volcano of the Cascade Range in southwestern Washington, erupted without warning on the morning of May 18, 1980, emitting a great column of condensed steam, heated gases, and ash from the summit crater. Within a few minutes the plume had risen to a height of 20 km (12 mi) and its contents were being carried eastward by stratospheric winds. The eruption was initiated by explosive demolition of the northern portion of the cone, concealed from this viewpoint. (J. G. Rosenbaum, U.S. Geological Survey.)

cated above the subduction zone between the Pacific plate and the North American plate.

CALDERAS

One of the most catastrophic of natural phenomena is a volcanic explosion so violent that it destroys the entire central portion of the volcano. Only a great central depression, named a *caldera*, remains. Although some of the upper part of the volcano is blown outward in fragments, most of it subsides into the ground beneath the volcano. Vast quantities of ash and dust are emitted and fill the atmosphere for many hundreds of square kilometers.

Krakatoa, a volcanic island in Indonesia, exploded in 1883, leaving a huge caldera. It is estimated that 75 cu km (18 cu mi) of rock disappeared during the explosion. Great seismic sea waves generated by the explosion killed many thousands of persons living on low coastal areas of Sumatra and Java. Another historic explosion was that of Katmai, on the Alaskan Peninsula, in 1912.

A caldera more than 3 km (2 mi) wide and 1000 m (3300 ft) deep was produced. The explosion was heard at Juneau, 1200 km (750 mi) distant, while at Kodiak, 160 km (100 mi) away, the ash formed a layer 25 cm (10 in.) deep.

A classic example of a caldera produced in prehistoric times is Crater Lake, Oregon (Figure 14.5). Mount Mazama, the former volcano, is estimated to have risen 1200 m (4000 ft) higher than the present rim. Valleys previously cut by streams and glaciers into the flanks of Mount Mazama were beheaded by the explosive subsidence of the central portion and now form distinctive notches in the rim (Figure 14.6). The event occurred about 6600 years ago.

FLOOD BASALTS AND SHIELD VOLCANOES

Geologists postulate that at various points beneath the lithosphere there occur *mantle plumes*, which are isolated columns of heated rock rising slowly within the astheno-

FIGURE 14.5 Crater Lake, Oregon, is surrounded by the high, steep wall of a great caldera. Wizard Island (center) is an almost perfect basaltic cinder cone with basalt lava flows; it was built on the floor of the caldera after the major explosive activity had ceased. (A. N. Strahler.)

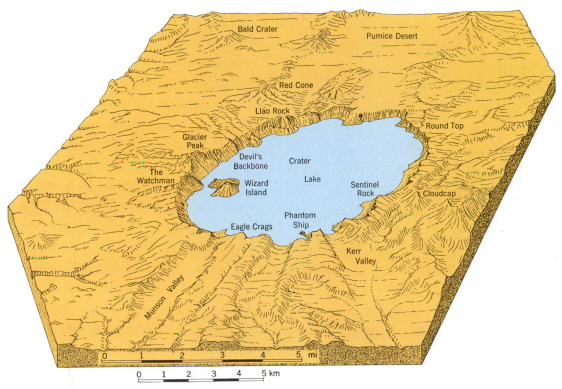

FIGURE 14.6 Crater Lake, Oregon, is an outstanding illustration of a caldera, now holding a lake. (Drawn by E. Raisz.)

sphere. Directly above a mantle plume, crustal basalt can be heated to the point of melting and produce a magma pocket. The site of magma is called a *hot spot*. Magma of basaltic composition makes its way through the overlying lithosphere to emerge at the surface as lava.

Where a mantle plume lies beneath a continental lithospheric plate, the hot spot may generate enormous volumes of basaltic lava that accumulate layer upon layer. The basalt may ultimately attain a thickness of thousands of meters and may cover thousands of square kilometers. These accumulations are called *flood basalts*. An important example is found in the Columbia Plateau region of southeastern Washington, northeastern Oregon, and westernmost Idaho (Figure 14.7); basalts of Cenozoic age cover an area of about 130,000 sq km (50,000 sq mi), about the same area as the state of New York.

Hot spots also form above mantle plumes in the oceanic lithosphere. The emerging basalt builds a class of volcanoes known as *shield volcanoes*. These are constructed on the deep ocean floor, far from the plate boundaries, and may be built high enough to rise above sea level as volcanic islands. As a lithospheric plate drifts slowly over a mantle plume beneath, a succession of shield volcanoes is formed. Thus a chain of volcanic islands comes into existence. Several chains of volcanic islands exist in the Pacific Ocean basin. Best known of the island chains is the Hawaiian group.

A few basaltic volcanoes also occur along the mid-oceanic ridge, where sea-floor spreading is in progress. Perhaps the outstanding example is Iceland, in the North

Atlantic Ocean. Iceland is constructed entirely of basalt flows superimposed on other basaltic rocks in the form of dikes and sills that entered the spreading rift at deeper levels. Mount Hekla, an active volcano on Iceland, is a shield volcano somewhat similar to those of Hawaii. Farther south along the Mid-Atlantic Ridge are other islands consisting of basaltic volcanoes: the Azores, Ascension Island, and Tristan da Cunha.

Shield volcanoes of the Hawaiian Islands are characterized by gently rising, smooth slopes that flatten near

FIGURE 14.7 Basalt lava flows exposed in cliffs bordering the Columbia River in Washington. Columnar jointing of the massive basalt is conspicuous in the arid climate. (A. N. Strahler.)

FIGURE 14.8 This air view of Mauna Loa, Hawaii, shows a chain of pit craters leading up to the great central depression at the summit. (U.S. Army Air Force.)

the top, producing a broad-topped volcano (Figures 14.8 and 14.9). Domes on the island of Hawaii rise to summit elevations 4000 m (13,000 ft) above sea level. Including the basal portion lying below sea level, they are more than twice that high. In width they range from 16 to 80 km (10 to 50 mi) at sea level and up to 160 km (100 mi) wide at the submerged base.

Explosive behavior and the emission of tephra are not as important in shield volcano eruptions as they are for composite cones built of felsic magmas. The basalt

lava in the Hawaiian domes is highly fluid and travels far down the low slopes, which do not usually exceed angles of 4 or 5°.

Lava domes have a wide, steep-sided central depression that may be 3 km (2 mi) or more wide and several hundred meters deep (Figures 14.8 and 14.9). These large depressions are a type of caldera produced by subsidence accompanying the removal of molten lava from beneath. Molten basalt is actually seen in the floors of deep pit craters that occur on the floor of the caldera or elsewhere over the surface of the lava dome (Figure 14.10). Most lava flows issue from fissures on the sides of the volcano.

FIGURE 14.9 Halemaumau fire pit, a steep-walled crater within the central depression (caldera) of Kilauea volcano, Hawaii. (A. N. Strahler.)

FIGURE 14.10 A fire fountain of molten basalt lava in the floor of an active fire pit east of the summit of Kilauea volcano. The fountain rises to a height of about 75 m (250 ft). A thin layer of recently congealed lava covers the magma pool. (John S. Shelton.)

FIGURE 14.11 A young cinder cone surrounded by rough-surfaced basalt lava flows. Lava Beds National Monument, northern California. (Alan H. Strahler.)

CINDER CONES

At points where frothy basalt magma is ejected under high pressure from a narrow vent, tephra accumulates in a steep-sided conical hill, called a *cinder cone* (Figure 14.11). Groups of cinder cones are common in areas where basaltic lava flows are emerging from fissures. The cones rarely grow to heights over a few hundred meters. The loose material absorbs heavy rain without permitting surface runoff. Erosion is delayed until weathering produces a soil that fills the interstices.

Lava flows sometimes issue from the same vent as a cinder cone; they may burst apart the side of the cone. Cinder cones may erupt in almost any conceivable topographic location—on ridges, on slopes, and in valleys (Figure 14.12). Cinder cones usually occur in groups, often many dozens in an area of a few tens of square kilometers.

ENVIRONMENTAL ASPECTS OF VOLCANIC ACTIVITY AND LANDFORMS

Eruptions of volcanoes and lava flows are environmental hazards of the severest sort, often taking a heavy toll of

FIGURE 14.12 A cinder cone with its lava flows has dammed a valley, making a lake. Farther downvalley, another lava mass has made a second dam. (Drawn by W. M. Davis.)

plant and animal life and the works of humans. What natural phenomenon can compare with the Mount Pelée disaster, in which thousands of lives were snuffed out in seconds? Perhaps only an earthquake or storm surge of a tropical cyclone is equally disastrous. The wholesale loss of life and the destruction of towns and cities are frequent in the history of peoples who live near active volcanoes. Loss occurs principally from sweeping clouds of incandescent gases that descend the volcano slopes like great avalanches; from lava flows whose relentless advance engulfs whole cities; from the descent of showers of ash, cinders, and bombs; from violent earthquakes associated with the volcanic activity. For habitations along low-lying coasts, there is the additional peril of great seismic sea waves, generated elsewhere by the explosive destruction of volcanoes.

In 1985, an explosive eruption of Ruiz Volcano in the Colombian Andes caused the rapid melting of ice and snow in the summit area. Mixing with volcanic ash, the water formed a variety of mudflow known as a *lahar*. Rushing downslope at speeds up to 145 km (90 mi) per hour, the lahar became channeled into a valley on the lower slopes, where it engulfed a town and killed more than 20,000 persons.

The surfaces of volcanoes and lava flows remain barren and sterile for long periods after their formation. Certain types of lava surfaces are extremely rough and difficult to traverse; the Spaniards who encountered such terrain in the southwestern United States named it "malpais" (bad ground). Most volcanic rocks in time produce highly fertile soils that are intensively cultivated.

Volcanoes are a valuable natural resource in terms of recreation and tourism. Few landscapes can rival in beauty the mountainous landscape of volcanic origin. National parks have been made of Mount Rainier, Mount Lassen, and Crater Lake in the Cascade Range. Hawaii Volcanoes National Park recognizes the natural beauty of Mauna Loa and Kilauea; their displays of molten lava are a living textbook of igneous processes.

GEOTHERMAL ENERGY RESOURCES

Geothermal energy is energy in the form of sensible heat that originates within the earth's crust and makes its way to the surface by conduction. Heat may be conducted upward through solid rock or carried up by circulating ground water that is heated at depth and makes its return to the surface. Concentrated geothermal heat sources are usually associated with igneous activity, but there also exist deep zones of heated rock and ground water that are not directly related to igneous activity. We will examine briefly several forms of intensified internal heat; together they comprise the available geothermal energy sources that are potentially useful to humans.

From observations made in deep mines and boreholes, we know that the temperature of rock increases steadily with depth. This rate of increase, called the *geothermal gradient*, averages 20 to 40C° per km (60 to 100F°) for at least the uppermost 5 km (3 mi). Although the rate of increase falls off quite rapidly with increasing depth, temperatures attain very high values in the upper mantle where rock is close to its melting point. Although the internal heat of the earth flows very slowly upward to the earth's surface, the quantity is extremely small. In one year, the amount of heat reaching the surface is just enough to melt an imaginary ice layer 6 mm (0.2 in.) thick. This small heat flow is insignificant compared with the amount of solar energy reaching the ground in one year.

It might seem simple enough to obtain all our energy needs by drilling deep holes at any desired location into the crust and letting the hot rock turn injected fresh water into steam, which we could use to generate electricity as our primary energy source. Unfortunately, at the depths usually required to furnish the needed heat intensity, crustal rock tends to close any cavity or opening by rupture and slow flowage; this phenomenon would either prevent the holes from being drilled, or would close them in short order. Generally, then we must look for *geothermal localities*, where special conditions have caused hot rock and hot ground water to lie within striking distance of conventional drilling methods.

At some widely separated geothermal localities over the globe, ground water reaches the surface in *hot springs* at temperatures not far below the boiling point of water, which is 100°C (212°F) at sea-level atmospheric pressure (Figure 14.13). At some of these same places, jetlike emissions of steam and hot water occur at intervals from small vents; these are *geysers* (Figure 14.13). The water that emerges from hot springs and geysers is largely ground water that has been heated in contact with hot rock and forced to the surface. In other words, this water is recycled surface water. Little, if any, is water that was originally held in rising bodies of magma.

A phenomenon related to hot springs and geysers is the *fumarole*, a jet of gases issuing from a small vent. Gas temperatures in fumaroles are extremely high, up to 320°C (650°F). Most of the gas (over 99%) is water. In other words, the fumarole emits largely superheated steam. Fumaroles are common in regions of current and recent volcanic activity.

The natural hot water and steam localities were the

FIGURE 14.13 Geothermal activity in Yellowstone National Park, Wyoming. (left) Mammoth Hot Springs. Terraces of siliceous sinter hold steaming pools of hot water. (right) Old Faithful Geyser. (A. N. Strahler.)

FIGURE 14.14 An electricity generating plant at The Geysers, California. Steam pipes in the foreground lead to the plant. After use in generating turbines, the steam is condensed in large cylindrical towers. (Pacific Gas & Electric Company.)

first type of geothermal energy source to be developed and at present account for nearly all production of electrical power. Wells are drilled to tap the hot water, which flashes into steam under reduced atmospheric pressure as it reaches the surface. The steam is separated from the water in large towers and fed into generating turbines to produce electricity (Figure 14.14). The hot water is usually released into surface stream flow, where it may create a pollution problem. The larger steam fields have sufficient energy to generate at least 15 megawatts of electric power, and a few can generate 200 megawatts or more.

Much greater energy sources than those we have just described lie in deeper zones of hot ground water, but these must be tapped by deep drilling. One region of deep, hot ground water, currently under investigation, is beneath the Imperial Valley of southern California. An area of 5000 sq km is involved, and extends over the border into Mexico in the Mexicali Valley. This region is tectonically active and has been interpreted as a zone of crustal spreading in which the lithospheric plate is being fractured. Rising basalt magma, found elsewhere in active spreading zones such as Iceland in the North Atlantic, may be responsible for the geothermal condition, but this interpretation is speculative. Test wells showed that a large reservoir of extremely hot ground water (260° to 370°C; 500° to 700°F) is present here. This water readily flashes into steam when penetrated by a drill hole. Steam pressure forces both steam and hot water to the surface, much like the action of a coffee percolator. Near Mexicali, Mexico, this resource has already been developed, with the daily flow of steam and water amounting to 2000 metric tons. The prospect of a large power development beneath Imperial Valley looks very good, while the salinity of the hot water is quite low. Presently developed Imperial Valley flash-steam sites have an estimated maximum total capacity of about 200 megawatts of electricity. Heat remaining in the water after it has been used to generate electricity could be used to distill the waste water and produce a substantial yield of irrigation water, a valuable commodity in this desert agricultural region.

In certain areas, the intrusion of magma has been sufficiently recent that solid igneous rock of a batholith is still hot in a depth range of perhaps 2 to 5 km (1 to 3 mi). At this depth, the rock is strongly compressed and contains little, if any, ground water. Rock in this zone may be as hot as 300°C (600°F) and could supply an enormous quantity of heat energy. The planned development of this resource includes drilling into the hot zone and then shattering the surrounding rock by hydrofracture —a method using water under pressure that is widely used in petroleum development. Surface water would be pumped down one well into the fracture zone and heated water pumped up another well. Experimental holes drilled in 1983 at Los Alamos in northern New Mexico reached a depth of 4400 m (13,000 ft), where the dry rock temperature is about 325°C (620°F). The dry granite rock was fractured by water injected under high pressure, and a pumping test was successful.

Attention is now being turned to calderas beneath which lie bodies of recently crystallized or partially molten magma. These sites are potential sources of enormous quantities of extractable heat at comparatively shallow depth. One example is the Yellowstone caldera, encompassing a large area of that national park in Wyoming. Another is the Long Valley caldera that lies on the eastern side of the Sierra Nevada in California; it includes the popular ski resort at Mammoth Lakes. The major caldera collapse here occurred over 700,000 years ago, but numerous earthquakes are being generated beneath the caldera surface, which has shown a measurable uplift. The roof of the magma chamber is placed at a depth of 5 to 7 km (3 to 5 mi), and a test well is planned to penetrate crystallized magma. In 1990, test drilling was being carried out under a small island within the Santorini (Thera) caldera in the Aegean Sea north of Crete. The eruption and collapse of the parent volcano of Santorini occurred around 1500 B.C.

The potential for U.S. electrical power generation from deep hot rock areas is estimated to be many times greater than for hot water areas. One estimate placed the potential output of the dry rock areas at 400,000 megawatts within a decade or two. (Compare this figure with a total U.S. geothermal production of 2200 megawatts in 1990.)

Finally, on the list of occurrences of geothermal energy, we come to an energy source trapped in thick continental shelf sediments of the continental margin. Recall from Chapter 13 that these sediments form a wedge, thickening toward the ocean basin. Rich petroleum deposits have been developed in the continental shelf wedge of the Gulf Coast region. Here, in a region extending from Texas to Louisiana, the strata in some areas have been found to hold hot water under high pressure and to contain dissolved natural gas. Subsurface bodies of this type are described as *geopressurized zones*. They are avoided in drilling for petroleum because the high pressure makes it difficult to control the well. These geopressurized areas should be explored, according to recent proposals, as sources of hot water for generating electricity. One estimate of the total energy resource in the Gulf area alone is a generating capability of between 30,000 and 115,000 megawatts of power. Development of this resource is probably not economically feasible at this time, but the ultimate potential is large.

LANDFORMS OF TECTONIC ACTIVITY

Our introduction to global plate tectonics in Chapter 13 brought out the essential distinction between two basically different expressions of tectonic activity. Along converging lithospheric plate boundaries, tectonic activity is basically that of compression, illustrated schematically in Figure 14.15. In subduction zones, sedimentary layers of the ocean floor are subject to compression within a trench as the descending plate forces them against the fixed plate. In arc–continent and continental collision, compression is of the severest kind, producing alpine structure with thrust faulting (see Figure 13.21).

In zones of rifting of continental plates, explained in Chapter 13, the brittle continental crust is pulled apart and yields by faulting. In the simple model shown in Figure 14.15, rifting is expressed in a pair of opposite-facing faults. The *fault block* between them moves down to form a depressed area. In the remainder of this chapter, we examine the landforms of folding and faulting and the phenomenon of earthquakes generated during the faulting process.

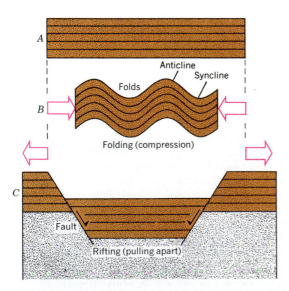

FIGURE 14.15 Folding occurs under compression; rifting occurs where the crust is pulled apart.

FORELAND FOLD BELTS

When plate collision begins to take place, wedges of strata of a passive continental margin come under strong forces of compression. The strata, which were originally more or less flat-lying, experience *folding*, as shown in Figure 14.15B. The wavelike undulations imposed on the strata consist of alternating archlike upfolds, called *anticlines*, and troughlike downfolds, called *synclines*. Thus the initial landform associated with an anticline is a broadly rounded mountain ridge; the landform corresponding to a syncline is an elongate, open valley.

Two well-known examples of open folds of comparatively young geologic age have long attracted the interest of geographers. One of these is the Jura Mountains of France and Switzerland. Figure 14.16 is a block diagram of a small portion of that fold belt. The strata are mostly limestone layers of Jurassic age and were capable of being deformed by bending with little brittle fracturing. Folding occurred in late Cenozoic (Miocene) time. Notice that each mountain crest is associated with the axis of an anticline, while each valley lies over the axis of a syncline. Some of the anticlinal arches have been partially removed by erosion processes. The rock structure can be seen clearly in the walls of the winding gorge of a major river that crosses the area. The Jura folds lie just to the north of the main collision orogen of the Alps. In this respect they are called *foreland folds*.

Our second example of a belt of foreland folds produced during continental collision is the Zagros Mountains of southwestern Iran. Seen from an orbiting space vehicle, the Zagros folds resemble an army of caterpillars crawling in parallel tracks (Figure 14.17). They were formed during late Cenozoic time when the Arabian plate smashed into the Eurasian plate, closing an ocean basin that formerly separated the two continental masses. Flat–lying strata on the passive margin of the Arabian plate were crumpled into open folds over a belt more than 200 km (125 mi) wide.

FAULTS AND FAULT LANDFORMS

A fault in the brittle rocks of the earth's crust is a result of sudden yielding under unequal stresses. Faulting is accompanied by a displacement along the plane of breakage, or *fault plane*. Faults are often of great hori-

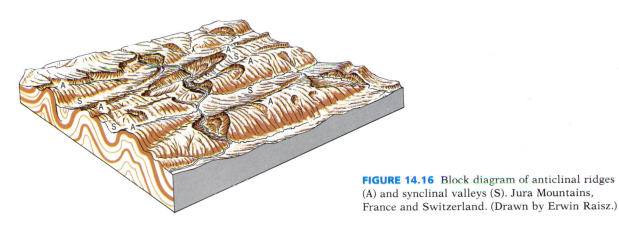

FIGURE 14.16 Block diagram of anticlinal ridges (A) and synclinal valleys (S). Jura Mountains, France and Switzerland. (Drawn by Erwin Raisz.)

zontal extent, so the surface trace, or *fault line*, can sometimes be followed along the ground for many kilometers. Little is known of what happens to faults at depth, but in all probability most extend down for at least several thousands of meters.

Faulting occurs in sudden slippage movements that generate earthquakes. A particular fault movement may result in a slippage of as little as a centimeter or as much as 15 m (50 ft). Successive movements may occur many years or decades apart, even several centuries apart, but accumulate into total displacements of hundreds or thousands of meters. In some places clearly recognizable sedimentary rock layers are offset on opposite sides of a fault, and the amount of displacement can be accurately measured.

According to the angle of inclination and relative direction of the displacement, four basic types of faults are recognized. A *normal fault* has a steep or nearly vertical fault plane (Figure 14.18A). Movement is predominantly in a vertical direction, so one side is raised, or upthrown, relative to the other side, which is downthrown. A nor-

mal fault results in a steep *fault scarp*, whose height is an approximate measure of the vertical element of displacement (Figure 14.19). Fault scarps range in height from a few meters to several hundred meters. Their length is measurable in kilometers; often they attain lengths of 100 km (60 mi) or more. Normal faulting is an expression of the pulling apart (extension) of the outer crust. Sliding on an inclined surface of the type indicated in Figure 14.18A must result in a horizontal separation of points situated on opposite sides of the fault.

Faults are rarely isolated features. Normal faults usually occur in multiple arrangements, commonly as a parallel series of faults. This gives rise to a belted landscape pattern. A narrow block dropped down between two normal faults is a *graben* (Figure 14.20 and Figure 13.27). A narrow block elevated between two normal faults is a *horst*. Grabens make conspicuous trenches, with straight, parallel walls. Horsts make blocklike plateaus or mountains, often with a flat top but steep, straight sides. An example of a very large graben, produced by continental rifting, is the Red Sea graben separating the Arabian

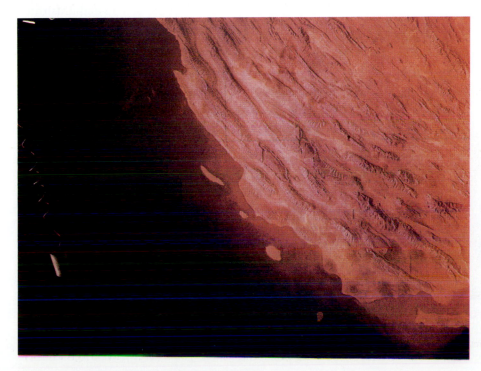

FIGURE 14.17 The Zagros Mountains of Iran, photographed by astronauts of the Gemini XII orbiting space vehicle in 1966. The view is toward the northwest and shows a portion of the Persian Gulf. The elongate ridges are anticlines, partly eroded by streams; they were elevated by folding which began late in the Cenozoic Era. (NASA 566-6383).

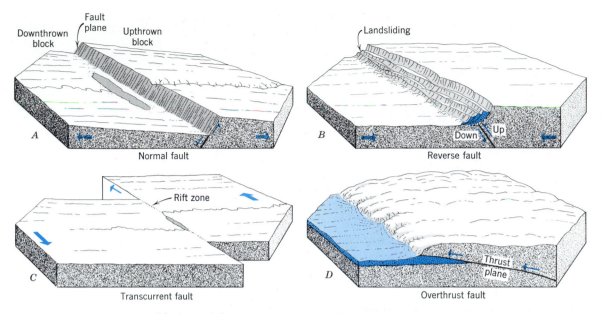

FIGURE 14.18 Four types of faults and their expression as landforms. (Drawn by A. N. Strahler.)

Peninsula from Africa (Figure 13.28). Two smaller grabens form the Gulf of Suez and the Gulf of Aqaba.

In a *reverse fault*, the inclination of the fault plane is such that one side rides up over the other and a crustal shortening occurs (Figure 14.18B). Reverse faults produce fault scarps similar to those of normal faults, but the possibility of landsliding is greater because an overhanging scarp tends to be formed. The San Fernando, California, earthquake of 1971 was generated by slippage on a reverse fault.

A *transcurrent fault* is unique in that the movement is predominantly in a horizontal direction (Figure 14.18C). On an ideal flat plain, no fault scarp would be produced. Instead, only a thin line would be traceable across the surface (Figure 14.21). Streams sometimes turn and follow the fault line for a short distance (Figure 14.22). Sometimes a narrow trench, or rift, marks the fault line.

The transform fault, described in Chapter 13 as one of the three kinds of lithospheric plate boundaries, is a special case of transcurrent fault. The San Andreas Fault, illustrated in Figures 14.21 and 14.23, is a transform fault and makes up a section of the active boundary between the Pacific plate and the North American plate. Many other transcurrent faults occur in California, but they are of lesser importance and do not mark plate boundaries.

The *low-angle overthrust fault* (Figure 14.18D) also involves predominantly horizontal movement, but the fault plane is nearly horizontal. One slice of rock rides over the adjacent ground surface. A thrust slice may be up to 50 km (30 mi) wide. The evolution of low-angle thrust faults was explained in Chapter 13 and illustrated in Figures 13.21 and 13.23.

THE RIFT VALLEY SYSTEM OF EAST AFRICA

The rifting of continental lithosphere that is the very first stage in splitting apart of a continent to form a new ocean basin is beautifully illustrated by the East African

FIGURE 14.19 This fault scarp was formed during the Hebgen Lake, Montana, earthquake of August 1959. In a single instant, a displacement of 6 m (19 ft) took place on a normal fault. (J. R. Stacy, U.S. Geological Survey.)

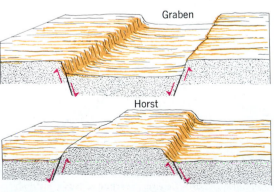

FIGURE 14.20 Graben and horst. (Drawn by A. N. Strahler.)

FIGURE 14.21 In this vertical air view, the nearly straight trace of the San Andreas Fault, a transcurrent fault, contrasts sharply with the sinuous lines of stream channels. Carrizo Plain, Kern County, California. (WESTERN GEOPHYSICAL COMPANY, a division of WESTERN ATLAS INTERNATIONAL, INC.)

FIGURE 14.23 The San Andreas Fault runs straight as an arrow across the Carrizo Plain in Kern County, California. You are looking in a northwesterly direction; the land on the left side of the fault line is moving away from you. (John S. Shelton.)

rift valley system. It has attracted the attention of geologists since the early 1900s. They gave the name *rift valley* to what is basically a graben, but with a more complex history that includes the building of volcanoes in the graben floor. Figure 14.24 is a sketch map of the East African rift valley system, which is a full 3000 km (1900 mi) long and extends from the Red Sea southward to the Zambezi River. The system consists of a number of grabenlike troughs, each a separate rift valley ranging in width from 30 to 60 km. As geologists had noted in earlier field surveys of this system, the rift valleys are like keystone blocks of a masonry arch that have slipped down between neighboring blocks because the arch has spread apart somewhat (Figure 14.25). Thus, the floors of the rift valleys are above the elevation of most of the African continental surface, even though some of the valley floors are occupied by long, deep lakes and by major rivers (Figure 14.26). The side of a rift valley may consist of multiple fault steps (Figure 14.27).

The rift-valley system consists of a number of domelike swells in the crust, the highest of which forms the Ethiopian Highlands on the north. Basalt lavas have risen from fissures in the floors of the rift valleys and from the flanks of the domes. Sediments, derived from the high plateaus that form the flanks of troughs, make thick fills in the floors of the valleys. Lake Victoria is flanked by two rift valleys, which join south of the lake. A single rift valley extends southward from this junction.

Two great composite volcanoes have been built close to the rift valley east of Lake Victoria. One is Mount Kilimanjaro, whose summit rises to over 6300 m (19,300 ft). The other, Mount Kenya, is only a little lower; it lies right on the equator. Geologists think the acidic magma that built these two great volcanoes came from thinned crust, domed upward by highly heated mantle rock that

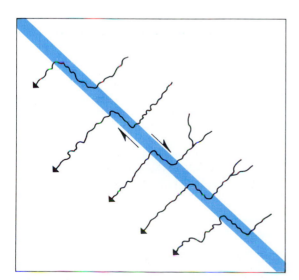

FIGURE 14.22 Schematic map of streams offset by long-continued movement along a transcurrent fault. (From A. N. Strahler, *Principles of Earth Science*, Harper & Row, Publishers. Copyright © by Arthur N. Strahler.)

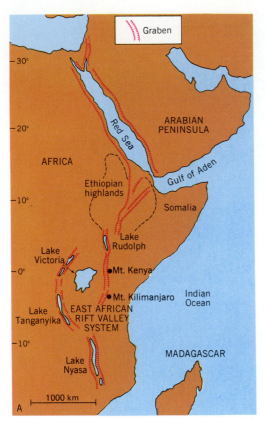

FIGURE 14.24 A sketch map of the East African rift valley system and the Red Sea to the north.

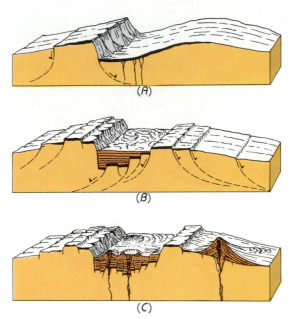

FIGURE 14.25 Development of a typical rift valley in East Africa. The diagrams are schematic and combine elements found in several localities. Width of the area shown is about 150 km. (A) Late Miocene and early Pliocene. Normal faulting has produced a tilted fault block on the left. Crust at right is deformed into a broad monocline with a cap of lava. (B) Late Pliocene. Renewed normal faulting has broken the valley floor into narrow blocks and raised the eastern side. The rift valley is now a graben structure. Lava flows have filled the valley floor. (C) Pleistocene and Holocene. After another episode of minor faulting, extrusive activity has built volcanoes in the rift valley and on the flank of the uplift. (Drawn by A. N. Strahler. Based on data of B. H. Baker, in *East Africa Rift System*, 1965, UNESCO Seminar, University College, Nairobi, p. 82.)

moved upward as the continental lithosphere was pulled apart.

EARTHQUAKES

The news media present detailed accounts of disastrous earthquakes. Nearly everyone has seen pictures of their destructive effects. Californians know about severe earthquakes from firsthand experience; but many other areas in North America have also experienced earthquakes, and a few of these have been severe. An *earthquake* is a motion of the ground surface, ranging from a faint tremor to a wild motion capable of shaking buildings apart and causing gaping fissures to open up in the ground.

The earthquake is a form of energy of wave motion transmitted through the surface layer of the earth in widening circles from a point of sudden energy release—the *earthquake focus*. Like ripples produced when a pebble is thrown into a quiet pond, these *seismic waves* travel outward in all directions, gradually losing energy. Earthquakes are produced by sudden movements along faults; commonly these are normal faults or transcurrent faults. The famed San Andreas Fault, pictured earlier, passes through the San Francisco Bay Area. The devastating earthquake of 1906 resulted from slippage along this transcurrent fault. The fault is 1000 km (600 mi) long and extends into southern California, passing about 60 km (40 mi) inland of the Los Angeles metropolitan area. Associated with the San Andreas Fault are several other important transcurrent faults, all capable of generating severe earthquakes.

Rock on both sides of an active fault is slowly bent over many years as tectonic forces are applied. Energy accumulates in the bent rock, just as it does in a bent crossbow. When a critical point is reached, the strain is relieved by slippage on the fault and a large quantity of energy is instantaneously released in the form of seismic waves. Slow bending of the rock takes place over many decades. In the case of a transcurrent fault, energy release causes offsetting of features that formerly crossed the fault in straight lines, for example, a roadway or fence (Figure 14.28). Faults of this type can also show a slow, steady displacement known as *fault creep*, which tends to reduce the accumulation of stored energy.

EARTHQUAKE ENERGY—THE RICHTER SCALE

The quantity of energy released during an earthquake at the place where the fault slippage originates is known as the *earthquake magnitude* (M). Rock rupture begins at a point on the fault plane where the frictional resistance is first overcome. This initial point is known as the *focus*; the point on the surface directly above the focus is the

FIGURE 14.26 A portion of Gregory's Rift in southern Kenya displayed in false-color remote-sensing imagery. The rift floor, about 40 km (25 mi) wide, is bounded by a series of fault steps leading up to forested highlands on either side. The red color, indicating green foliage, becomes deeper with increasing elevation. Fault scarps are most sharply defined on the eastern side of the valley. In the valley floor lies Lake Nalvasha (black); south of it are two volcanoes with prominent circular craters: Longonot (small) and Susuwa (large). (NASA ERTS 2188-07055, 29 July 1975. Reproduced by permission of Earth Observation Satellite Company, Lanham, Maryland, U.S.A.)

FIGURE 14.27 The rift valley wall in Ethiopia. Multiple fault scarps give the landscape a stepped appearance. (George Gerster/Comstock.)

FIGURE 14.28 During the San Francisco earthquake of 1906, this fence was offset 2.4 m (8 ft) by lateral movement along the San Andreas Fault near Woodville, California. (G. K. Gilbert, U.S. Geological Survey.)

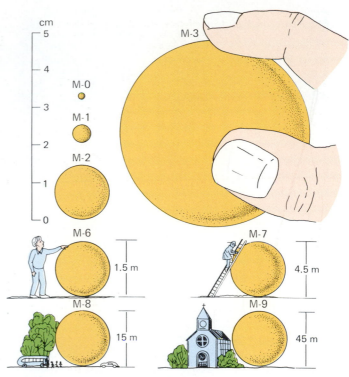

FIGURE 14.29 Energy release compared for various Richter magnitudes in terms of the relative volumes of spheres. The upper diagram shows spheres to their true scale relationships for magnitudes 0 through 3. The lower four diagrams give the diameters of spheres on the same relative scale as for magnitudes 0 through 3. (From A. N. Strahler, *The Earth Sciences*, 2nd ed., Harper & Row, Publishers. Copyright © by Arthur N. Strahler.)

epicenter. Actually, rupture spreads rapidly along the fault plane, so that slippage may extend over a distance measured in many kilometers, or even hundreds of kilometers.

In 1935 a leading seismologist, Charles F. Richter, devised a rating scale of earthquake magnitude. The *Richter scale* consists of numbers ranging from less than 0 (negative numbers) to more than 8.5. Values are given to the nearest one-tenth, thus: 2.5, 4.9, 6.2, 7.8, 8.5. There is neither a fixed maximum nor a minimum, but the highest-magnitude earthquakes thus far measured have been rated as 8.9 on the Richter scale. Earthquakes of magnitude 2.0 are the smallest normally detected by the human senses, but instruments can detect quakes as small as −3.0. The scale is logarithmic, which is to say that the amplitude of the recorded waves increases tenfold for each integer increase in Richter magnitude. An earthquake of magnitude 5.0 is ten times larger than one of magnitude 4.0. Figure 14.29 uses familiar objects to illustrate the energy increase associated with each number on the scale. Some information on the meaning of various magnitudes is given in Table 14.1 To convert the Richter numbers to tell the actual amount of energy release involves further calculation. For every increase of one integer of Richter magnitude, the quantity of energy increases by a factor of about 32 times. Thus, an earthquake of magnitude 7.0 releases about 32 times as much energy as one of magnitude 6.0. An earthquake of magnitude 8.0 releases about 1000 times as much energy as one of magnitude 6.0 (32 × 32 = 1024).

An atomic bomb blast, such as that at Bikini Atoll in 1946, yields about 10^{12} joules, equivalent to an earthquake of magnitude 5.0, which usually causes only minor damage in an urban area. At magnitude 8.9, the energy release is nearly 1 million times greater than from a single atomic bomb.

The total quantity of energy released by all earthquakes of the world in a single year is estimated to be from 10^{18} to 10^{19} joules. Most of this quantity is from a very few earthquakes of Richter magnitude greater than 7.0.

EARTHQUAKES AND PLATE TECTONICS

Seismic activity, the repeated occurrence of earthquakes, shows a close geographic relationship to lithospheric plate boundaries (Figure 14.30). The greatest intensity of seismic activity is found along converging plate boundaries where oceanic plates are undergoing subduction. Strong pressures build up at the downslanting contact of the two plates, and these are relieved by sudden fault slippages that generate earthquakes of large magnitude. This mechanism explains the great earthquakes experienced in Japan, Alaska, Chile, and other narrow zones close to trenches and volcanic arcs of the Pacific Ocean basin.

Transform boundaries that cut through the continental lithosphere are sites of intense seismic activity, with moderate to strong earthquakes. The most familiar example is the San Andreas Fault, which forms the transform boundary between the American plate and the Pacific plate in California.

Spreading boundaries are a third class of narrow zones of seismic activity related to lithospheric plates. Most of these boundaries are identified with the mid-oceanic ridge and its branches. Earthquakes are generated both along the ridge axis and on the transform faults that connect offset ends of the ridge, but they are mostly small earthquakes set off at shallow depths.

TABLE 14.1 Richter Magnitude and Energy Release

Magnitude, Richter scale	Energy release, joules	Comment
2.0	2.5×10^7	Smallest quake normally detected by humans.
2.5–3	10^8–10^9	Quake can be felt if it is nearby. About 100,000 shallow quakes of this magnitude per year.
4.5	10^{11}	Can cause local damage.
5.0	10^{12}	Energy release about equal to first atomic bomb, Alamogordo, N.M., 1945.
6.0	2.5×10^{13}	Destructive in a limited area. About 100 shallow quakes per year of this magnitude.
7.0	10^{15}	Rated a major earthquake above this magnitude. Quake can be recorded over whole earth. About 14 per year this great or greater.
8.25	6.0×10^{16}	San Francisco earthquake of 1906.
8.5	1.5×10^{17}	Chile, 1960; Alaska, 1964. Close to the maximum known.
8.9	8.8×10^{17}	Maximum ever recorded. Only two known: Colombia–Ecuador border, 1906; Japan, 1933.

Earthquakes also occur at scattered locations over the continental plates, far from active plate boundaries. In many cases, no active fault is visible and the geologic cause of the earthquake is obscure. Some large earthquakes of southern Asia are probably related to the continental suture between the Eurasian plate and the Arabian and Austral-Indian plates. Some isolated earthquakes of North America are associated with the rise of the crust following disappearance of the ice sheets of the last ice age (see Chapter 22).

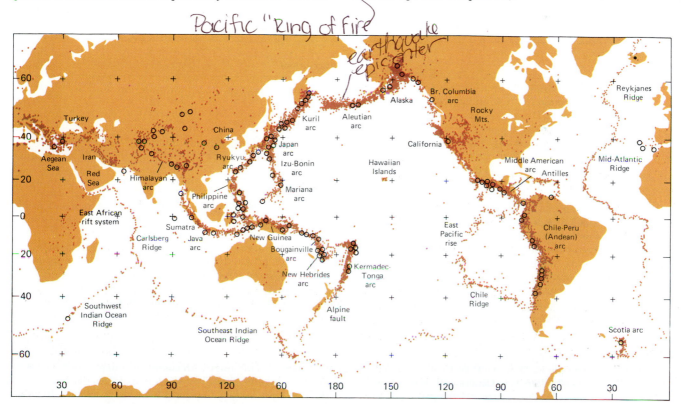

FIGURE 14.30 World map of earthquake epicenters 1961–67 and epicenters of great earthquakes. Epicenters of earthquakes originating at depths from 0 to 100 km are shown by color dots. Each dot represents a single epicenter or a cluster of epicenters. Color circles show epicenters of earthquakes of Richter magnitude 8.0 and greater, 1897–1976. (Epicenter data compiled from ESSA, Coast and Geodetic Survey, in J. Dorman and M. Barazangi, 1969, *Seismological Society of America, Bulletin*, vol. 50, No. 1, Plates 2 and 3. Epicenters of great earthquakes from W. Hamilton, 1979, *Professional Paper 1078*, U.S. Geological Survey, Figure 3. From A. N. Strahler, *The Earth Sciences*, 2nd ed., Harper & Row, Publishers. Copyright © by Arthur N. Strahler.)

EARTHQUAKES AS ENVIRONMENTAL HAZARDS

To gain an understanding of the environmental hazards that major earthquakes impose on humans and their urban and industrial structures and supply systems, we need to inquire into the nature of ground motions generated by earthquakes. First, however, it is important to know how the local intensity of ground shaking is measured and described.

EARTHQUAKE INTENSITY SCALES

The actual destructiveness of an earthquake depends upon factors other than the energy release given by Richter magnitude—for example, closeness to the epicenter and nature of the subsurface earth materials. An *earth-quake intensity scale*, designed to measure observed earth-shaking effects, is important in practical engineering aspects of seismology.

An intensity scale used extensively in the United States is the *modified Mercalli scale*. The original Mercalli scale was prepared by an Italian seismologist of that name in 1902, and was modified in 1956 by Charles Richter to apply to various types of building construction. The modified Mercalli scale recognizes 12 levels of intensity, designated by Roman numerals I through XII (Table 14.2). Each intensity is described in terms of phenomena that any person might experience. For example, at intensity IV, hanging objects swing, a vibration like that of a passing truck is felt, standing automobiles rock, and

TABLE 14.2   The Modified Mercalli Intensity Scale

Masonry types:	The quality of masonry, brick or otherwise, is specified by the following letter code:
Masonry A	Good workmanship, mortar, and design; reinforced, especially laterally, and bound together by using steel, concrete, etc.; designed to resist lateral forces.
Masonry B	Good workmanship and mortar; reinforced, but not designed in detail to resist lateral forces.
Masonry C	Ordinary workmanship and mortar; no extreme weaknesses like failing to tie in at corners, but neither reinforced nor designed against horizontal forces.
Masonry D	Weak materials, such as adobe; poor mortar; low standards of workmanship; weak horizontally.

Intensity value	Description
I.	Not felt. Marginal and long-period effects of large earthquakes.
II.	Felt by persons at rest, on upper floors, or favorably placed.
III.	Felt indoors. Hanging objects swing. Vibration like passing of light trucks. Duration estimated. May not be recognized as an earthquake.
IV.	Hanging objects swing. Vibration like passing of heavy trucks; or sensation of a jolt like a heavy ball striking the walls. Standing cars rock. Windows, dishes, doors rattle. Glasses clink. Crockery clashes. In the upper range of IV, wooden walls and frame creak.
V.	Felt outdoors; direction estimated. Sleepers wakened. Liquids disturbed, some spilled. Small unstable objects displaced or upset. Doors swing, close, open. Shutters, pictures move. Pendulum clocks stop, start, change rate.
VI.	Felt by all. Many frightened and run outdoors. Persons walk unsteadily. Windows, dishes, glassware broken. Knick-knacks, books, etc., off shelves. Pictures off walls. Furniture moved or overturned. Weak plaster and masonry D cracked. Small bells ring (church, school). Trees, bushes shaken visibly, or heard to rustle.
VII.	Difficult to stand. Noticed by drivers. Hanging objects quiver. Furniture broken. Damage to masonry D, including cracks. Weak chimneys broken at roof line. Fall of plaster, loose bricks, stones, tiles, cornices, also unbraced parapets and architectural ornaments. Some cracks in masonry C. Waves on ponds, water turbid with mud. Small slides and caving in along sand or gravel banks. Large bells ring. Concrete irrigation ditches damaged.
VIII.	Steering of cars affected. Damage to masonry C; partial collapse. Some damage to masonry B; none to masonry A. Fall of stucco and some masonry walls. Twisting, fall of chimneys, factory stacks, monuments, towers, elevated tanks. Frame houses moved on foundations if not bolted down; loose panel walls thrown out. Decayed piling broken off. Branches broken from trees. Changes in flow or temperature of springs and wells. Cracks in wet ground and on steep slopes.
IX.	General panic. Masonry D destroyed; masonry C heavily damaged, sometimes with complete collapse; masonry B seriously damaged. General damage to foundations. Frame structures, if not bolted, shifted off foundations. Frames cracked. Serious damage to reservoirs. Underground pipes broken. Conspicuous cracks in ground. In alluviated areas sand and mud ejected, earthquake fountains, sand craters.
X.	Most masonry and frame structures destroyed with their foundations. Some well-built wooden structures and bridges destroyed. Serious damage to dams, dikes, embankments. Large landslides. Water thrown on banks of canals, rivers, lakes, etc. Sand and mud shifted horizontally on beaches and flat land. Rails bent slightly.
XI.	Rails bent greatly. Underground pipelines completely out of service.
XII.	Damage nearly total. Large rock masses displaced. Lines of sight and level distorted. Objects thrown into the air.

Source: C. F. Richter (1958), *Elementary Seismology*, W. H. Freeman and Company, San Francisco, p. 136–38. Minor editorial changes, following B. A. Bolt (1978), *Earthquakes: A Primer*, W. H. Freeman and Company, San Francisco, Appendix C, p. 204–05.

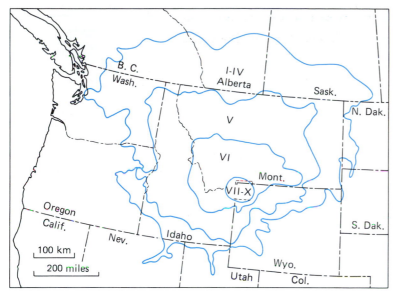

FIGURE 14.31 An isoseismal map of the Hebgen Lake earthquake of August 17, 1959. Roman numerals give intensity on the modified Mercalli scale. (U.S. Coast & Geodetic Survey. From A. N. Strahler, *The Earth Sciences*, 2nd ed., Harper & Row, Publishers. Copyright © by Arthur N. Strahler.)

windows and dishes rattle. Damage to various classes of masonry is used to establish criteria in the higher numbers of the scale. At an intensity of XII, damage to human-made structures is nearly total and large masses of rock are displaced. On the basis of reports gathered after an earthquake, maps can be prepared to show concentric zones of intensity. The numbered lines are *isoseismals* (Figure 14.31).

EARTHQUAKE GROUND MOTIONS

Special instruments are used to record the strong, damaging ground motions accompanying earthquakes. The strong motion seismograph records the amplitudes and accelerations of motions so large that they would cause an ordinary seismograph to "go off the scale." Portable instruments can be moved into an area where an earthquake is believed to be imminent. The recording mechanism is triggered by earthquake motions but otherwise remains inoperative.

An index of the severity of ground shaking during an earthquake is found in measurement of the peak acceleration—the rate of change of velocity. In simple language, the index measures how fast the ground motion speeds up or slows down. Units of acceleration are centimeters-per-second-per-second (cm/sec/sec or cm/sec^2), the same as for the acceleration of gravity, denoted by the symbol **g**. The value of **g** is 980 cm/sec^2. Earthquake acceleration is also given in proportional parts of **g**. Peak ground accelerations occur when the direction of motion reverses itself, producing a whiplash action that causes breakage in structural materials. Maximum accelerations are observed on solid bedrock, as compared with loose or soft sediments, and usually, the horizontal acceleration is greater than the vertical acceleration. High values of acceleration, in the range from 0.5 to 1.0 **g**, are associated with high velocities of ground motion. The combination of high values of horizontal velocity and acceleration easily topples unreinforced masonry

buildings (Figure 14.32). Duration of shaking is also important, since the longer the shaking lasts, the more likely is structural failure. Strong vertical accelerations sometimes produce the interesting effect of throwing objects, such as pebbles or boulders, into the air repeatedly "like peas on a drum." Even objects as heavy as a firetruck can be bounced repeatedly and displaced sideways as much as one or two meters. In the same way a frame building can be displaced from its foundation.

One of the major hazards from a severe earthquake is a secondary effect—collapse of the ground under the force of gravity because the ground shaking reduces the strength of the earth material on which heavy structures rest. Parts of many major cities, particularly port cities, have been built on naturally occurring bodies of soft,

FIGURE 14.32 Earthquake devastation in the downtown area of Managua, Nicaragua, December 1972. The number of persons killed in this disaster has been estimated as between 4000 and 6000, and property damage between $400 and $600 million. (Don Goode/Black Star.)

unconsolidated clay-rich sediment (such as the delta deposits of a river) or on filled areas in which large amounts of loose earth materials have been dumped to build up the land level. These water-saturated deposits often experience a change in property known as *liquefaction* when shaken by an earthquake. The material loses strength to the degree that it becomes a highly fluid mud, incapable of supporting buildings, which show severe tilting or collapse.

During the great San Francisco earthquake of 1906, the most severe shaking and structural damage occurred on areas bordering San Francisco Bay, where the buildings had been constructed on mud deposits or artificial fill. These areas showed effects associated with a Mercalli intensity of VII, VIII, or higher, whereas areas of the city where buildings rested on solid bedrock showed intensities lower than VII. The same pattern of destruction was evident in the Loma Prieta earthquake of 1989, described on later pages (Figure 14.33).

Liquefaction of clay-rich sediments during an earthquake can also lead to downhill flowage of the material under the force of gravity. These movements go under such names as landslides and earthflows; they are discussed in Chapter 15. For example, in the Alaskan Good Friday Earthquake of 1964, ground shaking at Anchorage set off the deep flowage of clay-rich sediments on which a section of the city was built. The flowage movement produced a large number of earth-blocks that settled down in a steplike arrangement, dislocating houses and streets and breaking water and gas mains (Figure 14.34). Violent ground shaking in this earthquake also set off great snowslides and rockslides in unpopulated mountain areas. Earthquake-triggered slides of this type have caused great loss of life in other parts of the world; we shall refer to them again in Chapter 15.

SEISMIC SEA WAVES

An important coastal hazard that is sometimes associated with a major earthquake is the *seismic sea wave*, or *tsunami*, as it is known to the Japanese. A train of these

FIGURE 14.34 Ground shaking during the Good Friday earthquake of March 27, 1964, at Anchorage, Alaska, set off large-scale slumping of unconsolidated sediments beneath these homes. (Gene Daniels/Black Star.)

waves is often generated in the ocean at a point near the earthquake source by a sudden movement of the seafloor. The waves travel over the ocean in ever-widening circles, but they are not perceptible at sea in deep water. When a wave arrives at a distant coastline, however, the effect is to cause a rise of water level. Wind-driven waves, superimposed on the heightened water level, allow the surf to attack places inland that are normally above the reach of waves.

The first evidence of a tsunami may, however, be the lowering of water level, causing a seaward withdrawal

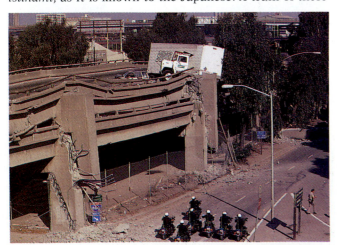

FIGURE 14.33 Collapsed section of the double-decked Nimitz Freeway (Interstate 880) in Oakland, California. Shaking by the Loma Prieta earthquake caused the upper level of the freeway to collapse upon the lower deck, crushing at least 39 persons in their cars. (Peter Menzel.)

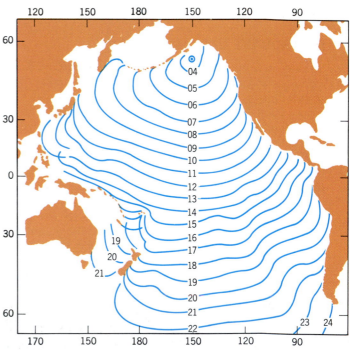

FIGURE 14.35 Map of the Pacific Ocean showing the locations of a tsunami wave front at one-hour intervals, GMT. The wave originated in the Gulf of Alaska as a result of the Good Friday earthquake of March 27, 1964. (After B. W. Wilson and A. Torum, 1968, U.S. Army Corps of Engineers, *Technical Memorandum No. 25*, Coastal Engineering Research Center, Washington, D.C., p. 38, Figure 27. From A. N. Strahler, *The Earth Sciences*, 2nd ed., Harper & Row, Publishers. Copyright © by Arthur N. Strahler.)

of the water line and exposing the floors of shallow bays. This is what seems to have happened at Lisbon, Portugal, on November 1, 1755, following an earthquake centered off the Portuguese coast. The sight of the exposed seafloor attracted a large number of townspeople, who were drowned when the following wave crest arrived. Several great catastrophes in recorded history seem to have been wrought by tsunamis—for example, the flooding of the Japanese coast in 1703, with a loss of more than 100,000 lives.

A particularly destructive train of seismic sea waves occurred in the Pacific Ocean on March 3, 1933, providing a good example of the phenomenon. The waves were produced by an earthquake centered beneath the ocean floor at a place 500 km northeast of Tokyo (lat. 39° N, long. 144° E). The wave train first reached the Japanese shore in one-half hour, Yokohama in 2 hours, Honolulu in 7.5 hours, San Francisco in 10.3 hours, and Iquique, Chile, in 22 hours. On exposed parts of the Japanese coast, where deep water lies close to shore, the waves rose as high as 10 m, causing destruction and death in low-lying areas.

The Good Friday Alaskan earthquake of 1964 released a tsunami that did great damage at a number of Pacific coastal points. In Alaska, Kodiak Island and Prince William Sound suffered severely from coastal flooding. San Francisco Bay experienced a water-level rise of about a meter that damaged small craft moored in marinas within the bay. Very severe effects were felt at Crescent City, in northern California, where damage amounted to about $10 million. Hilo, Hawaii, experienced a steep-walled water wave (a "bore") with a height of seven meters that rushed landward in a narrow estuary, causing great destruction. In Hilo alone, 83 persons died, and the total death toll for the Hawaiian Islands was over 170.

Following the 1964 tsunami, the Seismic Sea Wave Warning System (SSWWS) was set up, with headquarters in Honolulu. Seismologists at the headquarters issue warnings of possible destructive seismic sea waves, using the known depths of the ocean and the known instant of the earthquakes as a basis for computing the time the first waves will reach a given coast (Figure 14.35).

ASSESSING NATIONAL EARTHQUAKE HAZARDS

Geologists have intensified their field studies of active faults in order to map those areas where risk of earthquake damage is high. They are devising criteria by which to rate the level of risk that is incurred by the construction of urban housing and power plants—especially nuclear plants—on or close to faults. Each fault that is mapped is assigned a level of capability of producing an earthquake. One kind of field evidence is the determination of the geologic date when the last significant movement occurred; this is sometimes done by obtaining the age of the youngest bedrock or sediment that is displaced by the fault. Another approach is to set up special seismographs to determine if weak seismic

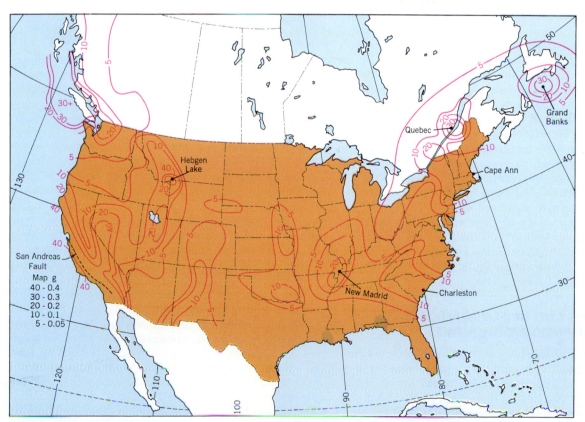

FIGURE 14.36 A seismic risk map of the 48 contiguous United States and southern Canada. The peak acceleration of gravity (g) to be expected is shown by isopleths numbered from 5 to 40. The key converts to values of g. (Based on data of the Building Seismic Safety Council and the Geological Survey of Canada.)

activity (background activity) can be detected along a fault that seems not to have undergone movement in recent geologic time.

Particular importance is being attached to determining the capability of faults close to nuclear power generating plants. The Nuclear Regulatory Commission defines a *capable fault* as one that has exhibited movement at or near the ground surface at least once within the past 35,000 years or movement of a recurring nature within the past 500,000 years.

Geologists and seismologists of the U.S. Geological Survey have prepared maps of the United States showing the level of seismic risk. One type of risk map tells the maximum intensity of ground shaking in terms of the value of the earth's surface gravitational acceleration (**g**) (Figure 14.36).

A study of this risk map shows some interesting things. First, the central and eastern portions of the United States, which lie in the geologically "stable" region of the North American lithospheric plate, show some areas of moderate earthquake risk, with probable peak accelerations exceeding 0.1 **g** in some fairly large areas. The fact is that several severe earthquakes have been recorded in this supposedly stable area. One was the great Charleston, South Carolina, earthquake of 1886. Another, and perhaps one of the greatest of historical record within the continental interior, was the New Madrid, Missouri, earthquake of 1811, which affected a large area of the Mississippi River floodplain. A series of violent shocks occurred over a period of many months, including three big ones that are estimated at Richter magnitudes 7.5, 7.3, and 7.8. It is said that the seismic waves rang church bells in Boston and that chimneys were felled and plaster walls cracked in Richmond, Virginia. In 1755, a severe quake centered near Cape Ann, Massachusetts, damaged buildings in Boston; its intensity is estimated as VIII on the modified Mercalli scale.

According to the U.S. Geological Survey, the eastern half of the United States has experienced more than 3500 earthquakes since 1700 and most of these produced effects in the range of III to V on the modified Mercalli scale. Earthquakes with epicenters in central and eastern North America are rarely associated with surface faults along which movement can be observed. The geologic circumstances associated with their occurrence are quite obscure in most cases. In contrast, earthquakes occurring in the western part of the continent are rather directly associated with visible faults and produce obvious surface changes.

EARTHQUAKES AND URBAN PLANNING— THE SAN FERNANDO EARTHQUAKE

About 6:00 A.M., on February 9, 1971, a violent earthquake shook the San Fernando Valley area, lying northwest of the city of Los Angeles. The quake lasted only about 60 seconds, but during this single minute 64 persons lost their lives. The toll in human lives and the severe structural effects of the San Fernando earthquake shocked the entire Los Angeles community into renewed awareness of the need for urban planning to minimize or forestall the damaging effects of a major earthquake.

Although the earthquake was not in the really severe

FIGURE 14.37 Severe structural damage and the collapse of buildings of the Veterans Administration Hospital in Sylmar, Los Angeles County, caused by the San Fernando earthquake of 1971. (J. Eyerman/Black Star.)

category according to the Richter scale (it measured 6.6, which is moderate in severity), local areas experienced a ground motion of acceleration as high as or higher than any previously measured in an earthquake. Fortunately, the ground shaking was of brief duration; had it persisted for a longer time, structural damage would have been much more severe than it was. Particularly disconcerting was the collapse of the Olive View Hospital in Sylmar, a new structure supposedly conforming with earthquake-resistant standards. The Veterans Hospital in Sylmar also suffered severe damage; several buildings collapsed and 42 persons were killed (Figure 14.37).

A crack produced in the Van Norman Dam prompted authorities to drain that reservoir to prevent dam collapse and disastrous flooding of a densely built-up area. The Sylmar Converter Station, one of the key elements in the electrical power transmission system of the Los Angeles area, was severely damaged. Collapse of a freeway overpass blocked the highway beneath, and freeway pavements were cracked and dislocated. Fortunately, the time of the quake was early morning, when most persons were at home and few were traveling the major arteries. Damage from the San Fernando earthquake of 1971 was estimated at $500 million.

Soon after the San Fernando disaster, the National Academy of Sciences and the National Academy of Engineering set up a joint panel of experts to study the earthquake effects and draw up recommendations. The panel concluded: "It is clear that existing building codes do not provide adequate damage control features. Such codes should be revised." The panel further recommended that public buildings, such as hospitals, schools, and buildings housing police and fire departments and other emergency services should be so constructed as to withstand the most severe shaking to be anticipated. Fortunately, most school buildings that were built after the severe Long Beach earthquake of the 1930s showed no structural damage. Many older school buildings, however, were rendered unfit for use.

The fault movement that set off the San Fernando earthquake was not on the great San Andreas Fault; rather, it was from an epicenter some 25 km (15 mi) from that great transcurrent fault. Instead it was generated by slip on a steep-angle reverse fault, such as illustrated in Block B of Figure 14.18. This kind of fault is typical of the Transverse Ranges of central California in a region from Los Angeles to Santa Barbara and signifies intense crustal compression in the crust of that area.

For residents of the Los Angeles area, a serious threat lies in the large number of active faults close at hand. Movements on these local faults have produced more than 40 damaging earthquakes since 1800, including the Long Beach earthquake of the 1920s and the San Fernando earthquake of 1971. In 1987, an earthquake of magnitude 6.1 struck the vicinity of Pasadena and Whittier, located within about 20 km (12 mi) of downtown Los Angeles. Known as the Whittier Narrows earthquake, it was generated along a local fault system that had not previously shown significant seismic activity. The brief but intense primary shock and aftershocks that followed damaged beyond repair many older structures built of unreinforced brick masonry. Although a slip along the San Andreas Fault, some 50 km (30 mi) to the north of this densely populated area, will release an enormously larger quantity of energy, its destructive effects in downtown Los Angeles will be somewhat moderated by the greater travel distance. Actually, the ground-shaking violence of a local earthquake of intensity 6.6 may be just as great as that from one of magnitude 8.3 on the distant San Andreas Fault, but we should not overlook the much longer duration of the San Andreas shaking and the enormously greater inhabited area it will reach.

HAZARDS OF THE SAN ANDREAS FAULT ZONE

Eight decades have passed since the great San Francisco earthquake of 1906 was generated by movement on the San Andreas Fault. Since that event, this sector of the fault has been locked; that is, devoid of a single displacement comparable to the 6-m (20-ft) offset in 1906. Figure 14.38 is a map showing the extent of the locked sections. The two lithospheric plates that meet along the fault have in the meantime been moving steadily with respect to one another and a large amount of unrelieved strain has already accumulated in the crustal rock on either

FIGURE 14.38 A sketch map of the San Andreas Fault showing the locked sections alternating with active sections of frequent small earthquakes and slow creeping motion. (Based on data of the U.S. Geological Survey.)

side of the fault. While the time of occurrence of the next great earthquake cannot now be predicted within a time window of even one or two decades, it is inevitable. As each decade passes, the probability of that event becomes greater.

The last major slip on the San Andreas Fault in the section closest to Los Angeles County occurred in 1857; it was the great Fort Tejon earthquake with a Richter magnitude now estimated to have been 8.3. Studies of the past history of the San Andreas Fault indicate that an earthquake like that of 1857 has a recurrence interval from 100 to 230 years, with an average of 140 years. Since 1979, geophysicists monitoring this locked sector of the fault have detected a number of physical changes that suggest increased possibility for a major earthquake.

Another locked sector of the San Andreas Fault system lies in the region of the San Gorgonio Pass and extends southeastward into the Imperial Valley. Here, the last great earthquake occurred some 250 to 300 years ago, so an earthquake of magnitude 8.0 or larger is considered even more imminent than for the Fort Tejon locked section. A recent estimate places at about 50 percent the likelihood that a very large earthquake will occur within the next 30 years somewhere along the southern California portion of the San Andreas Fault.

THE LOMA PRIETA EARTHQUAKE OF 1989

On October 17, 1989, the San Francisco Bay area was severely jolted by an earthquake—Richter magnitude of 7.1—epicentered about 80 km (50 mi) southeast of San Francisco at a point only 12 km (8 mi) from the city of Santa Cruz, on Monterey Bay. Displacement that caused the quake occurred deep beneath the surface on a 40 km (25 mi)-long segment of the San Andreas fault that had not slipped since the great San Francisco earthquake of 1906 (see Figure 14.38). The 1989 slippage amounted to about 1.8 m (6 ft) horizontally and 1.2 m (4 ft) vertically, but did not break the ground surface above it. The city of Santa Cruz suffered severe structural damage to older buildings, while numerous ground fissures (cracks) appeared on nearby mountain slopes, along with landslides that blocked roads.

Destructive ground shaking in the distant San Francisco Bay area proved surprisingly severe and caused great damage to buildings, bridges, and viaducts underlain by thick deposit of water-saturated, unconsolidated sand and mud. Landfills of such materials were especially hard hit. Severe shaking caused these materials to compact and to lose strength by the process we have referred to elsewhere in this chapter, and in Chapter 15 as spontaneous liquefaction. Here the ground motion was greatly amplified over that on adjacent solid bedrock, and heavy structures underwent sinking, tilting, and cracking. The same phenomenon caused major structural damage in the great 1906 earthquake, and those sensitive areas had been carefully mapped. The Marina area of San Francisco, in particular, again experienced severe damage similar to that in 1906. Quite unexpected was the collapse of the upper level of the Cypress section of Interstate 880 in Oakland (Figure 14.33),

and the collapse of a section of the nearby Bay Bridge. Altogether, 62 lives were lost in this earthquake, and the damage was estimated to be about 6 billion dollars.

In comparison, the 1906 earthquake took a toll of 700 lives and property damage equivalent to 20 billion 1987 dollars. This same section of the San Andreas fault has not slipped since 1909, when the lateral displacement was 6 m (20 ft). Geologic history indicates that the average rate of lateral motion between the two opposing lithospheric plates is 3.4 cm (1.34 in.) per year. Thus another slip of 6 m (20 ft) in the San Francisco area is likely to occur after a lapse of 176 years (i.e., about in the year 2080). A report issued by the U.S. Geological Survey contained this commentary:

The recurring theme on the Loma Prieta earthquake is that geologic conditions strongly influence damage. In other words, the geology determines where fault ruptures are likely to be, how hard the ground will shake, where landslides will occur, and where the ground will sink and crack. A parallel theme is that the pattern of damage from shaking and geologic effects observed in 1989 is very similar to that witnessed in 1906. Thus, many of the lessons taught by the 1906 shock have been forgotten or ignored. As philosopher-poet George Santayana aptly noted, "Those who cannot remember the past are prone to repeat it."*

---

*George Plafker and John P. Galloway, Eds., 1989, *Lessons Learned from the Loma Prieta, California, Earthquake of October 17, 1989*, U.S. Geological Survey Circular 1045, p. 48.

THE EARTH'S CRUST IN REVIEW

In three chapters we have made a survey of the composition, structure, and geologic activity of the earth's crust. One underlying concept is that of a rock cycle continuously in operation over some three billion years or more of geologic time. The recycling of crustal mineral matter has taken place through the mechanism of plate tectonics. The continents have gradually grown in extent through accumulation of felsic rock produced in subduction zones, and in this way, the first-order relief features of the globe have come into existence.

Active tectonic and volcanic belts are dominant environmental features of the continents. Cutting across the

belts of prevailing winds, high mountain chains induce orographic precipitation and, at the same time, create rainshadow deserts. Landforms produced by volcanic activity are important in these new mountain chains. Landforms produced by faulting are important in areas where continental rifting is occurring. These activities also pose environmental hazards for those human populations living nearby.

Another great class of landforms remains to be investigated—those formed by agents of erosion through processes of interaction of the land surface with the hydrosphere and atmosphere. These processes are the subject of the next five chapters.

LANDFORMS OF WEATHERING AND MASS WASTING

Now that we have completed a study of the earth's crust, its mineral composition, and its tectonic and volcanic landforms, we can focus on the shallow life layer itself. At this interface, the externally acting, solar-driven energy systems of the atmosphere and oceans mesh with the internally driven geologic system that has created and raised the continental masses. The geologic processes have caused varied rock types to become exposed to the surface environment. Our study of the interaction of these two great planetary systems began in Chapter 12 with a study of the chemical alteration of igneous rock and the production of sediment. This is a process essential to the rock transformation cycle and to growth of the continental crust. We will now look at the processes that shape the surface of the lands.

ting and drying, growth and decay of roots, and various other repetitive processes cause the soil and regolith to expand and contract in ceaseless daily and seasonal cycles.

The spontaneous downward movement of soil, regolith, and rock under the influence of gravity (but without the dynamic action of moving fluids) is included under the general term *mass wasting*. In Chapter 1 we emphasized the role of gravity as a pervasive environmental factor. All processes of the life layer take place in the earth's gravity field, and all particles of matter tend to respond to gravity. Movement to lower levels takes place when the internal strength of a mass of regolith, sediment, or rock declines to a critical point below which the force of gravity cannot be resisted. This failure of strength under the ever-present force of gravity takes

WEATHERING AND MASS WASTING

Weathering is the general term applied to the combined action of all processes causing rock to be disintegrated physically and decomposed chemically because of exposure at or near the earth's surface. The products of rock weathering tend to accumulate in a soft surface layer, called *regolith* (Figure 15.1). The regolith grades downward into solid, unaltered rock, known simply as *bedrock*. Regolith, in turn, provides the source of *sediment*, consisting of detached mineral particles transported and deposited in a fluid medium, which may be water, air, or glacial ice. Both regolith and sediment comprise parent materials for the formation of the true soil, a surface layer capable of supporting the growth of plants.

Weathering also plays a major role in denudation. Disintegration and decomposition of various kinds of hard bedrock greatly facilitate the erosion of the land surface by running water. Besides this function, weathering leads to a number of distinctive landforms, which we shall describe in this chapter. Although we have dealt with weathering in Chapter 12 as a one-way series of changes leading to chemical stability of minerals in the surface environment, it is important to keep in mind that the physical weathering processes continue to agitate and move soil and regolith. Frost action, alternate wet-

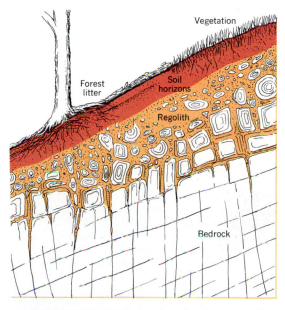

FIGURE 15.1 In a typical exposure cut into a hillside, the bedrock (broken into blocks by joints) grades upward into the regolith. Immediately beneath the surface is the soil, penetrated by roots of plants. (Drawn by A. N. Strahler.)

many forms and scales. We shall see that human activity is a major agent in causing or intensifying several forms of mass wasting.

THE WASTING OF SLOPES

As used in geomorphology, the term *slope* designates some small element or area of the land surface that is inclined from the horizontal. Thus we speak of "mountain slopes," "hillslopes," or "valley-side slopes" with reference to the inclined ground surfaces extending from divides and summits down to valley bottoms.

Slopes guide the flow of surface water under the influence of gravity. Slopes fit together to form drainage systems in which surface-water flow converges into stream channels; these, in turn, conduct the water and rock waste to the oceans to complete the hydrologic cycle. Natural processes have so completely provided the earth's land surfaces with slopes that perfectly horizontal or vertical surfaces are extremely rare.

Figure 15.2 shows a typical hillslope forming one wall of the valley of a small stream. Soil and regolith mantle the bedrock except in a few places where the bedrock is particularly hard and projects in the form of *outcrops*. *Residual regolith* is derived from the rock beneath and moves very slowly down the slope toward the stream. Beneath the valley bottom are layers of *transported regolith*—alluvium transported and deposited by the stream. This sediment had its source in regolith prepared on hillslopes many kilometers or tens of kilometers upstream. All terrestrial accumulations of sediment, whether deposited by streams, waves and currents, wind, or glacial ice, can be designated transported regolith in contrast to residual regolith.

PHYSICAL WEATHERING PROCESSES AND FORMS

Physical weathering is a broad term including several quite different physical processes that convert hard, massive bedrock into fragments, ranging in size from large blocks or boulders to fine sand and silt. In Chapter 12, we described the size grades of mineral particles produced by disintegration through physical weathering, as well as by grinding and crushing action of streams, waves and currents, wind, and glacial ice. Although we describe a number of different processes of physical weathering, it should be kept in mind that two or more processes may be acting simultaneously, along with chemical processes of decomposition.

Frost Action

One of the most important physical weathering processes is *frost action*, the repeated growth and melting of ice crystals in the pore spaces or natural fractures of bedrock. Frost action is, of course, limited to those mid-latitude and high-latitude climates with cold winters and to cold alpine climates at high altitudes.

Almost everywhere, bedrock is cut through by systems of fractures called *joints* (Figures 15.1 and 15.2). Rarely is igneous rock free of numerous joints that permit the entry of water. In sedimentary rocks the planes of stratification, or bedding planes, comprise a natural set of planes of weakness cut at right angles by sets of joints (Figures 15.3 and 15.4). Obviously, comparatively weak stresses can separate joint blocks, whereas strong stresses are required to make fresh fractures through solid rock. The process can be called *block separation*.

As coarse-grained igneous rock becomes weakened by chemical decomposition, water is able to penetrate the contact surfaces between mineral grains; here the water can freeze and exert strong forces to separate the grains. This form of breakup is termed *granular disintegration* (Figure 15.3). The product is a gravel or coarse sand in which each grain consists of a single mineral particle separated from others along the original crystal or grain boundaries.

The effects of frost action can be seen in all climates having a winter season with many alternations of freeze and thaw. Where bedrock is exposed on knolls and mountain summits, joint blocks are pried apart by water freezing in joint cracks. Under the most favorable conditions, seen on high mountain summits and in the arctic tundra,

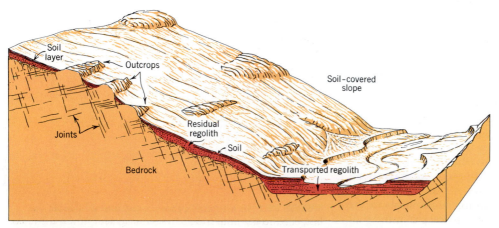

FIGURE 15.2 Soil, regolith, and outcrops on a hillslope. Alluvium lies in the floor of an adjacent stream valley. (Drawn by A. N. Strahler.)

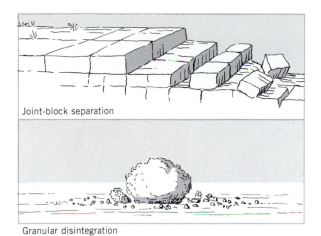

Joint-block separation

Granular disintegration

FIGURE 15.3 Two common forms of bedrock disintegration. (Drawn by A. N. Strahler.)

FIGURE 15.5 Frost-shattered blocks of quartzite on the summit of the Snowy Range, Wyoming, elevation 3700 m (12,000 ft). (A. N. Strahler.)

FIGURE 15.4 Jointing in sandstone resembles pavement blocks at Artists View, Catskill Mountains, New York. A tree has grown up between two joint blocks. (A. N. Strahler.)

FIGURE 15.6 Talus cones near the shore of Lake Louise in the Canadian Rockies. (A. N. Strahler.)

large angular rock fragments accumulate in a layer that completely blankets the bedrock beneath. The German name *felsenmeer* has been given to such expanses of broken rock (Figure 15.5).

Frost action on cliffs of bare rocks in high mountains detaches rock fragments that fall to the cliff base. Where production of fragments is rapid, they accumulate to form *talus slopes* (Figure 15.6). Later in this chapter, we explain how talus slopes are shaped by mass wasting processes.

In fine-textured soils and sediments composed mostly of silt and clay, freezing of soil water takes place in horizontal layers, or lenses. As these ice layers thicken, the overlying soil layer is heaved upward. Prolonged soil heaving can produce minor irregularities and small mounds on the soil surface. If a rock fragment lies on the surface, perpendicular ice needles grow beneath the fragment, raising it above the surface (Figure 15.7). The same process acting on a rock fragment below the soil surface will eventually bring the fragment to the surface.

Frost action is a dominant process in the arctic tundra environment, where it is a factor in the formation of a wide variety of unique soil structures and landforms. These we investigate in a special section later in this chapter.

FIGURE 15.7 Frost heaving above timberline. Needle ice, attached to the underside of a rock fragment, lifted the mass as the crystals lengthened. Below is the cavity in which the ice formed. (Mark A. Melton.)

Salt-Crystal Growth

Closely related to ice-crystal growth as a disruptive process is the *salt-crystal growth* in fractures and other openings in rock. This process operates extensively in dry climates. During long dry periods, water deep within the rock is drawn to the surface by capillary force. This water carries dissolved mineral salts. As evaporation takes place, minute salt crystals remain behind. The growth force of these crystals is capable of causing granular disintegration of the outer rock layer. The same process can be seen in action on building stones and concrete in cities. Deicing salt, spread on streets in winter, is remarkably effective in causing the disintegration of rock and concrete structures close to the ground.

Cliffs of sandstone are especially susceptible to disintegration by this activity. As Figure 15.8 shows, slow seepage of water takes place at the cliff base, above a dense, impervious layer of shale. Continual surface evaporation leaves the dissolved salts behind in pore spaces in the sandstone. The pressure of growing crystals disrupts the sandstone into scales and flakes. Sand grains thus released are swept away by wind gusts or by sheets of rainwater flowing down the cliff face. Recession of the cliff base eventually produces a niche, or shallow cave. In the southwestern United States, many such niches were occupied by Indians, who built masonry walls to enclose the natural openings. These cliff dwellings were protected not only from the elements, but also from attack by hostile Indians (Figure 15.9).

Wedging by Plant Roots

Plant roots, growing between joint blocks and along minute fractures between mineral grains, exert an expansive force tending to widen those openings. You may have seen a tree whose lower trunk and roots are firmly wedged between two great joint blocks of massive rock (Figure 15.4). Whether the tree has actually been able to spread the blocks farther apart or has merely occupied the available space is open to question. However, it is certain that pressure exerted by growth of tiny rootlets in joint fractures causes the loosening of countless small rock scales and grains. Heaving and cracking of concrete

FIGURE 15.9 The White House Ruin, a former Indian habitation, occupies a great niche in sandstone in the lower wall of Canyon de Chelly, Arizona. (Mark A. Melton.)

sidewalk slabs by root growth of nearby trees is commonplace evidence of the effectiveness of plants in contributing to physical weathering.

Temperature Change as a Cause of Rock Disruption

Another possible set of forces of rock disruption is temperature change. Rock-forming minerals expand when heated, but contract when cooled. Where rock surfaces are exposed daily to intense heating by direct solar rays, alternating with intense cooling by longwave radiation at night, the resulting expansion and contraction of mineral grains tends to break them apart. Given sufficient time, in which tens of thousands of daily cycles of expansion and contraction occur, the accumulated effect may be important as an agent of physical weathering. Although firsthand evidence is lacking, it seems likely that repeated temperature changes contribute to the breakup of rock already weakened by other agents of weathering. The intense heat of forest and brush fires is known to cause rapid flaking and scaling of exposed rock surfaces. The same effect is seen on stone blocks subjected to the intense heat of building fires.

Sheeting Structure and Exfoliation Domes

A curious but widespread process related to physical weathering results from *unloading*, the relief of confining pressure, as rock is brought nearer the earth's surface through the erosional removal of overlying rock. Rock formed at great depth beneath the earth's surface (particularly igneous and metamorphic rock) is in a slightly contracted state because of the tremendous load of the overlying rock. On being brought to the surface, the rock expands slightly and, in the process, thick shells of rock break free from the parent mass below. The new surfaces of fracture form a joint system described as *sheeting structure*. The structure shows best in massive rocks, such as granite, because in a closely jointed rock the expansion would be taken up among the blocks.

The rock sheets, or shells, produced by unloading gen-

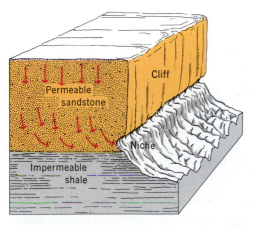

FIGURE 15.8 Seepage of water from the cliff base localizes the development of niches through rock weathering. (Drawn by A. N. Strahler.)

FIGURE 15.10 Exfoliation slabs of pink granite dip seaward on the Atlantic coast of Mt. Desert Island, Maine. (A. N. Strahler.)

erally parallel the ground surface and therefore tend to be inclined downward toward valley bottoms. On granite coasts, the shells are found to incline seaward at all points along the shore (Figure 15.10). Sheeting structure is well seen in quarries, where it greatly facilitates the removal of large stone blocks.

Where sheeting structure has formed over the top of a single large body of massive rock, an *exfoliation dome* is produced. These domes are among the largest of the landforms due primarily to weathering. In the Yosemite Valley region of California, where domes are spectacularly displayed, the individual rock shells may be as thick as 6 to 15 m (20 to 50 ft) (Figure 15.11).

Other varieties of large, smooth-sided rock domes lacking in shells are not true exfoliation domes; they are formed by granular disintegration of a single body of hard, coarse-grained intrusive igneous rock lacking in joints. Examples are the Sugar Loaf of Rio de Janeiro and Stone Mountain in Georgia (Figure 19.45). These smooth domes rise prominently above surrounding areas of weaker rock.

CHEMICAL WEATHERING PROCESSES AND FORMS

We investigated chemical weathering processes in Chapter 12 under the heading of chemical alteration. Recall that the dominant processes of chemical change affecting silicate minerals are oxidation, carbonic acid action, and hydrolysis. Feldspars and the mafic minerals—particularly olivine—are very susceptible to chemical decay. On the other hand, quartz is a highly stable mineral, almost immune to decay. One result of this fact is that, where felsic and mafic igneous rocks are exposed side by side, the felsic rock may show little alteration while the mafic rock may be in an advanced state of decomposition (Figure 15.12).

Decomposition by hydrolysis and oxidation changes strong igneous rock into very weak regolith. This change allows erosion to operate with great effectiveness wherever the regolith is exposed. Weakness of the regolith also makes it susceptible to natural forms of mass wasting.

FIGURE 15.11 Half Dome, Yosemite National Park, California, viewed from Glacier Point. The dome rises above a deep, glacially carved trough (left) that is a branch of Yosemite Valley. The rock exposed here is an igneous rock of granitic composition. Thick exfoliation shells cap Half Dome and its rounded side. The nearly vertical rock face of the dome was formed along a set of close parallel joints that cut through the rock mass. (A. N. Strahler.)

FIGURE 15.12 Mafic igneous rock undergoing decay in place to produce a thick regolith. White dikes of felsic rock are little affected. Sangre de Cristo Mountains, New Mexico. (A. N. Strahler.)

FIGURE 15.13 Spheroidal weathering in jointed coarse-grained granitic rock, Lake Gregory, San Bernardino Mountains, California. (A. N. Strahler.)

Spheroidal Weathering and Saprolite

As chemical alteration penetrates the bedrock, joint blocks are attacked from all sides. Decay progresses inward to make concentric shells of soft rock (Figure 15.13). This form of change is called *spheroidal weathering* and produces onionlike bodies from which thin layers can be broken away from a spherical core.

In warm, moist climates of the equatorial, tropical, and subtropical zones, hydrolysis and oxidation often result in the decay of igneous and metamorphic rocks to depths as much as 100 m (300 ft). Geologists who first studied this deep rock decay in the Southern Appalachians named the rotted layer *saprolite* (literally "rotten rock"). To the civil engineer, deeply weathered rock is a major hazard in constructing highways, dams, or other heavy structures. Although the saprolite is soft and can be removed by power shovels with little blasting, there is serious danger of failure of foundations under heavy loads. This regolith may also have undesirable plastic properties because of a high content of certain clay minerals, such as montmorillonite, which have the property of swelling during absorption of water.

The hydrolysis of granite is accompanied by granular disintegration, the grain-by-grain breakup of the rock. This process creates many interesting boulder and pinnacle forms by the rounding of angular joint blocks (Figure 15.14). These forms are particularly conspicuous in arid regions (Figure 15.15). There is ample moisture in most deserts for hydrolysis to act, given sufficient time. The products of granular disintegration of felsic igneous rock form a coarse desert gravel, called *grus*, which con-

sists largely of grains of quartz and partially decomposed feldspars.

Effects of Carbonic Acid Action

Carbonate rocks (limestone, marble) are particularly susceptible to the action of carbonic acid in rainwater and soil water. Mineral calcium carbonate (calcite) is dissolved, yielding calcium ions and bicarbonate ions. In regions of large water surplus, these ions are carried away in solution in the water of streams.

Carbonic acid reaction with limestone produces many interesting surface forms, mostly of small dimensions. Outcrops of limestone typically show cupping, rilling, grooving, and fluting in intricate designs (Figure 15.16). In a few places the scale of deep grooves and high, wall-like rock fins reaches proportions that prevent the passage of humans and animals.

In the wet low-latitude climates, mafic rock, particularly basaltic lava, undergoes rapid removal under attack by soil acids and produces landforms quite similar to those formed by carbonic acid on massive limestones

FIGURE 15.15 Large joint blocks of granite are gradually being rounded into huge boulders through grain-by-grain disintegration in a desert environment. Joshua Tree National Monument, California. (A. N. Strahler.)

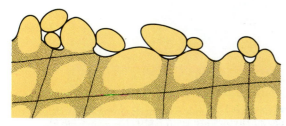

FIGURE 15.14 Stages in the development of egg-shaped boulders from rectangular joint blocks. (Drawn by W. M. Davis.)

FIGURE 15.16 This desert outcrop of massive white limestone shows rills, cups, and sharp ridges. Although the climate is dry, occasional rainshowers drench the rock surface, allowing carbonic acid to do its work. Charleson Range, Nevada. (John S. Shelton.)

Carbonic acid action also results in the removal of carbonate rock far below the surface, producing limestone caves and cavern systems. The movement of ground water is involved in this process, and we investigate it in Chapter 19.

MASS WASTING

Everywhere on the earth's surface, gravity pulls continually downward on all materials. Bedrock is usually so strong and well supported that it remains fixed in place but, where a mountain slope becomes too steep, bedrock masses break free, falling or sliding to new positions of rest. In cases involving huge masses of bedrock, the result can be catastrophic in loss of life and property in towns and villages in the path of the slide. Such slides are a major form of environmental hazard in mountainous regions. Because soil, regolith, and many forms of sediment are held together poorly, they are much more susceptible to being moved by the force of gravity than is bedrock. Abundant evidence shows that on most slopes at least a small amount of downhill movement is going on constantly. Although this motion is imperceptible, the regolith sometimes slides or flows rapidly.

in moist climates of higher latitudes. The effects of solution removal of basaltic lava are displayed in spectacular grooves, fins, and spires on the walls of deep alcoves in parts of the Hawaiian Islands (Figure 15.17).

Carbonic acid action is a major agent of denudation in regions of moist climate underlain by limestone. For example, in one study of a limestone valley in Pennsylvania, it was concluded that the ground surface is being lowered at an average of 0.3 m (1 ft) in 10,000 years through carbonic acid action alone. In moist climates, as we might expect, carbonate rocks underlie valleys and lowlands in contrast to other, less susceptible rock types that form adjacent ridges or uplands. Quite the reverse is true in dry climates, where limestone and dolomite are strongly resistant to weathering and form high ridges and plateaus. For example, the rim of the Grand Canyon and the surrounding high plateau are underlain by beds of dolomite (see Figures 12.14 and 19.5). Sandstone strata formed of quartz grains cemented by calcium carbonate are also highly resistant to weathering in a dry climate.

Taken altogether, the various kinds of downhill movements occurring under the pull of gravity, collectively called *mass wasting*, constitute an important process in denudation of the continental surfaces.

The processes of mass wasting and the landforms they produce are extremely varied and tend to grade one into another. We have selected only a few of the most important forms of mass wasting; they are named in Figure 15.18. Our table lists at the top three categories of information: (a) the kinds of earth materials involved; (b) the physical properties of those materials; and (c) the kinds of motion or movement that occur. The presence or absence of water, mixed in with the regolith or rock material, is a most important factor in determining how the moving mass behaves and how fast it travels downslope. Boxes in the body of the table contain the names of several forms of mass movements, arranged more or less according to composition (left to right) and speed

FIGURE 15.17 The steep walls of many narrow coastal ravines are deeply grooved by chemical weathering of the basaltic lava on the Napali Coast, Island of Hawaii. (Douglas Peebles.)

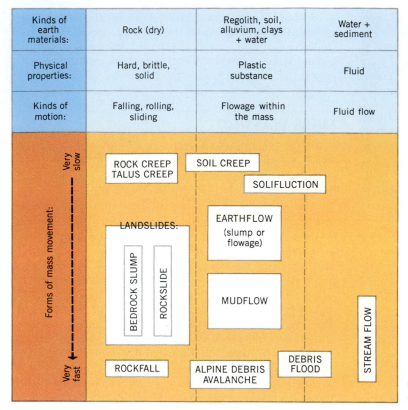

Kinds of earth materials:	Rock (dry)	Regolith, soil, alluvium, clays + water	Water + sediment
Physical properties:	Hard, brittle, solid	Plastic substance	Fluid
Kinds of motion:	Falling, rolling, sliding	Flowage within the mass	Fluid flow

Forms of mass movement: Very slow → Very fast

ROCK CREEP TALUS CREEP

SOIL CREEP

SOLIFLUCTION

LANDSLIDES:

BEDROCK SLUMP   ROCKSLIDE

EARTHFLOW (slump or flowage)

MUDFLOW

STREAM FLOW

ROCKFALL   ALPINE DEBRIS AVALANCHE   DEBRIS FLOOD

FIGURE 15.18 Processes and forms of mass wasting.

(top to bottom). As we describe each of these forms, you will find it helpful to refer back to this table for a better perspective on its relationship to the others.

Rockfall and Talus Cones

Steep rock cliffs created by action of glaciers or rapid uplift of a fault block shed rock masses or individual joint blocks that come to rest at the foot of the cliff; this is the process of *rockfall*, in which the rate of fall follows

the acceleration of gravity. Typically, in alpine environments, frost action is responsible for wedging the blocks free as water from melting snow refreezes in joints. Fragments set free in this manner accumulate in a distinctive landform, the *talus slope* (Figure 15.6). A talus slope, or *scree slope*, as it is often called, has a nearly constant slope angle, usually in the range of 34° to 36° from the horizontal.

Most cliffs are notched by narrow ravines that funnel the fragments into individual tracks, so as to produce

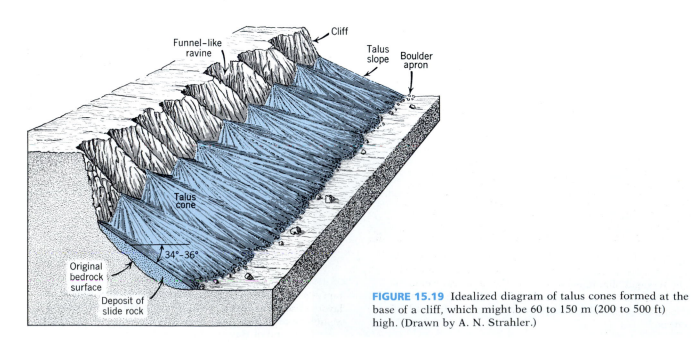

FIGURE 15.19 Idealized diagram of talus cones formed at the base of a cliff, which might be 60 to 150 m (200 to 500 ft) high. (Drawn by A. N. Strahler.)

talus cones arranged side by side along the cliff (Figure 15.19). Where a large range of sizes of particles is supplied, the larger pieces, because of their greater momentum and ease of rolling, travel to the base of the cone, whereas the tiny grains lodge in the apex. This process tends to sort the fragments by size, progressively finer from base to apex (Figure 15.19).

Most fresh talus slopes are unstable, so the disturbance created by walking across the slope or by the dropping of a large rock fragment from the cliff above will easily set off a sliding of the surface layers of particles. The upper limiting angle to which coarse, hard, well-sorted rock fragments will stand is termed the *angle of repose*.

Talus Creep and Rock Glaciers

In mountainous areas where glaciers deeply eroded the bedrock in the last ice age to leave high, steep rock cliffs, the production of talus cones has been rapid in many places. Locally, talus has accumulated over small masses of glacial ice that still remain in hollows (cirques) that were the starting points of former active glaciers (see Chapter 22 and Figure 22.8). In favorable locations, the rock fragments of the talus cones have developed a very slow down-slope motion, called *rock creep*, to produce curious tonguelike bodies that in many ways resemble glaciers of ice (Figure 15.20). The surface of these *rock glaciers*, which on close inspection consists of angular rock fragments, has developed wrinkles that are bowed outward in the direction of motion. The front of the rock glacier is steep and sharply defined.

Many rock glaciers contain ice that fills the spaces between rock fragments but cannot be seen at the surface. Movement of such rock glaciers may be controlled by slow flowage within the ice. Rock glaciers can be found in large numbers in the Alaska Range and in certain ranges of the Colorado Rockies.

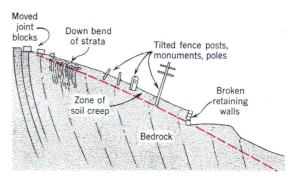

FIGURE 15.21 Evidences of the slow, downhill creep of soil and regolith. (After C. F. S. Sharpe.)

Soil Creep

On almost any steep, soil-covered slope, evidence can be found of extremely slow downhill movement of soil and regolith, a process called *soil creep*. Figure 15.21 shows some of the evidence that the process is going on. Joint blocks of distinctive rock types are found moved far downslope from the outcrop. In some layered rocks, such as shales or slates, edges of the strata seem to bend in the downhill direction. This is not true plastic bending, but is the result of slight movement on many small joint cracks (Figure 15.22). Fence posts and telephone poles lean downslope and even shift measurably out of line. Retaining walls of road cuts lean and break outward under pressure of soil creep from above.

What causes soil creep? Heating and cooling of the soil, growth of frost needles, alternate drying and wetting of the soil, trampling and burrowing by animals, and shaking by earthquakes all produce some disturbance of the soil and regolith. Because gravity exerts a downhill pull on every such rearrangement, the particles are moved slowly downslope.

Creep also affects rock masses enclosed in the soil or lying on bare bedrock. Huge boulders that have gradu-

FIGURE 15.20 Its wrinkled surface suggesting internal flowage, this great rock glacier descends a steep mountain slope in the Wrangel Mountains of Alaska. The talus fragments that form the moving tongue are shed from the side of Sourdough Mountain (light-colored slopes at upper right). (Steve McCutcheon.)

FIGURE 15.22 Slow downhill creep of regolith on this mountainside near Downieville, California, has caused vertical rock layers to seem to "bend over." (Copyright © 1988 by Mark A. Melton.)

ally crept down a mountain side in large numbers may accumulate at the mountain base to produce a boulder field.

Earthflows

In regions of humid climate, a mass of water-saturated soil, regolith, or weak clay or shale layers may move down a steep slope during a period of a few hours in the form of an *earthflow*. Figure 15.23 is a sketch of an earthflow showing how the material slumps away from the top, leaving a steplike terrace bounded by curved, wall-like scarp. The saturated material flows sluggishly to form a bulging toe.

Shallow earthflows, affecting only the soil and regolith, are common on sod-covered and forested slopes that have been saturated by heavy rains. An earthflow may affect a few square meters or it may cover an area of several hectares. If the bedrock of a mountainous region is rich in clay (shale or deeply weathered igneous rocks), earthflow sometimes involves millions of tons of bedrock (Figure 15.24).

Earthflows are a common cause of the blockage of highways and railroad lines, usually during periods of heavy rains. Usually the rate of flowage is slow, so the flows are not often a threat to life. Property damage to buildings, pavements, and utility lines is often large where construction has taken place on unstable soil slopes.

Examples of both large and small earthflows induced or aggravated by human activities are found in the Palos Verdes Hills of Los Angeles County, California. These movements occur in shales that tend to become plastic when water is added. The upper part of the earthflow undergoes a subsiding motion with backward rotation of the down-sinking mass, as illustrated in Figure 15.23. The interior and lower parts of the mass move by slow flowage, and a toe of extruded flowage material may be formed.

Largest of the earthflows in the Palos Verdes Hills area was the Portuguese Bend "landslide" that affected an area of about 160 hectares (400 acres). The total motion over a 3-year period was about 20 m (70 ft). Damage

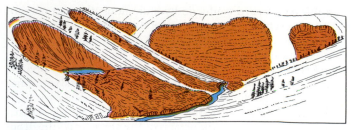

FIGURE 15.24 Earthflows in a mountainous region. (Drawn by W. M. Davis.)

to residential and other structures totaled some $10 million. The most interesting observation, from the point of view of assessing human impact on the environment, is that the slide has been attributed by geologists to infiltration of water from cesspools and from irrigation water applied to lawns and gardens. A discharge of over 115,000 liters (30,000 gallons) of water per day from some 150 homes is believed to have sufficiently weakened the shale beneath to start and sustain the flowage.

Earthflows Involving Quick Clays

One special form of earthflow has proved to be a major environmental hazard in parts of Norway and Sweden and along the St. Lawrence River and its tributaries in Quebec Province of Canada. In all these areas the flowage involves horizontally layered clays, sands, and silts of late Pleistocene age that form low, flat-topped terraces adjacent to rivers or lakes. Over a large area, which may be 600 to 900 m (2000 to 3000 ft) across, a layer of silt and sand 6 to 12 m (20 to 40 ft) thick begins to move toward the river, sliding on a layer of soft clay that has spontaneously turned into a near-liquid state. The moving mass also settles downward and breaks into steplike masses. Carrying along houses or farms, the layer ultimately reaches the river, into which it pours as a great, disordered mass of mud.

Figure 15.25 shows an earthflow of this type that occurred in 1898 in Quebec, along the Rivière Blanche. Beyond is the scar of a much older earthflow of the same type. The Rivière Blanche earthflow involved about 3

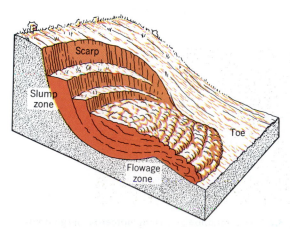

FIGURE 15.23 An earthflow with slump features well developed in the upper part. (Drawn by A. N. Strahler.)

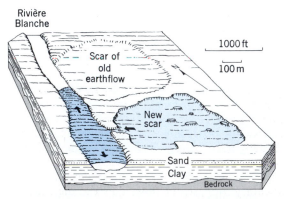

FIGURE 15.25 Block diagram of the 1898 earthflow near St. Thuribe, Quebec. (After C. F. S. Sharpe, *Landslides and Related Phenomena*, Columbia Univ. Press, New York.)

million cu m (3.5 million cu yd) of material and required 3 to 4 hours to move into the river through a narrow bottleneck passage.

Disastrous earthflows have occurred a number of times since the occupation of Quebec by Europeans. A particularly spectacular example was the Nicolet earthflow of 1955, which carried a large piece of the town into the Nicolet River. Fortunately only three lives were lost, but the damage to buildings and a bridge ran into the millions of dollars.

Clays that spontaneously change from a solid condition to a near-liquid condition are called *quick clays*. The process is called *spontaneous liquefaction*. A sudden shock or disturbance will often cause a layer of quick clay to begin to liquefy; once begun, the process cannot be stopped. A good example comes from the city of Anchorage, Alaska. Severe ground shaking by the Good Friday earthquake of 1964 set off the liquefaction of quick clays underlying an extensive flat area, or terrace, which was the site of a housing development. Clay flowage allowed the overlying sediment layer to subside and break into blocks (see Figure 14.34).

A study of quick clays in North America and Europe has led to an explanation of their strange behavior. The clays are marine glacial deposits, meaning that they are sediments laid down in shallow, saltwater bays during the Pleistocene Epoch. The thin, platelike particles of clay accumulated in a heterogeneous arrangement, often described as a "house of cards" structure. There is a very large proportion of water-filled void space between clay particles. It is thought that the original salt water saturating the clay acted as an electrolyte to bind the particles together, giving the clay layer strength. Some areas where quick clays occur have experienced a crustal uplift since they were laid down. The sediment deposit has been raised from immersion in salt water and the salt solution has been gradually replaced by fresh ground water. Now the clay is no longer bound by the electrolyte action and becomes sensitive. A mechanical shock causes the house of cards structure to collapse. Because a large volume of water is present (from 45 to 80% water content by volume), the mixture behaves as a liquid, with almost no strength remaining.

Today, many towns, cities, and farms occupy land underlain by quick clays, and an environmental hazard thus exists. Such areas require identification and mapping so that future land use is limited to farming with a dispersed resident population.

Mudflows

One of the most spectacular forms of mass wasting and one that is potentially a serious environmental hazard is the *mudflow*, a mud stream that travels down canyons in mountainous regions (Figure 15.26). In deserts, where vegetation does not protect the mountain soils, local thunderstorms produce rain much faster than it can be absorbed by the soil. As the water runs down the slopes it forms a watery mud, which flows down to the canyon floors. Following stream courses, the mud continues to flow until it becomes so thickened that it must stop. Large boulders, buoyed up in the mud, are carried with

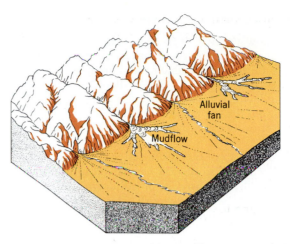

FIGURE 15.26 Thin, streamlike mudflows commonly issue from canyon mouths in arid regions, spreading out over the alluvial fan slopes at the base of a mountain range. (Drawn by A. N. Strahler.)

the flow (Figure 15.27). Roads, bridges, and houses in the canyon floor are engulfed and destroyed.

When a mudflow emerges from the canyon and spreads across a plain, property damage and loss of life can result. In desert regions, the plains lie close to the mountain ranges that supply irrigation water. These favored land surfaces are often densely populated and are sites of urban development, as well as intensive agriculture.

Mudflows also occur on the slopes of erupting volcanoes when heavy rains turn freshly fallen volcanic ash into mud. On high composite volcanoes, melting snow and glacial ice may also contribute large volumes of water to form mudflows. Known as lahars (Chapter 14) these fast-moving volcanic mudflows are funneled into valley bottoms, clogging stream channels and destroying roads, bridges, and houses. Immediately following the major eruption of Mount St. Helens on May 18, 1980, a lahar moved down the Toutle River, causing great destruction. Traveling a total distance of more than 120 km (75 mi), it eventually reached the Columbia River. Lahars from Mount St. Helens deposited a total of more than 50 million cubic meters of sediment into the lower Cowlitz and Columbia river channels.

FIGURE 15.27 This mudflow, carrying numerous large boulders, issued from a steep mountain canyon in the Wasatch Mountains, Utah. (Orlo E. Childs.)

Debris Floods and Avalanches

Mudflows show varying degrees of consistency, from a mixture about like the concrete that emerges from a mixing truck to thinner consistencies that are little different than the mixture in turbid stream floods. The watery type of mudflow is commonly called a *debris flood* in the western United States, particularly in southern California, where it occurs commonly and with disastrous effects.

In Los Angeles County, California, real estate development has been carried out on very steep hillsides and mountainsides by the process of bulldozing roads and homesites out of the deep regolith. The excavated regolith is pushed into adjacent embankments, where its instability poses a threat to slopes and stream channels below. When saturated by heavy winter rains, these embankments can give way, producing earthflows, mudflows, and debris floods that travel far down the canyon floors and spread out on lowland surfaces, burying streets and yards in bouldery mud. Many debris floods of this area are also produced by heavy rains falling on undisturbed mountain slopes denuded by the burning of vegetative cover in the preceding dry summer; some fires are set by humans, whether inadvertently or deliberately. Disturbance of slopes by construction practices is simply an added source of debris and serves to enhance what is already an important environmental hazard.

A related phenomenon of high, alpine mountain chains, where glacial erosion has produced extremely steep valley gradients and where large quantities of gla-cial rock rubble (moraine) and relict glaciers are perched precariously at high positions, is the *alpine debris avalanche*. The sudden rolling of a mixture of rock waste and glacial ice can produce a tongue of debris traveling down-valley at a speed little less than that of a freely falling body. A great disaster of this type occurred in 1970 in the high Andes of Peru. A severe earthquake (magnitude 7.7 on the Richter scale), which caused widespread death and destruction, set off the fall of a large snow cornice from a high peak, Huascarán. After a free fall of 3000 ft (900 m) the snow mass was partially melted by impact and incorporated a great quantity of loose rock to become a debris avalanche. Traveling down-valley at a speed calculated to have reached 480 km (300 mi) per hour, the avalanche wiped out the town of Yungay and several smaller villages. The death toll, which included earthquake casualties, was estimated in the thousands.

Landslides

The term *landslide* is used widely in a general sense to mean any downslope movement of a mass of regolith or bedrock under the influence of gravity. Many geologists and civil engineers refer to earthflows as landslides. We shall limit our use of the term to mean the rapid sliding of large rock masses beginning their descent as unit blocks, without internal flowage. Softening of clay minerals to produce a plastic mass, as in the case of an earthflow, is not involved in the behavior of a typical landslide. A landslide mass is rigid, rather than plastic in its behavior, traveling like a massive sled over a plane of slippage. In most cases, however, the sliding block breaks up into many smaller blocks, and these in turn may disintegrate into a rubble. In its final stages, the mass of rubble often shows a gross flowage motion.

Two basic forms of landslides are (1) rockslides and (2) slump blocks. A *rockslide* consists of a bedrock mass slipping on a sloping rock plane, as shown in Figure 15.28A. The plane of slippage is usually a fault plane, joint plane, or bedding plane. A *slump block* is a bedrock mass that moves down a curved slip surface, as shown in Figure 15.28B. As a slump block moves down, it also

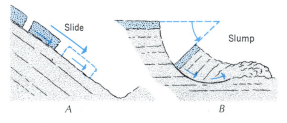

FIGURE 15.28 Landslides may involve *(A)* slip on a nearly plane surface or *(B)* slump with rotation on a curved plane.

FIGURE 15.29 This great landslide, locally known as the Hope Slide, descended from the high mountain summit at the upper left. The rock became pulverized into loose debris, which came to rest at the mountain base. Near Hope, British Columbia. (Mark A. Melton.)

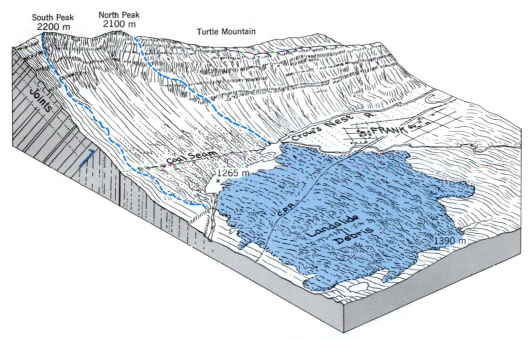

FIGURE 15.30 A classic example of an enormous, disastrous landslide is the Turtle Mountain slide, which took place at Frank, Alberta, in 1903. A huge mass of limestone slid from the face of Turtle Mountain between South and North peaks, descended to the valley, then continued up the low slope of the opposite valley side until it came to rest as a great sheet of bouldery rock debris. (Data from Canadian Geological Survey, Department of Mines. Drawn by A. N. Strahler.)

rotates on a horizontal axis, so the upper surface of the block becomes tilted toward the cliff that remains. A rockslide can travel a long distance down a mountainside, far from its original position, whereas a slump block remains close to its original position.

Rockslides

Wherever steep mountain slopes occur, there is a possibility of large, disastrous rockslides. In Switzerland, Norway, or the Canadian Rockies, for example, villages built on the floors of steep-sided valleys have been destroyed and their inhabitants killed by the sliding of millions of cubic yards of rock set loose without any warning (Figure 15.29).

The Turtle Mountain landslide of 1903 in Alberta, Canada, is shown in Figure 15.30. It involved the sliding of an enormous mass of limestone, its volume estimated at 27 million cu m (35 million cu yd), through a descent of 900 m (3000 ft). The debris buried part of the town of Frank, with a loss of 70 persons.

Many of the world's great rockslides have occurred in mountains deeply carved by glacial action in the Pleistocene Epoch. The dominant landforms are U-shaped troughs, abundant in the European Alps, the Cordilleran Range of North America, the Andes Range of South America, and the Himalayas of southern Asia.

Glacial troughs in coastal locations are typically inundated by the ocean, becoming steep-walled fiords (see Figure 22.14). The environmental hazard of rockslides is extremely high for towns and cities located on the floors of glacial troughs and at the heads of fiords. A rockslide entering the deep water of a fiord generates an enormous

wave of water that runs quickly to the head of the fiord, inundating and destroying habitations located on the low-lying trough floor.

In many places glacial troughs are occupied by deep trough lakes, which may be many kilometers long. A rockslide entering a trough lake sets off a great wave, traveling to both ends of the lake and riding some distance up the trough floor.

Slump Blocks

Slump blocks consist of great masses of bedrock sliding downward from a high cliff and, at the same time, rotating backward on a horizontal axis (Figure 15.31). Wherever massive sedimentary strata, usually sandstones or limestones, or lava beds, rest on weak clay or shale formations, a steep cliff tends to be formed by erosion. As the weak rock is eroded from the cliff base, the cap rock

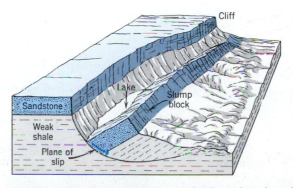

FIGURE 15.31 Slump blocks rotate backward as they slide from a cliff. (Drawn by A. N. Strahler.)

is undermined. When a point of failure is reached, a large block breaks off, slides down, and tilts back along a curving plane of slip. Slump blocks may be as much as 2 to 3 km (1 or 2 mi) long and 150 m (500 ft) thick. A single block appears as a ridge at the base of the cliff. A closed depression or lake basin may lie between the block and the cliff.

Slumping of large bedrock masses is rarely a threat to human life because of the slow rate of movement and the limited distance of travel. The environmental importance of slumping, in both bedrock and regolith, lies in the potential for disruption of such human-made struc-

tures as buildings and highways. If a house happens to lie over the upper part of an active earthflow where slump scarps are forming, it can be sheared into sections. Because slumping is difficult to stabilize, abandonment of the property may be the only economically feasible solution. Large bedrock slump blocks located along the base of a cliff should be identified and avoided in the construction of highways, dams, and power plants. Even though the slump block may appear to be stabilized and immobile, excavation of material at the base of the block may lead to reactivation of the block. The recognition of ancient earthflows and slump blocks is an important

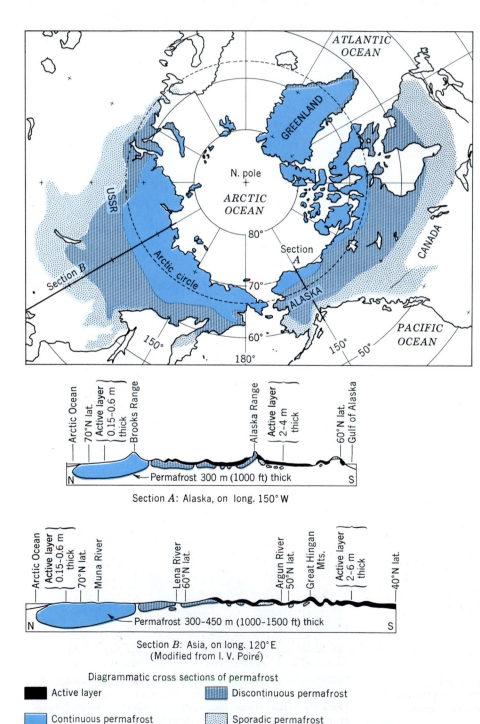

Section A: Alaska, on long. 150° W

Section B: Asia, on long. 120° E
(Modified from I. V. Poiré)

Diagrammatic cross sections of permafrost

▮ Active layer   ▦ Discontinuous permafrost

▮ Continuous permafrost   ░ Sporadic permafrost

FIGURE 15.32 Distribution of permafrost in the northern hemisphere and representative cross sections in Alaska and Asia. (From Robert F. Black, "Permafrost," Chapter 14 of P. D. Trask's *Applied Sedimentation*. Copyright © by John Wiley & Sons. Reprinted by permission of John Wiley & Sons, Inc.)

aspect of environmental protection because serious future economic losses can often be avoided in the planning stages of engineering projects.

GEOMORPHIC ASPECTS OF THE ARCTIC TUNDRA

The landscape of the treeless arctic tundra shows many distinctive effects of weathering and mass wasting under a regime of severe cold in which all soil water is solidly frozen throughout a winter of many months duration that is interrupted by a short summer season of surface thawing. Saturation of soil at this time leaves it vulnerable to mass wasting and water erosion, while the freezing of soil water that follows exerts a strong mechanical influence on the surface layer.

Arctic Permafrost

Perennially frozen ground, or *permafrost*, prevails over the tundra region and a wide bordering area of boreal forest climate. The surface zone of seasonal thaw, called the *active layer*, is from 0.6 to 4 m (2 to 14 ft) thick, depending on latitude and the nature of the ground. The distribution of permafrost in the northern hemisphere is shown in Figure 15.32. Three zones are recognized. Continuous permafrost, which extends without gaps or interruptions under all surface features, coincides largely with the tundra climate, but also includes a large part of the boreal forest climate in Siberia. Discontinuous permafrost, which occurs in patches separated by frost-free zones under lakes and rivers, occupies much of the boreal forest climate zone of North America and Eurasia. Sporadic occurrence of permafrost in small patches extends into the southern limits of the boreal forest climate.

Depth of the permafrost reaches 300 to 450 m (1000 to 1500 ft) in the continuous zone near lat. 70°N. Much of this permanent frost is an inheritance from more severe

FIGURE 15.34 Patterned ground near Barrow Point, Alaska. The rims of the ice-wedge polygons rise higher than the intervening ground, which contains small lakes. (William R. Farrand.)

conditions of the last ice age, but some permafrost bodies may be growing under existing climate conditions.

Features of the Active Surface Layer

In areas of coarse-textured regolith consisting of rock particles of a wide range of sizes, the annual cycle of thawing and freezing causes the coarsest fragments—pebbles and cobbles—to move horizontally and to become sorted out from the finer particles. This type of

FIGURE 15.33 Sorted circles of gravel form a system of netlike stone rings on this nearly flat land surface where water drainage is poor. The circles in the foreground are 3 to 4 m across; the gravel ridges are 20 to 30 cm high. Broggerhalvoya, western Spitsbergen, latitude 78°N. (Bernard Hallet, Quaternary Research Center, University of Washington, Seattle.)

FIGURE 15.35 Exposed by river erosion, this great wedge of solid ice fills a vertical crack in organic-rich floodplain silt along the Yukon River, near Galena in western Alaska. (Photograph No. PK 3161 by Troy L. Pewe.)

sorting produces ringlike arrangements of coarse fragments. Linked with adjacent rings, the gross pattern becomes netlike to form a system of *stone polygons* (also called "stone rings," or "stone nets") (Figure 15.33).

In fine-textured alluvium, such as that formed on river floodplains and deltaic plains in the arctic environment, ice has accumulated in vertical wedge-forms in deep cracks in the sediment. These *ice wedges* are interconnected into a system of polygons, called *ice-wedge polygons* (Figure 15.34). Ice wedges are thought to originate as shrinkage cracks, formed during extreme winter cold. During the spring melt, water enters the cracks and becomes frozen. Repeated cracking and the addition of new ice causes the ice wedge to thicken until it becomes as wide as 3 m (10 ft) and as deep as 30 m (30 ft) (Figure 15.35). Both stone polygons and ice-wedge polygons belong to a class of features called *patterned ground*. Another remarkable ice-formed feature of the arctic tundra is a conspicuous conical mound, called a *pingo* (Figure 15.36). The pingo has a core of ice and grows in height as more ice accumulates, forcing up the overlying sediment. In extreme cases, pingos reach a height of 50 m (165 ft) and a basal diameter as great as 600 m (2000 ft).

A special variety of earth flowage characteristic of arctic permafrost regions is *solifluction* (from Latin words meaning "soil" and "to flow"). It is active in early summer, when thawing has penetrated the upper few decimeters. At this time soil is fully saturated with water that cannot escape downward because of the impermeable frozen mass. Flowing almost imperceptibly, this saturated soil forms terraces and lobes that give a mountain slope a stepped appearance (Figure 15.37).

Environmental Problems of Permafrost

Environmental degradation of permafrost regions arises from surface changes made by humans. The undesirable consequences are usually related to the destruction or removal of an insulating surface cover, which may consist of a humus or peat layer in combination with living plants of the tundra or arctic forest. When this layer is

FIGURE 15.37 A solifluction lobe on Baffin Island in the arctic tundra. While bearing intact its cover of plants and soil, a bulging mass of water-saturated regolith has slowly moved downslope, overriding the ground surface below it. A backpack marks the base of the advancing lobe. (M. Church, University of British Columbia.)

scraped off, the summer thaw is extended to a greater depth, with the result that ice wedges and other ice bodies melt in the summer and waste downward. This activity is called *thermal erosion*. Meltwater mixes with silt and clay to form mud, which is then eroded and transported by water streams, leaving trenchlike morasses.

The consequences of disturbance of permafrost terrain became evident in World War II, when military bases, airfields, and highways were constructed hurriedly without regard for maintenance of the natural protective surface insulation. In extreme cases, scraped areas turned into mud-filled depressions and even into small lakes that expanded in area with successive seasons of thaw, engulfing nearby buildings. Engineering

FIGURE 15.36 This large pingo rises above the flat plain of the McKenzie River delta, N.W.T., Canada. (Steve McCutcheon.)

FIGURE 15.38 The elevated tunnel of sheet metal is a "utilidor" (utility corridor) in which steam and water lines are protected from freezing in the severe tundra winter. Notice that the buildings are mounted on posts and have small windows. Inuvik, Northwest Territories. (Mark A. Melton.)

FIGURE 15.39 The Ray Mine, a great open-pit copper mine near Teapot Mountain, Arizona. Because the copper minerals are disseminated through an enormous body of igneous rock, large volumes of rock must be excavated. (Mark A. Melton.)

practices now call for placing buildings on piles with an insulating air space below or for the deposition of a thick insulating pad of coarse gravel over the surface prior to construction. Steam and hot-water lines are placed above ground to prevent thaw of the permafrost layer (Figure 15.38).

Another serious engineering problem of arctic regions is the behavior of streams in winter. As the surfaces of streams and springs freeze over, the water beneath bursts out from place to place, freezing into huge accumulations of ice. Highways are thus made impassable.

The lessons of superimposing our technology on a highly sensitive natural environment were learned the hard way—by encountering unpleasant and costly effects that were not anticipated. The threat of environmental destruction will persist as oil continues to flow through the Trans-Alaska Pipeline. This facility carries hot oil from the northern shores of Alaska, across a permafrost landscape, to the port of Valdez on the south coast. Effects of this pipeline on permafrost and other elements of the environment were heavily debated and were the subject of intensive investigation. The possibility exists of damage to the ecosystem from spills caused by pipe breakage.

SCARIFICATION OF THE LAND

Industrial societies now possess enormous machine power and explosives capable of moving great masses of regolith and bedrock from one place to another. One such activity is the extraction of mineral resources. Another is the reorganization of terrain into suitable configurations for highway grades, airfields, building foundations, dams, canals, and various other large structures. Both activities involve removal of earth materials, which temporarily destroys preexisting ecosystems and habitats of plants and animals. The same activities include building up of new land on adjacent surfaces using those same earth materials. This process destroys ecosystems and habitats by burial. What distinguishes artificial forms of mass wasting from the natural forms is that machinery is used to raise earth materials against the force of grav-

ity. Explosives used in blasting can produce disruptive forces many times more powerful than the natural forces of physical weathering.

Scarification is a general term for excavations and other land disturbances produced for purposes of extracting mineral resources; it includes accumulation of waste matter as spoil or tailings. Among the forms of scarification are open-pit mines, strip mines, quarries for structural materials, borrow pits along highway grades, sand and gravel pits, clay pits, phosphate pits, scars from hydraulic mining, and stream gravel deposits reworked by dredging. Open-pit mining of low-grade copper ores is illustrated in Figure 15.39.

Scarification is on the increase. Demands for coal to meet energy requirements are on the rise. There are also increased demands for industrial minerals used in manufacturing and construction. At the same time, as the richer and more readily available mineral deposits are consumed, industry turns to poorer grades of ores and to less easily accessible coal deposits. As a result, the rate of scarification is further increased.

WEATHERING AND MASS WASTING IN REVIEW

In this chapter we have compared natural processes of wasting of the continental surfaces with changes of a similar nature induced by human activity. The processes of weathering and soil creep are for the most part slow-acting and produce effects that are visible only when accumulated over centuries. Natural mass-wasting processes also include catastrophic events. Indeed, some large landslides dwarf anything that humans have accomplished in equal time by earth-moving with machines and explosives. But machines work constantly with enormous quantities of energy, and the cumulative environmental damage and destruction continues to mount. Only now are the conflicts of interest beginning to emerge and to be squarely faced. Only now are we designing and implementing environmental protection measures necessary to control the spread of scarification and its many harmful side effects.

RUNOFF, STREAMS, AND GROUND WATER

The primary concern of this chapter is with the science of *hydrology*, which is a study of water as a complex but unified system on the earth. We began our investigation of hydrology in Chapter 9 by covering that phase of hydrology in which soil water is recharged by precipitation and returned directly to the atmosphere by evapotranspiration, or to be disposed of as runoff.

Recall that many soil-water budgets show a substantial water surplus, to be disposed of as runoff. Our investigation now deals with that water surplus and the paths it follows as subsurface water and surface water. There are two basic paths of escape for surplus water. First, surplus water may flow over the ground surface as runoff to lower levels. As it travels, the dispersed flow becomes collected into streams, which eventually conduct the runoff to the ocean. Second, surplus water may percolate through the soil, traveling downward under the force of gravity to become part of the underlying ground-water body. Following subterranean flow paths, this water emerges to become surface water, or it may emerge directly in the shore zone of the ocean. In this chapter, we trace both the surface and subsurface pathways of flow of surplus water. In so doing, we will complete the hydrologic cycle.

RUNOFF AS A VITAL RESOURCE

Surplus fresh water, as runoff, is a vital part of the environment of terrestrial life forms and of humans in particular. Surface water in the form of streams, rivers, ponds, and lakes constitutes one of the distinctive environments of plants and animals.

Our heavily industrialized society requires enormous supplies of fresh water for its sustained operation. Urban dwellers consume water in their homes at rates of 150 to 400 liters (40 to 100 gallons) per person per day. Enormous quantities of water are used for cooling purposes in air-conditioning units and power plants.

In view of projections based on existing rates of increase in water demands, we shall be hard put in the future to develop the needed supplies of pure fresh water. Water pollution also tends to increase as populations

grow and urbanization advances over broader areas. A disconcerting concept is that the available resource of pure fresh water is shrinking while demands are rising. Knowledge of hydrologic processes enables us to evaluate the total water resource, to plan for its management, and to protect it from pollution.

FORMS OF OVERLAND FLOW

Figure 16.1 shows what happens to rain falling on a hillslope that has its natural soil and plant cover intact. Some rain is caught and held above the ground on leaves and stems, a process called *interception*. Intercepted water may be returned directly to the atmosphere by evaporation. Precipitation reaching the soil surface enters the soil by infiltration; but when the rate of precipitation exceeds the rate of infiltration, water begins to accumulate as surface detention in minor surface depressions in the soil or behind small check dams of plant litter. The phenomenon is also called *depression storage*. Some of this water is evaporated directly into the atmosphere; the remainder eventually infiltrates the soil.

Runoff that flows down the slopes of the land in

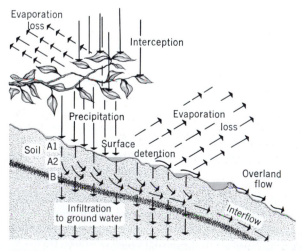

FIGURE 16.1 Precipitation, interception, and overland flow.

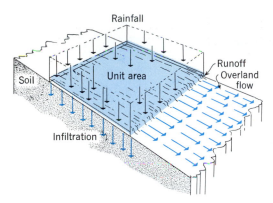

FIGURE 16.2 Rainfall, infiltration, and overland flow. (From A. N. Strahler, *The Earth Sciences*, 2nd ed., Harper & Row, Publishers. Copyright © by Arthur N. Strahler.)

broadly distributed sheets is referred to as *overland flow* (Figure 16.2). We distinguish overland flow from *stream flow (channel flow)*, in which the water occupies a narrow channel confined by lateral banks. Overland flow can take several forms. It may be a continuous thin film, called *sheet flow*, where the soil or rock surface is smooth. Flow may take the form of a series of tiny rivulets connecting one water-filled hollow with another where the ground is rough or pitted. On a grass-covered slope, overland flow is subdivided into countless tiny threads by water passing around the stems. Even in a heavy and prolonged rain, you might not notice overland flow in progress on a sloping lawn. On heavily forested slopes, overland flow may pass entirely concealed beneath a thick mat of decaying leaves. At the base of a hillslope, overland flow is disposed of by passing into the head of a gully, a stream channel, or a lake (Figure 16.3). In some cases, overland flow simply disappears into a porous layer of sandy soil.

Overland flow is measured in centimeters or inches of water per hour, just as for precipitation and infiltra-

tion. A simple formula expresses the rate at which overland flow will be produced by a given unit of ground surface as follows:

$$\text{Rate of production of overland flow} = \text{rate of precipitation} - \text{rate of infiltration}$$

For example, if the rate of infiltration became constant at a value of 1 cm/hr and the rate of rainfall was a steady 2 cm/hr (a heavy rain), the runoff would be produced at a rate of 1 cm/hr, assuming none to be returned to the atmosphere by evaporation.

INTERFLOW

Mature soils of certain of the major soil orders have a dense layer (B horizon) of clay accumulation, but the soil layer above it (A horizon) is sandy and loose. (These horizons are explained in Chapters 23 and 24.) When the infiltrating water reaches the B horizon, its downward passage is greatly impeded. The soil water then backs up into the more permeable A horizon and begins to move in the downslope direction, parallel with the ground surface (Figure 16.1). This lateral flow is called *interflow* (or *throughflow*). Eventually interflow reaches the base of the hillslope and seeps into a stream channel. Interflow is thus a pathway of runoff intermediate in position and speed of travel between overland flow and deeper ground-water flow. Interflow is also important where a permeable soil rests on a dense, impermeable bedrock mass.

DRAINAGE SYSTEMS

Overland flow, interflow, and out-seeping ground-water flow eventually contribute to a stream, which is a much faster, more concentrated form of runoff (Figure 16.4). We define a *stream* as a long, narrow body of flowing water occupying a trenchlike depression, or channel, and moving to lower levels under the force of gravity.

The total system of downslope water flow from the point of arrival at the ground surface comprises the *drainage system*. It consists of a branched network of stream channels, as well as the sloping ground surfaces that contribute overland flow and interflow to those

FIGURE 16.3 Overland flow taking the form of a thin sheet of water covers the nearly flat plain in the middle distance. This water is converging into stream flow in a narrow, steep-sided gully (left). The photograph was taken shortly after a summer thunderstorm had deluged the area. The locality, near Raton, New Mexico, shows steppe grassland vegetation. (Mark A. Melton.)

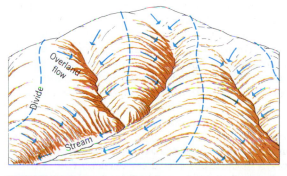

FIGURE 16.4 Overland flow from slopes in the headwater area of a stream system supplies water and rock debris to the smallest elements of the channel network.

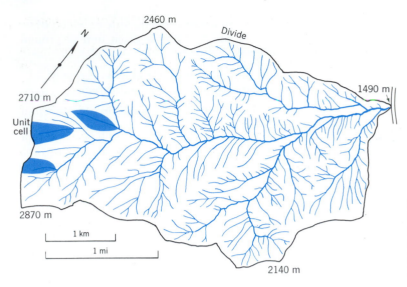

FIGURE 16.5 Channel network of the basin of Pole Canyon, Utah. (Data from U.S. Geological Survey and Mark A. Melton. From A. N. Strahler, *Planet Earth: Its Physical Systems Through Geologic Time,* Harper & Row, Publishers. Copyright © by Arthur N. Strahler.)

channels. The entire system is bounded by a drainage divide, outlining a more or less pear-shaped *drainage basin.* The ground slopes and channels are adjusted to dispose of, as efficiently as possible, the runoff and its contained load of mineral particles.

A typical stream network contributing to a single outlet is shown in Figure 16.5. Note that each fingertip tributary receives runoff from a small area of land surface surrounding the channel. This area may be regarded as the unit cell of the drainage system. The entire surface within the outer divide of the drainage basin constitutes the watershed for overland flow. A drainage system is a converging mechanism funneling and integrating the weaker and more diffuse forms of runoff into progressively deeper and more intense paths of activity.

STREAM CHANNEL GEOMETRY

A *stream channel* is a narrow trough, shaped by the forces of flowing water to be highly effective in moving the quantities of water and sediment supplied to the stream. Channels may be so narrow that a person can jump across them or as wide as 1.5 km (1 mi) for a great river—such as the Mississippi.

Hydraulic engineers who measure stream dimensions and flow rates have adopted a set of terms to describe channel geometry (Figure 16.6). Depth, *d*, in meters, is measured at any specified point in the stream as the vertical distance from surface to bed. Width, *w*, is the distance in meters across the stream from one water's edge to the other. The cross-sectional area, *A*, is the area in square meters of a vertical slice across the stream. The *wetted perimeter, P*, is the length of line of contact between the water and the channel, as measured from the cross section.

The rate of fall in altitude of the stream surface in the downstream direction is the *stream gradient;* it is often stated in meters per kilometer or feet per mile. A gradient of 5 m/km means that the stream surface undergoes a vertical drop of 5 m for each km of horizontal distance downstream. Slope can also be given in terms of percent

grade, a common practice in engineering. A grade of 3 percent, or 0.03, means that the stream drops 3 m for every 100 m of horizontal distance.

STREAM FLOW

As a stream flows under the influence of gravity, the water encounters resistance—a form of friction—with the channel walls. As a result, water close to the bed and banks moves slowly; that in the deepest and most centrally located zone flows fastest. Figure 16.6 indicates by arrows the speed of flow at various points in the stream. The single line of maximum velocity is located in midstream, where the channel is straight and symmetrical.

Our statement about velocity needs to be qualified. Actually in all but the most sluggish streams, the water

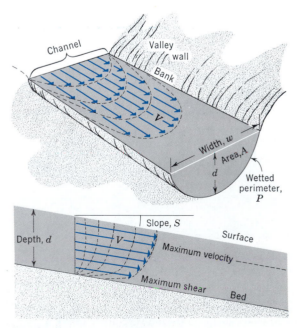

FIGURE 16.6 Stream flow within a channel is most rapid near the center and just below the water surface.

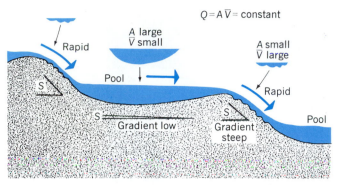

$$Q = A\bar{V} = constant$$

FIGURE 16.7 Schematic diagram of the relationship among cross-sectional area, mean velocity, and gradient. (From A. N. Strahler, *The Earth Sciences*, 2nd ed., Harper & Row, Publishers. Copyright © by Arthur N. Strahler.)

is affected by *turbulence*, a system of innumerable eddies that are continually forming and dissolving. A particular molecule of water, if we could keep track of it, would describe a highly irregular, corkscrew path as it is swept downstream. Motions include upward, downward, and sideward directions. Turbulence in streams is extremely important because of the upward elements of flow that lift and support fine particles of sediment. The murky, turbid appearance of streams in flood is ample evidence of turbulence, without which sediment would remain near the bed. Only if we measure the water velocity at a certain fixed point for a long period of time, say several minutes, will the average motion at the point be downstream and in a line parallel with the surface and bed. Average values are shown by the arrows in Figure 16.6.

Because the velocity at a given point in a stream differs greatly according to whether it is being measured close to the banks and bed or out in the middle line, a single value, the *mean velocity*, is needed. Mean velocity is computed for the entire cross section to express the activity of the stream as a whole.

STREAM DISCHARGE

A most important measure of stream flow is *discharge*, Q, defined as the volume of water passing through a given cross section of the stream in a given unit of time. Discharge is stated in cubic meters per second (abbreviated to cms); in English units, cubic feet per second (cfs). Discharge may be obtained by taking the mean velocity, *V*, and multiplying it by the cross-sectional area, *A*.

This relationship is stated by the important equation, Q = AV.

We realize that water will flow faster in a channel of steep gradient than one of gentle gradient because the component of gravity acting parallel with the bed is stronger for the steeper gradient. As shown in Figure 16.7, velocity increases quickly where a stream passes from a wide pool of low gradient to a steep stretch of rapids. As V increases, A must decrease; otherwise their product, AV, would not be held constant. In the pool, where velocity is low, A is correspondingly increased.

Stream channels differ in the amount of frictional resistance that the bed and banks offer to the flow of water. Resistance is large in a broad but shallow channel and is much less in a deep, narrow channel. The optimum channel would be semicircular in cross section. Streams are, however, usually required to have broad, shallow channels because of the load of mineral particles that must be carried or because the banks are weak and will not hold a steep attitude.

STREAM GAUGING

Information on daily discharges of streams is vital to a nation, not only as a measure of the surface-water resource, but also for the design of flood-protection structures and for the prediction of stream floods as they progress down a river system.

In the United States, the measurement of stream flow, or *stream gauging*, is under the jurisdiction of the U.S. Geological Survey. In cooperation with states and municipalities, that organization maintains over 11,000 gauging stations on principal streams and their tributaries. Figures on discharge are published by the Geological Survey in a series of Water-Supply Papers.

A stream-gauging station requires a device for measuring the height of the water surface, or *stream stage*. Simplest to install is a *staff gauge*, which is simply a graduated vertical scale attached to a post or bridge pier. This must be read directly by an observer whenever the stage is to be recorded. More useful is an automatic-recording gauge, which is mounted in a stilling tower built beside the river bank (Figure 16.8).

To measure stream discharge, one must determine both the area of cross section of the stream and the mean velocity. Velocity measurement requires a *current meter*. The meter is lowered into the stream at closely spaced intervals so the velocity can be read at a large number of points evenly distributed in a grid pattern through the

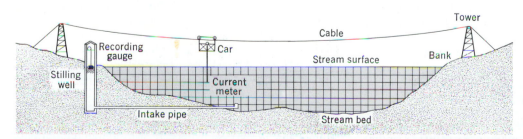

FIGURE 16.8 Idealized diagram of stream-gauging installation.

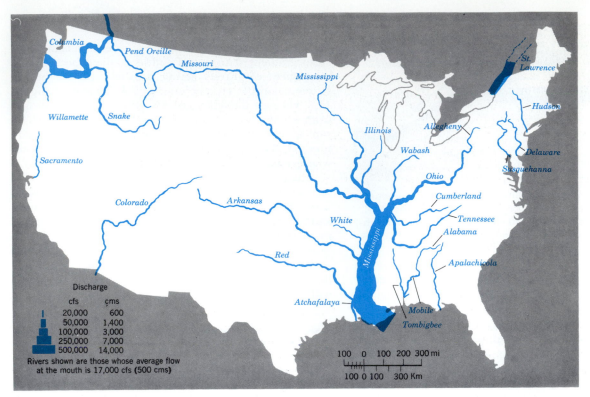

FIGURE 16.9 A flow map of the United States showing the relative magnitude of rivers by means of a broad line with width proportional to the mean annual water discharge. (U.S. Geological Survey.)

stream cross section (Figure 16.8). A bridge often serves as a convenient means of crossing over the stream; otherwise, a cable car or small boat is used. Revolving cups on the current meter turn at a rate proportional to current velocity. As the velocities are being measured from point to point, a profile of the riverbed is also being made by sounding the depth. From these readings, a profile is drawn and the cross-sectional area is measured from the profile. Mean velocity is computed by summing all individual velocity readings and dividing by the number of readings. Discharge can then be computed using the formula $Q = AV$.

Figure 16.9 is a map showing the relative discharge of major rivers of the United States. ("River" is a popular term applied to a large stream. The word "stream" is the scientific term designating channel flow of any magnitude of discharge.) The mighty Mississippi with its tributaries dwarfs all other North American rivers, although the MacKenzie, Columbia, and Yukon rivers, as well as the Great Lakes discharge through the St. Lawrence River, are also of major proportions. The Colorado River, a much smaller stream, crosses a vast semiarid and arid region in which little tributary flow is added to the snowmelt source high in the Rocky Mountains.

STREAM FLOW AND PRECIPITATION

It seems obvious that the discharge of a stream will increase in response to a period of heavy rainfall or snowmelt. The response is delayed, of course, but the length

of delay depends on a number of factors. The most important factor is the size of the drainage basin feeding the stream above the place where the gauge is located. The relationship between stream discharge and precipitation is best studied by means of a simple graph, called a *hydrograph*.

Figure 16.10 is a hydrograph for a drainage basin about 800 sq km (300 sq mi) in area, located in Ohio in a region of moist continental climate. The graph gives data for a 2-day summer storm. Rainfall is shown by a bar graph giving the precipitation (cm) in each 2-hour period. Also plotted on the graph (smooth line) is the discharge (cms) of Sugar Creek, the trunk stream of the drainage basin. The average total rainfall over the watershed of Sugar Creek was about 15 cm (6 in.); of this amount half passed down the stream within 3 days' time. Some rainfall was held in the soil as soil water, some evaporated, and some infiltrated to the water table to be held in long-term storage in the ground-water body.

Studying the rainfall and runoff graphs in Figure 16.10, we see that prior to the onset of the storm, Sugar Creek was carrying a small discharge. This flow, being supplied by the slow seepage of ground water into the channel, is termed *base flow*. After the heavy rainfall began, several hours elapsed before the stream gauge at the basin mouth began to show a rise in discharge. This interval, called the *lag time*, indicates that the branching system of channels was acting as a temporary reservoir. As the stage was rising, water was soaking into permeable bank materials, where it was being temporarily stored.

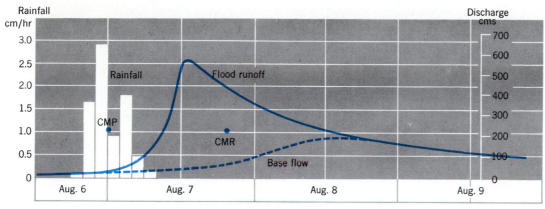

FIGURE 16.10 Four days of flow of Sugar Creek, Ohio. (After William G. Hoyt and Walter B. Langbein, *Floods;* copyright 1955 by Princeton University Press; Figs. 8 and 13, pp. 39 and 45. Reprinted by permission of Princeton University Press.)

Lag time is measured as the difference between center of mass of precipitation (CMP) and center of mass of runoff (CMR), as labeled in Figure 16.10. The peak flow of Sugar Creek was reached almost 24 hours after the rain began; the lag time was about 18 hours. Note also that the rate of decline in discharge was much slower than the rate of rise because overland flow was followed by contributions from interflow and, finally, from ground-water seepage.

In general, the larger a watershed, the longer is the lag time between peak rainfall and peak discharge and the more gradual is the rate of decline of discharge after the peak has passed. Notice that the flow of Sugar Creek showed a slow but distinct rise in the amount of discharge contributed by base flow.

BASE FLOW AND OVERLAND FLOW

In regions of moist climates, where the water table is high and normally intersects the important stream channels, the hydrographs of larger streams will show clearly the effects of three sources of water: overland flow, in-

terflow, and base flow. Figure 16.11 is a hydrograph of the Chattahoochee River, Georgia, a large stream draining a watershed of some 8700 sq km (3350 sq mi), much of it in the moist southern Appalachian Mountains. The sharp, abrupt fluctuations in discharge are produced by overland flow and interflow following rain periods of 1 to 3 days' duration. These are each similar to the hydrograph of Figure 16.10, except that they are shown in Figure 16.11 much compressed by the time scale.

After each rain period, the discharge falls off rapidly; but, if another storm occurs within a few days, the discharge rises to another peak. The enlarged inset graph shows details for the month of January. When a long period intervenes between storms, the discharge falls to a low value, the base flow, where it levels off.

Throughout the year the base flow, which represents ground-water inflow into the stream, undergoes a marked annual cycle. During the period of recharge (winter and early spring), water-table levels are raised and the rate of inflow into streams is increased. For the Chattahoochee River, the rate of base flow during January, February, March, and April holds uniform at about 110 cms (4000 cfs). The base flow begins to decline in spring,

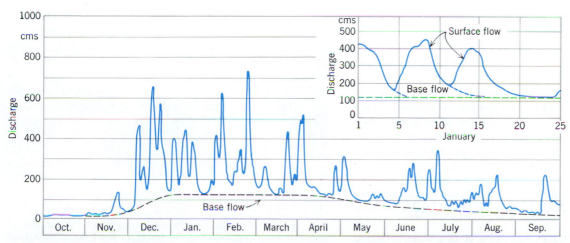

FIGURE 16.11 Flow peaks of the Chattahoochee River. (Data from U.S. Geological Survey in E. E. Foster, *Rainfall and Runoff*.)

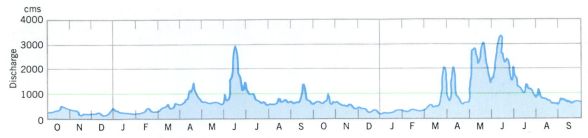

FIGURE 16.12 Discharge of the Missouri River. (Data from U.S. Geological Survey in E. E. Foster, *Rainfall and Runoff*.)

when heavy evapotranspiration losses reduce soil water and therefore cut off the recharge of ground water. The decline continues throughout the summer, reaching a low of about 30 cms (1000 cfs) by the end of October.

Finally, examine the hydrograph of the Missouri River at Omaha, Nebraska, for a 2-year period of record starting in October (Figure 16.12). This great river, draining 840,000 sq km (322,800 sq mi) of watershed, is a major tributary of the Mississippi River. Note that the discharge, ranging from 280 to 2800 cms (10,000 to 100,000 cfs), is many times greater than the discharges of smaller streams considered thus far. High rates of flow are chiefly from snowmelt, which occurs on the High Plains in spring and in the Rocky Mountain headwater areas in early summer. This event explains the sudden high discharges from April through June. During midwinter, when soil water is frozen and total precipitation is small over the watershed as a whole, the discharge rises little above the base flow. Ground-water recharge occurring in the spring raises summer levels of base flow to about 570 cms (20,000 cfs), or 2 to 3 times the winter base flow.

FLOODS

Everyone has seen enough news-media photos of river floods to have a good idea of the appearance of floodwaters and the havoc wrought by their erosive power and by the silt and clay they leave behind. Even so, it is not easy to define the term *flood*. Perhaps it is enough to say that a condition of flood exists when the discharge of a river cannot be accommodated within the margins of its normal channel, so the water spreads over adjoining ground on which crops or forests are able to flourish.

Most larger streams of moist climates have a *floodplain*, a belt of low, flat ground bordering the channel on one or both sides inundated by stream waters about once a year. This flood usually occurs in the season when abundant supplies of surface water combine with the effects of a high water table to supply more runoff than can stay within the channel. Such an annual inundation is considered a flood, even though its occurrence is expected and does not prevent the cultivation of crops after the flood has subsided. The seasonal inundation does not interfere with the growth of dense forests, which are widely distributed over low, marshy floodplains in all moist regions of the world. Still higher discharges of water, the rare and disastrous floods that may occur as in-

frequently as 3 to 5 decades, inundate ground lying well above the floodplain (Figure 16.13).

For practical purposes, the National Weather Service, which provides a flood-warning service, designates a particular stage of gauge height at a given place as the *flood stage;* this implies that the critical level has been reached above which *overbank flooding* may be expected to set in. Immediately at or below flood stage, the river may be described as being in the *bank-full stage*, the flow being entirely within the limits of the heavily scoured channel.

DOWNSTREAM PROGRESS OF A FLOOD WAVE

The rise of a river stage to its maximum height, or *flood crest*, followed by a gradual lowering of stage, is termed the *flood wave*. The flood wave is simply a large-sized rise and fall of river discharge of the type already analyzed in earlier paragraphs, and it follows the same principles. Figure 16.14*A* shows the downstream progress of a flood on the Chattooga-Savannah river system. On the Chattooga River near Clayton, Georgia, the flood peak, or crest, was reached quickly—1 day after the storm—and subsided quickly. On the Savannah River, 105 km (65

FIGURE 16.13 The city of Harrisburg, Pennsylvania, lies partly submerged beneath flood waters generated by Hurricane Agnes, June 1972. The Susquehanna River, seen in the distance, rose to nearly 5 m (16 ft) above flood stage, inundating the low river terrace on which the downtown portion of Harrisburg is built. (Department of the Army.)

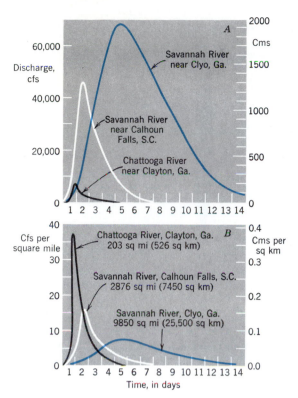

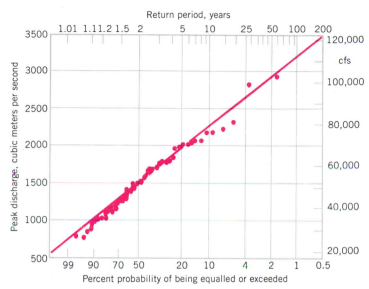

FIGURE 16.15 Flood frequency data for the Clearwater River at Kamaiah, Idaho. Each dot is a measured maximum yearly discharge in a 53-year record. (Data of U.S. Geological Survey from R. K. Linsley, M. A. Kohler, and J. L. Paulhus, *Hydrology for Engineers*, second edition, McGraw-Hill, New York, p. 347.)

FIGURE 16.14 The downstream progress of a flood wave. (After William G. Hoyt and Walter B. Langbein, *Floods;* copyright 1955 by Princeton University Press; Figs. 8 and 13, pp. 39 and 45. Reprinted by permission of Princeton University Press.)

mi) downstream at Calhoun Falls, South Carolina, the peak flow occurred a day later; but the discharge was much larger because of the larger area of watershed involved. Downstream another 153 km (95 mi), near Clyo, Georgia, the Savannah River crested 5 days after the initial storm, with a discharge of over 1700 cms (60,000 cfs). This set of three hydrographs shows that (1) the lag time in occurrence of the crest increases downstream; (2) the entire period of rise and fall of flood wave becomes longer downstream; and (3) the discharge increases greatly downstream as watershed area increases.

Figure 16.14*B* is a somewhat different presentation of the same flood data in that the discharge is given in terms of a common unit of area (cms/sq km), thus eliminating the effect of increase in discharge downstream and showing us only the shape, or form, of the flood crest.

FLOOD PREDICTION

The National Weather Service operates a River and Flood Forecasting Service through 85 offices located at strategic points along major river systems of the United States. Each office issues river and flood forecasts to the communities within the associated district, which is laid out to cover one or more large watersheds. Flood warnings are publicized by every possible means. Close cooperation is maintained with various agencies to plan the evacuation of threatened areas and the removal or protection of vulnerable property.

Hydrologists use sophisticated statistical methods to arrive at estimates of the probability that a flood of a given stage of discharge will occur in any given year. Estimates are, of course, based on gauge records of past years. Figure 16.15 is a graph on which the greatest discharge of each year of record is plotted as a dot. Discharge in units of cms is plotted on the vertical axis in equally spaced units (arithmetic scale). Numbers on the bottom horizontal scale tell the probability that the given discharge will be equaled or exceeded in a given year. For example, a discharge associated with the value of 20 percent is interpreted to mean "a discharge of this magnitude (2000 cms) can be expected to be equaled or exceeded in 20 out of 100 years." The numbers on the scale at the top of the graph tell the return period (or recurrence interval), which is simply the inverse value of the probability percentage on the bottom scale. In the example just cited, probability 20 percent, the return period is 5 years, meaning that on the average the stated discharge (2000 cms) can be expected to be equaled or exceeded once every 5 years. The horizontal scale on the graph has been adjusted to obtain the best possible fit of the plotted points to a straight line. The fitted line can be expressed by a mathematical equation; it is one of several estimating equations that might be applied to the same data in efforts to secure the best predictions.

A much simpler pictorial analysis is used in Figure 16.16 to suggest the likelihood that a given maximum stage will occur in a given calendar month of the year. Each bar divides all observed maximum stages of the calendar month into quartiles, or groups of 25 percent of the occurrences. The dots at top and bottom of the bar give the absolute maximum and minimum stages observed during the entire period of record.

The graph for the Mississippi River at Vicksburg illustrates a great river responding largely to spring floods so as to yield a simple annual cycle. All floods have occurred in the first 6 months of the year. The Colorado River at Austin, Texas, illustrates a river draining largely semiarid plains. Summer floods are produced directly by torrential rains from invading moist tropical air masses. Floods of the late summer and fall are often attributable to tropical storms (hurricanes) moving inland from the Gulf of Mexico. The Sacramento River at Red Bluff, California, has a winter flood season when rains are heavy, but a sharp dip to low stages in late summer, which is the very dry period in the Mediterranean climate. The flood expectancy graph for the Connecticut River at Hartford shows two seasons of floods. The more reliable of the two is the early spring, when snowmelt is rapid over the mountainous New England terrain. The second season is in the fall, when rare but heavy rainstorms, some of hurricane origin, bring exceptional high stages.

The regulation of floods of large rivers is described in Chapter 17 in connection with landforms of floodplains.

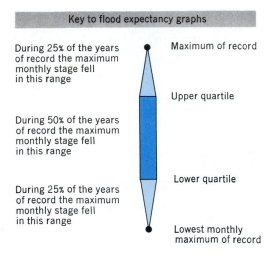

FIGURE 16.16 (Above and right) The highest stage that occurred in each month is given in terms of percentages on these graphs of four rivers. (National Weather Service.)

FLASH FLOODS AND THEIR PREDICTION

Known as *flash floods*, sudden and short-lived overbank discharges differ in important respects from the prolonged floods of large streams we have thus far described. Flash floods quickly follow torrential thunderstorm rains that deliver on the order of 20 to 30 cm (8 to 12 in.) in a 24–hour period over a small hilly or mountainous watershed within which ground slopes and stream channel gradients are steep. Three weather situations can be recognized as producing flash floods. One is a hurricane that has penetrated the continental interior; Camille, Agnes, and Amelia produced destructive flash floods. Another, typical of the central and eastern United States, is a slow-moving or stagnant weather front along which large amounts of moist tropical air are being imported. A third, found in the mountainous western states, is orographic in nature, typified by flash floods developed over California mountain ranges in the winter months.

Because of the small size of the affected watershed, the lag time of a flash flood is extremely short—perhaps only an hour or two—giving little or no time to warn inhabitants of the mainstream environs. The flood arrives as a swiftly moving wall of turbulent water, sweeping away buildings, vehicles, and retaining walls in its path. In heavily forested watersheds, as in the Appalachians, the water carries little bedload and is highly erosive. In arid western watersheds, great quantities of coarse rock debris are swept into the main channel and travel with the floodwater. An example is the Big Thompson River flash flood of July 1976, occurring in a watershed of the Colorado Front Range. (Channel erosion from this flood is pictured in Figure 17.8.) In Chapter 18, we will refer to flash floods in the desert (see Figure 18.8).

An additional hazard of the flash flood is the breaching of an earthfill dam across the main channel, suddenly releasing an enormous quantity of stored water. Just such an occurrence accompanied the great Johnstown Flood of May 1889, in the Conemaugh Valley of the Allegheny Plateau, in western Pennsylvania; it drowned more than 2100 persons out of a town population of 30,000. A similar flash flood in 1977 hit the same valley, taking 76 lives—this in spite of extensive flood control measures that had been implemented between the two floods.

The National Weather Service operates flash flood warning systems at some 55 locations throughout the United States. Included in the instrumentation of each key observing point is an electronic river level sensor which immediately signals to an alarm station that a critical water level for flash flooding has been reached. The alarm station releases various forms of warning devices, such as sirens and lights. Quick action is of the essence, because many flash floods are generated during night hours and lag time is so very brief. In recent decades, about 200 lives have been lost annually in floods, and flash floods accounted for 80% of those deaths.

HYDROLOGIC EFFECTS OF URBANIZATION

Hydrology of a watershed is altered in two ways by urbanization. First, an increasing percentage of the surface is rendered impervious to infiltration by the construction of roofs, driveways, walks, pavements, and parking lots. It has been estimated that in residential areas, for a lot size of 1400 sq m (15,000 sq ft), the impervious area amounts to about 25 percent; for a lot size of 560 sq m (6000 sq ft), the impervious area amounts to about 80 percent.

An increase in proportion of impervious surface reduces infiltration and increases overland flow from the urbanized area. An important result is to increase the frequency and height of flood peaks during heavy storms. There is also a reduction of recharge to the ground-water body beneath; this reduction, in turn, decreases the base-flow contribution to channels in the area. Thus the full range of stream discharges, from low stages in dry periods to flood stages, is made greater by urbanization.

A second change caused by urbanization is the introduction of storm sewers that allow storm runoff from

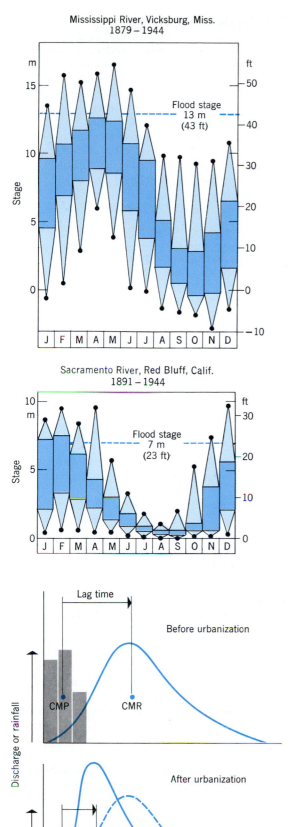

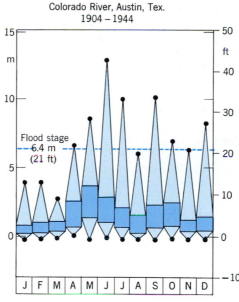

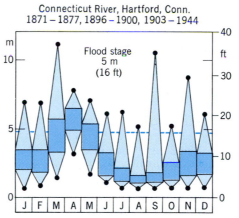

paved areas to be taken directly to stream channels for discharge. Runoff travel time to channels is shortened at the same time that the proportion of runoff is increased by expansion in impervious surfaces. The two changes together conspire to reduce the lag time, as shown by the schematic hydrographs in Figure 16.17.

Many rapidly expanding suburban communities are soon finding that certain low-lying residential areas, formerly free of flooding, are being subjected to inundation by overbank flooding of a nearby stream. The need for careful terrain study and land-use planning is obvious in such cases to protect the unwary home buyer from locating in a flood-prone neighborhood.

One partial solution to the problem created by storm sewering is to return water runoff to the ground-water body by means of infiltrating basins. This program has

FIGURE 16.17 (left) Schematic hydrographs showing the effect of urbanization on lag time and peak discharge. Points CMP and CMR are centers of mass of rainfall and runoff, respectively, as in Figure 16.10. (After L. B. Leopold, U.S. Geological Survey, Circular 554.)

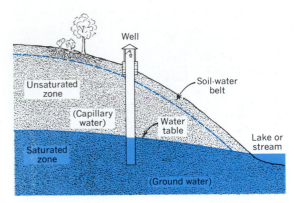

FIGURE 16.18 Zones of subsurface water.

The true position of the water table is shown by the level of standing water in a well drilled or dug to some depth below the water table. Where wells are numerous in an area, the position of the water table can be mapped in detail by plotting the water heights and noting the trends in elevation from one well to the other. The water table is highest under the highest areas of surface—hilltops and divides—but descends toward the valleys, where it appears at the surface close to streams, lakes, or marshes (Figure 16.19). The reason for such a configuration of the water table is that water percolating down through the unsaturated zone tends to raise the water table, whereas seepage into streams, swamps, and lakes tends to draw off ground water and lower its level.

been adopted on Long Island, in New York, where infiltration rates are high in sandy glacial materials. Another method of disposal of storm runoff is by recharge wells. At Orlando, Florida, storm runoff enters wells that penetrate cavernous limestone. The capacity of the system to absorb runoff without clogging appears to be adequate. In Fresno, California, a large number of gravel-packed wells of 76-cm (30-in.) diameter receive runoff from streets. The system has proved successful in disposing of storm drainage.

GROUND WATER

Ground water is that part of the subsurface water that fully saturates the pore spaces of the rock mass or soil. The ground water occupies the *saturated zone* (Figure 16.18). Above it is the *unsaturated zone,* in which water does not fully saturate the pores. The upper surface of the saturated zone is the *water table.* Water is held in the unsaturated zone by capillary force in tiny films adhering to the mineral surfaces. The unsaturated zone may be absent or very shallow where the saturated zone is at or close to the surface in low, flat regions.

At the base of the unsaturated zone is the *capillary fringe,* a thin layer in which the water has been drawn upward from the ground-water table through capillary force. The action is much like the rise of kerosene in a lamp wick or of water in a blotter. Water in the capillary fringe largely fills the soil pores and is thus continuous with the ground-water body. Thickness of the capillary fringe depends on the soil texture because capillary rise is higher when the openings are smaller. In a silty material, the capillary fringe may be 1 m (3 ft) thick, but as little as 1 cm thick in coarse sand or fine gravel with large pore spaces. (See Table 12.2.)

GROUND-WATER MOVEMENT

Because ground water moves extremely slowly, a difference in water-table level, or *hydraulic head,* is build up and maintained between areas of high elevation and those of low elevation. In periods of water surplus accompanying abnormally high precipitation, this head is increased by a rise in the water-table elevation under divide areas; in periods of water deficit, occasioned by drought, the water table falls (Figure 16.19).

In humid climates having a strong seasonal cycle of precipitation, and in climates in which soil water is frozen for several months of the year, a seasonal cycle of rise and fall of the water table is produced. This cycle is illustrated by a graph of water-table fluctuations in an observation well on Cape Cod, Massachusetts (Figure 16.20). Percolating water reaches the water table in abundant quantities from late winter to late spring, a phenomenon known as *ground-water recharge.* This recharge period, which reflects the period of water surplus in the soil–water budget, causes a seasonal rise in the water table. Recharge declines to very small amounts from midsummer to early winter, and during this time the water table steadily declines as water moves to lower levels under the force of gravity. The effect of a major drought shows in the graph by a downward trend of the average elevation in 1965 and 1966.

The subsurface phase of the hydrologic cycle is completed when the ground water emerges in places where the water table intersects the ground surface. Such places are the channels of streams and the floors of marshes and lakes. By slow seepage and spring flow, the water enters fast enough to balance the rate at which water enters the ground-water table by percolation from above.

Figure 16.21 shows paths of flow of ground water. Flow takes paths curved concavely upward. Water enter-

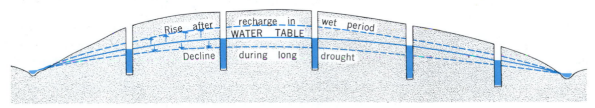

FIGURE 16.19 The water table conforms roughly to surface topography.

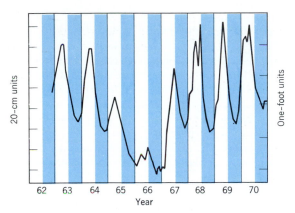

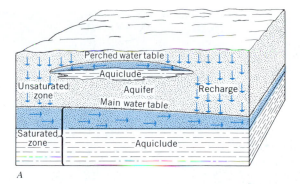

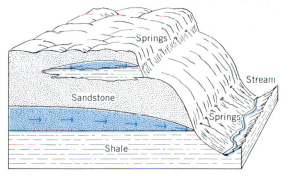

FIGURE 16.20 Hydrograph of an observation well on Cape Cod, Massachusetts, showing the characteristic annual cycle of rise and fall of the water table. (Data from U.S. Geological Survey.)

FIGURE 16.22 Ground water in horizontal strata. (*A*) An unconfined aquifer with an overlying aquiclude and a perched water table. (*B*) Springs controlled by the upper surfaces of aquicludes. (From A. N. Strahler, *Physical Geology*, Harper & Row, Publishers. Copyright © by Arthur N. Strahler.)

ing the hillside midway between divide and stream flows rather directly. Close to the divide point on the water table, however, the flow lines go almost straight down to great depths, from which they recurve upward to points under the streams. Progress along these deep paths is incredibly slow; that near the surface is much faster. The most rapid flow is close to the place of discharge in the stream, where the arrows are shown to converge.

GROUND WATER IN STRATA

Sedimentary strata often exert a strong control over the storage and movement of ground water. Clean, well-sorted sand—such as that found in beaches, dunes, or stream alluvium—can hold ground water in an amount equal to about one-third its bulk volume. The *porosity* of this material is high. Ground water moves freely through such coarse sediments, whose *permeability* is high. Coarse sandstone strata may retain relatively high porosity and permeability, becoming *aquifers* that both store ground water in substantial quantities and permit it to move freely within the strata. Other stratified permeable materials that form good aquifers are volcanic tephra and some kinds of frothy (scoriaceous) lavas. Limestone that has been rendered cavernous by the action of carbonic acid may also form an excellent aquifer. In contrast, beds of clay and shale have low permeability and resist the

movement of ground water; they are known as *aquicludes*. Dense lava flows and sills may also serve as aquicludes.

Where strata lie nearly horizontal or are gently dipping, the ground-water flow paths may be quite different from the simple patterns shown in Figure 16.21. Suppose for example, that the region is one of sedimentary strata with beds of sandstone alternating with beds of shale (Figure 16.22*A*). The main water table is shown to lie within the sandstone aquifer. The upper portion of the saturated zone occupies the lower part of the aquifer. The shale aquiclude is also in the saturated zone, but because of low permeability the water it holds is, for all practical purposes, stagnant. Ground water moves rather rapidly through the aquifer. The direction of this motion must necessarily be almost horizontal and is in the direction of downward slope (gradient) of the water table.

Figure 16.22 also shows a lenslike body of shale within the upper part of the sandstone formation. Acting as an aquiclude, this lens blocks the downward percolation of water through the unsaturated zone. As a result, a lenslike body of ground water collects above the shale lens. The upper surface of this ground-water lens is known as a *perched water table*. Ground water in this lens moves toward the edges of the lens, from which it trickles down through the lower part of the aquifer and eventually reaches the main water table.

The aquifer shown in Figure 16.22 is described as an *unconfined aquifer*, because the water table is free to receive recharge from above and can rise or fall freely

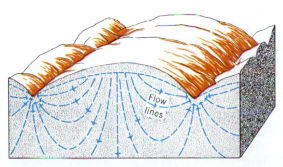

FIGURE 16.21 Theoretical paths of ground-water movement under divides and valleys. (Data from M. K. Hubbert. Drawn by A. N. Strahler.)

FIGURE 16.23 Thousand Springs, Idaho, emerges from the north side of the Snake River Canyon, opposite the mouth of the Salmon River. The spring issues from a highly permeable layer of scoriaceous basalt. A cliff of columnar basalt can be seen high above the falls. (Jack Williams.)

within the aquifer in response to changes in the amount of recharge received.

Diagram B of Figure 16.22 shows how aquicludes can influence the way in which ground water emerges at the surface. Using the same geological structure as in Diagram A, we have shown a deep river valley cut through the strata. The shale lens is now exposed in the upper wall of the valley. Ground water emerges from the perched water table in a line of springs. A *spring* is any flow of ground water emerging naturally from the solid surface of the earth. (Springs may also emerge from the floors of streams, lakes, or the ocean.) A second line of springs is formed in the lower wall of the valley, where the main water table intersects the surface. Springs of emerging ground water occur under a wide variety of geological conditions and we have shown only one simple case.

Springs are usually insignificant features, going unnoticed because of a concealing cover of vegetation. Many small springs cease to flow in the summer season. Some large springs yield copious, sustained flows of water. An impressive example is the Thousand Springs of the Snake River Canyon in Idaho, where a large volume of flow emerges from highly permeable layers of scoriaceous basalt (Figure 16.23).

Where strata are gently inclined, conditions may be favorable for the development of *artesian flow*, the spontaneous upward movement of ground water in a well or spring. Figure 16.24 shows a dipping *confined aquifer* of

sandstone, capped by an aquiclude of shale. The aquifer receives water along its exposed outcrop at a relatively high position and moves to lower levels in the aquifer. Because of hydrostatic pressure, water can rise in wells to a level higher than the upper surface of the aquifer. The limit of rise is the *hydrostatic pressure surface*, shown by a sloping line in Figure 16.24. The elevation of this surface declines with increasing distance from the intake area because energy is expended in the flow of ground water through the aquifer. As the diagram shows, a flowing *artesian well* occurs where the ground surface and wellhead lie below the pressure surface. Where the wellhead lies above the pressure surface, the water will rise only to the level of the pressure surface; it will not emerge as a flowing well. Natural artesian springs can occur where faulting or solution removal has produced a passageway for water to make its way upward through the confining aquiclude.

The word "artesian" is taken from a French word *artésien*, meaning "pertaining to the province of Artois." In about 1750 in this northernmost province of France, deep wells were first drilled to reach confined aquifers and obtain artesian flow. Since then, many regions of sedimentary strata have furnished large amounts of artesian water. One of the most extensive areas producing artesian ground water is the Great Plains region, covering parts of Kansas, Nebraska, and eastern Colorado and extending northward into the region of the Black Hills of South Dakota. Here the Dakota sandstone under-

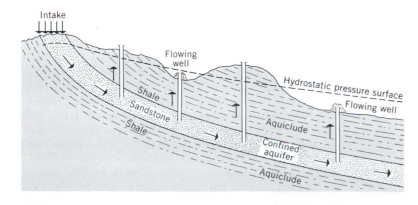

FIGURE 16.24 Schematic cross section of an artesian flow system. (From A. N. Strahler, *Physical Geology*, Harper & Row, Publishers. Copyright © by Arthur N. Strahler.)

lies the entire area and has furnished artesian water in many places (see Figure 16.27). Numerous irrigation wells were drilled into the Dakota formation early in this century. Artesian water was withdrawn at a rate that greatly exceeded the rate of recharge. Consequently, few of these wells produce surface flow today and most must be pumped to furnish water.

GROUND-WATER WITHDRAWAL

Withdrawal of ground water for a wide range of human uses has begun to make serious impacts on the environment in many places. The drilling of vast numbers of wells, from which water is forced out in great volumes by powerful pumps, has profoundly altered nature's balance of ground-water recharge and discharge. Increased urban populations and industrial developments require larger water supplies, needs that cannot always be met from construction of new surface-water reservoirs.

In agricultural lands of the semiarid and desert climates, heavy dependence is placed on irrigation water from pumped wells, especially because many of the major river systems have already been fully developed for irrigation from surface supplies. Wells can be drilled within the limits of a given agricultural or industrial property and can provide immediate supplies of water without any need to construct expensive canals or aqueducts.

Formerly, the small well needed to supply domestic and livestock needs of a home or farmstead was actually dug by hand as a large cylindrical hole, lined with masonry where required. By contrast, the modern well put down to supply irrigation and industrial water is drilled by powerful machinery that may bore a hole 40 cm (16 in.) or more in diameter to depths of 300 m (1000 ft) or more. Drilled wells, often called *tube wells*, are sealed off by metal casings that exclude impure near-surface water and prevent clogging of the tube by caving the walls. Near the lower end of the hole, where it enters the aqui-

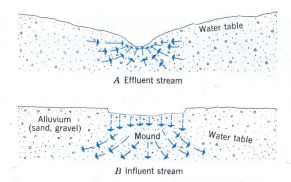

FIGURE 16.26 Effluent and influent streams. (From A. N. Strahler, *The Earth Sciences*, 2nd ed., Harper & Row, Publishers. Copyright © by Arthur N. Strahler.)

fer, the casing is perforated to admit the water. The yields of single wells range from as low as a few liters per day in a domestic well to many millions of liters per day for large, deep industrial or irrigation wells.

As water is pumped from a well, the level of water in the well drops. At the same time, the surrounding water table is lowered in the shape of a conical surface, termed the *cone of depression*. The difference in height between the cone base and the original water table is the *drawdown* (Figure 16.25). By producing a steeper gradient of the water table, the flow of ground water toward the well is also increased, so the well will yield more water. This increase holds only for a limited amount of drawdown, beyond which the yield fails to increase. The cone of depression may extend as far out as 16 km (10 mi) or more from a well where heavy pumping is continued. Where many wells are in operation, their intersecting cones produce a general lowering of the water table.

In areas of moist climates with a large annual water surplus, natural recharge is by general percolation over the entire ground area surrounding the well. Here the prospects of achieving a balance of recharge and withdrawal are highly favorable through the control of pumping. An important recycling measure is the return of waste waters or stream waters to the ground-water table by means of *recharge wells*, in which water flows down rather than up.

Depletion often greatly exceeds the rate at which the ground water of the area is recharged by percolation from rain or from the beds of streams. In an arid region, much of the ground water for irrigation is from wells driven into thick alluvial sands and gravels. Recharge of these deposits depends on the seasonal flows of water from streams originating high in adjacent mountain ranges.

Streams in a dry climate, flowing on plains underlain by sand and gravel, lose water by seepage through the channel floor. This water recharges the ground-water body below and causes the water table to be raised in the form of a mound (Figure 16.26A). Streams of this type are called *influent streams*. In contrast, in moist climates where the water table is high and slopes toward the channels, streams receive flow by out-seepage of ground water; these are *effluent streams* (Figure 16.26B).

In dry climates, particularly, the extraction of ground

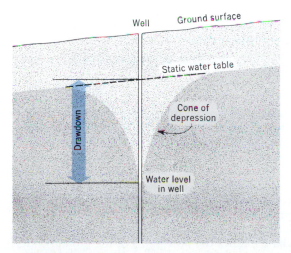

FIGURE 16.25 Drawdown and cone of depression in a pumped well.

FIGURE 16.27 The perfect circles of green laid out in orderly patterns are thriving plots of food and forage crops—corn, wheat, beans, alfalfa, or grass—irrigated by walking sprinkler systems. Called centerpivot irrigation, the system consists of a long pipe mounted on wheeled supports. The pipe is held fast to a center point where the water is injected and creeps slowly round and round in a huge circle emitting water from sprinkler heads along the pipe. Water is supplied from deep wells close by. (Courtesy of Valmont Industries, Incorporated, Valley, Nebraska.)

water by pumping can greatly exceed the recharge by stream flow. Cones of depression deepen and widen; deeper wells and more powerful pumps are then required. Overdrafts of water accumulate, and the result is exhaustion of a natural resource not renewable except by long lapses of time. Figure 16.27 shows an area in the High Plains of the central United States, east of the Rockies. Pumped wells distribute ground water through sprinkler systems that continuously revolve around a center point.

ENVIRONMENTAL PROBLEMS RELATED TO GROUND-WATER WITHDRAWAL

POLLUTION OF GROUND WATER

Disposal of solid wastes poses a major environmental problem in heavily populated areas of North America because the advanced industrial economy produces an endless source of liquid waste (sewage) and solid waste (garbage and trash). Traditionally, the solid wastes were trucked to the municipal dump and burned there in continually smoldering fires that emitted foul smoke and gases. The partially consumed residual waste was later buried under a cover of compacted earth.

In recent decades, there has been a major improvement in solid-waste disposal methods. One method is high-temperature incineration. Another is the *sanitary landfill* method in which waste is not allowed to burn. Instead, the waste is continually covered by a protective layer of sand or clay available at the landfill site. The waste is thus buried in the unsaturated zone. Here it is subject to reaction with percolating rainwater that has infiltrated the ground surface. This water picks up a variety of ions from the waste body and carries these down as *leachate* to the water table. Once in the water table, the leachate follows the flow paths of the ground water.

As shown in Figure 16.28, a mounding of the water

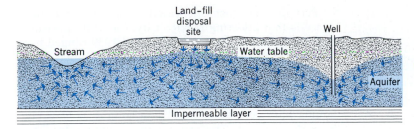

FIGURE 16.28 Leachate from a waste disposal site moves toward a supply well (right) and a stream (left). (From A. N. Strahler, *Planet Earth: Its Physical Systems Through Geologic Time*, Harper & Row, Publishers. Copyright © by Arthur N. Strahler.)

table develops beneath the disposal site. Loose soil of the disposal area facilitates infiltration of precipitation, while lack of vegetation reduces the evapotranspiration. Consequently, the recharge here is greater than elsewhere, and the mound is maintained. After leachate has moved vertically down by gravity percolation to the water table mound, it moves radially outward from the mound to surrounding lower points on the water table. As shown in Figure 16.28, a supply well with its cone of depression draws ground water from the surrounding area. Linkage between outward flow from the waste disposal site and inward flow to the well can bring leachate into the well, polluting the ground water supply. Pollution of supply wells by partially treated effluent infiltrating the ground at sewage disposal plants can occur in a basically similar manner.

An important step in guarding against this form of pollution is to place a monitor well (or several monitor wells) on a line between the disposal site and the well. Chemical tests for presence of leachate are made regularly, while the slope of the water table can also be determined. Movement of leachate toward the supply well may be blocked by placement of a recharge well (or wells), building a freshwater accumulation (actually an inverted cone) that will oppose the movement of the leachate.

It is not necessary for a ground-water mound to be present for pollutants to travel to distant points. Where the water table has a pronounced slope, as it generally does everywhere except near the summit of a broad ground-water divide, leachate or any pollutant introduced at a given point migrates as a pollution plume along the flow paths of the ground water.

Another potential source of pollution of ground-water supplies is from highways and streets, through spillage of chemicals and from deicing salt applied during the winter months. Spillage of large volumes of liquids from tank trucks and tank cars as a result of highway and railroad accidents poses a serious threat because a large slug of pollutant can be injected into the ground-water recharge system. The commonest serious pollutants are automobile fuel and heating oil, but many toxic industrial chemicals are also transported in tank trucks and cars. Leakage of fuels from underground storage tanks, used in all gasoline stations, is a related source of possible pollution.

SALTWATER INTRUSION

A serious consequence of sustained, heavy ground-water withdrawal in coastal zones is that wells near the shore eventually draw salt ground water and must be abandoned. To understand how this happens, we must examine the relationship between salt and fresh ground water.

Figure 16.29 shows an idealized diagram of the relationship in an island or a narrow peninsula. The body of fresh ground water takes the shape of a gigantic lens with convex faces, except that the upper surface has a broad curvature whereas the lower surface, in contact with the salt ground water, bulges deeply downward. Because fresh water is less dense than salt, we can think of this freshwater lens as floating on salt water, pushing it down much as the hull of an ocean liner pushes aside the surrounding water. The ratio of densities of fresh water to salt water is 40 to 41. Hence, if the water table is, say, 10 m above sea level, the bottom of the freshwater lens will be located 400 m below sea level, or 40 times as deep as the water table is high with respect to sea level. The fresh ground water extends seaward some distance beyond the shoreline. Although the salt ground water is stagnant, the fresh water travels in the curved paths shown by arrows in Figure 16.29.

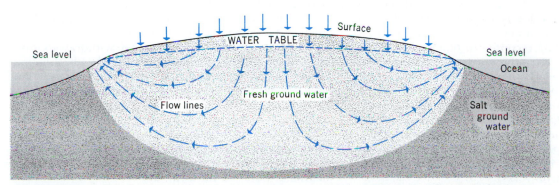

FIGURE 16.29 Freshwater and saltwater relationships in an island or peninsula. (After G. Parker.)

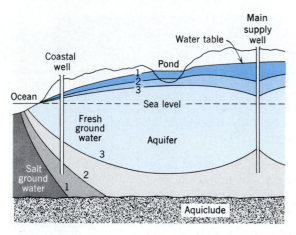

FIGURE 16.30 Schematic cross section showing lowering of water table and pond levels, and intrusion of salt ground water, as a result of ground-water withdrawal. (Vertical scale is greatly exaggerated above sea level.) (From A. N. Strahler and A. H. Strahler, *Environmental Geoscience,* Hamilton Publ. Co. Copyright © by Arthur N. Strahler.)

In coastal regions, fresh ground-water supplies are vulnerable to contamination by *saltwater intrusion*. As shown in Figure 16.30, pumping of a well has not only caused the water table to decline, but has also caused the interface between salt ground water and fresh ground water to move inland as an invading wedge. Ultimately, the salt water is drawn into the well and the water is no longer fit for consumption. Notice that in Figure 16.30 a small coastal well also suffered salt contamination. In the case of an island or narrow peninsula, where salt ground water lies beneath a freshwater lens, salt water is easily drawn directly upward to contaminate a well. If pumping is stopped after salt contamination has occurred the salt water will be gradually pushed back to its original limits. To hasten this process and to prevent further salt intrusion, fresh water can be pumped down into recharge wells located between the contaminated well and the shore.

LAND SUBSIDENCE FROM GROUND-WATER WITHDRAWAL

Another serious environmental effect of excessive ground-water withdrawal is that of subsidence of the ground surface. Several localities have been affected in California, where ground water has been pumped from basins filled with alluvial (stream and lake-deposited) sediments. Water-table levels in these basins dropped over 30 m (100 ft), with a maximum drop of 120 to 150 m (400 to 500 ft) being recorded in one locality in the San Joaquin Valley. Here the maximum ground subsidence was about 3 m (10 ft) in a 35-year period, and caused damage to wells in the area.

Another important area of ground subsidence accompanying water withdrawal is beneath Houston, Texas, where the ground surface subsided from 0.3 to 1.0 m (1 to 3 ft) in a metropolitan area 50 km (30 mi) across. Damage resulted to buildings, pavements, airport runways, and flood-control works.

Perhaps the most celebrated case of ground subsidence is that affecting Mexico City. Carefully measured ground subsidence between 1891 and 1959 ranged from 4 to 7 m (13 to 23 ft). Subsidence began at a much earlier date as a result of withdrawal of ground water from an aquifer system beneath the city and has caused many serious engineering problems. Clay beds overlying the aquifer have contracted greatly in volume as water has been drained out. The combat the ground subsidence, recharge wells were drilled to inject water into the aquifer. In addition, new water supplies from sources outside the city area were developed to replace local ground-water use.

Venice, Italy, has suffered severe damage to ancient buildings as a result of flooding during times of storms on the adjacent Adriatic Sea. Flooding results because the land on which Venice was built has been gradually subsiding. Venice was built in the Eleventh Century A.D. on low-lying islands in a coastal lagoon, sheltered from the ocean by a barrier beach. The area is underlain by some 1000 m (3000 ft) of layers of sand, gravel, clay, and silt, with some layers of peat. Compaction of the soft layers has been going on gradually for centuries under the heavy load of city buildings. However, ground-water withdrawal, greatly accelerated in recent decades, has aggravated the condition.

By 1980, the amount of sinking had amounted to about 20 cm (8 in.). While this last figure does not seem large, the total effect (that includes a rise in the Adriatic sea level) was to greatly increase the frequency of flooding during winter storms. Sea level is normally raised by the effects of coastal storms, and high tides occurring at the same time may add to the rise in water level. The problem of flooding during storms is aggravated by the fact that the canals of Venice receive raw sewage, so that the flood water is contaminated.

Most of the subsidence in recent decades is attributed to withdrawals of large amounts of ground water from industrial wells at Porto Marghera, the modern port of Venice, located a few kilometers distant on the mainland shore. Although this pumping was greatly curtailed, reducing the rate of subsidence to a very small natural rate (about 1 mm per year), the threat of flooding with its attendant damage to churches and other buildings of great historical value remained undiminished. Hopes of flood control have been pinned to the construction of seawalls and floodgates on the barrier beach that lies between Venice and the open ocean.

By the early 1980s, the port city of Bangkok, Thailand, had become the world's most rapidly sinking city as a result of massive ground-water withdrawals from soft marine sediments beneath the city. By 1980, the rate of subsidence had reached 14 cm (6 in.) per year and major floods were frequent. Some reduction in the rate of ground-water withdrawal during the 1980s achieved a modest decrease in the rate of subsidence.

Land subsidence also occurs when large volumes of petroleum are withdrawn from oil pools beneath the surface. The principle is essentially the same as in the case of ground-water withdrawal.

LAKES AND PONDS

Lakes are integral parts of drainage systems and participate in runoff of water in the hydrologic cycle. Lakes are of major environmental importance to humans in many ways. They represent large bodies of fresh water in storage; they support ecosystems that provide food for humans. Today, the recreational value of lakes is assuming increasing importance.

Where lakes are not naturally present in the valley bottoms of drainage systems, they can be created as needed by placing dams across the stream channels. Many regions that formerly had almost no natural lakes are now abundantly supplied. As you travel by airplane across such a region the glint of sunlight from hundreds of artificial lakes will catch your eye. Some are small ponds made to serve ranches and farms; others cover hundreds of square kilometers. Obviously, an abundance of new lakes represents a major environmental modification and has far-reaching consequences.

The term *lake* includes a very wide range of kinds of water bodies. Lakes have in common only the requirement that they have an upper water surface exposed to the atmosphere and no appreciable gradient with respect to a level surface of reference. Ponds (small, usually shallow water bodies), marshes, and swamps with standing water can be included.

Lake water may be fresh or saline, and we may have some difficulty in deciding whether a body of salt water adjacent to the open ocean is to be classed as a lake or an extension of the sea. A practical criterion rules that a coastal water body is not a lake if it is subject to influx of salt water from the ocean. Lake surfaces may, however, lie below sea level; an example is the Dead Sea with surface elevation of −396 m (−1300 ft). The largest of all lakes, the Caspian Sea, has a surface elevation of −25 m (−80 ft). Significantly, both of these large below-sea-level lakes are saline.

Basins occupied by lakes show a wide range of origins as well as a vast range in dimensions. Basins are created by geologic processes and it should not be surprising that there are lakes produced by every category of geologic process.

An important point about lakes in general is that they are for the most part short-lived features in terms of geologic time. Lakes disappear from the scene by one of two processes, or a combination of both. First, lakes that have

FIGURE 16.32 A freshwater pond on Cape Cod, Massachusetts. Hygrophytic plants occupy a zone close to the water's edge. Pine forest in the background is on higher, well-drained ground. (A. N. Strahler.)

stream channel outlets will be gradually drained as the outlet channels are eroded to lower levels. Where a strong bedrock threshold underlies the outlet, erosion will be slow but nevertheless certain. Second, lakes accumulate inorganic sediment carried by streams entering the lake and organic matter produced by plants within the lake.

Lakes also disappear by excessive evaporation accompanying climatic changes. Many former lakes of the southwestern United States flourished in moister periods of glacial advance during the Pleistocene Epoch, but today are greatly shrunken or have disappeared entirely under the present arid regime (see Chapter 22). A special case of lake disappearance is that of the lowering of the water table by excessive withdrawal of ground water.

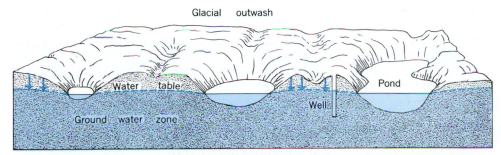

FIGURE 16.31 Water table ponds in glacial deposits, largely sand, on Cape Cod, Massachusetts. (From A. N. Strahler, *A Geologist's View of Cape Cod*, Copyright © by Arthur N. Strahler. Reproduced by permission of Doubleday and Co.)

In moist climates, water level of lakes and ponds coincides closely with the water table in the surrounding area. Seepage of ground water, as well as direct runoff of precipitation, maintains these free-water surfaces permanently throughout the year. Examples of such freshwater ponds are found widely distributed in North America and Europe, where plains of glacial sand and gravel contain natural pits and hollows left by the melting of stagnant ice masses (see Chapter 22). Figure 16.31 is a block diagram showing small freshwater ponds on Cape Cod. The surface elevation of these ponds coincides closely with the level of the surrounding water table.

Many former freshwater water-table ponds have become partially or entirely filled by the organic matter from growth and decay of water-loving plants (Figure 16.32). The result is a bog with a surface close to the water table. (Bog succession is described in Chapter 26; see Figure 26.2).

Freshwater marshes and swamps, in which water stands at or close to the ground surface over broad areas, represent the appearance of the water table at the surface. Such areas of poor surface drainage have a variety of origins. For example, the broad, shallow freshwater swamps of the Atlantic and Gulf coastal plain represent regions only recently emerged from the sea. Other marshes are created by the shifting of river channels on floodplains.

WATER BALANCE OF LAKES AND RESERVOIRS

In this chapter our theme is a hydrologic one, related to the flow of water in the hydrologic cycle. Consequently, it is appropriate here to look into the water balance of lakes. A lake represents a simple open material-flow system with respect to the mass balance of the water itself. A lake has an upper surface boundary exposed to the atmosphere and a lower surface boundary in contact with a solid mineral surface. Water may enter and leave through both of these boundary surfaces. Incoming streams and overland flow from ground surfaces draining into the lake represent point and line sources of water input, while an overflow channel represents an output point. Consequently, we can set up a fairly simple water-balance equation (Figure 16.33). Let the letter I stand for input of water; O, for output of water. The units used may be volume or mass of water per unit time. Units of volume per unit of time are identical with discharge, Q,

in stream flow and are the more useful choice, since input and output from stream channels will be gauged in terms of discharge. By means of subscript letters we can differentiate the sources of input and output as follows: r represents runoff by stream flow and overland flow; g represents subsurface flow as ground-water movement; p represents direct fall of precipitation upon the lake surface; e represents evaporation from the free-water surface. Finally, let G stand for the net change in storage of water in the lake. The equation can now be written as

$$G = (I_r + I_p + I_g) - (O_r + O_e + O_g)$$
$$\text{input} \qquad\qquad \text{output}$$

When the system is in steady state, there will be no change in storage, so that the term G becomes zero and drops out. Then

$$I_r + I_p + I_g = O_r + O_e + O_g$$
$$\text{input} \qquad\qquad \text{output}$$

Actual measurement of each term for a given lake is a most difficult technical problem. Stream gauges at input and output points can give reasonably reliable data on the major elements of surface flow in the system but the increment from direct overland flow is not subject to direct measurement. Ground-water flow is next to impossible to measure directly and must be estimated from a knowledge of local gradients of the water table and permeability of the aquifer. Precipitation can be directly measured with a rain gauge. Evaporation from the free-water surface is not easy to measure, since it depends on water temperature, atmospheric temperature and humidity, barometric pressure, and wind speed. Formulas taking into account these variables are available for estimating evaporation from lakes.

Direct measurement of evaporation from a free-water surface makes use of the *evaporating pan*, which is simply a circular container 1.2 to 1.8 m (4 to 6 ft) in diameter and 25 cm (10 in.) or more in depth. Water is added as required and the amount of surface lowering due to evaporation measured with a gauge. Evaporation data are collected by the National Weather Service at many observing stations. The evaporation from the pan is generally greater than from a lake or reservoir, so that the pan readings require correction by a reduction factor ranging from 60 to 80 percent.

Figure 16.34 is a map of the United States showing the average annual evaporation from a free-water surface as measured by the pan method. Notice the high values in the hot desert regions of the southwestern United States in contrast to low values in the cool northeastern region. Listed below are some representative figures giving annual total evaporation for actual reservoir surfaces:

	Cm	Inches
Gardiner, Maine	61	24
Birmingham, Alabama	109	43
El Paso, Texas	180	71
Tucson, Arizona	152	60
San Juan, Puerto Rico	140	55
Gatun, Canal Zone	122	48
Atbara, Sudan	315	124

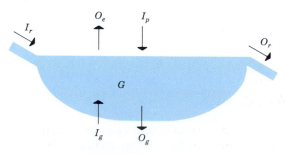

FIGURE 16.33 Schematic diagram of the water balance of a lake. (Terms are defined in text.)

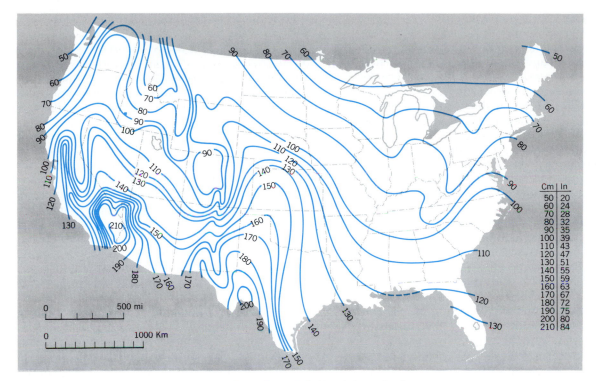

FIGURE 16.34. Annual evaporation from the surface of shallow lakes. The figures give water depth in centimeters. (Data of NOAA, National Weather Service.)

SALINE LAKES AND SALT FLATS

Lakes with no surface outlet are characteristic of arid regions. Here, on the average year after year, the rate of water loss by evaporation balances the rate of stream inflow. If the rate of inflow should increase, the lake level will rise. At the same time the lake surface will increase in area, allowing a greater rate of evaporation. A new balance can then be achieved. Since dissolved solids are brought into the lake by streams—usually ones that head in distant highlands where a water surplus exists—and there is no surface outlet, the solids accumulate with resultant increase in salinity of the water. Salinity, or degree of "saltiness," refers to the abundance of certain common ions in the water. Eventually, salinity levels reach a point where salts are precipitated in the solid state.

Evaporation control is a subject of major importance in conserving the water supplies in the reservoir behind a dam, particularly where the reservoir is situated in a region of arid climate. This situation occurs where an exotic river is dammed. (An *exotic river* is one that is sustained in its flow across an arid region through runoff derived from a distant region of water surplus.) The Colo-rado River in Arizona is a good example. A large reservoir, such as Lake Mead behind Hoover Dam, presents an enormous water surface exposed to intense evaporation.

Because the input of stream water is finite, a reservoir may be designed with such a large capacity that it will never completely fill in an arid climate, because there is a point at which annual evaporation equals the annual input.

Poorly drained, shallow basins (playas) accumulate highly soluble salts. These form sterile, white salt flats (see Figures 12.15 and 18.14). On rare occasions these flats are covered by a shallow layer of water, brought by flooding streams heading in adjacent highlands. A number of desert salts are of economic value and have been profitably extracted. An example well known to most persons is borax (sodium borate), widely used as a water-softening agent. In shallow coastal estuaries in the desert climate, sea salt is commercially harvested by allowing it to evaporate in shallow basins. One well-known salt source of this kind is the Rann of Kutch, a coastal lowland in the tropical desert of westernmost India, close to Pakistan. Here the evaporation of shallow water of the Arabian Sea has long provided a major source of salt, the only product of value in the region.

THE ARAL SEA—A DYING SALINE LAKE

Here is an example of how a change in rate of input of water results in a change in storage of a large saline lake in a region of dry climate. The Aral Sea, located to the east of the Caspian Sea in the southcentral Soviet Union, was once the world's fourth largest lake. Its area, depth, and volume were essentially stable in the late 1800s, continuing until about 1960. Its fishing industry supplied about one-tenth of the total catch of the Soviet Union.

The Aral is sustained by runoff from two rivers—Syr Dar'ya and Amy Dar'ya—both nourished by meltwaters in high mountain watersheds to the southeast. The lake has no surface outlet, and in this semidesert climate, with its large annual water deficit, requires that the output be entirely from surface evaporation.

Shrinkage of the Aral Sea by desiccation (net water loss) set in during the 1970s as a result of two causes: one natural, the other human-induced. A series of dry years reduced the river inflow, but mainly the reduction in river input resulted from greatly increased water withdrawal for irrigation. Most of the water diversion has been through the Karakum Canal, the largest and longest in the Soviet Union. Water diversions through this canal increased by a factor of 14 between 1956 and 1986, as a government project for greatly increased cotton production came into operation.

As water diversions increased, the surface level of the Aral was sharply lowered and its area reduced. In the period 1960 to 1987, lake volume had decreased by 66% and salinity of the lake water had increased from 10% to 27%. Thus the shoreline receded rapidly, exposing the former lake bed, which became encrusted with salts. The flourishing fishing port of Muynak became a ghost town, now 50 km (30 mi) from the new shoreline. Strong winds now blow salt particles and mineral dusts in great clouds southwestward over the irrigated cotton fields and westward over grazing pastures. These salts—particularly the sodium chloride and sodium sulfate components—are highly toxic to plants.

Largely because of increased water salinity, 20 out of 24 native fish species disappeared from the Aral, and the commercial catch has been reduced to zero. Ecosystems of the river deltas have been severely impacted. What were once islands of great biological diversity have been greatly degraded, with the disappearance of many water-loving native plants, including shrubs and grasses. Many species of animals have disappeared from the deltas, so that commercial hunting and trapping have almost ceased.

Concurrent with the contraction of the Aral Sea, the local climate seems to have become generally warmer, but with a greater seasonal temperature range, while the growing season has shortened. To what extent this change is directly caused by the lake shrinkage is not known. Ground-water levels have fallen by several meters, and this has caused reductions in ground-water withdrawal in adjacent irrigated areas.

Although not inevitable, the fate of the Aral Sea appears grim. Reduction of inflow to near-zero values is foreseen, and salinity of the remaining lake will eventually be about the same as in the open ocean. Public outcry against the prevailing political policy has increased sharply. Several schemes have been proposed to reverse the lake shrinkage. They include the possibility of massive long-distance importation of water from the north by southward diversions of rivers that now drain northward into the Arctic Ocean. Help, if there is any, may come too late to save the Aral Sea.

Data sources: P.M. Micklin, 1988, *Science*, vol. 241, p. 1170–76; J. Perera, 1988, *New Scientist*, vol. 120:1640, p. 20; *Geographical Magazine*, 1990, vol. 57, no. 3, p. 13.

DESERT IRRIGATION AND SALINIZATION

Human interaction with the tropical desert environment is as old as civilization itself. Two of the earliest sites of civilization—Egypt and Mesopotamia—lie in the tropical deserts. The key to successful human occupation of the deserts lies in the availability of large supplies of water from nondesert sources. This is a concept so familiar to all that it scarcely needs to be stated. For Egypt and Mesopotamia, the water sources of ancient times were exotic rivers deriving their flow from regions having a water surplus and flowing across the desert region because of geologic events and controls having nothing to do with climate.

A look at a global population map shows that population density is less than 1 person per square km (2 persons per sq mi) over nearly all the area of the tropical deserts. Only where exotic streams cross the desert does the population density rise sharply. Valleys of the Nile, the Tigris and Euphrates, and the Indus are striking examples in the Old World. But in the coastal desert of Peru, we also find a substantial population long dependent on exotic streams fed from the Andes range and crossing the coastal desert to reach the Pacific Ocean.

Can we increase our production of food by expanding agriculture into the tropical deserts? Making the desert bloom is a romantic concept fostered on the American scene for generations by bureaucrats, politicians, and land developers. Were these promoters of vast irrigation schemes working in the long-term public interest? Only in recent decades have the undesirable environmental impacts of desert irrigation come to the forefront. But we could have read the modern scenario in the history of the rise and fall of the Mesopotamian civilization.

Irrigation systems in arid lands divert the discharge of a large river, such as the Nile, Indus, Jordan, or Colorado, into a distributary system that allows the water to infiltrate the soil of areas under crop cultivation. Ultimately, such irrigation projects suffer from two undesirable side effects: salinization and waterlogging of the soil.

The irrigated area is subject to very heavy soil-water losses through evapotranspiration. Salts contained in the irrigation water remain in the soil and increase in concentration. This process is called *salinization*. (Areas of salinization show as white surfaces on the remote sensing imagery of Figure A.II.2.) Ultimately, when salinity of the soil reaches the limit of tolerance of the plants, the land must be abandoned. Prevention or cure of salinization may be possible by flushing the soil salts downward to lower levels by the use of more water. This remedy requires greater water use than for crop growth alone.

Infiltration of large volumes of water causes a rise in the water table and may, in time, bring the zone of saturation close to the surface. This phenomenon is called *waterlogging*. Crops cannot grow in perpetually saturated soils. Furthermore, when the water table rises to the point that upward movement under capillary action can bring water to the surface, evaporation is increased and salinization is intensified.

One of the largest of the modern irrigation projects affected adversely by salinization and waterlogging lies within the basin of the lower Indus River in Pakistan. Here the annual rate of rise of the water table has averaged about 0.3 m (1 ft), while an estimated 36,000 hectares (90,000 acres) goes out of production annually because of excessive salinity. The government of Pakistan has embarked upon a major program of recovery of waterlogged salinized land. By means of an elaborate system of drains, salty water is to be extracted from the affected land and conveyed to the sea by means of a conduit following the east bank of the Indus River. The project is due for completion early in the 1990s.

Other agricultural areas of major salinization include the Euphrates Valley in Syria, the Nile Delta of Egypt, and the wheat belt of Western Australia. In the United States, major heavily salinized agricultural areas are found in the San Joaquin and Imperial valleys of California, while numerous affected cropland areas in a wide range of sizes are identifiable throughout the entire semiarid and arid regions of the western United States.

The question of expanding irrigation agriculture in the southwestern United States was studied by the Committee on Arid Lands of the American Association for the Advancement of Science. In its 1972 report, this body recommended that additional large-scale importation of irrigation water to the American Southwest from distant sources should be made only where there are compelling reasons to do so. One such reason is to augment rapidly failing ground-water supplies in districts already under irrigation. A second is to arrest the progress of salinization in areas already under irrigation. In short, the committee agreed that additional water should be imported only to prevent social and economic disruption in established irrigated areas. The lesson is that the search for new regions in which to expand agriculture should be directed to other, more favorable environments where the scales are not so heavily weighted by enormous evaporative water losses.

CHEMICAL SOURCES OF WATER POLLUTION

In this chapter we have examined the flow of water itself, largely from the hydrologist's point of view. Streams, lakes, bogs, and marshes are specialized habitats of plants and animals; their ecosystems are particularly sensitive to changes induced by humans in the water balance and in water chemistry. Not only does our industrial society make radical physical changes in water flow by construction of engineering works (dams, irrigation systems, canals, dredged channels), but we also pollute and contaminate our surface waters with a large variety of wastes. Some of these wastes are in the form of ions in solution. The subject of water quality is appropriate to discuss in this chapter. Introduction of mineral sediment into surface waters as a result of land disturbances will be discussed in Chapter 17.

Chemical pollution by direct disposal into streams and lakes of wastes generated in industrial plants is a phenomenon well known to the general public and can be seen firsthand in almost any industrial community in the United States. Direct outfall of sewage, whether raw (untreated) or partially treated, is another form of direct pollution of streams and lakes.

In urban and suburban areas, pollutant matter entering streams and lakes includes deicing salt, lawn conditioners (lime and fertilizers), and sewage effluent. In agricultural regions, important sources of pollutants are fertilizers and the body wastes of livestock. Major sources of water pollution are associated with the mining and processing of mineral deposits. The possibility of contamination of radioactive substances released from nuclear processing plants also exists.

Among the common chemical pollutants of both surface water and ground water are sulfate, nitrate, phosphate, chloride, sodium, and calcium ions. Sulfate ions enter runoff by fallout from polluted urban air and as sewage effluent. Important sources of nitrate ions are fertilizers and sewage effluent. Excessive concentrations of nitrate in water supplies are highly toxic to humans but, at the same time, the removal of nitrate is difficult and expensive. Phosphate ions are contributed in part by fertilizers and by detergents in sewage effluent. Phosphate and nitrate are plant nutrients and can lead to excessive growth of algae in streams and lakes. This process is called *eutrophication*. Chloride and sodium ions are contributed both by fallout from polluted air and by deicing salts used on highways. Instances are recorded in which water supply wells close to highways have become polluted from deicing salts.

A particular form of chemical pollution of surface water goes under the name of *acid mine drainage*. It is an important form of environmental degradation in parts of Appalachia where abandoned coal mines and strip mine workings are concentrated. Ground water emerging from abandoned mines and seeping from strip mine spoil banks is charged with sulfuric acid and various salts of metals, particularly of iron. The sulfuric acid is formed by the reaction of water with iron sulfides, particularly the mineral pyrite (Fe_2S), which is a common constituent of coal seams. Acid of this origin in stream waters can have adverse effects on animal life. In sufficient concentrations, it is lethal to certain species of fish and has at times caused massive fish kills. The acid waters also cause corrosion to boats and piers. Government sources have estimated that nearly 10,000 km (6000 mi) of streams in this country, together with almost 100 sq km (40 sq mi) of reservoirs and lakes are seriously affected by acid mine drainage. One particularly undesirable by-product of acid mine drainage is precipitation of iron to form slimy red and yellow deposits in stream channels.

Sulfuric acid may also be produced from the drainage of mines from which sulfide ores are being extracted and from tailings produced in plants where the ores are processed. Such chemical pollution also includes salts of various toxic metals, among them zinc, lead, arsenic, copper, and aluminum. Yet another source of chemical pollution of streams is by radium derived from the tailings of uranium ore processing plants.

Toxic metals, among them mercury, along with pesticides and a host of other industrial chemicals, are introduced into streams and lakes in quantities that are locally damaging or lethal to plant and animal communities. In addition, sewage introduces live bacteria and viruses that are classed as biological pollutants; these pose a threat to the health of humans.

Thermal pollution is a term applied generally to the discharge of heat into the environment from the combustion of fuels and from nuclear energy conversion into electric power. Thermal pollution of water takes the form of heavy discharges of heated water locally into streams, estuaries, and lakes. The thermal environmental impact may thus be quite drastic in a small area.

THE RUNOFF ENERGY SYSTEM

Runoff of fresh water on and beneath the land surface is powered by gravity; it is an open energy system in which work is done as the water moves to lower levels under a hydraulic gradient and encounters frictional resistance. Figure 16.35 shows the principal energy storages and transformations that make up the system. The total open system of runoff has as its energy input the quantity of available potential energy present in precipitation that has arrived at the land surface. This water mass occupies an elevated position above a base level of flow (in this case, the ocean surface).

Two subsystems are shown. The surface water subsystem consists of overland flow and stream flow; the ground-water subsystem consists of subsurface water moving under a hydraulic gradient, whether the paths of motion be directed downward or upward.

In the surface-water subsystem potential energy is initially in storage at the ground surface, where it is transformed into kinetic energy as overland flow and stream flow move to lower levels. Kinetic energy is temporarily stored in the water flow, just as a moving automobile holds a store of kinetic energy. Flowage is met by internal friction of turbulent water motions and by friction along the boundary of the flowing water body with the ground beneath and the air above. Thus kinetic energy is transformed into sensible heat, which is temporarily held in storage and is then lost to the atmosphere by conduction, or is transformed to longwave radiation. Surface-water evaporation provides another energy output to the atmosphere as latent heat. Some kinetic energy of the flowing water passes directly out of the system where the trunk stream enters the ocean.

The ground-water subsystem receives potential energy during recharge, and this also undergoes transformation into kinetic energy. As some of the ground-water flow emerges at the land surface in springs, it adds to the kinetic energy of surface flow. Also shown is transfer of sensible heat from the ground-water subsystem to the surface-water subsystem during the seepage and springflow process. Otherwise, ground water flows to the inter-

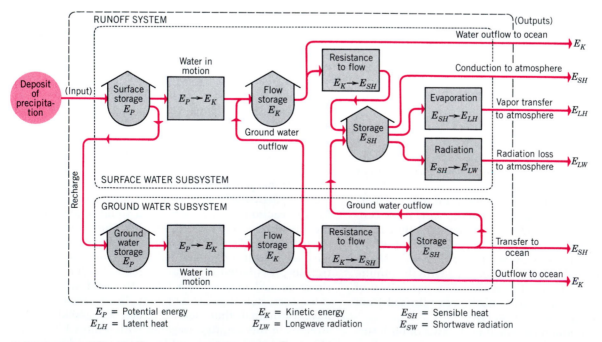

E_P = Potential energy E_K = Kinetic energy E_{SH} = Sensible heat
E_{LH} = Latent heat E_{LW} = Longwave radiation E_{SW} = Shortwave radiation

FIGURE 16.35 The runoff system as an open energy flow system.

face of fresh water and salt water beneath the coastline and emerges through the sea floor, transporting both sensible heat and kinetic energy of motion out of the system.

We realize, of course, that the initial supply of potential energy for the runoff system is furnished by solar energy, which evaporates water from the ocean surface, carries it high into the atmosphere as water vapor, and allows it to fall as precipitation. All this is part of the total hydrologic energy system.

FRESH WATER AS A NATURAL RESOURCE

Fresh water is a basic natural resource essential to our varied and intense agricultural and industrial activities. Runoff held in reservoirs behind dams provides water supplies for great urban centers, such as New York City and Los Angeles; diverted from large rivers, it provides irrigation water for highly productive lowlands in arid lands, such as the Imperial Valley of California and the Nile Valley of Egypt. To these uses of runoff are added hydroelectric power, where the gradient of a river is steep, or routes of inland navigation, where the gradient is gentle.

Unlike ground water, which represents a large water storage body, fresh surface water in the liquid state is stored only in small quantities. (An exception is the Great Lakes system.) Referring back to Figure 9.5, note that the quantity of available ground water is about 20 times as large as that stored in freshwater lakes, while the water held in streams is only about 1/100 that in lakes. Because of small natural storage capacities, surface water can be drawn only at a rate comparable with its annual renewal through precipitation. Dams are built to develop useful storage capacity for surplus runoff that would otherwise escape to the sea; but, once the reservoir has been filled, water use must be scaled to match the natural supply rate averaged over the year. The development of surface-water supplies brings on many environmental changes, both physical and biological, and these must be taken into account in planning for future water developments.

Studies of the total U.S. water resource have led to the conclusion that for the nation as a whole, there is an adequate freshwater supply for at least several decades. For individual regions of the country, however, we can anticipate severe shortages that can be met only by the transfer of water from regions or surplus. The enormous concentration of persons and industry into a small area (e.g., in Los Angeles County and the New York City region), requires the development of surface-water facilities in distant uplands, along with elaborate water-transfer systems using expensive and vulnerable aqueducts. It is only by continual and costly expansion of these facilities that shortages can be avoided. Compounding distribution problems is water pollution. If the quality of our freshwater supplies continues to be degraded, we must resort more and more to expensive water treatment procedures necessary to keep the water usable.

In recent years, much emphasis has been placed on alternatives to massive water transfers to meet rising demands. The basic alternative is a reduction in water use, particularly reductions in wasteful use in urban systems and in the huge demands for irrigation in arid lands. It is obvious that problems of water-resource management involve a broad spectrum of economic, political, and cultural factors.

C H A P T E R 17

LANDFORMS MADE BY RUNNING WATER

Geomorphology deals largely with the action of *fluid agents* that erode, transport, and deposit mineral and organic matter. The four fluid agents are: (1) running water in surface and underground flow systems; (2) waves, acting with currents in oceans and lakes; (3) glacial ice, moving sluggishly in great masses; and (4) wind, blowing over the ground.

Of the four agents, three are forms of water. Consequently, the science of hydrology is inseparably interwoven with geomorphology. One might not be far wrong in saying simply that the hydrologist is preoccupied with "where water goes," the geomorphologist with "what water does." Hydrology concerns itself with the hydrologic cycle in an attempt to calculate the water balance and measure rates of flow of water in all parts of that cycle (Chapter 16). Geomorphology concerns itself with geologic work that the water in motion performs on the land.

We have used denudation to mean the total action of all processes by which the exposed rocks of the continents are worn away and the resulting sediments are transported to the sea by the fluid agents. Denudation is thus an overall lowering of the land surface; it tends toward reducing the continents to nearly featureless sea-level surfaces and, ultimately, through wave action, to submarine surfaces. If denudation had not been repeatedly counteracted by crustal uplifts throughout geologic time, it would have eliminated all terrestrial environments.

An important point that emerges as we look back through geologic time is that terrestrial life environments have been in constant change, even as plants and animals have undergone their evolutionary development. The varied denudation processes have produced, maintained, and changed a wide variety of landforms, which have been the habitats for evolving life forms. In turn, the life forms have become adapted to those habitats and have diversified to a degree that matches the diversity of the landforms themselves.

Geomorphic and hydrologic systems have long been subjected to radical modification by the works of humans. Agriculture has for centuries altered the surface properties of areas of subcontinental size. Agriculture has modified the action of running water and the water

balance, to say nothing of having radically changed the character of the soil. Urbanization is an even more radical alteration, seriously upsetting hydrologic processes. Engineering and mining activities, such as strip-mining and the construction of highways, dams, and canals, not only upset the hydrologic systems but can completely destroy or submerge entire assemblages of landforms. Of the four fluid agents of landform sculpture, only glaciers of ice have so far successfully resisted changes of activity imposed by humans.

Invariably, our attempts to control the action of running water, waves, and currents produce unpredicted and undesirable side effects, some of which are physical, others ecological. An important reason to study the fluid agents is to predict the consequences of such changes and to plan wisely for management of the environment.

FLUVIAL PROCESSES AND LANDFORMS

Landforms shaped by running water are conveniently described as *fluvial landforms* to distinguish them from landforms made by the other fluid agents—glacial ice, wind, and waves. Fluvial landforms are shaped by the *fluvial processes* of overland flow and stream flow. Weathering and the slower forms of mass wasting, such as soil creep, operate hand in hand with overland flow and cannot be separated from the fluvial processes.

Fluvial landforms and the fluvial processes dominate the continental land surfaces the world over. Throughout geologic history, glacial ice has been present only in comparatively small global areas located in the polar zones and in high mountains. Landforms made by wind action occupy only trivially small parts of the continental surfaces, and landforms made by waves and currents are restricted to a very narrow contact zone between oceans and continents. In terms of area, the fluvial landforms are dominant in the environment of terrestrial life and are the major source areas of human food resources through the practice of agriculture. Almost all lands in crop cultivation and almost all grazing lands have been shaped by fluvial processes.

Fluvial processes perform the geologic activities of

FIGURE 17.1 Erosional and depositional landforms. (Drawn by A. N. Strahler.)

erosion, transportation, and deposition. Consequently, there are two major groups of fluvial landforms: erosional landforms and depositional landforms (see Figure 14.1). Valleys are formed where rock is eroded away by fluvial agents. Between the valleys are ridges, hills, or mountain summits representing unconsumed parts of the crustal block. All such sequential landforms shaped by the progressive removal of the bedrock mass are *erosional landforms*.

Fragments of soil, regolith, and bedrock that are removed from the parent mass are transported by the fluid agents and deposited elsewhere as sediment to make an entirely different set of surface features, the *depositional landforms*. Figure 17.1 illustrates the two groups of landforms. The ravine, canyon, peak, spur, and col are erosional landforms; the fan, built of coarse sediment below the mouth of the ravine, is a depositional landform. The floodplain, built of alluvium transported by a stream, is also a depositional landform.

NORMAL AND ACCELERATED SLOPE EROSION

Fluvial action starts on the uplands of drainage basins. Overland flow, by exerting a dragging force over the soil surface, picks up particles of mineral matter ranging in size from fine colloidal clay to coarse sand or gravel, depending on the speed of the flow and the degree to which the particles are bound by plant rootlets or held down by a mat of leaves. Added to this solid matter is dissolved mineral matter in the form of ions produced by acid reactions or direct solution. This slow removal of soil is part of the natural geologic process of landmass denudation; it is both inevitable and universal. Under stable natural conditions, the erosion rate in a moist climate is slow enough that a soil with distinct horizons is formed and maintained, enabling plant communities to maintain themselves in a stable equilibrium. Soil scientists refer to this state of activity as the *geologic norm*.

By contrast, the rate of soil erosion may be enormously speeded up by human activities or by rare natural events to result in a state of *accelerated erosion*, in which the soil is removed much faster than it can be formed. This condition comes about most commonly from a forced change in the plant cover and in the physical state of the ground surface and uppermost soil horizons. The destruction of vegetation by the clearing of land for cultivation or by forest fires sets the stage for a series of drastic changes. The interception of rain by fo-

FIGURE 17.2 A large raindrop (above) lands on a wet soil surface, producing a miniature crater (below). Grains of clay and silt are thrown into the air, and the soil surface is disturbed. (Official U.S. Navy photograph.)

liage is ended; protection afforded by a ground cover of fallen leaves and stems is removed. Consequently, the rain falls directly on the mineral soil.

The direct force of a falling raindrop (Figure 17.2) causes a geyserlike splashing in which soil particles are lifted and then dropped into new positions, a process termed *splash erosion*. It is estimated that a torrential rainstorm has the ability to disturb as much as 225 metric tons of soil per hectare (100 tons per acre). On a sloping ground surface, splash erosion shifts the soil slowly downhill. A more important effect is to cause the soil surface to become much less able to infiltrate water because the natural soil openings become sealed by particles shifted by raindrop splash. Reduced infiltration permits a much greater proportion of overland flow to occur from a given amount of rain. Increased overland flow intensifies the rate of soil removal.

Another effect of the destruction of vegetation is to reduce greatly the resistance of the soil surface to the force of erosion under overland flow. On a slope covered by grass sod, even a deep layer of overland flow causes little soil erosion because the energy of the moving water is dissipated in friction with the grass stems, which are tough and elastic. On a heavily forested slope, countless check dams made by leaves, twigs, roots, and fallen tree trunks take up the force of overland flow. Without such vegetation cover the eroding force is applied directly to the bare soil surface, easily dislodging the grains and sweeping them downslope.

CHANGES IN INFILTRATION CAPACITY

Use of the land surface by humans for purposes of crop cultivation and the grazing of domesticated animals can have a profound effect on the infiltration of precipitation through the soil surface. We can imagine the soil surface to be a fine sieve that receives rainfall of a given intensity and is capable of transmitting water downward at a given rate (see Figure 16.2). Rainfall intensity is usually stated for short time periods, for example, cm per 10 min or cm/hr. Infiltration is also stated as depth per unit of time. When rainfall intensity exceeds infiltration rate, the excess water leaves the surface as overland flow. Overland flow may also be stated in intensity units, as water depth per unit of time. The following equation then applies:

$$P - I = R_0$$

where P is precipitation rate (intensity)
       I is infiltration rate,
and R_0 is overland flow,

all stated as cm/hr or in./hr. Evaporation is assumed to be zero during the rainfall period. Obviously infiltration rate will equal precipitation rate until the limit of the infiltration rate, or *infiltration capacity*, is reached.

Now, it is an important fact about soils that their infiltration capacity is usually great at the start of a rain which has been preceded by a dry spell, but drops rapidly as the rain continues to fall and to soak into the soil. After 2 or 3 hours the infiltration capacity becomes almost constant. This change with time is shown in the graphs of Figure 17.3. In the first quarter-hour or so, infiltration capacity drops sharply; then the curve flattens rapidly. One reason for the high starting value and its rapid drop is that the larger soil openings that allow water to flow downward rapidly become clogged by particles brought from above, or tend to close up as the colloidal clays take up water and swell. From this effect we can easily reason that a sandy soil with little or no clay will not undergo a great drop in infiltration capacity, but will continue to let the water through indefinitely at a generous rate. In contrast, a clay-rich soil is quickly sealed to the point that it allows only a very slow rate of infiltration. This principle is illustrated by Figure 17.3A, which shows the infiltration curves of two soils—one sandy, one rich in clay. A sandy soil may be able to infiltrate water of a heavy, long-continued rain without any overland flow occurring, whereas the clay soil must divert much of the rain into overland flow.

Certain forms of disturbance of soils to which humans contribute tend to decrease the infiltration capacity and increase the amount of surface runoff. Trampling by livestock will tamp the porous soil into a dense, hard layer (Figure 17.3B). Forest cutting followed by crop cultivation tends to leave the soil exposed so that raindrop impact quickly seals the soil pores (Figure 17.3C). Fires,

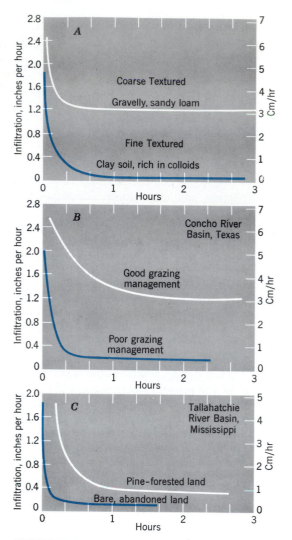

FIGURE 17.3 Infiltration rates vary greatly according to soil texture and land use. (Data from Sherman and Musgrave; Foster.)

caused by lightning or set by humans, in destroying the protective vegetation and surface litter, also expose the soil to raindrop impact and cause reduced infiltration capacity. It is little wonder, then, that humans have, through unwise farming and grazing practices, radically changed the original proportions of infiltration to runoff. As a result of reduced infiltration, severe erosion damage has occurred in many areas.

LAND USE AND SEDIMENT YIELD

We can get a good appreciation of the contrast between normal and accelerated erosion rates by comparing the quantity of sediment derived from cultivated surfaces with that derived from naturally forested or reforested surfaces within a given region in which climate, soil, and topography are fairly uniform. *Sediment yield* is a technical term for the quantity of sediment removed by overland flow from a unit area of ground surface in a given unit of time. Yearly sediment yield is stated in metric tons per hectare, or tons per acre.

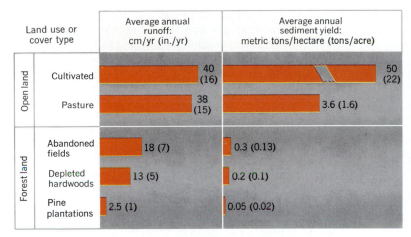

Land use or cover type		Average annual runoff: cm/yr (in./yr)	Average annual sediment yield: metric tons/hectare (tons/acre)
Open land	Cultivated	40 (16)	50 (22)
	Pasture	38 (15)	3.6 (1.6)
Forest land	Abandoned fields	18 (7)	0.3 (0.13)
	Depleted hardwoods	13 (5)	0.2 (0.1)
	Pine plantations	2.5 (1)	0.05 (0.02)

FIGURE 17.4 The bar graph shows that both runoff and sediment yield are much greater for open land than for land covered by shrubs and forest. Cultivated land has an enormous sediment yield, as compared with any of the other types. (Data of S. J. Ursic, Department of Agriculture.)

Figure 17.4 gives data on annual average sediment yield and runoff by overland flow from several types of upland surfaces in northern Mississippi. Notice that both surface runoff and sediment yield decrease greatly with increased effectiveness of the protective vegetative cover. Sediment yield from cultivated land undergoing accelerated erosion is over 10 times greater than that from pasture and about 1000 times greater than from pine plantation land. The reforested land has a sediment-yield rate representing the geologic norm of soil erosion for this region; it is about the same as for mature pine and hardwood forests that have not experienced cultivation.

The distinction between normal and accelerated slope erosion applies to regions in which the water balance shows an annual surplus. Under a midlatitude semiarid climate with summer drought, the natural plant cover consists of short-grass prairie. Although sparse and providing rather poor ground cover of plant litter, the grass cover is strong enough that the geologic norm of erosion can be sustained. In these semiarid environments, however, the natural equilibrium is highly sensitive to upset. Depletion of plant cover by fires or the grazing of herds of domesticated animals can easily set off rapid erosion. These sensitive, marginal environments require cautious use because they lack the potential to recover rapidly from accelerated erosion once it has set in.

Erosion at a very high rate by overland flow is actually a natural geologic process in certain favorable localities in semiarid and arid lands; it takes the form of *badlands*. One well-known area of badlands is the Big Badlands of South Dakota, along the White River. Badlands are underlain by clay formations, easily eroded by overland flow. Erosion rates are too fast to permit plants to take hold, and no soil can develop. A maze of small stream channels is developed; ground slopes are very steep (Figure 17.5). Badlands such as these are self-sustaining and have been in existence at one place or another on continents throughout much of geologic time.

FORMS OF ACCELERATED SOIL EROSION

Humid regions of substantial water surplus, for which the natural plant cover is forest or prairie grasslands, experience accelerated soil erosion when humans expend enough energy to remove the plant cover and keep the land barren by annual cultivation. With fossil fuels to power machines of plant and soil destruction, humans have easily overwhelmed the restorative forces of nature

FIGURE 17.5 Badlands at Zabriskie Point, Death Valley National Monument, California. (Alan H. Strahler.)

FIGURE 17.6 Shoestring rills on a steep bank of mine wastes illustrate the susceptibility of artificial accumulations of disturbed soil and regolith to accelerated erosion. Sediment from this source forms unwanted alluvial deposits in the nearest stream bed. Oatman, Arizona. (A. N. Strahler.)

over vast expanses of continental surfaces. We now consider the consequences of these activities.

When a plot of ground is first cleared of forest and plowed for cultivation, little erosion will occur until the action of rain splash has broken down the soil aggregates and sealed the larger openings. Overland flow then begins to remove the soil in rather uniform thin layers, a process termed *sheet erosion*. Because of seasonal cultivation, the effects of sheet erosion are often little noticed until the upper horizons of the soil are removed or greatly thinned. Reaching the base of the slope, where the surface slope is sharply reduced to meet the valley bottom, soil particles come to rest and accumulate in a thickening layer of *colluvium*. This sediment deposit, too, has a sheetlike distribution and may be little noticed, except where it can be seen that fence posts or tree trunks are being slowly buried.

Material that continues to be carried by overland flow to reach a stream in a valley axis is then carried farther downvalley and may accumulate as alluvium in layers on the valley floor. *Alluvium* is sediment deposited by a stream; it may consist of gravel, sand, silt, or clay.

Deposition of alluvium results in the burial of fertile floodplain soils under infertile, sandy layers. This form of valley sedimentation chokes the channels of small streams, causing the water to flood broadly over the valley bottoms.

Where slopes are steep, runoff from torrential rains produces a more intense activity, *rill erosion*, in which innumerable closely spaced channels are scored into the soil and regolith (Figure 17.6). If these rills are not destroyed by soil tillage, they may soon begin to integrate into still larger channels, called *gullies*. Gullies are steep-walled, canyonlike trenches whose upper ends grow progressively upslope (Figure 17.7). Ultimately, a rugged, barren topography, like the badland forms of the dry climates, results from accelerated soil erosion that is allowed to proceed unchecked.

The natural soil, with its well-developed horizons, is a nonrenewable natural resource. The rate of soil formation is extremely slow in comparison with the rate of its destruction once accelerated erosion has begun and is allowed to go unchecked. Soil erosion as a potentially disastrous form of environmental degradation was brought to public attention decades ago in the United States.

Curative measures developed by the Soil Conservation Service have proved effective in stopping accelerated soil erosion and in permitting the return to slow erosion rates approaching the geologic norm. These measures include the construction of terraces to eliminate steep slopes, permanent restoration of overly steep slope belts to dense vegetative cover, and the healing of gullies by placing check dams in the gully floors.

GEOLOGIC WORK OF STREAMS

The geologic work of streams consists of three closely interrelated activities: erosion, transportation, and deposition. *Stream erosion* is the progressive removal of mineral material from the floor and sides of the channel, whether bedrock or regolith. *Stream transportation* consists of movement of the eroded particles by dragging along the bed, by suspension in the body of the stream,

FIGURE 17.7 Deep branching gullies have carved up an overgrazed pasture near Shawnee, Oklahoma. Contour terracing and check dams have been emplaced in an effort to halt the headward growth of the gullies. (Mark A. Melton.)

or in solution. *Stream deposition* is the accumulation of transported particles on the streambed and floodplain, or in the floor of a standing body of water into which the stream empties. Deposition is synonymous with sedimentation. Obviously, erosion cannot occur without some transportation taking place, and the transported particles must eventually come to rest. Erosion, transportation, and deposition are simply three phases of a single activity.

STREAM EROSION

Streams erode in various ways, depending on the nature of the channel materials and the tools with which the current is armed. The force of the flowing water alone, exerting impact and a dragging action on the bed and banks, can erode poorly consolidated alluvial materials, such as gravel, sand, silt, and clay. This erosion process, called *hydraulic action*, is capable of excavating enormous quantities of unconsolidated materials in a short time (Figure 17.8). The undermining of the banks causes large masses of alluvium to slump into the river, where the particles are quickly separated and become part of the stream's load. This process of *bank caving* is an important source of sediment during high river stages.

Where rock particles carried by the swift current strike against bedrock channel walls, chips of rock are detached. The rolling of cobbles and boulders over the streambed will further crush and grind smaller grains to produce an assortment of grain sizes. These processes of mechanical wear constitute *abrasion*; it is the principal means of erosion in bedrock too strong to be affected by simple hydraulic action.

Finally, the chemical processes of rock weathering—acid reactions and solution—are effective in the removal of rock from the stream channel and may be referred to

FIGURE 17.9 These potholes in lava bedrock attest to the abrasion that takes place on the bed of a swift mountain stream. McCloud River, California. (Alan H. Strahler.)

as *corrosion*. The effects of corrosion are conspicuous in limestone, which develops cupped and fluted surfaces.

One interesting form produced by stream abrasion is the *pothole*, a cylindrical hole carved into the hard bedrock of a swiftly moving stream (Figure 17.9). Often a spherical or discus-shaped stone is found in the pothole and is apparently the tool, or grinder, with which the pothole was deepened. A spiraling flow of water in the pothole rotates the grinder at the base of the hole, boring gradually into the rock. Many other features of abrasion, such as plunge pools, chutes, and troughs lend variety to the rock channel of a mountain stream.

STREAM TRANSPORTATION

The solid matter carried by a stream is the *stream load*; it is carried in three forms. Dissolved matter is transported invisibly in the form of chemical ions. All streams carry

FIGURE 17.8 Rapid bank caving undermined this cabin on Big Thomson Creek, Colorado, during a disastrous summer flood in 1976. Alluvium was flushed out from the channel floor in the foreground leaving only scoured bedrock and a litter of large boulders. (David C. Shelton, Colorado Geological Survey.)

FIGURE 17.10 In this vertical air view of the gorge of the Salt River, Arizona, the turbid storm runoff of a small tributary contrasts sharply with the clear water of the main stream. (Mark A. Melton.)

FIGURE 17.11 Carrying a heavy charge of suspended load, this turbulent mountain stream is beginning to cut into a bouldery layer carried and deposited as bed load in a higher stage of flow. Rincon Creek, Santa Barbara and Ventura counties, California. (A. N. Strahler.)

some dissolved salts resulting from mineral alteration. Clay and silt particles are carried in *turbulent suspension*; that is, they are held up in the water by the upward elements of flow in turbulent eddies in the stream. This fraction of the transported matter is the *suspended load* (Figure 17.10). Sand, gravel, and cobbles move as *bed load* close to the channel floor by rolling or sliding and an occasional low leap (Figure 17.11).

The load carried by a stream varies enormously in the total quantity present and the size of the fragments, depending on the discharge and stage of the river. In flood, when velocities of 6 m (20 ft) per second or more

are produced in large rivers, the water is turbid with suspended load. Large boulders may be moving over the streambed where the river gradient is steep.

CHANNEL CHANGES IN FLOOD

We think of a stream in flood as changing largely through increase in height of water surface, which causes channel overflow and inundation of the adjoining floodplain. Because of the turbidity of the water, we cannot see important changes taking place on the streambed; but these can be determined by sounding the river depth while stream-gauging measurements are being taken (Figure 17.12). These changes apply to a channel formed in thick alluvial materials, not in bedrock. At first the bed may be built up by large amounts of bed load supplied to the stream during the first phases of heavy runoff. This upbuilding is soon reversed, however, and the bed is actively deepened by scour as stream stage rises. Thus, in the period of highest stage, the streambed is at its lowest elevation. When the discharge then starts to decline, the level of the stream surface drops and the bed is built back up by the deposition of the alluvium. In the example shown in Figure 17.12, a layer of alluvium about 3 m (10 ft) thick was reworked, that is, moved about in the complete cycle of rising and falling stages.

Alternate deepening by scour and shallowing by deposition of load are responses to changes in the ability of the stream to transport its load. The maximum load of debris that can be carried by a stream is a measure of *stream capacity*. Load is stated as weight of material moved through a given stream cross section in a given unit of time; common units are metric tons per day. Total load includes both the bed load and the suspended load.

Where a stream is flowing in a channel of hard bedrock, it may not be able to pick up enough alluvial material to supply its full capacity for bed load. Bedrock channels are typical of streams occupying deep gorges and having steep gradients. When a flood occurs in a bedrock channel, the channel cannot be quickly deepened in response. In an alluvial river, where thick layers of silt, sand, and gravel underlie the channel, the rising river

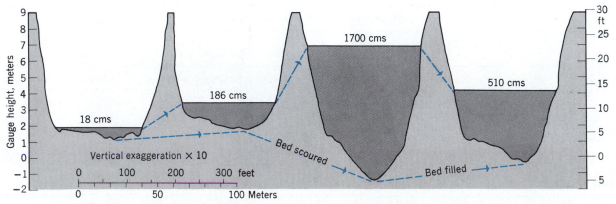

FIGURE 17.12 Changes in channel form of the San Juan River near Bluff, Utah, during the progress of a flood. (After Leopold and Maddock, U.S. Geological Survey.)

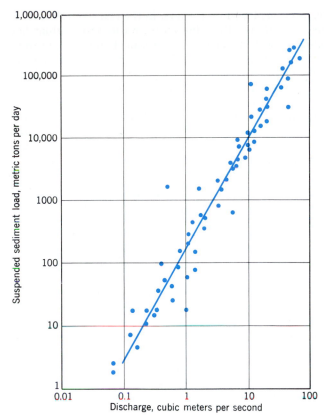

FIGURE 17.13 Increase of suspended sediment load with increase in discharge. Both scales are logarithmic. The dots show how individual observations are scattered about the line of best fit. Data are for the Powder River, Arvada, Wyoming. (After L. B. Leopold and T. Maddock, U.S. Geological Survey, Professional Paper 252.)

for bed load is increased from 8 to 16 times. It is small wonder, then, that most of the conspicuous changes in a stream channel, including sidewise shifting of the course, occur in flood stage. Few important changes occur in low stages.

Suspended load also increases greatly as discharge increases and stage rises. Increased turbulence is the cause. Figure 17.13 gives an example of the relationship between suspended load and discharge, measured at a single gauging station. The scales are logarithmic and show that the suspended load in flood stages may be over 10,000 times as great as in the lowest stages. The line fitted to the plotted points shows that a 10-fold increase in discharge brings a 100-fold increase in suspended load. Suspended load may be derived from erosion by overland flow on watershed surfaces and from bank caving.

When the flood crest has passed and the discharge begins to decrease, the capacity to transport load also declines. Therefore some of the particles that are in motion must come to rest on the bed in the form of sand and gravel bars. First the largest cobbles will cease to roll, then the pebbles and gravel, then the sand. Particles of fine sand and silt, carried in suspension, can no longer be sustained and settle to the bed. Clay particles of colloidal size continue to travel far downstream with the flood wave. In this way the falling stream adjusts to its decreasing capacity. When restored to low stage, the water may become clear.

easily picks up and sets in motion all the material that it is capable of moving. In other words, the increasing capacity of the stream for load is easily satisfied.

Capacity for bed load increases sharply with stream velocity because the swifter the current, the more intense the turbulence and the stronger the dragging force against the bed. Capacity to move bed load goes up approximately as the third to fourth power of the velocity. Thus, when stream velocity is doubled in flood, capacity

SUSPENDED SEDIMENT LOAD OF LARGE RIVERS

The great rivers of the world show an enormous range in the quantity of suspended sediment load and in sediment yield. Data for seven selected major rivers (Table 17.1) reveal some interesting relationships among load, climate, and land–surface properties.

The Yellow River (Huanghe) of China heads the world list in annual suspended sediment load, while its sediment yield is one of the highest known for a large river basin. The explanation lies in a high soil-erosion rate on intensively cultivated upland surfaces of wind-deposited silt (loess) in Shansi (Shanxi) province. (See Chapter 21

TABLE 17.1 Suspended Sediment Loads and Sediment Yields of Selected Large Rivers

River	Drainage Area (sq km × 10³)	Average Discharge (cu m/sec)	Average Annual Sediment Load (metric tons × 10³)	Average Annual Sediment Yields (metric tons/sq km)
Yellow (Huanghe), China	715	1.6	1,900,000	2600
Ganges, India	960	12	1,500,000	1400
Colorado, U.S.A.	640	0.17	140,000	380
Mississippi, U.S.A.	3200	19	310,000	97
Amazon, Brazil	6100	190	360,000	60
Congo, Congo	4000	42	65,000	16
Yenisei, USSR	2500	18	11,000	4

Data source: J. N. Holeman, "The sediment yield of major rivers of the world," *Water Resources Research*, vol. 4, no. 4, pp. 737–47.

Note: Data rounded to two digits.

and Figure 21.20). Much of the drainage area is in a semi-arid climate having dry winters; vegetation is sparse, and the runoff from heavy summer rains entrains a large amount of sediment.

The Ganges River derives its heavy sediment load from steep mountain slopes of the Himalayas and from intensively cultivated lowlands, all subjected to torrential rains of the tropical wet monsoon season. The Colorado River represents an exotic stream, deriving its runoff largely from snowmelt and precipitation on high mountain watersheds of the Rockies; but most of its suspended sediment is from tributaries in the semiarid plateau lands through which it passes.

The sediment load of the Mississippi River comes largely from subhumid and semiarid grassland watersheds of its great western tributary, the Missouri River. The Amazon River, a colossus in discharge and basin area, has a very low sediment yield because, like the Congo River, much of its basin lies within a wet equatorial climate where the land surface bears a highly protective rainforest. The Yenisei River of Siberia has a remarkably low sediment load and sediment yield for its vast drainage area, most of which is in the needleleaf forest, or taiga, of the subarctic and high midlatitude zones.

Although it is difficult to assess the importance of human-induced soil disturbance on the sediment load of major rivers, most investigators are generally agreed that cultivation has greatly increased the sediment load of rivers of eastern and southeastern Asia, Europe, and North America. The increase due to human activities is thought to be greater, by a factor of $2\frac{1}{4}$ than the geologic norm for the entire world land area. For the more strongly affected river basins, the factor may be 10 or more times larger than the geologic norm.

The sediment load carried by a large river is of considerable importance in planning for the construction of large water-storage dams and canal systems for irriga-tion. Sediment will be trapped in the reservoir behind a dam, eventually filling the entire basin and ending the useful life of the reservoir as a storage body. At the same time, depriving the river of its sediment in the lower course below the dam may cause serious upsets in river activity. Resulting deep scour of the bed and lowering of river level may render irrigation systems inoperable. In designing for canal systems, the forms of artificial channels must be adjusted to the size and quantity of sediment carried by the water; otherwise obstruction by deposition or abnormal scour may follow.

CONCEPT OF EQUILIBRIUM AND THE GRADED STREAM

A stream system, fully developed within its drainage basin of contributing valleyside slopes, has undergone a long period of adjustment of its geometry so it can discharge, through the trunk exit, not only the surplus water produced by the basin but also the solid load with which the channels are supplied. A purely hydraulic system could operate without a gradient because accumulated surplus water can generate its own surface slope and is capable of flowage on a horizontal surface. The transport of bed load requires a gradient, and it is in response to this requirement that a stream channel system has adjusted its gradient and has achieved an average steady state of operation, year in and year out and from decade to decade. In this condition, the stream is referred to as a *graded stream*; it is considered to have achieved an equilibrium in its state of operation.

As an idealized concept we consider that in an equilibrium condition, the supply of load furnished a stream from its watershed exactly matches the stream's capacity for transport. In nature, the balance between load and a stream's capacity exists only as an average condition over periods of many years. As already explained,

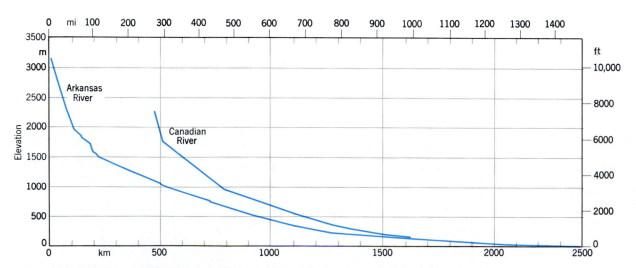

FIGURE 17.14 Longitudinal profiles of the Arkansas and Canadian rivers. The middle and lower parts of the profiles are for the most part smoothly graded, whereas the poorly graded upper parts reflect rock inequalities and glacial modifications within the Rocky Mountains. (From Gannett, U.S. Geological Survey.)

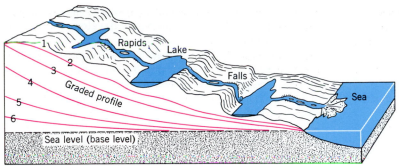

streams scour their channels in flood and deposit load when in the falling stage. Thus in terms of conditions of the moment, a stream is rarely in equilibrium; but over long periods of time, the graded stream maintains its level by restoring those channel deposits temporarily removed by the excessive energy of flood flows.

The channel gradient of a graded stream diminishes in steepness in the downstream direction. The *longitudinal profile* of a stream is shown in Figure 17.14 by a graph in which altitude (vertical scale) is plotted against downstream distance (horizontal scale). One important cause of the downstream decrease in gradient is that, as discharge increases and channel cross section becomes larger, the stream becomes more efficient in its operation. Frictional resistance is disproportionately less for

a larger channel, so the same work can be accomplished on a lower gradient. In some cases, also, the average size of particles of the bed load diminishes in the downstream direction; the finer load can be carried on a lower gradient.

EVOLUTION OF A GRADED STREAM

It will be helpful in developing the concept of a graded stream system to investigate the changes that will take place along a stretch of stream that is initially poorly adjusted to the transport of its load. An ungraded channel is illustrated in Figure 17.15. The starting profile is imagined to be brought about by crustal uplift in a series

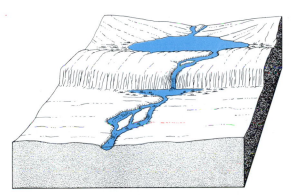

(A) Stream established on a land surface dominated by landforms of recent crustal activity.

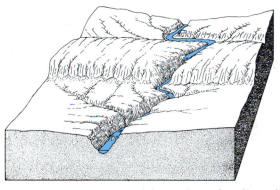

(B) Gradation in progress; lakes and marshes drained; deepening gorge; tributary valleys extending.

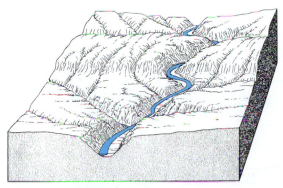

(C) Graded profile attained; beginning of floodplain development; valley widening in progress.

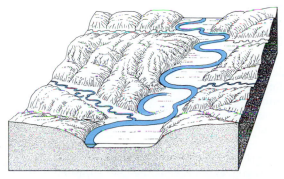

(D) Floodplain widened to accommodate meanders; floodplains extended up tributary valleys.

FIGURE 17.16 Evolution of a stream and its valley. (Drawn by E. Raisz.)

FIGURE 17.17 The Grand Canyon of the Yellowstone River has been carved in volcanic rock (rhyolite) of the Yellowstone Plateau. Through the chemical action of rising hot-water (hydrothermal) solutions, the rock has received its pastel shades of color. Yellowstone National Park, Wyoming. (A. N. Strahler.)

of fault steps, bringing to view a surface that was formerly beneath the ocean and exposing it to fluvial processes for the first time. Runoff collects in shallow depressions, which fill and overflow from higher to lower levels. In this way a throughgoing channel originates and begins to conduct runoff to the sea.

Figure 17.16 illustrates the gradation process in a series of block diagrams. Over *waterfalls* and *rapids*, which are simply steep-gradient portions of the channel, flow velocity is greatly increased and abrasion of bedrock is most intense (Block A). As a result, the falls are cut back and the rapids trenched, while the ponded reaches are filled by sediment and lowered in level as the outlets are cut down. In time the lakes disappear and the falls are transformed into rapids. Erosion of rapids reduces the gradient to one more closely approximating the average gradient of the entire system (Block B). At the same time, branches of the stream system are being extended into the landmass, carving out a drainage basin and transforming the original landscape into a fluvial landform system.

In the early stages of gradation and extension, the capacity of the stream exceeds the load supplied to it, so little or no alluvium accumulates in the channels. Abrasion continues to deepen the channels, with the result that they come to occupy steep-walled *gorges* or *canyons* (Figure 17.17). Weathering and mass wasting of these rock walls contributes an increasing supply of rock de-

bris to the channels. Debris shed from land surfaces contributing overland flow to the newly developed branches is also on the increase.

We can anticipate a great decrease in stream capacity for bed load, resulting from the gradual reduction in the channel gradient. This decrease will be converging on the increasing load with which it is being supplied. There will come a point in time at which the supply of load exactly matches the capacity of the stream to transport it. At this point the stream has achieved the graded condition and possesses a *graded profile*, descending smoothly and uniformly in the downstream direction (Figure 17.15). After attaining this state of balance, the stream continues to cut sidewise into its banks on the outside of bends. This *lateral cutting* does not appreciably alter the gradient and therefore does not materially affect the equilibrium.

The first indication that a stream has attained a graded condition is the beginning of floodplain development. On the outside of a bend, the channel shifts laterally into a curve of larger radius and thus undercuts the valley wall. On the inside of the bend, alluvium accumulates as a *point-bar deposit*. Widening of the bar deposit produces a crescentic piece of low ground, which is the first stage in floodplain development. This stage is illustrated in Block C of Figure 17.16. As lateral cutting continues, the floodplain strips are widened and the channel develops sweeping bends, called *alluvial meanders* (Block D). The floodplain is then widened into a continuous belt of flat land between steep valley walls.

Floodplain development reduces the frequency with which channel scour attacks and undermines the adjacent valley wall. Weathering, mass wasting, and overland flow then act to reduce the steepness of the valleyside slopes (Figure 17.18). As a result, in a moist climate, the gorgelike aspect of the valley gradually disappears and eventually gives way to an open valley with soil-covered slopes protected by a dense plant cover.

With the passage of long spans of time, the graded profile is slowly lowered throughout its entire length, as shown by numbered profiles in Figure 17.15. At its mouth, the stream surface merges smoothly with sea level. The bottom of the stream channel lies below sea level for some distance upstream, a condition that may persist many tens of kilometers inland in the case of a large river. Sea level, projected inland beneath the conti-

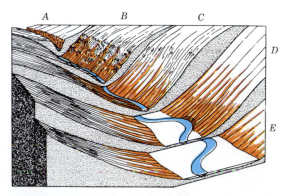

FIGURE 17.18 After a stream has become graded, its valley walls become more gentle in slope and the bedrock is covered by soil and weathered rock. (Drawn by W. M. Davis.)

nent as an imaginary surface, is the *base level* of stream activity. It is the theoretical lower limit to which the profile of the stream surface can be reduced. Because the overall rate of denudation diminishes gradually with time, later stages of reduction of the graded profile and the contributing slopes of the drainage basin become immeasurably prolonged. Although base level represents the theoretical limit to which the stream profile can be reduced, the achievement of that limit would require an infinitely long span of time. In reality, crustal movements and changes of sea level repeatedly disturb the relationship between stream profile and base level. We shall return to this subject at the end of this chapter.

ENVIRONMENTAL SIGNIFICANCE OF GORGES AND WATERFALLS

Deep gorges and canyons of major rivers exert environmental controls in a variety of ways. There is little or no room for roads or railroads between the stream and valley sides, so roadbeds must be cut or blasted at great expense and hazard from the sheer rock walls. Maintenance is expensive because of flooding or undercutting by the stream and the sliding and falling of rock, which can damage or bury the roadbed. But a gorge may afford the only feasible passage through a mountain range. A deep canyon is also a barrier to movement across it and may require the construction of expensive bridges.

Although small waterfalls are common features of alpine mountains carved by glacial erosion (Chapter 22), large waterfalls on major rivers are comparatively rare the world over. New river channels resulting from flow

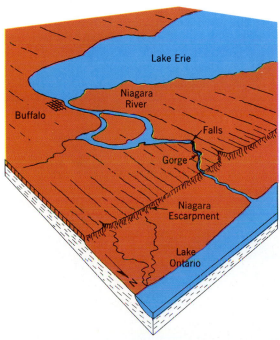

FIGURE 17.19 A bird's-eye view of the Niagara River, with its falls and gorge carved in strata of the Niagara Escarpment. View is toward the southwest from a point over Lake Ontario. (Based on a sketch by G. K. Gilbert, 1896. From A. N. Strahler, *The Earth Sciences,* 2nd ed., Harper & Row, Publishers. Copyright © by Arthur N. Strahler.)

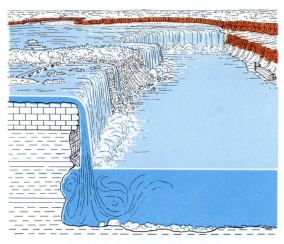

FIGURE 17.20 Niagara Falls is formed where the river passes over the eroded edge of a massive limestone layer. Continual undermining of weak shales at the base keeps the fall steep. (Drawn by E. Raisz.)

diversions caused by ice sheets of the Pleistocene Epoch provide one class of falls and rapids of large discharge. Certainly the preeminent example is Niagara Falls. Overflow of Lake Erie into Lake Ontario happened to be situated over a gently inclined layer of limestone, beneath which lies easily eroded shale (Figure 17.19). As Figure 17.20 shows, the fall is maintained by continual undermining of the limestone by erosion in the plunge-pool at the base of the fall. The height of the falls is now 52 m (170 ft), and its discharge is about 17,000 cms (200,000 cfs). The drop of Niagara Falls is utilized for the production of hydroelectric power by the Niagara Power Project, in which water is withdrawn upstream from the falls and carried in tunnels to generating plants located 6 km (4 mi) downstream from the falls. Capability of this project is 2400 megawatts of power.

Great rapids and waterfalls on rivers draining the rifted belt of East Africa are of special interest because of their close relationship to plate tectonics (see Figures 14.24 and 14.25). Deep lakes within the fault basins feed falls and rapids, such as the Murchison Falls, located on the Victoria Nile near its junction with the outlet of Lake Albert. Far to the south, Victoria Falls on the Zambezi River marks the zone of descent of that river over a great escarpment (Figure 17.21).

Most large rivers of steep gradient do not possess falls, and it is necessary to build dams to create artificially the vertical drop necessary for turbine operation. An example is the Hoover Dam, behind which lies Lake Mead, occupying the canyon of the Colorado River. With a dam height of 220 m (726 ft), the generating plant of Hoover Dam is capable of producing 1345 megawatts of power, about half as much as the Niagara Project.

We should not lose sight of the aesthetic and recreational values of gorges, rapids, and waterfalls of major rivers. The Grand Canyon of the Colorado River, probably more than any single product of fluvial processes, highlights the scenic value of a great river gorge. Against the advantages in obtaining hydroelectric power and fresh water for urban supplies and irrigation by construction of large dams, we must weigh the permanent

FIGURE 17.21 Victoria Falls is located on the Zambesi River, which forms the border between Zambia and Zimbabwe in southern Africa. The water falls 128 m (420 ft) to the bottom of a deep, narrow gorge excavated along a fault line. (Harm J. deBlij.)

loss of some large segments of our finest natural scenery, along with the destruction of ecosystems adapted to the river environment. It is small wonder, then, that new dam projects are meeting with stiff opposition from concerned citizen groups who fear that the harmful environmental impacts of such structures far outweigh their future benefits.

AGGRADATION AND ALLUVIAL TERRACES

A graded stream, delicately adjusted to its supply of water and rock waste from upstream sources, is highly sensitive to changes in those inputs. Changes in climate and in surface characteristics of the watershed bring changes in discharge and load at downstream points, and these changes in turn require channel readjustments.

Consider first the effect of an increase in bed load beyond the capacity of the stream. At the point on a channel where the excess load is introduced, the coarse sediment accumulates on the streambed in the form of bars of sand, gravel, and pebbles. These deposits raise the elevation of the streambed. The process is called *aggradation;* it is a form of sediment deposition. As more bed materials accumulate, the stream channel gradient is increased; the increased flow velocity enables bed materials to be dragged downstream and spread over the channel floor at progressively more distant downstream reaches.

Aggradation typically changes the channel cross section from one of narrow and deep form to a wide, shallow cross section. Because bars are continually being formed, the flow is divided into multiple threads, and these rejoin and subdivide repeatedly to give a typical *braided channel* (Figure 17.22). The coarse channel deposits spread across the former floodplain, burying fine-textured alluvium with coarse material. We have already seen that alluvium accumulates as a result of accelerated soil erosion.

How is aggradation induced in a stream system by

FIGURE 17.22 This braided river channel in the Yukon Territory carries meltwater from a glacier terminus, visible in the distance. Notice that an alluvial fan (left) has been built out into the valley floor by a tributary stream. (John S. Shelton.)

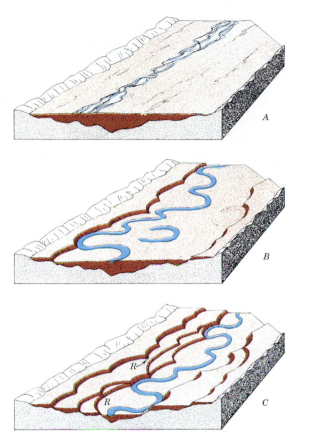

FIGURE 17.23 Alluvial terraces form when a graded stream slowly cuts away the alluvial fill in its valley. (Drawn by A. N. Strahler.)

natural processes? One natural cause of aggradation that has been of major importance in stream systems of North America and Eurasia is glaciation during the Pleistocene Epoch (Chapter 22). The advance of a valley glacier results in the input of a large quantity of coarse rock debris at the head of a drainage system. Valley aggradation was widespread in a broad zone marginal to the great ice sheets of the Pleistocene Epoch; the accumulated alluvium filled most valleys to depths of many tens of meters. Figure 17.23A shows a valley filled in this manner by an aggrading stream. The case could represent any one of a large number of valleys in New England or the Middle West.

Suppose, next, that the source of bed load is cut off or greatly diminished. In the case illustrated in Figure 17.23, the ice sheets have disappeared from the headwater areas and, with them, the supplies of coarse rock debris. Reforestation of the landscape has restored a protective cover to valley walls and hillslopes of the region, holding back coarse mineral particles from entrainment in overland flow. Now the streams have copious water discharges but little bed load. In other words, they are operating below capacity. The result is channel scour and deepening. The channel form becomes deeper and narrower. Gradually, the stream profile level is lowered, a process called *degradation*. Because the stream is very close to being in the graded condition at all times, its dominant activity is lateral cutting by growth of meander bends, as shown in Block B of Figure 17.23. The valley alluvium is gradually excavated and carried downstream, but it cannot all be removed because the channel encounters hard bedrock in many places. Consequently, as shown in Block C, steplike surfaces remain on both sides of the valley. The treads of these steps are *alluvial terraces*. As each terrace is cut, the width of the next one above it is reduced.

All older terraces would be destroyed were it not that bedrock of the valley wall here and there projects through the alluvium and protects higher terraces. In Block C of Figure 17.23, the valley alluvium has been largely removed; but some *rock-defended terraces* remain on the valley sides, protected from stream attack by the rock that outcrops at the points labeled R. Notice that the scarps separating terraces are curved in broad arcs concave toward the valley. The curvature is easily explained as the result of cutting of the scarps by curved meander bends.

Alluvial terraces have always attracted occupation by humans because of their advantages over both the valley-bottom floodplains, which are subject to annual flooding, and the hillslopes beyond, which may be too steep and rocky to cultivate. Besides, the soils of terraces are easily tilled and made prime agricultural land (Figure 17.24). Towns are easily laid out on the flat ground of a terrace, and roads and railroads are easily run along the terrace surfaces parallel with the river.

FIGURE 17.24 Alluvial terraces of the Rakaia River gorge on the South Island of New Zealand. The flat terrace surface in the foreground is used as pasture for sheep. Two higher terrace levels can be seen at the left. (F. Kenneth Hare.)

AGGRADATION INDUCED BY HUMAN ACTIVITY

Channel aggradation is a common form of environmental destruction brought on by human activities. Accelerated soil erosion following cultivation, lumbering, and forest fires is the most widespread source of sediment for valley aggradation. Aggradation of channels has also been a serious form of environmental degradation in coal-mining regions, along with water pollution (acid mine drainage), mentioned in Chapter 16. Throughout the Appalachian coalfields, channel aggradation is widespread because of the large supplies of coarse sediment from mine wastes. Strip mining has enormously increased the aggradation of valley bottoms because of the vast expanses of broken rock available to entrainment by runoff (see Figures 19.18 and 19.19).

Urbanization and highway construction are also major sources of excessive sediment, causing channel aggradation. Large earth-moving projects are involved in creating highway grades and in preparing sites for industrial plants and housing developments. Although these surfaces are eventually stabilized, they are vulnerable to erosion for periods ranging from months to years. The regrading involved in these projects often diverts overland flow into different flow paths, further upsetting the activity of streams in the area.

Mining, urbanization, and highway construction not only cause drastic increases in bed load, which cause channel aggradation close to the source, but also increase the suspended load of the same streams. Suspended load travels downstream and is eventually deposited in lakes, reservoirs, and estuaries far from the source areas. This sediment is particularly damaging to the bottom environments of aquatic life. The accumulation of fine sediment reduces the capacity of reservoirs, limiting the useful life of a large reservoir to perhaps a century or so. Excess sediment also results in rapid filling of tidal estuaries, requiring increased dredging of channels.

LANDFORMS OF FLOODPLAINS

An *alluvial river* is one that flows on a thick accumulation of alluvial deposits constructed by the river itself in earlier stages of its activity. A characteristic of an alluvial river is that it experiences overbank floods with a frequency ranging between annual and biennial occurrence during the season of large water surplus over the watershed. Overbank flooding of an alluvial river normally inundates part or all of a floodplain that is bounded on either side by rising slopes, called *bluffs*.

Typical landforms of an alluvial river and its floodplain are illustrated in Figure 17.25. Dominating the floodplain is the meandering river channel itself, plus abandoned reaches of former channel (Figure 17.26). Meanders develop narrow necks, which are cut through, shortening the river course and leaving a meander loop abandoned. This event is called a *cutoff*. It is followed quickly by deposition of silt and sand across the ends of the abandoned channel, producing an *oxbow lake*. The oxbow lake is gradually filled in with fine sediment

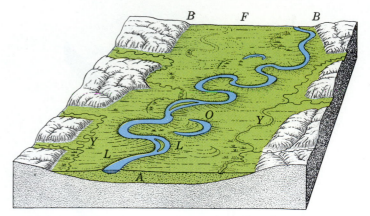

FIGURE 17.25 Floodplain landforms of an alluvial river. L = natural levee; O = oxbow lake; Y = yazoo stream; A = alluvium; B = bluffs; F = floodplain. (Drawn by E. Raisz.)

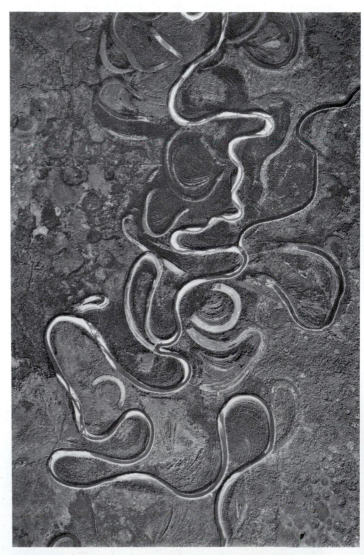

FIGURE 17.26 This vertical air photograph, taken from an altitude of about 6 km (20,000 ft) shows meanders, cutoffs, oxbow lakes and swamps, and floodplain of the Hay River in Alberta. (Department of Energy, Mines and Resources, Canada.)

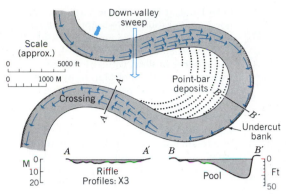

FIGURE 17.27 An idealized map and cross-profiles of a meander bend of a large alluvial river, such as the lower Mississippi River. Small arrows show position of the swiftest current. (From A. N. Strahler, *The Earth Sciences*, 2nd ed., Harper & Row, Publishers. Copyright © by Arthur N. Strahler.)

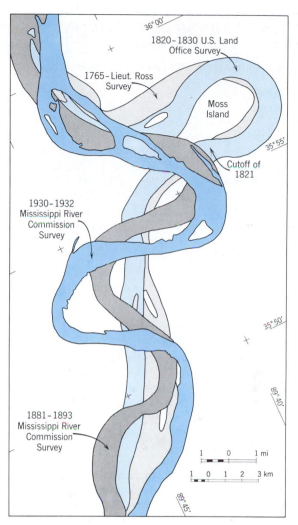

FIGURE 17.28 A series of four surveys of the Mississippi River shows considerable changes in the position of the channel and the form of the meander bends. Note that one meander cutoff has occurred (1821) and new bends are being formed. (After U.S. Army Corps of Engineers.)

brought in during high floods and with organic matter produced by aquatic plants. Eventually the oxbows are converted into swamps, but their identity is retained indefinitely.

The growth of alluvial meanders leaves distinctive marks on the floodplain. As shown in Figure 17.27, the channel in a meander bend is deep close to the outside or undercut bank, which yields by caving and allows the bend to grow in radius. Here the channel has its greatest depth, forming a pool. As flow passes from one bend to the next the threads of swiftest current cross the channel diagonally in a zone known as the "crossing." This element of the channel, which is shallow and has many shifting bars, constitutes a riffle. Thus pools and riffles occur in alternation, corresponding with each meander bend. The meander bend not only grows laterally, but also shifts slowly downvalley in a migratory movement known as *downvalley sweep*. Figure 17.28 shows both meander growth and sweep on the Mississippi River over a century's time. The combined effect of meander growth and sweep gives to the point-bar deposits nested arcuate patterns consisting of bars (embankments of bed material) and swales (troughs between bars) (Figure 17.29). *Bar-and-swale* topography thus produced is clearly visible in Figure 17.26.

FIGURE 17.29 An oxbow lake of the Snake River, near Roberts, Idaho. The Snake River can be seen at the upper left. At both ends, the lake is blocked by hygrophytic (swamp) forest. The pattern of nested arcs within the former meander bend are sandbars and intervening low troughs (swales), showing the stages in growth of the meander. (John S. Shelton.)

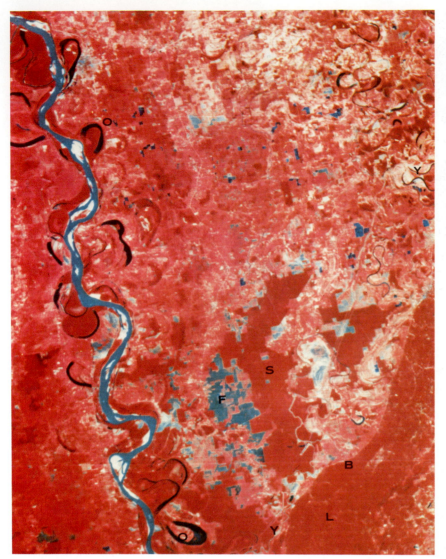

(left) Landsat image of the Yazoo Basin, north of Vicksburg, Mississippi, taken in late August 1973. Blue represents water surfaces, such as the Mississippi River (left) and patches of flood water *(F)* remaining from the great flood of March–April 1973. Oxbow lakes *(O)* appear as dark blue crescents. Forested land forms solid red patches on backswamp areas *(S)* of the floodplain. The eastern bluffs *(B)* of the floodplain show as a sharp line, with forested upland underlain by loess *(L)* at the right. The Yazoo River *(Y)* follows the eastern edge of the floodplain. Sand bars in the Mississippi River channel show as pure white. Other white areas are cultivated fields with a large proportion of bare soil. Highways and railroad embankments form thin white lines. (NASA.)

(above) An abandoned section of the Mississippi River near Mound Bayou, Louisiana. Soil color differences in the cultivated area reveal old bars and swales. (A. N. Strahler.)

(left) Cultivated natural levees (right) and forested backswamp (background) of the Mississippi River, just south of New Orleans, Louisiana. An artificial levee runs close to the river channel (Orlo E. Childs.)

FIGURE 17.30 The Mississippi River floodplain.

FIGURE 17.31 This river floodplain in Bangladesh is largely under water during a 1973 flood. Villages occupy the higher ground of the natural levees bordering the river channel. (A. Moldvay/Photo Researchers.)

During periods of overbank flooding, when the entire floodplain is inundated, water spreads from the main channel over adjacent floodplain deposits. As the current rapidly slackens, sand and silt are deposited in a zone adjacent to the channel. The result is an accumulation known as a *natural levee*. Because deposition is heavier closest to the channel and decreases away from the channel, the levee surface slopes away from the channel (Figures 17.25 and 17.30). In times of overbank flooding, the higher ground of the natural levees is often revealed by a line of trees and buildings on either side of the channel (Figure 17.31). Between the levees and the bluffs is lower ground, the *backswamp* (Figures 17.25 and 17.30).

After a flood wave has passed, the water from the backswamps drains back into the river by means of smaller streams that flow downvalley, parallel with the main channel (see Figure 17.25). These parallel streams are prevented from joining the large stream because of the natural levees. At some downvalley point where the main channel is undercutting the floodplain bluffs, the parallel tributary is forced into a junction with the main channel.

A famous example is the Tallahatchie–Yazoo River system in Mississippi. For a distance of about 280 km (175 mi), this stream flows parallel with the Mississippi River, to make its junction at Vicksburg, where the big river is cutting into the eastern bluffs of the floodplain (Figure 17.30). Many years ago, a leading geographer selected the Tallahatchie–Yazoo as the type example of a floodplain stream with its junction deferred far downvalley by the presence of a natural levee; he named it a *yazoo stream*. This quaint term is still widely used.

Overbank flooding results not only in the deposition of a thin layer of silt on the floodplain, but also brings an infusion of dissolved mineral substances that enter the soil. As a result of the resupply of nutrients, floodplain soils retain their remarkable fertility in regions of soil–water surplus from which these substances are normally leached away. Important areas of alluvial soils in the Mississippi River floodplain and the lowland of the Ganges River maintain a high fertility by this process.

ALLUVIAL RIVERS AND HUMAN OCCUPATION

Alluvial rivers attracted human habitations long before the dawn of recorded history. Early civilizations arose in the period 4000 to 2000 B.C. in alluvial valleys of the Nile River in Egypt, the Tigris and Euphrates rivers in Mesopotamia, the Indus River in what is now Pakistan, and the Huanghe (Yellow River) in China. Fertile and easily cultivated alluvial soils, situated close to rivers of reliable flow, from which irrigation water was easily lifted or diverted, led to intense utilization of these fluvial zones. With respect to human culture, the alluvial river and its floodplain are sometimes referred to by geographers as the *riverine environment*. Today about half the world's population lives in southern and southeastern Asia; most of these persons are small farmers who cultivate alluvial soils of seven great river floodplains.

To the agricultural advantages of alluvial zones is added the value of the rivers themselves as arteries of transportation. Navigability of large alluvial rivers led to the growth of towns and cities, many situated at the outsides of meander bends where deep water lies close to the bank, or at points on the floodplain bluffs where the river is close by (e.g., Memphis, Tennessee, and Vicksburg, Mississippi).

FLOOD ABATEMENT MEASURES

In the face of repeated disastrous floods, vast sums of money have been spent on a variety of measures to reduce flood hazards on the floodplains of alluvial rivers. The economic, social, and political aspects of flood abatement are beyond the scope of physical geography; we shall review only the engineering principles applied to the problem. Two basic forms of regulation are to (1) detain and delay runoff by various means on the ground surfaces and in smaller tributaries of the watershed, and (2) modify the lower reaches of the river where floodplain inundation is expected.

The first form of regulation aims at treatment of watershed slopes, usually by reforestation or planting of other vegetative cover to increase the amount of infiltration and reduce the rate of overland flow. This type of treatment, together with construction of many flood-storage dams in the valley bottoms, can greatly reduce the flood crests and allow the discharge to pass into the main stream over a longer period of time.

Under the second type of flood control, designed to protect the floodplain areas directly, two quite different theories can be practiced. First, the building of *artificial levees*, parallel with the river channel on both sides can function to contain the overbank flow and prevent inundation of the adjacent floodplain. Artificial levees, also known as *dikes*, are broad embankments built of earth; they must be high enough to contain the greatest floods, or they will be breached rapidly by great gaps at the points where water spills over. Under the control of the Mississippi River Commission, which was created in 1879, a vast system of levees was built along the Mississippi River in the expectation of containing all floods. Levees have been continuously improved and now total more than 4000 km (2500 mi) in length and in places are as high as 10 m (30 ft).

Because of natural and artificial levees, the river channel is slowly built up to a higher level than the floodplain; this makes these "bottomlands," as they are called, subject to repeated inundations and consequent heavy loss of life and property. Floodplains of many of the world's great rivers present serious problems of this type. In China in 1887, the Huanghe (Yellow River) inundated an area of 130,000 sq km (50,000 sq mi), causing the direct death of 1 million people and the indirect death of a still greater number through ensuing famine.

The second theory, practiced in more recent decades on the Mississippi River by the U.S. Army Corps of Engineers, is to shorten the river course by cutting channels directly across the great meander loops to provide a more direct river flow. Shortening has the effect of increasing the river slope, which in turn increases the

mean velocity. Greater velocity enables a given flood discharge to be moved through a channel of smaller cross-sectional area; the flood stage is correspondingly reduced. Channel improvement, begun in the 1930s, initially had a measurable effect in reducing flood crests along the lower Mississippi, and the levees were not in such great danger of being overtopped. New meander bends grew rapidly, however, and they proved difficult to control.

Certain parts of the floodplain are also set aside as temporary basins into which the river is to be diverted according to plan to reduce the flood crest. In the delta region, floodways are designed to conduct flow from river channels to the ocean by alternate routes.

The flood abatement measures we have described constitute a structural approach to the problem of coping with flood hazards. An alternative lies in a nonstructural approach that recognizes floods as natural phenomena and accepts the fact that even the most elaborate and costly engineering structures may not prevent flood damage. Nonstructural measures include zoning the floodplain for uses that can accommodate occasional floods (e.g., agricultural use and recreational use) without exorbitant financial loss and danger to human life. Under wise floodplain zoning, new urban housing and industry are kept out of flood-prone areas.

Unfortunately, the history of settlement of our nation has been such that flood-prone areas have attracted towns and cities and the industrial plants and highways that go with them. Cheap river transport was, of course, a potent factor in encouraging the early growth of these urban concentrations close to riverbanks on low floodplain sites. A river not only supplied large amounts of water needed for domestic use and industry, but also served as a convenient flow system by means of which sewage and industrial wastes could be disposed of cheaply.

Continued urban and industrial development on floodplains of major rivers began to be questioned as early as the 1930s. In 1937 the *Engineering News-Record* posed this question: "Is it sound economics to let such property be damaged year after year, to rescue and take care of the occupants, to spend millions for their local protection, when a slight shift of location would assure safety?" A new, broader outlook on the relationship of humans to floods was presented in 1945 by a geographer, Professor Gilbert White, in a treatise titled *Human Adjustment to Floods*. Since that time, geographers have played a leading role in analyzing problems of flood hazards and in suggesting broad policies for human adaptation to floods.

STREAM REJUVENATION AND ENTRENCHED MEANDERS

A graded stream experiences a major disruption in activity when the crust beneath it is raised with respect to sea level. During and following the crustal uplift, the stream undergoes channel degradation in an attempt to reestablish grade at a lower level. This process, termed *rejuvenation*, begins as a series of rapids near the mouth, where the flow passes from the former mouth down to the lowered sea level. The rapids quickly shift upstream, and soon the entire stream valley is being trenched to form a new valley (Figure 17.32).

If rejuvenation affects an alluvial river with a floodplain, the effect is to give a steep-walled inner gorge. On either side lies the former floodplain, now a flat terrace

high above river level (Figure 17.33). This feature is called a *rock terrace* to distinguish it from an alluvial terrace.

Rejuvenation may cause the meanders to become impressed into the bedrock and give the inner gorge a meandering pattern. These sinuous bends are termed *entrenched meanders* to distinguish them from the floodplain meanders of an alluvial river (Figures 17.34 and 17.35).

Although entrenched meanders are not free to shift about as floodplain meanders, they can enlarge slowly so as to produce cutoffs. Cutoff of an entrenched meander leaves a high, round hill largely surrounded by the deep abandoned river channel and separated from the valley wall by the shortened river course (see rear part of Figure 17.34). As you might guess, such hills formed ideal natu-

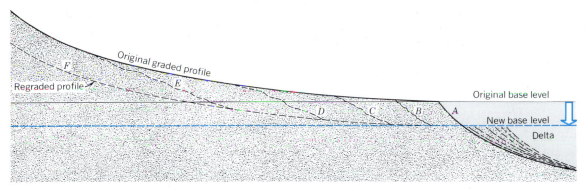

FIGURE 17.32 A drop in base level brings on rejuvenation and regrading of the stream profile, starting at point A and progressing upstream. (From A. N. Strahler, *The Earth Sciences*, 2nd ed., Harper & Row, Publishers. Copyright © by Arthur N. Strahler.)

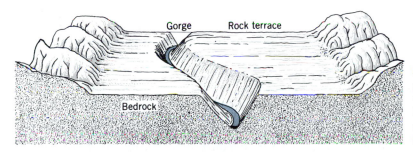

FIGURE 17.33 Following rejuvenation, a winding gorge has been carved into a former floodplain, which has become a high rock terrace. (From A. N. Strahler, *The Earth Sciences*, 2nd ed., Harper & Row, Publishers. Copyright © by Arthur N. Strahler.)

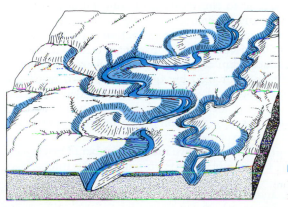

FIGURE 17.34 Rejuvenation of a meandering stream has produced entrenched meanders. One meander neck has been cut through, forming a natural bridge. (Drawn by E. Raisz.)

FIGURE 17.35 The Goosenecks of the San Juan River in Utah are deeply entrenched river meanders in horizontal sedimentary strata. The canyon, carved in sandstones and limestones of Pennsylvanian age, is about 370 m (1300 ft) deep. (Breck Kent/Earth Scenes.)

FIGURE 17.36 Rainbow Bridge was formed by the cutoff of an entrenched meander of Bridge Creek, a tributary to the Colorado River in Glen Canyon, southern Utah. The arch of massive sandstone rises 81 m (267 ft) above the narrow creekbed beneath it. The arch at its summit is 13 m (42 ft) thick. Weathering of the sandstone has played the major role in shaping the bridge, following the cutoff. The surface of Lake Powell (foreground) reaches to the base of the arch, making it accessible by boat. (A. N. Strahler.)

ral fortifications. Many European fortresses of the Middle Ages were built on such cutoff meander spurs. A good example is Verdun, near the Meuse River.

Under unusual circumstances, where the bedrock includes a strong, massive sandstone formation, meander cutoff leaves a *natural bridge* formed by the narrow meander neck. One well-known example is Rainbow Bridge near Navajo Mountain, in southeastern Utah (Figure 17.36). Other fine examples can be seen in Natural Bridges National Monument at White Canyon in San Juan County, Utah.

FLUVIAL PROCESSES IN REVIEW

Running water as a fluid agent of denudation dominates the sculpturing of the continents. We have examined fluvial systems in which overland flow becomes organized into channel systems, converging the flow into larger and larger streams. As streams erode, transport, and deposit rock material, they shape a variety of secondary landforms. Gorges and canyons dominate the earlier stages in valley evolution. These transient forms finally give way to broad floodplains on which meandering rivers move on low gradients. More subtle landforms, such as the faint natural levees, now dominate the fluvial landscape. Overbank flooding and widespread inundation bring major environmental impact to these flat alluvial lands, where they are intensively occupied by humans. Efforts to control flooding by engineering works often fail, despite enormous inputs of money. Doubts are now raised as to the wisdom of continuing to protect lands naturally subject to river inundations.

The impact of humans on the action of running water is severe in agricultural and urban areas. Soil erosion is easily induced, but controlled only with great difficulty. Channels are damaged by aggradation from upstream sources of sediment. Reservoirs impounded by great dams are being steadily filled with sediment, and we have good reason to question the desirability of building more of these expensive works.

In this chapter, we have made a number of brief references to the role of climate in varying the intensity of fluvial processes. In the next chapter, we turn to a systematic analysis of the way denudation is influenced by climate.

DENUDATION AND CLIMATE

In the last two chapters we examined the processes of weathering, mass wasting, overland flow, and stream flow. These processes act in concert and cannot be considered in isolation. Fluvial landforms reflect the simultaneous action of all the contributing processes. In this chapter we turn to a wider outlook on fluvial denudation, broadening our focus to cover an entire region of subcontinental dimensions and emphasizing the landscape changes that take place over long spans of time.

The concept of a *cycle of denudation* has been strongly rooted in geographic thought, ever since its introduction in the late 1800s by a leading American geographer, William Morris Davis. The denudation cycle, as envisioned by Davis, has its initial stage at a time when the crust of a continent is rapidly elevated by internal earth forces. Previously, this portion of crust may have been the floor of a shallow inland sea or of a shallow continental shelf. Once raised high above sea level as a landmass, the fluvial agents begin their attack. Streams come into existence and evolve through the stages we examined in the last chapter, eventually becoming graded in their trunk sections. In a stage of youth, branching tributaries carve up the elevated landmass into innumerable erosional landforms. Davis envisioned that when the landscape assumes a rugged, mountainous aspect, it has entered the stage of maturity.

As time passes, land surfaces are lowered in altitude and the slopes decline in steepness. As relief diminishes, thick regolith mantles the hillslopes and divides and the landscape assumes a subdued aspect—a stage of late maturity. Eventually relief diminishes to the point that only a low, undulating land surface remains. In this stage of old age, the land surface has become a *peneplain*. In coining this word, Professor Davis had in mind two words: "penultimate" and "plain." Realizing that erosion processes must act with increasing slowness in the latter stages of the denudation cycle, Davis postulated that the cycle would be terminated through disruption by a geologic event in which the land surface is abruptly lifted or submerged. If the peneplain is uplifted, a new cycle will be initiated.

We no longer use Davis's stage names—youth, maturity, old age—because they imply a biologic model of growth stages in individuals of an animal species. We do, however, incorporate many of Davis's concepts into a useful physical model of landmass denudation that takes into account the new paradigm of plate tectonics.

CONCEPT OF THE AVAILABLE LANDMASS

That part of the continental crust lying above sea level, which is the base level for erosion by streams, constitutes the *available landmass*. It is a mass of bedrock subject to denudation and available for ultimate removal. As denudation takes place, however, the crust gradually and spontaneously rises, so the total available landmass is actually of a much greater thickness than appears at any one instant. To understand what happens, we must investigate a principle of geology.

Figure 18.1 is a schematic diagram showing the earth's outer shell of hard, brittle rock—the lithosphere—resting on a plastic rock layer—the asthenosphere. An important geologic principle, established for many decades, is that the lithosphere literally floats on the plastic asthenosphere, much as a layer of sea ice or an iceberg floats on denser seawater. In the case of an iceberg, if you were to remove the part showing above water, the iceberg would rise and bring more ice above

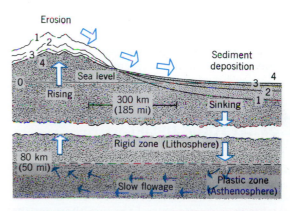

FIGURE 18.1 Isostatic compensation for landmass erosion and sediment deposition. (From A. N. Strahler, *The Earth Sciences*, 2nd ed., Harper & Row, Publishers. Copyright © by Arthur N. Strahler.)

water level. If you placed an added load of ice on top of the iceberg, it would sink lower in the water and come to a new position of rest. This flotation principle, applied to the lithosphere, is called *isostasy*. The word is derived from Greek words: *isos*, "equal," and *stasis*, "standing."

As Figure 18.1 shows, when bedrock is eroded from a continent (left), the lithosphere rises beneath it; but the amount of rise is somewhat less than the thickness of material removed. When the eroded sediment accumulates in layers (right), the added load causes the lithosphere to sink lower; but the amount of sinking is somewhat less than the thickness of sediment accumulating. A rising or sinking of the crust in response to denudation or sediment accumulation is referred to as *isostatic compensation*. In the plastic zone below, a lateral movement of mass by very slow flowage accompanies the rising and sinking of the lithosphere. This activity occurs at a depth averaging about 80 km (50 mi), where rock is close to its melting point and has little strength to resist flowage.

Because of slow crustal uplift due to isostasy, the available landmass is actually several times greater than the mass lying above base level at any one instant. As a satisfactory working figure, we can say that the removal of 5000 m of rock would be accompanied by a compensating crustal uplift of 4000 m, so the net lowering of the land surface would amount to only 1000 m. In setting up a model denudation cycle, the net lowering figure must be used.

TECTONIC UPLIFT AND PLATE TECTONICS

A geologic phenomenon quite different from crustal rise accompanying denudation is a rapid upward movement associated with tectonic activity related to lithospheric plate interactions. As subduction continues without interruption for tens of millions of years, the continental margin undergoes spasmodic uplift. This happens because magma rising from the downsinking plate is added to the overlying continental lithosphere, increasing its thickness. During arc–continent and continent–continent collisions, thickening of the continental lithosphere may occur rapidly. Telescoping of continental shelf strata during arc–continent collision is one mode of rapid crustal thickening. During continent–continent collision (such as that which produced the Tibetan plateau), subduction can double the thickness of continental lithosphere and produce a high plateau. Thus, orogeny works in different ways to induce tectonic uplift of continents.

RATES OF TECTONIC UPLIFT

Our model of denudation begins with an event of tectonic uplift. It is reasonable to assume that the rate of tectonic uplift is far more rapid than the rate of fluvial denudation. Based on direct geologic evidence, rates of recent tectonic uplift in various parts of the world are on the order of 4 to 12 m per 1000 years (15 to 40 ft/1000 yr). Using a rate of 6 m/1000 yr (20 ft/1000 yr), a mountain summit might rise from sea level to an elevation of 6 km

(20,000 ft) in about 1 million years, assuming none of the mass to have been lost by erosional reduction during the uplift. Even with erosion in progress during uplift, a full-sized mountain mass could be created in a time span vastly shorter than the time required for its reduction to a low plain by denudation.

RATES OF DENUDATION

Geologists and hydrologists have made a number of attempts to estimate rates of continental denudation. One method is to take available data on the load of streams, combining both suspended load and solid load. Another form of data consists of the measured accumulations of sediment in reservoirs. As we showed in Chapter 17, sediment yield varies enormously from place to place, depending on climate, vegetation cover, and parent material of the soil. Relief is another major factor; some areas of high relief and steep slopes have enormously high rates of sediment yield compared with undulating plains under a cover of forest. We shall investigate these regional differences later in this chapter, in connection with the influences of climate on geomorphic processes.

For our model denudation cycle, which presupposes a moist climate with a substantial water surplus, we shall use an initial denudation rate comparable to that observed today in high mountains: 1 to 1.5 m/1000 yr (3 to 5 ft/1000 yr). Notice that this rate is only a small fraction of the assumed rate of tectonic uplift.

We reason that denudation rates are strongly influenced by average summit elevation of the landmass. The higher a mountain block rises, the steeper will be the gradients, on the average, of streams carving into that mass. Both intensity of stream abrasion and the ability of a stream to transport load (stream capacity) increase strongly as channel gradients increase. Valley wall slopes are also steeper, on the average, where stream gradients are steeper, so mass wasting and erosion by overland flow will also be more intense. Thus, as gradients diminish with time, the rate of denudation will diminish steadily from the high initial value we have assumed.

We anticipate that, by the time the landmass has been lowered to an average summit elevation of about 300 m (1000 ft), the denudation rate will have decreased to a value comparable to that found today for the eastern United States or the Amazon River basin: about 4 cm/1000 yr (0.2 ft/1000 yr). This rate is corrected for greatly increased erosion in the United States caused by human activities.

A MODEL DENUDATION SYSTEM

Using the assumptions made thus far about rates of tectonic uplift, isostatic compensation, and denudation, we can devise a model of the denudation process and from it perhaps obtain some idea of the order of magnitude of time required to reduce the landmass to a peneplain. First, we image that a large crustal mass is raised as a single block from a previous position close to sea level.

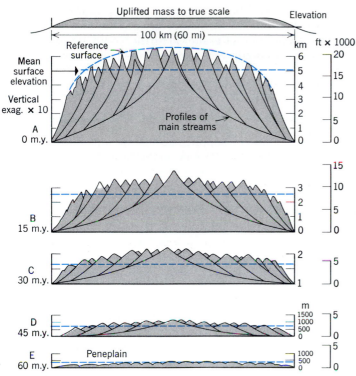

FIGURE 18.2 Schematic diagram of landmass denudation. In this model, the average surface elevation is reduced by one-half every 15 million years. (From A. N. Strahler, *The Earth Sciences*, 2nd ed., Harper & Row, Publishers. Copyright © by Arthur N. Strahler.)

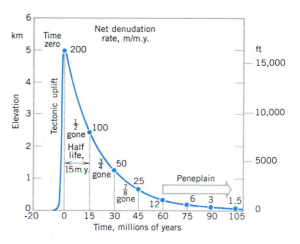

FIGURE 18.3 Graph of decrease in average surface elevation with time, as shown in Figure 18.2. (From A. N. Strahler, *The Earth Sciences*, 2nd ed., Harper & Row, Publishers. Copyright © by Arthur N. Strahler.)

An arbitrary width of about 100 km (60 mi) is assigned to the uplifted block because this dimension is about the order of magnitude of width of a number of present-day mountain ranges, including the European Alps, Carpathians, Pyrenees, Caucasus, Alaska Range, Sierra Nevada, Northern Cascades, Rocky Mountains, and Appalachians. Length of the uplifted block is not important for landform analysis; a segment some tens of kilometers will do. As Figure 18.2 shows, the uplifted block has a very broadly domed summit and steep sides; it is bordered by low areas, at or below sea level, that serve as receiving sites for sediment carried out from the mountain block by streams.

The initial surface of reference, a plain close to sea level, is raised rapidly to an elevation of 6 km (20,000 ft); see Figure 18.2. Although this initial surface is carved up by streams during uplift, its position in space is shown by a dashed line in cross section (A); it is labeled "reference surface." The rapid tectonic uplift is shown on a graph of elevation versus time, Figure 18.3, by a steeply rising curve. We have allowed 5 million years (5 m.y.) for uplift, but most of this occurs within a 2-m.y. period. Uplift tapers off in rate and ceases at a point labeled "Time-zero" on the horizontal scale of the graph.

Denudation in progress during uplift increased sharply in intensity as average elevation increased so that, at time-zero, the mass has already been carved into a maze of steep-walled gorges, organized into a fluvial system of steep-gradient streams. The profile of the rugged mountain mass and the main stream system are sug-

gested in greatly exaggerated scale in Stage A of Figure 18.2. The average elevation of the eroded surface lies at 5 km (16,400 ft). Thus about 1 km (3300 ft) of rock has been removed during uplift.

Starting at time-zero, an average denudation rate of 1 m/1000 yr is assumed for the entire surface. An additional assumption is that isostatic compensation occurs continuously so that the net lowering rate is only one-fifth of the denudation rate, or 0.2 m (20 cm)/1000 yr. With decreasing elevation, the rate of denudation itself diminishes in a constant ratio such that one-half the available landmass is removed in each 15-m.y. period. We may call this time unit the *half-life* of the available mass. After a lapse of 15 m.y. (1 half-life), the average elevation will have been reduced to 2.5 km (8200 ft), shown as Stage B in Figure 18.2. Now the net rate of lowering will have fallen to 0.1 m/1000 yr. At the end of 30 m.y. (2 half-lives), average elevation and denudation rate are again reduced by one-half: to one-quarter of the initial values (Stage C). Now three-quarters of the mass is gone and the average altitude is down to 1.25 km (4100 ft).

The program of decrease in elevation of landmass with time, described here and illustrated in Figure 18.3, is known as *exponential decay*. This is a common program in natural and physical science. For example, it applies to the rate at which a radioactive element decays spontaneously by radioactivity. Exponential decay is exactly the inverse of a program of exponential increase seen in an annual percentage growth rate, such as a savings account at compound interest, a gross national product (GNP), or a pecentage annual population increase. The exponential decay rate, as applied to the denudation of a landmass, simply reflects the principle that the energy available to perform the work of denudation is proportionate to the height of a mass above sea level. Therefore, as average elevation declines, the intensity of change is proportionately diminished (stages B through D).

In Stage E of our model denudation program, an average elevation of about 300 m (1000 ft) is reached after about 60 m.y. The land surface can now be described as a peneplain. No exact definition of a peneplain exists in terms of average elevation or relief. We can only say that hillslopes are very gentle, divides are very broadly rounded, and the region has the aspect of an undulating plain.

LANDMASS REJUVENATION

Our simple model of denudation needs an important modification to bring it closer to reality, as judged from the evidence of geologic history. We have assumed that isostatic compensation occurs uniformly and constantly as denudation proceeds, but this is probably not the case. It is more reasonable to suppose that isostatic compensation occurs spasmodically. Because the lithosphere is a massive, strong layer, it resists being lifted until a certain minimum thickness of rock has been removed. Then, beyond the critical point of strength, any further removal triggers a rapid uplift, restoring the equilibrium. Figure 18.4 is a graph showing the rapid rise in mean surface elevation in response to spasmodic landmass uplifts, each of which is followed by accelerated denudation. Notice that each of the two uplifts shown by peaks in the graph resulted in an approximate doubling of the available landmass.

Uplifts of the continental shields to produce rejuvenations may be caused in other ways related to the movements of lithospheric plates.

A series of block diagrams, Figure 18.5, shows the evolution of the landscape following uplift. Block A shows the elevated peneplain still intact because the small area shown lies in an inland location, far from the coastline where changes begin. What follows is *landmass rejuvenation*. (We explained stream rejuvenation in Chapter 17.) The trenching of previously graded streams begins near their mouths and progresses up the valleys, finally reaching the area of our block diagram. A system of narrow, steep-walled gorges spreads throughout the landmass (Block B). The peneplain is now dismembered into isolated upland patches. The landscape is in the stage that Professor Davis regarded as "youthful." Steep valley-wall slopes continue to increase in height and number; they gradually replace the last peneplain remnants. In the stage shown in Block C, an "early-mature stage," the landscape is extremely rugged (Figure 18.6). Ridges are sharp-crested; little or no flat land can be

found anywhere. Relief has reached its maximum in this stage. By this time, the principal streams have again attained the condition of grade, so further lowering of the valley floors proceeds very slowly. Floodplains begin to appear.

As time passes, relief diminishes over the area, and divides are reduced in elevation and become broadly rounded (Block D). Regolith accumulates in greater thickness on divides and hillslopes. Erosion rates become slower, and sediment supplied to the streams becomes smaller in quantity and finer in texture. Floodplains

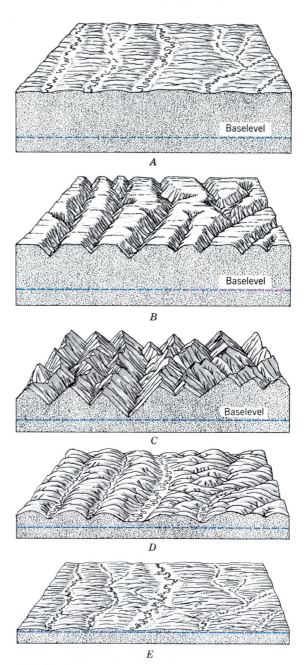

FIGURE 18.5 Stages in the denudation cycle caused by uplift of a peneplain and a rejuvenation of the stream system. Another uplift of the peneplain, shown in Block E, will bring a return to conditions shown in Block A and the cycle will be repeated. (Drawn by A. N. Strahler.)

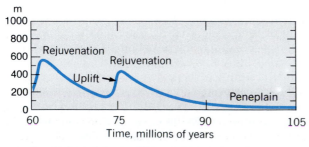

FIGURE 18.4 Repeated rapid uplifts because of isostatic compensation cause successive rejuvenations.

FIGURE 18.6 Sharp ridge crests and steep slopes characterize this portion of the Sawtooth Mountains of south central Idaho. Here, a great batholith of granite provides a landmass of uniform composition and structure. (John S. Shelton.)

widen and the major streams take on the characteristics of alluvial rivers. The subdued landscape shown in Block D would, in Professor Davis's language, be in the "late mature stage." Then, with the passage of many tens of millions of years, a peneplain is again formed with respect to the stable base level (Block E). This event marks the "old stage" in the denudation cycle. At this point in time, another isostatic uplift is triggered off and the peneplain is uplifted, bringing conditions back to those shown in Block A.

The landform evolution we have described is placed in the setting of a moist climate having a substantial annual water surplus and little or no soil-water shortage at any time of year. Streams receive important water contributions from interflow and ground-water seepage, so the larger streams usually are sustained in flow throughout the entire year. These conditions are typical of moist subtropical and continental climates of midlatitudes, as well as the wet equatorial and trade-wind littoral climates. Associated with these climates is the production of thick regolith by chemical weathering beneath a forest cover. The well-protected hillslopes yield little coarse sediment, even during periods of heavy rainfall. Thus the streams carry most of their load as suspended fine sediment and as dissolved matter (ions). Although some of the silt accumulates on floodplains during overbank floods, most of the suspended sediment is carried to the ocean, where it becomes marine sediment.

CLIMATE AND LANDSCAPE REGIONS

For decades, some geomorphologists have harbored a vision of a world system of several basic landscape types, each type uniquely dictated by climate and each clearly different from the others in the assemblage of landforms present. World maps have been drawn to show *climogenetic regions* (or *morphogenetic regions*). These maps appear to be closely correlated with climate systems and are little more than world climate maps with new sets of labels. If one were to travel through all parts of a single climogenetic region, the landscape would actually prove greatly different from place to place and from continent to continent.

Obviously, climate is not the sole factor determining landform characteristics. The climogenetic concept omits from consideration the factor of geologic control acting through differences in bedrock composition and structure and through the geologic history of the region in terms of volcanic and tectonic activity. In short, although climate is a potent control over landform development, as we have already established, it is but one of two basic controls of landform development; the other is geology. We shall develop the factor of geologic control in Chapter 19.

CLIMATE-PROCESS SYSTEMS

The role of climate in shaping landforms can be expressed through recognition of *climate-process systems*. Each system comprises a unique combination of levels of intensity of several basic denudation processes. The major process categories are weathering, mass wasting, fluvial processes, and wind action. The action of waves and currents in shaping coastlines is usually omitted from consideration because it affects only a thin line bordering the continents. Glacial processes are, of course, under direct climate control, but the landforms made beneath glacial ice by erosion and deposition are best considered separately. (Chapter 22 covers this subject.) Processes of weathering and mass wasting acting in the cold environment close to bodies of glacial ice comprise a *periglacial* climate-process system.

In Chapter 15 we examined the periglacial climate-process system in our review of landforms of the arctic tundra where permafrost prevails. We turn now to two other great climate-process systems. One is found in the desert environment of tropical and midlatitude regions, the other in the savanna environment where the wet–dry tropical climate prevails.

FIGURE 18.7 A roadside soil exposure showing rocklike slabs of caliche (calcrete). Pecos Plains near Vaughan, New Mexico. (A. N. Strahler.)

FIGURE 18.9 A bar of bouldery rock fragments (right) built during a brief desert flood. Gonzales Pass Canyon, Arizona. (Copyright © by Mark A. Melton.)

THE DESERT CLIMATE-PROCESS SYSTEM

Geomorphic Processes of the Desert Environment

A common denominator of desert landscape evolution exists through a unique set of weathering processes, fluvial processes, and the work of wind. Mechanical weathering by salt-crystal growth is a dominant and pervasive process. Rock disruption by thermal expansion and contraction during the diurnal cycle of heating and cooling is also a universal process, but its effectiveness is questionable. Although chemical weathering is considered relatively less important in deserts than it is in moister climates, evidence is clear that hydrolysis and oxidation affect the silicate minerals wherever they are exposed throughout the desert environment. Surfaces of pebbles, boulders, and rock outcrops become darkened by thin films of clay minerals and oxides of iron and manganese. This black, iridescent coating is called *desert varnish*. In the absence of below-freezing temperatures,

frost action can be considered ineffective in the warmer desert regions.

Evaporation of soil water and ground water, brought toward the surface of soil and bedrock by capillary film movement, results in a variety of rocklike crusts. The surface layer of porous sandstone becomes indurated into a hard crust by carbonate deposition. Where the crust is broken away, crumbling of the softer rock inside through salt-crystal growth leads to the formation of deep pits and hollows, and these may be enlarged into shallow caves (Chapter 15).

Within desert soils (which belong to the soil order of Aridisols) precipitation of calcium carbonate below the surface layer commonly forms a hard, rocklike layer called the petrocalcic horizon (Figure 18.7). Where the overlying soil horizons have been removed by erosion, the petrocalcic horizon is exposed as a hard rocklike crust. Geographers refer to this layer as *calcrete* (also *caliche*) to distinguish it from crusts of iron oxides or

FIGURE 18.8 A flash flood has filled this desert channel with raging, turbid waters. A distant thunderstorm produced the runoff. Coconino Plateau, Arizona. (A. N. Strahler.)

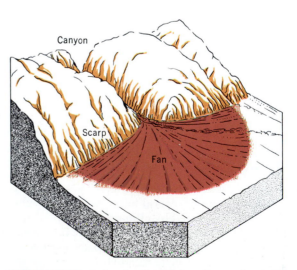

FIGURE 18.10 A simple alluvial fan. (Drawn by A. N. Strahler.)

FIGURE 18.11 Great alluvial fans extend out upon the floor of Death Valley. The canyons from which they originate have carved deeply into a great uplifted fault block. (Mark A. Melton.)

silica found in the savanna environment. As calcrete crusts are eroded, they give rise to stepped landforms resembling the cliffs, mesas, and buttes of eroded sedimentary strata. Crusts formed of gypsum are also common.

Fluvial Processes in an Arid Climate

The general appearance of a desert landscape is strikingly different from that of a humid region, reflecting differences in both vegetation and landforms. Torrential rainfalls occur in dry as well as in moist climates, and most landforms of desert regions are formed by running water. (Areas blanketed by dune sands are an important exception.) A particular locality in a dry desert may experience heavy rain only once in several years; but, when it does fall, stream channels carry water and perform important work as agents of erosion, transportation, and deposition. Although running water is a rather rare phenomenon in dry deserts, it works with spectacular effectiveness on the few occasions when it does act. This is explained by the meagerness of vegetation in dry deserts. The few small plants that survive offer little or no protec-

tion to soil or bedrock. Without a thick plant cover to protect the ground and hold back the swift downslope flow of water, large quantities of sediment are swept into the streams. A dry channel is transformed in a few minutes into a raging flood of muddy water, heavily charged with rock fragments (Figure 18.8). The bed load left behind when the flood has passed consists of coarse, bouldery debris, heaped into massive channel bars (Figure 18.9).

Recall from Chapter 16 that, in a dry climate, streams flowing across thick accumulations of permeable alluvium are of the influent class (see Figure 16.26). Stream water infiltrates the bed and contributes to a groundwater mound beneath. Because of loss of discharge through influent seepage and direct surface evaporation, streams of arid lands usually show braided channels and are actively engaged in aggradation. These streams are often short and disappear on alluvial deposits or end on shallow, dry lake floors.

Alluvial Fans

A very common desert landform built by braided, aggrading streams is the *alluvial fan*, a low cone of alluvial sands and gravels resembling in outline an open Japanese fan (Figure 18.10). The apex, or central point of the fan, lies at the mouth of a canyon or ravine. The fan is built out on an adjacent plain. Alluvial fans are of many sizes; some desert fans are many kilometers across (Figure 18.11).

Fan-building streams carry heavy bed loads of coarse rock waste from a mountain or upland region. The braided channel shifts constantly, but its position is firmly fixed at the canyon mouth. The lower part of the channel, below the apex, sweeps back and forth. This activity accounts for the semicircular fan form and the downward slope in all radial directions from the apex.

Large, complex alluvial fans also include mudflows (see Figure 15.26). As shown in Figure 18.12, mud layers are interbedded with sand and gravel layers. Water infiltrates the fan at its head, making its way to lower levels along sand layers that serve as aquifers. The mudflow layers act as barriers, or aquicludes. Ground water trapped beneath an aquiclude is under hydraulic pressure from water accumulated higher in the fan apex. When a well is drilled into the lower slopes of the fan,

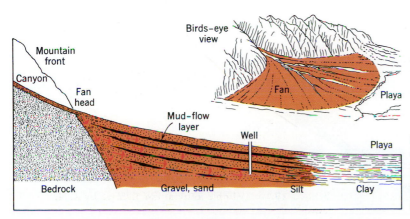

FIGURE 18.12 An idealized cross section of an alluvial fan showing mudflow layers (aquicludes) interbedded with sand layers (aquifers). (From A. N. Strahler, *Planet Earth: Its Physical Systems Through Geologic Time,* Harper & Row, Publishers. Copyright © by Arthur N. Strahler.)

water rises spontaneously as artesian flow (see Chapter 16).

Alluvial fans are the dominant class of ground-water reservoirs in the southwestern United States. Sustained heavy pumping of these reserves for irrigation has lowered the water table severely in many fan areas. The natural rate of recharge is extremely slow in comparison. However, efforts are made to increase this recharge by means of water-spreading structures and infiltrating basins on the fan surfaces.

As explained in Chapter 16, a serious side effect of excessive ground-water withdrawal is that of subsidence of the ground surface. Important examples of subsidence are found in alluvial valleys filled to great depth with alluvial fan and lake-deposited sediments.

Landscape Evolution in the Basin and Range Province

Where tectonic activity has been active in an area of continental desert, the assemblage of fluvial landforms is particularly diverse. The Basin and Range Province of the western United States is such an area; it includes large parts of Nevada and Utah, southeastern California, southern Arizona and New Mexico, and adjacent parts of Mexico. Between uplifted and tilted crustal blocks are down-dropped tectonic basins.

Keep in mind that geologic history plays a major role in determining the overall appearance and relief of the desert landscape. The Basin and Range structure of the mountainous desert of the American Southwest, associ-

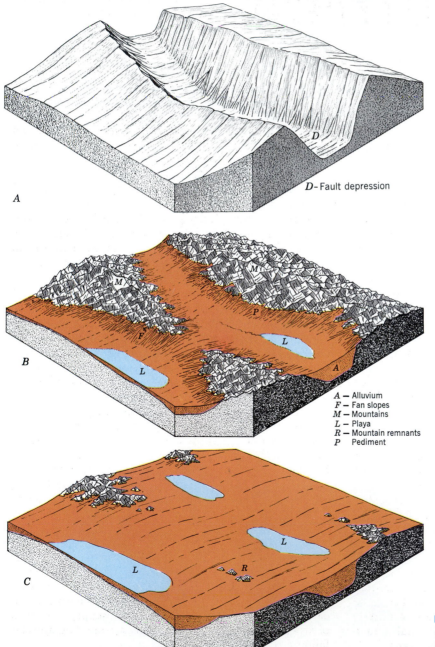

D– Fault depression

A — Alluvium
F — Fan slopes
M — Mountains
L — Playa
R — Mountain remnants
P   Pediment

FIGURE 18.13 Stages in the denudation of mountain blocks in a desert environment. (Drawn by A. N. Strahler.)

FIGURE 18.14 Racetrack Playa, a flat, white plain, is surrounded by alluvial fans and rugged mountains. This desert valley lies in the northern part of the Panamint Range, not far west of Death Valley, California. In the distance rises the steep eastern face of the Inyo Mountains, a great fault block. (John S. Shelton.)

ated with continental rifting and block faulting, is almost unique as a geologic type. The largest of the world's tropical deserts—those of Africa and Australia—occupy stable continental shields that are fragments of the ancient subcontinent of Gondwana. A large variety of rocks and structures is exposed in these shields, while some large areas have covers of relatively undisturbed sedimentary strata. Other deserts are found in the intermontane basins of alpine belts, where plate boundaries have experienced collision in recent geologic time. Deserts of Iran, Afghanistan, and Pakistan fall into this category. The coastal desert of Peru and Chile owes its position and narrow width to an active subduction zone along which the Nazca plate is descending beneath the American plate.

Block A of Figure 18.13 shows two lifted fault blocks with a down-dropped block between. Although denudation acts on the uplifted blocks as they are being raised, we have shown them as very little modified at the time tectonic activity has ceased. Our model landscape thus has a very great local relief; that is, there is a very great elevation difference between block summits and the nearby basin floors. This elevation difference might be on the order of 2 to 3 km (6,500 to 10,000 ft). At first, the faces of the fault block are extremely steep. They are scored with deep ravines. Many cones of talus have been built along the base of each block.

Block B shows a later stage of denudation in which streams have carved up the mountain blocks to produce a rugged terrain of deep, steep-walled canyons and high, sharp divides. Figure 18.11 shows this type of landscape as it appears today along the western side of Death Valley, California. Large alluvial fans have been built side by side; the fan deposits form a continuous apron extending far out into the basin.

In a central position in the basins lie *playas*, the flat floors of lakes that are dry much of the time or contain only shallow lakes (Figure 18.14). Fine silts and clays lie under the playa surfaces, along with layers of soluble salts. Salt lakes of playas represent the terminal sites of stream flow. In many cases, large alluvial fans, built across the entire basin, serve as barriers between individual depressions in the basin floors. Because evaporation greatly exceeds precipitation, these lakes do not rise above the limits of the closed depressions in which they lie.

As the mountain masses are lowered in height and reduced in extent, the fan slopes encroach on the mountain bases. Close to the receding mountain front, erosion processes develop a sloping rock floor called a *pediment* (Figures 18.13 and 18.15). As the remaining mountains continue to shrink in extent, the pediments widen to form broad rock platforms thinly veneered with alluvium. In the deserts of the southwestern United States and northern Mexico, the graded slope extending from mountain base to playa is called a *bajada*. The term includes both alluvial fan and pediment surfaces.

Block C of Figure 18.13 shows an advanced stage of denudation in the desert environment. The once lofty mountains are reduced to small remnants, but these retain their steep slopes and sharp ridge crests. A small mountain mass, isolated from the main body of the range and completely surrounded by pediment or fan surfaces, is an *inselberg* (German term meaning "island mountain").

Throughout the denudation stages we have described, fluvial processes are limited to local transport of rock particles from a mountain range to the nearest basin. The basin floor, which receives all the sediment, must be gradually rising in elevation as the mountains are diminishing in elevation. Because there is no outflow to the sea, the concept of a base level of denudation has no meaning and a peneplain does not represent the ultimate stage of denudation, as it does in the humid environment. Each arid basin becomes a closed system so far as mass transport is involved. Only the hydrologic system is open, with water entering as precipitation and leaving as evaporation. The desert land surface produced in an advanced stage of fluvial activity is an undulating plain, called a *pediplain*. As an idealized landscape, it consists of areas of pediment merging with surrounding alluvial fan and playa surfaces.

FIGURE 18.15 A pediment formerly carved in granite at the foot of a desert mountain slope is now seen in profile as a straight, sloping line. Later erosion has exposed the bedrock, and weathering has formed the joint blocks into boulders (intermediate distance). Little Dragoon Mountains, near Adams Peak, about 20 km (12 mi) northeast of Benson, Arizona. (Copyright © 1991 by Mark A. Melton.)

HILLSLOPE EVOLUTION IN HUMID AND ARID CLIMATES

In describing the evolution of landscapes in both humid and arid environments, we have indicated two quite different modes of change in the forms of land slopes. Collectively, all land surfaces contributing runoff and sediment to stream channels are referred to as *hillslopes*, whether these be land surfaces of high mountains or low hills, or the walls of canyons. An individual element of a hillslope begins at a point on a drainage divide, or watershed, and follows the path of overland flow to the nearest stream channel. We visualize this hillslope element as a narrow strip to which a unit width is assigned—one meter or one foot, for example. A *hillslope profile* is a plot of elevation versus horizontal distance, following the midline of the hillslope element.

Hillslope profiles show a great variety of shapes. The meaning of each shape and the way it changes with time are subject to interpretation. Alternative possibilities are strongly debated among geomorphologists. One difficulty in drawing firm conclusions about the evolution of hillslopes is that they change so very slowly in terms of the human life span. Systematic, evolutionary changes must be inferred rather than observed.

We present two contrasting models of hillslope evolution, one for moist climates in low and middle latitudes, the other for arid climates in tropical and midlatitude zones. These models agree rather well with slopes observed in regions underlain by bedrock that is uniform in resistance to weathering and erosion. For a simple model, we can assume that the bedrock is a plutonic mass, such as granite or diorite, and that it is fractured into small blocks of uniform dimensions by three sets of uniformly spaced joints. Our ideal rock mass is therefore homogeneous in its physical and chemical composition.

Figure 18.16 is a schematic diagram of hillslope evolution as denudation progresses in a humid climate (left) and in an arid climate (right). The hillslope profile ends in a stream channel, which is assumed to be undergoing a continuous lowering of elevation. For the humid climate, the profile is shown with a rounded summit. Here the effects of rain-splash erosion and soil creep are dominant and tend to produce an upwardly convex profile. The remainder of the profile is straight to the point of junction with the channel. On this straight, steep slope, denudation proceeds by a combination of processes, including overland flow and soil creep. Material is removed both as sediment and as dissolved solids (ions). Rapid channel lowering ensures the removal of all material brought down the slope and maintains the steepness of the profile. For some time, the hillslope may be uniformly lowered, as shown by profiles a and b, which are parallel lines. As channel lowering slows with time, the rounded summit is expanded in area, while the lower portion of the profile declines in angle (profiles c and d). Relief is diminishing with time. Now an up-concavity begins to form at the base of the slope (profile e). In some cases a deposit of colluvium forms here. The slope profile is now S-shaped in outline and can be described as a sigmoid profile. The sigmoid form persists to the peneplain stage (profile f).

In the arid environment (Figure 18.16, right) the initial profile form is assumed to be steep and straight from the base to a sharp-edged summit. Throughout all stages

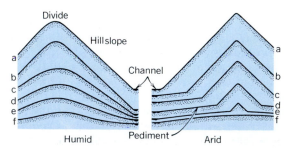

FIGURE 18.16 Hillslope evolution with decreasing relief in humid and arid climates. (From A. N. Strahler, *The Earth Sciences*, 2nd ed., Harper & Row, Publishers. Copyright © by Arthur N. Strahler.)

of denudation, the angle of this steep profile element remains constant; but the slope base retreats toward the divide, as shown in successive profiles. The slope is mantled with a thin layer of coarse, bouldery regolith and bedrock outcrops are numerous. From the channel, a pediment begins to form at an early stage (profile b). The pediment widens as the base of the steep element retreats. The sharp break between pediment and mountain front is a characteristic feature of the desert of the southwestern United States and northern Mexico (Figure 18.15). Profile e of Figure 18.16 represents an inselberg surrounded by a broad pediment. The steep slopes disappear entirely when the inselberg is removed, as shown in profile f; now only a pediment remains.

In summarizing the two idealized programs of slope evolution, we can make this generalization: Hillslopes evolving in humid climates follow a program of *reclining retreat*, in which the entire profile between divide and channel diminishes in angle with time. Slopes evolving in an arid climate show *parallel retreat*, in which the initial slope angle is retained but the base of the steep element retreats from the stream channel and is replaced by a pediment.

Keep in mind that the concepts we have presented are generalizations and that differences of opinion exist as to many details of slope evolution. Few large areas of the continents are underlain by homogeneous rock, as our model requires. Highly varied bedrock structures are usually present locally, and these features exert strong controls over slope form, as we will explain in Chapter 19. Rejuvenation can be expected to produce complex slope profiles in humid climates, as a new generation of steeper slopes replaces an older generation of gentler slopes. There is also the possibility that a climate change has occurred so that, for example, slope forms of an earlier arid environment are undergoing modification in a humid environment.

Our slope models do not apply closely to landscapes found in the tropical wet-dry climates or to landscapes of the arctic tundra climate. This observation leads to the conclusion that several models may be needed to encompass the full range of denudational landscapes found over the globe.

THE CLIMATE-PROCESS SYSTEM OF THE SAVANNA ENVIRONMENT

Geomorphic processes of the savanna environment, dominated by the wet-dry tropical climate, are closely tied to soil-forming processes. This includes not only processes acting under present climates but also processes that may have acted during long time spans in the late Cenozoic Era, in which a moister climate may have prevailed. Large areas of savanna landscape lie on stable shield areas of the former subcontinent of Gondwana (Chapter 13). Bedrock of these shield areas is extremely complex in rock composition and structure. Exposures of plutonic igneous masses (batholiths) are numerous, while metamorphic belts of great complexity are also abundant. Some parts of the shields bear sedimentary covers; elsewhere are expanses of ancient flood basalts. Thus the

landforms of the savanna regions are to a large degree structurally controlled, making difficult the recognition of distinctive landforms governed solely by processes unique to the wet-dry tropical climate.

The nature and intensity of weathering processes acting under the savanna soil-water regime are poorly understood. Chemical weathering during the wet season is considered by some investigators to be a very important process affecting igneous and metamorphic rocks rich in aluminosilicate minerals. The intensity of erosion by running water is also considered to be extremely high because of the rather poor protection afforded the soil surface by savanna vegetation, particularly where it has been burned over prior to the rains and heavily grazed and trampled by large herds of wild herbivores or cattle. Sediment loads of streams—both suspended load and bed load—are very high during periods of peak flow. In contrast, stream discharges fall to low values or to zero during the long dry season, when broad braided channels of sand and silt are exposed to view.

Two landscape features of the savanna region deserve notice: (1) prominent, isolated rock knobs and (2) bench-like upland surfaces. The rock knobs are of a special type, called *bornhardts;* they are found throughout the savanna region of East Africa and South Africa. Typically, a bornhardt is a steep-sided knob of granite or similar plutonic rock having a rounded summit and often showing exfoliation shells (Figure 18.17). Surrounding the bornhardt is a more or less level land surface, or plain, underlain by regolith. Some geomorphologists consider this type of plain to be a pediplain formed by processes similar to pediment-producing processes acting in deserts. Figure 18.18 is a schematic cross section showing one interpretation of the origin of the rock knob. Deep chemical weathering has taken place beneath the deep regolith surrounding the knob, where the bedrock is closely jointed or consists of more susceptible rock. Related to bornhardts are *tors*, which consist of rounded joint blocks. Heaps of rounded boulders rising above a

FIGURE 18.17 The rounded rock hills at left and right are bornhardts. This savanna woodland scene is in the Transvaal province of the Union of South Africa. (G. Douglas, courtesy of SATOUR.)

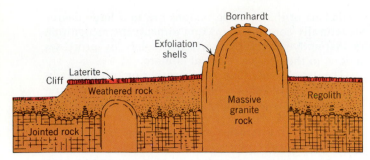

FIGURE 18.18 Schematic diagram of a bornhardt with its surrounding plain bearing a laterite crust. (Drawn by A. N. Strahler.)

FIGURE 18.19 Laterite, formed by hardening of plinthite, is being quarried for building stone in this scene from the state of Orissa in India. (Henry D. Foth. Used by permission.)

plain are a common landform of the savanna landscape of the Indian peninsula.

To imply that bornhardts and tors are unique to the savanna environment, as some authors have done, would be inaccurate. Similar landforms can be found in a wide range of climates, including the desert and the moist midlatitude climates. For example, the rock tors of the Dartmoor region of southwestern England bear a close resemblance to the tors of Africa.

The second landscape feature, and one that is undoubtedly associated with the tropical environment, is a stepping or benching of the land surface caused by the presence of a layer or layers of rocklike material derived from the soil.

As we explain in Chapter 24, a distinctive soil horizon known as plinthite is characteristic of very old soils of the order Oxisols. Plinthite consists of dense sesquioxides of iron and aluminum that are residual products of weathering. Plinthite can harden irreversibly into a rocklike layer, becoming *laterite*, when the soil is exposed to repeated wetting and drying in the tropical wet-dry climate. An alternate term for laterite is *ferricrete*, which recognizes the presence of iron oxide as the cementing material.

Laterite represents a plinthite horizon that has become exposed to the surface environment through erosional removal of overlying soil horizons of an Oxisol. The same process is carried out by humans in Southeast Asia, where plinthite is quarried and cut into building blocks that harden into enduring laterite blocks (Figure 18.19). Figure 18.18 shows a laterite layer forming a cap layer or crust over the plain surrounding the bornhardt. Where streams have carved valleys into the laterite layer, it forms a small cliff. Laterite crusts are not necessarily related to bornhardts or other landforms and occur widely over the uplands of savanna regions. Two, three, or more laterite crusts may be formed in succession at lower levels, as denudation proceeds. The benched landscape resembles in some ways the landscape of plateaus, cliffs, and mesas developed in flat-lying sedimentary strata of covered shields (Chapter 19). Elsewhere in tropical environments, crusts consist of weathered materials of soil or regolith cemented by silica; this material is called *silcrete* to distinguish it from ferricrete.

A general hypothesis for the development of crusts

of ferricrete and silcrete is that they originated as soil horizons during long periods of moister climate representing a former geographic expansion of the wet monsoon climate into what is today the region of wet-dry tropical climate. Exposure of the soil horizon and its conversion into a laterite crust accompanied a change to the existing climate, as the belt of wet equatorial climate contracted and withdrew into lower latitudes.

LANDSCAPE AND CLIMATE

This chapter has emphasized the role of climate in shaping landforms, but at the same time we have pointed out that caution is in order lest we neglect the role of geology in determining the total landscape of a given area. Tectonics plays a major role along with climate, as we noted in the great difference between the desert landscape of the Basin and Range region and that of the stable Gondwana shield areas of desert southern Africa and Australia.

History, both of climate and geology, must be taken into account if we are to arrive at a full understanding of landscape. Climates of the past have left their imprint on the landscape, as we found in the case of the laterite crusts of the tropical savanna environment. Geologic history has left its indelible imprint on the bedrock mass from which landforms are sculptured by the agents of denudation. To arrive at a full picture of the variety of landforms that a landscape can contain, we turn in the next chapter to consider how variations in bedrock composition and structure can control the distinctive shapes of hills and valleys and of ridges and escarpments. From the generalizations that characterize the broad and distant view of landscapes as expressions of climate-process systems and tectonic history, we shall shift our field of view into a much higher level of magnification that brings out the fine details of a landscape.

LANDFORMS AND ROCK STRUCTURE

Fluvial denudation acts on any landmass exposed to the atmosphere, but those landmasses differ greatly from place to place in rock composition and rock structure. As you might expect, landforms of fluvial denudation can take on a wide variety of shapes and patterns, expressing the complexity of the crustal rock from which they are carved.

Recall from the previous chapter that the mass of rock lying above base level constitutes the landmass. In theory, denudation cannot reduce the land surface below the base level, which is the inland extension of sea level. But the earth's crust can be lifted higher above sea level by means of tectonic movements. Crustal rise can increase the available landmass. Through repeated crustal uplifts and repeated episodes of denudation, deep-seated rocks and structures appear at the surface. In this way the root structures of mountain belts come to be exposed, along with batholiths and other plutons.

In Chapter 14 we classified all landforms as being either initial or sequential in origin. As a landmass is first brought into existence by tectonic activity or volcanic activity, its surface configuration is made up of initial landforms. Initial landforms produced directly by volcanic activity, folding, and faulting were described in Chapter 14. Denudation soon converts the initial landforms into sequential landforms. In this chapter our interest is in the erosional class of sequential landforms, controlled in shape, size, and arrangement by the underlying rock stucture.

Denudation acts more rapidly on the weaker rock types, lowering them to produce valley floors and leaving the stronger rocks to stand out in bold relief as ridges and uplands. This is a subject we will pursue further in the present chapter. Rock structure controls the placement of streams and the shapes and heights of the intervening divides. A distinctive assemblage of landforms and stream patterns is developed for each of the major types of crustal structures. The habitats of plants and animals are varied by these landforms. Humans occupy these habitats and exploit them for agricultural lands, communication lines, and urban development. In all these activities a distinctive set of constraints and opportunities is imposed by each structural type.

EROSIONAL LANDFORMS AND CONTINENTAL HISTORY

The erosional landforms we describe in this chapter have been developed on the stable crust of the continental lithosphere, far removed from active plate boundaries where tectonic and volcanic activities are taking place. We will examine geologic relics of past arc–continent and continent–continent collisions and of deep-seated tectonic and intrusive activity that once took place over subduction zones. These continent-building activities started early in Precambrian time, at least 3 billion years ago, and persisted through the Paleozoic and Mesozoic eras as one orogeny after another left its mark.

Since the close of the Cretaceous Period, about -65 m.y., the stable continental shields have experienced fluvial denudation, punctuated by occasional epeirogenic downwarping that brought marine submergence to large interior regions, and by upwarping that raised the submerged areas to become land again. True, some minor faulting has locally affected the shields in the Cenozoic Era, but this activity is scarcely perceptible in the landscape. (We must make an exception of areas of major continental rifting—East Africa and the American Basin and Range province.) There has also been sporadic and isolated volcanic activity in hot spots within the shields, as the continental lithosphere has drifted over mantle plumes. These isolated volcanic features have been treated in the previous chapter; they include flood basalts and some isolated groups of volcanoes, such as the San Francisco Peaks of northern Arizona.

ROCK STRUCTURE AS A LANDFORM CONTROL

As denudation takes place, landscape features develop in close conformity with patterns of bedrock composition and structure. Figure 19.1 shows four types of sedimentary rock, together with a mass of much older igneous rock on which the sediments were deposited. The diagram shows the usual landform habit of each rock type,

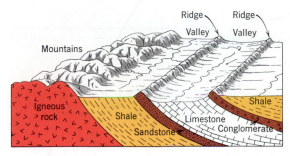

FIGURE 19.1 Many landscape features originate through the slow erosional removal of weaker rock, leaving the more resistant rock standing as ridges or mountains. (Drawn by A. N. Strahler.)

whether to form valleys or mountains. The cross section shows conventional rock symbols used by geologists. These rock strata have been strongly tilted and deeply eroded.

Shale is a weak rock and is reduced by fluvial action to form the lowest valley floors of the region. Limestone, easily dissolved by carbonic acid action, also forms valleys in humid climates. In arid climates, on the other hand, limestone is highly resistant and usually forms high, prominent landforms. Sandstone is usually a resistant rock; it forms ridges and uplands. As a group, the felsic igneous rocks are resistant to denudation. Occurring as plutons, they typically form uplands rising above adjacent sedimentary strata.

The metamorphic rocks are, as a group, more resistant to denudation than their sedimentary parent types. As shown in Figure 19.40, however, there are conspicuous landform differences among the types of metamorphic rocks.

DIP AND STRIKE

Natural layers and planes of weakness are characteristic of the structure of each kind of rock. A system of geometry is needed to measure and describe the attitude of

these natural planes and to indicate them on maps. Examples of such planes are the bedding layers of sedimentary strata, the sides of a dike, and the joint sets in granite. Rarely are these planes truly horizontal or vertical.

The acute angle formed between a natural rock plane and an imaginary horizontal plane is the *dip*. The amount of dip is stated in degrees ranging from 0° for a horizontal plane to 90° for a vertical plane. Figure 19.2 shows the dip angle for an outcropping layer of sandstone, against which a horizontal water surface is at rest. The compass direction of the line of intersection between the inclined rock plane and an imaginary horizontal plane is the *strike*. In Figure 19.2, the strike is north.

STRUCTURAL GROUPS OF LANDMASSES

Landmasses fall into three major groups, and each of these in turn contains two or more distinctive types. As major groups, we can recognize the following:

Group A. Undisturbed sedimentary strata. These are thick covers of sedimentary rocks overlying ancient shield rocks of continental lithosphere. The strata are most commonly of marine origin deposited on the floors of continental shelves of passive continental margins and shallow inland seas; they have been brought above sea level by crustal rise and are now in the process of undergoing fluvial denudation. Because no significant amount of bending or faulting has affected these strata, their attitude is nearly horizontal over very large areas.

Group B. Disturbed structures of tectonic activity that has long since ceased. These landmasses show strongly the effects of bending and breaking of the crust by tectonic processes. Sedimentary strata showing folding and faulting are included in this group. Metamorphic rocks produced in root zones of ancient collision belts are another type.

Group C. Eroded igneous masses. Exposed large plutons—bodies of intruded igneous rock—are one important type of igneous mass. A quite different type includes extinct composite and shield volcanoes undergoing deep erosion.

Figure 19.3 illustrates eight landmass types within the three major groups. Within Group A, two important types are coastal plains and horizontal strata. Group B includes deeply eroded folds, domes, fault blocks, and metamorphic belts. Group C includes exposed plutons and eroded volcanoes.

A description of each landmass type will give you insight into the meaning of diverse landforms of the continents and the ways in which these landforms affect the environment and provide a variety of mineral resources. Each of our national parks and wilderness areas owes its distinctive scenery to the influence of a particular landmass type. These protected natural areas are a national resource on which no dollar value can be placed; their diversity depends largely on the kinds of rocks and structures out of which their landforms have been carved by fluvial denudation.

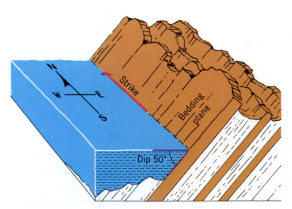

FIGURE 19.2 Strike and dip. (Drawn by A. N. Strahler.)

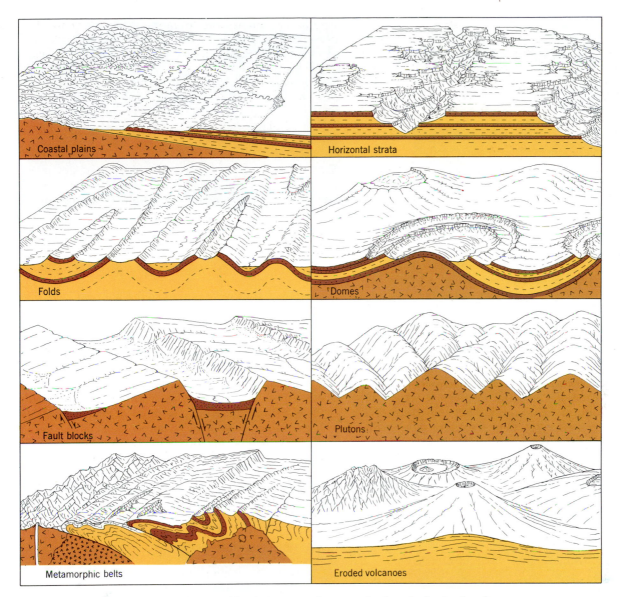

FIGURE 19.3 Several distinctive types of landmasses can be recognized on the basis of rock composition and structure. (Drawn by A. N. Strahler.)

UNDISTURBED SEDIMENTARY STRATA

Undisturbed sedimentary strata of more or less horizontal attitude cover large parts of the continental shields (see Figure 13.8). Although there is no set limit to the age of strata in these covers, almost all are Paleozoic, Mesozoic, and Cenozoic ages. Rarely are Precambrian strata found undisturbed by tectonic activity. The undisturbed sedimentary strata fall into two major classes: interior shield covers and coastal plains.

INTERIOR SHIELD COVERS

Extensive areas of the continental shields are covered by thick sequences of nearly horizontal sedimentary strata.

At various times in the 600 million years following the end of Precambrian time, these strata were deposited in shallow inland seas. Following crustal uplift, with little disturbance other than minor warping or faulting, these areas became continental surfaces undergoing denudation.

In arid climates, where vegetation is spare and the action of overland flow is especially effective, sharply defined landforms develop on horizontal sedimentary strata (Figure 19.4). The dominant landform is a sheer rock wall, or *cliff*. At the base of the cliff is an inclined slope. This slope flattens out to make a bench, terminated at the outer edge by the cliff of the next lower set of forms. These forms are wonderfully displayed in the walls of the great canyons of the Colorado Plateau region (Figure 19.5).

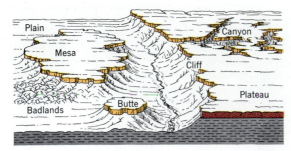

FIGURE 19.4 In arid climates, a distinctive set of landforms develops in flat-lying sedimentary formations. (Drawn by A. N. Strahler.)

FIGURE 19.6 A butte of sandstone not far from Ship Rock, northwestern New Mexico. Weak shales form the base of the butte. (A. N. Strahler.)

In these arid regions, erosion strips away successive rock layers, leaving behind *plateaus* capped by hard rock layers. Cliffs retreat as near-perpendicular surfaces because the weak shale formations exposed at the cliff base are rapidly washed away by storm runoff. When undermined, the rock in the upper cliff face repeatedly breaks away along vertical joint fractures.

Cliff retreat produces *mesas*, table-topped plateaus bordered on all sides by cliffs (Figure 19.4). Mesas represent the remnants of a formerly extensive layer of resistant rock. As a mesa is reduced in area by retreat of the rimming cliffs, it maintains its flat top. Before its complete consumption, the final landform is a small steep-sided hill known as a *butte* (Figure 19.6).

Areas underlain by weak shales or clays erode rapidly and may be entirely barren of plant cover. The weak material is dissected by a network of small streams to produce badlands, a landscape of miniature valleys and

divides. (Badlands were explained in Chapter 17; see Figure 17.5).

Regions of horizontal strata have branching stream networks formed into a *dendritic drainage pattern* (Figure 19.7). The smaller streams in this pattern take a variety of directions (Figure 19.8). This treelike branching pattern is also found in areas of exposed batholiths (see Figure 19.43).

Shield covers may include areas of flood basalts, in which erosional landforms resemble those of horizontal strata. One such occurrence is the Columbia Plateau region of eastern Washington and Oregon, described in Chapter 14 (see Figure 14.7). Another is the Deccan Plateau of western India. In both areas, basalt lavas cover thousands of square kilometers and are several thousand

FIGURE 19.5 The Grand Canyon of the Colorado River, Arizona, seen from the South Rim. (A. N. Strahler.)

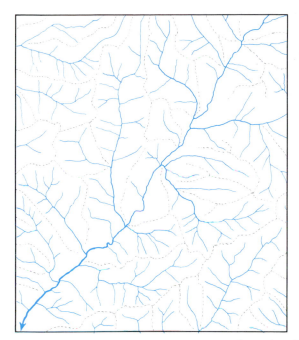

FIGURE 19.7 A dendritic drainage pattern formed on horizontal strata. Divides are shown in dotted line.

meters thick. Interbedded with the lavas are lake and stream deposits of sands, gravels, and clays. Canyons, mesas, and buttes carved from lava flows are quite similar to those formed of sedimentary strata.

FIGURE 19.8 In this Gemini VII space photo, canyons carved into horizontal limestone strata show a dendritic drainage pattern. The area is part of the Hadramawt Plateau, near the southern coast of the Arabian peninsula. At the upper left is a broad dry valley floored with sandy alluvium. Width of the area is about 160 km (400 mi). (NASA No. S65-64010.)

LIMESTONE CAVERNS

The reaction of carbonic acid, formed by atmospheric carbon dioxide dissolved in soil water and ground water, with limestone (calcium carbonate), was mentioned in Chapter 12 as one of the processes of chemical alteration. In Chapter 15, we described the effects of carbonic acid action on exposed surfaces of limestone and, in some special circumstances, on basaltic lavas.

Caverns are interconnected subterranean cavities in limestone, formed by carbonic acid carried in circulating ground water. Cavern systems in limestone can be thought of as a kind of underground landform consisting of relief features exposed to the atmosphere as well as to ground water and running water. Surface openings leading down into caverns are also an important class of landform.

Extensive cavern systems are found in horizontal limestone strata, as well as folded and faulted limestone strata. Most persons are familiar with the names of famous caverns, such as Mammoth Cave or Carlsbad Caverns. Millions of Americans have visited these famous tourist attractions. Figure 19.9 suggests how caverns may develop. In the upper diagram the action of carbonic acid is shown to be particularly concentrated in the saturated zone just below the water table. Products of solution are carried along in the ground-water flow paths to emerge in streams.

Limestone caverns consist of passageways and rooms forming highly complex systems. Patterns of joints in the limestone exert a strong control over the network of passages (Figure 19.10). Variations in composition among individual limestone beds will also help to control spacing of passages.

After cavern development has been in progress for a long period of time the land surface above is deeply pocked with depressions, which are parts of the old cavern system lying close to the surface. These depressions are termed *sinks*, or *sinkholes*. They may be partly filled with clay soil and can hold ponds or marshes, or may be cultivated if the soil is well drained (Figure 19.11). Some sinks are gaping holes or fissures leading down to open caverns beneath.

In a second stage, shown in Figure 19.9, the stream has deepened its valley and the water table has been lowered accordingly. New caverns are being formed at the lower level, while those previously formed are now in the unsaturated zone, or zone of aeration.

One result of the lowering of the water table as the main stream continues to entrench its valley is that new levels of cavern systems are added beneath the older, higher levels. Figure 19.12 shows three such levels in the Mammoth Cave area of Kentucky. Vertical shafts connect these levels with surface sinkholes.

In the second stage, percolating water is now exposed to the air on ceilings and walls of the caverns. The carbonic acid reaction is reversible, so that under favorable conditions that permit some of the dissolved carbon dioxide in the solution to escape into the air as a gas, calcite is precipitated as a mineral deposit. Calcium carbonate precipitated on the inner surfaces of caverns

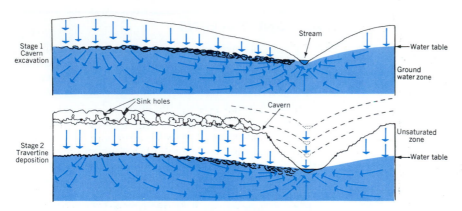

FIGURE 19.9 Cavern development in the ground-water zone, followed by travertine deposition in the unsaturated zone. (From A. N. Strahler, *The Earth Sciences*, 2nd ed., Harper & Row, Publishers. Copyright © by Arthur N. Strahler.)

occurs as travertine, a banded form of calcite. Encrustations forming where water drips from cave ceilings constitute *dripstone*, while encrustations made in moving water of pools and streams on cave floors form *flowstone*. Dripstone and flowstone accumulate as elaborate encrustations, known popularly as "formations," that give many caves their great beauty.

Certain scenic features of caverns are illustrated in Figure 19.13. From ceiling points where the slow drip of water takes place, spikelike forms known as *stalactites* are built downward. From points on the cavern floor upon which there falls a steady drip of water, postlike columns termed *stalagmites* are built upward. Stalactites and stalagmites may join into columns, and the columns forming under a single joint crack may fuse into solid walls. Growth below a joint crack may produce a drip curtain. Blocks of limestone fallen from the ceiling pond the runoff along the cavern floor, making pools in which travertine terraces are formed.

The environmental importance of caverns is felt in several ways. Throughout the early development of the human species, caverns were an important habitation. We find the skeletal remains of humans, together with their implements and cave drawings, preserved through the centuries in caverns in many parts of the world. Today caverns are being used as storage facilities, living quarters, and factories.

Caverns have provided some valuable deposits of guano, the excrement of birds or bats, which is rich in nitrates. Guano deposits have been used in the manufacture of fertilizers and explosives. Bat guano was taken from Mammoth Cave for making gunpowder during the War of 1812.

FIGURE 19.10 This map of a portion of Anvil Cave in Morgan County, Alabama, was made by the Huntsville Grotto of the National Speleological Society. The passageways lie within a nearly horizontal limestone layer overlain by massive sandstone. Maximum height from floor to ceiling ranges from 8 to 10 m (25 to 30 ft). (Courtesy of W. W. Varnedo and the Geological Survey of Alabama.)

FIGURE 19.11 A shallow sinkhole on the Kaibab Plateau, Arizona. Clay impedes the drainage of water, allowing a pond to form. A needleleaf forest of spruce is seen in the distance. (A. N. Strahler.)

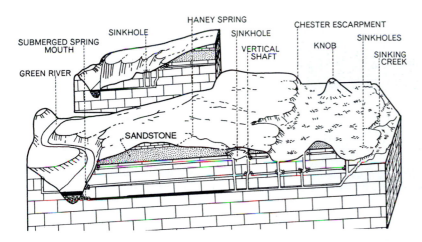

FIGURE 19.12 Block diagram showing the typical relationship between caverns and strata in the Mammoth Cave region of southcentral Kentucky. The lowermost level is occupied by the underground Echo River, its lower course completely submerged. (From W. B. White, R. A. Watson, E. R. Pohl, and R. Brucker, 1970, The Central Kentucky Karst, *Geographical Review*, vol. 60, p. 109. Copyright © by the American Geographical Society. Used by permission.)

KARST LANDSCAPES

Where limestone solution is very active, we find a landscape with many unique landforms. This is especially true along the Dalmatian coastal area of Yugoslavia, where the landscape is called *karst*. The term is used by geographers for the topography of any limestone area where sinkholes are numerous and small surface streams are nonexistent.

Development of a karst landscape is shown in Figure 19.14. In an early stage, funnellike sinkholes are numerous. Later, the caverns collapse, leaving open, flat-floored valleys. Some important western-hemisphere regions of karst topography are the Mammoth Cave region of Kentucky, the Yucatan Peninsula, and parts of Cuba and Puerto Rico.

Karst landscapes exhibit a wide variety of forms. One common type, known as *cockpit karst* (also "cone karst"), consists of numerous closely-set hills with rounded summits; between them are closed depressions with sinkholes. A good example is the sinkhole plain of central Kentucky, lying south of Mammoth Cave and the Chester Escarpment (see Figure 19.12). Cockpit karst is also strongly developed in Puerto Rico. A second example of a distinctive karst landscape is *tower karst*, perhaps best displayed in southern China and West Malaysia. There, steep-sided conical limestone hills 100 to 500 m (300 to 1500 ft) high dominate the scenery (Figure 19.15). A form of tower karst also occurs in Puerto Rico, where the isolated limestone knobs are called "mogotes."

FIGURE 19.14 Features of a karst landscape. (A) Rainfall enters the cavern system through sinkholes in the limestone. (B) Extensive collapse of caverns reveals surface streams flowing on shale beds beneath the limestone. The flat-floored valleys can be cultivated. (Drawn by E. Raisz.)

FIGURE 19.13 Travertine deposits in Carlsbad Caverns include stalactites (slender rods hanging from the ceiling) and sturdy columns. (Mark A. Melton.)

FIGURE 19.15 Tower karst near Guilin (Kweilin), Guanxi Province, in southern China at lat. 25°S. White limestone can be seen exposed in the nearly vertical sides of the towers. Flooding of the alluvial plain surrounding the towers occurs in the summer monsoon. (Bruno Barbey/Magnum Photos.)

ENVIRONMENTAL AND RESOURCE ASPECTS OF SHIELD COVERS

Horizontal strata of shield covers contain many minerals and rocks of economic value. Building stone, such as the Bedford limestone in Indiana or the Berea sandstone in Ohio, is a valuable resource. Limestone is quarried for use in the manufacture of portland cement or as flux in iron smelting. Some important deposits of lead, zinc, and iron ores occur in sedimentary rocks. For example, the lead and zinc mines of the Tristate district (Missouri, Kansas, and Oklahoma) are in horizontal limestones. Uranium ores are important in strata of the Colorado Plateau.

Perhaps the greatest mineral resources of sedimentary strata of the shield covers are bituminous coal and lignite, described in Chapter 12. As the map of North American coal fields shows (Figure 19.16), large reserves of coal lie close to the surface in areas of horizontal strata in the Great Plains region west of the Mississippi, in the northern High Plains still farther west, in sedimentary basins of the Rocky Mountains, and in the Colorado Plateau near the common corner of the states of Utah, Arizona, Colorado, and New Mexico. The great demands for this coal in decades to come make it likely that large areas will be strip mined and that power-generating plants of large capacity will be built close to the supplies of coal (Figure 19.17). The future role of fossil fuels—coal

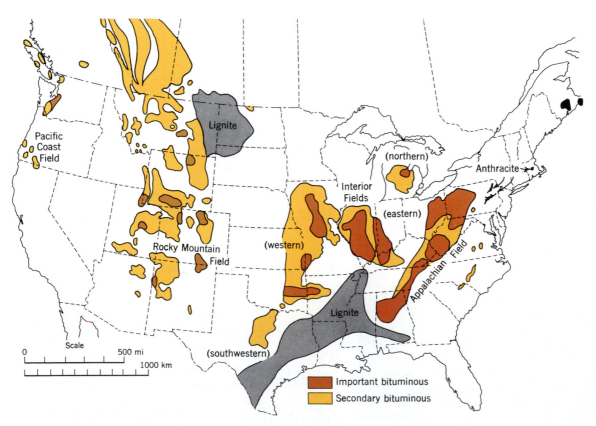

FIGURE 19.16 Coal fields of the United States and southern Canada. The lignite area of the southern states lies within the coastal plain. Anthracite fields are in folded strata.

FIGURE 19.17 Strip mining of a great coal seam at the Antelope Mine, Douglas, Wyoming. Overburden is being removed by the huge machine in the distance, while the exposed coal is being removed by machines in the foreground. (Kristin Finnegan/Allstock.)

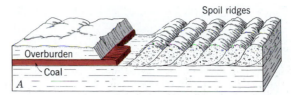

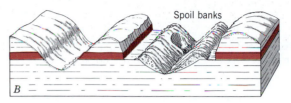

FIGURE 19.18 Area strip mining *(A)* and contour strip mining *(B)*. (Drawn by A. N. Strahler.)

and lignite, petroleum and natural gas—has in recent years come under serious question because the combustion of these fuels plays a major part in causing the rather steady observed increase in atmospheric carbon dioxide. That this combustion, along with that of the burning of forests and woodlands, will play a leading role in global warming is widely accepted by scientists. Environmentalists also are greatly concerned with the land scarification and air pollution that such developments can bring.

The major form of land scarification both today and in years to come is associated with the mining of coal. For deep-lying coal seams, vertical shafts are driven down from the surface to reach the coal, which is mined by extension of horizontal *drifts* and rooms into the face of the seam. In mountainous terrain, drifts can be driven directly into the seam where it is exposed on the mountainside. Subterranean mining of coal is often followed by damage to the underlying land surface through mine collapse leading to subsidence, which can be considered a form of mass wasting. One example comes from the city of Scranton, Pennsylvania. Here the collapse of

abandoned anthracite mine workings has repeatedly caused settling and fracturing of the ground, damaging streets and houses. Near Hanna, Wyoming, the cave-in of abandoned shallow coal mines has produced many deep pits, while underground burning of the coal seam has added to progressive collapse.

Where the coal seams lie close to the surface or actually outcrop along hillsides, the *strip mining* method is used. Earth-moving equipment removes the covering strata (overburden) to bare the coal, which is lifted out by power shovels. There are two kinds of strip mining, each adapted to the given relationship between ground surface and coal seam. *Area strip mining* is used in regions of nearly flat land surface under which the coal seam lies horizontally (Figure 19.18A). After the first trench is made and the coal removed, a parallel trench is made, the overburden of which is piled as a spoil ridge into the first trench. Thus the entire seam is gradually uncovered, and a series of parallel spoil ridges remains. Phosphate beds are mined extensively by the area strip mining method in Florida, and the method is also used for mining clay layers. The *contour strip mining* method is used where a coal seam outcrops along a steep hillside (Figure 19.18B). The coal is uncovered as far back into the hillside as possible, and the overburden is dumped on the downhill side. The result is a bench bounded on one side by a steep rock cliff, or "high wall," and on the other side by a ridge of loose spoil with a steep outer slope leading down into the valley bottom. The benches form sinuous patterns following the plan of the outcrop (Figure 19.19).

Spoil ridges left by strip mining are highly susceptible to rapid erosion under torrential rainfall, yielding large amounts of sediment. This material is carried down-valley, where it accumulates in valley bottoms (Chapter 17). On steep mountain slopes of the Appalachian area, mass wasting in the form of earthflows and mudflows affects the unstable spoil banks, while the presence of sulfur-bearing minerals in the coal seams gives rise to acid water drainage, severely impacting the streams that drain the area.

FIGURE 19.19 Contour strip mining near Lynch, Kentucky. A highway makes use of the winding bench at the base of the high rock wall. (Billy Davis/Black Star.)

COASTAL PLAINS

Coastal plains are undisturbed accumulations of marine strata lapping over passive continental margins. Refer to Figure 13.12 for the geologic interpretation of coastal plain strata. Where continental and oceanic lithosphere are in contact within a single lithospheric plate, sediment accumulates in a continental shelf wedge. The margin of the continental plate subsides because of isostatic compensation (Chapter 18) as the sediment load accumulates. Water depth over the continental shelf remains shallow, but the sediment wedge can reach thicknesses of thousands of meters over a span of 50 to 100 million years. Passive continental margins with thick continental shelf deposits occur on continents enclosing the North and South Atlantic ocean basins. (The world maps of lithospheric plates in Figure 13.17 show that the continental margins are well inside the lithospheric plates.) As these continental margins experienced episodes of epeirogenic downsinking, they became submerged inland extensions of the continental shelves. During such submergences, marine clay, marl (lime mud), and sand were deposited in thin, wedge-shaped layers. Epeirogenic uplift of the continental margin has, in many places, brought a broad zone of these marine strata above sea level as coastal plains.

Block A of Figure 19.20 shows a recently emerged coastal plain. It is a smooth, almost featureless surface, sloping gently toward the coastline. Streams that drain the older landmass extend their courses across the plain to reach the sea. These are known as *consequent streams* because they follow the initial slope of the new land surface. (The radial streams that originate on the slopes of a new volcano also qualify as consequent streams by this same definition.) The diagram shows that parallel lines of uplifted beach ridges formed a partial barrier to the

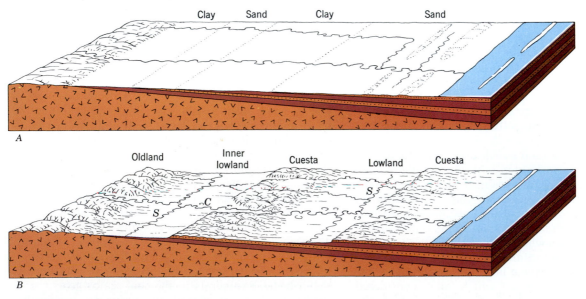

FIGURE 19.20 Development of a broad coastal plain. (A) Early stage; plain recently emerged. (B) Advanced stage; cuestas and lowlands developed. S = subsequent stream; C = consequent stream. (Drawn by A. N. Strahler.)

consequent streams, causing one of them to turn parallel with the shoreline and join the other.

In an advanced stage of coastal-plain erosion, a new series of streams and topographic features has developed (Block B of Figure 19.20). Where more easily eroded strata (usually clay, marl, or shale) are exposed, denudation is rapid, making *lowlands*. Between them rise broad belts of hills called *cuestas*. Cuestas are commonly underlain by sand, sandstone, or limestone. The lowland lying between the area of older rock (*oldland*) and the first cuesta is called the *inner lowland*.

Streams that develop along the trend of the lowlands, parallel with the shoreline, are of a class known as *subsequent streams*. They take their positions along zones of weak rock and therefore follow closely the patterns of rock structure. Subsequent streams occur in many regions, and we shall mention them again in the discussion of domes and folds. The drainage lines on a fully dissected coastal plain combine to form a *trellis drainage pattern*. In this pattern, the subsequent streams trend at about right angles to the consequent streams.

COASTAL PLAIN OF NORTH AMERICA

The coastal plain of eastern North America is a major geographic region, ranging in width from 160 to 500 km (100 to 300 mi) and extending for 3000 km (2000 mi) along the Atlantic and Gulf coasts. The coastal plain starts at Long Island, New York, as a partly submerged cuesta, and widens rapidly southward so as to include much of New Jersey, Delaware, Maryland, and Virginia (Figure 19.21). Throughout this portion, the coastal plain has but one cuesta. The inner lowland is a continuous broad valley developed on weak clay strata.

In Alabama and Mississippi, the coastal plain is fully dissected. Cuestas and lowlands run in belts roughly parallel with the coast (Figure 19.22). The cuestas are underlain by sandy formations and support pine forests. Limestone forms lowlands such as the Black Belt in Alabama, named for its dark, fertile soils.

ENVIRONMENTAL AND RESOURCE ASPECTS OF COASTAL PLAINS

Lowlands of the broad coastal plain of the eastern United States show intensive agricultural development because of the natural fertility of soils derived from carbonate strata. Sandy cuestas, bearing soils poor in plant nutrients, provide valuable pine forests in the southern United States. Transportation tends to follow the lowlands and to connect the larger cities located there. For example, important highways and railroads connect New York City with Trenton, Philadelphia, Baltimore, Washington, and Richmond, all of which are situated in the inner lowland (Figure 19.21). The inner limit of coastal plain strata is known as the Fall Line, because here the major rivers have developed falls and rapids on the resistant oldland rocks. Cities grew at the Fall Line, where the rivers become navigable tidal estuaries. Thus geology has played a part in the location of the nuclei of Megalopolis.

The seaward dip of sedimentary strata in a coastal plain provides a structure favorable to the development of artesian water supplies. Water penetrates deeply into sandy strata of a cuesta, overlain by aquicludes of shale or clay. When a well is drilled into the sand formation considerably seaward of its surface exposure, water under hydraulic pressure rises toward the surface. Artesian water in large quantities is available in many parts of

FIGURE 19.21 The coastal plain of the Atlantic seaboard states shows little cuesta development except in New Jersey. The inner limit of the coastal plain is marked by a series of fall-line cities. (After A. K. Lobeck.)

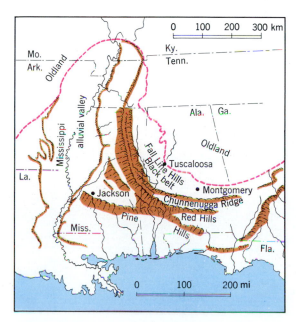

FIGURE 19.22 The Alabama–Mississippi coastal plain is belted by a series of sandy cuestas and shale or marl lowlands. (After A. K. Lobeck.)

the Atlantic and Gulf coastal plains, although it is no longer sufficient to supply the demands of densely populated and industrialized communities.

The Gulf Coastal Plain of the United States contained an enormous accumulation of petroleum and natural gas, much of which has already been withdrawn.

Other mineral deposits of economic importance in coastal plains include sulfur, occurring in the coastal plain of Louisiana and Texas; phosphate beds, found in Florida; and clays, used in the manufacture of pottery, tile, and brick in New Jersey and the Carolinas.

SEDIMENTARY DOMES

The second structural class of landmasses found within the continental lithospheric plates consists of disturbed structures of past tectonic activity. The class includes two structural types in which deformation is simple: sedimentary domes and open folds. A third structural type includes belts of intensely deformed strata showing tight folding and thrust faulting. These strata have been largely transformed into metamorphic rock.

The *sedimentary dome* is a circular or oval tectonic structure in which strata have been raised into a domed shape (Figure 19.23). Sedimentary domes occur in various places within the covered shield areas of the continents. A few seem to have been forced up by magma intruding at depth within a sequence of strata. More commonly, ancient shield rock lies beneath the strata of the dome, which represents a tectonic uplift of a thick layer of crust. Domes of the central Rocky Mountain region have been explained as lying above steep reverse faults that fail to emerge at the present land surface.

Erosion features of a large sedimentary dome are illustrated in the two block diagrams of Figure 19.24. The diameter of the dome is on the order of 50 to 100 km (30 to 60 mi). Strata are first removed from the summit region of the dome, exposing older strata beneath. Eroded edges of steeply dipping strata form sharp-crested sawtooth ridges called *hogbacks* (Figures 19.25 and 19.26).

When the last of the strata has been removed, the ancient shield rock is exposed in the central core of the dome. Here a mountainous terrain develops on plutons lacking in layered structure.

Figure 19.26 is an idealized block diagram to show the relationship of landforms to dip of strata. In the foreground are hogbacks formed on steeply dipping sandstone layers. Weak shale forms narrow valleys between the hogbacks and between exposed shield rock and a hogback. Subsequent streams occupy the valleys of weak rock. Toward the central part of the block diagram, the angle of dip of the strata is shown to lessen, so the hogback form gives way to broader ridges of cuesta form. The cuesta has a steep, clifflike face on one side and a gentle, descending surface on the opposite side. Cuestas on coastal plains may show this asymmetrical form if the strata underlying the cuesta are well indurated. Cuestas are found on flanks of sedimentary domes where the dip is low. Toward the rear of the diagram, the strata are shown to flatten in attitude and become nearly horizontal. Here, erosion produces plateaus, cliffs, and mesas.

The stream network on a deeply eroded dome shows dominant subsequent streams forming a circular system, called an *annular drainage pattern* (Figure 19.27). The shorter tributaries make a radial arrangement. The total

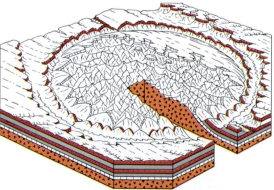

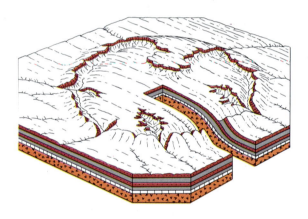

FIGURE 19.23 A small, nearly circular sedimentary dome near Sundance, Wyoming. (John S. Shelton.)

FIGURE 19.24 Erosion of strata from the summit of a sedimentary dome structure. (A) Strata partially removed, forming an encircling hogback ridge. (B) Strata removed from center of dome, revealing a core of much older igneous or metamorphic rock. (Drawn by A. N. Strahler.)

FIGURE 19.25 Hogbacks of sandstone lie along the eastern base of the Colorado Front Range, which is a large domed structure. The view is southward toward the city of Boulder. The rugged forested terrain in the distance is developed on igneous and metamorphic rock exposed in the core of the uplift. (John S. Shelton.)

pattern resembles a trellis pattern bent into a circular form.

Important accumulations of petroleum and natural gas occur in domes of sedimentary strata. Many small domes of the Rocky Mountains have been important petroleum producers, for example, the Rock Springs Dome and Teapot Dome in Wyoming. Oil has accumulated in sandstone reservoirs overlain by shale cap rock.

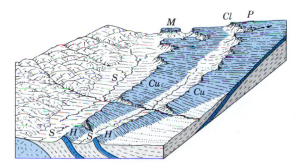

FIGURE 19.26 Hogbacks gradually merge into cuestas and the cuestas into plateaus and mesas, where the dip of the strata becomes less from one place to another. S = subsequent stream; H = hogback ridge; Cu = cuesta; M = mesa; Cl = cliff; P = plateau. (Drawn by W. M. Davis.)

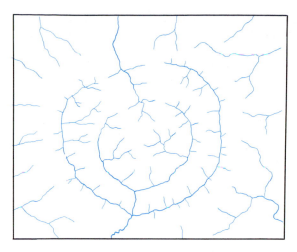

FIGURE 19.27 The drainage pattern on an eroded dome combines annular and radial elements.

THE BLACK HILLS DOME

A classic example of a large sedimentary dome is the Black Hills dome of western South Dakota and eastern Wyoming (Figure 19.28). Valleys that encircle this dome are ideal locations for railroads and highways, so it is natural that towns and cities should have grown in these valleys. One valley in particular, the Red Valley, is continuously developed around the entire dome and has been called the Race Track because of its shape. It is underlain by a weak shale, which is easily washed away. The cities of Rapid City, Spearfish, and Sturgis are located in the Red Valley. On the outer side of the Red Valley is a high, sharp hogback of Dakota sandstone, known simply as Hogback Ridge. It rises some 150 m (500 ft) above the level of the Red Valley. Farther out toward the margins of the dome the strata are less steeply inclined and form a series of cuestas. Artesian water is obtained from wells drilled in the surrounding plain.

The eastern central part of the Black Hills consists of a mountainous core of intrusive and metamorphic rocks. These mountains are richly forested, whereas the intervening valleys are beautiful open parks. Thus the region is attractive as a summer resort area. Harney Park, elevation 2207 m (7242 ft), is the highest peak of the core. In the northern part of the central core, in the vicinity of Lead and Deadwood, are valuable ore deposits. At Lead is the fabulous Homestake Mine, formerly one of the world's richest gold-producing mines.

The western central part of the Black Hills consists of a limestone plateau deeply carved by streams. The original dome had a flattened summit. The limestone plateau represents one of the last remaining sedimentary rock layers to be stripped from the core of the dome.

LANDFORMS OF ERODED FOLD BELTS

According to the model of mountain making developed in Chapter 13, strata of the continental margins are

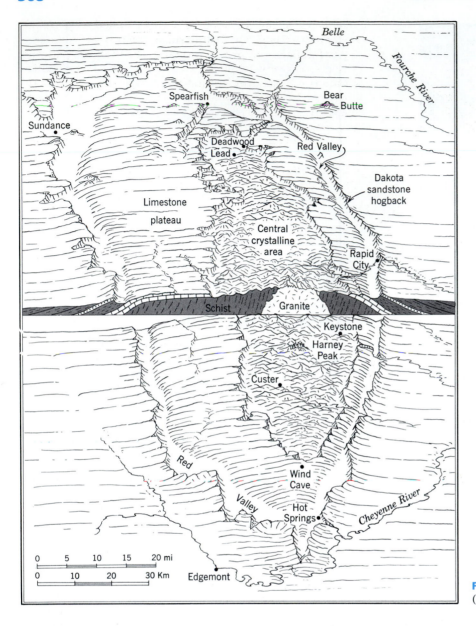

FIGURE 19.28 The Black Hills dome. (Drawn by A. N. Strahler.)

deformed into folds along narrow belts during arc–continent collision (see Figure 13.22). These are foreland folds, illustrated in Figures 14.16 and 14.17. They are initial landforms of tectonic origin. Long after folding has ceased, denudation exposes the mountain roots, bringing the folded strata into relief. Alternating hard and soft rock layers give the crust a strong grain, like the grain in a wooden plank. Denudation removes the weaker bands of the grain, allowing the harder bands to stand out in relief, much as surfwood is etched by sand and waves.

Parallel, wavelike folds pass through a series of erosion stages, illustrated in Figure 19.29. A downfold of strata is known as *syncline;* an upfold is known as an *anticline.* In an early stage of the development of folds, anticlines are identical with the mountains, or ridges; synclines are identical with the valleys.

Block A of Figure 19.29 shows erosion of anticlines occurring during the last stages of folding. Synclines are being filled with alluvial fan materials swept down from the adjacent anticlines. After folding has ceased, the upper layers of soft, unconsolidated rock are removed until, as shown in Block B, a hard, well-cemented sandstone layer is exposed, reflecting the full amplitude of the folding. Coinciding with synclines are *synclinal valleys;* coinciding with anticlines are *anticlinal mountains.*

Streams that drain the steep flanks of the anticlines quickly cut deep ravines, exposing the underlying layers. Breaching spreads rapidly to the crest of the anticline, where a long, narrow valley is opened out along the summit (Block B). This valley is occupied by a subsequent stream excavating a belt of weak rock; it is an *anticlinal valley* because it lies on the center line of the anticline. As this valley grows in length, depth, and breadth, it replaces the original anticlinal mountain. A reversal of topography thus occurs. The synclinal valley, which originally contained the major stream, is now shrunken between the growing anticlinal valleys on either side. The anticlinal valleys are now rapidly deepened because the core of weak rock is exposed to attack. Eventually the

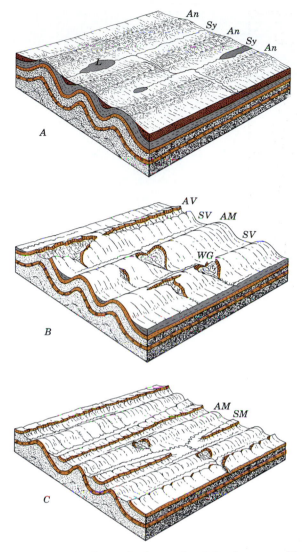

FIGURE 19.29 Stages in the erosional development of folded strata. (A) While folding is still in progress, erosion cuts down the anticlines; alluvium fills the synclines, keeping relief low. An = anticline; Sy = syncline; L = lake. (B) Long after folding has ceased, erosion exposes a highly resistant layer of sandstone or quartzite. AV = anticlinal valley; SV = synclinal valley; WG = watergap. (C) Continued erosion partly removes the resistant formation but reveals another below it. AM = anticlinal mountain; SM = synclinal mountain. (Drawn by A. N. Strahler.)

syncline becomes a mountain ridge, called a *synclinal mountain* (Block C). At this stage, the original topography has been completely reversed. The drainage pattern is a trellis type, similar in most respects to the trellis pattern of a mature coastal plain, but different in that the major subsequent streams are more closely spaced and their tributaries are shorter (Figure 19.30).

Here and there in a belt of folds, a major stream crosses several folds at nearly right angles, passing through the sharply defined ridges by narrow *watergaps*. These streams may have existed prior to the folding and were able to maintain themselves as the folds were formed. Streams that maintain their courses across a rising barrier, whether an anticline or an uplifted fault block, are called *antecedent streams*.

As the dissection of the folded strata progresses, there is a continuous change in the form and position of the various types of ridges and valleys. In time the synclinal ridges are completely removed by erosion. Meanwhile, new ridges are appearing in the centers of the anticlinal valleys. These form as a result of the uncovering of still older resistant strata that were folded along with the rest. The new ridges, which may be thought of as second-generation anticlinal mountains, grow in height as the weak rock is stripped from both sides. Two such anticlinal mountains are shown in Block C of Figure 19.29.

ZIGZAG RIDGES AND PLUNGING FOLDS

The folds illustrated in Figure 19.29 are continuous and level crested; they produce ridges that are approximately parallel and continue for great distances. In some fold regions, however, the fold crests rise or descend from place to place. Folds of this form are called *plunging folds*. When eroded, they give rise to a zigzag line of ridges (Figure 19.31).

The topographic form of a plunging syncline differs from that of an anticline which plunges in the same direction, when both are dissected. Figure 19.31 illustrates landforms of plunging folds. The plunging syncline is represented by a ridge with a slightly concave summit but steeply descending cliffs on the end and sides. Along the direction of plunge of the fold center line, or axis, this ridge develops an increasing concavity, then separates into two diverging ridges. The plunging anticline

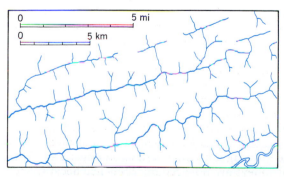

FIGURE 19.30 A trellis drainage pattern on folds.

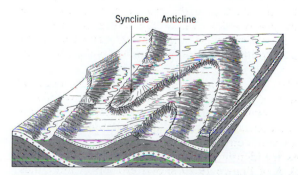

FIGURE 19.31 Plunging folds give zigzag ridges when deeply eroded. (Drawn by E. Raisz.)

FIGURE 19.32 Like an old wooden plank deeply etched by drifting sand to reveal the grain, the surface of south-central Pennsylvania shows zigzag ridges formed by bands of hard quartzite. Strata were crumpled into folds during a collision of continents that took place over 200 million years ago to produce the Appalachians. In this image, obtained by a Landsat orbiting satellite, the ridges may appear as trenches; if so, try inverting the page to see the relief as it actually is. The area shown is about 160 km (100 mi) in width. (NASA.)

FIGURE 19.33 Radar image of a portion of the Folded Appalachians in south–central Pennsylvania. Zigzag ridges in the vicinity of Hollidaysburg include two plunging anticlines and two plunging synclines. At the lower right are the entrenched meanders of the Raystown Branch of the Juniata River. The area shown is about 40 km (25 mi) wide. (SAR image by courtesy of Intera Technologies Corporation, Calgary, Alberta, Canada.)

has a smoothly rounded summit that descends gradually to the level of the valley in the direction of plunge.

Zigzag ridges formed on plunging folds are remarkably well developed in central Pennsylvania. They are clearly shown on the remote sensing image (Figure 19.32). (For most viewers, the ridges will appear correctly as raised features if the page is inverted, placing north toward the bottom.) The series of watergaps of the Susquehanna River, shown in geologic detail in Figure 19.34, can be identified in the eastern part of the frame. Figure 19.33 is a radar image of a portion of the area shown in Figure 19.32; it lies about 4 cm (1.5 in.) from the top and 2.5 cm (1.0 in.) from the left. Using the block diagram of Figure 19.31 as a guide, you can easily identify three conspicuous plunging anticlines and three plunging synclines on the radar image.

ENVIRONMENTAL AND RESOURCE ASPECTS OF FOLD REGIONS

Some of the environmental and resource aspects of dissected fold regions are illustrated by the Appalachians of south-central and eastern Pennsylvania (Figure 19.34). The ridges, of resistant sandstones and conglomerates, rise boldly to heights of 600 m (2000 ft) above broad lowlands underlain by weak shales and limestones. Major highways run in the valleys, crossing from one valley to another through the watergaps of streams that have cut through the ridges. Important cities are situated near the watergaps of major streams. An example is Harrisburg, located where the Susquehanna River issues from a series of watergaps. Where no watergaps are conveniently located, roads must climb in long, steep grades over the ridge crests. The ridges are heavily forested; the valleys are rich agricultural belts.

In various fold regions of the world, an important resource is anthracite, or hard coal. This coal occurs in strata that have been tightly folded. Pressure has converted the coal from bituminous into anthracite. Because of extensive erosion, all coal has been removed except that which lies in the central parts of synclines. Figure 19.35 shows coal-bearing synclines in Pennsylvania. The coal seams dip steeply; workings penetrate deeply to

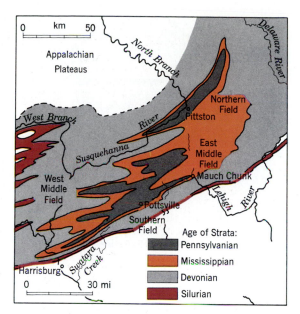

FIGURE 19.35 Anthracite coal basins of central Pennsylvania correspond with areas of Pennsylvania strata, downfolded into long synclinal troughs.

reach the coal that lies in the bottoms of the synclines. Seams near the surface are worked by strip mining. Although workable anthracite seams are now exhausted in the Pennsylvania field, environmental damage persists through continued subsidence and cracking of the ground surface above abandoned workings.

Many broad, gentle anticlinal folds, as well as sedimentary domes, form important traps for the accumulation of petroleum. The principle is shown in Figure 19.36. Oil migrates in permeable sandstone beds to reach the anticlinal crest or dome summit, where it is trapped by an impervious cap rock of shale. Many of the oil and gas pools of western Pennsylvania, where petroleum production first succeeded, are on low anticlines. Many elongate domes of elliptical outline, resembling anticlines plunging down at both ends, form important oil-bearing structures in the large sedimentary basins of the middle Rocky Mountains.

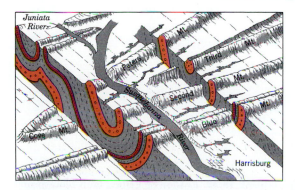

FIGURE 19.34 A great synclinal fold, involving three resistant quartzite formations and thick intervening shales, has been eroded to form bold ridges through which the Susquehanna River has cut a series of watergaps. (Based on a drawing by A. K. Lobeck.)

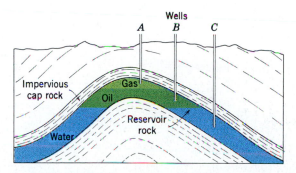

FIGURE 19.36 Idealized cross section of an oil pool in an anticline or dome in sedimentary strata. Well A will draw gas; well B will draw oil; and well C will draw water. The cap rock is shale; the reservoir rock is sandstone. (From A. N. Strahler, *Planet Earth: Its Physical Systems Through Geologic Time*, Harper & Row, Publishers. Copyright © by Arthur N. Strahler.)

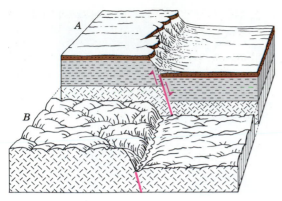

FIGURE 19.37 (A) Fault scarp. (B) Fault-line scarp. (From A. N. Strahler, *The Earth Sciences*, 2nd ed., Harper & Row, Publishers. Copyright © by Arthur N. Strahler.)

FIGURE 19.38 Fault-line scarp, MacDonald Lake, near Great Slave Lake, Northwest Territories, Canada. (Canadian Forces Photograph No. 5120-105R.)

EROSIONAL DEVELOPMENT OF FAULT SCARPS AND BLOCKS

Landforms shaped from normal faults throughout the erosion period that follows their formation are illustrated in Figure 19.37. At the rear part of Block A is the original fault scarp, produced directly by crustal movement. Although eroded by running water, the scarp base is straight. A few stream-cut canyons notch the scarp; some talus cones and alluvial fans have been built along the scarp base.

Block B shows a scarp resulting entirely through erosion; it is a *fault-line scarp*. The scarp appears because a weak shale formation has been stripped from both sides of the fault, revealing a basement of resistant igneous rocks upon which the shale was once deposited. Fault-line scarps are common features of the continental shields (Figure 19.38). These scarps have persisted hundreds of millions of years since the faults ceased to be active because the fault planes penetrated deeply into the crust.

As continental rifting begins to occur on a vast scale within a continental lithospheric plate, enormous crustal blocks are raised or depressed by faulting. This rifting process has dominated the landscape of the Basin and Range Province of the southwestern United States and

northern Mexico. In Chapter 18, we used large fault blocks as the geologic base upon which to illustrate the progress of fluvial denudation in an arid climate. In the Basin and Range Province, mountain blocks of two basic types can be recognized. A tilted block has one steep side (the fault scarp) and one gentle side. A lifted block is a form of horst; it has a flattened summit and two steep boundary fault scarps.

Figure 19.39 shows some erosional features of a large, tilted fault block. The freshly uplifted block has a steep mountain face that is rapidly dissected into deep canyons. The upper mountain face reclines in angle as it is removed, while rock debris accumulates in the form of alluvial fans adjacent to the fault block. Vestiges of the fault plane are preserved as *triangular facets*, the snubbed ends of ridges between canyon mouths.

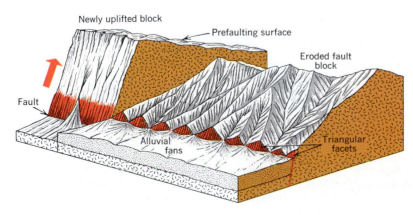

FIGURE 19.39 Erosion of a tilted fault block produces a rugged mountain range. A line of triangular facets at the mountain base marks the position of the original fault plane. (Drawn by A. N. Strahler.)

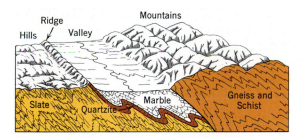

FIGURE 19.40 Metamorphic rocks tend to form elongate, parallel belts of valleys and mountains. (Drawn by A. N. Strahler.)

mentary strata because of the great production that has come from pools of this type.

Fault scarps and fault-line scarps can form imposing topographic barriers across which it is difficult to build roads and railroads. The great Hurricane Ledge of southern Utah is a feature of this type, in places a steep wall 760 m (2500 ft) high.

Grabens may be so large as to form broad lowlands. An illustration is the Rhine graben of Western Germany. Here a belt of rich agricultural land, 32 km (20 mi) wide and 240 km (150 mi) long, lies between the Vosges and Black Forest ranges, both of which are block mountains faulted up in contrast with the downdropped Rhine graben block.

ENVIRONMENTAL AND RESOURCE ASPECTS OF FAULTS AND BLOCK MOUNTAINS

Faults are of environmental and economic significance in several ways. Fault planes are usually zones along which the rock has been pulverized, or at least considerably fractured. This breakage has the effect of permitting ore-forming solutions to rise along fault planes. Many important ore deposits lie in fault planes or in rocks that faults have broken across.

Another related phenomenon is the easy rise of ground water along fault planes. Springs, both cold and hot, are commonly situated along fault lines. They occur along the base of a fault-block mountain. Examples are Arrowhead Springs along the base of the San Bernardino Range and Palm Springs along the foot of the San Jacinto Mountains, both in southern California.

Petroleum, too, finds its way along fault planes where the rocks have been rendered permeable by crushing, or it becomes trapped in porous beds that have been faulted against impervious shale beds. Some of the most intensive searches for oil center about areas of faulted sedi-

METAMORPHIC BELTS

Where strata have been tightly folded and altered into metamorphic rocks in the orogens of arc–continent and continent–continent collisions, denudation develops a landscape with a strong grain of ridges and valleys paralleling the strike of the folds and thrust faults. These features lack the sharpness and parallelism of ridges and valleys in belts of open folds, but each belt clearly reflects the greater or lesser resistance to denudation offered by each type of metamorphic rock.

Figure 19.40 shows the control of metamorphic rocks over landforms in the humid climate of eastern North America, where metamorphic rocks of the Older Appalachians form a narrow zone continuous from Georgia to the Maritime Provinces. Marble tends to form open valleys; slate and schist make belts of medium to strong relief; quartzite usually stands out boldly and may produce conspicuous narrow hogback ridges. Most metamorphic belts have been broken by overthrust faults that run parallel with the belt and often separate one rock type from another. Subsequent stream valleys occupying these fault lines accentuate the grain of the topography.

FIGURE 19.41 The Labrador fold belt in Quebec and Newfoundland, displayed by Landsat remote-sensing imagery. This region was reduced to a peneplain (perhaps repeatedly), which has since been etched by fluvial and glacial erosion to reveal complex folds of metamorphic rocks derived from Precambrian sedimentary strata. Lakes, occupying ice-scoured rock basins, appear in black. (NASA 1483-15013, October 4, 1973.)

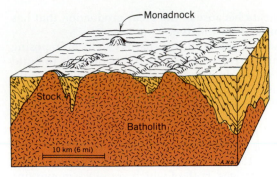

FIGURE 19.42 Batholiths appear at the land surface only after long-continued erosion has removed thousands of meters of overlying rocks. Small projections of the granite intrusion (known as stocks) appear first and are surrounded by older rock. (Drawn by A. N. Strahler.)

Much of New England, particularly the Taconic and Green Mountains, illustrates these principles well. The larger valleys trend north and south and are underlain by marble. These are flanked by ridges of gneiss, schist, slate, or quartzite. The highlands of the Hudson and of northern New Jersey continue this belted pattern southward, where it joins the Blue Ridge. Near Harpers Ferry, Maryland, quartzite ridges rise prominently above broad valley belts of schist.

Vast areas of the exposed continental shields show complex folding of ancient metamorphic rocks of Precambrian age (Figure 19.41). These structures are interpreted as the remains of numerous continental collisions that occurred during early growth stages of the major shield areas.

EXPOSED BATHOLITHS

The third structural group of landmasses includes exposed plutons originating as batholiths of igneous rock that solidified under a thick cover of overlying rocks (Figure 19.42). Most batholiths are composed of felsic igneous rock, which is commonly granite or a closely related rock type. Batholiths are produced along active lithospheric plate boundaries, but details are obscure and the origin of the igneous material is subject to highly varied interpretations. One hypothesis considers granite to be derived from thick masses of metamorphic rock—schists and slates, for example. Melting and recrystallization in tectonic zones converted the parent materials into felsic igneous rock.

FIGURE 19.44 Map of Cretaceous batholiths of western North America. The word "Batholith" has been omitted from each label. (From C. O. Dunbar and K. M. Waage, *Historical Geology*, 3rd ed., p. 376, Figure 16.7. Copyright © 1969 by John Wiley & Sons. Reprinted by permission of John Wiley & Sons, Inc.)

FIGURE 19.43 This dendritic drainage pattern is developed on the dissected Idaho batholith.

FIGURE 19.45 Stone Mountain, Georgia, is a striking erosion remnant, or *monadnock*, about 2.4 km (1.5 mi) long and rising 193 m (650 ft) above the surrounding Piedmont peneplain surface. The rock is granite, almost entirely free of joints, and has been rounded into a smooth dome by weathering processes. (Landis Aerial Photo.)

Where the rock mass of the batholith is quite uniform and free of strong faults, it is eroded into a maze of canyons and drainage divides that follow no predominant trend (see Figure 18.6). The drainage pattern is dendritic and resembles that developed on horizontal sedimentary strata (Figure 19.43).

Exposed batholiths occur widely over the continental lithosphere. Good examples are found throughout the Cordilleran ranges of western North America. During the Cretaceous Period, large batholiths were formed from Lower (Baja) California to Alaska (Figure 19.44). One of these rock masses most familiar to Americans is the Si-

erra Batholith; it comprises the bulk of the Sierra Nevada fault block in California. Another example is the Idaho Batholith, exposed over an area of about 40,000 sq km (16,000 sq mi). Largest of the group is the Coast Range Batholith of British Columbia. An example of a small body of massive granite, intruded into older metamorphic rocks of the Piedmont upland of Georgia, is Stone Mountain (Figure 19.45).

The name *monadnock* has been given to an isolated mountain or hill, such as Stone Mountain, rising conspicuously above a peneplain. A monadnock develops because the rock within the monadnock is much more resis-

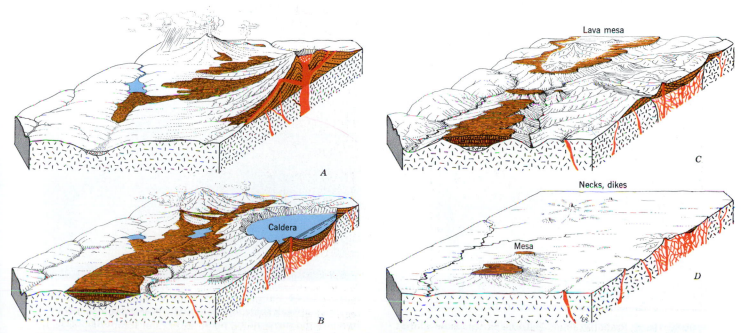

FIGURE 19.46 Stages in the erosional development of volcanoes and lava flows. (Drawn by E. Raisz.)

FIGURE 19.47 Mount Shasta in the Cascade Range, northern California, is a dissected composite volcano. It has been carved into by streams and small alpine glaciers. Part way up the right-hand side is a more recent subsidiary volcanic cone, called Shastina, with a sharp crater rim. (John S. Shelton.)

tant to denudation processes than the bedrock of the surrounding region (Figure 19.45). The name, which was first used nearly a century ago by William M. Davis, is taken from Mount Monadnock in southern New Hampshire. Monadnocks of quartzite of ridgelike form are common in areas of metamorphic rock (Figure 19.40).

EROSIONAL LANDFORMS OF VOLCANOES AND LAVA FLOWS

Figure 19.46 shows successive stages in the erosion of volcanoes, lava flows, and a caldera. In Block A, active volcanoes are in the process of building. These are initial landforms. Lava flows issuing from the volcanoes have

spread down into a stream valley, following the downward grade of the valley and forming a lake behind the lava dam.

In Block B, some changes have taken place, the most conspicuous of which is the destruction of the largest volcano to produce a caldera. A lake occupies the caldera, and a small cone has been built inside. One of the other volcanoes, formed earlier, has become extinct. It has been dissected by streams, losing the original conical form. Smaller, neighboring volcanoes are still active, and the contrast in form is marked.

In Block C of Figure 19.46, all volcanoes are extinct and have been deeply eroded. The caldera lake has been drained and the rim worn to a low, circular ridge. The

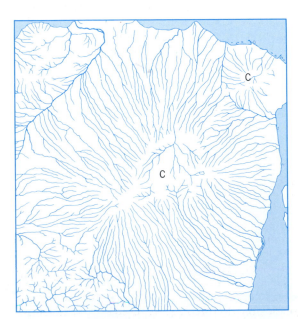

FIGURE 19.48 Radial drainage patterns of volcanoes in the East Indies. The letter C shows the location of a crater.

FIGURE 19.49 Ship Rock, New Mexico, is a volcanic neck enclosed by a weak shale formation. The peak rises 520 m (1700 ft) above the surrounding plain. At the right, wall-like dikes extend far out from the central peak. (John S. Shelton.)

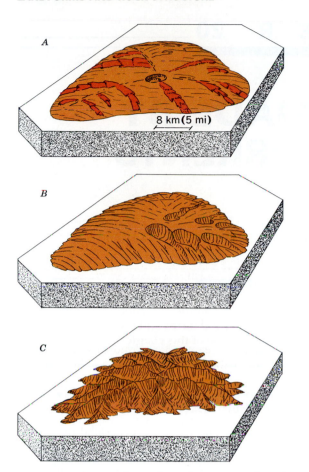

FIGURE 19.51 Waimaea Canyon, over 760 m (2500 ft) deep, has been eroded into the flank of an extinct shield volcano on Kauai, Hawaii. Basaltic lava flows are exposed. (A. N. Strahler.)

FIGURE 19.50 Lava domes in various stages of erosion make up the Hawaiian Islands. (Drawn by A. N. Strahler.) (A) Initial dome with central depression and fresh flows issuing from radial fissure lines. (B) Early stage of erosion, with deeply eroded valley heads. (C) Advanced erosion stage, with steep slopes and mountainous relief.

lava flows that formerly flowed down stream valleys have been able to resist erosion far better than the rock of the surrounding area and have come to stand as platforms (mesas) high above the general level of the region.

An example of a dissected volcano is Mount Shasta in the Cascade Range (Figure 19.47). A smaller subsidiary cone of more recent date, named Shastina, is attached to the side.

The drainage pattern of streams on a volcanic cone is of necessity a *radial pattern*, shown in Figure 19.48. It is often possible to recognize volcanoes from a drainage map alone because of the perfection of the radial pattern. Where a well-formed crater exists, small streams flow from the crater rim toward the bottom of the crater. Here the water is absorbed in the permeable layers of ash within the cone or is conducted outward by means of a single gap in the crater rim. This inward drainage, described as centripetal, often adds to the certainty of interpreting a volcano from a drainage pattern.

Block D of Figure 19.46 shows an advanced stage of erosion of volcanoes. There remains now only a small sharp peak, or *volcanic neck*, representing the solidified lava in the pipe, or neck, of the volcano. Radiating from

this are wall-like dikes, formed of lava that previously filled fractures around the base of the volcano. Perhaps the finest illustration of a volcanic neck with radial dikes is Ship Rock, New Mexico (Figure 19.49). Because the central neck and radial dikes extend to great depths in the rock below the base of the volcano, they may persist as landforms long after the cone and its associated flows have been removed.

Shield volcanoes show erosion features quite different from those of the composite volcanoes. Figure 19.50 shows the stages of erosion of Hawaiian shield volcanoes, beginning in Diagram A with the active volcano and its central depression. Radial streams cut deep canyons into the flanks of the extinct shield volcano, and these canyons are opened out into deep, steep-walled amphitheaters. Eventually, as Diagram C shows, the original surface of the shield volcano has been entirely obliterated, and a rugged mountain mass made up of sharp-crested divides and deep canyons remains (Figure 19.51).

LANDFORMS AND ROCK STRUCTURE

In this chapter we have emphasized the role of bedrock geology in controlling the configuration of landforms shaped by fluvial denudation. In our previous chapter we balanced the role of tectonics against climate in the development of landscapes. Each climate-process system leaves its distinctive marks on the landscape, but rock structure shows its effects clearly over a wide range of climatic environments. Open folds display their zigzag ridge crests in the moist climate of Pennsylvania just as surely and clearly as do open folds in the semiarid climate of Colorado or Texas. If our analysis of landforms and rock structure has been successful, it can lead only to the conclusion that geology is in equal partnership with climate to form a dual control system over landscapes of fluvial denudation.

CHAPTER 20

LANDFORMS MADE BY WAVES AND CURRENTS

The continental shoreline, where the salt water of the ocean contacts fresh water and the solid mineral base of the continents, is a complex environmental zone of great importance. Humans have occupied the shore zone for a number of reasons. First, there are food resources of shellfish, finfish, and waterfowl to be had in the shallow waters and estuaries. Second, the shoreline is a base from which ships embark to seek marine food resources farther from land and to transport people and goods between the continents; in time of war the shoreline is a critical barrier to be defended from invading forces arriving by sea. Third, the coastal zone is a recreational facility, with its sea breezes, bathing beaches, and opportunities for surfing, skin diving, sport fishing, and boating.

Along with its opportunities, the shore zone imposes restraints and hazards on humans and their structures. Some coasts are rocky and cliffed; they provide little or no shelter in the form of harbors. Along other coasts the enormous energy of storm waves can cut back the shore, undermining buildings and roads. High water levels in time of storm can cause inundation of low-lying areas and bring the force of breaking waves to bear on ground many meters above the normal levels reached by seawater. For centuries, humanity has been at war with the sea, building fortresslike walls to keep out the sea and even forcing the sea to give up coastal land so that more crops for food and forage could be produced.

Long segments of our coastlines face environmental degradation and destruction as urbanization of the coastal zone demands more land and expanded port facilities. The pressures of a growing population to seek its holiday pleasures along the shore threaten to destroy the very benefits the shore offers. Management of the environment of the continental shorelines requires a knowledge of the natural forms and processes of this sensitive zone; the purpose of this chapter is to provide that basic knowledge.

WAVES IN DEEP WATER

Waves of the oceans and lakes are generated by winds. Energy is transferred from the atmosphere to the water surface by a rather complex mechanism involving both friction of the moving air on the water surface and direct wind pressure.

Wind-generated ocean waves belong to a type known as *oscillatory* waves because the wave form travels through the water and causes an oscillatory water motion. A simple terminology applied to waves is illustrated in Figure 20.1. Wave height is the vertical distance between trough and crest. Wave length is the horizontal distance from trough to trough, or crest to crest.

In the oscillatory wave, a tiny particle, such as a drop of water or a small floating object, completes one vertical circle, or wave orbit, with the passage of each wave length (Figure 20.2). Particles move forward on the wave crest, backward in the wave trough. At the sea surface, the orbit is of the same diameter as the wave height, but dies out rapidly with depth. The water particles return

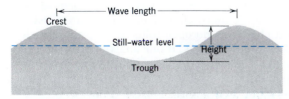

FIGURE 20.1 Terminology of water waves.

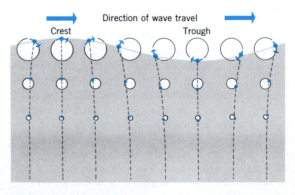

FIGURE 20.2 Orbital motion in deep-water waves of relatively low height.

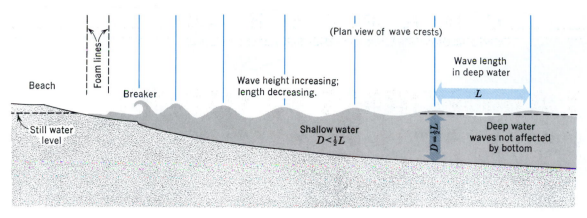

FIGURE 20.3 As waves enter shallow water, the form changes until breaking occurs.

to the same starting point at the completion of each orbit. In ideal waves of this type, there is no net motion of water in the direction of the wind.

SHOALING WAVES AND BREAKERS

Most shore zones have a fairly smooth, sloping bottom extending offshore into deeper water. As a train of waves enters progressively shallower water depths, there comes a point at which the orbital motion of the waves encounters interference with the bottom. As a general rule this critical depth is about equal to one-half the wave length (Figure 20.3).

As the waves continue to travel shoreward, the wave length decreases while the wave height increases, as shown in Figure 20.3. Consequently, the wave is steepened and becomes unstable. Rather suddenly the crest of the wave moves forward and the wave is transformed into a *breaker*, which then collapses (Figure 20.4). The turbulent water mass then rides up the beach as the *swash*, or *uprush*. This powerful surge causes a landward movement of sand and gravel on the beach. When the force of the swash has been spent against the slope of the beach, the return flow, or *backwash*, pours down the beach; but much of the water disappears by infiltration into the porous beach sand. Sand and gravel are swept seaward by the backwash.

MARINE EROSION OF COASTS

Throughout this chapter we will use the term *shoreline* to mean the shifting line of contact between water and land. The broader term *coastline*, or simply *coast*, refers

to a zone in which coastal processes operate or have a strong influence. The coastline includes the shallow water zone in which waves perform their work, as well as beaches and cliffs shaped by waves and coastal dunes.

The forward thrust of water produced by storm breakers is a powerful eroding agent along those coasts where the land is high close to the shore (Figure 20.5). The impact of storm waves carves a steep wall, or *marine cliff*, into a bedrock mass. Wave erosion is extremely slow in hard bedrock, so environmental change by natural processes is comparatively unimportant along rocky coasts. On the other hand, wave erosion is an important environmental concern along coasts made up of soft sedimentary strata, regolith, alluvium, glacial deposits, or sand dunes.

In a few places along the coasts of North America and northern Europe, marine cliffs are being rapidly eroded back in weak glacial deposits. For example, the Atlantic Ocean (eastern) shore of Cape Cod has a 24-km (15-mi) stretch of marine cliff ranging in height from 20 to 50 m (60 to 170 ft) carved into unconsolidated glacial sands and gravels. This feature is best described as a *marine scarp* because the loose sand holds a slope angle of about 30° to 35° from the horizontal and is by no means a sheer wall (Figure 20.6). On Cape Cod the rate of shoreline retreat has averaged nearly 1 m (3 ft) per year for the past century, and a single winter storm will cut the scarp back a distance of several meters. Shoreline retreat is called *retrogradation*.

Some details of a marine cliff, or *sea cliff*, formed in hard bedrock are shown in Figure 20.7. A deep basal indentation, the *wave-cut notch*, marks the line of most intense wave erosion. The waves find points of weakness in the bedrock, penetrating deeply to form *crevices* and *sea caves*. More resistant rock masses project seaward

FIGURE 20.4 A breaking wave. (Drawn by W. M. Davis.)

FIGURE 20.5 Breaking storm waves generated over the North Atlantic Ocean are attacking the base of this rock cliff at Land's End, West Cornwall, England. Joint blocks dislodged by wave pressure leave behind a sheer rock wall. (John Serafin/Peter Arnold, Inc.)

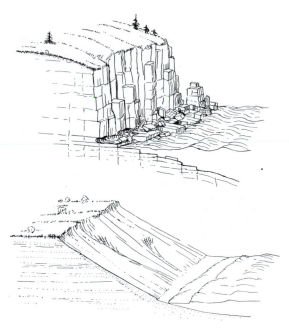

FIGURE 20.6 The nearly vertical cliff (above) is typical of the granite coast of Maine, while the gentler marine scarp of sand and gravel (below) is found on Cape Cod and other localities where glacial sediments are being eroded. (Drawn by A. N. Strahler.)

and are cut through to make picturesque *arches* (Figure 20.8). After an arch collapses, the remaining rock column forms a *stack*, which is ultimately leveled (Figure 20.7).

As the sea cliff retreats landward, continued wave abrasion forms an *abrasion platform*. This sloping rock floor continues to be eroded and widened by abrasion beneath the breakers. If a beach is present, it is little more than a thin layer of gravel and cobblestones.

Retreat of a coastline may take place so rapidly that a stream emptying into the sea is unable to lower its channel fast enough to maintain a sea-level junction. The result is a *hanging valley*, in which the stream channel ends abruptly near the top of the cliff and the valley cross-profile is abruptly truncated (Figure 20.9). Where bedrock is weak, the marine cliff may be unstable. Yielding occurs spontaneously, and mass movements take the form of earthflows and slump blocks (Figure 20.10).

Sea cliffs are features of spectacular beauty as well as habitats for many forms of life, including sea mammals and shore birds. Only in recent years has the need to preserve these cliffed coasts in their natural state been appreciated fully. Although the strength of these rocky features resists alteration by humans, the cliff line is vulnerable to heavy use for summer homes, motels, and restaurants. Intensive use not only destroys the pristine scenery, but also adds pollution.

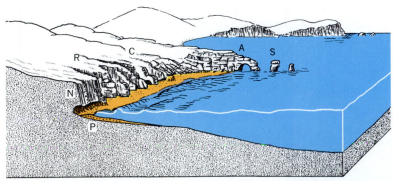

FIGURE 20.7 Landforms of sea cliffs. A = arch; S = stack; C = cave; N = notch; P = abrasion platform. (Drawn by E. Raisz.)

FIGURE 20.8 Rock arches have been carved from this wave-cut cliff of horizontal sandstone strata on the Lake Superior shore. (Orlo E. Childs.)

BEACHES

Sediment enters the shore zone of breaking waves from a number of possible sources. Sediment may be derived directly from a marine cliff or scarp that is being actively eroded. Sediment entering the ocean or a lake from the mouth of a stream is another major source of supply.

Storm waves may scour the offshore zone, dragging sand and gravel landward to reach the breaker zone.

Whatever the origin of the sediment, it is shaped by swash and backwash into a wedge-shaped sediment deposit, familiar to everyone as a *beach* (Figure 20.11). Sediment composing beaches ranges from fine sand to cobblestones several centimeters in diameter. Within a given stretch of beach, the sediment is usually quite well sorted into a particular size grade. Thus there are beaches of fine sand, of coarse sand, of gravel, or of cobbles. As a rule, beaches composed of fine sand are broad and have a very gentle seaward slope, whereas beaches of coarse sand or gravel are quite steep. Beaches formed of cobblestones are very steep and show a high crest, or ridge form. Particles of silt and clay do not form beaches, but instead are easily carried away from the shoreline in suspension in currents, to settle out in deep water.

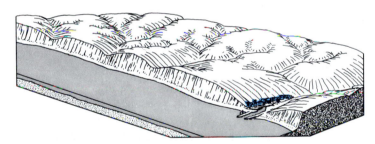

FIGURE 20.9 Hanging valleys appear as notches in a marine cliff. The large stream at the right has been able to maintain an accordant junction with the sea level and provides a small harbor. (Drawn by W. M. Davis.)

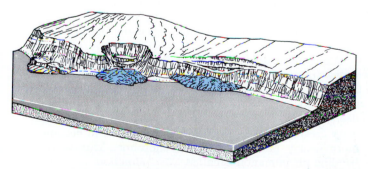

FIGURE 20.10 Coastal landslides in weak sedimentary strata are the result of oversteepening of a marine cliff by wave attack. (Drawn by W. M. Davis.)

FIGURE 20.11 A curving pocket beach on the island of Kauai, Hawaii. The foaming water of a collapsed breaker is approaching the beach, where it will form the landward swash. Receding before it is the backwash, leaving a thin water film on the sand surface. (A. N. Strahler.)

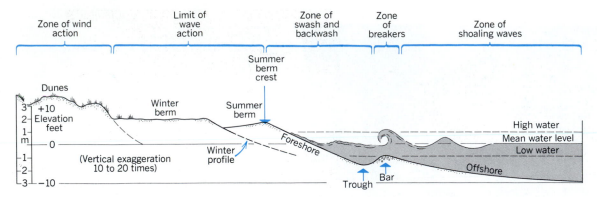

FIGURE 20.12 Typical forms and zones of a sand beach in the midlatitude zone. (From A. N. Strahler, *The Earth Sciences*, 2nd ed., Harper & Row, Publishers. Copyright © by Arthur N. Strahler.)

THE BEACH PROFILE

A broad sand beach with its bordering shallow-water zone of breaking waves, on the one side, and a zone of wind action with dune development forming a border on the landward side represents a succession of unique life habitats in which each assemblage of plant and animal forms is adapted to a different environment.

Figure 20.12 is an idealized profile across a typical broad beach developed by exposure to waves of the open ocean in a midlatitude location experiencing a strong contrast between wave action of summer and winter seasons. The profile shows summer conditions, in which waves are of low height and comparatively low levels of energy.

During the summer, accumulation of sand takes place, building a *summer berm*, which is a benchlike structure. A higher *winter berm* lies behind the summer berm. The winter berm can be cut back deeply by a storm, but is rebuilt following the storm. The *foreshore*

is the sloping beach face in the zone of swash and back-wash. Beneath the breaker zone is a low underwater bar, called an *offshore bar*. It lies in the *offshore* region of the beach, in the zone of shoaling waves and below the level of low tide. Actually, there are many variations in beach-profile forms from place to place and season to season, depending on wave form, wave energy, and the composition of the beach.

Beaches can experience a widening, or out-building process called *progradation*, as shown in Figure 20.13A. The sand that is added to the beach may come from deeper water, or it may be brought along the shore from another part of the coast. A common indication of progradation is the presence of numerous parallel *beach ridges*, each of which represents a former berm crest. Beaches can also experience a narrowing, or back-cutting retrogradation, shown in Figure 20.13B. In this case the sand may be moved to deeper water in the offshore zone, or it may be transported along the shoreline to another part of the coast.

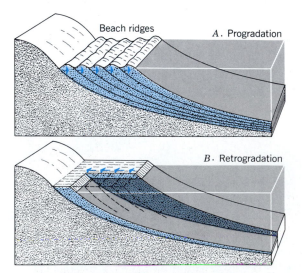

FIGURE 20.13 Progradation and retrogradation. (From A. N. Strahler, *The Earth Sciences*, 2nd ed., Harper & Row, Publishers. Copyright © by Arthur N. Strahler.)

LITTORAL DRIFT

In an idealized situation in which waves approach a straight shoreline, their crests parallel with that line, a given wave breaks at the same instant at all points and the swash rides up the beach at right angles to the shoreline. The backwash returns along the same line. Consequently, particles move up and down the beach slope along a fixed line.

Along most shorelines most of the time, however, waves approach the coast at an oblique angle (see Figure 20.16). As these waves travel toward a shoreline over a gently shoaling ocean bottom, they undergo a gradual decrease in velocity of forward travel. As a result, the wave crests become curved in plan and tend to become more nearly parallel with the shoreline. This wave-bending phenomenon is called *wave refraction*.

Despite some refraction, however, the wave crest arrives in the breaker zone with an oblique approach to the shoreline. The swash of the breaker then rides obliquely up the foreshore, as shown in Figure 20.14. As

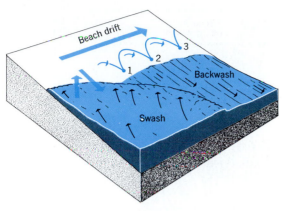

FIGURE 20.14 Beach drift of sand, caused by an oblique approach of the swash. (Drawn by A. N. Strahler.)

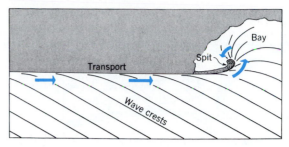

FIGURE 20.16 Along a straight coast, littoral drift carries sand in one direction to reach the mouth of a bay, where it is formed into a sandspit. (From A. N. Strahler, *The Earth Sciences*, 2nd ed., Harper & Row, Publishers. Copyright © by Arthur N. Strahler.)

a result the sand, pebbles, and cobbles are moved obliquely up the slope. After the swash has spent its energy, the backwash flows down the slope of the beach, being controlled by the pull of gravity, which moves it in the most direct downhill direction. Now the particles are dragged directly seaward and come to rest at positions to one side of the starting points. Because wave fronts approach consistently from the same direction on a particular day, this movement is repeated many times. Individual rock particles thus travel a considerable distance along the shore. Multiplied many thousands of times to include the numberless particles of the beach, this form of mass transport, called *beach drift*, is a major process in shoreline development.

Rarely is beach drift not taking place, in one direction or the other, along a marine shoreline. Usually, a given stretch of shoreline is subjected to a dominant direction of wave approach throughout a given season of the year, or throughout the entire year. Consequently, beach drift can be assigned a single direction of net transport as the seasonal or yearly average.

A process related to beach drifting is *longshore drift*. When waves approach a shoreline under the influence of strong winds, the water level is slightly raised near shore by a slow shoreward drift of water. An excess of water, which must escape, is thus pushed shoreward. A *longshore current* is set up parallel to shore in a direction away from the wind (Figure 20.15). When wave and wind conditions are favorable, this current is capable of mov-

ing sand along the bottom in the breaker zone in a direction parallel to the shore.

Both beach drifting and longshore drifting move particles in the same direction for a given set of onshore winds and oblique wave fronts and therefore supplement each other's influence in sediment transportation. The combined transport by beach and longshore drift is termed *littoral drift*. Let us now apply the principle to the evolution of beach deposits.

Littoral drift along a straight section of shore is illustrated in Figure 20.16. Where an embayment occurs,

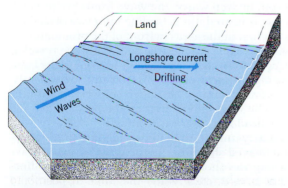

FIGURE 20.15 Longshore current drifting. (Drawn by A. N. Strahler.)

FIGURE 20.17 A white sandspit, growing in a direction toward the observer, leaves only a narrow inlet (foreground) for tidal currents. Sandy shoals lie in the bay at the right. Martha's Vineyard, Massachusetts. (Donald W. Lovejoy.)

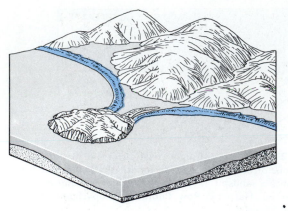

FIGURE 20.18 Two tombolos have connected this cliffed island with the mainland. (Drawn by W. M. Davis.)

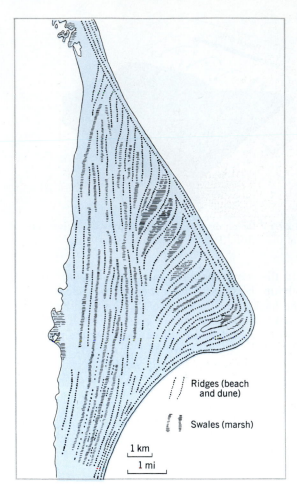

FIGURE 20.20 Sketch map of Cape Canaveral, Florida, as it was in 1910, before human-induced modification. Ridges near the shore (*right*) are beach ridges; those farther inland are dune ridges built on older beach ridges. (After Douglas Johnson, 1919.)

drift continues along the line of the straight shore, with the result that an embankment of sediment is constructed along that line. This narrow beach deposit, extending out into open waters, is called a *sandspit*, or simply a *spit* (Figure 20.17). Because of wave refraction in shallow water of the bay, sediment is carried around the spit end by littoral drift. Thus the spit develops a landward curvature. A sandspit may grow in length and become extended across the entire mouth of a bay, forming a *baymouth bar*. (See Figure 20.32*D*.) The word *bar* refers to any long, narrow sand embankment formed by wave action. Littoral drift from an island may form a *tombolo*, which is a sand bar connecting the island with the mainland (Figure 20.18).

Where littoral drift converges from opposite directions upon a given point on a shoreline, sediment accumulates in the form of a *cuspate bar* (Figure 20.19). When progradation continues over a long period of time, a much larger feature, called a *cuspate foreland*, is constructed (Figure 20.20). It consists of many beach ridges separated from one another by narrow belts of low, marshy ground, called *swales*. Wind action may transform the sandy beach ridges into dunes, but the original plan of the ridges continues to be prominent.

WAVE REFRACTION ON AN EMBAYED COAST

On a coastline with promontories and bays, such as that shown in Figure 20.21, wave refraction concentrates wave energy upon the promontories. Successive positions of a wave crest are shown by lines numbered 1, 2, 3, and so on. In deep water, the wave fronts are parallel. As the shore is neared, the retarding influence of shallow water is felt first in the areas in front of the promontories. Shallowing of water reduces the speed of wave travel at those places; but in the deeper water in front of the bays, the retarding action has not yet occurred. Consequently, the wave front is refracted in rough conformity with the shoreline. The waves break first on the promontory and on the bay head last, as shown in Figure 20.21.

Particularly important in understanding the development of embayed shorelines is the distribution of wave energy along the shore. On Figure 20.21, dashed lines (lettered a, b, c, d, etc.) divide the wave at position 1 into equal parts, which may be taken to include equal amounts of energy. Along the headlands, the energy is

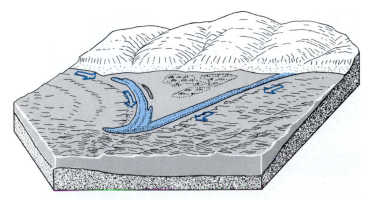

FIGURE 20.19 This cuspate bar, which has enclosed a triangular lagoon, receives drifted beach materials from both sides. (Drawn by E. Raisz.)

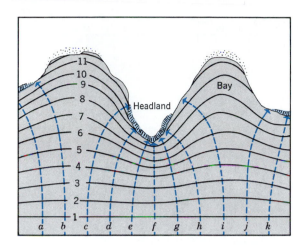

FIGURE 20.21 Wave refraction on an embayed coast.

FIGURE 20.23 A series of cliffed headlands on the Big Sur coast of northern California. A pocket beach lies in the foreground. Above the cliff is a marine terrace, partly buried in alluvium. (Alan H. Strahler.)

concentrated into a short piece of shoreline; along the bays, it is spread over a much greater length of shoreline. Consequently, the breaking waves act as powerful erosional agents on the promontories, but are relatively weak and ineffective at the bay heads. The important principle is that headlands and promontories are rapidly eroded back. The process tends to produce a simple, broadly curving shoreline as a stable form.

Concentrated wave attack on promontories produces a high marine cliff and a broad abrasion platform. Waves breaking obliquely on the bay shores cause littoral drift of the detritus toward the bay heads, as shown in Figure 20.22. This sediment accumulation becomes a crescent-shaped *pocket beach* (Figure 20.23).

LITTORAL DRIFT AND SHORE PROTECTION

Along stretches of shoreline affected by retrogradation, the beach may be seriously depleted or even entirely destroyed. When this occurs, cutting back of the coast can be rapid in weak materials, destroying valuable shore property. Protective engineering structures designed for direct resistance to frontal wave attack, such as seawalls, are prone to failure and are extremely expensive (Figure

20.24). In some circumstances, a successful alternative strategy is to install structures that will cause progradation, building a broad protective beach. The principle here is that the excess energy of storm waves will be dissipated in reworking the beach deposits. Cutting back of the beach in a single storm will be restored by beach-building between storms.

Progradation requires that sediment moving as littoral drift be trapped by the placement of baffles across the path of transport. To achieve this result, groins are installed at close intervals along the beach. A *groin* is simply a wall or embankment built at right angles to the shoreline; it may be constructed of huge rock masses, reinforced concrete, or wooden pilings. Figure 20.25 shows the shoreline changes induced by groins. Sand accumulates on the updrift side of the groin, developing a curved shoreline. On the downdrift side of the groin, the beach will be depleted because of the cutting off of the normal supply of drift sand. The result may be harmful retrogradation and cutting back of the coast. For this reason groins must be closely spaced so the trapping ef-

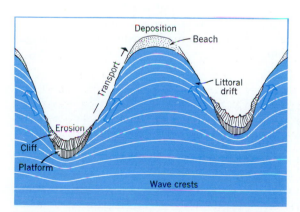

FIGURE 20.22 Littoral drift moves sediment from promontories to bay heads. (From A. N. Strahler, *The Earth Sciences*, 2nd ed., Harper & Row, Publishers. Copyright © by Arthur N. Strahler.)

FIGURE 20.24 Concrete seawalls of different styles protect pieces of shorefront property along this stretch of Lake Michigan shoreline of northern Indiana. (Ned L. Reglein.)

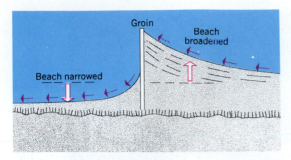

FIGURE 20.25 Construction of a groin causes marked changes in configuration of the sand beach. (From A. N. Strahler, *Planet Earth: Its Physical Systems Through Geologic Time*, Harper & Row, Publishers. Copyright © by Arthur N. Strahler.)

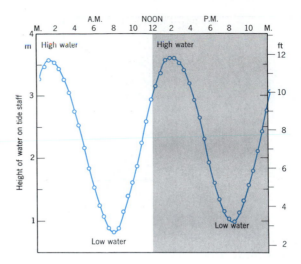

FIGURE 20.27 Height of water at Boston Harbor measured every half hour.

fect of one groin will extend to the next (Figure 20.26). Ideally, when the groins have trapped the maximum quantity of sediment, beach drift will be restored to its original rate for the shoreline as a whole.

In some instances the source of beach sand is from the mouth of a river. Construction of dams far upstream on the river may drastically reduce the sediment load and therefore also cut off the source of sand for littoral drift. Retrogradation may then occur on a long stretch of shoreline. The Mediterranean shoreline of the Nile Delta has suffered retrogradation because of reduction in sediment supply following dam construction far up the Nile.

TIDAL CURRENTS

Most marine coastlines are influenced by the *ocean tide*, a rhythmic rise and fall of sea level under the influence of changing attractive forces of moon and sun on the rotating earth. Where tides are great, the effects of changing water level and the currents set in motion are of major importance in shaping coastal landforms.

The tidal rise and fall of water level is graphically represented by the *tide curve*. We can make half-hourly observations of the position of water level against a tide staff (measuring stick) attached to a pier or sea wall. We then plot the changes of water level and draw the tide curve. Figure 20.27 is a tide curve for Boston Harbor covering a day's time. The water reached a maximum

height, or *high water*, of about 4 m (12 ft), then fell to a minimum height, or *low water*, of about 1 m (3 ft). This occurred about 6¼ hours after high water. A second high water occurred about 12½ hours after the previous high water, completing a single tidal cycle. In this example the *range of tide*, or difference between heights of successive high and low waters, is about 3 m (9 ft).

The rising tide sets in motion currents of water, known as *tidal currents*, in bays and estuaries. The relationships between tidal currents and the tide curve are shown in Figure 20.28. When the tide begins to fall, an *ebb current* sets in. This flow ceases about the time the tide is at its lowest point. As the tide begins to rise, a landward current, the *flood current*, begins to flow.

Tidal currents may be active in the lower reaches of large rivers that have experienced partial drowning because of recent sea-level rise. The tidal section of the river is a form of estuary, in which fresh water and salt water are mixed. River discharge augments the ebb current but opposes the flood current, so the ebb current attains a higher speed than the flood current. Thus a floating object would travel farther seaward than landward with each tidal cycle and would eventually reach the open ocean. This process provides a natural flushing action that carries human-made wastes, such as sewage, to the ocean.

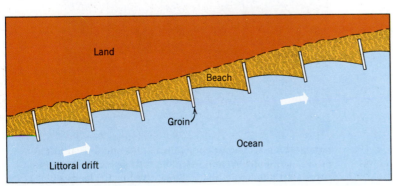

FIGURE 20.26 A system of multiple groins can successfully maintain a broad beach.

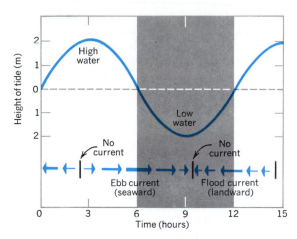

FIGURE 20.28 The ebb current flows seaward as the tide level falls; the flood current flows landward as the tide level rises.

TIDAL POWER

Tidal power makes use of the tidal rise and fall of ocean level, an unending source of energy. To harness the power of the tides, a bay must be located along a coast subject to a large range of tide. An ideal location is to be found on the Atlantic Coast of Maine, New Brunswick, and Nova Scotia, around the Bay of Fundy. Here the average tide range exceeds 10 m (30 ft). The rugged coast near the boundary of Maine and Nova Scotia is deeply indented by narrow bays, with constricted places where the bays empty into the ocean. These constrictions can be closed off by dams and the tidal flow passed through conduits connecting the bay and the ocean. Turbines placed in the conduits can be turned by the swift tidal current.

Plans for a tidal power plant at Passamaquoddy Bay, at the entrance to the Bay of Fundy, were developed and modified over a long span of years. Work on a small part of the total project was begun in 1935 and suspended in 1937. Two important tidal power plants have since come into operation. One is the La Rance plant on the Brittany coast of France. Opened in 1966, the maximum generating capacity of the plant, which has 24 conduits, is 240 megawatts when the turbines are running at full speed. The average output for the year is only 60 megawatts, which is about the output of one small nuclear power station. A second tidal power plant is operating in the Soviet Union in an inlet of the White Sea.

In 1977 a Canadian study board reviewed the potential for development of tidal power in the Bay of Fundy and concluded that a major project is economically feasible. A tidal generating plant intended as a pilot project began operation in 1984 at Annapolis Royal, Nova Scotia, on the south side of the Bay. The plant uses the ebb flow through a dam placed across the inlet of a long narrow bay; it generates about 20 megawatts of electricity. Two other such narrow bays are being studied as sites for much larger tidal generating plants. One of these is to be located in Minas Basin, at the extreme eastern end of the Bay of Fundy, where the highest tidal range occurs. The maximum capacity of this plant would be

about 5000 megawatts; its average production about 1500 megawatts. This is equivalent to the electricity that can be generated annually from 23 million barrels of crude oil.

In Great Britain, studies have been made of the feasibility of installing a number of tidal generating plants offshore of its larger shallow bays, such as The Wash and the Humber estuary on the east coast, and Morecambe Bay on the west. Long, low barrages (dams) would be constructed some 10 km (6 mi) offshore with numerous sluicegates, so as to allow turbines to generate electricity during both rising and falling tides. Also being studied are more conventional tidal power plants, such as those planned for the Bay of Fundy, for installation in the narrow Severn and Mersey estuaries.

The potential for tidal power development on a global scale is not great. For the world the annual tidal energy potentially available by exploitation of all suitable coastal sites is only about 1 percent of the total energy potentially available through hydropower development. For the United States, the developable tidal energy resources are about two-thirds as great as the hydropower resources.

TIDAL CURRENT DEPOSITS

Ebb and flood currents generated by tides perform several important functions along a shoreline. For one, the currents that flow in and out of bays through narrow inlets are very swift and can scour an inlet strongly. Scour keeps the inlet open despite the tendency of shore drifting processes to close it with sand.

Also, tidal currents carry much fine silt and clay in suspension, derived from streams that enter the bays or from bottom muds agitated by storm wave action against the outer shoreline. This fine sediment settles to the floors of the bays and estuaries, where it accumulates in layers and gradually fills the bays. Much organic matter is present in this sediment.

FIGURE 20.29 Salt marsh of *Spartina* grass forms a yellowish-green border zone around this small tidal inlet on Cape Cod, Massachusetts. At the right, the salt-marsh peat has been removed to make a boat landing. (A. N. Strahler.)

In time, tidal sediments fill the bays and produce *mud flats*, which are barren expanses of silt and clay exposed at low tide but covered at high tide. Next, a growth of salt-tolerant plants takes hold on the mud flat. The plant stems entrap more sediment and the flat is built up to approximately the level of high tide, becoming a *salt marsh* (Figure 20.29). A thick layer of peat is eventually formed at the surface. Tidal currents maintain their flow through the salt marsh by means of a highly complex network of sinuous tidal streams (Figure 20.30).

Salt marsh is important to geographers because it is land that can be drained and made agriculturally productive. The salt marsh is first cut off from the sea by the construction of an embankment of earth (a dike), in which gates are installed to allow the freshwater drainage of the land to exit during flood flow. Gradually, the salt water is excluded and soil water of the diked land

FIGURE 20.30 Salt marsh with sinuous tidal channels (right) lies behind a protective barrier beach (left) on outer Cape Cod. A broad tidal inlet allows ebb and flood tide currents to drain and flood the marsh with each tidal cycle. A tidal delta can be seen on the landward side of the inlet. Sand dunes (foredunes) cover the barrier beach at the lower left. (John S. Shelton.)

A. Ria coast

B. Fiord coast

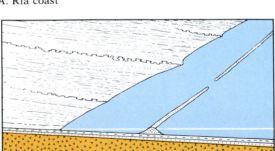

C. Barrier-island coast

D. Delta coast

E. Volcano coast (left); F. Coral-reef coast (right)

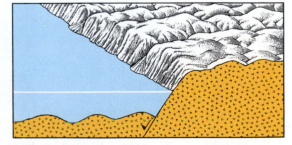

G. Fault coast

FIGURE 20.31 Seven common kinds of coastlines are illustrated here. These examples have been selected to illustrate a wide range in coastal features. (Drawn by A. N. Strahler.)

becomes fresh. Such diked lands are intensively developed in Holland (where they are called *polders*).

Over many decades, the surface of reclaimed salt marsh subsides because of the compaction of underlying peat layers and may come to lie well below mean sea level. The threat of flooding by salt water, when storm waves breach the dikes, hangs constantly over the inhabitants of such low areas. The reclamation of salt marsh by dike construction and drainage was practiced by New World settlers in New England and Nova Scotia. In the industrial era, large expanses of salt marsh have been destroyed by landfill, a practice only recently inhibited by new legislation.

COMMON KINDS OF COASTLINES

There are many different kinds of coastlines, each kind unique because of the distinctive landmass against which the ocean water has come to rest. One group of coastlines derives its qualities from *submergence*, the

partial drowning of a coast by a rise of sea level or a sinking of the crust. Another group derives its qualities from *emergence*, the exposure of submarine landforms by a falling of sea level or a rising of the crust. Another group of coastlines results when new land is built out into the ocean by volcanoes and lava flows, by the growth of river deltas, or by the growth of coral reefs.

A few important types of coastlines are illustrated in Figure 20.31. The *ria coast* (A) is a deeply embayed coast resulting from submergence of a landmass dissected by streams. This coast has many offshore islands. A *fiord coast* (B) is deeply indented by steep-walled fiords, which are submerged glacial troughs (Chapter 22). The *barrier-island coast* (C) is associated with a recently emerged coastal plain. The offshore slope is very gentle, and a barrier island of sand is usually thrown up by wave action at some distance offshore. Large rivers build elaborate deltas, producing *delta coasts* (D). The *volcano coast* (E) is formed by the eruption of volcanoes and lava flows, partly constructed below water level. Reef-building corals create new land and make a *coral-reef coast* (F).

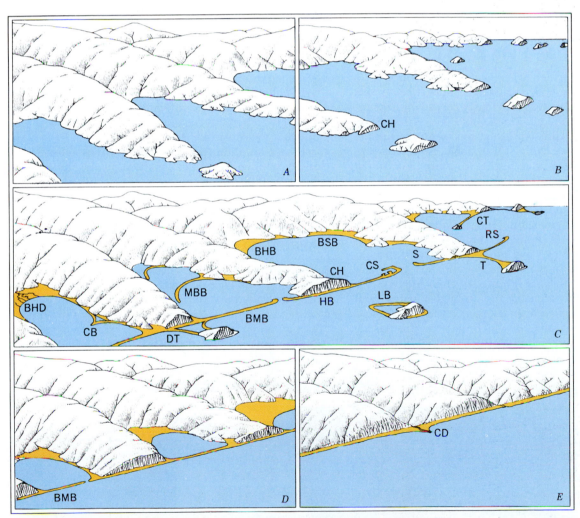

FIGURE 20.32 Stages in the evolution of a ria coastline. (Drawn by A. N. Strahler.) T = tombolo; S = spit; RS = recurved spit; CS = complex spit; CT = complex tombolo; LB = looped bar; CH = cliffed headland; DT = double tombolo; HB = headland beach; BMB = baymouth bar; MBB = midbay bar; CB = cuspate bar; BHB = bayhead beach; BSB = bayside beach; BHD = bayhead delta; L = lagoon; I = inlet; CD = cuspate delta.

FIGURE 20.33 Golden Gate is the narrow, steep-walled entrance to San Francisco Bay, a branching arm of the sea that receives river flow from the Sierra Nevada far inland. This photo was taken in 1949, when the Golden Gate Bridge was twelve years old. (Ned L. Reglein.)

Down-faulting of the coastal margin of a continent can allow the shoreline to come to rest against a fault scarp, producing a *fault coast* (G).

DEVELOPMENT OF A RIA COAST

The ria coast is formed when a rise of sea level or a crustal sinking (or both) brings the shoreline to rest against the sides of valleys previously carved by streams. This event is illustrated in Frame A of Figure 20.32. Soon wave attack forms cliffs on the exposed seaward sides of the islands and headlands (Frame B). Sediment produced by wave action then begins to accumulate in the form of beaches along the cliffed headlands and at the heads of bays. This sediment is carried by littoral drift and is built into sandspits across the bay mouths (baymouth bars) and as connecting links (tombolos) between islands and mainland (Frame C). Finally, all outlying islands are planed off by wave action and a nearly straight shoreline develops in which the sea cliffs are fully connected by baymouth bars (Frame D). Now the bays are sealed off from the open ocean, although narrow tidal inlets may persist, kept open by tidal currents. Frame E shows a much later stage in which the coastline has receded beyond the inner limits of the original bays.

The influence of ria coastlines on human activity has been strong throughout the ages. The deep embayments of the ria shoreline make splendid natural harbors. One of the finest of these is San Francisco Bay (Figure 20.33). Much of the ria coastline of Scandinavia, France, and the British Isles is provided with such harbor facilities. Consequently, these peoples have a strong tradition of fishing, shipbuilding, ocean commerce, and marine activity generally. Mountainous relief of ria and fiord coasts made agriculture difficult or impossible, forcing the people to turn to the sea for a livelihood. New England and the Maritime Provinces of Canada have a ria coastline with abundant good harbors. The influence of this environment was to foster the same development of fishing, whaling, ocean commerce, and shipbuilding seen in the British Isles and Scandinavian countries.

BARRIER-ISLAND COASTS

In contrast to ria and fiord coasts, with their bold relief and deeply embayed outlines, we find low-lying coasts from which the land slopes gently beneath the sea. The coastal plain of the Atlantic and Gulf coasts of the United States presents a particularly fine example of such a gently sloping surface. A coastal plain is a belt of rela-

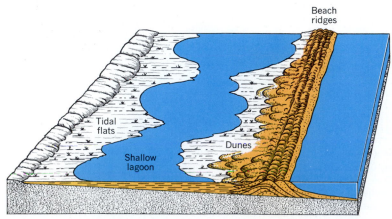

FIGURE 20.34 A barrier island is separated from the mainland by a wide lagoon. Sediments fill the lagoon, while dune ridges advance over the tidal flats. (From A. N. Strahler, *The Earth Sciences*, 2nd ed., Harper & Row, Publishers. Copyright © by Arthur N. Strahler.)

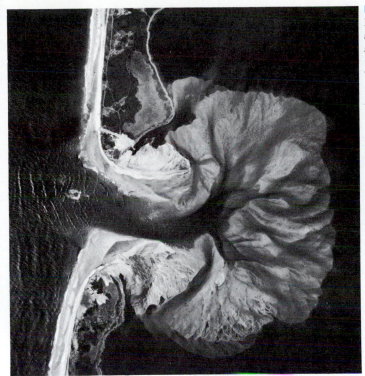

FIGURE 20.35 East Moriches Inlet was cut through Fire Island, a barrier island off the Long Island shoreline, during a severe storm in March 1931. This aerial photograph, taken a few days after the breach occurred, shows the underwater tidal delta being built out into the lagoon (right) by currents. The entire area shown is about 1.6 km (1 mi) long. North is to the right; the open Atlantic Ocean is on the left. (U.S. Army Air Forces Photograph.)

FIGURE 20.36 The Gulf Coast of Texas is dominated by its offshore barrier island.

FIGURE 20.37 This Landsat image shows some details of the Texas barrier-island coast. Near the top of the frame is Corpus Christi Bay (circular shoreline), with the city of that name built on the southwestern shore. A ship channel can be traced diagonally across the lagoon to a pass (Aransas Pass) in the barrier island. Farther south, the barrier island (Padre Island) has an inner dune belt (white). Other dunes on the mainland appear as small white spots. Irrigated farmland near Corpus Christi appears as a pattern of dark squares. Ranchland lies to the south. (NASA Landsat 1092–16314, October 1972.)

(left) Seen from Apollo 9 spacecraft, the white barrier beach of sand—the Outer Banks—projects sharply seaward in two points: Cape Hatteras (H) and Cape Lookout (L). Pamlico Sound (P) lies between the barrier and the mainland shore, which is deeply embayed as a result of postglacial submergence. Extensive areas of salt marsh (S) make up the low coastal zone. Through two inlets (I) between Hatteras and Lookout sediment is being carried seaward in plumes of lighter color. The Gulf Stream boundary (G) is sharply defined. The area shown is about 160 km (100 mi) across (NASA AS9–20–3128.)

(above) A close look at the barrier beach on Cape Hatteras. A dune ridge, well protected by grass cover, lies to the left of the beach. The view is north; the Atlantic Ocean is at the right. (Mark A. Melton.)

(above) Near Cape Hatteras this stretch of barrier beach has been swept over by storm swash, topping the dune barrier and carrying sand in sheets toward the bay at the left. (Mark A. Melton.)

FIGURE 20.38 The barrier-island coast of North Carolina.

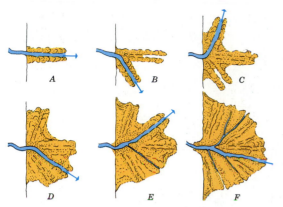

FIGURE 20.39 Stages in the formation of a simple delta. (After G. K. Gilbert.)

tively young sedimentary strata, formerly accumulated beneath the sea as deposits on the continental shelf (see Figure 19.20). Emergence as a result of repeated crustal uplifts has characterized this coastal plain during the latter part of the Cenozoic Era and into recent time.

Along much of the Atlantic and Gulf coast, a *barrier island* exists; it is a low ridge of sand built by waves and further increased in height by the growth of sand dunes (Figure 20.34). Behind the barrier island lies a *lagoon*, which is a broad expanse of shallow water, often several kilometers wide and in places filled largely with tidal deposits.

A characteristic feature of most barrier islands is the presence of gaps, known as *tidal inlets*. Through these gaps, strong currents flow alternately seaward and landward as the tide rises and falls. In heavy storms, the barrier may be breached by new inlets (Figure 20.35). Tidal currents will subsequently tend to keep a new inlet open, but it may be closed by shore drifting of sand.

Shallow water results in generally poor natural harbors along barrier-island shorelines. The lagoon itself may serve as a harbor if channels and dock areas are dredged to sufficient depths. Ships enter and leave through one of the passes in the barrier island, but artificial seawalls and jetties are required to confine the cur-

rent and keep sufficient channel depth. In some cases, major port cities are located where a large river empties into the lagoon. The lower courses of large rivers provide tidal channels that may be dredged to accommodate large vessels and thus make seaports of cities many kilometers inland.

One of the finest examples of a barrier island and lagoon is along the Gulf Coast of Texas (Figures 20.36 and 20.37). Here the island is unbroken for as much as 160 km (100 mi) at a stretch, and passes are few. The lagoon is 8 to 16 km (5 to 10 mi) wide. Galveston is built on the barrier island adjacent to an inlet connecting Galveston Bay with the sea. Most other Texas ports, however, are located on the mainland shore. Corpus Christi, Rockport, Texas City, Port Lavaca, and other ports are located along the shores of river embayments.

Barrier islands of the coast of North Carolina are characterized by prominent cuspate headlands—Cape Hatteras and Cape Lookout—and by a large lagoon—Pamlico Sound (Figure 20.38). The inner shoreline is deeply embayed, indicating recent submergence of the coastal plain. The barrier beach has been moving landward through the process of *overwash*, in which storm swash travels across the entire barrier and sweeps sand into the lagoon. This barrier beach is a subject of intense environmental controversy. Those who seek to develop the beach as a recreation facility and to protect private property from destruction are opposed by conservationists and National Park Service officials, who advocate that natural processes should be allowed to proceed unchecked.

DELTA COASTS

The deposit of clay, silt, and sand made by a stream where it flows into a body of standing water is known as a *delta* (Figure 20.39). Deposition is caused by rapid reduction in velocity of the current as it pushes out into the standing water. Typically, the river channel divides and subdivides into lesser channels called *distributaries*. The coarse particles settle out first, forming foreset beds. Silts and clays continue out farthest and come to rest as bottomset beds (Figure 20.40). Contact of fresh with salt

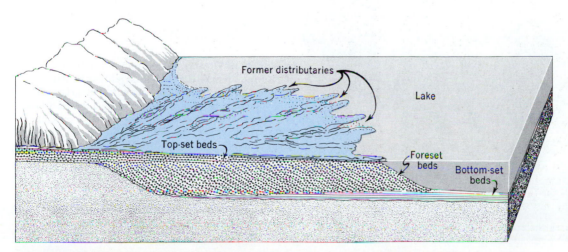

FIGURE 20.40 Structure of a simple delta shown in a vertical section. (After G. K. Gilbert.)

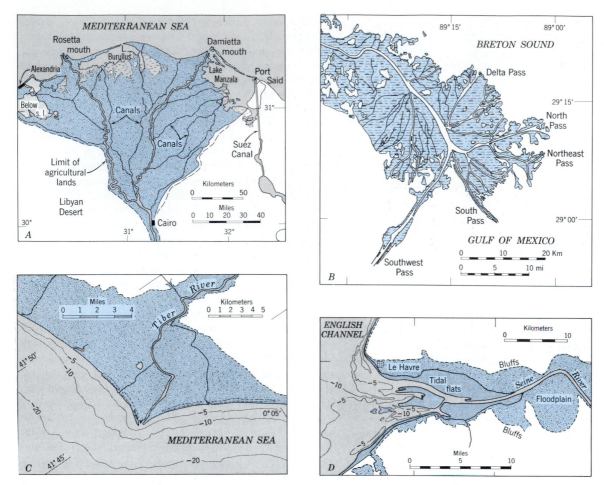

FIGURE 20.41 Deltas. (A) The Nile delta has an arcuate shoreline and is triangular in plan. (B) The Mississippi delta is of the branching, bird-foot type with long passes. (C) The Tiber delta on the Italian coast is pointed, or cuspate, because of strong wave and current action. (D) The Seine delta is filling in a narrow estuary.

FIGURE 20.42 In this photograph, taken by astronauts from an orbiting Gemini satellite, the Nile delta appears in true color as a dark blue-green triangle bounded by pale yellowish-brown desert areas. Two major distributaries (N) of the Nile River end in prominent cusps: the Damietta mouth (D) and the Rosetta mouth (R). Alexandria (A) is just within view in the foreground. Sand carried by littoral drift from the river mouths has accumulated in a broad barrier beach (B), separated from the delta plain by an open lagoon (L). The Suez Canal (C) connects the Mediterranean Sea (left) with the Gulf of Suez (right). Beyond lies the Sinai Peninsula. (NASA S–65–34776.)

FIGURE 20.43 The Mississippi delta recorded from Landsat orbiting satellite and presented in false-color imagery. The natural levees of the bird-foot delta appear as lacelike filaments in a great pool of turbid river water. To the left (west and north) of the active modern delta are the remains of older deltas. The Chandeleur Islands, a system of arcuate barrier islands, have been built of material from the subsided remains of these older deltas. New Orleans can be seen at the upper left, occupying the natural levees of the Mississippi River and the southern shore of Lake Pontchartrain. (NASA Landsat 1177–16023, January 1973.)

water causes the finest clays to clot (flocculate) into larger aggregates, which settle to the seafloor.

Deltas show a variety of shapes, both because of the configuration of the coastline and because of wave action. The Nile delta, whose resemblance to the Greek letter "delta" suggested the name for this type of landform, has distributaries that branch out in a radial arrangement (Figure 20.41A). Because of its broadly curving shoreline, causing it to resemble in outline an

alluvial fan, this type may be described as an *arcuate delta* (Figure 20.42). The Mississippi delta presents a very different sort of picture (Figure 20.41B). It is a *bird-foot delta*, with long, projecting fingers growing far out into the water at the ends of each distributary (Figure 20.43).

Where wave attack is vigorous, the sediment brought out by the stream is spread along the shore in both directions from the river mouth, giving a pointed delta with curved sides. Because of its resemblance to a sharp tooth, this type is called a *cuspate delta* (Figure 20.41C). Where a river empties into a long, narrow estuary, the delta is confined to the shape of the estuary (Figure 20.41D). This type can be called an *estuarine delta*.

Deltas of large rivers have been of environmental importance from earliest historical times because their extensive flat fertile lands support dense agricultural populations. Important coastal cities, linking ocean and river traffic, are often situated on or near deltas. Examples are Alexandria on the Nile, Calcutta on the Ganges-Brahmaputra, Amsterdam and Rotterdam on the Rhine, Shanghai on the Changjiang (Yangtze), Marseilles on the Rhône, and New Orleans on the Mississippi.

Delta growth is often rapid, ranging from about 3 m (10 ft) per year for the Nile to 60 m (200 ft) per year for the Po and Mississippi rivers. Some cities and towns that were at river mouths several hundred years ago are today several kilometers inland. An important engineering problem is to keep an open channel for ocean-going vessels that have to enter the delta distributaries to reach port. The mouths of the Mississippi River delta distributaries, known as passes, have been extended by the construction of jetties (Figure 20.43). The narrowed stream is forced to move faster, scouring a deep channel.

CORAL-REEF COASTS

Coral-reef coasts are unique in that the addition of new land is made by organisms: corals and algae. Growing together, these organisms secrete rocklike deposits of mineral carbonate, called *coral reefs*. As coral colonies die, new ones are built on them, accumulating as limestone. Coral fragments are torn free by wave attack, and the pulverized fragments accumulate as sand beaches.

Coral-reef coasts occur in warm, tropical and equatorial waters between the limits lat. 30°N and 25°S. Water temperatures above 20°C (68°F) are necessary for dense reef-coral growth. Reef corals live near the water surface. Water must be free of suspended sediment and well aerated for vigorous coral growth. For this reason corals

FIGURE 20.44 The Island of Moorea and its fringing coral reef, Society Islands, South Pacific Ocean. The island is a deeply dissected volcano with a history of submergence. (Jack Fields/Photo Researchers.)

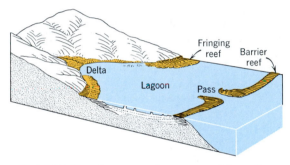

FIGURE 20.45 A barrier reef is separated from the mainland by a shallow lagoon. (Drawn by W. M. Davis.)

FIGURE 20.46 Atolls of the Banda Sea near the Celebes Islands in Indonesia. The location is about lat. 5° S. The atolls appear in this Landsat image as pale blue loops, partly obscured by cumuliform clouds. Islands, reddish in color because of the rain forest cover (green vegetation appears red on this false-color image), are almost completely obscured under cloud cover, but their fringing reefs can be discerned at a few points. The area shown is about 120 km (75 mi) across. (NASA ERTS E–1414–01221, September 1973.)

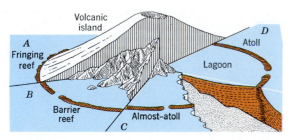

FIGURE 20.47 The subsidence theory of barrier-reef and atoll development is shown in four stages, beginning with a fringing reef attached to a volcanic island and ending with a circular reef. (Drawn by W. M. Davis.)

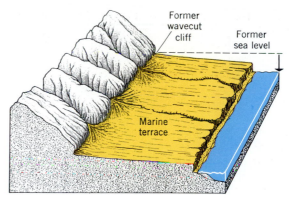

FIGURE 20.48 A raised shoreline becomes a cliff parallel with the newer, lower shoreline. The former abrasion platform is now a marine terrace. (Drawn by A. N. Strahler.)

thrive in positions exposed to wave attack from the open sea. Because muddy water prevents coral growth, reefs are missing opposite the mouths of muddy streams. Coral reefs are remarkably flat on top. They are exposed at low tide and covered at high tide.

Three general types of coral reefs may be recognized: (1) fringing reefs, (2) barrier reefs, and (3) atolls. *Fringing reefs* are built as platforms attached to shore (Figure 20.44). They are widest in front of headlands where wave attack is strongest, and the corals receive clean water with abundant food supply.

Barrier reefs lie out from shore and are separated from the mainland by a lagoon (Figure 20.45). Narrow gaps occur at intervals in barrier reefs. Through these openings, excess water from breaking waves is returned from the lagoon to the open sea.

Atolls are more or less circular coral reefs enclosing a lagoon, but without any land inside (Figure 20.46). On large atolls, parts of the reef have been built up by wave action and wind to form low island chains, connected by the reef. Most atolls have been built on a foundation of volcanic rock. These foundations are thought to have been basaltic volcanoes, built on the deep ocean floor. According to Charles Darwin's *subsidence theory*, the extinct volcanoes slowly subsided after being planed off by wave erosion. At the same time the fringing coral reef continued to build upward (Figure 20.47).

The environmental aspects of atoll islands are unique in some respects. For example, there is no rock other than coral limestone, composed of calcium carbonate.

This means that trees requiring other minerals, such as silica, cannot be cultivated without the aid of fertilizers or some outside source of rock from a larger island composed of volcanic or other igneous rock.

The palm tree is native to atoll islands because it thrives on brackish water, and the seed, or palm nut, is distributed widely by floating from one island to another. Native inhabitants cultivated the coconut palm to provide food, clothing, fibers, and building materials. Fresh water is scarce on small atoll islands. Rainfall must be caught in open vessels or catchment basins and carefully conserved. Fish and other marine animals are an important part of the human diet on atoll islands. Calm waters of the lagoon make a good place for fishing and for beaching canoes.

RAISED SHORELINES AND MARINE TERRACES

The active life of a shoreline is sometimes cut short by a rapid rise of the coast. When this event occurs, a *raised shoreline* is formed. The marine cliff and abrasion platform are abruptly raised above the level of wave action. The former abrasion platform has now become a *marine terrace* (Figure 20.48). Fluvial denudation begins to de-

FIGURE 20.49 Marine terraces on the western slope of San Clemente Island, off the southern California coast. More than twenty terraces have been identified in this series; the highest has an elevation of about 400 m (1300 ft). (John S. Shelton.)

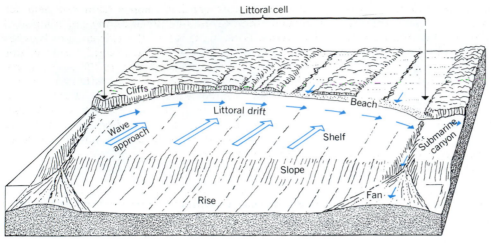

FIGURE 20.50 Pictorial diagram of a littoral cell typical of the coast of southern California. (Based on data of D. L. Inman and B. M. Brush, *Science*, vol. 181, p. 26. (From A. N. Strahler, *Physical Geology*, Harper & Row, Publishers. Copyright © by Arthur N. Strahler.)

stroy the terrace, and it may also undergo partial burial under alluvial fan deposits.

Marine terraces along mountainous coasts are important to humans as strips of flat ground extending for tens of kilometers parallel with the shoreline. Highways and railroads follow these terraces, and they are excellent sites for coastal towns and cities. Agriculture makes use of the flat terrace surfaces where the soil is good.

Raised shorelines are common along the continental and island coasts of the Pacific Ocean because here tectonic processes are active along coastal mountain ranges and island chains. Repeated uplifts result in a series of raised shorelines in a steplike arrangement. Fine examples of these multiple marine terraces are seen on the western slope of San Clemente Island, off the California coast (Figure 20.49).

THE LITTORAL CELL AS A MATERIAL FLOW SYSTEM

A unifying concept applying to shore processes and landforms of continental coastlines is that of the *littoral cell*, a distinct coastal compartment that represents an open material flow system. The system consists basically of an input of sediment from the land, transport and storage of sediment along the shore and within the zone of shoaling

waves, and an output of sediment to deep water of the ocean basin.

A block diagram, Figure 20.50, shows a typical littoral cell on the Pacific coast of North America, where marine cliffs are present along much of the shoreline. The prevailing direction of littoral drift is from left to right, as indicated by the arrows. The cell is defined as the stretch of coastline between two cliffed promontories along which the shoreline is concave in plan toward the ocean. Wave attack upon the cliff supplies sediment for littoral drift. Sediment in storage begins to appear as a narrow beach some distance downdrift of the promontory. Farther along the coast, streams enter the cell, bringing large quantities of fluvial sediment and causing the beach to become greatly thickened and widened. Ultimately, the drifting sediment is carried into a submarine canyon in the continental shelf and is swept by turbidity currents into deep water where it may accumulate as a submarine fan.

The littoral cell includes the shallow offshore region of the shelf, on which sediment is also stored and moved about. In winter, much of the beach sediment is temporarily moved into the offshore zone, to be returned to the beach in summer. This annual cycle is typical of beaches in middle latitudes. Not shown in the diagram is a circuit in which beach sand is carried inland by onshore winds

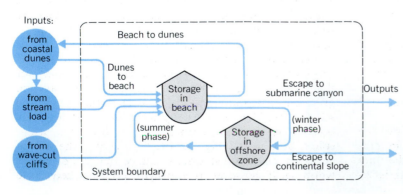

Schematic diagram of the littoral cell as an open material flow system. (From A. N. Strahler, *Physical Geology*, Harper & Row, Publishers. Copyright © by Arthur N. Strahler.)

FIGURE 20.51

to form coastal dunes, from which the sand can be returned to the beach by offshore winds, or by streams, or during high tides and surf associated with large storms.

Figure 20.51 is a schematic diagram of the littoral cell as an open material flow system. Three system inputs and three outputs are shown, along with two forms of sediment storage.

It has been estimated that on a world basis the major source of sediment for littoral cells is from streams (about 95%), with very little derived from cliff erosion. Many coasts, such as barrier-island and delta coasts, are completely lacking in bedrock cliffs. Because of the dependence of most beaches upon stream sediment for their nourishment and maintenance, the construction of river dams poses a threat to the existence of beaches in many parts of the world. As explained earlier in this chapter, when sediment is trapped in the reservoir behind a dam, beaches in the downdrift direction from the river mouth become depleted and may disappear entirely. This unwanted effect is particularly well demonstrated by littoral cells of the California coast, where most of the major streams have been dammed to serve as water storage reservoirs.

RISING SEA LEVEL AND COASTAL INUNDATION—A THREAT FROM GLOBAL WARMING

Few aspects of the negative consequences of a strong global warming have been more dramatically portrayed by the news media than a predicted rise in sea level of several meters, inundating low-lying coasts and their great cities. The threat is real, but like other possible impacts of global warming, there is uncertainty both as to the time schedule it will follow and the magnitude of the effect.

A review of the scientific principles involved here will be helpful. We are discussing eustatic changes in sea level. *Eustatic* is an adjective denoting worldwide sea-level change; it can be distinguished from local relative changes in sea level caused by the uplift or downsinking of the earth's crust—tectonic effects, that is. Eustatic change can result from either (a) a change in water volume of the world ocean or (b) a change in the water-holding capacity of the ocean basins. The second possibility relates to plate tectonics and is on a time scale longer than that involved with global warming in our time.

Climate change has its effect primarily through changes in water volume of the oceans, and this process, in turn, can operate in two ways. First is a volume change resulting from water temperature change. Global warming will raise the temperature of the uppermost layer of the oceans—at most a few hundred meters thick—and the resulting thermal expansion of that water will cause a substantial eustatic rise. The second way is the wasting (or growth) of glacial ice accumulations on the continents. Global warming, by increasing the rate of melting, tends to cause mountain glaciers and ice sheets to shrink in volume, and that meltwater increases the volume of the world ocean. Keep in mind, however, that global warming, by increasing precipitation, will also promote the growth of glaciers, and we must consider the ratio of the opposed processes of melting and growth. Large swings in world sea level took place throughout the past 3 million years as ice sheets cyclically expanded and contracted. We defer to Chapter 22 further discussion of the mechanism of possible sea-level rise by ice melting.

Has a slow eustatic rise of sea level accompanied the overall global temperature rise that started in the mid-1880s? Tide gauges and their records provide the only evidence available, and interpretation of the data is extremely difficult for a number of reasons. Crustal rise or sinking is taking place locally along many coasts where the gauges are located. Cyclic changes in ocean currents cause local mean sea-level changes recorded by individual gauges. Scientific attempts to take these and other factors into account yield the conclusion that mean sea level has indeed been rising. For the 80-year period 1900–1980 the total eustatic rise comes to about 80 mm (3 in.), and about half of this is attributed to thermal expansion. The average rate of rise has been about 1.2 mm/yr through the early and middle 1900s. Recently, the data seem to be showing a doubling of that rate to about 2.4 mm/yr. Although this accelerating rate of rise is in harmony with the expected effect of global warming, the same reasons for caution are valid as those for questioning the proposition that the warming is induced by increased emissions of greenhouse gases through human activity.

Projections of the above rates into the next century are as widely divergent as the global warming rates we presented in Chapter 4. Computer modeling evaluated by the National Academy of Sciences in 1989 showed estimates of sea level increase by the year 2100 from as small as 0.3 m (1 ft) to as large as 2 m (7 ft).

Using a rise of 1 m (3 ft) by the middle or late 2000s, various national and international agencies have estimated the impacts, both physical and financial, on major coastal cities of the world. Large river deltas, along with reclaimed tidal estuaries and marshes and the cities built on them, would face catastrophic damage through storm surf and storm surges in concert with high tides. Costs of raising protective barriers to meet this threat are enormously high, and their effectiveness has been questioned. Along cliffed coasts under direct attack by waves, retrogradation will be intensified, resulting in massive losses of shorefront properties.

One note on the positive side comes from biologists specializing in the ecology of coral reefs. They point out that reef-building corals will intensify their rates of growth to maintain the correct level in relation to the sea surface. Their increased production of calcium carbonate, withdrawing carbon dioxide from the atmosphere at a rate as high as 4 to 9 percent of the rate of emission, could substantially reduce the rate of global warming. If, however, reef growth could not keep pace with the rate of rise of sea level, there could result the wholesale extermination of living reefs.

In Chapter 22, we investigate the role that ice sheets may play in causing a global change of sea level. Is it possible that the enormous Antarctic Ice Sheet may grow—rather than shrink—during global warming, and that sea level will fall rather than rise?

Data sources: Stanley S. Jacobs, 1986–87, *Oceanus*, vol. 29:4, pp. 50–54; *New Scientist*, 1990, vol. 125:1702, p. 36.

COASTAL LANDFORMS IN REVIEW

Variety is the keyword in describing the landforms produced by waves, currents, and organisms along the world's shorelines. We have dealt with only a few representative examples of the kinds of coastlines to be found on our planet. Each kind of coastal zone offers a different habitat for life forms, and each offers a different situation to confront humans as development and exploitation of coastal resources continues.

In past centuries the coastal zone was used principally as a source of food or as a place from which to embark on the ocean in search of food or to trade with peoples in other lands. Today the situation is changing. The coastal zone is now in great demand as a recreation zone, and its economic value lies mostly in the worth of the waterfront land as real estate. Marinas spring up at every available sheltered spot, and recreational boating and sport fishing dwarf the shrinking commercial-fishing activity. Luxury condominiums rise on filled lands where the tidal flat and salt marsh once supported a complex marine ecosystem. Industry, too, presses for its share of the coastal zone for uses as sites of nuclear power plants and oil refineries.

Wise decisions on coastal-zone management often depend partly on accurate knowledge of the operation of coastal processes, such as wave erosion and the littoral drift of sediment. Shoreline changes engineered to check erosion have often set off unwanted retrogradation or progradation elsewhere along the coast, or have resulted in the filling of channels and harbors. These changes also profoundly affect the shallow-water ecosystems.

And now, as climate change caused by the increase in atmospheric carbon dioxide looms as inevitable, the predicted global warming seems certain to raise sea level worldwide, threatening densely inhabited low-lying coasts with inundation.

Transportation and deposition of sand by wind is an important process in shaping coastal landforms. We have made references in this chapter to coastal sand dunes derived from beach sand. In the next chapter we shall investigate the transport of sand by wind and the shaping of dune forms. In so doing we complete the linkage between wind action and wave action in controlling coastal environments.

LANDFORMS MADE BY WIND

Wind blowing over the solid surface of the lands is another active agent of landform development. Ordinarily, wind is not strong enough to dislodge mineral matter from the surfaces of hard rock, or of moist, clay-rich soils, or of soils bound by a dense plant cover. Instead, the action of wind in eroding and transporting sediment is limited to land surfaces where small mineral and organic particles are in the loose state. Such areas are typically deserts and semiarid lands (steppes). An exception is the coastal environment, where beaches provide abundant supplies of loose sand, even where the climate is moist and the land surface inland from the coast is well protected by a plant cover.

Landforms shaped and sustained by wind erosion and deposition represent distinctive life environments, often highly specialized with respect to the communities of animals and plants they support. In climates with barely sufficient soil water, there is a contest between wind action and the growth of plants that tends to stabilize landforms and protect them from wind action. We shall find precarious balances in the ecosystems of certain marginal climatic zones. These balances are not only altered by natural changes in climate, but are easily upset by human activities, often with serious consequences. To understand these environmental changes, we need to acquire a working knowledge of the physical processes of wind action on the land surfaces.

EROSION BY WIND

Wind performs two kinds of erosional work. Loose particles lying on the ground surface may be lifted into the air or rolled along the ground. This process is *deflation*. Where the wind drives sand and dust particles against an exposed rock or soil surface, causing it to be worn away by the impact of the particles, the process is *wind abrasion*. Abrasion requires cutting tools carried by the wind; deflation is accompanied by air currents alone.

Deflation acts wherever the ground surface is thoroughly dried out and is littered with small, loose particles of soil or regolith. Dry river courses, beaches, and areas of recently formed glacial deposits are highly sus-

ceptible to deflation. In dry climates, almost the entire ground surface is subject to deflation because the soil or rock is largely bare. Wind is selective in its deflational action. The finest particles, those of clay and silt sizes, are lifted most easily and raised high into the air. Sand grains are moved only by moderately strong winds and travel close to the ground. Gravel fragments and rounded pebbles can be rolled over flat ground by strong winds, but they do not travel far. They become easily lodged in hollows or between other large grains. Consequently, where a mixture of sizes of loose particles is present on the ground, the finer sizes are removed and the coarser particles remain behind.

A landform produced by deflation is a shallow depression called a *blowout*, or *deflation hollow*. This depression may be from a few meters to a kilometer or more in diameter, but it is usually only a few meters deep. Blowouts form in plains regions in dry climates. Any small depression in the surface of the plain, particularly where the grass cover is broken through, may develop into a blowout. Rains fill the depression, creating a shallow pond or lake. As the water evaporates, the mud bottom dries out and cracks, forming small scales or pellets of dried mud that are lifted out by the wind. In grazing lands, animals trample the margins of the depression into a mass of mud, breaking down the protective grass-root structure and facilitating removal when dry. In this way the depression is enlarged (Figure 21.1). Blowouts are also found on rock surfaces where the rock is being disintegrated by weathering.

In the great desert of the southwestern United States and northern Mexico, the floors of intermontane basins are vulnerable to deflation. The flat floors of shallow playas have in some places been reduced by deflation as much as several meters over areas of many square kilometers.

In many areas of eroded playa deposits, wind abrasion has modified water-carved gully systems into U-shaped troughs separated by sculptured ridges of playa sediment. Known as *yardangs*, these long, tapered ridges typically develop an overhanging prow facing into the wind (Figure 21.2). Here sandblasting constantly undermines the upwind point of the yardang. The term "yardang" was used for these features by natives of the Lop

FIGURE 21.1 This blowout hollow on the plains of Nebraska contains a remnant column of the original material, thus providing a natural meter stick for the depth of material removed by deflation. (N. H. Darton, U.S. Geological Survey.)

Nur region in the Tarim Basin of westernmost China. It was there in 1900 that a distinguished explorer of central Asia, Sven Hedin, found the yardangs and reported their existence to the western world. Yardangs also occur in great numbers in the Western Desert of Egypt and the Lut Desert of Iran. A small group of yardangs can be found at the margin of Rogers Dry Lake in the Mojave Desert of California.

Where rainbeat, overland flow, and deflation have been active for a long period on the gently sloping surface of a desert alluvial fan or alluvial terrace, rock fragments ranging in size from pebbles to small boulders become concentrated into a surface layer known as *desert pavement* (Figure 21.3). As overland flow from torrential downpours carries away the fine particles, the remaining large fragments become closely fitted together, concealing the smaller pebbles—grains of sand, silt, and clay—that lie beneath. The pavement acts as an armor that effectively protects the finer particles from further removal by overland flow and deflation. The pavement is easily disturbed by the wheels of trucks and motorcycles, exposing the finer particles and allowing severe water erosion and deflation to occur.

DUST STORMS

Strong, turbulent winds, blowing over desert surfaces, lift great quantities of fine dust into the air, forming a dense, high cloud called a *dust storm*. In semiarid grass-

FIGURE 21.2 Two sharp-crested yardangs viewed end-on, the prevailing wind direction being away from the photographer. These yardangs have been carved by wind abrasion (sandblast action) into consolidated deposits of fine gravel, sand, silt, and clay. The trough between the two yardangs is floored by a gravel layer. Margin of Rogers Lake (a dry playa), Mojave Desert, California. (Copyright © 1991 by Mark A. Melton.)

FIGURE 21.3 A desert pavement in western Arizona. Fragments of many sizes litter the flat ground surface. The rock surfaces are darkened by a coating known as "desert varnish." Below, a closeup of the same surface shows that the pebbles are fitted closely together, protecting the sand layer below. (A. N. Strahler.)

FIGURE 21.4 An approaching dust storm. The wall of dust represents a rapidly moving cold front. Coconino Plateau, Arizona. (D. L. Babenroth.)

lands a dust storm is generated where ground surfaces have been stripped of protective vegetative cover by cultivation or grazing. A dust storm approaches as a dark cloud extending from the ground surface to heights of several kilometers (Figure 21.4). Within the dust cloud, there is deep gloom or even total darkness. Visibility is cut to a few meters, and a fine choking dust penetrates everywhere.

It has been estimated that as much as 1000 metric tons of dust may be suspended in a cubic kilometer of air (4000 tons/cu mi). On this basis, a large dust storm can carry more than 100 million metric tons of dust— enough to make a hill 30 m (100 ft) high and 3 km (2 mi) across the base. Dust travels long distances in the air. Dust from a single desert storm is often traceable as far as 4000 km (2500 mi).

Human activities in the very dry, hot deserts have contributed measurably to the raising of dust clouds. In the desert of northwest India and Pakistan (the Thar Desert bordering the Indus River), the continued trampling of fine-textured soils by the hooves of grazing animals and by human feet produces a blanket of dusty hot air that hangs over the region for long periods and extends to a height of 9 km (30,000 ft).

In other deserts, such as those of North Africa and the southwestern United States, ground surfaces in the natural state contribute comparatively little dust because of the presence of desert pavements and sheets of coarse sand. This protective layer is easily destroyed by wheeled vehicles, exposing finer-textured materials and allowing deflation to raise dust clouds. The disturbance of large expanses of North African desert by tank battles

during World War II caused great dust clouds; dust from this source was identified as far away as the Caribbean region. Dust clouds, visible on Landsat images, have been noted in recent years in the Mojave Desert of southern California. They result from various types of ground disturbances, including the grinding action of wheels of off-road vehicles, such as trail motorcycles and four-wheel-drive recreation vehicles.

SAND TRANSPORT BY WIND

The true desert *sandstorm* is a low cloud of moving sand that rises usually only a few centimeters and at most 2 m (6 ft) above the ground. It consists of sand particles driven by a strong wind. Those who have experienced sandstorms report that the head and shoulders of a person standing upright may be entirely above the limits of the sand cloud. The reason the sand does not rise higher is that the individual particles are engaged in a leaping motion, termed *saltation* (Figure 21.5). Grains describe a curved trajectory and strike the ground with considerable force, but at a low angle. The impact with other grains causes the sand grains to rebound into the air. At the same time, the surface layer of sand grains creeps downwind as the result of the constant impact of the bouncing sand grains.

The erosional effect of sand in saltation is thus concentrated on surfaces exposed less than a meter (3 ft) above the flat ground surface. Fence posts and telephone poles on wind-swept sandy plains are quickly cut through at the base unless a protective metal sheathing or heap of large stones is placed around the base.

SAND DUNES

A *dune* is an accumulation of sand shaped by the wind and capable of movement over the underlying ground surface. Dunes that are bare of vegetation and constantly changing form under wind currents are *live dunes*, or active dunes. They may be inactive, as *fixed dunes*, covered by vegetation that has taken root and prevented further shifting of the sand.

Most dune sands are composed of the mineral quartz. Quartz grains become highly spherical in shape through abrasion (see Figure 12.9). Rarely, dunes are composed of tephra (volcanic sand), of shell fragments that accumulate in beaches, of grains of gypsum, or of heavy minerals such as magnetite.

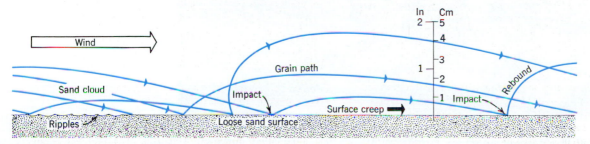

FIGURE 21.5 Sand particles travel in a series of long leaps. (After R. A. Bagnold.)

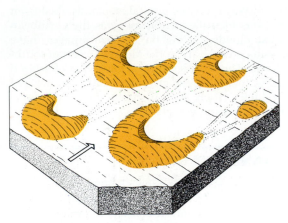

FIGURE 21.6 Barchans, or crescentic dunes. Arrow indicates prevailing wind direction. (Drawn by A. N. Strahler.)

FIGURE 21.7 A single, isolated crescent dune. Prevailing wind direction is from right to left. Obviously, the dune has migrated across the dry stream channels of the desert surface since the last water flood occurred. Salton Sea region, California. (John S. Shelton.)

One common type of live sand dune is an isolated heap of free sand called a *crescent dune*, or *barchan*. As the name suggests, this dune has a crescentic outline; the points of the crescent are directed downwind (Figure 21.6). On the windward side of the crest, the sand slope is gentle and smoothly rounded. On the lee side of the dune, within the crescent, is a steep curving dune slope, the *slip face*. This face maintains an angle of about 35° from the horizontal (Figure 21.7). Sand grains slide down the steep face after being blown free of the sharp crest. When a strong wind is blowing, the flying sand makes a perceptible cloud at the crest.

Crescent dunes rest on a flat, pebble-covered ground surface. The sand heap may originate as a drift in the lee of some obstacle, such as a small hill, rock, or clump of brush. Once a sufficient mass of sand has formed, it begins to move downward, taking the form of a crescent dune. For this reason the dunes are usually arranged in chains extending downwind from the sand source.

Where sand is so abundant that it completely covers the ground, dunes take the form of wavelike ridges separated by troughlike furrows. The dunes are called *transverse dunes* because, like ocean waves, their crests trend at right angles to the direction of wind (Figure 21.8). The entire area may be called a *sand sea* because it resembles a storm-tossed sea suddenly frozen to immobility. The sand ridges have sharp crests and are symmetrical, the gentle slope being on the windward, the steep slip face on the lee side. Deep depressions may lie between the dune ridges. Sand seas require enormous quantities of sand, often derived from the weathering of a sandstone formation underlying the ground surface or from adjacent alluvial plains. Still other transverse dune belts form adjacent to beaches that supply abundant sand and have strong onshore winds (see Figure 21.11).

Another group of dunes belongs to a family in which the curve of the dune crest is bowed convexly downwind, the opposite of the curvature of crests in the crescent dune. These dunes are parabolic in outline and are

FIGURE 21.8 Transverse dunes of a sand sea, near Yuma, Arizona. Isolated crescent dunes can be seen at the right. The view is eastward; prevailing winds are northerly. (John S. Shelton.)

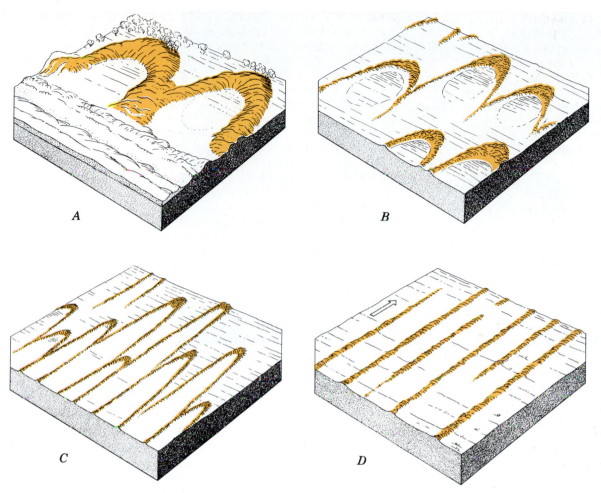

FIGURE 21.9 Four types of dunes. Types A, B, and C are of the parabolic class. The prevailing wind direction, shown by an arrow in Block D, is the same for all four examples. (Drawn by A. N. Strahler.) A. Coastal blowout dunes. B. Parabolic dunes on a semiarid plain. C. Parabolic dunes drawn out into hairpin dunes. D. Longitudinal dune ridges on a desert plain.

FIGURE 21.10 This coastal blowout dune is advancing over a forest, with the slip face gradually burying the tree trunks. Lake Michigan coast, near Michigan City, Indiana. (Barry Voight.)

classed as *parabolic dunes.* A common representative of this class is the *coastal blowout dune,* formed adjacent to beaches. Here large supplies of sand are available and are blown landward by prevailing winds (Figure 21.9A). A saucer-shaped depression is formed by deflation; the sand is heaped in a curving ridge resembling a horseshoe in plan. On the landward side is a steep slip face that advances over the lower ground and buries forests, killing the trees (Figure 21.10). Coastal blowout dunes are well displayed along the southern and eastern shores of Lake Michigan. Dunes of the southern shore have been protected for public use as the Indiana Dunes State Park.

On semiarid plains, where vegetation is sparse and winds are strong, groups of *parabolic blowout dunes* develop to the lee of shallow deflation hollows (Figure 21.9B). Sand is caught by low bushes and accumulates in a broad, low ridge. These dunes have no steep slip faces and may remain relatively immobile. In some cases the dune ridge migrates downwind, drawing the parabola into a long, narrow form with parallel sides resembling a hairpin in outline (Figure 21.9C). These *hairpin dunes,* stabilized by vegetation, are seen in Figure 21.11.

Another class of dunes is described as longitudinal because the dune ridges run parallel with the wind direc-

FIGURE 21.11 Long, narrow hairpin dunes, stabilized by vegetation, extend landward from a beach with abundant sand supply. Where the plant cover has been breached, the loose sand has moved freely and resumed its landward advance. San Luis Obispo Bay, California. (John S. Shelton.)

tion. On desert plains and plateaus, where sand supply is meager but winds are strong from one direction, *longitudinal dunes* are formed (Figure 21.9D). These are usually only a few meters high, but may be several kilometers long. In some areas, the longitudinal dune is produced by extreme development of the hairpin dune, such that the parallel side ridges become the dominant form.

Longitudinal dunes in close association with parabolic dunes are abundant over a vast area in the Navajo and Hopi lands of northeastern Arizona and northwestern New Mexico. In this high plateau within the semiarid (steppe) climate, the troughs between the sand ridges, as well as the flanks of the ridges, are covered with shrub vegetation. Longitudinal dunes of the tropical deserts in Africa, Arabian Peninsula, and central Australia occur in

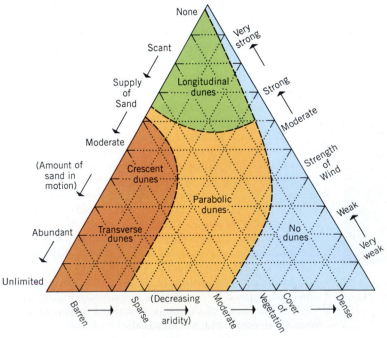

FIGURE 21.12 A schematic diagram of the relationships of the four common dune types of western North America to factors of sand supply, wind strength, and density of vegetative cover. (Adapted from John T. Hack, *The Geographical Review*, vol. 31, Figure 19, p. 260.)

FIGURE 21.13 Longitudinal sand dunes in the southern Arabian Peninsula trend northeast–southwest, parallel with the prevailing northeast winds. In the lower right-hand part of the frame is a dissected plateau of the flat-lying sedimentary strata. (NASA Landsat 1186-06381, January 1973.)

extreme desert environments, where plants are few, or not visibly present (see Figure 21.13).

From what has been given in earlier paragraphs as a description of the environment in which each major kind of dune is usually found, we can attempt to organize that information along the lines of a system of three major factors simultaneously influencing dune form: (1) supply of sand—whether scant or abundant; (2) density of vegetative cover—whether sparse or dense; and (3) strength of wind—whether weak or strong. Figure 21.12 is a triangular graph on which relative values of the three variable

factors can be simultaneously plotted. When this is done, each dune type falls into a different location on the graph, and areas or domains can be roughly drawn for each. In this case, the data analysis applies only to common dune types of the semiarid to desert climate regions of the southwestern United States and northern Mexico. Wind strength is assumed to be that of the predominant wind direction.

DUNES OF THE SAHARA AND ARABIAN DESERTS

Wind is a major agent of landscape development in the Sahara Desert. Enormous quantities of reddish dune sand have been derived from the weathering of older sandstone formations. The sand is formed into a great expanse of free sand dunes, called an *erg*. Elsewhere there are vast flat-surfaced sheets of sand, armored against further deflation by a layer of pebbles. A surface of this kind is called a *reg*.

Dunes of the great ergs are similar to the sand seas of transverse dunes we have already described. Longitudinal dune ridges, oriented parallel with dominant winds, occupy vast areas in tropical Africa and the Arabian peninsula (Figure 21.13).

Some of the Saharan dunes are complex forms, not represented in the western hemisphere. One of these is the *seif* dune, or sword dune, a huge tapering sand ridge whose crestline rises and falls in alternate peaks and saddles. The side slopes are indented by crescentic slip faces. Seif dunes may be tens of kilometers long. Another Saharan type is the *star dune* (heaped dune), a large hill of sand whose base resembles a many-pointed star when seen from above (Figure 21.14). Radial ridges of sand rise toward the dune center, culminating in sharp peaks as high as 100 m (300 ft) or more above the base. Star dunes remain fixed in position and have served for centuries as reliable landmarks for desert travelers.

Permanent human habitations are few and far between in the Sahara. They are localized in low places where ground water lies close to the surface and can be reached by dug wells. With fresh water assured, a

FIGURE 21.14 Star dunes of reddish-yellow sand cover much of the area in this Gemini VII photo of the Sahara Desert in western Algeria. The upper bluish zone is a shallow lake. Water and sediment are brought to the lake basin by a major stream, Wadi Saura, entering from the lower right. Bedrock is exposed in narrow hogback ridges representing a deeply eroded fold structure in sedimentary rocks. The area shown is about 50 km (30 mi) across. (NASA No. S65-6380.)

FIGURE 21.15 Looking down on the Algerian oasis of Souf, in the heart of the Sahara Desert, we see small groves of date palms planted on the floors of hollows in a sea of dune sand. (George Gerster/Rapho-Photo Researchers.)

productive oasis can thrive, with its groves of date palms, some citrus trees, and small plots of grains and vegetables (Figure 21.15). Often the oasis is surrounded by sand dunes, and these must be prevented from invading the oasis through the construction of barriers of brush on the dune crests. Modern tube wells, some of them tapping deep artesian sources of ground water, have been installed in recent decades at favorable locations.

COASTAL DUNES

Landward of sand beaches we usually find a narrow belt of dunes in the form of irregularly shaped hills and depressions; these are the *foredunes*. They normally bear a cover of beachgrass and a few other species of plants capable of survival in the severe environment. Dunes de-

FIGURE 21.16 Beachgrass thriving on coastal foredunes has trapped drifting sand to produce a ridge (left) that rises sharply above the barren surface of a blowout depression (right). Oregon Dunes National Recreation Area. (A. N. Strahler.)

veloped under a partial cover of plants are called *phytogenic dunes*.

On coastal foredunes, the cover of beachgrass and other small plants, sparse as it seems to be, acts as a baffle to trap sand moving landward from the adjacent beach (Figure 21.16). As a result, the foredune ridge is built up as a barrier rising several meters above high tide level. For example, dune summits of the Landes coast of France reach heights of 90 m (300 ft) and span a belt 10 km (6 mi) wide.

The swash of storm waves cuts away the upper part of the beach, and the dune barrier is attacked. Between storms the beach is rebuilt and, in due time, the dune ridge is also restored if a plant cover is maintained. The coastal blowout dune is a normal feature of this process. Thus the foredunes form a protective zone for tidal lands lying on the landward side of a barrier island.

If the plant cover of the dune ridge is depleted by vehicular and foot traffic or by the bulldozing of sites for approach roads and buildings, a blowout will rapidly develop. The new cavity may extend as a trench across the dune ridge. With the onset of a storm with high water levels, swash is funneled through the gap and spreads out on the tidal marsh or tidal lagoon behind the ridge. This activity, called overwash, was mentioned in Chapter 20. Sand swept through the gap is spread over the tidal deposits.

Breaching of dune barriers along the North Sea Coast has caused severe losses in lives and property. These barriers are part of the system to exclude seawater from reclaimed polders and fenlands. Protection of the dunes of the Netherlands coast assumes vital importance in view of the loss of life and property that a series of storm breaches can bring.

Another form of environmental damage related to dunes is the rapid downwind movement of sand when the dune status is changed from one of fixed, plant-controlled forms to that of active dunes of free sand. When plant cover is depleted, wind rapidly reshapes the dunes to produce crests and slip faces, as shown in Figure

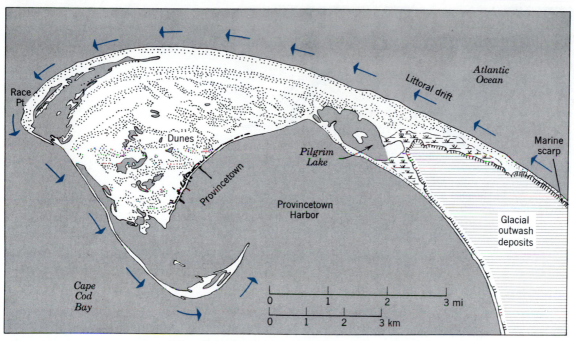

FIGURE 21.17 A sketch map of the Provincelands of Cape Cod, Massachusetts, as it appeared in the 1880s. Dunes show as a stippled pattern. Color arrows show the direction of littoral drift of sand. (Drawn by A. N. Strahler.)

21.10. The free sand slopes advance on forests, roads, buildings, and agricultural lands. In the Landes region of coastal dunes on the southwestern coast of France, landward dune advance has overwhelmed houses and churches and even caused entire towns to be abandoned.

A striking case of human interference with a dune environment is that of the Provincelands of Cape Cod, located at the northern tip of that peninsula, making up the fist of the armlike outline of the Cape (Figure 21.17). The Provincelands has been constructed of beach sand built into a succession of beach ridges, and these have been modified in form and increased in height by dune building. When the first settlers arrived in Provincetown, a city now occupying the south shore of the Provincelands, the dunes were naturally stabilized by grasses and other small plants covering the dune summits and by pitch pine and other forest trees on lower surfaces and in low swales between dune ridges, although dunes were probably active then, as now, on ridges close to the northern shore. Inhabitants grazed their livestock on the dune summits and rapidly cut the forests for fuel, with the result that the dunes were activated and began to move southward.

By 1725 Provincetown was being overwhelmed by drifting sand. Some buildings were partially buried and some had to be abandoned. Sand was carted away from the streets in large volumes. By the early 1800s the major dune ridges were moving southward at a rate estimated to be 27 m (90 ft) per year. Fortunately, beachgrass plantings, begun in 1825, and the rigorous enforcement of laws forbidding grazing and tree cutting, resulted finally in the stabilization of all but the northernmost dunes. The area is today a part of the Cape Cod National Sea-

shore. Park authorities have made extensive new plantings of dune grass and have restricted vehicular traffic in an attempt to bring further control to sand movement.

LOESS

Wind-deposited silt, called *loess*, is a variety of sediment lacking horizontal stratification. Instead, because of slight shrinkage during compaction, loess has a tendency to break away along vertical planes. This structure, called *cleavage*, is similar to vertical jointing in sedimentary rocks. Because of cleavage, deep natural exposures of loess in the upper walls of stream valleys tend to form vertical cliffs. Cliff retreat occurs by the collapse of loess columns. A perpendicular highway cut made in loess is remarkably stable and may resist erosion for decades (Figure 21.18).

The thickest deposits of loess are in northern China, where a layer over 30 m (100 ft) thick is common and a maximum of 90 m (300 ft) has been measured. Loess covers many hundreds of square kilometers and appears to have been derived as dust from the interior of Asia. Loess deposits are also of major importance in the United States, central and eastern Europe, central Asia, and Argentina.

In the United States, thick loess deposits lie in the Missouri–Mississippi valley (Figure 21.19). Much of the prairie plains region of Indiana, Illinois, Iowa, Missouri, Nebraska, and Kansas is underlain by a loess layer ranging in thickness from 1 to 30 m (3 to 100 ft). Extensive deposits also occur along the lands bordering the lower Mississippi River floodplain on its east side, throughout

FIGURE 21.18 (left) This thick layer of loess near Vicksburg, Mississippi, shows no horizontal layering. The material has excellent cohesion and can stand unsupported in vertical walls for decades on end. (right) A closeup of the same cliff shows that the material is a soft but cohesive silt. (Orlo E. Childs.) •

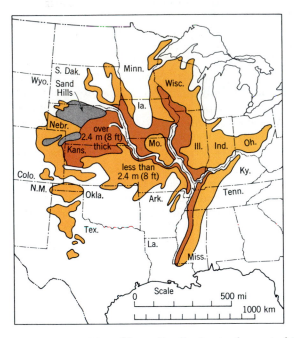

FIGURE 21.19 Map of loess distribution in the central United States. (Data from Map of Pleistocene Eolian Deposits of the United States, Geological Society of America, 1952.)

Tennessee and Mississippi. Other important loess deposits are the Palouse region of southeast Washington, western Idaho, and northeast Oregon.

The American and European loess deposits are directly related to the continental glaciers of the Pleistocene Epoch. At the time when the ice covered much of North America and Europe, a generally dry winter climate prevailed in the land bordering the ice sheets. Strong winds blew southward and eastward over the bare ground, picking up silt from the floodplains of braided streams that discharged the meltwater from the ice. This dust settled on the ground between streams, gradually building up to produce a smooth, level ground surface. The loess is particularly thick along the eastern sides of the valleys because of prevailing westerly winds, and it is well exposed along the bluffs of most streams flowing through the region today.

Loess is of major importance in world agricultural resources. Loess forms the parent matter of black and brown soils suited to the cultivation of grains. The highly productive plains of southern Russia, the Argentine Pampa, and the rich wheat-producing region of north China are underlain by loess. In the United States, corn is extensively cultivated on the loess plains in such states as Iowa and Illinois, where rainfall is sufficient; wheat

FIGURE 21.20 A succession of contour terraces, cut into thick loess and covered by tree plantings, is designed to prevent deep gullying and loss of soil. Arched entrances to cave dwellings can be seen at lower left and lower right. Grain fields on the flat valley floor occupy surfaces of thick sediment trapped behind dams. Shanxi Province, near Xi'an (Sian), People's Republic of China. (Alan H. Strahler.)

is grown farther west on loess plains of Kansas and Nebraska and in the Palouse region of eastern Washington.

Loess forms vertical walls along valley sides and is able to resist sliding or flowage. Because it is easily excavated, loess has been widely used for cave dwellings both in China and in central Europe.

The thick loess deposit covering a large area of northeastern China in the province of Shanxi (Shansi) and adjacent provinces poses a difficult problem of severe soil erosion. Although the loess is capable of standing in vertical walls, it also succumbs to deep gullying during the period of torrential summer rains. From the steep walls of these great scars, fine sediment is swept into streams and carried into tributaries of the Huanghe (Hwang Ho, Yellow River). An intensive program of slope stabilization has been implemented by the Chinese government by using artificial contour terraces, seen in Figure 21.20, in combination with tree planting. Valley bottoms have been dammed so as to trap the silt to form flat patches of land suitable for cultivation.

DEFLATION INDUCED BY HUMAN ACTIVITY—THE DUST BOWL

Cultivation of vast areas of short-grass prairie under a climate of substantial seasonal water shortage is a practice that invites deflation of soil surfaces. Much of the Great Plains region of the United States is such a marginal region. In past centuries these plains have experienced many dust storms generated by turbulent winds. Strong cold fronts frequently sweep over this area, lifting dust high into the troposphere at times when soil water is low.

On the Great Plains of Kansas, Oklahoma, and Texas, deflation and soil drifting reached disastrous proportions during a series of drought years in the middle 1930s, following a great expansion of wheat cultivation. These former grasslands are underlain by soils having a thick, loose upper horizon. During the drought, a sequence of exceptionally intense dust storms occurred. Within their formidable black clouds, visibility declined to nighttime darkness, even at noonday. The area affected became known as the Dust Bowl. Many centimeters of soil were removed from fields and transported out of the region as suspended dust, while the coarser silt and sand particles accumulated in drifts along fence lines and around buildings (Figure 21.21). The combination of environmental degradation and repeated crop failures caused widespread abandonment of farms and a general exodus of farm families.

Among geographers who have studied the Dust Bowl phenomenon, there is a difference of opinion as to how great a role soil cultivation and livestock grazing played in inducing deflation. The drought was a natural event over which humans had no control, but it seems reasonable that the natural grassland would have sustained far less soil loss and drifting if it had not been destroyed by the plow.

Although we cannot prevent cyclic occurrences of drought over the Great Plains, measures can be taken to minimize the deflation and soil drifting that occur in periods of dry soil conditions. Improved farming practices include the use of listed furrows (deeply carved furrows) that act as traps to soil movement. Stubble mulching reduces deflation when land is lying fallow. Tree belts may have significant effect in reducing the intensity of wind stress at ground level.

The problem area associated with the Dust Bowl lies primarily within the easternmost region of the semiarid subclass of the dry midlatitude climate (9s), or steppe climate (see Chapter 11 and Figure 8.10). The eastern boundary of that climate runs nearly north–south close to the 98th meridian of longitude. The area of dry farming of grains (especially winter wheat) extends to the western limit of the High Plains and includes northern Texas and the Oklahoma Panhandle, eastern Colorado, and the western halves of Kansas, Nebraska, and the Dakotas. It is generally agreed by agronomists that rainfall in this belt is insufficient, on the average, to support sustained agriculture without irrigation, and that devastating droughts such as that of the mid-1930s will be repeated. Another drought occurred during the mid-1950s. Crop irrigation has greatly increased in the area (and farther east), using ground water pumped largely from the deep Ogalalla aquifer. Volume pumped and area under irrigation from this source increased nearly sixfold in the period from 1950 to 1980. Ultimately, this water source will decline sharply. In view of the gloomy future for agriculture in the semiarid region, the proposal has been made by ecologists that it be returned to uncultivated shortgrass prairie, restoring it to what has already been dubbed a "Buffalo Common."

FIGURE 21.21 A typical scene in the Dust Bowl during the late 1930s. Drifts of sand and silt have accumulated around abandoned farm buildings and a fence in Dallam County, Texas. (Soil Conservation Service, U.S. Department of Agriculture.)

WIND AS AN ENVIRONMENTAL AGENT IN THE LIFE LAYER

This chapter and the one before it have had much in common through the mutual mechanism of transfer of momentum and kinetic energy from the wind to a liquid or solid surface over which it blows. While wind-driven waves are the dominant agent in shaping shorelines, the action of wind in transporting beach sand landward to form coastal dunes can exert a measure of control over breaking waves, because sediment carried inland beyond the reach of waves of normal energy subtracts mass from the beach zone, and in so doing slows progradation of beaches. On the other hand, dune accumulations absorb the exceptional high-energy impact of storm waves and so inhibit retrogradation and the overwash of barrier beaches. Both breaking waves and wind can carry sediment upgrade to higher positions, something that gravity flow of runoff cannot accomplish.

One final process remains to be examined on the list of active agents of erosion, transportation, and deposition: glacial ice. One unique feature of the action of glaciers is that they are quite insensitive to human activities. This quality probably derives from their vast bulk and sluggish motion. No one has yet tried to build a dam to stop the flow of a great glacier, as we have done so many times to halt the flow of a large river. So we turn from considering processes and forms that are extremely sensitive to human impact to a geologic process that imposes its environmental qualities upon us and accepts no degree of control in return.

GLACIAL LANDFORMS AND THE ICE AGE

Glacial ice has played a dominant role in shaping landforms of large areas in midlatitude and subarctic zones. Glacial ice also exists today in two great accumulations of continental dimensions and in many smaller masses in high mountains. In this sense, glacial ice is an environmental agent of the present, as well as of the past, and is itself a landform. Glacial ice of Greenland and Antarctica strongly influences the radiation and heat balances of the globe. Moreover, these enormous ice accumulations represent water in storage in the solid state; they constitute a major component of the global water balance. Changes in ice storage can have profound effects on the position of sea level with respect to the continents.

Coastal environments of today have evolved with a rising sea level following the melting of ice sheets of the last ice advance in the Pleistocene Epoch, or Ice Age. When we examine the evidence of former extent of those great ice sheets, we need to keep in mind that the evolution of modern humans as an animal species occurred during a series of climatic changes that placed many forms of environmental stress on all terrestrial plants and animals in the midlatitude zone. There were also important climatic effects extending into the tropical zone and even into the equatorial zone.

GLACIERS

Most of us know ice only as a brittle, crystalline solid because we are accustomed to seeing it only in small quantities. Where a great thickness of ice exists—100 m (300 ft) or more—the ice at the bottom behaves as a plastic material. The ice will then slowly flow in such a way as to spread out the mass over a larger area, or to cause it to move downhill, as the case may be. This behavior characterizes a *glacier*, defined as any large natural accumulation of land ice affected by present or past motion.

Conditions necessary for the accumulation of glacial ice are simply that snowfall of the winter shall, on the average, exceed the amount of ablation of snow that occurs in summer. The term *ablation* includes both evaporation and the melting of snow and ice. Thus each year

a layer of snow is added to what has already accumulated. As the snow compacts by surface melting and refreezing, it turns into a granular ice; then it is compressed by overlying layers into hard crystalline ice. When the ice becomes so thick that the lower layers become plastic, outward or downhill flow commences and an active glacier has come into being.

At sufficiently high altitudes, even in the tropical and equatorial zones, glaciers form because air temperature is low enough and mountains receive heavy orographic precipitation. Glaciers that form in high mountains are characteristically long and narrow because they occupy former stream valleys. These mountain glaciers bring the plastic ice from collecting grounds high on the range down to lower altitudes and, consequently, to warmer temperatures. Here the ice disappears by ablation. These *alpine glaciers* are a distinctive type (Figure 22.1).

In arctic and polar zones, prevailing temperatures are low enough that ice can accumulate over broad areas, wherever uplands exist to intercept heavy snowfall. As a result, the uplands become buried under enormous plates of ice whose thickness may reach several thousand meters. An ice mass limited to high summit areas of mountains or plateaus is referred to as an *icecap*. During glacial periods, an icecap may expand greatly, spreading over surrounding lowlands and enveloping all landforms it encounters. This extensive type of ice layer is called a *continental glacier* or an *ice sheet*.

ALPINE GLACIERS

Figure 22.2 illustrates a number of features of alpine glaciers. The center illustration is of a simple glacier occupying a sloping valley between steep rock walls. Snow is collected at the upper end in a bowl-shaped depression, the *cirque*. The upper end lies in the *zone of accumulation*. Layers of snow in the process of compaction and recrystallization constitute the *firn* (or *névé*). The smooth *firn field* is slightly concave-up in profile. Flowage in the glacial ice beneath carries the excess ice out of the cirque, downvalley. An abrupt steepening of the grade forms a *rock step*, over which the rate of ice flow is accelerated and produces deep *crevasses* (gaping fractures),

FIGURE 22.1 This large alpine glacier in Alaska displays two medial moraines, which can be traced upvalley to glacier junctions. The ice is heavily crevassed where it passes over an icefall. Can you find evidence that this glacier was formerly much thicker than today? (Charles Moore/Black Star.)

Glaciers

which form an *ice fall*. The lower part of the glacier lies in the *zone of ablation*. Here the rate of ice wastage is rapid and old ice is exposed at the glacier surface, which is extremely rough and deeply crevassed. The *glacier terminus* is heavily charged with rock debris. The lower portion of the glacier has an upwardly convex cross-profile, shown in the upper-right drawing of Figure 22.2.

The uppermost layer of a glacier, about 60 m (200 ft) in thickness, is brittle and fractures readily into crevasses, whereas the ice beneath behaves as a plastic sub-

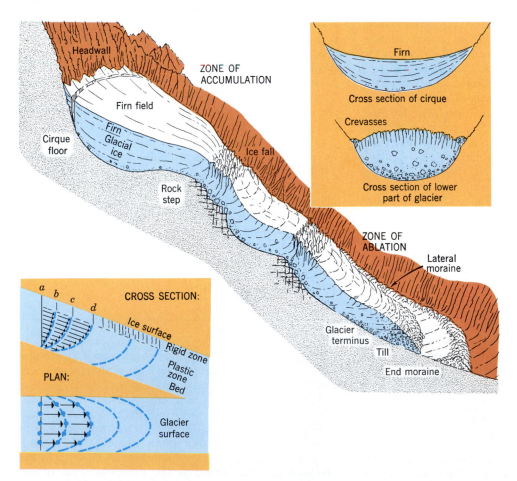

FIGURE 22.2 Structure and flowage of a simple alpine glacier. (Drawn by A. N. Strahler.)

stance and moves by slow flowage (Figure 22.2, lower left). A line of stakes placed across the glacier surface would be gradually deformed into a parabolic curve by glacier flow, indicating that rate of movement is fastest in the center and diminishes toward the sides. Rate of continual flowage of glaciers is very slow, indeed, amounting to a few centimeters per day for large ice sheets and the more sluggish alpine glaciers, but ranging up to several meters per day for an active alpine glacier.

A simple glacier readily establishes a dynamic equilibrium in which the rate of accumulation at the upper end balances the rate of ablation at the lower end. Equilibrium is easily upset by changes in the average annual rates of nourishment or ablation.

Besides showing the continual slow internal flowage we have described, some glaciers experience episodes of very rapid movement, described as *surges*. When surging occurs, the glacier may develop a sinuous pattern of movement, not unlike the open bends of an alluvial river (Figure 22.3). A surging glacier (sometimes called a ''galloping glacier'' in the news media) may travel downvalley several kilometers in a few months.

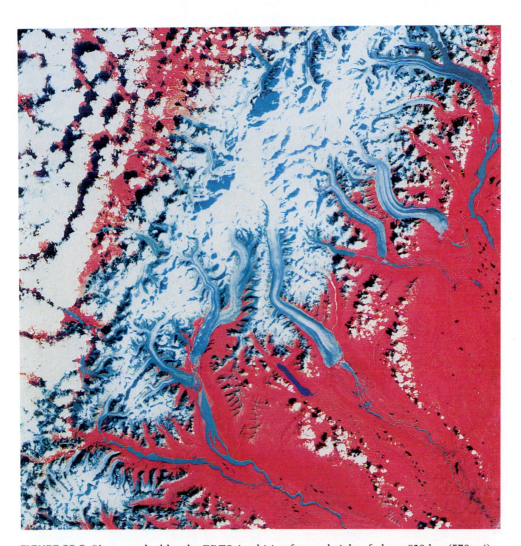

FIGURE 22.3 Photographed by the ERTS-1 orbiting from a height of about 920 km (570 mi), glaciers of the Alaska Range in south-central Alaska appear as blue curving bands, emerging from high collecting grounds on a snow-covered mountain axis trending from northeast to southwest across the center of the photo. The puffy white patches at the left and right are clouds. Darker lines running down the length of a glacier are medial moraines, formed of rock debris on the ice surface. Where these moraines have been distorted into a sinuous pattern, the glacier has experienced a rapid downvalley surge at rates up to 1.2 m (4 ft) per hour. Those glaciers with smooth moraines, paralleling the banks, are experiencing very slow, uniform flow throughout their entire length. The group of high mountain peaks in the upper central area includes Mount McKinley, highest point in North America. This is a false-color image created by combining data of three narrow spectral bands, each assigned a primary color for printing. The area shown is about 105 km (65 mi) across; north is toward the top. (NASA EROS Data Center, No. 81033210205A2.)

A GLACIER AS A FLOW SYSTEM OF MATTER AND ENERGY

Glacier equilibrium can be interpreted through an open material flow system. As shown in the cross section (Figure 22.4), matter in the form of solid precipitation enters

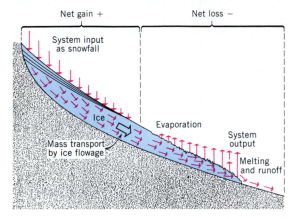

FIGURE 22.4 Schematic diagram of a glacier as a flow system of matter and energy. (Redrawn from A. N. Strahler, *Physical Geology*, Harper & Row, Publishers. Copyright © 1981 by Arthur N. Strahler.)

the system through the surface of the zone of accumulation. Downvalley flow carries the glacier ice to the zone of ablation, where it leaves the system by direct evaporation (sublimation) or as meltwater of runoff. As Figure 22.5 shows, matter enters directly into storage in the solid state, undergoes either of two changes of state, and leaves the system in both vapor and liquid states. Glacial equilibrium requires the flow system to be in a steady state, in which rate of precipitation input (P) balances the sum of the output rates of evaporation (E) and runoff (R), while the quantity of matter in storage (G) remains constant.

The energy system of a glacier can be represented by two subsystems, shown in Figure 22.6. The gravity flow subsystem receives an input of potential energy (E_P) in the form of the mass of snow deposited at the high elevation of the firn surface. Under the force of gravity, the glacier ice flows to lower levels and the potential energy is transformed into kinetic energy (E_K) of ice motion. The moving ice represents kinetic energy in storage. Because of internal resistance to flow as well as friction at the glacier boundary, kinetic energy of motion is transformed into sensible heat of friction (E_{SH}), which is conducted out of the ice, leaving the system. The thermal subsystem receives its input by absorption of solar shortwave energy (E_{SW}) and longwave energy (E_{LW}) and by con-

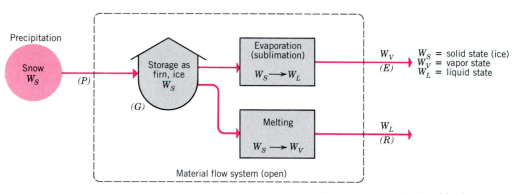

FIGURE 22.5 Flow diagram of a glacier as an open material flow system. (A. N. Strahler.)

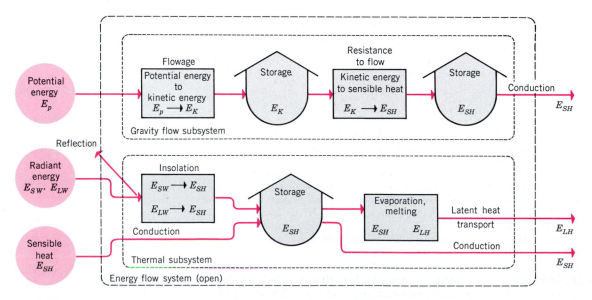

FIGURE 22.6 A glacier as an open energy flow system. (A. N. Strahler.)

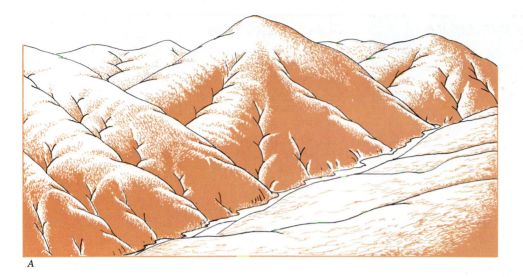

A

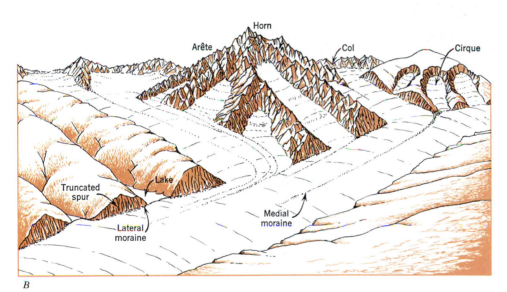

B

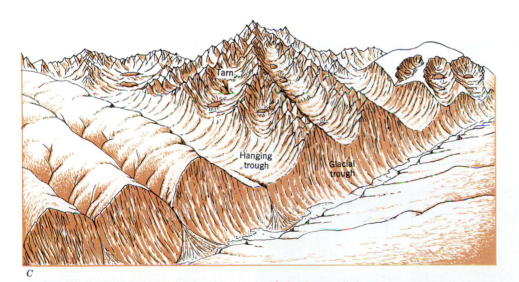

C

FIGURE 22.7 Landforms produced by alpine glaciers. (A) Before glaciation sets in, the region has smoothly rounded divides and narrow, V-shaped stream valleys. (B) After glaciation has been in progress for thousands of years, new erosional forms are developed. (C) With the disappearance of the ice, a system of glacial troughs is exposed. (Drawn by A. N. Strahler.)

FIGURE 22.8 The bowl shaped hollow (center and right) high on the mountain side is a cirque, formerly occupied and excavated by a small alpine glacier. The steep cirque headwall, carved from massive granitic rock, has shed talus blocks that partly cover the cirque floor. Sundance Mountain, Colorado Rockies. (Copyright © 1990 by Mark A. Melton.)

duction of sensible heat into the glacier ice from the overlying atmosphere. In summer, rain supplies sensible heat to the ice surface, causing melting. As melting occurs, sensible heat is transformed into latent heat of fusion (E_{LH}), which leaves the system with the outgoing meltwater. Evaporation from the ice surface allows latent heat of sublimation to be passed directly into the atmosphere. Sensible heat may also flow out of the system by conduction. Although separate storages of sensible heat are shown in the two subsystems, they are one and the same storage.

GLACIAL EROSION

Glacial ice is usually heavily charged with rock fragments, ranging from pulverized rock flour to huge angular boulders of fresh rock. Some of this material is derived from the rock floor on which the ice moves. In alpine glaciers, rock debris is also derived from material that slides or falls from valley walls. Glaciers are capable of great erosive work. *Glacial abrasion* is erosion caused by ice-held rock fragments that scrape and grind against the bedrock. By *plucking*, the moving ice lifts out blocks of bedrock that have been loosened by the freezing of water in joint fractures. Rock debris obtained by erosion must eventually be left stranded at the lower end of a glacier when the ice disappears by ablation. Both erosion and deposition result in distinctive landforms.

LANDFORMS MADE BY ALPINE GLACIERS

Landforms made by alpine glaciers are shown in a series of diagrams in Figure 22.7. Previously unglaciated mountains are attacked and modified by glaciers, after which the glaciers disappear and the remaining landforms are exposed to view. Block A shows a region sculptured entirely by weathering, mass wasting, and streams. The mountains have a smooth, full-bodied appearance, with rather rounded divides. Soil and regolith are thick. Imagine now that a climatic change results in the accumulation of snow in the heads of most of the valleys high on the mountain sides.

An early stage of glaciation is shown at the right side of Block B, where snow is collecting and cirques are being carved by the outward motion of the ice and by intensive frost shattering of the rock near the masses of compacted snow. Glaciers fill the valleys and are integrated into a system of tributaries that feed a trunk glacier just

FIGURE 22.9 The Mer de Glace, a glacier in the French Alps, seen from an aerial tramway. The glacier is greatly shrunken in both thickness and width as compared with its full dimensions two centuries ago, during the Little Ice Age. A massive lateral moraine runs along the left side of the glacier. To the right and left are alpine horns. (Paul W. Tappan.)

FIGURE 22.10 A U-shaped glacial trough in the Beartooth Plateau, near Red Lodge, Montana. Talus cones have been built out from the steep trough walls. (Alan H. Strahler.)

as in a stream system. Tributary glaciers join the main glacier with smooth, accordant junctions; but, as we shall see later, the rock floors of their channels are strongly discordant in level.

Vigorous freezing and thawing of meltwater from snows lodged in crevices high on the walls of a cirque shatters the bare rock into angular fragments; these fall or creep down upon the snowfield and are incorporated into the glacier. Frost shattering also affects the rock walls against which the ice rests (Figure 22.8). The cirques thus grow steadily larger. Their rough, steep walls soon replace the smooth, rounded slopes of the original mountain mass (Figure 22.7, Block B, center).

Where two cirque walls intersect from opposite sides, a jagged, knifelike ridge, called an *arête*, results. Where three or more cirques grow together, a sharp-pointed peak is formed by the intersection of the arêtes. The name *horn* is applied to such peaks (Figure 22.9). One of the best known is the striking Matterhorn of the Swiss Alps. Where opposed cirques have intersected deeply, a pass or notch, called a *col*, is formed.

Glacier flow constantly deepens and widens its channels so that after the ice has finally disappeared there remains a deep, steep-walled *glacial trough*, characterized by a relatively straight or direct course and by the U-shape of its cross-profile (Figure 22.10). Tributary glaciers also carve U-shaped troughs. But they are smaller in cross section, with floors lying high above the floor level of the main trough; these are called *hanging troughs*. Streams, which later occupy the abandoned trough systems, form scenic waterfalls and cascades where they pass down from the lip of a hanging trough to the floor of the main trough. These streams quickly cut a small V-shaped notch in the trough bottom.

Valley spurs that formerly extended down to the main stream before glaciation occurred have been beveled off by ice abrasion and are termed *truncated spurs* (Block B of Figure 22.7). The bedrock is not always evenly excavated under a glacier, so that floors of troughs and cirques may contain *rock basins* and *rock steps*. Cirques and upper parts of troughs thus are occupied by small lakes, called *tarns* (Figure 22.7, Block C). The major

troughs usually contain large, elongate *trough lakes*, sometimes referred to as *finger lakes*. Landslides are numerous because glaciation leaves oversteepened trough walls. In glaciated countries such as Switzerland and Norway, slides are a major type of natural disaster because towns and cities lie in the trough floors, where they are readily destroyed by debris avalanches, rockfalls, and landslides (Chapter 15).

Debris may be carried by an alpine glacier within the ice or it may be dragged along between the ice and the valley wall as a *lateral moraine* (Figures 22.2 and 22.9). Where two ice streams join, this marginal debris is dragged along to form a *medial moraine*, riding on the ice in midstream (Figures 22.7 and 22.9). At the terminus of a glacier, debris accumulates in a heap known as a *terminal moraine*, or *end moraine*. This heap is usually in the form of a curved embankment lying across the valley floor and bending upvalley along each wall of the trough to merge with the lateral moraines (Figures 22.2, 22.11,

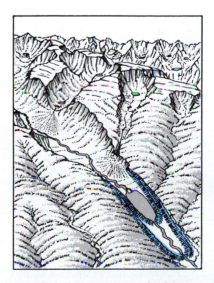

FIGURE 22.11 Moraines of a former valley glacier appear as curved embankments marking successive positions of the ice margins. (After W. M. Davis.)

FIGURE 22.12 A terminal moraine, shaped like the bow of a great canoe, lies at the mouth of a deep glacial trough on the east face of the Sierra Nevada. Cirques can be seen in the distance. Lee Vining, California. (John S. Shelton.)

and 22.12). As the end of the glacier recedes, scattered debris is left behind. Successive halts in ice retreat produce successive moraines, termed *recessional moraines*.

GLACIAL TROUGHS AND FIORDS

Many large glacial troughs now are nearly flat-floored because aggrading streams that issued from the receding ice front were heavily laden with rock fragments. The deposit of alluvium extending downvalley from a melting glacier is the *valley train*. Figure 22.13 shows a comparison between a trough with little or no fill (B) and another with an alluvial-filled bottom (C).

When the floor of a trough open to the sea lies below sea level, the seawater will enter as the ice front recedes, producing a narrow bay known as a *fiord* (Figure 22.13D). Fiords may originate either by submergence of the coast or by glacial erosion to a depth below sea level. Most fiords are explained in the second way. When floating, from three-fourths to nine-tenths of a body of glacial ice lies below water level. Therefore a glacier several hundred meters thick can erode to considerable depth below sea level before the buoyancy of the water reduces its erosive power where it enters the open water.

Fiords are observed to be opening up today along the Alaskan coast, where some glaciers are melting back rap-

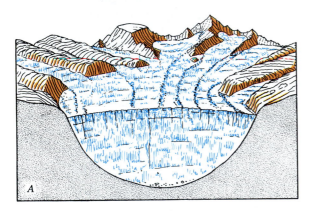

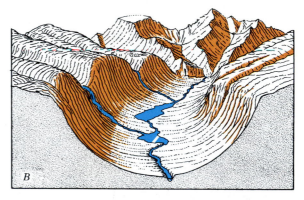

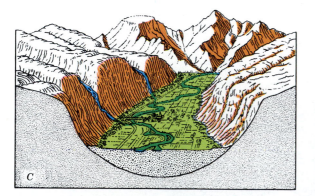

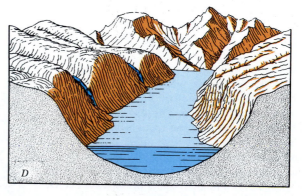

FIGURE 22.13 Development of a glacial trough. (Drawn by E. Raisz.) (A) During maximum glaciation, the U-shaped trough is filled by ice to the level of the small tributaries. (B) After glaciation, the trough floor may be occupied by a stream and lakes. (C) If the main stream is heavily loaded, it may fill the trough with alluvium. (D) Should the glacial trough have been deepened below sea level, it will be occupied by an arm of the sea, or fiord.

idly and the fiord waters are being extended along the troughs. Fiords are found largely along mountainous coasts in lat. 50° to 70°N and S (Figure 22.14).

ENVIRONMENTAL ASPECTS OF ALPINE GLACIATION

In past centuries the high, rugged terrain produced by alpine glaciation was sparsely populated and formed impassable barriers between settlements or between nations. This difficulty of access protected the alpine terrain from the impact of humans and has left a legacy of wilderness areas of striking scenic beauty. Mountainsides steepened by glaciation make ideal ski slopes. Ski resorts have mushroomed by the dozens in glaciated mountains, and this form of recreation is now a major industry. In the summer, these same areas are invaded by tens of thousands of campers and hikers. Rock climbing draws hundreds more to scale the near-vertical rock walls of glacial troughs, cirques, and horns.

The floors of glacial troughs have served for centuries as major lines of access deep into the heart of glaciated alpine ranges. For example, in the Italian Alps, several major flat-floored glacial troughs extend from the northern plain of Italy deep into the Alps. The principal Alpine passes between Italy on the south and Switzerland and Austria on the north lie at the heads of these troughs. An example is the Brenner Pass, located at the head of the Adige trough.

ICE SHEETS OF THE PRESENT

Two enormous accumulations of glacial ice are the Greenland and Antarctic ice sheets. These are huge plates of ice, over 3000 m (10,000 ft) thick in the central areas, resting on landmasses of subcontinental size. The Greenland Ice Sheet has an area of 1,740,000 sq km (670,000 sq mi) and occupies about seven-eighths of the entire island of Greenland (Figure 22.15). Only a narrow, mountainous coastal strip of land is exposed.

The Antarctic Ice Sheet covers 13 million sq km (5 million sq mi) (Figure 22.16). A significant point of difference between the two ice sheets is their position with reference to the poles. The antarctic ice rests almost squarely on the south pole, whereas the Greenland Ice Sheet is considerably offset from the north pole, with its center at about lat. 75°N. This position illustrates a fundamental principle: A large area of high land is essen-

FIGURE 22.14 Geirangerfjord, Norway, is a deeply carved glacial trough occupied by an arm of the sea. (M. Desjardins/Photo Researchers.)

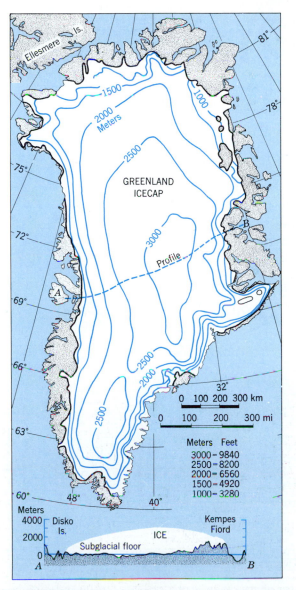

FIGURE 22.15 Generalized map of Greenland. (After R. E. Flint, *Glacial and Pleistocene Geology.*)

tial to the accumulation of a great ice sheet. No land exists near the north pole; ice accumulates there only as sea ice.

The surface of the Greenland Ice Sheet has the form of a very broad, smooth dome. From a high point at an altitude of about 3000 m (10,000 ft), there is a gradual slope outward in all directions. The rock floor of the ice sheet lies near sea level under the central region, but is higher near the edges. Accumulating snows add layers of ice to the surface, while at great depth, the plastic ice flows slowly outward toward the edges. At the outer edge of the sheet, the ice thins down to a few hundred meters. Continual loss through ablation keeps the position of the ice margin relatively steady where it is bordered by a coastal belt of land. Elsewhere the ice extends in long tongues, called *outlet glaciers*, to reach the sea at the heads of fiords. From the floating glacier edge, huge masses of ice break off and drift out to open sea with tidal currents to become icebergs.

Ice thickness in Antarctica is even greater than that of Greenland (Figure 22.16). For example, on Marie Byrd Land, an ice thickness of 4000 m (13,000 ft) was measured. Here the rock floor lies 2000 m (6500 ft) below sea level.

An important glacial feature of Antarctica is the presence of great plates of partially floating glacial ice, called *ice shelves* (Figure 22.16). The largest of these is the Ross Ice Shelf, with an area of about 520,000 sq km (200,000 sq mi) and a surface height averaging about 70 m (225 ft) above the sea. Ice shelves are fed by the ice sheet, but they also accumulate new ice through the compaction of snow.

SEA ICE

The phenomenon of floating sea ice began to assume major scientific importance during and following World War II, when submarine operations began to be extended in the Arctic Ocean and floating ice masses were occupied as scientific observing stations. Discovery of large

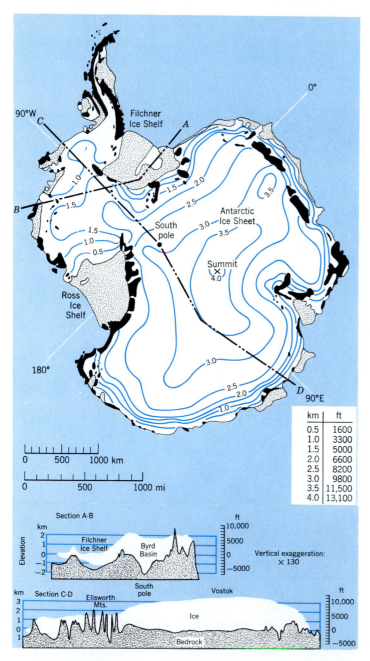

FIGURE 22.16 The Antarctic Ice Sheet and its ice shelves. (Based on data from American Geophysical Union. From A. N. Strahler, *The Earth Sciences*, 2nd ed., Harper & Row, Publishers. Copyright © by Arthur N. Strahler.)

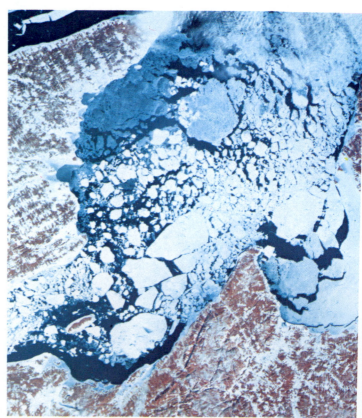

FIGURE 22.17 Ice floes along the shoreline of Prince Edward Island, Nova Scotia, in mid-January. The snow-covered ice plates appear pure white; open leads are solid blue. (NASA Landsat image 1180–14314.)

petroleum deposits beneath the Arctic Slope of Alaska and Canada has recently added a new interest in sea ice as a serious environmental hazard in the development of those oil resources.

The oceanographer distinguishes *sea ice*, formed by direct freezing of ocean water, from icebergs and ice islands, which are bodies of land ice broken free from tide-level glaciers and continental ice shelves. Aside from differences in origin, a major difference between sea ice and floating masses of land ice is in thickness. Sea ice, which begins to form when the surface water is cooled to temperatures of about −2°C (28½°F), is limited in thickness to about 5 m (15 ft).

Pack ice is the name given to ice that completely covers the sea surface (Figure 22.17). Under the forces of winds and currents, pack ice breaks up into individual patches, termed *ice floes*. The narrow strips of open water between such floes are *leads*. Where ice floes are forcibly brought together by winds, the ice margins buckle and turn upward into pressure ridges resembling walls or irregular hummocks. These rugged features made travel on foot across the sea ice a terrible ordeal for early explorers of the polar sea. The surface zone of sea ice is composed of fresh water because the salt has been excluded in the process of freezing. The Arctic Ocean, which is surrounded by landmasses, is largely covered by pack ice throughout the year, although open leads are numerous in the summer (Figure 22.18). The relatively warm North Atlantic drift maintains an ice-free zone off the northern coast of Norway.

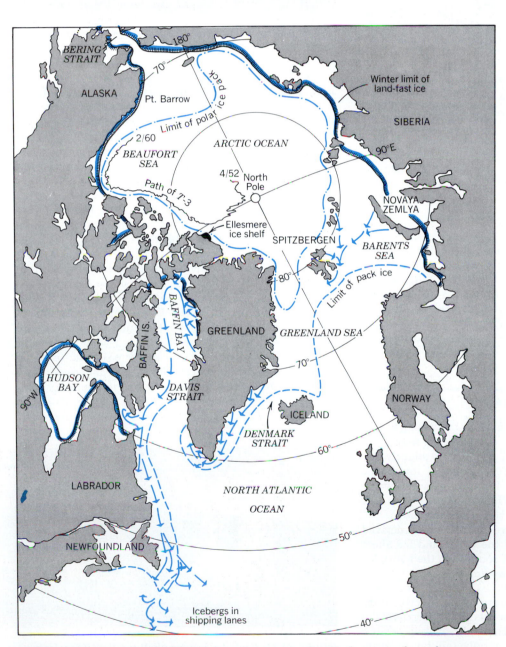

FIGURE 22.18 Sea ice of the Arctic Ocean. Common tracks of icebergs are shown by arrows. A tilted homolographic projection is used for this map. (Based on data from the National Research Council. From A. N. Strahler, *The Earth Sciences*, 2nd ed., Harper & Row, Publishers. Copyright © by Arthur N. Strahler.)

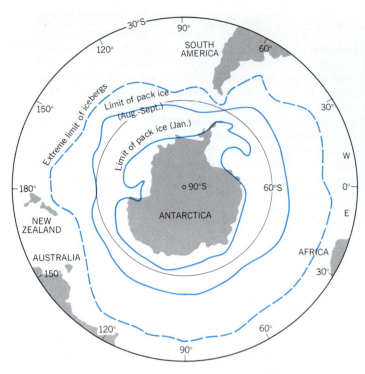

FIGURE 22.19 Sea ice of the antarctic region. (Data from the National Academy of Sciences and American Geographical Society. From A. N. Strahler, *The Earth Sciences*, 2nd ed., Harper & Row, Publishers. Copyright © by Arthur N. Strahler.)

The situation is quite different in the antarctic region, where a vast open ocean bounds the sea ice zone on the equatorward margin (Figure 22.19). Because the ice floes can drift freely north into warmer waters, the antarctic ice pack does not spread far beyond lat. 60°S in the cold season. In March, close to the end of the warm season, the ice margin shrinks to a narrow zone bordering the Antarctic continent.

ICEBERGS AND ICE ISLANDS

Icebergs, formed by the breaking off, or *calving*, of blocks from a valley glacier or tongue of an icecap, may be as thick as several hundred meters. Being only slightly less dense than seawater, the iceberg floats very low in the water, about five-sixths of its bulk lying below water level (Figure 22.20). The ice is fresh, of course, because it is formed of compacted and recrystallized snow.

In the northern hemisphere, icebergs are derived largely from glacier tongues of the Greenland Ice Sheet (Figure 22.18). They drift slowly south with the Labrador and Greenland currents and may find their way into the North Atlantic in the vicinity of the Grand Banks of Newfoundland. Icebergs of the antarctic are distinctly different. Whereas those of the North Atlantic are irregular in shape and present rather peaked outlines above water, the antarctic icebergs are commonly tabular in form, with flat tops and steep, clifflike sides (Figure 22.21). This

FIGURE 22.20 An iceberg of the North Atlantic Ocean. (James Holland/Black Star.)

is because tabular bergs are parts of ice shelves. In dimensions, a large tabular berg of the antarctic may be many kilometers broad and over 600 m (2000 ft) thick, with an ice wall rising 60 to 90 m (200 to 300 ft) above sea level.

Somewhat related in origin to the tabular bergs of the antarctic are *ice islands* of the Arctic Ocean. These huge plates of floating ice may be 30 km (20 mi) across and have an area of 800 sq km (300 sq mi). The bordering ice cliff, 6 to 10 m (20 to 30 ft) above the surrounding pack ice, indicates an ice thickness of 60 m (200 ft) or more.

The few ice islands known are derived from a shelf of land-fast glacial ice attached to Ellesmere Island, about lat. 83°N (Figure 22.18). The ice islands move slowly with the water drift of the Arctic Ocean, and a charting of their tracks reveals much about circulation in that ocean (see path of T-3 in Figure 22.18). Ice islands serve as permanent and sturdy platforms for scientific research stations, from which observations of oceanography, meteorology, and geophysics can be carried out over long periods.

FIGURE 22.21 A tabular iceberg off Antarctica. It is a fragment of an ice shelf. (Wolfgang Kaehler/AllStock.)

ICE SHEETS AND GLOBAL WARMING

Holding 91% of the earth's ice, the Antarctic Ice Sheet has the potential—if it were to melt entirely—of raising the earth's mean sea level by about 60 m (200 ft). (Crustal sinking as a result of the added load would reduce this to about 40 m.) To this we would want to add the mass of the Greenland Ice Sheet, which holds most of the remaining volume of land ice.

As far back as the 1970s, climatologist John Mercer had pointed out that a portion of the Antarctic Ice Sheet lying west of the Transantarctic Mountains was in a potentially unstable condition, and that it might surge rapidly and pour into the ocean, causing a rapid sea level rise of about 5 m (16 ft). That theory depends heavily on the role of the large ice shelves of western Antarctica, which merge on their landward sides with the ice sheet (see Figure 22.16). Because the ice shelves are afloat, their detachment and/or melting away would not change sea level. They do, however, play a part in the scenario by tending to hold back the downslope flow of ice streams in the ice sheet. The reason for this holding effect is that the shelf ice is in contact with the bottom (grounded) at a number of points. If the floating ice shelf becomes greatly thinned by melting on the underside, losing these contact points, it may move out to sea, so that the unsupported land ice can quickly surge forward

to enter the sea. Thus far there are no indications that this activity has begun. On the other hand, if Mercer's prediction should materialize, it could result in a sea level rise as much as 5 m (16 ft)—a disaster of monumental proportions.

Running exactly counter to this worst-case scenario is another brought forward strongly in 1989 by scientists long active in glaciological research—among them geophysicist Charles R. Bentley. They reason that although global warming will accelerate glacier melting, increased precipitation over the ice sheets will produce a net growth in their thickness. (Increased precipitation during global warming is covered in Chapter 6.) Measurements made by radar on satellites have shown that the Greenland Ice Sheet has increased in thickness by 23 cm (9 in.) in the past decade. Growth in bulk of the Antarctic ice sheet, by itself, could cause a lowering of global sea level. In balance with total melting, ice sheet growth would result in only a modest rate of rise of sea level—perhaps no faster than what has been documented over the past decade.

Data sources: Oceanus, 1986–87, vol. 29:4, p. 50–54; Science News, 1989, vol. 136, p. 397; New Scientist, 1989, vol. 124:1689, p. 62–65; Science, 1989, vol. 246, p. 1563; New Scientist, 1990, vol. 125:1702, p. 36.

THE ICE AGE

The period of growth and outward spreading of great ice sheets is known as a *glaciation.* We can safely assume that a glaciation is associated with a general cooling of average air temperatures over the regions where the ice sheets originated. At the same time, ample snowfall must have persisted over the growth areas to allow the ice masses to grow in volume. The opposite kind of change—a shrinkage of ice sheets in depth and volume—would result in the receding of the ice margins toward the central highland areas and eventual disappearance of the ice sheets. This period is called a *deglaciation.* Following a deglaciation, but preceding the next glaciation, is a period in which a mild climate prevails; it is called an *interglaciation.*

A succession of alternating glaciations and interglaciations, spanning a total period on the order of 1 to 10 million years (m.y.) or more, constitutes an *ice age.* Throughout the past 2½ to 3 m.y., the earth has been experiencing an ice age that has long been known simply as *The Ice Age*—a title adopted by the naturalist Louis Agassiz in the 1830s. Our present position in time is within an interglaciation, following a deglaciation that set in quite rapidly about 15,000 years ago. In the preceding glaciation, called the *Wisconsinan Glaciation,* ice sheets covered much of North America and Europe and parts of northern Asia and southern South America. The

maximum ice advance of the Wisconsinan Glaciation was reached about 18,000 years ago.

Figures 22.22 and 22.23 show the extent to which North America and Europe were covered at the maximum known spread of the last advance of the ice. Canada was engulfed by the vast Laurentide Ice Sheet. It spread south into the United States, covering most of the land lying north of the Missouri and Ohio rivers, as well as northern Pennsylvania and all of New York and New England. Alpine glaciers of the Cordilleran ranges coalesced into a single ice sheet that spread to the Pacific shores and met the Laurentide sheet on the east.

In Europe, the Scandinavian Ice Sheet centered on the Baltic Sea, covering the Scandinavian countries. It spread south into central Germany and far eastward to cover much of Russia. In north-central Siberia, large icecaps formed over the northern Ural Mountains and highland areas farther east. At one time ice from these centers grew into a large ice sheet covering much of central Siberia. The British Isles were almost covered by a small ice sheet that had several centers on highland areas and spread outward to coalesce with the Scandinavian Ice Sheet. The Alps at the same time were heavily inundated by enlarged alpine glaciers.

South America, too, had an ice sheet. This grew from icecaps on the southern Andes Range south of about latitude 40°S, spreading westward to the Pacific shore, as well as eastward to cover a broad belt of Patagonia. It

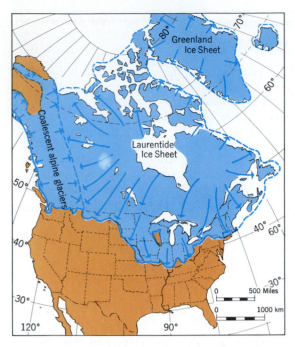

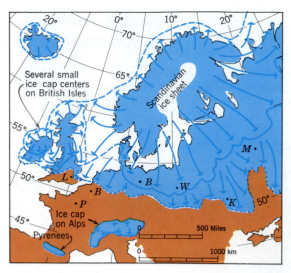

FIGURE 22.23 The Scandinavian Ice Sheet dominated northern Europe during the Pleistocene glaciations. The solid line shows limits of ice in the last glacial stage; the dotted line on land shows maximum extent at any time. (After R. F. Flint.)

FIGURE 22.22 Pleistocene ice sheets of North America at their maximum spread reached as far south as the present Ohio and Missouri rivers. (After R. F. Flint.)

covered all of Tierra del Fuego, the southern tip of the continent. The South Island of New Zealand, which today has a high spine of alpine mountains with small relict glaciers, developed a massive icecap in late Pleistocene time. All high mountain areas of the world underwent greatly intensified alpine glaciation at the time of maximum ice-sheet advance. Today only small alpine glaciers remain and, in less favorable locations, the glaciers are entirely gone.

Maximum southern extent of ice in each of the last four glaciations in the north-central United States is shown in Figure 22.24. The ice fronts advanced southward in great lobes. Notice that an area in southwestern Wisconsin escaped inundation by Pleistocene ice sheets. Known as the Driftless Area, it was apparently bypassed by glacial lobes moving on either side. Landforms made by the last ice advance and recession are very fresh in appearance and show little modification by erosion processes. It is to these deposits and landforms that we now turn our attention.

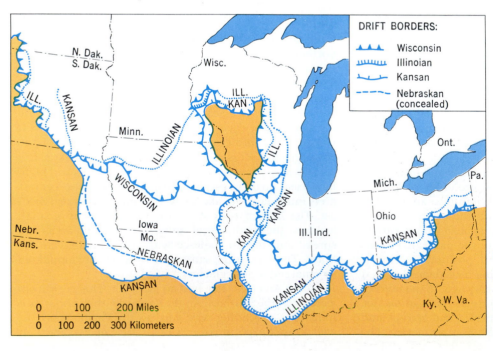

FIGURE 22.24 Drift borders in the north-central United States. In each glacial stage, the ice sheet reached a different line of maximum advance. (After R. F. Flint, *Glacial and Pleistocene Geology.*)

EROSION BY ICE SHEETS

Like alpine glaciers, ice sheets proved to be highly effective eroding agents. The slowly moving ice scraped and ground away much solid bedrock. Left behind were smoothly rounded rock masses bearing countless minute abrasion marks. Scratches, called *striations*, trend in the general direction of ice movement (Figures 22.25 and 22.26); but variations in ice direction from time to time resulted in intersecting lines. Certain kinds of rock were susceptible to deep grooving (Figure 22.26).

Where a strong, sharp-pointed piece of rock was held by the ice and dragged over the bedrock surface, the moving point produced a series of curved cracks fitted together along the line of ice movement. These *chatter marks*, and closely related *crescentic gouges*, whose curvature is the opposite, are good indicators of the direction of ice movement (Figure 22.26). Some very hard rocks acquired highly polished surfaces from the rubbing of fine clay particles against the rock. The evidences of ice erosion described here are common throughout the eastern United States and Canada. They may be seen on almost any exposed hard rock surface.

Bearing abrasion marks is a type of conspicuous knob of solid bedrock that has been shaped by the moving ice (Figure 22.27). One side, that from which the ice was approaching, is characteristically smoothly rounded and shows a striated and grooved surface; this is the stoss side. The other, or lee side, where the ice plucked out angular joint blocks, is irregular, blocky, and steeper than the stoss side. The quaint name *roches moutonnées*, which probably originated in the European Alps, has been applied to such glaciated rock knobs.

Vastly more important than the minor abrasion forms are enormous excavations that the ice sheets made in some localities of weak bedrock. In many places, the ice current was accentuated by the presence of a valley paralleling the direction of ice flow. Under such conditions the ice sheet behaved much as a valley glacier, scooping out a U-shaped trough (Figure 22.28). Fine examples are the Finger Lakes of western New York State. Here a set of former stream valleys lay parallel to southward spread of the ice, which scooped out a series of deep troughs. Blocked at the north ends by glacial debris, the basins now hold elongated lakes. Many hundreds of lake basins were created in a similar manner all over the glaciated portion of North America and Europe. Countless small lakes of Minnesota, Canada, and Finland occupy rock basins scooped out by ice action. Irregular debris deposits left by the ice are also important in causing lake basins.

FIGURE 22.25 Glacial striations and shallow grooves on a bedrock surface heavily abraded by Pleistocene alpine glaciers in the Boundary Ranges of British Columbia, near the Alaska border. (J. M. Ryder, University of British Columbia.)

DEPOSITS LEFT BY ICE SHEETS

The term *glacial drift* includes all varieties of rock debris deposited in close association with glaciers. Drift is of

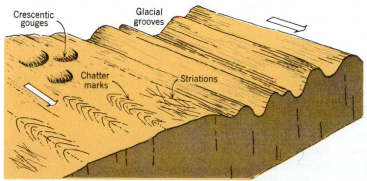

FIGURE 22.26 Features of ice abrasion. Chatter marks and crescentic gouges; glacial grooves paralleling the ice flow (arrow). (From A. N. Strahler, *The Earth Sciences*, 2nd ed., Harper & Row, Publishers. Copyright © by Arthur N. Strahler.)

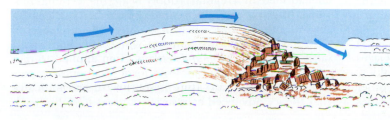

FIGURE 22.27 A glacially abraded rock knob. (From A. N. Strahler, *The Earth Sciences*, 2nd ed., Harper & Row, Publishers. Copyright © by Arthur N. Strahler.)

FIGURE 22.28 A Pleistocene continental ice sheet carved this finger-lake trough in granite on Mount Desert Island, Maine. Beyond is the smoothly rounded ice-abraded profile of Mount Sargent. (A. N. Strahler.)

two major types: (1) *Stratified drift* consists of layers of sorted and stratified clays, silts, sands, or gravels deposited by meltwater streams or in bodies of water adjacent to the ice. (2) *Till* is a heterogeneous mixture of rock fragments ranging in size from clay to boulders, deposited directly from the ice without water transport.

Over those parts of the United States formerly covered by Pleistocene ice sheets, glacial drift averages from 6 m (20 ft) thick over mountainous terrain, such as New England, to 15 m (50 ft) and more thick over the lowlands of the north-central United States. Over Iowa, drift is from 45 to 60 m (150 to 200 ft) thick; over Illinois, it averages more than 30 m (100 ft) thick. In some places, where deep stream valleys existed prior to glacial advance, as in Ohio, drift is a few hundred meters deep.

To understand the form and composition of deposits left by ice sheets, one must consider the conditions prevailing at the time of existence of the ice. Block A of Figure 22.29 shows a region partly covered by an ice sheet with a stationary front edge. This condition occurred when the rate of ice ablation balanced the amount of ice brought forward by the spreading of the ice sheet. Although the Pleistocene ice fronts advanced and receded in many minor and major fluctuations, there were long periods when the front was essentially stable and thick deposits of drift accumulated.

The transportational work of an ice sheet is like that of a huge conveyor belt. Anything carried on the belt is dumped off at the end and, if not constantly removed, will pile up in increasing quantity. Rock fragments brought within the ice are deposited at the edge as the ice evaporates or melts. There is no possibility of return transportation.

The advancing ice of a valley glacier or a continental ice sheet picks up rock fragments in a great range of sizes from the bedrock surface over which it moves. Boulders the size of houses may be moved by the ice. At the other extreme, vast numbers of particles of silt and clay grades

are produced by ice abrasion performed by rock fragments held fast in the moving ice and serving as cutting tools. All these fragments tend to be very angular, as would be expected of hard bedrock crushed and ground under high pressures beneath a thick ice layer. As the glacial ice melts in a stagnant marginal zone, the rock particles it holds are lowered to the solid surface beneath, where they form a layer of debris (Figure 22.30). This *ablational till* shows no sorting and often consists of a mixture of sand and silt, with many angular pebbles and boulders. Beneath this residual layer there may be a basal layer of dense *lodgment till*, consisting of clay-rich debris previously dragged forward beneath the moving ice.

As parent matter for soil, glacial till has a unique combination of characteristics. It has an abundance of clay particles, which may be in part clay minerals derived from older regolith; or they may be finely crushed minerals, capable of being altered into clay minerals during a long period of chemical weathering following disappearance of the ice. The clay fraction in dense till impedes the movement of soil water, so water drainage is often very slow. Some ablational tills of sandy composition, on the other hand, are of loose texture and drain rapidly. The presence of cobbles and boulders, called *erratics*, can make soil tillage difficult or even impossible. Because glacial tills range greatly in age and composition, each occurrence must be evaluated separately in terms of physical and chemical properties.

Glacial till that accumulated at the immediate ice edge formed an irregular, rubbly heap, the terminal moraine. After the ice disappeared (Figure 22.29, Block B), the moraine became a belt of knobby hills interspersed with basinlike hollows, some of which now hold small lakes. The name *knob-and-kettle* is often applied to moraine belts. Terminal moraines form great curving patterns; the convex curvature is southward and indicates that the ice advanced as a series of great *ice lobes*, each

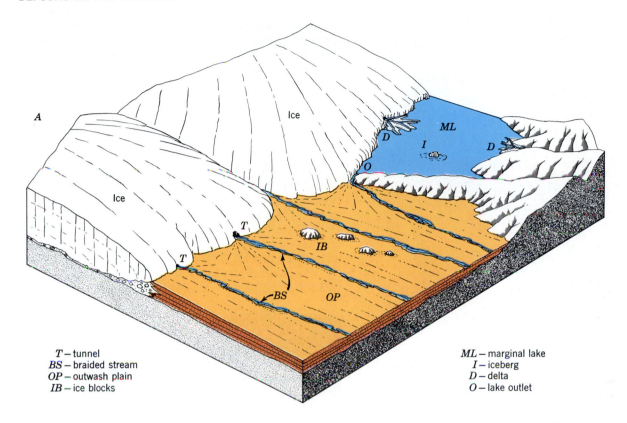

A

T – tunnel
BS – braided stream
OP – outwash plain
IB – ice blocks

ML – marginal lake
I – iceberg
D – delta
O – lake outlet

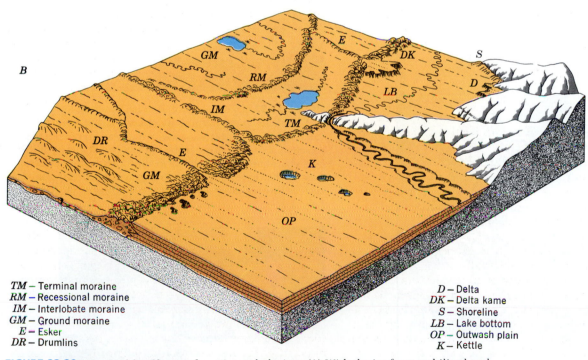

B

TM – Terminal moraine
RM – Recessional moraine
IM – Interlobate moraine
GM – Ground moraine
E – Esker
DR – Drumlins

D – Delta
DK – Delta kame
S – Shoreline
LB – Lake bottom
OP – Outwash plain
K – Kettle

FIGURE 22.29 Marginal landforms of continental glaciers. (A) With the ice front stabilized and the ice in a wasting, stagnant condition, various depositional features are built by meltwater. (B) The ice has wasted completely away, exposing a variety of new landforms made under the ice. (Drawn by A. N. Strahler.)

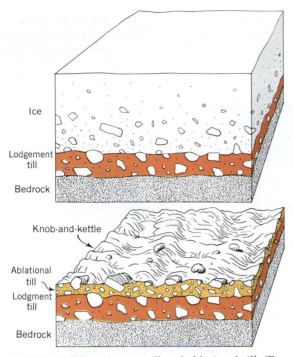

FIGURE 22.30 Lodgment till and ablational till. (From *A Geologist's View of Cape Cod* by A. N. Strahler. Copyright © 1966 by Arthur N. Strahler. Reproduced by permission of Doubleday and Company, Inc.)

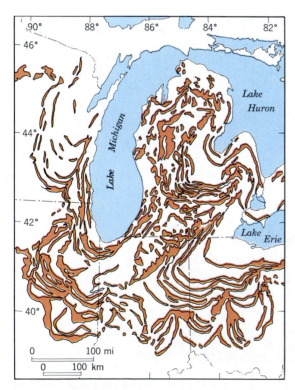

FIGURE 22.31 Moraine belts of the north-central United States have a festooned pattern left by ice lobes. (After R. F. Flint and others, *Glacial Map of North America.*)

with a curved front (Figure 22.31). Where two lobes come together, the moraines curve back and fuse together into a single moraine pointed northward. This deposit is an *interlobate moraine*. In its general recession accompanying disappearance, the ice front paused for some time along a number of lines, causing morainal belts similar to the terminal moraine belt to be formed. These belts are recessional moraines (Figures 22.29 and 22.31). They run roughly parallel with the terminal moraine but are often thin and discontinuous.

An entirely different class of sediment derived from glacial ice is *glaciofluvial sediment*, transported from stagnant ice masses by streams of meltwater and deposited in broad, shallow channels as a form of coarse alluvium (Figure 22.29A). As layer upon layer of sand accumulates near the ice margin, an *outwash deposit* is formed (Figure 22.32). Following disappearance of the ice, the outwash deposit remains as an *outwash plain*, which may be pitted with deep kettles (Figure 22.29B).

FIGURE 22.32 Thick layers of outwash sands and gravels such as these on the north shore of Long Island were excavated in great quantities for use in highway and building construction. The dark layer at the top is a bed of glacial till left by a glacial advance. Boulders in the foreground are glacial erratics that have rolled down from the till bed. Exposures such as this are rarely seen today in this region. (A. K. Lobeck.)

FIGURE 22.33 An esker, such as this one in Maine, is a valuable source of sand and gravel. Cobbles and boulders must first be screened out. (A. N. Strahler.)

Large streams issued from tunnels in the ice, particularly when the ice—for many kilometers back from the front—became stagnant, without forward movement. Tunnels then developed throughout the ice mass, serving to carry off the meltwater. Now that the ice has disappeared, the former ice tunnel is represented by a long, sinuous ridge known as an *esker*. The esker is the deposit of sand and gravel formerly laid on the floor of the ice tunnel (Figure 22.33). All the ice has melted away, only the streambed deposit remains, forming a ridge (Figure 22.29B). Eskers are often many kilometers long; a few are more than 150 km (100 mi) long.

Another common glacial form is the *drumlin*, a smoothly rounded, oval hill resembling the bowl of an inverted teaspoon. It consists of glacial till (Figure 22.34). Drumlins invariably lie in a zone behind the terminal moraine. They commonly occur in groups or swarms, which may number in the hundreds. The long axis of each drumlin parallels the direction of ice movement, and the drumlins thus point toward the terminal moraines and serve as indicators of direction of ice movement. Drumlins were formed under moving ice by a plastering action in which layer upon layer of bouldery clay was spread on the drumlin.

Between moraines, the surface overridden by the ice is overspread by a continuous cover of glacial till. This layer cover is often inconspicuous because it forms no prominent landscape feature. The till may be thick and may obscure or entirely bury the hills and valleys that existed before glaciation. Where thick and smoothly spread, the deposit forms a level *till plain*. Plains of this origin are widespread throughout the Middle West.

Between the ice front and rising ground, valleys that may have opened out northward were blocked by ice. Under such conditions, *marginal glacial lakes* formed along the ice front (Figure 22.29A). These lakes overflowed along the lowest available channel between the ice and the rising ground slope, or over some low pass along a divide. Streams of meltwater from the ice built sand deposits, called *glacial deltas*, into these marginal lakes. When the ice disappeared, the lake drained away, exposing layers of fine clay and silt accumulated on the bottom. These fine-grained strata, which had settled out from suspension in turbid lake waters, are called *glaciolacustrine sediments* and are a variety of stratified drift. Sediment commonly shows banding, with alternating dark and light layers, termed *varves* (Figure 22.35). An individual varve consists of a light-colored silt layer beneath a dark layer of fine clay. The silt layers are interpreted as summer deposits from highly turbid lake water; the clay layers, as stillwater deposits formed in winter when the lakes were sealed over by an ice cover. Glacial lake plains are extremely flat, with meandering streams and extensive areas of marshland.

FIGURE 22.35 Samples of varved clays from glacial lake beds near New York City. The columns measure about 10 cm (4 in.) from top to bottom of the photograph. (C. A. Reeds, American Museum of Natural History.)

FIGURE 22.34 This small drumlin, located south of Sodus, New York, shows a tapered form from upper right to lower left, indicating that the ice moved in that direction (north to south). (Ward's Natural Science Establishment, Inc., Rochester, N.Y.)

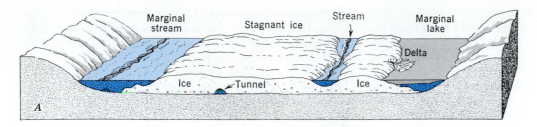

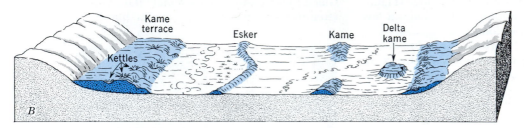

FIGURE 22.36 Kames may originate as stream or lake deposits laid between a stagnant ice mass and valley sides. (Drawn by A. N. Strahler.)

Deltas, built with a flat top at what was formerly the lake level, are now curiously isolated, flat-topped landforms known as *delta kames* (Figure 22.29). Delta and stream channel deposits built between a stagnant ice mass and the wall of a valley are now *kame terraces*, whose steep scarps are ice-contact slopes (Figure 22.36). Kame terraces are difficult to distinguish from the uppermost member of a series of alluvial terraces, but most kames have undrained depressions or pits produced by the melting of enclosed ice blocks. Built of very well-washed and sorted sands and gravels, kames commonly show the dipping stratification planes characteristic of deltas.

ENVIRONMENTAL AND RESOURCE ASPECTS OF GLACIAL DEPOSITS

Throughout Europe and North America landforms associated with the ice are of major environmental importance and the deposits constitute a natural resource as well. Agricultural influences of glaciation are both favorable and unfavorable, depending on preglacial topography and whether the ice eroded or deposited heavily.

In hilly or mountainous regions, such as New England, the glacial till is thinly distributed and extremely stony. Soils developed on glacial deposits of the northern United States and in Canada are acid and low in fertility. Extensive bogs are floored by bog soils unsuited to agriculture unless transformed by water-drainage systems. Early settlers found cultivation of till difficult because of countless boulders and cobbles in the soil. Till accumulations on steep mountain slopes are subject to mass movements in the form of earthflows. Clays in the till become weakened after absorbing water from melting snows and spring rains. Where slopes have been oversteepened by excavation for highways, movement of till is a common phenomenon.

The steep slopes, the irregularity of knob-and-kettle topography, and the abundance of boulders along mo-

raine belts conspired to prevent crop cultivation but invited use as pasture. These same features, however, make moraine belts extremely desirable as suburban residential areas. Pleasing landscapes of hills, depressions, and small lakes made ideal locations for larger estates.

Flat till plains, outwash plains, and lake plains, on the other hand, comprise some of the most productive agricultural land in the world. Fertile soils formed on these till plains and on lake plains bordering the Great Lakes. We must not lose sight of the fact that in the prairie areas, wind-deposited silt (loess) forms a blanket over clay-rich till and sandy outwash. This upland loess formed an ideal parent material for the development of grain-producing soils, but has proved highly vulnerable to accelerated soil erosion.

Stratified drift deposits are of great economic value. The sands and gravels of outwash plains, delta kames, and eskers provide the aggregate necessary for concrete and the base courses beneath highway pavements. The purest sands may be used for molds, which are needed for metal castings.

Stratified drift, where thick, forms an excellent aquifer and is a major source of ground-water supplies. Deep accumulations of stratified sands in preglacial valleys are capable of yielding ground water in adequate quantities for municipal and industrial uses. Water development of this type is widespread in Ohio, Pennsylvania, and New York.

THE LATE-CENOZOIC ICE AGE

The Ice Age occurred during the last 3 m.y. of the Cenozoic Era. As shown in the table of geologic time, Table 13.1, the Cenozoic Era has seven epochs. The Ice Age falls within the last three epochs: Pliocene, Pleistocene, and Holocene. These three epochs comprise only a small fraction—about eight percent—of the total duration of the Cenozoic Era.

During the first half of this century, most geologists

TABLE 22.1 **Classic Names for North American Glaciations, Based on Continental Evidence (ca 1950)**

Glaciations	Interglaciations
Wisconsinan	
	Sangamonian
Illinoian	
	Yarmouthian
Kansan	
	Aftonian
Nebraskan	

associated the Ice Age with the Pleistocene Epoch, which was then thought to have begun about 1 million years ago (-1 m.y.). In other words, they identified the boundary between the Pliocene and Pleistocene epochs as the starting point of the Ice Age. That supposition has now been totally invalidated by new evidence from deep-sea sediments showing that many glaciations and interglaciations occurred in late Pliocene time, and that the Ice Age probably commenced as far back in time as -3 m.y. For this reason, we can rename the Ice Age the *Late-Cenozoic Ice Age*, leaving the date of its onset to be established when more evidence becomes available.

From the middle 1800s until about 1950, the record of glaciations, deglaciations, and interglaciations was almost entirely interpreted from continental deposits. During this early period of research there emerged a history of four distinct North American glaciations in the Pleistocene Epoch. Names of the four glaciations and interglaciations of North America are given in Table 22.1. These names are used for map boundaries shown in Figure 22.24. At present, opinion is divided as to whether this classic system of names should be retained or abandoned.

Unraveling the history of glaciations and interglaciations was first attempted solely on the basis of evidence from glacial and related deposits and landforms exposed on the continental surfaces. Because ice-sheet advance and recession were evidently not synchronous over all parts of even one continent, the correlation of events proved extremely difficult. One form of evidence has come from interpretation of layered deposits (science of stratigraphy). The till of one glaciation is typically followed by a loess layer associated with deglaciation, then by formation of an ancient soil (paleosol) and by deposition of organic matter such as bog peat indicative of an interglaciation with its mild climate. Older tills show varying degrees of chemical alteration by weathering. Landforms of earlier glaciations, where not buried under new glacial deposits, show increasing degrees of modification by mass wasting and fluvial erosion with increasing age.

RADIOCARBON DATING

Starting in the early 1950s, the *radiocarbon dating method* of establishing the absolute age of sample materials became an important tool of late-Pleistocene and Holocene research. The method was used to make age

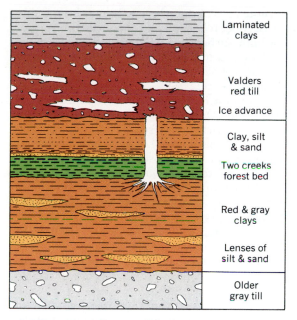

FIGURE 22.37 The Two Creeks forest bed, exposed near Manitowoc, Wisconsin, was developed in a substage of mild climate, but was overridden by ice advance in the Valderan Substage that followed. (After J. L. Hough, 1958, *Geology of the Great Lakes*, p. 102, Figure 31, Univ. of Illinois Press, Urbana.)

determinations on such carbon-bearing substances as charcoal, invertebrate shells, wood, and peat derived from archaeological sites and glacial deposits. An example of the use of the radiocarbon method was to determine the ages of tree trunks felled by the rapid advance of the ice sheet at a locality in Wisconsin. A forest had grown up in a mild period (a minor interglaciation) in late Wisconsinan time. The advancing ice broke off the tree trunks and incorporated them into dense red clay of a moraine. Figure 22.37 shows the succession of layers at this locality. Radiocarbon analysis of the logs gave an age close to $-12,000$ years. This was a record of the final advance of ice in the Wisconsinan Glaciation; it was almost immediately followed by the final ice recession from that region, marking the end of the Pleistocene Epoch and the beginning of the Holocene Epoch.

The radiocarbon method was supplemented and corrected by the record of tree rings (science of dendrochronology), which gives an absolute age chronology back to about -7500 years. Unfortunately, the radiocarbon method is reasonably accurate only for ages no greater than about $-40,000$ years, which is only a trivial fraction of the duration of the Pleistocene Epoch.

OXYGEN–ISOTOPE CHRONOLOGY OF DEEP-SEA CORES

A great scientific breakthrough in the study of late-Cenozoic glacial history came in the 1960s. First, it became possible to measure the absolute age of lava rock and certain types of water-laid sediments by means of paleomagnetism. The earth's magnetic field has undergone many sudden reversals of polarity in Cenozoic time

In figure: Laminated clays / Valders red till / Ice advance / Clay, silt & sand / Two creeks forest bed / Red & gray clays / Lenses of silt & sand / Older gray till

and the absolute ages of these reversals have been firmly established. Second, it became possible to take long sample cores of undisturbed fine-textured sediments of the deep ocean floor and to determine by magnetic polarity reversals the age of sediment layers at various control points within each core.

The cores contain materials that have settled out of the overlying ocean. One kind of material is fine clay of inorganic mineral particles. A second kind of material consists of the tests (hard parts) of microorganisms that lived (as plankton), in the near-surface zone of the overlying ocean. It was soon recognized that the relative abundances of certain species of microorganisms are indicators of the former temperature of the surface water layer—whether cold or warm—within a range of about 6C° (11F°). By studying the percentage compositions of species, investigators have been able to identify cycles of alternating colder and warmer global climate. They have placed the cycles in an absolute framework of age in years before present.

The most important advance in interpretation of deep-sea cores was the development of a method of determining the *oxygen–isotope ratio*. Without giving a lengthy explanation of the method, we simply note that the carbonate mineral matter of tests contains atoms of oxygen. (The typical composition of one class of tests is calcium carbonate, $CaCO_3$.) The common form of oxygen is an isotope known as oxygen-16 (O^{16}), but there are other, less common isotopes of oxygen, one of which is oxygen-18 (O^{18}). The oxygen–isotope ratio is the ratio of O^{18} to O^{16} written as O^{18}/O^{16}. For reasons we cannot develop in depth here, the growth of ice sheets during the onset of a glaciation causes the oxygen–isotope ratio in cores to increase, whereas a melting of ice sheets and return of the meltwater to the ocean results in a decrease in that ratio.

Figure 22.38 is a curve drawn to show the fluctuations in oxygen–isotope ratio within a typical deep-sea core taken from the subantarctic ocean in a location between Africa, Australia, and Antarctica. This kind of curve is now called a *paleoglaciation curve*. The record shows 13 *isotope–ratio stages* over a time span of nearly 500,000 years. Also shown is a curve of inferred sea-surface temperatures based on ratios of species of microorganisms

in the core. Smoothing these curves so as to leave only the major peaks and valleys suggests that there were 5 glaciations in the 450,000 years of record, for an average of 1 glaciation about every 90,000 years. Isotope–ratio stages have been established and numbered back to about -2 m.y., and are recognizable even to -3 m.y. Within the 2 m.y. span there are 41 isotope–ratio stages. It now appears that glaciations may have been occurring as far back as -3 m.y., or well into the Pliocene Epoch, and that the total number of glaciations in late-Cenozoic time may have been more than 30, spaced at intervals of about 90,000 years. Based on examination of the composition of deep-sea sediments in the longer cores taken in the Southern Ocean it seems probable that the Antarctic Ice Sheet began to build between -11 and -14 m.y., reaching its approximate present dimensions by about -3.5 m.y.

The discovery of a long sequence of late-Cenozoic glaciations brought to an end the identification of the Ice Age as matched in time with the Pleistocene Epoch. The revised Late-Cenozoic Ice Age has endured $3\frac{1}{2}$ m.y. and perhaps as long as 4 m.y. It may continue for several or many more millions of years, depending on the persistence of the combination of factors that bring about an ice age.

It is known that a great ice age occurred in late Paleozoic time and endured for at least 100 m.y. There is evidence that this ice age, which saw the presence of a great ice sheet on the Gondwana subcontinent of Pangaea through Carboniferous and Permian times, actually began much earlier in Africa. Ice ages that are known to have occurred in the Silurian and Devonian periods in what is now northern Africa may have been continuous through the entire Paleozoic Era—a time span of some 200 m.y. Paleomagnetic evidence shows that the Gondwana subcontinent was in a polar location during this entire time. We will return to this subject in discussing the causes of ice ages.

ORIGIN OF THE GREAT LAKES

The Great Lakes of the United States and Canada are truly remarkable environmental features. By an accident

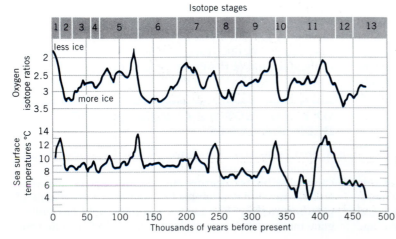

FIGURE 22.38 Curve of oxygen-isotope ratios, or paleoglaciation curve (*above*), and curve of inferred summer sea-surface temperatures from deep-sea cores in the subantarctic ocean (*below*). (Based on data of J. D. Hays, J. Imbrie, and N. J. Shackleton, *Science*, vol. 194, p. 1130, Figure 9.)

of nature—the Late-Cenozoic glaciation—these huge freshwater bodies happen to lie in a midlatitude continental area that has proved enormously productive both agriculturally and industrially. Taken together, the surface area of the Great Lakes is about 246,000 sq km (95,000 sq mi), much greater than any single freshwater lake in the world. Other large freshwater lakes of the world lie in less favorable environments from the human standpoint: Great Bear and Great Slave lakes in the far north of Canada; lakes Victoria and Tanganyika in equatorial Africa; and Lake Baikal in Siberia.

In preglacial time, lowlands existed where the Great Lakes now stand. These lowlands were occupied by ma-jor steams. Repeated ice advances of at least four glaciations extensively and deeply eroded weak rocks of the former lowlands and carried the debris south to form blocking lines of moraines.

Stages in Great Lakes evolution are shown in the series of six maps in Figure 22.39. These maps are simplified and show only a few representative stages. Map A shows the earliest lakes beginning to form as the ice front receded. Lakes Chicago and Maumee were marginal glacial lakes, ponded between the ice front and higher ground to the south. Both lakes overflowed southward by streams draining into the Mississippi River system.

Map B shows continued retreat and the diversion of

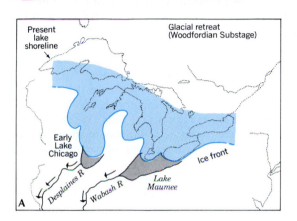

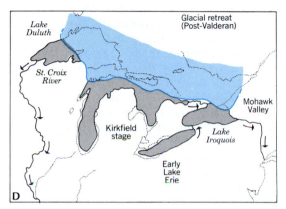

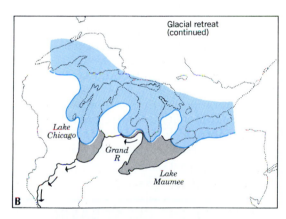

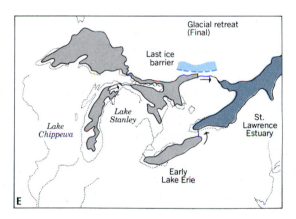

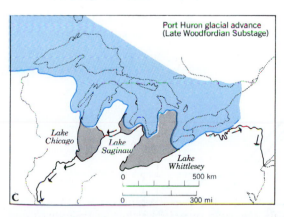

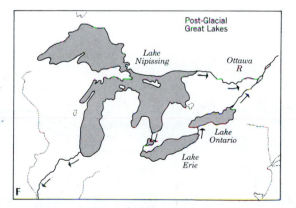

FIGURE 22.39 Six stages in evolution of the Great Lakes. (After J. L. Hough, 1958, *Geology of the Great Lakes*, Univ. of Illinois Press, p. 284–96. © 1958 by the Board of Trustees of the University of Illinois. From A. N. Strahler, *The Earth Sciences*, 2nd ed., Harper & Row, Publishers. Copyright © by Arthur N. Strahler.)

one lake into another by a marginal stream following the ice front. Map C catches the action at a point when drainage was established eastward along the ice front to enter the Hudson River system by way of the Mohawk valley. In Map D, final ice recession was under way and part of ancestral Lake Superior had opened up, draining south into the Mississippi system while the other lakes drained east to the Hudson system.

Map E shows the very last of the ice disappearing in Ontario and opening an outlet along what is now the Ottawa River. This outlet led directly into the St. Lawrence valley, which was then an estuary of salt water. Map F shows a stage of maximum extent of the Great Lakes. Crustal tilting caused the Ottawa River outlet to be abandoned, and lowering of lake levels caused abandonment of the drainage of Lake Michigan into the Mississippi system.

One result of this complex history of changing lake levels and areal extents is that today there are broad marginal zones of lake plains along the shores of Lakes Michigan, Huron, Erie, and Ontario. These plains are intensively developed as agricultural lands and have absorbed the urban expansion of such major lake cities as Chicago, Toledo, Detroit, Cleveland, Toronto, Buffalo, and Erie. Many serious environmental problems beset the Great Lakes. One reason is that the water supplies for a large number of industrial cities are drawn from the same lakes into which are fed, in return, enormous quantities of pollutant wastes.

ALPINE AND PERIGLACIAL ICE-AGE ENVIRONMENTS

Evidence of lowered global air temperatures during glaciations comes from both terrestrial and marine sources. One line of evidence is the lowered elevation of the *snowline*, or lower limit of snow accumulations lasting through the entire year. Figure 22.40 is a meridional profile along the Cordilleran ranges and Andes from 65°N to 60°S latitude, which shows the altitude of the snowline of today as compared with that during a glaciation. These snowlines are smoothed lines, evening out local differences in altitude due to varying local influences. In North America, the amount of lowering was about 1500 m (5000 ft). In the Andes at low latitudes the snowline was

lowered about 600 m (2000 ft). Similar figures apply on other continents. The lowering of snowline is interpreted as representing a drop of mean annual global air temperature of between 5 and 7C° (9 and 13F°).

Lowering of the snowline must have been accompanied by a lowering of all life zones stratified according to altitude. This change would have required plants and animals to migrate to lower elevations or to lower latitudes to remain in the same thermal environments. A reverse effect would be caused by the rising of snowline during deglaciation.

The southward spread of ice sheets brought with it an advance zone of frost-controlled climate similar to that found today in coastal lands bordering the Greenland Ice Sheet, on arctic islands of northern Canada, and along the Alaskan arctic coast. The term *periglacial* describes this environmental zone, which would have had a tundra landscape with some areas of barren soils and, elsewhere, a tundra plant cover (see Chapter 15). Ground ice formed ice-wedges and frost polygons, while spontaneous flowage of saturated sediments occurred in summers as mudflows and solifluction lobes. Today we find these features of the periglacial environment as relict forms in soils and alluvium in a zone bordering the former ice limits.

Other climate zones were pushed toward the equator, as lowered air temperatures set in and shifts occurred in zones of global atmospheric circulation. The upper-air westerlies extended their zone of influence into lower latitudes, as Hadley cell circulation weakened and the subtropical high-pressure belt migrated equatorward. With this migration the tropical deserts moved into lower latitudes and may have produced desert conditions as far equatorward as 10° to 15° latitude. With lowered atmospheric temperatures, it seems likely that resultant reduction in atmospheric moisture would have reduced convectional activity and precipitation in the equatorial belt. All of these climatic shifts affected plants and animals, forcing them to migrate with the shifting climate zones in order to survive.

PLUVIAL LAKES OF THE GREAT BASIN

Among the nonglacial phenomena associated with the late-Cenozoic glaciations were changes in the water bal-

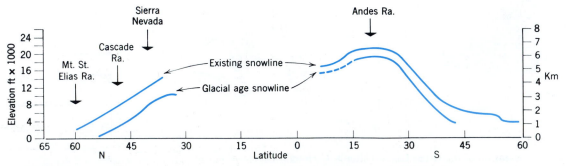

FIGURE 22.40 Generalized meridional profile of the present snowline and lowered Pleistocene snowline of maximum glaciations. (Data of R. F. Flint, *Glacial and Pleistocene Geology*, John Wiley & Sons, New York, p. 47, Figure 4.1.)

ances of closed basins of the Basin and Range Province. As we explained in Chapter 16 on the subject of inland saline lakes, this region has a great excess of potential evaporation over precipitation. This arid condition results today in total absence of lake water in most of these basins, in occasional stands of shallow water in others, and in a few instances such as Great Salt Lake, Utah, and Pyramid Lake, Nevada, in permanent lakes of high salinity. Runoff reaching these lakes comes by way of streams receiving discharge from neighboring mountain ranges where a water surplus occurs at high elevations. Obviously, there is a delicate equilibrium among evaporation, inflow of streams, and storage in the water balance of those closed basins presently holding water.

During glaciations, the water balance changed in favor of small water surpluses, with the result that water occupied a large number of the intermountain basins, bringing into existence a large number of *pluvial lakes.* The word "pluvial" suggests an increase in precipitation during glaciations as the cause of the lakes. Evaporation would also have been less under a climate of lower air temperatures. We know that alpine glaciers of certain of the neighboring higher ranges, such as the Wasatch Mountains and Sierra Nevada, made major advances during glaciations to reach low altitudes, showing the effects of greater ice accumulation and reduced ablation.

Figure 22.41 is a map showing pluvial lakes as they were during maximum extent during the Wisconsinan Glaciation. Altogether, there were about 120 pluvial lakes in existence then. Some overflowed into others and

probably held fresh water at times. Largest of the pluvial lakes was Lake Bonneville, an expansion of the present-day Great Salt Lake in Utah. It reached an areal extent of 52,000 sq km (20,000 sq mi), about the same as Lake Michigan, and for a time overflowed northward into the Snake River. Its maximum depth was 330 m (1000 ft). Abandoned shorelines of Lake Bonneville can be seen today along the mountain slopes against which the lake waters rested.

Expansion, contraction, and changing salinities of the pluvial lakes constituted great swings in environmental conditions affecting ecosystems of the basins. A remarkable example of adaptation of animals to changing environments is seen in the case of the desert pupfish. There are today some 20 populations of these tiny fish surviving in isolated spring-fed streams and tiny pools in Death Valley, California. This tectonic basin, which lies below sea level and is one of the hottest surface environments on earth, was occupied by pluvial Lake Manly (Figure 22.41). As lake waters disappeared, the fish were forced into a few remaining spring localities and became isolated from one another. Their tolerance to a wide range of temperatures is quite phenomenal. Blue-green algae provide the fish with food.

ICE-AGE CHANGES OF SEA LEVEL

Important changes in sea level were associated with the Late-Cenozoic Ice Age. We know that the volume of ice

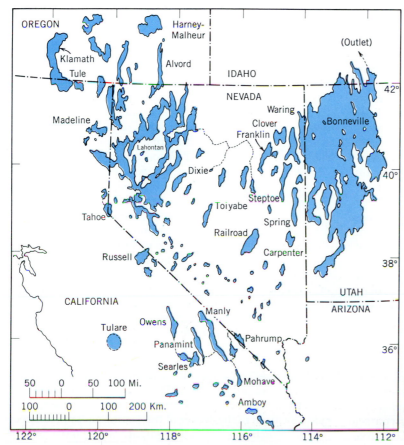

FIGURE 22.41 Pluvial lakes of the western United States. The dotted lines are overflow channels. (Based on a map by R. F. Flint, *Glacial and Pleistocene Geology,* John Wiley & Sons, New York, p. 227, Figure 13.2.)

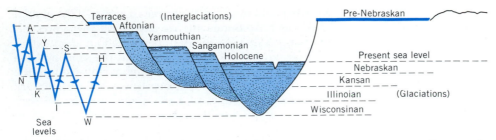

FIGURE 22.42 Schematic diagram of development of alluvial terraces during oscillations of sea level, superimposed on a general crustal rise. Actual deposits would be fragmentary. (From A. N. Strahler, *Physical Geology*, Harper & Row, Publishers. Copyright © by Arthur N. Strahler.)

held on Antarctica is such that, if it melted, together with all other glacial ice, there would result a sea-level rise of about 60 m (200 ft). Imagine the large coastal zones that would be inundated by such a change, if it were to occur in the near future. Major coastal urban centers of North America and Europe would be drowned and a large land area removed from agricultural production. The extent to which this inundation might occur within the next century or two because of global warming brought on by the steady increase in atmospheric carbon dioxide is a matter of great concern to environmental planners, and we have discussed it in Chapter 20.

On the other hand, the growth of ice sheets withdrew vast quantities of water from the oceans, with a resulting decline in sea level. We now have good evidence on which to base a reconstruction of the sea-level changes accompanying the closing glacial event within the Wisconsinan Glaciation. This evidence comes from a variety of organic and sedimentary materials brought up from the floor of the Atlantic continental shelf. These materials include shells and coralline algae whose growth habitats are known to lie close to sea level. Salt-marsh peat is also a valuable indicator of sea level. The samples can be dated by the radiocarbon method.

Plotting the present depths of these samples against age, a generalized curve of sea level can be drawn. About 35,000 years before present, sea level stood near its present position, for this was a time of ice recession within the Wisconsinan Glaciation—a sort of mini-interglaciation (Farmdalian substage). As the final ice advance set in, sea level declined and reached a low point of −60 to −80 m (−200 to −260 ft) at about −18,000 years. At this time a broad zone of the continental shelf was exposed and the shoreline lay some 100 to 200 km (60 to 125 mi) east of its present position. Remains of freshwater plants show that this exposed shelf was a richly vegetated landscape; animal remains show that it supported land animals, such as elephants (mastodons and mammoths). Although glacial ice stood not far away at the time, the climate was not as severe as one might suppose, being essentially like that of the subarctic lands of Canada, and at times not much different than that of northern New England today.

One effect of lowered sea level during glaciations was on streams entering the sea. The mouths of these streams were extended in length to reach the more distant shoreline. Lowered stream baselevel causes the lower courses of the streams to degrade their channels and to carve

trenches into the continental shelves. Trenching and valley widening also progressed upstream into what is now the continental mainland. As sea level rose, the same streams were forced to aggrade and to fill their valleys with alluvium. This succession of events was repeated with each glaciation and interglaciation, but with an added effect—that of gradually rising continental crust. The combined effect is illustrated schematically in Figure 22.42. Each successive trenching was carried to a lower level, but each filling ceased at a lower level than the previous filling. The result is a set of stepped alluvial terraces. Keep in mind that this diagram is highly idealized and that you would not expect to find a complete succession of terraces in any one locality. Nevertheless, major streams draining the Atlantic and Gulf coasts of North America show at least partial development of terraces of this type. (Aggradation and degradation in headwater areas near the ice margin showed a different pattern of events, as we explained in Chapter 17.)

ICE SHEETS AND CRUSTAL REBOUND

The bedrock surface beneath the Greenland and Antarctic ice sheets has been depressed under the load of the ice layer. Actually, some parts of the continental crustal surface under these ice sheets are now below sea level. We might suppose that the same kind of crustal depression occurred under the Pleistocene ice sheets. If so, the disappearance of the ice sheets would be followed by a crustal rise, called *postglacial rebound*.

There is excellent evidence that crustal rebound is in progress over the former European and North American ice sheet centers. At numerous points along the coasts of the Scandinavian Peninsula and Finland there are ancient beaches and wave-cut benches now high above the present sea level. Some of these elevated shoreline markers can be dated through various artifacts associated with them and by the radiocarbon method. In this way the rate of crustal uplift in postglacial time can be estimated. Precise measurements of elevations of surveying reference points also show clearly that uplift is in progress. Figure 22.43 is a map of the Baltic region showing lines of equal rates of uplift. The maximum rate is in the northern end of the Gulf of Bothnia, where it is about 1 m (3 ft) per century. Not surprisingly, this is the approximate location of the center of the Scandinavian Ice Sheet where the ice was thickest (compare with Figure 22.23).

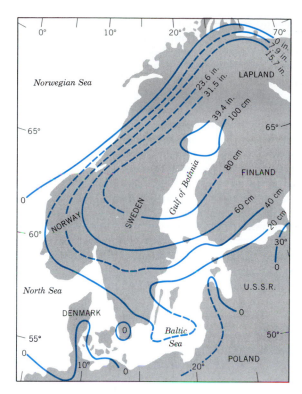

FIGURE 22.43 Present rate of uplift of the Baltic region is shown here by lines of equal uplift in centimeters per century. (Based on data by B. Gutenberg in J. A. Jacobs, R. D. Russell, and J. T. Wilson, *Physics and Geology*, New York, McGraw-Hill, p. 98, Figure 4–5. From A. N. Strahler, *The Earth Sciences*, 2nd ed., Harper & Row, Publishers. Copyright © by Arthur N. Strahler.)

FIGURE 22.44 Upraised marine shorelines bordering the shoreline of Hudson Bay. Snow bands lie in swales between beach ridges. (Canadian Government Department of Energy, Mines, and Resources; National Air Photo Library.)

In North America the postglacial crustal rebound is nicely documented by a great succession of uplifted beach ridges bordering the present coastlines of Hudson Bay and the Arctic Ocean (Figure 22.44). Each beach ridge consists of boulders rearranged by breaking ocean waves into a low embankment paralleling the water line. The Laurentide Ice Sheet, which was centered about over Hudson Bay, disappeared between −6000 and −8000 years. The crust in this area has risen 70 to 100 m (200 to 300 ft) since the ice disappeared, for an average rate of uplift on the order of 1 m (3 ft) per century. This rate compares favorably with the maximum rate of uplift in the Baltic region.

CAUSES OF ICE AGES

Few geological puzzles have proved as intriguing and as difficult to solve as the cause or causes of multiple glaciations and interglaciations. Despite scientific advances on many fronts in the earth sciences and an enormous gain in our knowledge of how the earth works through plate tectonics, the debate as to what causes glaciations continues unabated in its fervor and in the wide range of ideas proposed.

We must approach this puzzle on two quite different levels. First is an overall set of conditions that seems to have favored the onset and endurance of an ice age—the alternating growth and disappearance of large ice sheets on the continents over a time span as long as 2 to 3 million years. In this broad view of the problem, we need to take into account ancient ice ages, such as the late-Paleozoic glaciation so well documented on the Gondwana continent. The second level is that of the immediate, or forcing causes that are responsible for actually precipitating a glaciation at a particular point in time, for causing a deglaciation to follow, and for this cycle to be repeated. The questions are these: Why are there cycles of glaciation and deglaciation? What controls the length of these cycles? What causes them to be initiated or triggered? In answering these questions we may find ourselves considering several quite different triggering mechanisms and favorable factors acting together to cause glaciation. A meaningful inquiry into the causes and factors involved in glaciations and interglaciations requires a good working knowledge of atmospheric and oceanic sciences. This is because climate change is the key to the glacial cycles.

FUNDAMENTAL CAUSES OF AN ICE AGE

Four fundamental factors could tend to cause an entire ice age lasting 2 to 3 million years or longer: (1) A favorable position of the continents with respect to the polar regions. (2) A withdrawal of oceans from land surfaces accompanying widespread continental uplift. (3) A sustained period of increased volcanic activity. (4) A sustained period of diminished intensity of solar energy reaching the earth.

Of these four fundamental factors, the first three are geologic in nature. The late-Paleozoic ice age saw a great ice sheet on the subcontinent of Gondwana. The continent itself was located at polar latitudes and included the south pole. Paleomagnetic evidence confirms the slow drifting of Gondwana over the polar region. As early as the Silurian Period, glaciation occurred in western Africa, when the pole was located in that area. By late Paleozoic time, the pole path was contained within what was then the single continental nucleus of southern Africa, southern South America, Antarctica, India, and Australia. This ice age ended when breakup of Gondwana was followed by the movement of most of its fragments to lower latitudes. Only Antarctica ended up in the south polar position. Look next at the north polar region in late Paleozoic time, as shown in Figure 13.30, Map A. In Permian time only the northern tip of the Eurasian continent projected into the polar zone. As the continental plates rifted apart and the Atlantic basin opened up, North America moved westward and poleward to a position opposite Eurasia, while Greenland took up a position between North America and Europe. The effect of these plate motions was to bring an enormous landmass area to a high latitude and to surround a polar ocean with land.

The reasons why a large land area at high latitude favors the onset of an ice age are fairly simple. First, the landmasses on which ice sheets can grow become located in a cold climate zone where the rate of snowfall can greatly exceed ablation. Second and perhaps less obvious, the presence of the landmasses blocks the poleward flow of warm ocean currents that might otherwise tend to make the polar climate relatively mild. At present, the polar sea (Arctic Ocean) is connected with the Atlantic and Pacific oceans only by narrow straits. The polar ocean with its year-round cover of sea ice maintains a cold climate, which extends well into the fringes of the bordering landmasses. To summarize, the most important single factor in making an ice age possible is a geologic event—the arrival of a large expanse of continental surface at a polar position. There was, however, a long lapse of time in the late Cenozoic Era between the time the landmasses surrounded the north pole and the first ice sheets began to form. Clearly, other factors brought on the start of the ice sheet growth.

The second factor on our list is a general withdrawal of shallow seas from the continental margins in early Cenozoic time. In late Cretaceous time—near the close of the Mesozoic Era—shallow continental seas were extensive. By Miocene time, however, the seas had retreated to the edges of continents. This lowering of sea level relative to the continents is usually attributed to a general rise of the continental crust. On the other hand, the withdrawal of oceans may have been caused by an increase in the average depth of the ocean basins as the rate of seafloor spreading decreased following a high rate of spreading in the Cretaceous Period.

Why would a relative lowering of sea level and a reduction in the surface expanse of ocean water tend to promote an ice age? The reasons are not by any means simple, because many different effects are involved. Large landmasses tend to have colder winters (but warmer summers) than adjacent oceans in the same latitude zone. An increase in surface elevation of a continent would tend to lower the average temperature near the ground surface, but the amount of uplift over continental interiors may have been too small to be of consequence.

Some geologists have argued that an increase in tectonic and volcanic activity in the late Cenozoic caused the rise of lofty mountain ranges capable of trapping large amounts of snowfall, and thus made possible the first growth of icecaps. Others argue that with more or less continuous plate subduction going on throughout geologic time, there were high mountain ranges on active continental margins more or less continually at one place or another. Thus the growth of alpine mountains and volcanic chains is not in itself a cause of continental glaciation.

VOLCANIC ACTIVITY AND ICE AGES

Our third factor, also geologic, is an increase in the intensity of volcanic activity on a global basis over a sustained time span. The link between volcanic activity and glaciation is through the emission of vast amounts of extremely fine volcanic dust during explosive volcanic eruptions, particularly the eruption of composite volcanoes and caldera explosions. The dust rapidly reaches the upper atmosphere and spreads over the entire globe within months or a year or two following an eruption. The dust particles, in turn, cause a climate change that may be global in scope.

Following the eruption of the volcano Krakatoa in 1883, a veil of volcanic dust formed in the stratosphere and spread into high latitudes. In Europe and North America, the dust veil brought on a large number of extremely brilliant sunsets and attracted a great deal of attention. Solar observatories recorded a 20 percent drop in the intensity of solar energy reaching the earth's surface in the first year following the explosion. For each of the next 3 years the reduction was about 10 percent. Eventually most of the dust settled out into the lower atmosphere and the blocking effect disappeared. Some atmospheric scientists agree that the climatic effect of a stratospheric dust veil is to cause a cooling of the average atmospheric temperature near the earth's surface. If we accept this conclusion, we can then examine the geologic record with regard to the intensity of volcanic activity before and during the Late-Cenozoic Ice Age.

A record of past volcanic activity can be obtained from deep-sea cores. A major volcanic eruption is recorded as a thin layer of volcanic ash consisting of minute shards of volcanic glass. Some indication of the general intensity of volcanic activity can be had by determining the number of ash layers per thousand years. Some research scientists claim the evidence points to greatly increased volcanic activity in the Pleistocene

Epoch. One study of deep-sea cores led to the conclusion that volcanic activity in the Pleistocene was at a level four times higher than the average for the 20-million-year span covered by the cores. There was also a moderately high level of volcanic activity in mid-Pliocene time and a number of minor peaks in the Miocene Epoch. These and other attempts to relate volcanic activity to the cause of an ice age have been strongly disputed by other scientists.

Our fourth possible cause of an ice age is in the realm of astronomy and we will refer to it only briefly. The hypothesis states that our sun, as it travels through space in a rotary motion with the rest of the Milky Way galaxy, occasionally passes through a cold interstellar dust cloud. As the dust is drawn into the sun's surface it is highly heated and increases the sun's brightness. Thus the sun's energy output is increased for a time. The effect is to increase the input of energy into the earth's atmosphere, and this may supply the added precipitation needed for ice sheets to form and grow. Perhaps, on the other hand, the increased solar input would raise the average air temperature—an opposite effect.

CAUSES OF GLACIATIONS AND INTERGLACIATIONS

We turn next to the timing and triggering mechanisms that may have been responsible for cycles of glaciation and interglaciation known to have been repeated at least several times, and perhaps as many as 20 to 30 times. A general lowering of air temperatures on a global scale toward the close of the Cenozoic Era is fairly well documented. This overall cooling is the result of one or more of the fundamental ice-age causes we have already reviewed. As we have seen, reduced temperatures during glaciations generally are recorded on the continents in a lowering of the snowline. And, of course, we have seen that the general cooling is well documented by evidence from deep-sea cores.

Here is a short list of hypotheses of cyclic glaciations currently being debated: (1) Triggering by bursts of vol-

canic activity; (2) control by astronomical cycles of the earth's tilt and orbital motion; (3) control by changes in arctic sea-ice cover; (4) control by changes in the albedo of the land and water surfaces, largely from presence or absence of snow cover.

Under the first hypothesis the principle is the same as the general effect of increased volcanic dust on global air temperature. Some investigators have tried to correlate specific events of increased volcanic activity with the onset of specific glaciations. The validity of any such correlation is much in doubt.

The control of cycles of glaciations and interglaciations by astronomical cycles is a strong contender for acceptance at the present time, despite its being discredited by a number of capable scientists. The *astronomical hypothesis*, as it has often been called, has been under consideration on and off for about 40 years. It is based on facts about the motions of the earth in its orbit around the sun. These facts are well established by astronomical observations and are not the subject of debate. Two factors are involved here: (1) the changing distance between earth and sun; (2) the changing angle of tilt of the earth's axis of rotation. As to the first factor, the distance that separates the earth and sun at summer solstice, June 21, undergoes a cyclic variation. A single cycle lasts 21,000 years, during which the earth–sun distance may vary from 1 to 5 percent greater than the average distance to 1 to 5 percent less than the average distance. The axial tilt (now 23½°) undergoes a 40,000-year cycle of change in which the tilt angle may be increased to as much as about 24° and decreased to as little as 22°. If we pick a point on the earth at, say, lat. 65°N, we can calculate the total cycle of change in the intensity of incoming solar radiation, or insolation, on a day in summer (June 21). This cycle of incoming solar radiation combines the cycle of changing earth–sun distance and the cycle of axial tilt. Figure 22.45 is a graph showing the cyclic changes in summer daily insolation at lat. 65°N. This graph has been named the *Milankovitch curve*, after the astronomer who calculated the variations and published the results in 1938; he proposed that these cycles of insolation have controlled the glacial cycles.

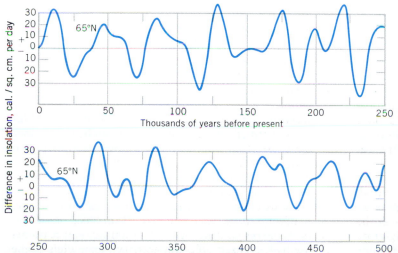

FIGURE 22.45 The Milankovitch curve, showing fluctuations in summer daily insolation at lat. 65°N over the past 500,000 years, based upon calculations made by A. D. Vernekar, 1968. The zero value represents the present solstice insolation for lat. 65°N. (Data from W. Broecker and J. van Donk, *Reviews of Geophysics*, vol. 8, p. 190, Figure 10. From A. N. Strahler, *Physical Geology*, Harper & Row, Publishers. Copyright © by Arthur N. Strahler.)

In its simplest form, the Milankovitch hypothesis tells us that a northern-hemisphere glaciation should coincide with each major low point in the insolation curve, whereas an interglaciation should coincide with each major high point. Actually, the glacial events might lag behind the insolation peaks and valleys. The greatest variations in 1 cycle represent a difference of about 5 percent in the total quantity of daily insolation at a given latitude. The capability of this amount of variation to produce the glacial cycle is a subject of continued debate.

The Milankovitch curve was taken up again in the mid-1960s by Wallace S. Broecker, a geochemist, and his collaborators at the Lamont-Doherty Geological Observatory of Columbia University. They noted that the high insolation peaks represent a cycle averaging about 80,000 to 90,000 years. Each high insolation peak was thought to be associated with a rapid deglaciation. During intervening long periods of reduced insolation, ice sheets would grow in volume. These investigators claimed to have found close correlations between the insolation cycle and cycles of both sea-level changes and oxygen–isotope ratios over the past 140,000 years. As you might expect, Broecker's interpretations had both supporters and detractors.

About 10 years later, the insolation hypothesis based on the Milankovitch curve was revived by a team of scientists working on major global climate changes and their causes. Using oxygen–isotope ratios and percentages of indicator microorganisms in two deep-sea cores, James D. Hayes of Columbia University and John Imbrie of Brown University found strong correlations between global climate changes and the insolation cycle. So impressive did they find this evidence that they claimed to be virtually certain the insolation cycle is responsible for glaciations and interglaciations. Of course, their data were strongly challenged, and their opponents declared with equal assurance that no such relationship could be derived from these data.

In 1990, Broecker and a colleague, George H. Denton, described an updated version of their theory of glaciations and deglaciations based on the Milankovitch curve. Of the insolation cycles that comprise that curve, they emphasize a 100,000-yr cycle as dominant, although both the 21,000-yr and 40,000-yr cycles superimpose lesser cycles of ice increase and decrease. Ice volume increases gradually through each major cycle, then falls dramatically at its close. The northern and southern hemispheres appear to be synchronized in this cycle, with the snowline falling and rising simultaneously in both hemispheres. Oxygen–isotope ratios measured in ice-sheet cores have confirmed the synchronous fluctuations.

New evidence now came into play. Analysis of air bubbles trapped in ice-sheet cores was examined for carbon dioxide content, and this was found to track closely the temperature curve indicated by the oxygen–isotope ratios. A significant finding was that the carbon dioxide peaks appear shortly in advance of the temperature peaks, and this suggests that the former is in control. This finding led Broecker and Denton to look to the deep-ocean circulation for an explanation. Today, under an interglacial regime, warm surface water of tropical origin flows north in the eastern North Atlantic Ocean to reach the subarctic zone. There, the warm water gives off an enormous amount of heat to the atmosphere. Chilled in the process, the highly saline surface water becomes much denser and sinks to great depth. It then travels over great distances through the Indian and Pacific oceans, emerging near the surface off the western coast of North America, Based on evidence of deep-sea cores, this "conveyor" system appears to have been disrupted during glaciations, so that its warming effect was not in operation.

Surface winds control the motions (currents) of the warm shallow layer of the oceans, so that changes in the global atmospheric circulation are implicated in the "flipping" of the conveyor between its on and off modes. We have already shown in Chapter 5 how global warming by increased concentration of the atmospheric greenhouse gases can be expected to intensify the Hadley cells and circulation of the westerlies. Changes in atmospheric moisture and cloud cover are also involved (Chapter 6). Somewhere in this schedule of complex atmospheric changes may lie the cause of flipping on and off of the oceanic conveyor, which in turn may terminate or initiate a glaciation.

We bring to a close this overview of the causes of ice ages and glaciations leaving unmentioned a number of interesting hypotheses. The entire subject is so complex as to be difficult for even a research scientist to grasp fully. Interactions between the atmosphere, the oceans, and the continental surfaces (including the ice sheets) are numerous and closely interrelated. Many threads of cause and effect cross and recross these earth realms. Changes that occur in one realm are fed back to the other realms in a most complex manner.

HOLOCENE CLIMATE CYCLES

The elapsed time span of about 10,000 years since the Wisconsinan Glaciation ended comprises the Holocene Epoch; it began with the rapid warming of ocean surface temperatures. Continental climate zones shifted rapidly poleward. Soil-forming processes began to act on new parent matter of glacial deposits in midlatitudes. Plants became reestablished in glaciated areas in a succession of climate stages. The first of these was known as the *Boreal stage*. Boreal refers to the present subarctic region where needleleaf forests dominate the vegetation. The history of climate and vegetation throughout the Holocene time has been interpreted through a study of spores and pollens found in layered order from bottom to top in postglacial bogs. (This study is called palynology.) Plants can be identified and ages of samples can be determined. A dominant tree was spruce. Interpretation of pollens indicates that the Boreal stage in midlatitudes had a vegetation similar to that now found in the region of boreal forest climate.

There followed a general warming of climate until the *Atlantic stage* was reached about 8000 years ago (−8000 years). Lasting for about 3000 years, the Atlantic stage had average air temperatures somewhat higher than those of today—perhaps on the order of 2.5C°

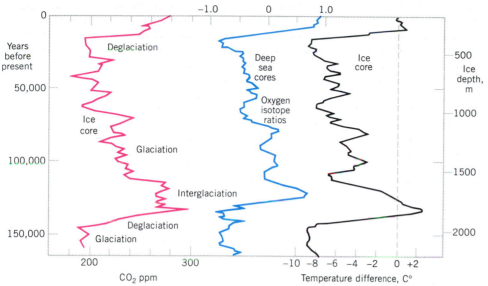

FIGURE 22.46 Records of atmospheric carbon dioxide abundance and calculated surface temperatures from an ice core drilled at Vostok station in Antarctica. Also shown is a corresponding record of oxygen–isotope ratios of deep-sea cores. (Data of C. Lorius, N. I. Barkov, J. Jouzel, Y. S. Korotkevich, V. M. Kotlyakov, and D. Raynaud, 1988, *EOS*, vol. 69, no. 26, p. 691, Figure 1. Copyright © by The American Geophysical Union. Used by permission.)

higher. We call such a period a climatic optimum with reference to the midlatitude zone of North America and Europe. There followed a period of temperatures below average, the *Subboreal stage*, in which alpine glaciers showed a period of readvance. In this stage, which spanned the age range −5000 to −2000 years, sea level—drawn far down from the level during glaciation—had returned to a position close to that of the present, and coastal submergence of the continents was largely completed.

The past 2000 years, from the time of Christ to the present, show climatic cycles on a finer scale than those we have described as Holocene climatic stages. This refinement in detail of climatic fluctuations is a consequence of the availability of historical records and of more detailed evidence generally. A secondary climatic optimum occurred in the period A.D. 1000 to 1200 years (−1000 to −800). This warm episode was followed by the *Little Ice Age* (A.D. 1450–1850; −550 to −150 years). During this time valley glaciers made new advances to lower levels. In the process, the ice overrode nearby forests and thus left a mark of its maximum extent.

An independent record of climate fluctuations through the Wisconsinan Glaciation and Holocene Epoch is available from oxygen–isotope ratios measured in ice layers of the Greenland and Antarctic ice sheets. Air temperature at the time of ice formation influences the O^{18}/O^{16} ratio. A lowering of air temperatures results in a decreasing proportion of O^{18} in the ice molecules. Consequently, a given snow or ice layer carries with it a permanent record of average atmospheric temperature prevailing during the year in which it was formed. It is not possible to assign a specific temperature value to a particular isotope ratio, but temperature fluctuations from warmer to colder periods and vice versa can be

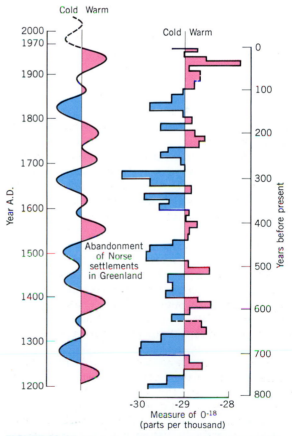

FIGURE 22.47 Variations in the oxygen–isotope ratio within an ice core from the Greenland Ice Sheet during the past 800 years. (From S. J. Johnson, et al., Climatic oscillations 1200–2000 A.D. Reprinted from *Nature*, vol. 227, No. 5257, p. 483. Copyright © 1970 Macmillan Journals Limited.)

readily recognized from a succession of samples taken at intervals along an ice core.

Figure 22.46 shows ice core data from the Antarctic Ice Sheet at Vostok station, operated by Soviet scientists. A record going back more than 150,000 years is presented for abundance of atmospheric carbon dioxide (left) and for atmospheric temperature relative to a zero value that is the present mean annual temperature of the station (right). Also shown (center) is a composite record of the oxygen–isotope ratios derived from sediments in deep-sea cores. The general correlation between carbon dioxide content and temperature difference is striking throughout the entire period of record, although it is not certain whether the increase in carbon dioxide preceded or followed the two major periods of warming during the two prominent deglaciations.

.Figure 22.47 shows ice-core data for the past 800 years, going back to the year A.D. 1200. It illustrates the short-period cycles of temperature change for the portion of the Holocene Epoch within range of documentation by human historical records. At the left is a smooth curve fitted to the isotope data by a mathematical procedure known as Fourier analysis. The particular analysis was designed to reveal cycles of a medium range of time periods. We notice at once that the smooth curve shows the warming trend of the first half of the present century, which is often attributed to a documented increase in carbon dioxide from fuel combustion (see Figure 4.24). A cooling trend that has set in since about 1940 in the northern hemisphere is also clear.

When we look back into earlier centuries, we find a series of cycles of temperature variation of approximately the same amplitude as the latest variations. Cycles with periods of 78 to 181 years have been recognized by Fourier analysis of these isotope data. The important point is that air-temperature fluctuations of the past 800 years are of the same order of magnitude that we observe in our century. In other words, the recent trends of warming and cooling are within the normal range found throughout the record. What some scientists have interpreted as human-induced changes in global atmospheric temperature may, instead, be caused by natural variations in the earth's radiation balance.

In Chapter 4 we examined conflicting predictions as to the role of humans in causing global climate change. Cycles of glaciation and interglaciation demonstrate the power of natural forces to make drastic swings from cold to warm climates. Lesser climatic cycles of the Holocene Epoch also occurred through natural causes. Only following the Industrial Revolution do we recognize possible linkages between global air-temperature change and combustion of hydrocarbon fuels on a massive scale. There is general agreement that increased carbon dioxide and other greenhouse gases tend to cause a rise in average temperatures and that an increase in input of industrial dusts reaching the upper atmosphere tends to lower average temperatures near the earth's surface. But we do not know as yet to what extent observed changes in global temperatures are parts of a natural cycle and to what extent they are influenced by the impacts of an industrial society. Whether humanity is hastening or delaying the onset of another glaciation will continue to be debated.

LANDFORMS IN REVIEW

The group of eight chapters we have now completed has reviewed geomorphic processes that operate on the surface of the continents. The great variety and complexity of landforms we have described is not difficult to comprehend when each agent of denudation is examined separately. Landforms produced by glaciers, waves, and wind are localized in distinctive environments—alpine and arctic regions, coastlines, and dry regions, respectively. The landforms of fluvial denudation are the most widespread and complex features of the continental landscape. Fluvial denudation integrates weathering, mass wasting, and fluvial processes into highly complex systems of denudation responding to climate controls.

Human influence on landforms is most strongly felt on surfaces of fluvial denudation because of the severity of surface changes caused by agriculture and urbanization. Landforms shaped by wind and by waves and currents are also highly sensitive to changes induced by human activity. Only glaciers maintain their integrity and are thus far undisturbed by human activity. Perhaps even this last realm of Nature's superiority will eventually fall prey to human interference through climate changes induced by industrial activity.

In the series of five chapters that now follows, we turn increasingly to the processes and forms of the biosphere and its living matter. The soil layer that lies at the interface between atmosphere and lithosphere contains not only mineral matter, but also organic matter that includes living plants and animals. Every major topic of physical geography we have thus far studied will be brought to bear on the science of the soil layer. Chapters that follow turn to the dynamic processes and forms of life to be found in ecosystems. Of these, we will concentrate on biogeography, which includes kinds and classes of natural vegetation of the lands.

THE SOIL LAYER

The soil is the very heart of the life layer on the lands. The soil layer is a place in which plant nutrients are produced and held. As we emphasized in Chapter 9, the soil layer also holds water in storage for plants to use. In Chapter 8 we explained that the role of climate is to vary the input of water and heat into the soil. This same heat energy and water are responsible for the breakup and chemical change of rock to produce the parent mineral matter of the soil (Chapter 12). That mineral matter is the source of many plant nutrients.

But climate acting on rock cannot make a soil layer capable of sustaining a rich plant cover. Plants themselves, together with many forms of animal life, play a major role in determining the qualities of the soil layer. Those qualities have evolved through centuries by the interaction of organic processes with physical and chemical soil processes. The organic processes include the synthesis of organic compounds, and these are eventually added to the body of the soil. Plants use the mineral nutrients to build complex organic molecules. Upon the death of plant tissues, these nutrients are released and reenter the soil, where they are reused by living plants. The concept of nutrient cycling by plants will emerge in Chapter 25 as one of the keys to understanding the development of the soil layer.

THE DYNAMIC SOIL

The soil is a dynamic layer in the sense that many complex physical and chemical activities are going on simultaneously within it. Because climate and plant cover vary greatly from place to place over the globe, the combined effects of soil-forming activities are expressed differently from place to place. Anyone can observe that the pale gray soil beneath a spruce forest in Maine is quite different in composition and structure from the dark brown soil beneath the prairie farmlands of Iowa. But, in each of these localities, the soil has reached a physical and chemical state related to the climate controls and soil-forming processes prevailing there.

The geographer is keenly interested in the differences in soils from place to place. The capability of a given soil type to furnish food crops largely determines which areas of the globe support the bulk of the human population. Despite changes in population distribution made possible by technology and industrialization, most of the world's inhabitants still live where the soil furnishes them food. Many of the same humans die prematurely because the soil does not furnish adequate nutrition for all who need it.

The substance of the soil exists in all three states—solid, liquid, and gas. The solid portion consists of both inorganic (mineral) and organic substances. The liquid present in the soil is a complex solution capable of engaging in a multitude of important chemical reactions. Gases present in the open pores of the soil consist not only of the atmospheric gases, but also of gases liberated by biological activity and chemical reactions within the soil. Soil science, formally called *pedology*, is obviously a highly complex body of knowledge. We shall do no more than cover a few of the high points of this science.

THE NATURE OF SOIL

Thus far, we have used the word "soil" without giving it a precise definition. *Soil*, as the term is used in soil science, is a natural surface layer containing living matter and supporting or capable of supporting plants. Substance of the soil includes both inorganic (mineral) matter and organic matter, the latter both living and dead. Living matter in the soil consists not only of plant roots, but of many kinds of organisms, including microorganisms. The upper limit of the soil is air or shallow water. Horizontal limits of the soil may be deep water or barren areas of rock or ice.

The lower limit of the soil is often difficult to define in simple and exact terms. Nonsoil below the soil may be bedrock or any form of regolith and sediment devoid of living roots and other signs of biologic activity. In contrast, soil shows evidences of biologic activity.

Soils usually show *soil horizons*, which are distinctive horizontal layers set apart from other soil zones or layers by differences in physical and chemical composition, organic content, structure, or a combination of those prop-

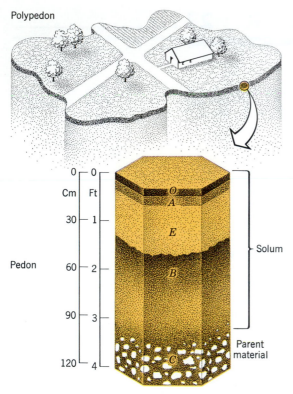

FIGURE 23.1 Concept of the pedon and polypedon. (Drawn by John Balbalis.)

erties. Soil horizons are developed by the interactions, through time, of climate, living organisms, and the configuration of the land surface (relief).

The word "soil" is used by civil engineers and geologists to mean any surface layer of unconsolidated mineral matter of low strength as compared with strong, hard bedrock. Our definition of soil excludes surficial materials that do not support the growth of plants.

Although most classes of soil—those with horizons—require a long span of time in which to develop, a layer capable of supporting plants can come into existence very rapidly. An example would be an accumulation of silt on a river floodplain. Here the parent matter may include organic matter and nutrients formed elsewhere and transported to the new location. Generally, the transformation of raw parent matter into a soil with horizons requires 1 to 2 centuries at the least. The time required for a soil to reach an equilibrium state with the environment is usually estimated in thousands of years.

CONCEPT OF THE PEDON*

Modern soil science makes use of the concept of the *polypedon*, which is the smallest distinctive division of the soil of a given area. A unique single set of properties

*Throughout this chapter and the next, numerous phrases and sentences have been taken verbatim from the following source: Soil Survey Staff, *Soil Taxonomy*, Soil Conservation Service, U.S. Dept. of Agriculture, Agriculture Handbook No. 436, Government Printing Office, Washington, D.C.

applies to the polypedon, and this set differs from that applying to adjacent polypedons. The polypedon is conceived in terms of space geometry as being composed of pedons. A *pedon* is a soil column extending down from the surface to reach a lower limit in some form of regolith or bedrock. As Figure 23.1 shows, soil scientists often visualize a pedon as a 6-sided (hexagonal) column. The surface area of a single pedon ranges from 1 to 10 sq m (10 to 100 sq ft). The *soil profile* is the display of horizons on one face of the pedon. Obviously, the same soil profile is displayed on all 6 faces of the pedon. In practice, a soil scientist digs a deep pit, exposing a soil profile in the side of the pit.

Figure 23.1 shows a number of soil horizons. Most horizons are visibly set apart on the basis of color or texture. Mineral soil horizons are designated by a set of capital letters and numeral subscripts, starting with A at the top. In Figure 23.1 we see A, E, B, and C horizons. An organic horizon, designated by the letter O, lies on the A horizon (see also Figure 23.8).

The *soil solum* consists of the A, E, and B horizons of the soil profile; these are the dynamic and distinctive layers of the soil. The C horizon, by contrast, is the parent material. The soil solum occupies the zone in which living plant roots exert control on the soil horizons; the C horizon lies below that level of root activity.

TABLE 23.1 **Grade Sizes of Sediment Particles, U.S. Department of Agriculture System**

Grade Name	Diameter Limits
	2.0 mm
Very coarse sand	
	1.0 mm
Coarse sand	
	0.5 mm
Medium sand	
	0.25 mm
Fine sand	
	0.10 mm
Very fine sand	
	0.05 mm
Silt	
	0.002 mm (2 microns)
Clay noncolloidal	2 to 0.01 microns
colloidal	Below 0.01 microns

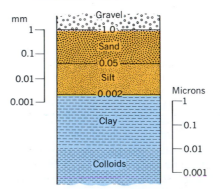

FIGURE 23.2 Size grades used in the description of soil textures. (U.S. Department of Agriculture system.)

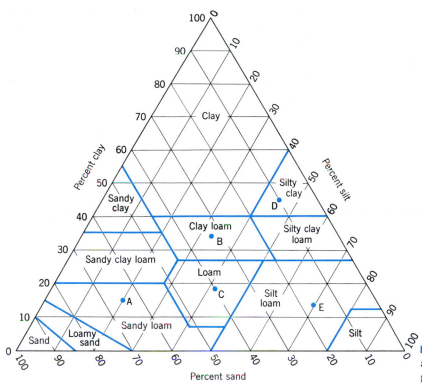

FIGURE 23.3 Soil-texture classes shown as areas bounded by heavy lines on a triangular graph. (U.S. Department of Agriculture.)

SOIL COLOR

For persons unfamiliar with soil science, color—easily perceived at a distance—is the most obvious property of a soil. The dark brown to black color of the soil is conspicuous to the traveler in farmlands of Iowa and Nebraska; the pervasive red color of soil in the Piedmont upland of Georgia does not escape notice. Certain color relationships are quite simple. Black color usually indicates the presence of abundant organic matter (humus); red color usually indicates the presence of sesquioxide of iron (hematite). The soil color may in some areas be inherited from the parent matter; but, more generally, it is a property generated by the soil-forming processes.

Description of soil color has been placed on an objective basis through the use of books of standard (Munsell) colors adapted to the needs of soil science. Three measurable variables determine color. One is the hue, or dominant color of the pure spectrum, depending on wavelength (See Figure 3.3). A second variable is value, the degree of darkness or lightness of the color. The third is chroma, the purity or strength of the spectral color. By using standard color books, the observer can express soil color as a letter–numeral code, telling hue, value, and chroma.

SOIL-TEXTURE CLASSES

A scale of grade sizes of mineral particles was explained in Chapter 12. The same grade scale is used in describing soil texture. Table 12.2, giving details of this scale, is reproduced here as Table 23.1. Figure 23.2 summarizes the main divisions of this grade scale. *Soil-texture classes* are based on varying proportions of sand, silt, and clay,

expressed as percentages. The standard system of classes used by the U.S. Department of Agriculture is shown in a triangular diagram, Figure 23.3. Percentages of all three components are shown simultaneously. The corners of the triangle represent 100 percent of each of the three grades of particles—sand, silt, or clay. *Loam* is a mixture in which no one of the three grades dominates over the other two. Loams therefore appear in the central region of the triangle. A particular soil whose components give it a position at Point A in the triangle has 65 percent sand, 20 percent silt, and 15 percent clay; it falls into a texture class known as *sandy loam*. Another soil, whose texture is represented by Point B, has $33\frac{1}{3}$ percent sand, $33\frac{1}{3}$ percent silt, and $33\frac{1}{3}$ percent clay; it falls into the class of a *clay loam*. Figure 23.4 gives five examples of soil textures; their positions are shown on Figure 23.3.

Texture is important because it largely determines

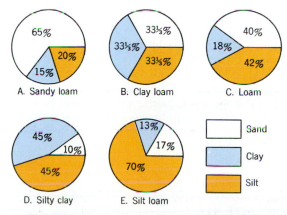

FIGURE 23.4 Typical compositions of five soil-texture classes. These examples are shown as lettered points on Figure 23.3. (U.S. Department of Agriculture.)

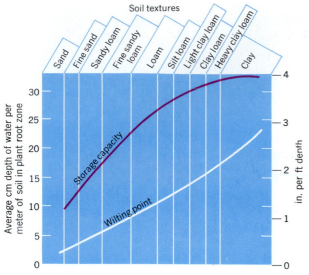

FIGURE 23.5 Storage capacity and wilting point vary according to soil texture.

the water retention and transmission properties of the soil. Sand may drain too rapidly; in a clay soil, the individual pore spaces are too small for adequate drainage. Where clay and silt content is high, root penetration may be difficult.

Recall from Chapter 9 that storage capacity (field capacity) of a soil is its capacity to hold water against the pull of gravity. Figure 23.5 shows how storage capacity varies with soil texture. Pure sand holds the least water, while pure clay holds the most. Loams hold intermediate amounts. Sand transmits the water downward most rapidly, clay most slowly. When planning the quantity of irrigation water to be applied, these factors must be taken into account. Sand reaches its full capacity very rapidly, and added water is wasted. Clay-rich loams take up water very slowly and, if irrigation is too rapid, water will be lost by surface runoff. By the same token, sandy soils require more frequent watering than clay-rich soils. The organic content of a soil also strongly affects its water-holding capacity. The intermediate loam textures are generally best as agricultural soils because they drain well, but also have favorable water-retention properties.

Soil texture is largely an inherited feature of a given soil and depends on the composition of the parent matter. Some kinds of parent matter furnish a large spread in particle sizes; others yield mostly sand or mostly clay.

Agricultural soil scientists also use a measure of soil-water storage termed the *wilting point*. Soil water in an amount less than the value at the wilting point cannot be absorbed by plants rapidly enough to meet their needs. At this point, the foliage of plants not adapted to drought will wilt. As Figure 23.5 shows, the wilting point depends on texture.

SOIL CONSISTENCE

Soil consistence refers to the quality of stickiness of wet soil and to the plasticity of moist soil, as well as the degree of coherence or hardness of the soil when it holds

small amounts of moisture or is in the dry state. Stickiness of a wet soil is evaluated by pressing a quantity of soil between thumb and finger, then separating the digits and observing the extent to which soil adheres to the skin. Plasticity is evaluated by rolling a small amount of wet soil into a rod shape. If plasticity is high, the soil can be rolled into a thin wire. Consistence when a soil is dry is expressed by various levels of hardness, ranging from loose (noncoherent) to extremely hard. Cementation, which usually occurs in a particular horizon, is not affected by wetting and ranges from weakly cemented material, easily broken in the hands, to an indurated condition resembling hard rock. Cementation is caused by the accumulation of mineral substances, such as calcium carbonate, silica, or iron oxides.

SOIL STRUCTURE

Soil structure refers to the presence of aggregations (lumps or clusters) of soil particles. Each aggregate is separated from adjoining aggregates by natural surfaces of weakness (cracks). In some instances, the aggregates are coated with surface films of material that help keep them apart. In other cases, aggregates are simply held together by forces of internal cohesion. An individual natural soil aggregate is called a *ped*. (An aggregate caused by breakage during plowing is a *clod*.) Soil structure is described in terms of the shape, size, and durability of peds.

Four primary types of soil structure are recognized: platy, prismatic, blocky, and spheroidal. These are illustrated in Figure 23.6. Platy soil structure consists of plates—thin flat pieces—in a horizontal position. In prismatic structure, peds are formed into vertical columns, often flat-sided, which may be 0.5 to 10 cm (0.2 to 4 in.) across. Blocky structure consists of angular, equidimensional peds with flattened surfaces that fit the surfaces of adjacent peds. Spheroidal structure consists of peds more or less rounded in outline with surfaces that do not fit those of adjacent peds. In the granular variety of spheroidal structure, shown in Figure 23.6, the peds are small and the soil is very porous.

A description of soil structure includes not only the form of the peds, but also their sizes (fine, medium, coarse) and their degree of durability (weak, strong). Soil structure is a physical property of great agricultural importance because it influences the ease with which water will penetrate a dry soil, the susceptibility of the soil to erosion, and the ease of cultivation.

Related to soil structure is the presence of thin films or coatings, called *cutans* (skins), on soil peds or individual coarse mineral grains. Cutans of clay (*argillans*) are clay skins coating the outer surfaces of peds or sand grains (Figure 23.7). These coatings consist of clay particles carried down through the soil by infiltrating water. Other cutan types include thin coatings of oxides of iron or manganese on mineral grains, and films of organic matter.

Also important in soil structure is the nature of the void spaces between peds. Size and degree of interconnection of voids are important in determining the ease with which water and air move through the soil.

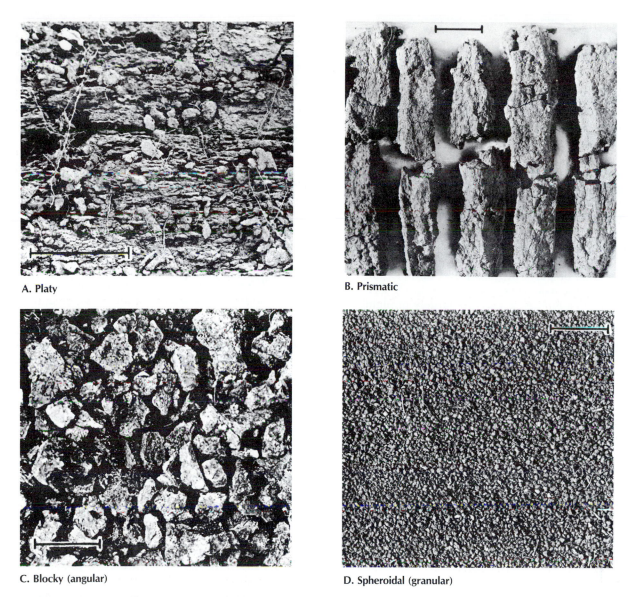

A. Platy

B. Prismatic

C. Blocky (angular)

D. Spheroidal (granular)

FIGURE 23.6 Four basic soil structures are illustrated. The black bar on each photograph represents 2.5 cm (1 in.). (Division of Soil Survey, U.S. Dept. of Agriculture.)

FIGURE 23.7 Magnified view of clay skins (cutans) coating and connecting grains of quartz sand. The white bar is 2 mm long. (Soil Conservation Service.)

SOIL HORIZONS

Soil horizons range greatly in thickness and distinctness. Figure 23.8 lists and describes the most important soil horizons. In some cases the upper and lower boundaries of a horizon are sharply defined; in others, the boundaries are gradual or diffuse. (Examine the photographs of soil profiles in Figure 24.4.) In general, soil horizons are of two classes: organic horizons and mineral horizons.

Organic horizons, designated by the capital letter O, overlie the mineral horizons and are formed of accumulations of organic matter derived from plants and animals (Figure 23.8). Typically, the uppermost organic horizon, designated as O_i, consists of vegetative matter in original forms recognizable with the unaided eye. Beneath the O_i horizon lies the O_a horizon, which consists of altered remains of parts of plants and animals not recognizable with the naked eye. Material of the O_a horizon is referred to as *humus;* it consists largely of plant tissues partly oxidized by consumer organisms (Chapter 25). The process by which the O_a horizon is produced has been called *humification*.

Mineral horizons consist predominantly of inorganic mineral matter, of which two basic groups are recognized: (1) skeletal minerals and (2) clay minerals and related weathering products. The *skeletal minerals*, mostly in the form of particles of sand and silt grades, comprise the bulk of most soils. Skeletal materials may consist of individual grains of a single mineral—quartz, for example—or grains that are aggregates of several minerals. The clay minerals and related alteration products are those described in Chapter 12. They form the

mineral soil fraction most important in soil-forming processes, in the development of horizons, and in determining the natural fertility of soils. The clay minerals have special physical and chemical properties because of their colloidal size and because of the platelike shape of the individual clay particles.

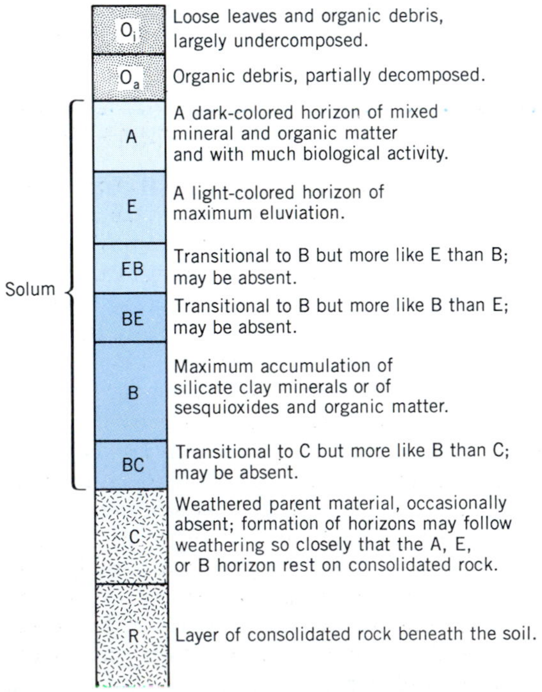

Mineral horizons are designated by the letters A, E, and B, with various subdivisions designated by adding subscript lowercase letters. Our emphasis in this description of the properties of the A, E, and B horizons is on soils of moist climates formed under forest cover. Our description would not apply closely to soils of semiarid and arid climates. Mineral horizons have less than 20 percent organic matter when no clay is present, less than 30 percent organic matter when the mineral fraction consists of 50 percent or more clay.

The *A horizon* is usually rich in finely divided organic matter and is therefore usually darker than the E horizon below. The *E horizon* is characterized by loss of clay minerals and of oxides of iron and aluminum. A concentration of quartz grains of sand or coarse silt grade usually remains, and the horizon is often pale.

The *B horizon* typically shows a gain of mineral matter, which may come from the A and E horizons above. High concentrations of clay minerals, oxides of iron and aluminum, and organic matter (humus) are often found in the B horizon. Thus the B horizon is typically less friable than the A horizons; it may be dense and tough, and cementation may also occur.

The *C horizon*, beneath the B horizon, is a mineral layer of regolith or sediment (but not bedrock), little affected by biologic activity. The C horizon is not part of the soil solum and is described simply as the layer of parent material. The C horizon is, however, affected by physical and chemical processes. An example is the accumulation of calcium carbonate in dry climates, which causes cementation in some soils. In these environments, the C horizon may also show accumulations of silica or soluble salts. Bedrock underlying the C horizon (or the B horizon when the C horizon is missing) is designated the *R horizon*.

To indicate special characteristics of a soil horizon, certain lowercase letters are used after the capital letters A, E, or B. A few examples are listed below. (We will not use these symbols on later pages, but will describe the horizon in words and apply a name to it.)

b	Buried horizon
f	Soil frozen
h	Accumulation of humus
k	Accumulation of carbonate matter
p	Horizon disturbed by plowing
q	Accumulation of silica
t	Accumulation of translocated clay
x	Brittle layer (fragipan)

FIGURE 23.8 Designations of horizons of a hypothetical soil profile that might represent a forest soil in a cool, moist climate. (Soil Conservation Service, U.S. Department of Agriculture.)

THE SOIL SOLUTION

Both air and water combine to form the *soil solution*, which comprises the environment for chemical reactions affecting the solid fraction of the soil. The soil atmosphere consists basically of air that enters pore spaces in the soil, diffusing into all interconnected openings. Fluctuations in barometric pressure are believed capable of inducing soil air to move alternately inward and outward, resulting in some degree of circulation.

Three of the atmospheric gases present in soil air play active roles in soil processes: molecular oxygen (O_2), molecular nitrogen (N_2), and carbon dioxide (CO_2). These active roles require that the gases be dissolved in water. As gas molecules dissolved in soil water, neither nitrogen nor oxygen is directly involved in chemical reactions affecting clay minerals and carbonate minerals in the soil. Carbon dioxide, on the other hand, is of major importance in direct reactions because it combines with soil water to form a weak solution of carbonic acid. Complex organic acids, produced during the decomposition of organic matter, are also important reagents in the soil solution. An acid serves as the active agent in attacking the bonded atoms of the crystal structure of clay minerals.

IONS

To understand chemical activity in the soil solution, we must refer to the activity of ions. An *ion* is an atom or group of atoms bearing an electrical charge. When certain compounds are dissolved in water, the atoms become separated as ions. For example, ordinary table salt is the compound sodium chloride, consisting of atoms of sodium (Na) and atoms of chlorine (Cl) in a one-to-one ratio. Thus the chemical formula for sodium chloride is NaCl. In its solid state, sodium chloride is a crystalline substance formed of atoms of sodium bonded weakly to adjacent atoms of chlorine. When placed in water, NaCl dissolves, meaning that the Na and Cl atoms separate and move freely among the water molecules. Separation from the crystal structure results in a sodium atom having a single positive electrical charge; it is now an ion, indicated by the symbol Na^+. A chlorine atom assumes a negative charge, becoming a chlorine ion: Cl^-.

Chemists refer to a positively charged ion as a *cation*, to a negatively charged ion as an *anion*. Some kinds of ions consist of two different kinds of atoms joined together. For example, the ammonium ion consists of one atom of nitrogen (N) joined with four atoms of hydrogen (H) with the formula NH_4^+; it is a cation. The sulfate ion is another example, with one atom of sulfur (S) joined to four atoms of oxygen (O), with the formula SO_4^{--}; it is an anion. Notice that the sulfate ion has two negative charges. Some kinds of ions have a single charge, others a double charge, and others a triple charge.

Ions important in soils include the following:

Cations

H^+	Hydrogen
Al^{+++}	Aluminum
$Al(OH)^{++}$	Hydroxyl aluminum
Ca^{++}	Calcium
Mg^{++}	Magnesium
K^+	Potassium
Na^+	Sodium
NH_4^+	Ammonium

Anions

Cl^-	Chlorine
SO_4^{--}	Sulfate
OH^-	Hydroxide
HCO_3^-	Bicarbonate
NO_3^-	Nitrate

The soil solution contains several varieties of ions derived from precipitation. Rainwater holds sea salts, suspended mineral particles, and various pollutants. When carried down to earth as rain, sea salts contribute all the ions that are present in seawater. Most of these ions are chlorine (Cl^-) and sodium (Na^+). Ions of magnesium (Mg^{++}), sulfate (SO_4^{--}), calcium (Ca^{++}), and potassium (K^+) are contributed in minor amounts. Mineral dusts lifted from the ground into the atmosphere and carried upward in turbulent wind account for many of the potassium, calcium, and magnesium ions found in rainwater.

Sulfate ions (SO_4^{--}) are usually present in rainwater. They come largely from sulfate particles and gaseous sulfur compounds injected into the atmosphere by fossil-fuel combustion, forest fires, volcanoes, and biologic activity. Also present are nitrate ions (NO_3^-) and ammonium ions (NH_4^+) produced from gaseous forms of nitrogen introduced into the atmosphere from a variety of sources, including the combustion of fuels, the decay of organic matter, and fertilizers. Phosphate ions (PO_4^{---}) are also present in rainwater, but in much smaller amounts than ammonium and nitrate.

An important point about the pollutant ions is that they may form acids in the soil solution. Sulfate ions form sulfuric acid (H_2SO_4); nitrate ions form nitric acid (HNO_3). In certain regions over which washout of heavily polluted air occurs, these acids cause many undesirable impacts on biologic activity and may to some extent be causing changes in the normal soil-forming processes.

SOIL COLLOIDS AND CATION EXCHANGE

Clay mineral particles of colloidal dimensions are chemically active in the soil because of their great surface area. A clay particle can be pictured as a very thin, platelike object with flat, parallel upper and lower surfaces. (Figure 12.10 shows electron microscope photographs of colloidal clay particles.) The crystalline structure of the clay minerals is such that the atoms are arranged in systematic repeating geometric patterns, called *crystal lattices*. (The same is true of all crystalline minerals.) For the clay minerals, the lattice structure takes the form of flat, parallel *lattice layers* of extreme thinness. For this reason the clay minerals are referred to as *layer silicates*.

The chemical bonds that hold together the atoms within each lattice layer are strong, whereas the bonds between layers are weak. Because of this structure, water molecules and various free ions can penetrate between the layers of the clay mineral, leading to its chemical alteration and to its physical disruption.

The lattice layer structure of clay minerals is such that oxygen atoms, which are negatively charged, are nearest the upper and lower surfaces. This condition is indicated in Figure 23.9 by minus signs on the clay parti-

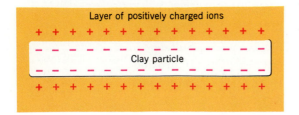

FIGURE 23.9 Schematic diagram of a thin, flat colloidal particle with negative surface charges and a layer of positively charged ions (cations) held to the surface.

cle. As a result, positive ions, or cations, will be attracted to the clay particle surface and held there by electrostatic attraction. Cations of hydrogen (H^+), aluminum (Al^{+++}), sodium (Na^+), potassium (K^+), calcium (Ca^{++}), and magnesium (Mg^{++}) are commonly present in soil solutions, and all may be found on clay particle surfaces. In many soil reactions, these cations replace one another in the process of *cation exchange.*

Cation exchange is governed by a replacement order, indicating which ion is capable of replacing another. This is a kind of seniority system in which the ion of a given rank can take over the position of ions of lower seniority. The aluminum ion can displace any of the other metallic ions, so it occupies the top position on the list. In order of replacement ability, cations of hydroxyl aluminum, calcium, magnesium, potassium, and sodium follow.

The capacity of a given quantity of soil to hold and exchange cations is called the *cation-exchange capacity (CEC)* and is a general indicator of the degree of chemical activity of a soil. Capacity is indicated in a unit known as the *milliequivalent,* which is a measure of the ratio of weight of ions to weight of soil. The exact definition of this unit is not important here, but its relative magnitude does concern us. The exchange capacities of various soil colloids are given in Table 23.2. Among the clay minerals, both vermiculite and montmorillonite have high CEC. Illite has intermediate CEC, while kaolinite has a low value. Even lower is the CEC of the sesquioxides of iron and aluminum. Particles of feldspar and quartz have nearly zero CEC.

The continued chemical weathering of a soil over a very long period of time tends to bring a shift in clay mineral composition. In earlier stages of weathering, the

TABLE 23.2 Cation-Exchange Capacities of Various Soil Colloids

Material	CEC
Organic matter (humus colloids)	150–500
Clay minerals	
Vermiculite	100–150
Montmorillonite	80–150
Illite	10–40
Kaolinite	3–15
Hydrous sesquioxides of aluminum and iron	4
Feldspar, quartz	1–2

content of high CEC minerals—vermiculite and montmorillonite—may be comparatively large, so the soil as a whole has a high CEC. In advanced stages, the minerals of high CEC are removed or altered in favor of minerals of low CEC—kaolinite and the sesquioxides of aluminum and iron. As a result, total CEC gradually declines to low values. Generally, when CEC falls below 10 in the B horizon the soil is classed as having low CEC. Soils with high CEC usually have a high capacity to store plant nutrients (base cations) and are potentially fertile soils (if the soil is not strongly acid in chemical balance).

SOIL ACIDITY AND ALKALINITY

The various soil cations capable of being readily exchanged on colloidal particles belong to two general classes. One class, important plant nutrients, consists of the *base cations* (or simply, *bases*). The base cations most important in soils are the following:

Calcium	Ca^{++}
Magnesium	Mg^{++}
Potassium	K^+
Sodium	Na^+

When base cations comprise the large majority of cations held by soil colloids, the soil is in a condition described as *alkaline.*

The other class consists of *acid-generating cations.* Three acid-generating cations are important in soils. One is the aluminum ion, Al^{+++}; it is associated with extreme acidity. The second is the hydroxyl aluminum ion, $Al(OH)^{++}$, associated with a moderate degree of acidity. The third is the hydrogen ion, H^+, forming about 10 percent of the acid-generating ions in acid soils. These ions must be exchangeable, that is, free to change places with other ions on the surfaces of colloids. Usually present are many Al^{+++} and H^+ ions too tightly bound to be readily exchangeable. A soil is described as *acid* when the total numbers of readily exchangeable acid-generating cations comprise from 5 to 60 percent of the total cation exchange capacity; the larger the percentage, the greater the degree of acidity.

The range of alkalinity or acidity of a soil is measured in terms of a number known as the *pH* of the soil solution. (The pH is a measure of the concentration of hydrogen ions; it is the logarithm to the base 10 of the reciprocal of the weight in grams of hydrogen ions per liter of water. Consequently, the smaller the pH number, the greater the hydrogen ion concentration). A pH of 7.0 is neutral in this scale; values below 5 represent a strongly acid soil solution; values above 10 represent a strongly alkaline soil solution. Table 23.3 shows a classification of soils according to acidity and alkalinity. For agricultural soils, this quality is very important because certain crops require near-neutral values of pH and cannot thrive on acid soils. Plants differ considerably in their preference for soil acidity or alkalinity, and this is an important factor in the distribution of plant types.

As Table 23.3 shows, agricultural soils with pH under about 6 require the application of lime for the successful

TABLE 23.3 Soil Acidity and Alkalinity

pH	4.0	4.5	5.0	5.5	6.0	6.5	6.7	7.0	8.0	9.0	10.0	11.0
Acidity	Very strongly acid		Strongly acid	Moderately acid	Slightly acid			Neutral	Weakly alkaline	Alkaline	Strongly alkaline	Excessively alkaline
Lime Requirements	Lime needed except for crops requiring acid soil		Lime needed for all but acid-tolerant crops		Lime generally not required			No lime needed				
Occurrence	Rare	Frequent	Very common in cultivated soils of humid climates						Common in sub-humid and arid climates		Limited areas in deserts	

Data source: C. E. Millar, L. M. Turk, and H. D. Foth, *Fundamentals of Soil Science,* 3rd edition, John Wiley & Sons, New York. See Chart 4.

cultivation of many crops. *Lime,* as the word is used in agriculture, may be either calcium oxide (CaO) or calcium carbonate ($CaCO_3$). Nearly all lime used in agriculture is natural limestone of calcium carbonate composition. It is ground into powder and spread over fields. Lime eliminates acid-producing ions from soil colloids, replacing them with calcium ions. After the soil pH has been raised to the desired level, fertilizers rich in nutrients (nitrogen, phosphorus, potassium) must also be added to the soil because these nutrients are deficient in most acid soils.

BASE STATUS OF SOILS

As in many forms of human society, soils are stratified into "status" levels and are classified into major groups on that basis. Status in soils is determined by the *percentage base saturation (PBS),* defined as the percentage of exchangeable base cations with respect to the total cation exchange capacity of the soil. A value of 35 percent has been used by soil scientists as a dividing number separating one class of soils as *high base status* (PBS greater than 35%) from those of another class of *low base status* (PBS less than 35%). Soils of high base status have high natural fertility for food crops; those of low base status are naturally low in fertility and require special treatment and the application of chemicals to correct the deficiency. *Base status of soils* thus has enormous impact on human food resources and on the possibilities for future expansion of agricultural food production into areas not now under cultivation.

SOIL-TEMPERATURE REGIMES

Soil temperature, which we introduced in Chapter 4, is an important factor in determining the characteristics of a soil. Temperature acts as a control over biologic activity and influences the intensity of chemical processes affecting the clay minerals. Below the freezing point, 0°C (32°F), there is no biologic activity; chemical processes affecting minerals are inactive. Between 0°C and 5°C (32 and 41°F), root growth of most plants and germination

of most seeds is impossible; but water can move through the soil and carbonic acid activity may be important. A horizon as cold as 5°C (41°F) acts as a thermal barrier to the roots of most plants. The germination of seeds of many low-latitude plants requires a soil temperature of 24°C (75°F) or higher.

At any moment the temperature within a soil varies from horizon to horizon. The temperature near the surface experiences both a diurnal and an annual cycle (Chapter 4). The range of these cycles may be small or large according to latitude and the degree of continentality in the thermal regime. (See Chapter 8, Thermal Regimes, and Figure 8.2). The annual cycle will be almost imperceptible in the wet equatorial climate (1), but very large in moist continental (10) and boreal forest (11) climates.

Each pedon has a characteristic *soil-temperature regime* that can be measured and described. For purposes of classifying soils, the temperature regime can be described by the mean annual soil temperature and the average seasonal fluctuations from that mean.

Each pedon has a mean annual temperature that is essentially the same in all horizons at all depths. Some representative examples are as follows:

Location	Mean Annual Soil Temperature	
	°C	(°F)
Irkutsk, U.S.S.R.	−2	(28)
Bozeman, Montana	6	(43)
Urbana, Illinois	10	(50)
Colombo, Sri Lanka	27	(81)

Table 23.4 gives the names and temperature boundaries of six soil temperature regimes. Two criteria are specified: (1) Mean annual temperature, T, and (2) difference between mean of summer (or warm season) and mean of winter (or cold season), Ts − Tw, at a depth of 50 cm (20 in.). (The notation $0° < T < 8°$ is read "mean annual temperature greater than 0°C but less than 8°C.")

The soil temperature regimes are used as a means of classifying certain subclasses of soils. In so doing, a pre-

TABLE 23.4 Soil Temperature Regimes

Name of Regime	Mean Annual Soil Temperature, °C (T)	Difference Between Mean Temperature (°C) of Warm Season and of Cold Season (Ts − Tw)
Pergelic	T < 0°	—
Cryic	0° < T < 8°	—
Frigid	T < 8°	> 5°
Mesic	8° < T < 15°	> 5°
Thermic	15° < T < 22°	> 5°
Hyperthermic	T < 22°	> 5°

Equivalents:	°C	°F	5°C = 9°F
	0	32	
	8	47	
	15	59	
	22	72	

Data source: Soil Survey Staff, *Soil Taxonomy*, Agriculture Handbook No. 436, Government Printing Office, Washington, D.C., pp. 62–63.

fix is added to the soil name as follows:

Regime	Prefix used
Cryic	Cry-, Cryo-
Frigid	Bor-
Thermic	Trop-

We will make use of these prefixes in Chapter 24.

SOIL-WATER REGIMES

Soil-water regimes, explained in Chapter 9, are used by American soil scientists in the modern system of soil classification. (In publications of the Soil Survey Staff of the U.S. Department of Agriculture, "soil moisture" is used instead of "soil water.") The Thornthwaite method of calculating the soil-water budget has been adopted, with minor simplifications, for the purpose of estimating the actual soil-water conditions. These estimates do not apply under special conditions. For example, in a dry climate the soil-water budget calls for virtually zero soil-water storage in all months. In some low-lying places in deserts, however, the ground-water table is close to the surface, supplying the roots of plants with adequate water during much of the year.

An important criterion in determining the soil-water regime is the distinction between a dry soil and a moist soil. As a soil loses water by evaporation, the remaining films of capillary water clinging to grain contacts shrink. As they shrink, the films set up increasing resistance to being drawn up into the tiny rootlets of plants. The measure of this resistance is the tension (force) of the water films in units of *bars*. One bar is about equal to the standard pressure of the atmosphere (approximately 1000 millibars). When the drying out of the soil reaches a point that the soil-water tension is 15 bars or greater, the remaining soil water is considered to be unavailable to plants. Thus 15 bars is the dividing point between a dry condition and a moist condition.

The definition of dry and moist conditions is applied to the *soil-water control section*. The upper boundary of this control section is defined as the depth to which 2.5 cm (1 in.) of water, applied to the surface of a dry soil, will moisten the soil. "Moisten" means to cause the tension to fall to a value below 15 bars. The lower boundary of the control section is the depth to which the application of 7.5 cm (3 in.) of water at the soil surface will moisten a dry soil. In soils of coarse loam grade the soil-water control section has a depth of 20 cm (8 in.) at the upper boundary and 60 cm (24 in.) at the lower boundary. The depth is strongly influenced by soil texture class, being shallower for fine textures and deeper for coarse textures.

Five soil-water regimes are recognized and given names under the modern U.S. system of soil classification:

Aquic regime (L. *aqua*, "water"). The soil is saturated with water most of the time because the ground-water table lies at or close to the surface throughout most of the year. The aquic regime is found in bogs, marshes, and swamps. It occurs independently of the various soil-water regimes associated with well-drained upland locations.

Udic regime (L. *udus*, "humid"). The soil is not dry in any part of the profile as long as 90 days per year. The udic regime is found in the moist climates (6h, 8h, 10h, 11h), as we have defined them in Chapter 9. The soil-water budget shows little or no water deficiency (D) in the growing season (summer), and there is a seasonal water surplus that causes water to move through the soil at some part of the year. For soil-water regimes with a water surplus in all months (Climates 1, 2, 6p, 8p, 10p, 11p) the name is modified to *perudic*. Under a perudic regime, even in the dry season moisture tension rarely becomes as great as 1 bar.

Ustic regime (L. *ustus*, "burnt," implying dryness). The soil has a moderate quantity of water in storage during the season when conditions are favorable to plant growth and the soil water is not frozen, but the soil is dry for 90 or more cumulative days in most years. The ustic regime is associated with the semiarid subtype(s) of the dry climates (4s, 5s, 9s) and with the tropical wet-dry climate (3).

Aridic (torric) regime (L. *aridus*, "dry," and *torridus*, "hot and dry"). The two names for this regime are used in different categories of the soil classification system. For warm soils the soil is never moist in some or all parts of the profile for as long as 90 consecutive days. The aridic (torric) regime applies to the semidesert (sd) and desert (d) subtypes of the dry climates (4, 5, 9), in a wide latitude range from tropical zone to midlatitudes.

Xeric regime (Gr. *xeros*, "dry"). This regime applies to areas of Mediterranean climate (7s, 7sd, 7sh) with a long, dry summer and a rainy winter. The soil is dry in all parts of the profile for 45 or more consecutive days in the dry summer season, but moist in all parts for 45 or more consecutive days in the moist winter season.

In the modern soil classification system, prefixes taken from the names of the soil-water regimes are used to designate one of the major classes of soils and many of the subclasses. The prefixes are as follows:

Regime name	Prefix
Aquic	Aqu-
Udic	Ud-
Ustic	Ust-
Aridic	Aridi-
Torric	Torr-
Xeric	Xer-

The soil-water regime is a powerful controlling factor in determining the properties of soils and the nature of soil-forming processes. The quantity of soil water in storage determines the rate of primary production by plants and general organic activity within the soil. In the udic regime, water moving through the soil is capable of removing ions (leaching). In the ustic regime, limited water storage is associated with the accumulation of calcium carbonate. In the aridic regime, soluble salts tend to accumulate in the soil. In the aquic regime, water saturation prevents oxygen from entering the soil. It would not be an understatement to say that the combined influence of soil-temperature regime and soil-water regime shapes almost the entire complex of chemical and biological properties of a soil, assuming only that the parent matter originally contained a wide spectrum of silicate minerals.

LANDFORM AND SOIL

Configuration of the ground surface as a factor in soil formation can be included in the single word, *landform*. Landform includes the varying steepness of the ground surface, or *slope*, as well as the compass orientation—the *aspect*—of an element of ground surface. Another landform property is the *relief*, or average elevation difference between adjacent high and low points (hilltops versus valley bottoms). Strong relief and steep slopes are combined in many areas of hills and mountains. Small relief and gentle slopes are combined in low plains and flat uplands.

Figure 23.10 shows how landform influences the thickness of A, E, and B horizons of the soil solum. Taking the profile in an undulating upland surface (left) as the normal condition, the flat upland shows thickened horizons because removal of the soil surface by erosion is much slower on very gentle slopes. In the hilly region with steep slopes, erosion removes the upper horizons more rapidly and the profile is reduced in thickness. The soil profile beneath the poorly drained meadow is entirely different in character. The combined A and B horizons are rich in organic matter, and a gray layer (glei horizon) is formed beneath the organic layer, where oxygen is deficient. In the adjacent bog a thick upper layer of peat is usually present.

Aspect of the ground surface influences both the soil temperature and the soil-water regimes. In midlatitudes, where the sun's rays strike at an intermediate angle between the horizon and zenith, slopes facing north (northern hemisphere) receive greatly reduced insolation. Here soil temperatures are cooler than average, and the soil-water regime is pushed in the direction of a moister climate.

BIOLOGIC PROCESSES IN SOIL FORMATION

The total role of biologic processes in soil formation includes the presence and activities of living plants and animals as well as their nonliving organic products. Living plants contribute to soil formation in two basic

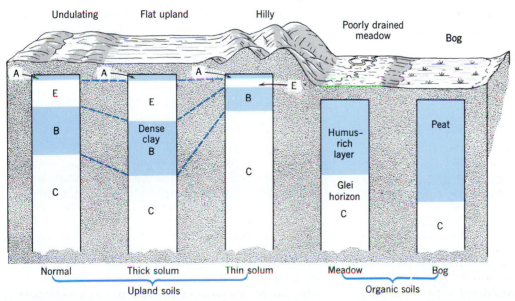

FIGURE 23.10 Relief and slope strongly influence the thickness and composition of the soil profile. (After U.S. Dept. of Agriculture, *Yearbook of American Agriculture*.)

ways. First is the production of organic matter—the biomass—both above the soil as stems and leaves and within the soil as roots. This primary production, discussed in Chapter 25, provides the raw material of organic matter in the O horizon and in lower horizons. The decomposer organisms process this raw material, reducing it to humus and ultimately to its initial components—carbon dioxide and water. Second is the recycling of nutrients from the soil to plant structures above ground and their return to the soil in dead plant tissues. Nutrient recycling is a mechanism by which nutrients are prevented from escaping through the leaching action of surplus soil water moving downward through the soil. In our discussion of the global soil types, Chapter 24, this process will receive special attention.

Animals living in the soil, or entering and leaving the soil by means of excavated passageways, span a wide range in species and individual dimensions. The total roles of animals in biologic processes of soil cannot be overestimated in soils with sufficient heat and moisture to support large animal populations. For example, earthworms continually rework the soil not only by burrowing, but also by passing the soil through their intestinal tracts. They ingest large amounts of decaying leaf matter, carrying it down from the surface and incorporating it into the mineral soil horizons. The granular structure of the darkened A horizon derives its quality from this activity. Many forms of insect larvae perform a similar function. Small tubular soil openings are also formed by many species of burrowing insects. Large openings are made by larger animals—moles, gophers, rabbits, badgers, prairie dogs, and many other species. The growth of roots followed by decay leaves tubular openings in the soil.

In moist climates, the evolution of a soil from parent mineral matter is accompanied by increasing plant growth and changes in plant species. We will examine this evolution process—called plant succession—in Chapter 26. Various close relationships between ecosystems and soil characteristics will become evident in descriptions of the widely varied classes of soils.

Human activity is also a potent agent in influencing the physical and chemical nature of the soil. Large areas of agricultural soils have been tilled and fertilized for centuries. Both structure and composition of these agricultural soils have undergone profound changes and can now be recognized as distinct soil classes of importance equal to natural soils. The modern system of soil classification presented in the next chapter includes soil types produced by or greatly modified by human activities.

REVIEW OF BASIC PEDOGENIC PROCESSES

Facts and concepts developed in this chapter have prepared the way for a brief review of several basic soil-forming processes, or *pedogenic processes*.

Pedogenic processes can be classified into four groups: (1) addition of material to the soil body; (2) losses from the soil body; (3) translocations of materials within the soil body; (4) transformations of material within the soil body.

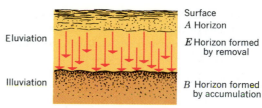

FIGURE 23.11 Eluviation and illuviation, leading to the formation of E and B horizons.

Additions of materials to the soil body are covered by the general term *soil enrichment*. Inorganic enrichment can come as sediment added to the soil surface by overland flow (colluvium) or overbank stream flooding (alluvium), and by wind (loess or volcanic ash). Another form of enrichment is from the organic litter of plants growing in the soil. This material accumulates in the O horizon and produces finely divided humus that is carried down into the mineral horizons.

Losses of material from the soil body include removal of surface material by soil erosion and by *leaching*, the

FIGURE 23.12 A thin layer of wind-deposited silt and dune sand (pale brown layer at top) has buried a Spodosol profile on outer Cape Cod. The pale grayish E horizon overlies a reddish B horizon, or spodic horizon. (A. N. Strahler.)

downward washing out and removal by surplus soil water percolating through the soil.

Translocation of materials within the soil takes place in a number of quite different ways, each with a different cause and often uniquely related to a particular soil-water regime. Two simultaneous processes of downward translocation are eluviation and illuviation, typical of the udic soil-water regime. *Eluviation* consists of the downward transport of fine particles, particularly the colloids (both mineral and organic), carrying them out of an uppermost, or A, mineral soil horizon, as shown in Figure 23.11. Eluviation leaves behind coarse skeletal mineral grains, forming the E horizon. In cool moist climates, the E horizon contains largely quartz in sand or coarse silt grade sizes. *Silication* is a term applied to this increase in proportion of silica because it remains behind while other materials are removed.

Illuviation is the accumulation of materials in a lower horizon, brought down from a higher horizon (Figure 23.11). Typically, illuviation forms the B horizon. The materials that accumulate may be clay particles, organic particles (humus), or sesquioxides of iron and aluminum.

The effects of both eluviation and illuviation can be seen most vividly in exposures of sandy soils in the cool, moist climate of New England and southeastern Canada (Figure 23.12).

The translocation of calcium carbonate is another important process. Removal of calcium carbonate, or *decalcification*, takes place as carbonic acid reacts with carbonate mineral matter. The soluble products are carried down into a lower horizon. Their accumulation constitutes *calcification;* it may take place in the B horizon or in the C horizon below the soil solum. Precipitation of soluble salts and the reverse process of removal of salts are *salinization* and *desalinization*, respectively.

Transformations within the soil body affect both inorganic and organic materials. Decomposition of primary to secondary minerals is one such transformation, already described in detail in Chapter 12. By synthesis, new minerals and organic compounds can be formed from the products of decomposition. Decomposition also affects organic materials. Humification, the process of transformation of plant tissues into humus, may be followed by total disappearance of the organic matter as water and carbon dioxide by respiration.

Horizonation, the degree of development of soil horizons, is the result of complex combinations of the pedogenic processes already listed. In the next chapter, as each class and subclass of soil is described, we will have an opportunity to explain how horizons vary from place to place under the complex interaction of the various soil-forming factors—parent material, temperature and soil-water regimes, biologic activity, and time.

PEDOLOGY AS A COMPLEX SCIENCE

Our scientific approach to pedology began with a description of the materials—mineral and organic—of which the true soil layer, or solum, is formed. We followed this with an examination of the structuring of these components into a soil profile. Next came a review of the chemistry of soils, with emphasis on the role of ions in the soil solution. Then climate was introduced through the influences of temperature and the soil–water budget, and geomorphology through the influence of landform. Finally, all these topics came together in an overview of the soil-forming processes—a dynamic view stressing changes that occur in soils through time.

A GEOGRAPHY OF SOILS

What we have yet to accomplish in the next chapter is to classify all of the known varieties of soils of the globe into kinds or classes, giving them a set of names. The final effort will be to do what strongly interests geographers, which is to plot the distribution of the kinds of soils over the entire land surface of the globe, and then to associate the patterns of that distribution with climate patterns. Fortunately, the system of climate we have stressed—one based on the soil-water balance—meshes beautifully with the global system of soil types. Two special fields of science—hydrology and pedology—now come together in our next chapter.

C H A P T E R 24

WORLD SOILS

From the standpoint of physical geography, the most important aspect of pedology is the classification of soils into major classes and subclasses recognized in terms of their areal distribution over the earth's land surfaces. Geographers are particularly interested in the ways climate, parent materials, and landform are linked with the distribution of types of soils. Geographers are also interested in the kinds of natural vegetation associated with each of the major soil types. The geography of soils is thus an essential ingredient in establishing the quality of environmental regions of the globe—important because soil fertility, along with availability of fresh water, is the basic measure of the potential of an environmental region to produce food for the human race.

EARLIER SOIL CLASSIFICATIONS

The founder of modern theories of soil origin and classification was V. V. Dokuchaiev, a Russian geologist. His studies between 1882 and 1900 led him to the concept that soil is an independent body whose character is determined primarily by climate and vegetation. A Russian follower of Dokuchaiev, K. D. Glinka, expanded the concepts of horizons in the soil profile.

For the development of modern soil science in the United States during the 1920s and 1930s, much of the credit rests with C. F. Marbut, who served for many years as chief of the Soil Survey Division of the U.S. Department of Agriculture (USDA). Marbut became acquainted with Russian pedological views, adapted them to conditions in the United States, and created a comprehensive system of soil classification. Built on his work, but also departing from it, was the 1938 USDA system, used with a number of modifications for the next 25 years.

THE COMPREHENSIVE SOIL CLASSIFICATION SYSTEM

In the early 1950s, a United States national cooperative effort was launched to develop a completely new scheme of soil classification. Soil scientists of the Soil Conserva-

tion Service, the faculties of land-grant universities, and many soil scientists of other nations participated in the new developments. After progressing through a succession of stages over a period of several years, the new scheme was ready for presentation by American pedologists to the Seventh International Congress of Soil Science in 1960. Known at the time as the Seventh Approximation (because it was the seventh in the series of revisions), the new system was prepared by the Soil Survey Staff of the Soil Conservation Service and was published in 1960.

Many modifications and refinements were made in the decade that followed. The system was named the *Comprehensive Soil Classification System* (CSCS). The system defines its classes strictly in terms of the morphology and composition of the soils, that is, in terms of the soil characteristics themselves. Moreover, the definitions are made as nearly quantitative as possible. Every effort was made to use definitions in terms of features that can be observed or inferred so that subjective or arbitrary decisions as to classification of a given soil will be avoided.

The CSCS recognizes and gives equal importance to classes of soils deriving their characteristics from human activities, such as long-continued cultivation and applications of lime and fertilizers and the accumulation of agricultural wastes. Because human occupancy and agricultural exploitation of large expanses of soils have gone on for centuries in various parts of the world, recognition of such modified soils is desirable and realistic in a classification system.

The CSCS terminology uses a large number of newly coined words. The new terms give the system a major advantage because the syllables were selected to convey precisely the intended meaning with respect to the properties or genetic factors relating to the soil class.

THE SOIL TAXONOMY

The classification system of the CSCS is known as the *Soil Taxonomy*; it is based on a hierarchy of six categories, or levels, of classification. They are listed here with the numbers of classes recognized within each category:

Orders	10
Suborders	47
Great groups	185
Subgroups	1000 (approx.)
Families	5000 (approx.)
Series	10,000 (approx.)

Numbers given for the lowest three categories refer only to soils of the United States. When soils of all continents are eventually included, the numbers of at least the two lowest categories will be greatly increased. On the other hand, numbers given for the first three categories are expected to remain approximately the same because they are designed to provide for classification of all known soils.

We will concentrate on the uppermost two levels of the soil taxonomy—orders and suborders—for the purpose of understanding the major differences of soils from region to region. We emphasize those units of greatest geographic extent and of greatest importance in understanding the patterns of natural vegetation—forest, grasslands, desert—and the ways major soil units reflect the varying intensity of the soil-forming processes we studied in Chapter 23.

DIAGNOSTIC HORIZONS FOR CLASSIFICATION

To understand how the 10 soil orders are differentiated requires a knowledge of several *diagnostic horizons*. The diagnostic horizons are freshly named and rigorously defined as an integral part of the CSCS; they are not simply relics of older soil classification systems.

The basic definition of a soil horizon is the same as given in Chapter 23, but requires precise restatement when applied to the CSCS. A *soil horizon* is a layer that is approximately parallel with the soil surface and has a set of properties that have been produced by soil-forming processes. The properties of a soil horizon are not like those of the layer above or below it. A soil horizon is differentiated from those adjacent to it partly by characteristics that can be seen or measured in the field, such as color, structure, texture, consistence, and the presence or absence of carbonates. In some cases, however, laboratory measurements are required to supplement field observations.

We consider first the diagnostic horizons of mineral soils—those with only a small proportion (less than 20%) of organic carbon by weight. Diagnostic criteria of organic soils are separately considered; they define the kind of organic material present.

Soil horizons fall into two major groups: epipedons and subsurface diagnostic horizons. An *epipedon* (Gr. *epi*, "over" or "upon") is simply a horizon that forms at the surface. It is not, however, identical in meaning with the A horizon, which we defined in Chapter 23. The subsurface horizons originate below the soil surface. In many cases they are identified as parts of the A horizon or as the E or B horizon. The brief descriptions of selected diagnostic horizons that follow are intended to give a general picture of the horizons rather than to supply rigorous definitions; exact definitions, required in soil classification, are extremely detailed and precise and are often lengthy.

Epipedons

Mollic epipedon (L. *mollis*, "soft"). A relatively thick, dark-colored surface horizon. The dark color is due to presence of organic matter (humus) derived from roots or carried underground by animals. The horizon is usually rich in base cations—calcium, magnesium, and potassium—so the base saturation (PBS) is over 50 percent and often much higher. Structure is usually granular or blocky, with the peds loosely organized when the soil is dry.

Umbric epipedon (L. *umbra*, "shade," hence "dark"). A dark surface horizon resembling the mollic epipedon, but with PBS less than 50 percent.

Histic epipedon (Gr. *histos*, "tissue"). A thin surface horizon of peat, usually formed in a wet place. The horizon is saturated with water for 30 consecutive days or more during the year (unless the soil has been artificially drained). Very thick accumulations of peat are associated with organic soils; they are not classed as epipedons.

Ochric epipedon (Gr. *ochros*, "pale"). A surface horizon that is light in color and contains less than 1 percent organic matter. Also included are surface horizons too thin, too dry, or too hard to qualify as any one of the previously listed epipedons.

Plaggen epipedon (Ger. *plaggen*, "sod"). A human-made surface layer greater than 50 cm (20 in.) thick. It has been produced by long-continued manuring, incorporating sod or other livestock bedding materials into the soil. This horizon is common in western Europe, but rare elsewhere.

Subsurface Diagnostic Horizons

Argillic horizon (L. *argilla*, "white clay"). An illuvial horizon (usually the B horizon) in which layer-lattice clay minerals have accumulated by illuviation (see Chapter 23). Clay cutans, called "argillans," are usually present. The argillic horizon normally forms beneath an eluvial (E) horizon.

Agric horizon (L. *ager*, "field"). An illuvial horizon formed under cultivation and containing significant amounts of illuvial silt, clay, and humus. Plowing facilitates the down-washing of these materials, which accumulate immediately beneath the plow layer.

Natric horizon (L. *natrium*, "sodium"). Although similar to an argillic horizon, a natric horizon has prismatic structure and a high proportion of Na$^+$, amounting to 15 percent or more of the CEC.

Calcic horizon. A horizon of accumulation of calcium carbonate or magnesium carbonate.

Petrocalcic horizon. A hardened calcic horizon that does not break apart when soaked in water.

Gypsic horizon. A horizon of accumulation of hydrous calcium sulfate (gypsum).

Plinthite (Gr. *plinthos*, "brick"). Iron-rich concentrations, usually in the form of dark red mottles, present in deeper horizons and capable of hardening into rocklike material with repeated wetting and drying.
Salic horizon. Horizon enriched by soluble salts.
Albic horizon (L. *albus*, "white"). A pale, often sandy horizon from which clay and free iron oxides have been removed. Commonly it is the E horizon, overlying a spodic horizon.
Spodic horizon (Gr. *spodos*, "wood ashes"). A horizon containing precipitated amorphous materials composed of organic matter and sesquioxides of aluminum, with or without iron. The spodic horizon is found mostly in quartz sands and is formed partly by illuviation. It normally underlies an E horizon.
Cambic horizon (L. *cambiare*, "to exchange"). An altered horizon with texture as fine as or finer than very fine sand that has lost sesquioxides or bases, including carbonates, through leaching. Although considered a B horizon, it has accumulated little clay, as compared with an argillic horizon; it lacks the dark color and organic-matter content of a histic, a mollic, or an umbric epipedon.
Oxic horizon (Fr. *oxide*, "oxide"). A highly weathered horizon at least 30 cm (12 in.) thick, rich in clays and sesquioxides of low CEC (16 or less). There remain few or no primary minerals capable of releasing base cations. Oxic horizons are very old and seldom found outside equatorial, tropical, and subtropical latitude zones.

Other Horizons or Layers of Diagnostic Value

Duripan (L. *durus*, "hard"). A dense, hard subsurface horizon cemented by silica that does not soften during prolonged soaking.
Fragipan (L. *fragilis*, "brittle"). Dense, moderately brittle layer, often mottled in color. Weak cementation may be due to close packing and binding by clay. A fragipan tends to retard downward movement of water.

Diagnostic Materials of Organic Soils

Fibric soil materials (L. *fibra*, "fiber"). Organic matter consisting of fibers readily identifiable as to botanical origin. An example is *Sphagnum* peat from bogs in cold climates.
Hemic soil materials (Gr. *hemi*, "half"). Organic matter of intermediate degree of composition between fibric and sapric materials.
Sapric soil materials (Gr. *sapros*, "rotten"). Highly decomposed organic matter; denser than fibric and hemic materials, with little content of identifiable fibers.

THE SOIL ORDERS

Every polypedon falls within one, and only one, of the 10 *soil orders*. Each order has its unique criteria, so selected that the criteria for a given order exclude members of all other orders. Criteria may include (1) gross composition, whether organic or mineral or both (e.g., percent clay or percent organic matter); or (2) degree of development of horizons; or (3) presence or absence of certain diagnostic horizons; or (4) degree of weathering of the soil minerals, expressed as cation-exchange capacity (CEC) or as percent base saturation (PBS).

Those soil scientists who devised the Soil Taxonomy were well aware that a polypedon derives its properties from some unique combination of soil-forming processes and influences—climate and living organisms acting on parent materials over time and influenced by landform. A working soil taxonomy cannot, however, be constructed in terms of the soil-forming processes—on a genetic basis, that is—because this information is not always available. Moreover, the origin of a given soil feature is often subject to differences of interpretation in terms of the processes involved. This is why the Soil Taxonomy of the CSCS is based on observable features of the soil profile and additional facts that can be determined by laboratory analysis of soil samples.

As geographers, we cannot rest easily with an alphabetical listing of the 10 soil orders (as is the practice of the Soil Survey Staff). Instead, a natural grouping and arrangement of orders will prove both meaningful and helpful. The three major classes we use are not, however, recognized by the CSCS. A brief descriptive statement introduces each soil order in the following sequence:

Soils with poorly developed horizons or no horizons and capable of further mineral alteration:

Entisols   Soils lacking horizons.

Inceptisols Soils having weakly developed horizons and containing weatherable minerals.

Soils with a large proportion of organic matter.

Histosols Soils with a thick upper layer of organic matter.

Soils with well-developed horizons or with fully weathered minerals, resulting from long-continued adjustment to prevailing soil-temperature and soil-water regimes:

Oxisols   Very old, highly weathered soils of low latitudes, with an oxic horizon and low CEC.

Ultisols   Soils of mesic and warmer soil-temperature regimes, with an argillic horizon and low base status (PBS < 35%).

Vertisols Soils of subtropical and tropical zones with high clay content, developing deep, wide cracks when dry and showing evidence of movement between aggregates.

Alfisols   Soils of humid and subhumid climates, with high base status (PBS > 35%) and an argillic horizon (B horizon).

Spodosols Soils with a spodic horizon (B horizon), an albic horizon (E horizon) with low CEC, and lacking carbonate minerals.

Mollisols   Soils chiefly of midlatitudes, with a mollic epipedon and very high base status, associated with subhumid and semiarid soil-water regimes.

Aridisols   Soils of dry climates, with or without argillic horizons, and with accumulations of carbonates or soluble salts.

The names of the soil orders combine a set of *formative elements* with the syllable *sol*, meaning "soil" (Table 24.1).

Within each soil order are *suborders* ranging in number from as few as two to as many as seven. Suborders are defined in various ways. The basis for defining suborders within one order may be quite different from that used for defining suborders within another order. For example, within some orders the list of suborders begins with soils of wet places and moves in sequence to well-drained and increasingly dry environments. Either soil-temperature regime or soil-water regime, or both, may be used to define suborders within certain orders. In still other orders, suborders are defined in terms of organic content or mineral content. We will select for description here those suborders that are widespread, usually over well-drained upland areas, and having close association with important vegetation classes.

The formative elements used for suborders are given in a table in Appendix V. These elements form prefixes, attached to the formative elements of the order names. For example, a *Boralf* is a suborder of the Alfisols. The element *bor* connotes its occurrence in a cold (boreal) climate; *alf* is the formative element in Alfisol.

Throughout the remainder of this chapter, we shall take up each soil order in turn, giving a description of the order followed by those soil orders that are widespread in occurrence and important to an understanding of native vegetation and agricultural uses. Figure 24.1 is a schematic diagram of the soil orders and major suborders. A world map of soils, Figure 24.2, shows the world distribution of soil orders and a number of the most important suborders. Bear in mind that, over large areas (particularly in low latitudes), the map boundaries are not well substantiated by field observations. For some orders, such as Aridisols, Mollisols, Spodosols, Ultisols, and Oxisols, the world pattern strongly reflects climate controls. A single color is assigned to each of these orders. Within the Alfisols, each of four suborders is represented by a different color. The colors show that each suborder is strongly related to climate and latitude zone. This treatment helps to simplify the map and to accentuate the relationship between soils and climate; it carries forward our objective as geographers to recognize important global environmental patterns.

Figure 24.3 is a map of soil orders and suborders of the United States and southern Canada. In this region, map boundaries are well established by detailed field observations. Note that the legend of this map uses a different system of designation of subtypes from that used in the world map. Figure 24.4 shows typical profiles of 12 selected soil suborders for 8 of the soil orders.

ENTISOLS

Entisols have in common the combination of a mineral soil and the absence of distinct pedogenic horizons that would persist after normal plowing. (The syllable *ent* in Entisol has no root meaning, but can be associated with "recent.") Entisols are soils in the sense that they support plants, but they may be found in any climate and under any vegetation. Lack of distinct horizons in the Entisols may be the result of a parent material, such as quartz sand, in which horizons do not readily form; or it may be the result of lack of time for horizons to form in recent deposits of volcanic ash or alluvium, or on actively eroding slopes, or in soils recently disturbed by plowing to a depth of one meter or more.

Entisols occur throughout the full global range from equatorial to arctic latitude zones. Entisols of the subarctic zone and tropical deserts (along with arctic areas of Inceptisols) are the poorest of all soils from the standpoint of potential agricultural productivity. In contrast, Entisols and Inceptisols of floodplains and deltaic plains in warm and moist climates are among the most highly productive agricultural soils in the world because of their favorable texture, ample nutrient content, and large soil-water storage. Dense agricultural populations in central China and in the Ganges–Brahmaputra plain of India and Bangladesh bear witness to this fact.

Of the five suborders of Entisols, three warrant brief mention. *Fluvents* (*fluv*: "fluvial") are formed on recent stream alluvium. They may be agriculturally productive

TABLE 24.1 Formative Elements in Names of Soil Orders

Name of Order	Formative Element	Derivation of Formative Element	Pronunciation of Formative Element
Entisol	ent	Meaningless syllable	recent
Inceptisol	ept	L. *inceptum*, beginning	inept
Histosol	ist	Gr. *histos*, tissue	histology
Oxisols	ox	Fr. *oxide*, oxide	ox
Ultisol	ult	L. *ultimus*, last	ultimate
Vertisol	ert	L. *verto*, turn	invert
Alfisol	alf	Meaningless syllable	alfalfa
Spodosol	od	Gr. *spodos*, wood ash	odd
Mollisol	oll	L. *mollis*, soft	mollify
Aridisol	id	L. *aridus*, dry	arid

because of favorable texture and nutrient value inherited from soils that supplied the sediment. *Orthents* (*orth*: "common" or "ordinary") occupy large areas of glacially abraded surfaces in both arctic and alpine environments. *Psamments*: (*psamm*: "sand") are derived from sand-textured parent matter that was deposited as dune sand or beach sand. Numerous large patches of Psamments are found in deserts, surrounded by the Aridisols. A noteworthy occurrence of Psamments in the United States is the Sand Hills Region of Nebraska, a patch of steppe grasslands underlain by dune sands of a dune field that was active in Pleistocene time.

INCEPTISOLS

Inceptisols are uniquely defined by a combination of the following properties: (1) Soil water is available to plants more than half the year for more than 3 consecutive months during a warm season. The soil-water regime is mostly udic or perudic. (2) One or more pedogenic horizons formed by alteration or concentration of matter are present, but without accumulation of translocated materials other than carbonate minerals or amorphous (noncrystalline) silica. (3) Soil textures are finer than loamy sand. (4) The soil contains some weatherable minerals.

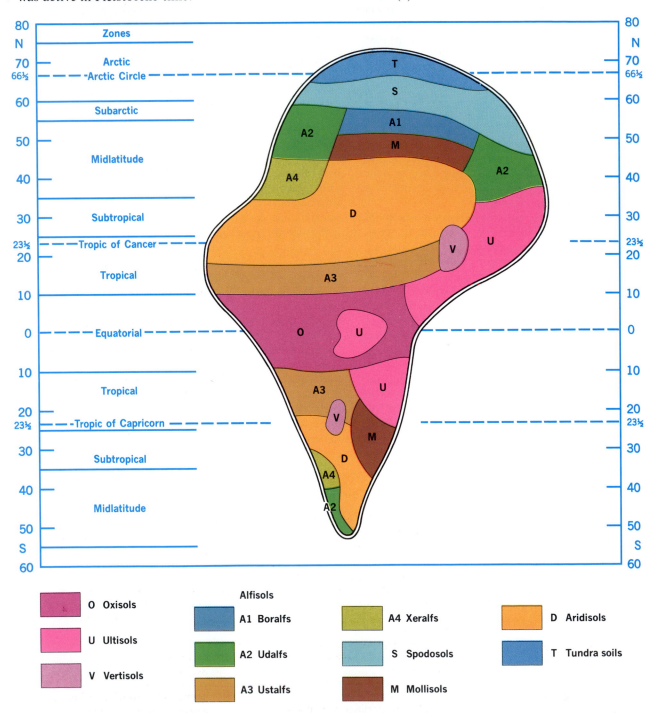

O Oxisols	
U Ultisols	
V Vertisols	

Alfisols

A1 Boralfs	**A4** Xeralfs
A2 Udalfs	**S** Spodosols
A3 Ustalfs	**M** Mollisols

D Aridisols	
T Tundra soils	

FIGURE 24.1 Schematic diagram of the soil orders and major suborders on an imaginary supercontinent.

A **Alfisols**
  A1 *Boralfs*
    A1a with Histosols
    A1b with Spodosols
  A2 *Udalfs*
    A2a with Aqualfs
    A2b with Aquolls
    A2c with Hapludults
    A2d with Ochrepts
    A2e with Troporthents
    A2f with Udorthents
  A3 *Ustalfs*
    A3a with Tropepts
    A3b with Troporthents
    A3c with Tropustults
    A3d with Usterts
    A3e with Ustochrepts
    A3f with Ustolls
    A3g with Ustorthents
    A3h with Ustox
    A3j Plinthustalfs with
        Ustorthents
  A4 *Xeralfs*
    A4a with Xerochrepts
    A4b with Xerorthents
    A4c with Xerults

D **Aridisols**
  D1 *Aridisols, undifferentiated*
    D1a with Orthents
    D1b with Psamments
    D1c with Ustalfs
  D2 *Argids*
    D2a with Fluvents
    D2b with Torriorthents

E **Entisols**
  E1 *Aquents*
    E1a Haplaquents with
        Udifluvents
    E1b Psammaquents with
        Haplaquents
    E1c Tropaquents with
        Hydraquents
  E2 *Orthents*
    E2a Cryorthents
    E2b Cryorthents with Orthods
    E2c Torriorthents with
        Aridisols
    E2d Torriorthents with Ustalfs
    E2e Xerorthents with Xeralfs
  E3 *Psamments*
    E3a with Aridisols
    E3b with Orthox
    E3c with Torriorthents
    E3d with Ustalfs
    E3e with Ustox
    E3f of shifting sands
    E3g Ustipsamments with
        Ustolls

H **Histosols**
  H1 *Histosols, undifferentiated*
    H1a with Aquods
    H1b with Boralfs
    H1c with Cryaquepts

I **Inceptisols**
  I1 *Andepts*
    I1a Dystrandepts with
        Ochrepts

I2 *Aquepts*
    I2a Cryaquepts with Orthents
    I2b Halaquepts with Salorthids
    I2c Haplaquepts with
        Humaquepts
    I2d Haplaquepts with
        Ochraqualfs
    I2e Humaquepts with
        Psamments
    I2f Tropaquents with
        Hydraquents
    I2g Tropaquepts with
        Plinthaquults
    I2h Tropaquepts with
        Tropaquents
    I2j Tropaquepts with
        Tropudults

I3 *Ochrepts*
    I3a Dystrochrepts with
        Fragiochrepts
    I3b Dystrochrepts with Orthox
    I3c Xerochrepts with Xerolls

I4 *Tropepts*
    I4a with Ustalfs
    I4b with Tropudults
    I4c with Ustox

I5 *Umbrepts*
    I5a with Aqualfs

M **Mollisols**
  M1 *Albolls*
    M1a with Aquepts
  M2 *Borolls*
    M2a with Aquolls
    M2b with Orthids
    M2c with Torriorthents
  M3 *Rendolls*
    M3a with Usterts
  M4 *Udolls*
    M4a with Aquolls
    M4b with Eutrochrepts
    M4c with Humaquepts
  M5 *Ustolls*
    M5a with Argialbolls
    M5b with Ustalfs
    M5c with Usterts
    M5d with Ustochrepts
  M6 *Xerolls*
    M6a with Xerorthents

O **Oxisols**
  O1 *Orthox*
    O1a with Plinthaquults
    O1b with Tropudults
  O2 *Ustox*
    O2a with Plinthaquults
    O2b with Tropustults
    O2c with Ustalfs

S **Spodosols**
  S1 *Spodosols, undifferentiated*
    S1a cryic regimes, with Boralfs
    S1b cryic regimes, with
        Histosols
  S2 *Aquods*
    S2a Haplaquods with
        Quartzipsamments
  S3 *Humods*
    S3a with Hapludalfs
  S4 *Orthods*
    S4a Haplorthods with Boralfs

U **Ultisols**
  U1 *Aquults*
    U1a Ochraquults with Udults
    U1b Plinthaquults with Orthox
    U1c Plinthaquults with
        Plinthaquox
    U1d Plinthaquults with
        Tropaquepts
  U2 *Humults*
    U2a with Umbrepts
  U3 *Udults*
    U3a with Andepts
    U3b with Dystrochrepts
    U3c with Udalfs
    U3d Hapludults with
        Dystrochrepts
    U3e Rhodudults with Udalfs
    U3f Tropudults with Aquults
    U3g Tropudults with
        Hydraquents
    U3h Tropudults with Orthox
    U3j Tropudults with Tropepts
    U3k Tropudults with Tropudalfs
  U4 *Ustults*
    U4a with Ustochrepts
    U4b Plinthustults with
        Ustorthents
    U4c Rhodustults with Ustalfs
    U4d Tropustults with
        Tropaquepts
    U4e Tropustults with Ustalfs

V **Vertisols**
  V1 *Uderts*
    V1a with Usterts
  V2 *Usterts*
    V2a with Tropaquepts
    V2b Tropofluvents
    V2c with Ustalfs

X **Soils in areas with mountains**
  X1 Cryic great groups of Entisols, Inceptisols, and Spodosols.
  X2 Boralfs and cryic great groups of Entisols and Inceptisols.
  X3 Udic great groups of Alfisols, Entisols, and Ultisols; Inceptisols.
  X4 Ustic great groups of Alfisols, Inceptisols, Mollisols, and Ultisols.
  X5 Xeric great groups of Alfisols, Entisols, Inceptisols, Mollisols, and Ultisols.
  X6 Torric great groups of Entisols; Aridisols.
  X7 Histic and cryic great groups of Alfisols, Entisols, Inceptisols, and Mollisols; ustic great groups of Ultisols; cryic great groups of Spodosols.
  X8 Aridisols; torric and cryic great groups of Entisols, and cryic great groups of Spodosols and Inceptisols.

Z **Miscellaneous**
  Z1 Ice sheets
  Z2 Rugged mountains, mostly devoid of soil (includes glaciers, permanent snowfields, and in some places, areas of soil.)

(Legend for Figure 24.2, page 464.) (See next page).

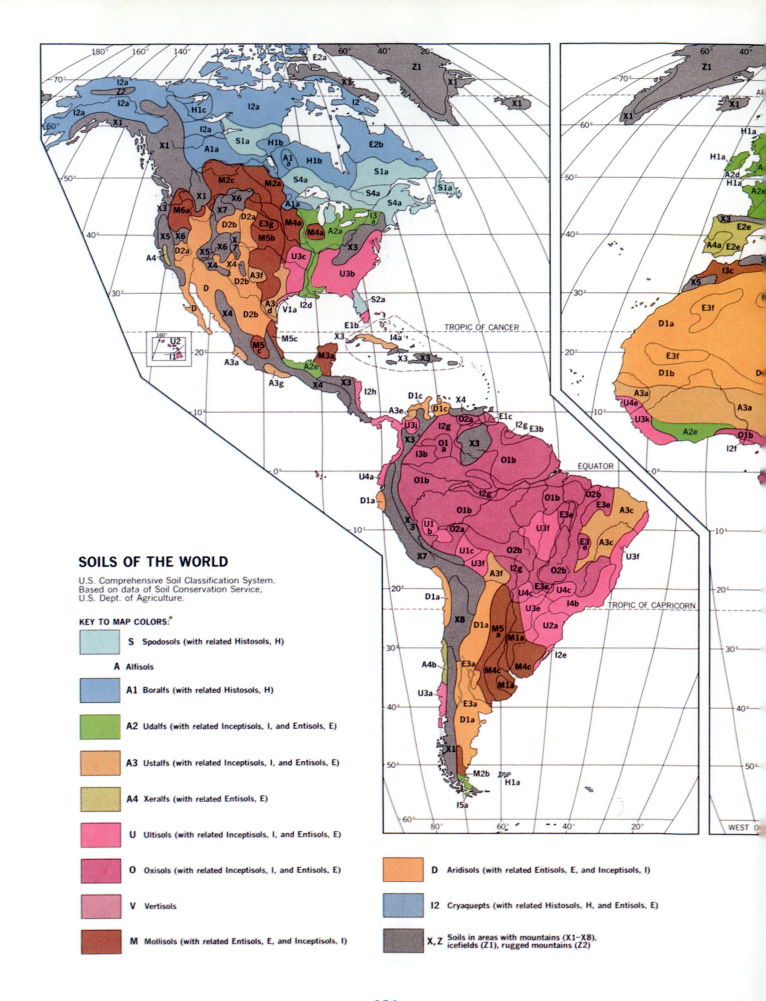

SOILS OF THE WORLD

U.S. Comprehensive Soil Classification System.
Based on data of Soil Conservation Service,
U.S. Dept. of Agriculture.

KEY TO MAP COLORS:

S Spodosols (with related Histosols, H)

A Alfisols

A1 Boralfs (with related Histosols, H)

A2 Udalfs (with related Inceptisols, I, and Entisols, E)

A3 Ustalfs (with related Inceptisols, I, and Entisols, E)

A4 Xeralfs (with related Entisols, E)

U Ultisols (with related Inceptisols, I, and Entisols, E)

O Oxisols (with related Inceptisols, I, and Entisols, E)

V Vertisols

M Mollisols (with related Entisols, E, and Inceptisols, I)

D Aridisols (with related Entisols, E, and Inceptisols, I)

I2 Cryaquepts (with related Histosols, H, and Entisols, E)

X, Z Soils in areas with mountains (X1–X8), icefields (Z1), rugged mountains (Z2)

TROPIC OF CANCER

EQUATOR

TROPIC OF CAPRICORN

WEST

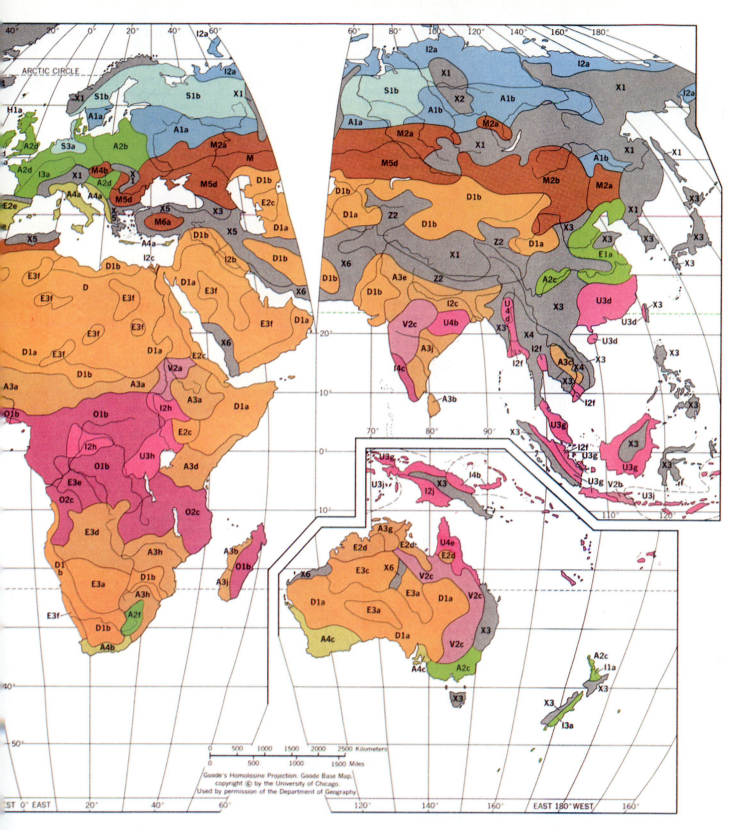

FIGURE 24.2 World map of soils according to the U.S. Comprehensive Soil Classification System. (U.S. Department of Agriculture.)

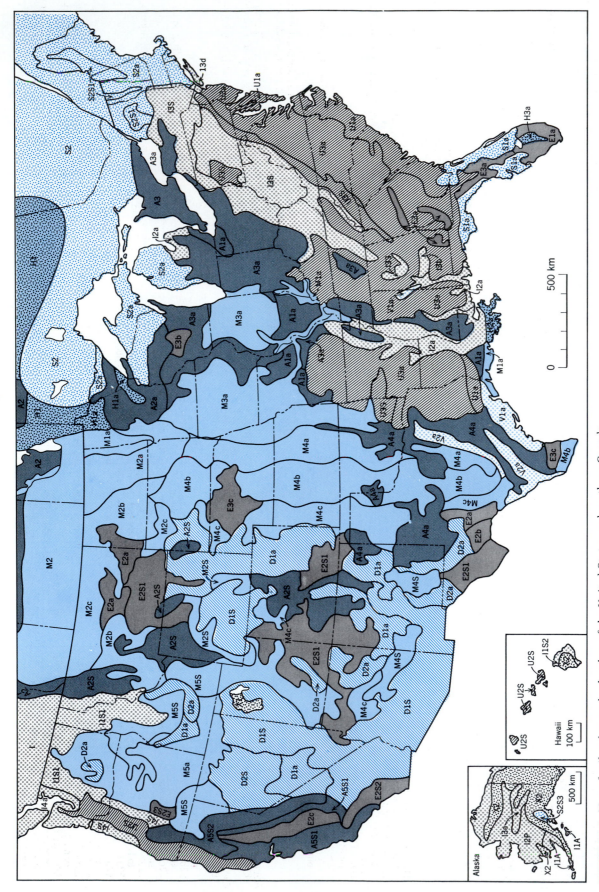

FIGURE 24.3 Map of soil orders and suborders of the United States and southern Canada. Legend on facing page. (Soil Conservation Service, U.S. Department of Agriculture.)

ALFISOLS

AQUALFS
A1a—Aqualfs with Udalfs, Haplaquepts, Udolls; gently sloping.

BORALFS
A2a—Boralfs with Udipsamments and Histosols; gently and moderately sloping.
A2S—Cryoboralfs with Borolls, Cryochrepts, Cryorthods, and rock outcrops; steep.

UDALFS
A3a—Udalfs with Aqualfs, Aquolls, Rendolls, Udolls, and Udults; gently or moderately sloping.

USTALFS
A4a—Ustalfs with Ustochrepts, Ustolls, Usterts, Ustipsamments, and Ustorthents; gently or moderately sloping.

XERALFS
A5S1—Xeralfs with Xerolls, Xerorthents, and Xererts; moderately sloping to steep.
A5S2—Ultic and lithic subgroups of Haploxeralfs with Andepts, Xerults, Xerolls, and Xerochrepts; steep.

ARIDISOLS

ARGIDS
D1a—Argids with Orthids, Orthents, Psamments, and Ustolls; gently and moderately sloping.
D1S—Argids with Orthids, gently sloping; and Torriorthents, gently sloping to steep.

ORTHIDS
D2a—Orthids with Argids, Orthents, and Xerolls; gently or moderately sloping.
D2S—Orthids, gently sloping to steep, with Argids, gently sloping; lithic subgroups of Torriorthents and Xerorthents, both steep.

ENTISOLS

AQUENTS
E1a—Aquents with Quartzipsamments, Aquepts, Aquolls, and Aquods; gently sloping.

ORTHENTS
E2a—Torriorthents, steep, with borollic subgroups of Aridisols; Usterts and aridic and vertic subgroups of Borolls; gently or moderately sloping.
E2b—Torriorthents with Torrerts; gently or moderately sloping.
E2c—Xerorthents with Xeralfs, Orthids, and Argids, gently sloping.
E2S1—Torriorthents; steep, and Argids, Torrifluvents, Ustolls, and Borolls; gently sloping.
E2S2—Xerorthents with Xeralfs and Xerolls; steep.
E2S3—Cryorthents with Cryosamments and Cryandepts; gently sloping to steep.

PSAMMENTS
E3a—Quartzipsamments with Aquults and Udults; gently or moderately sloping.
E3b—Udipsamments with Aquolls and Udalfs; gently or moderately sloping.
E3c—Ustipsamments with Ustalfs and Aquolls; gently or moderately sloping.

HISTOSOLS

HISTOSOLS
H1a—Hemists with Psammaquents and Udipsamments; gently sloping.
H2a—Hemists and Saprists with Fluvaquents and Haplaquepts; gently sloping.
H3a—Fibrists, Hemists, and Saprists with Psammaquents; gently sloping.

INCEPTISOLS

ANDEPTS
I1a—Cryandepts with Cryaquepts, Histosols, and rock land; gently or moderately sloping.
I1S1—Cryandepts with Cryochrepts, Cryumbrepts, and Cryorthods; steep.
I1S2—Andepts with Tropepts, Ustolls, and Tropofolists; moderately sloping to steep.

AQUEPTS
I2a—Haplaquepts with Aqualfs, Aquolls, Udalfs, and Fluvaquents; gently sloping.
I2P—Cryaquepts with cryic great groups of Orthents, Histosols, and Ochrepts; gently sloping to steep.

OCHREPTS
I3a—Cryochrepts with cryic great groups of Aquepts, Histosols, and Orthods; gently or moderately sloping.
I3b—Eutrochrepts with Uderts; gently sloping.
I3c—Fragiochrepts with Fragiaquepts, gently or moderately sloping; and Dystrochrepts, steep.
I3d—Dystrochrepts with Udipsamments and Haplorthods; gently sloping.
I3S—Dystrochrepts, steep, with Udalfs and Udults; gently or moderately sloping.

UMBREPTS
I4a—Haplumbrepts with Aquepts and Orthods; gently or moderately sloping.
I4S—Haplumbrepts and Orthods; steep, with Xerolls and Andepts; gently sloping.

MOLLISOLS

AQUOLLS
M1a—Aquolls with Udalfs, Fluvents, Udipsamments, Ustipsamments, Aquepts, Eutrochrepts, and Borolls; gently sloping.

BOROLLS
M2a—Udic subgroups of Borolls with Aquolls and Ustorthents; gently sloping.
M2b—Typic subgroups of Borolls with Ustipsamments, Ustorthents, and Boralfs; gently sloping.
M2c—Aridic subgroups of Borolls with Borollic subgroups of Argids and Orthids, and Torriorthents; gently sloping.
M2S—Borolls with Boralfs, Argids, Torriorthents, and Ustolls; moderately sloping or steep.

UDOLLS
M3a—Udolls, with Aquolls, Udalfs, Aqualfs, Fluvents, Psamments, Ustorthents, Aquepts, and Albolls; gently or moderately sloping.

USTOLLS
M4a—Udic subgroups of Ustolls with Orthents, Ustochrepts, Usterts, Aquents, Fluvents, and Udolls; gently or moderately sloping.
M4b—Typic subgroups of Ustolls with Ustalfs, Ustipsamments, Ustorthents, Ustochrepts, Aquolls, and Usterts; gently or moderately sloping.
M4c—Aridic subgroups of Ustolls with Ustalfs, Orthids, Ustipsamments, Ustorthents, Ustochrepts, Torriorthents, Borolls, Ustolls, and Usterts; gently or moderately sloping.
M4S—Ustolls with Argids and Torriorthents; moderately sloping or steep.

XEROLLS
M5a—Xerolls with Argids, Orthids, Fluvents, Cryoboralfs, Cryoborolls, and Xerorthents; gently or moderately sloping.
M5S—Xerolls with Cryoboralfs, Xeralfs, Xerorthents, and Xererts; moderately sloping or steep.

SPODOSOLS

AQUODS
S1a—Aquods with Psammaquents, Aquolls, Humods, and Aquults; gently sloping.

ORTHODS
S2a—Orthods with Boralfs, Aquents, Orthents, Psamments, Histosols, Aquepts, Fragiochrepts, and Dystrochrepts; gently or moderately sloping.
S2S1—Orthods with Histosols, Aquents, and Aquepts; moderately sloping or steep.
S2S2—Cryorthods with Histosols; moderately sloping or steep.
S2S3—Cryorthods with Histosols, Andepts and Aquepts; gently sloping to steep.

ULTISOLS

AQUULTS
U1a—Aquults with Aquents, Histosols, Quartzipsamments, and Udults; gently sloping.

HUMULTS
U2S—Humults with Andepts, Tropepts, Xerolls, Ustolls, Orthox, Torrox, and rock land; gently sloping to steep.

UDULTS
U3a—Udults with Udalfs, Fluvents, Aquents, Quartzipsamments, Aquepts, Dystrochrepts, and Aquults; gently or moderately sloping.
U3S—Udults with Dystrochrepts; moderately sloping or steep.

VERTISOLS

UDERTS
V1a—Uderts with Aqualfs, Eutrochrepts, Aquolls, and Ustolls; gently sloping.

USTERTS
V2a—Usterts with Aqualfs, Orthids, Udifluvents, Aquolls, Ustolls, and Torrerts; gently sloping.

Areas with little soil

X1—Salt flats.
X2—Rock land (plus permanent snow fields and glaciers).

Slope classes

Gently sloping—Slopes mainly less than 10 percent, including nearly level.
Moderately sloping—Slopes mainly between 10 and 25 percent.
Steep—Slopes mainly steeper than 25 percent.

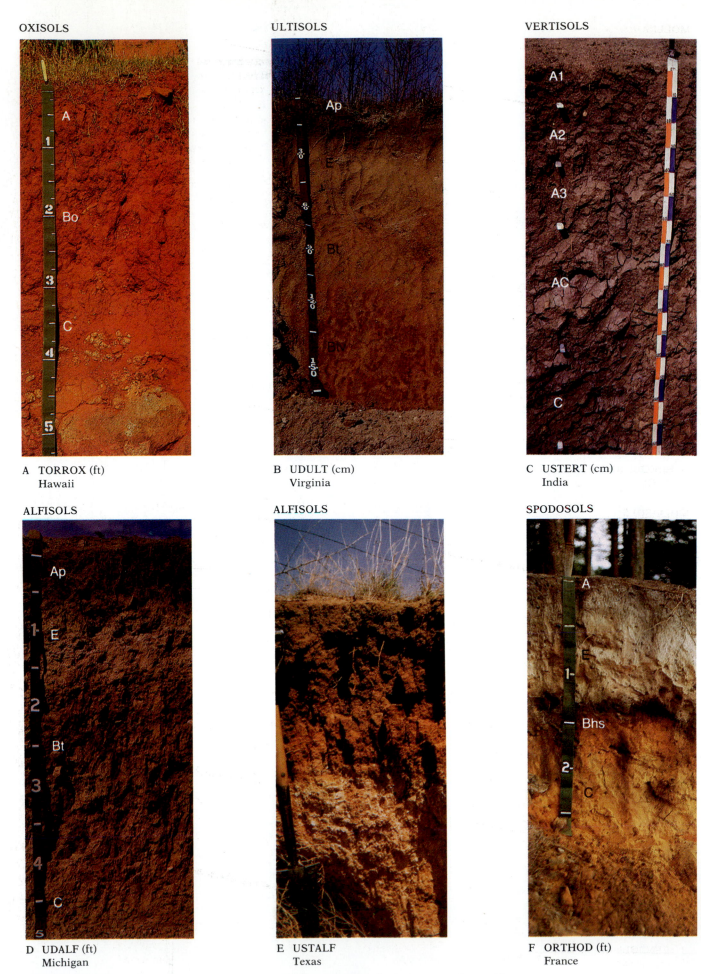

OXISOLS

A

1

Bo

2

3

C

4

5

A TORROX (ft)
Hawaii

ULTISOLS

Ap

30

E

60

Bt

90

120

150

B UDULT (cm)
Virginia

VERTISOLS

A1

A2

A3

AC

C

C USTERT (cm)
India

ALFISOLS

Ap

1

E

2

Bt

3

4

C

5

D UDALF (ft)
Michigan

ALFISOLS

E USTALF
Texas

SPODOSOLS

A

E

1

Bhs

2

C

F ORTHOD (ft)
France

FIGURE 24.4 Soil profiles of several soil orders. (Panels *A* through *K* by Henry D. Foth. Used by permission. Panel *L* from U.S. Department of Agriculture, Soil Conservation Service.)

MOLLISOLS

G BOROLL (cm)
USSR

MOLLISOLS

H UDOLL (ft)
Argentina

MOLLISOLS

I USTOLL
Colorado

MOLLISOLS

J RENDOLL (ft)
Argentina

ARIDISOLS

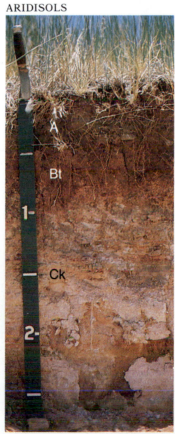

K ARGID (ft)
Colorado

HISTOSOLS

L FIBRIST (ft)
Minnesota

469

FIGURE 24.5 Profile of a tundra soil (Cryaquept) in northern Yukon Territory, Canada. Permafrost (perennially frozen soil water) appears at a depth of between 40 and 60 cm. (Henry D. Foth. Used by permission.)

FIGURE 24.6 Profile of a Histosol (suborder Saprist) seen in the wall of a pit cut into a bog near Belle Glade, Florida. Water is being pumped from the floor of the pit. (Henry D. Foth. Used by permission.)

(5) The clay fraction of the soil has a moderate to high CEC.

Inceptisols are found in a wide range of latitudes having the requisite soil-water regimes. One area of widespread occurrence is in the region of tundra climate (12). Inceptisols are also found in high mountains associated with alpine tundra climate; such locations are found at all latitudes, ranging down to the equatorial zone.

The Inceptisols are mostly found on relatively young geomorphic surfaces, for example, surfaces shaped by glacial ice of the last great ice advance (Wisconsinan Glaciation, Chapter 22). Because of their youth, these Inceptisols usually lack distinct horizons. Under the cold tundra climate with its permafrost, chemical decomposition of minerals is much inhibited.

Inceptisols also occur on recently accumulated alluvial sediments of floodplains where sedimentation is no longer active and pedogenic horizons have developed. These occurrences are shown in the world soils map as ranging from equatorial to midlatitude zones. Examples are floodplains of the Mississippi, Amazon, Congo, and Ganges–Brahmaputra rivers and deltaic plains of the Nile, Irrawaddy, and Mekong rivers.

Of the five suborders of Inceptisols, one in particular deserves mention. The *Aquepts* (*aqu:* "water") are Inceptisols of wet places (bogs and marshes). When the prefix *cry* is added to signify the cryic temperature regime, the great group of *Cryaquepts* is indicated. Cryaquepts are noteworthy for their widespread occurrence in the arctic and subarctic climate zones of the northern hemisphere where they are commonly called *tundra soils* (Figure 24.5) and are characterized by a surface layer of peat (a histic epipedon).

HISTOSOLS

Histosols are unique in having a very high content of organic matter in the upper 80 cm (32 in.); (see Figure 24.4L). More than half this thickness has from 20 to 30

percent of organic matter or more, or the organic horizon rests on bedrock or rock rubble. Most Histosols are peats or mucks consisting of more or less decomposed plant remains that accumulated in water; but some have formed from forest litter or moss or both in a cool, continuously humid climate, but are freely drained.

Histosols of cool climates are commonly acid in reaction and low in plant nutrients. Important areas of such Histosols are in the Northwest Territories of Canada and in an area lying to the south of Hudson Bay in Manitoba, Ontario, and Quebec. Histosols are found in lower latitudes as well, where conditions of poor drainage have favored thick accumulations of plant matter (Figure 24.6).

Histosols that are mucks (fine black materials of sticky consistency) are agriculturally valuable in midlatitudes, where they occur in the beds of countless former lakes in the glaciated zone. After appropriate drainage and application of lime and fertilizers, these mucks are

FIGURE 24.7 Garden crops cultivated on a Histosol of a former glacial lake bed, northern New Jersey. Special drainage is required to remove excess soil water. (A. N. Strahler.)

FIGURE 24.8 This peat bog in Connemara, Ireland, has been trenched to reveal the Histosol profile. Piles of peat blocks, seen in the foreground, will be dried for use as fuel. (Copyright © 1988 by Mark A. Melton.)

highly productive for truck-garden vegetables (Figure 24.7). Peat bogs are extensively used for cultivation of cranberries (cranberry bogs); *Sphagnum* peat is dried and baled for sale as a mulch for use on suburban lawns and shrubbery beds. Dried peat from bogs has been used for centuries in Europe as a low-grade fuel (Figure 24.8).

OXISOLS

Oxisols are unique through the combination of three properties: (1) extreme weathering of most minerals, other than quartz, to sesquioxides of aluminum and iron and to kaolinite; (2) very low CEC of the clay fraction; and (3) a loamy or clayey texture (sandy loam or finer). An oxic horizon is usually present within 2 m (6 ft) of the surface (Figures 24.4A and 24.16B). Plinthite may be present in the deeper parts of the profile (Figure 24.9).

Oxisols characteristically develop in equatorial, tropical, and subtropical zones on land surfaces that have been stable over long periods of time. Most of these surfaces are of early Pleistocene age or much older (i.e., older than 2 million years). During soil development, the climate must be moist. In some areas, however, the Oxisols now occupy seasonally dry environments (wet-dry tropical climate) because a climate change has occurred since the time of their formation. Recall from Chapter 18 that under favorable conditions, plinthite on exposure hardens irreversibly into laterite, a rocklike material (Figure 18.21).

Oxisols of low latitudes lack distinct horizons, except for darkened surface layers, even though they are formed in strongly weathered parent material (Figure 24.10). Red, yellow, and yellowish-brown colors are normal to the well-drained Oxisols because of the presence of iron sesquioxides. Supplies of plant nutrients are commonly low. When phosphorus is added as a fertilizer, the

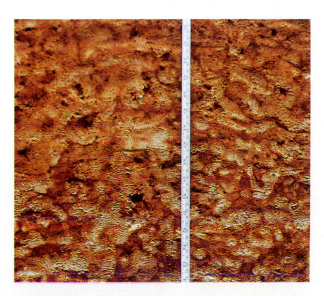

FIGURE 24.9 Red mottles characterize this plinthite horizon in a Udult profile (order Ultisols) in North Carolina. The area shown is about 50 cm (20 in.) wide. (Soil Conservation Service).

FIGURE 24.10 An Oxisol (suborder Torrox) in Hawaii. This soil has developed by deep chemical weathering of basalt, residual boulders of which can be seen in place near the base of the exposure. Sugarcane is being cultivated here. (Henry D. Foth. Used by permission.)

soil has the capacity to fix the phosphorus into forms unavailable to plants. The clay fraction consists largely of silicate clay minerals, but with an important proportion of iron and aluminum sesquioxides. Despite the large clay fraction, however, the soils are friable (easily broken apart) and are easily penetrated by water and roots.

The world map of soils (Figure 24.2) shows Oxisols to be dominant in vast areas of equatorial and tropical South America and Africa. These are regions of wet equatorial climate (1) and wet-dry tropical climate (3). (See world climate map, Figure 8.9.)

Oxisols and Ultisols of low latitudes were used under systems of shifting cultivation prior to the advent of modern technology; a large proportion of these soils is still used in that way. The levels of plant nutrients under natural conditions are so low that the yields obtainable under hand tillage are very low, especially after a garden patch has been used for a year or two. Substantial use of lime, fertilizers, and other industrial inputs is necessary for high yields and sustained production.

ULTISOLS

Ultisols are unique through the combination of three properties: (1) There is an argillic horizon, (2) the supply of exchangeable bases (CEC) is low, particularly in the lower horizons; and (3) the mean annual soil temperature is greater than 8°C (47°F). A point of importance is that the percentage of base saturation (PBS) diminishes rapidly with depth. The highest PBS is within the upper few centimeters, reflecting the recycling of base cations by plants or the additions of fertilizers.

Ultisols characteristically form under forest vegetation in climates with a slight to pronounced seasonal soil-water deficit alternating with a surplus (udic and ustic regimes). Thus, at the season of soil-water surplus, some water passes through the soil into the substratum, allowing leaching to occur. As a result, bases not held in plant tissue are depleted. The B horizon of well-drained Ultisols is characteristically red or yellowish-brown in color due to concentration of sesquioxides of iron (Figures 24.4B and 24.11).

The world soils map (Figure 24.2) shows a large area of dominantly Ultisols in the southeastern United States and another in southern China. There are also important areas of Ultisols in Bolivia, southern Brazil, western and central Africa, India, Burma, the East Indies, and northeastern Australia.

Reference to the world climate map, Figure 8.9, shows that the climate in areas of Ultisols ranges from the moist subtropical climate (6) in the subtropical zone to the wet-dry tropical climate (3) and monsoon and trade-wind littoral climate (2) in the tropical zone. These climates have large water surpluses in one season, although they cover a large spread in temperature characteristics. The land surfaces in these areas have been subjected to prolonged erosion and weathering, so igneous bedrock beneath is often deeply decayed, forming thick saprolite.

FIGURE 24.11 Ultisol profile in North Carolina. The thin, pale layer at the top is an E horizon, showing the effects of removal of materials by eluviation. Near the base of the thick, reddish B horizon is a mottled zone of plinthite. (Soil Conservation Service.)

Ultisols of low latitudes were largely used under systems of shifting cultivation prior to the development of modern technology, as they still are where the benefits of modern technology and capital are not available. The soils are low enough in plant nutrients so that growth of cultivated plants is poor without the general use of lime and fertilizers. Given essential industrial products and adequate management skills, however, the Ultisols can be highly productive.

VERTISOLS

Vertisols are uniquely defined through a combination of the following properties: (1) a high content of clay (montmorillonite) that shrinks and swells greatly with changes in soil-water storage; (2) deep wide cracks at some season; and (3) evidences of soil movement in the form of slickensides, gilgae, or structural aggregates tilted between horizontal and vertical. Horizons are normally weakly expressed and may not be apparent (Figure 24.4C).

The evidences of movement, just mentioned, require description. *Slickensides* are soil surfaces with grooves or striations (scratch marks) made by movement of one mass of soil against another while the soil is in a moist, plastic state (Figure 24.12). *Gilgae* are very small relief features that may be knobs and basins or narrow ridges

FIGURE 24.12 A Vertisol (suborder Ustert) in India. Slicken-sided surfaces have been exposed in this trench, giving evidence of soil movement. (Henry D. Foth. Used by permission.)

with valleys between. The height of these features is seldom over 1 m (3 ft). Heaving of the soil causes tilting of various objects supported by the soil, including fence posts and sidewalk slabs.

Vertisols typically form under grass or savanna vegetation in subtropical and tropical climates with a moderate or pronounced seasonal soil-water deficit. These climates include dry tropical climate, semiarid subtype (4s), and wet-dry tropical climate (3). At some season the soils dry out enough that the soil becomes deeply cracked (Figure 24.13). When rains come, some masses of surface soil drop into the cracks before they close, so the soil slowly "swallows itself." Vertisols the world over have had more names than any other kind of soil. Examples are *black cotton soils*, *tropical black clays*, and *regur*.

Vertisols are found only in the latitude range of the tropical and subtropical zones. The U.S. soils map, Figure 24.3, shows three narrow belts of Vertisols on the coastal plain of Texas. One large tropical area is in the

Deccan region of India, where the Vertisols are formed on weathered basalts. Other areas shown in the world map are the Sudan region of east-central Africa and a large north-south belt in eastern Australia.

Vertisols are very high in exchangeable bases such as calcium and magnesium; most are near neutral in pH; and most have intermediate amounts of organic matter. The soils retain large amounts of water because of their fine texture, but much of this soil water is held by the montmorillonite clay and is not available to plants. Where soil cultivation depends on human or animal power, agricultural yields are low. Problems of using these soils are especially difficult; for example, the soil becomes highly plastic and difficult to till when wet. Many areas of Vertisols have thus been left in grass or savanna, which may provide grazing for cattle. With the use of modern technology and its energy resources, however, production of food and fiber from Vertisols could be substantial.

ALFISOLS

Alfisols are uniquely defined by the following combination of properties: (1) a gray, brownish, or reddish horizon (ochric epipedon) not darkened by humus close to the surface; (2) a horizon of clay accumulation (argillic horizon); (3) a medium to high percentage base saturation (PBS); and (4) soil water available to plants more than half the year or more than 3 consecutive months during a warm season. The agrillic horizon is a B horizon (Figure 24.4D). It is enriched by accumulated silicate clay minerals and is moderately saturated with exchangeable bases, such as calcium and magnesium. The E horizon above it is marked by some loss of bases, silicate clay minerals, and sesquioxides.

On the world soils map, (Figure 24.2), dominant areas of Alfisols are shown in central North America, Europe, central Siberia, north China, and southern Australia. These areas are identified with the moist continental climate (10), the marine west-coast climate (8), and the Mediterranean climate (7); see the world climate map, Figure 8.9. Other important areas of Alfisols shown on the map are located in eastern Brazil; western, eastern, and southern Africa; western Madagascar; northern Australia; eastern India; and Indochina. These occurrences are in areas of wet-dry tropical climate (3) and the semiarid subtypes of the tropical and subtropical dry climates (4s, 5s).

On the whole, Alfisols are agriculturally fairly productive under simple management because soil water is usually adequate in one season and bases are not depleted. In western Europe and China, Alfisols, along with the Inceptisols occurring within the same regions, have supported dense populations for centuries.

Of the five suborders of Alfisols, four deserve recognition because of their widespread occurrence on upland surfaces. Each of the four represents a different climate and soil-water regime and is given separate recognition on our world soil diagram and map (Figures 24.1 and 24.2).

FIGURE 24.13 Soil cracks in Vertisol in Texas. Locally this soil is known as the Houston black clay. The crop growing here is cotton. (Soil Conservation Service.)

FIGURE 24.14 Ustalf profile in Texas. The pale upper layer is an ochric epipedon. The argillic horizon of darker color sets in at about 25 cm (10 in.). (Soil Conservation Service.)

Boralfs (*bor:* "boreal") are Alfisols of cold forest lands of North America and Eurasia. They have a gray surface horizon, a brownish subsoil, and are associated with a mean annual temperature less than 8°C (47°F).

Udalfs (*ud:* "humid") are Alfisols of the midlatitude zone and are closely associated with the moist continental climate (10) in North America, Europe, and eastern Asia (Figure 24.4D). They form under the udic soil-water regime. A deciduous forest was the natural vegetation on Udalfs, but today these soils are intensively farmed. They are highly productive soils when moderate amounts of lime and fertilizers are applied.

Ustalfs (*ust:* L. *ustus,* "burnt") are brownish to reddish Alfisols of the warmer climates under the ustic soil-water regime. They are brownish to reddish throughout the profile (Figures 24.4E and 24.14). Ustalfs range from the subtropical zone to the equator and are found associated with the wet-dry tropical climate (3) in Southeast Asia, Africa, Australia, and South America. In Africa, Ustalfs may owe their high base status to the constant rain of fine dust carried by prevailing winds from the adjacent tropical desert. Ustalfs of north India and Pakistan are highly productive under irrigation and are major producers of wheat.

Xeralfs are Alfisols of xeric soil-water regime, found in the Mediterranean climate (7), with its cool moist win-

FIGURE 24.15 Xeralf profile in California. The tape measure rests on the top of the dense argillic horizon (B horizon). (Soil Conservation Service.)

ters and dry warm summers. The Xeralfs are typically brownish or reddish in color (Figure 24.15). Good examples are found in coastal and inland valleys of central and southern California. These soils have a high natural fertility and support cattle-grazing grasslands and the cultivation of grape, citrus, and avocado.

SPODOSOLS

Spodosols have a unique property: a spodic horizon (B horizon) of accumulation (illuviation) of dark amorphous materials. The amorphous materials contain organic material, compounds of aluminum, and commonly iron, all brought downward by eluviation from an overlying A horizon (Figures 24.16A and 24.4F). In undisturbed Spodosols, a bleached gray to white E horizon overlies the B horizon. This is an albic horizon (Figure 24.17). Spodosols are strongly acid, low in plant nutrients, such as the base cations of calcium and magnesium, and generally low in humus. Spodosols usually have sand texture and small water-holding capacities.

Spodosols, with some exceptions, were formed under forest cover in a midlatitude climate that is both cool and moist. In comparing the world distribution of Spodosols with that of climates (compare Figure 24.2 with Figure 8.9), the correspondence with the boreal forest climate (11) is striking. Spodosols also occupy northerly portions of the moist continental climate (10). Correspondence with needleleaf (boreal) forest is also striking (see Figure 27.6).

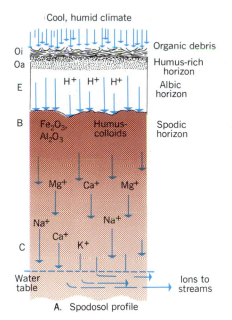

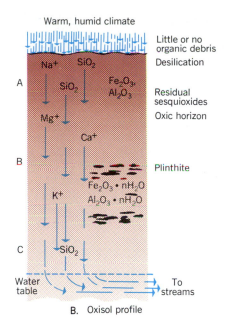

FIGURE 24.16 Schematic composition diagrams of (A) Spodosol profile and (B) Oxisol profile. Both develop in soil-water regimes with a large water surplus.

As is the case of the northern Inceptisols, regions of Spodosols are largely those that experienced the most recent continental glaciation; the soils are therefore very young. Nevertheless, Spodosols can be recognized as occurring in latitudes down to the equatorial zone. For example, soils of much of northern and central Florida are classed as Spodosols.

Spodosols of middle and high latitudes are naturally poor soils in terms of agricultural productivity. Because they are acid, the application of lime is a necessity. Because they are poor in the nutrients required by most crops, the application of fertilizers is required. With proper management and the input of required industrial products, the Spodosols can be highly productive. Another unfavorable factor is shortness of the growing season in the more northerly areas of Spodosols.

MOLLISOLS

Mollisols are uniquely defined through a combination of the following properties: (1) a mollic epipedon, which is a very dark brown to black surface horizon that is more than one-third of the combined thickness of the A and B horizons or is more than 25 cm (10 in.) thick and has a loose structure or soft consistence when dry (Figures 24.4G–J and 24.18); (2) a dominance of calcium among the extractable base cations in the A and B horizons; (3) a dominance of crystalline clay minerals of moderate or high cation exchange capacity (CEC); and (4) less than 30 percent of clay in some horizon if the soil has deep wide cracks at some season.

Mollisols typically form under grass in climates with a moderate to pronounced seasonal soil-water deficiency. A few form in marshes or on marls (compacted muds of calcium carbonate composition) in moist climates. Comparison of Mollisol distribution on the world soils map (Figure 24.2) with distribution of climates in Figure 8.9 shows that the Mollisols are closely associated with the

FIGURE 24.17 A Spodosol (suborder Orthod), developed in sandy parent material, in Quebec, Canada. The strongly leached albic horizon (E) appears almost pure white in contrast to the reddish-brown spodic horizon beneath it. (Henry D. Foth. Used by permission.)

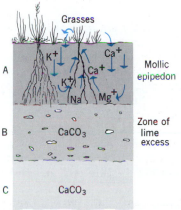

FIGURE 24.18 Schematic diagram of a Mollisol profile.

semiarid subtype of the dry midlatitude climate (9s), with the subhumid subtype and adjacent parts of the moist continental climate (10sh, 10h), and with the subhumid subtype of the moist subtropical climate (6sh). In North America, Mollisols dominate the Great Plains region, the Columbia Plateau, and the northern Great Basin. Smaller areas exist along the Gulf coastal plain of Mexico and the Yucatan Peninsula. In South America, a large area of Mollisols covers the Pampas of Argentina and Uruguay. In Eurasia, a great belt of Mollisols stretches from Rumania eastward across the steppes of Russia, Siberia, and Mongolia, into Manchuria.

Because of their well-developed granular texture and high base saturation, Mollisols are among the naturally most fertile soils in the world. They now produce the bulk of the grain that moves in commercial trade channels. Most of these soils have not been widely used for crop production except during the last century. Prior to that time they were used mainly for grazing by nomadic herds. The Mollisols have favorable characteristics for growing cereals in large-scale farming and are relatively easy to manage. Production varies considerably from one year to the next because seasonal rainfall is variable and soil-water storage is the limiting factor in productivity.

Of the seven suborders of Mollisols, four deserve special attention because of their widespread occurrence on well-drained upland surfaces. Be sure to refer to Figure 24.3 for the extent of each suborder in the United States and southern Canada.

Borolls are Mollisols of the cold-winter semiarid plains (steppes) (Figure 24.4G). *Udolls*, which are Mollisols of the udic soil-water regime, have brownish hues throughout and no horizon of accumulation of calcium carbonate (Figure 24.4H). Udolls are associated with the former tall-grass prairie and occupy large parts of Iowa, Illinois, and Missouri, well known as the corn belt.

Ustolls, Mollisols of the ustic soil-water regime, have a horizon of accumulation of soft, powdery calcium carbonate starting at a depth of about 50 to 100 cm (20 to 40 in.) (Figure 24.4I). A petrocalcic horizon may be formed within 1 m (3 ft) of the soil surface. In the southwestern United States, this material is known as *caliche* (Figure 18.7). Ustolls are associated with the short-grass prairies of the High Plains of South Dakota, Nebraska, Kansas, Oklahoma, and Texas (Figure 24.19). Their capacity to produce high yields of wheat is legendary.

The *Xerolls* are found in areas of the xeric soil-water regime under the Mediterranean climate of inland areas that have cool moist winters and rainless hot summers. Xerolls are brownish to reddish throughout the profile. They occupy a large area in parts of eastern Washington and Oregon, southern Idaho, and the northern parts of California and Nevada. In the Palouse Region, centering on the common meeting point of Washington, Oregon, and Idaho, Xerolls are a highly productive soil, derived from a parent layer of wind-deposited silt. Wheat and potatoes are important products of the Palouse soil.

Two other suborders of Mollisols are not shown on our maps. One is the *Aquolls*, formed in the aquic soil-water regime in places where the soil is seasonally saturated. Aquolls can be found in the Red River valley of northern Minnesota and North Dakota. The other suborder is the *Rendolls*, formed on highly calcareous parent matter, usually under a forest cover. It has very dark A and B horizons and a thick, dense C horizon of calcium carbonate (Figure 24.4J).

ARIDISOLS

Aridisols are uniquely defined through a combination of the following properties: (1) a lack of water available for

FIGURE 24.19 Exposed in this riverbank is a Ustoll developed on thick loess, which shows vertical columnar parting. (A. N. Strahler.)

FIGURE 24.20 A salic horizon, appearing as a white layer, lies close to the surface in this Orthid profile in the Nevada desert. The scale is marked in feet; 1 ft = 30 cm. (Soil Conservation Service.)

FIGURE 24.21 This gray desert soil, an Aridisol, has proved highly productive when cultivated and irrigated. The locality is near Palm Springs, California, in the Coachella Valley. (Ned L. Reglein.)

most plants for very extended periods; (2) one or more pedogenic horizons; (3) a surface horizon or horizons not significantly darkened by humus; and (4) absence of deep wide cracks. The Aridisols have practically no available soil water most of the time that the soil is warm enough for plant growth, for example, higher than 5°C (41°F). Soil water, in most years, is not available to plants for as long as 90 consecutive days when the soil temperature is above 8°C (47°F). Carbonate accumulations are often large at depth in the profile (Figure 24.4K).

Aridisols form under conditions that preclude entry of much water into the soil—either extremely scanty rainfall or slight rainfall that for one reason or another does not enter the soil. The vegetation consists of scattered plants, ephemeral grasses and forbs, cacti, and shrubs, all of which are well adapted to sustained drought.

The world soils map (Figure 24.2) shows that the Aridisols occupy regions on the climate map (Figure 8.9) of semidesert and desert subtypes of the dry tropical, subtropical, and midlatitude climates (4sd, 4d, 5sd, 5d, 9sd, 9d) and the Mediterranean climate (7s, 7sd). These same arid regions include large areas of Entisols, occupying areas of dune sand (Psamments) or rocky desert floor.

The Aridisols have two suborders. The suborder of *Argids* is illustrated by the profile in Figure 24.2K. Argids have an argillic horizon (labeled "Bt" on the profile) in which clay minerals have accumulated by illuviation. The Argids have formed largely on surfaces of late Pleistocene (Wisconsinan) age or older and, consequently, have developed throughout one or more of the Pluvial (moist) periods (Chapter 22). Illuviation of the B horizon is thought to have taken place during these moist periods. The other suborder, *Orthids*, are Aridisols without an argillic horizon. One variety of Orthids—the Salorthids—is characterized by having a salic horizon of salt accumulation. This horizon appears as a white layer in Figure 24.20.

Most Aridisols are used, as they have been through the ages, for nomadic grazing. This use is dictated by the limited rainfall, which is inadequate for crops without irrigation. Locally, where water supplies from exotic streams or ground water permit, Aridisols can be highly productive for a wide variety of crops under irrigation (Figure 24.21).

SOILS AND ALTITUDE

As we explained in Chapters 4 and 6, an increase in altitude brings lower air temperatures and increased orographic precipitation. Correspondingly, the soil-temperature and soil-water regimes undergo a change. The altitude effect is particularly conspicuous on the flanks of isolated mountain ranges within the dry midlatitude climate (9). Figure 24.22 shows a schematic soil profile, changing gradually with increased elevation as an observer traces soil horizons from a semiarid basin floor to the summit of a high mountain range—in this case the Bighorn Mountains in Wyoming. At the lowest elevation (left), the soil profile is that of an Aridisol (an Argid). With increasing elevation the profile becomes that of a Ustoll, which grades into the profile of a Udoll under cool moist conditions above 2300 m (7500 ft). On the summit uplands, about 2400 m (8000 ft), conditions are like those of the boreal forest climate. Here the soil profile is that of a Spodosol, formed under needleleaf forest.

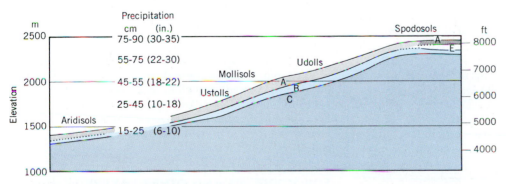

FIGURE 24.22 This schematic diagram shows the gradation of soils from a semiarid basin (left) to a cool moist climate (right) as one ascends the west slope of the Bighorn Mountains, Wyoming. (After J. Thorp.)

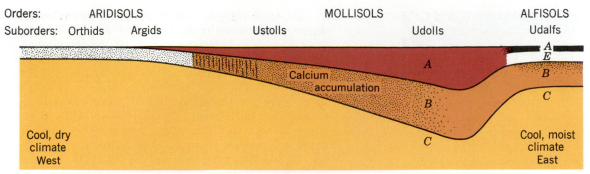

FIGURE 24.23 A schematic diagram of the changing soil profile from a cool dry desert on the west to a cool moist climate on the east. (After C. E. Millar, L. M. Turk, and H. D. Foth, *Fundamentals of Soil Science*, John Wiley and Sons, New York.)

Notice that a similar series of profile intergradations can be found in a west-to-east traverse across the United States, as shown in Figure 24.23, starting in the northern desert of the Great Basin and crossing the Midwest.

THE CANADIAN SYSTEM OF SOIL CLASSIFICATION*

The formation in 1940 of the National Soil Survey Committee of Canada was a milestone in the development of soil classification and of pedology generally in Canada. Prior to that time, the 1938 USDA system was used in Canada, but although the Canadian experience showed that the concept of zonal soils was useful in the western plains, it proved less applicable in eastern Canada where parent materials and relief factors had a dominant influence on soil properties and development in many areas.

Canadian pedologists observed closely the evolution of the U.S. Comprehensive Soil Classification System (CSCS) during the 1950s and 1960s and ultimately adopted several important features of that system. Nevertheless, the special needs of a workable Canadian national system required that a completely independent classification system be established. Because Canada lies entirely in a latitude zone poleward of the 40th parallel, there was no need to incorporate those orders found only in lower latitudes. Furthermore, the vast expanse of Canadian territory lying within the boreal forest and tundra climates necessitated the recognition at the highest taxonomic level (the order) of soils of cold regions that appear only as suborders and even as great groups in the CSCS—for example, the Cryaquents and Cryorthents that occupy much of the soils map of northern Canada above the 50th parallel (Figure 24.2).

The overall philosophy of the Canadian system is pragmatic: the aim is to organize the knowledge of soils in a reasonable and usable way. The system is a natural or taxonomic one in which the classes (taxa) are based upon properties of the soils themselves and not upon interpretations of the soils for various uses. Thus the taxa are concepts based upon generalization of properties of real bodies of soils rather than idealized concepts of the kinds of soils that would result from the action of presumed genetic processes. In this respect, the philosophy agrees with that used in the CSCS. Although taxa in the Canadian system are defined on the basis of actual soil properties that can be observed and measured, the system has a genetic bias in that properties or combinations of properties that reflect genesis are favored in distinguishing among the higher taxa. Thus the soils brought together under a single soil order are seen as the product of a similar set of dominant soil-forming processes resulting from broadly similar climatic conditions.

The Canadian system recognizes the pedon as the basic unit of soils; it is defined as in the CSCS. Major mineral horizons of the soil (A,B,C) are defined in much the same way as in the U.S. system. Thus the Canadian system of soil taxonomy is more closely related to the U.S. system than to any other. Both are hierarchical and the taxa are defined on the basis of measurable soil properties. However, they differ in several respects. The Canadian system is designed to classify only the soils that occur in Canada and is not a comprehensive system. The U.S. system includes the suborder, a taxon not recognized in the Canadian system. Because 90 percent of the area of Canada is not likely to be cultivated, the Canadian system does not recognize as diagnostic those horizons strongly affected by plowing and application of soil conditioners and fertilizers.

*Throughout this section numerous sentences and phrases are taken verbatim or paraphrased from the following work: Canada Soil Survey Committee, *The Canadian System of Soil Classification*, Research Branch, Canada Department of Agriculture, Publication 1646, 1978. Table 24.2 and Figure 24.24 are also compiled from this source.

SOIL HORIZONS AND OTHER LAYERS

The definitions of classes in the Canadian system are based mainly on the kinds, degrees of development, and the sequence of soil horizons and other layers in pedons.

The major mineral horizons are A, B, and C. The major organic horizons are L, F, and H, which are mainly forest litter at various stages of decomposition, and O, which is derived mainly from bog, marsh, or swamp vegetation. Subdivisions of horizons are labeled by adding lowercase suffixes to the major horizon symbols; for example, Ah or Ae.

Besides the horizons, nonsoil layers are recognized. Two such layers are R, rock, and W, water. Lower mineral layers not affected by pedogenic processes are also identified. In organic soils, layers are described as *tiers*.

The principal mineral horizons, A, B, and C, are defined as follows:

A   Mineral horizon formed at or near the surface in the zone of leaching or eluviation of materials in solution or suspension, or of maximum *in situ* accumulation of organic matter or both.

B   Mineral horizon characterized by enrichment in organic matter, sesquioxides, or clay; or by the development of soil structure; or by change of color denoting hydrolysis, reduction, or oxidation.

C   Mineral horizon comparatively unaffected by the pedogenic processes operative in A and B horizons. The processes of gleying and the accumulation of calcium and magnesium and more soluble salts can occur in this horizon.

Lowercase suffixes, used to designate subdivisions of horizons, are shown in Table 24.2.

SOIL ORDERS OF THE CANADIAN SYSTEM

Nine soil orders make up the highest taxon of the Canadian System of Soil Classification. Listed in alphabetical order, they are as follows:

Brunisolic	Gleysolic	Podzolic
Chernozemic	Luvisolic	Regosolic
Cryosolic	Organic	Solonetzic

Table 24.3 lists the great groups within each order.

BRUNISOLIC ORDER

The central concept of the *Brunisolic order* is that of soils under forest having brownish-colored Bm horizons. Most Brunisolic soils are well to imperfectly drained. They occur in a wide range of climatic and vegetative environments, including boreal forest; mixed forest, shrubs, and grass; and heath and tundra. As compared with the Chernozemic soils, the Brunisolic soils show a weak B horizon of accumulation attributable to their moister environ-

TABLE 24.2 Subhorizons and Organic Horizons of the Canadian System of Soil Classification

Subhorizons; Lowercase Suffixes

b   Buried soil horizon.

c   Cemented (irreversible) pedogenic horizon.

ca Horizon of secondary carbonate enrichment in which the concentration of lime exceeds that in the unenriched parent material.

e   Horizon characterized by the eluviation of clay, Fe, Al, or organic matter alone or in combination.

f   Horizon enriched with amorphous material, principally Al and Fe combined with organic matter; reddish near upper boundary, becoming yellower at depth.

g   Horizon characterized by gray colors, or prominent mottling, or both, indicative of permanent or intense reduction.

h   Horizon enriched with organic matter.

j   Used as a modifier of suffixes e, f, g, n, and t to denote an expression of, but failure to meet, the specified limits of the suffix it modifies.

k   Denotes presence of carbonate as indicated by visible effervescence when dilute HCl is added.

m   Horizon slightly altered by hydrolysis, oxidation, or solution, or all three to give a change in color or structure, or both.

n   Horizon in which the ratio of exchangeable Ca to exchangeable Na is 10 or less. It must also have the following distinctive morphological characteristics: prismatic or columnar structure, dark coatings on ped surfaces, and hard to very hard consistence when dry.

p   Horizon disturbed by human activities such as cultivation, logging, and habitation.

s   Horizon of salts, including gypsum, which may be detected as crystals or veins or as surface crusts of salt crystals.

sa Horizon with secondary enrichment of salts more soluble than Ca and Mg carbonates; the concentration of salts exceeds that in the unenriched parent material.

t   Illuvial horizon enriched with silicate clay.

u   Horizon that is markedly disrupted by physical or faunal processes other than cryoturbation.

x   Horizon of fragipan character. (A fragipan is a loamy subsurface horizon of high bulk density and very low organic matter content. When dry it is hard and seems to be cemented.)

y   Horizon affected by cryoturbation as manifested by disrupted or broken horizons, incorporation of materials from other horizons, and mechanical sorting.

z   A frozen layer.

Organic Horizons

O   Organic horizon developed mainly from mosses, rushes, and woody materials.

L   Organic horizon characterized by an accumulation of organic matter derived mainly from leaves, twigs and woody materials in which the organic structures are easily discernible.

F   Same as L, above, except that original structures are difficult to recognize.

H   Organic horizon characterized by decomposed organic matter in which the original structures are indiscernible.

TABLE 24.3 **Great Groups of the Canadian Soil Classification System**

Order	Great Group
Brunisolic	Melanic Brunisol
	Eutric Brunisol
	Sombric Brunisol
	Dystric Brunisol
Chernozemic	Brown
	Dark Brown
	Black
	Dark Gray
Cryosolic	Turbic Cryosol
	Static Cryosol
	Organic Cryosol
Gleysolic	Humic Gleysol
	Gleysol
	Luvic Gleysol
Luvisolic	Gray Brown Luvisol
	Gray Luvisol
Organic	Fibrisol
	Mesisol
	Humisol
	Folisol
Podzolic	Humic Podzol
	Ferro-Humic Podzol
	Humo-Ferric Podzol
Regosolic	Regosol
	Humic Regosol
Solonetzic	Solentz
	Solodized Solonetz
	Solod

ment. Brunisolic soils lack the diagnostic podzolic B horizon of the Podzolic soils, in which accumulation in the B horizon is strongly developed. The Melanic Brunisol shown in profile in Figure 24.24 can be found in the St. Lawrence Lowlands, surrounded by Podzolic soils (Figure 24.25).

CHERNOZEMIC ORDER

The general concept of the *Chernozemic order* is that of well to imperfectly drained soils having surface horizons darkened by the accumulation of organic matter from the decomposition of xerophytic or mesophytic grasses and forms representative of grassland communities or of grassland-forest communities with associated shrubs and forbs. The major area of Chernozemic soils is the cool, subarid to subhumid Interior Plains of western Canada. Most Chernozemic soils are frozen during some period each winter and the soil is dry at some period each summer. The mean annual temperature is higher than 0°C and usually less than 5.5°C. The associated climate is typically the semiarid (steppe) variety of the dry mid-latitude climate (Figure 24.25).

Essential to the definition of soils of the Chernozemic order is that they must have an A horizon (typically, Ah) in which organic matter has accumulated and they must meet several other requirements. The A horizon is at least 10 cm thick; its color is dark brown to black. It usually has sufficiently good structure that it is neither massive and hard nor single-grained when dry. The profile shown in Figure 24.24 is that of the Orthic subgroup of the Brown great group; it shows a Bm horizon that is typically of prismatic structure. The C horizon is one of

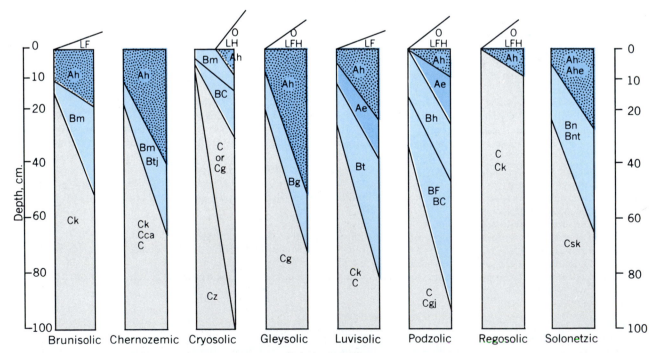

FIGURE 24.24 Representative schematic profiles of eight of the nine orders of the Canadian system of soil classification. Slanting lines show the range in depth and thickness of each horizon. (The horizon planes are actually approximately horizontal within the pedon.) See Table 24.2 for explanation of symbols. (From Canada Soil Survey Committee, Research Branch, Canada Department of Agriculture, 1978.)

lime accumulation (Cca). Clearly, the Chernozemic soils can be closely correlated with the Mollisols of the CSCS.

CRYOSOLIC ORDER

Soils of the *Cryosolic order* occupy much of the northern third of Canada where permafrost remains close to the surface of both mineral and organic deposits. Cryosolic soils predominate north of the tree line, are common in the subarctic forest area in fine textured soils, and extend into the boreal forest in some organic materials and into some alpine areas of mountainous regions. Cryoturbation (intense disturbance by freeze–thaw activity) of these soils is common, and it may be indicated by patterned ground features such as sorted and nonsorted nets, circles, polygons, stripes, and earth hummocks.

Cryosolic soils are found in either mineral or organic materials that have permafrost either within 1 m of the surface or within 2 m if more than one-third of the pedon has been strongly cryoturbated, as indicated by disrupted, mixed, or broken horizons. The profile shown in Figure 24.24 is that of the Orthic subgroup of the Static Cryosol great group. Note the presence of organic L, H, and O surface horizons and the thin Ah horizon. The Cryosolic soils are closely correlated with the Cryaquepts of the CSCS.

GLEYSOLIC ORDER

Soils of the *Gleysolic order* have features indicative of periodic or prolonged saturation with water and reducing conditions. They occur commonly in patchy association with other soils in the landscape. Gleysolic soils are usually associated with either a high ground-water table at some period of the year or temporary saturation above a relatively impermeable layer. Some Gleysolic soils

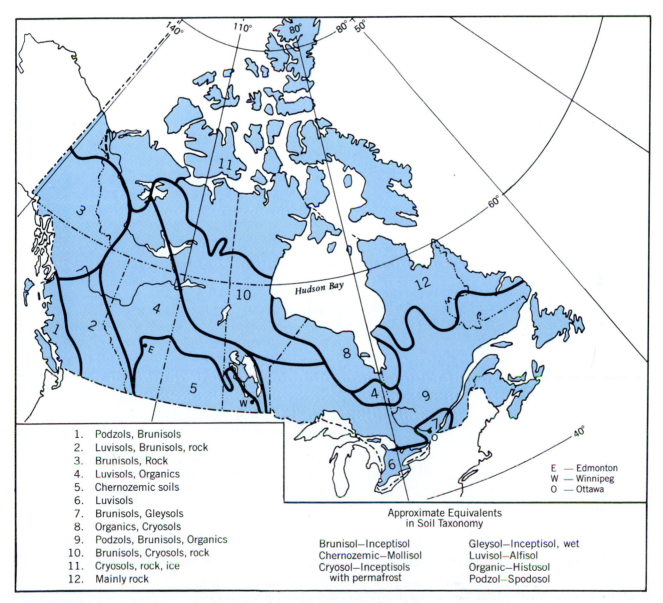

1. Podzols, Brunisols
2. Luvisols, Brunisols, rock
3. Brunisols, Rock
4. Luvisols, Organics
5. Chernozemic soils
6. Luvisols
7. Brunisols, Gleysols
8. Organics, Cryosols
9. Podzols, Brunisols, Organics
10. Brunisols, Cryosols, rock
11. Cryosols, rock, ice
12. Mainly rock

E — Edmonton
W — Winnipeg
O — Ottawa

Approximate Equivalents in Soil Taxonomy

Brunisol—Inceptisol
Chernozemic—Mollisol
Cryosol—Inceptisols with permafrost

Gleysol—Inceptisol, wet
Luvisol—Alfisol
Organic—Histosol
Podzol—Spodosol

FIGURE 24.25 Generalized map of soil regions of Canada. (Courtesy of Land Resource Research Institute, Agriculture Canada.) (Illustration is taken from *Fundamentals of Soil Science*, 7th ed., by Henry D. Foth [Wiley, 1984].)

may be submerged under shallow water throughout the year. Vegetation is, of course, hygrophytic. The profile shown in Figure 24.24 is that of the Gleysol great group. It has a thick Ah horizon. The underlying Bg horizon is grayish and shows mottling typical of reducing conditions.

LUVISOLIC ORDER

Soils of the *Luvisolic order* generally have light-colored, eluvial horizons (Ae) and they have illuvial B horizons (Bt) in which silicate clay has accumulated. These soils develop characteristically in well to imperfectly drained sites, in sandy loam to clay base-saturated parent materials under frost vegetation in subhumid to humid, mild to very cold climates. The genesis of Luvisolic soils is thought to involve the suspension of clay in the soil solution near the soil surface, downward movement of the suspended clay with the soil solution, and deposition of the translocated clay at a depth where downward motion of the soil solution ceases or becomes very slow. The representative profile shown in Figure 24.24 is that of the Orthic subgroup of the Gray Brown Luvisol great group.

Luvisolic soils occur from the southern extremity of Ontario to the zone of permafrost and from Newfoundland to British Columbia. The largest area of these soils are Gray Luvisols occurring in the central to northern Interior Plains under deciduous, mixed, and coniferous forest. In this location they appear to correlate with the Boralfs of the Alfisol order in the CSCS. Gray-Brown Luvisolic soils of southern Ontario would correlate with the suborder of Udalfs.

ORGANIC ORDER

Soils of the *Organic order* are composed largely of organic materials. They include most of the soils commonly known as peat, muck, or bog soils. Organic soils contain 17 percent or more organic carbon (30 percent organic matter) by weight. Most Organic soils are saturated with water for prolonged periods. They occur widely in poorly and very poorly drained depressions and level areas in regions of subhumid to perhumid climate and are derived from vegetation that grows in such sites. However, one group of Organic soils consists of leaf litter overlying rock or fragmental material; soils of this group may occur on steep slopes and rarely be saturated with water. (No profile of the Organic soils is shown in Figure 24.24.) Organic soils can be correlated with the Histosols of the CSCS.

PODZOLIC ORDER

Soils of the *Podzolic order* have B horizons in which the dominant accumulation product is amorphous material composed mainly of humified organic matter in varying degrees with Al and Fe. Typically Podzolic soils occur in coarse to medium textured, acid parent materials, under forest and heath vegetation in cool to very cold humid to perhumid climates. Podzolic soils can usually be recognized readily in the field. Generally they have organic surface horizons that are commonly L, F, and H. Most Podzolic soils have a reddish brown to black B horizon (Bh) with an abrupt upper boundary. The profile shown in Figure 24.24 is that of the Orthic subgroup of the Humic Podzol great group.

The Podzolic soils correspond closely with the Spodosols (Orthods) of the CSCS.

REGOSOLIC ORDER

Regosolic soils have weakly developed horizons. The lack of development of genetic horizons may be due to any number of factors: youthfulness of the parent material, recent alluvium; instability of the material, colluvium on slopes subject to mass wasting; nature of the material, nearly pure quartz sand; climate, dry cold conditions. Regosolic soils are generally rapidly to imperfectly drained. They occur in a wide range of vegetation and climates. The profile shown in Figure 24.24 is that of the Orthic subgroup of the Regosol great group. It has only a thin humic A horizon (Ah) and a surface horizon of organic materials.

Regosolic soils correspond with the Entisols of the CSCS.

SOLONETZIC ORDER

Soils of the *Solonetzic order* have B horizons that are very hard when dry and swell to a sticky mass of very low permeability when wet. Typically the Solonetzic B horizon has prismatic or columnar macrostructure that breaks into hard to extremely hard, blocky peds with dark coatings. Solonetzic soils occur on saline parent materials in some areas of the semiarid to subhumid Interior Plains in association with Chernozemic soils and to a lesser extent with Luvisolic and Gleysolic soils. Most Solonetzic soils are associated with a vegetative cover of grasses and forbs. The profile shown in Figure 24.24 is that of the Brown subgroup of the Solonetz great group.

Solonetzic soils are thought to have developed from parent materials that were more or less uniformly salinized with salts high in sodium. Leaching of salts by descending rainwater presumably results in deflocculation of the sodium-saturated colloids. The peptized colloids are apparently carried downward and deposited in the B horizon. Further leaching results in depletion of alkali cations in the A horizon, which becomes acidic, and a platy Ahe horizon usually develops. The underlying Solonetzic B horizon (Bn, Bnt) usually consists of darkly stained, fused, intact columnar peds. This stage is followed by the structural breakdown of the upper part of the B horizon and eventually to its complete destruction in the most advanced stage, known as *solodization*. Solonetzic soils are correlated with the suborder of Argids in the order of Aridisols under the CSCS.

GLOBAL SOILS IN REVIEW

In two chapters devoted to the basic elements of soil science, we have been limited to important concepts of soil formation and to a brief description of the soil orders and suborders. One dominant theme has been the role of climate in shaping the characteristics of the soil. This theme has made good use of a knowledge of the global heat and water balances and, more specifically, of the soil-water balance. Soils associated with water-surplus budgets are strikingly different in structure, composition, and fertility from those associated with budgets featuring a large soil-water shortage. Surprisingly, the most fertile soils from the standpoint of plant nutrients are those of the semiarid climates—Mollisols, for example. A great surplus of water in a warm climate is not conducive to producing rich agricultural soils because, under such conditions, the nutrient bases are leached from the soil and the clay minerals are altered to varieties with low capacity to hold base cations—as in the Oxisols, for example.

We have stressed too that soil character is also shaped by organisms living in or on the soil. Plant tissues supply humus that profoundly influences the soil profile. Moreover, plants recycle nutrient bases and thus delay or prevent their escape from the soil.

In the next three chapters, we turn to a study of the biosphere. First, we investigate the flow of energy and matter within the biosphere. Second, we develop concepts of biogeography—a geographic approach to ecosystems and their environments. Third, we examine the global distribution of vegetation of the lands, looking for distributional patterns linked to soil types and controlled by global climate patterns.

ENERGY FLOWS AND MATERIAL CYCLES IN THE BIOSPHERE

All living organisms of the earth, together with the environments with which they interact, constitute the *biosphere*. Organisms, whether belonging to the plant or animal kingdom, also interact with each other. The study of these interactions—in the form of exchanges of matter, energy, and stimuli of various sorts—between life forms and the environment is the science of *ecology*, very broadly defined. The total assemblage of components entering into the interactions of a group of organisms is known as an ecological system, or more simply, an *ecosystem*. The root *eco* comes from a Greek word connoting a house in the sense of household, which implies that a family lives together and interacts within a functional physical structure.

To the geographer, ecosystems are part of the physical composition of the life layer. A forest, for example, is not only a living community of organisms; it is also a collection of physical objects on the land surface. A forest is physically very different from an expanse of prairie grassland. Plant geographers are aware of these place-to-place differences in the physical aspect of vegetation. They attempt to categorize the varied types; they map the global distribution of vegetation forms.

Geographers also view ecosystems as natural resource systems. Food, fiber, fuel, and structural material are products of ecosystems; they represent organic compounds placed in storage by organisms through the expenditure of energy derived ultimately from the sun. Geographers are interested in the influence of climate on ecosystem productivity. Our basic understanding of global climates and the global range of soil-water budgets will prove most useful in explaining the global pattern of ecosystems on the lands.

Ecosystems have inputs of matter and energy that are used to build biological structures, reproduce, and maintain necessary internal energy levels. Matter and energy are also exported from an ecosystem. An ecosystem tends to achieve a balance of the various processes and activities within it. For the most part, these balances are quite sensitive and can be easily upset or destroyed.

THE ECOSYSTEM AND THE FOOD CHAIN

As an example of an ecosystem, consider a salt marsh (Figure 25.1). A variety of organisms are present: algae and aquatic plants, microorganisms, insects, snails, and crayfish, as well as such larger organisms as fishes, birds, shrews, mice, and rats. Inorganic components will be found as well: water, air, clay particles and organic sediment, inorganic nutrients, trace elements, and light energy. Energy transformations in the ecosystem occur by means of a series of steps or levels, referred to as a *food chain*.

The plants and algae in this chain are the *primary producers*. They use light energy to convert carbon dioxide and water into carbohydrates (long chains of sugar molecules) and eventually into other biochemical molecules needed for the support of life. This process of energy conversion is called photosynthesis. Organisms engaged in photosynthesis form the base of the food chain.

At the next level are the *primary consumers* (the snails, insects, and fishes); they live by feeding on the producers. At a still higher level are the *secondary consumers* (the mammals and birds); they feed on the primary consumers. As in many ecosystems, still higher levels of feeding are evident: the marsh hawks and owls. The *decomposers* feed on detritus, or decaying organic matter, derived from all levels. They are mostly microscopic organisms (microorganisms) and bacteria.

The food chain is really an energy flow system, tracing the path of solar energy through the ecosystem. Solar energy is stored by one class of organisms, the primary producers, in the chemical products of photosynthesis. As these organisms are eaten and digested by consumers, chemical energy is released. This chemical energy is used to power new biochemical reactions, which again produce stored chemical energy in the bodies of the consumers.

At each level of energy transformation, energy is lost as waste heat. In addition, much of the energy input to each organism must be used in respiration. Respiration can be thought of as the burning of fuel to keep the organism operating: it will be discussed in more detail in the

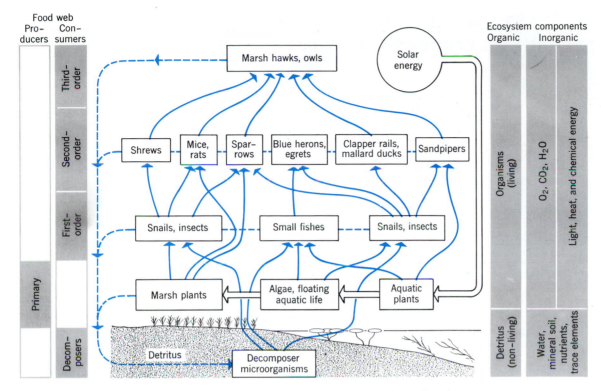

FIGURE 25.1 Flow diagram of a salt-marsh ecosystem in winter. The arrows show how energy flows from the sun to producers, consumers, and decomposers. (Food chain after R. L. Smith, *Ecology and Field Biology*, Harper and Row, New York.)

next section. Energy expended in respiration is used for bodily maintenance and cannot be stored for use by other organisms higher up in the food chain. This means that, generally, both the numbers of organisms and their total amount of living tissue must decrease drastically up the food chain.

PHOTOSYNTHESIS AND RESPIRATION

Stated in the simplest possible terms, *photosynthesis* is the production of carbohydrate. *Carbohydrate* is a general term for a class of organic compounds consisting of the elements carbon, hydrogen, and oxygen. Carbohydrate molecules are composed of chains of carbon molecules with hydrogen (H) atoms and hydroxyl (OH) atompairs attached to their sides. We can symbolize a single carbon atom with its attached hydrogen atom and hydroxyl molecules as —CHOH—. The leading and trailing dashes indicate that the unit is just one portion of a longer chain.

Photosynthesis of carbohydrate requires a series of complex biochemical reactions using water (H_2O) and carbon dioxide (CO_2) as well as light energy. A simplified chemical reaction for photosynthesis can be written as follows:

$$H_2O + CO_2 + \text{light energy} \rightarrow \text{—CHOH—} + O_2$$

Oxygen in the form of gas molecules (O_2) is a by-product of photosynthesis.

Respiration is the process opposite to photosynthesis, in which carbohydrate is broken down and combined with oxygen to yield carbon dioxide and water. The overall reaction is as follows:

$$\text{—CHOH—} + O_2 \rightarrow CO_2 + H_2O + \text{chemical energy}$$

As in the case of photosynthesis, the actual reactions are far from simple. The chemical energy released is stored in many types of energy-carrying molecules and used later to synthesize all the biological molecules necessary to sustain life.

At this point, it is helpful to link photosynthesis and respiration in a continuous cycle involving both the primary producer and the decomposer. Figure 25.2 shows one closed loop for hydrogen (H), one for carbon (C), and two loops for oxygen (O). We are not taking into account that there are two atoms of hydrogen in each molecule of water and carbohydrate, or that there are two atoms of oxygen in each molecule of carbon dioxide and oxygen gas. Only the flow pattern counts in this representation.

A good place to start is the soil, from which water is drawn up into the body of a living plant. In the green leaves of the plant, photosynthesis takes place while light energy is absorbed by the leaf cells. Carbon dioxide is being brought in from the atmosphere at this point. Here oxygen is liberated and begins its atmospheric cycle. The plant tissue then dies and falls to the ground, where it is acted on by the decomposer. Through respiration, oxygen is taken out of the atmosphere or soil air and combined with the decomposing carbohydrate. Energy is

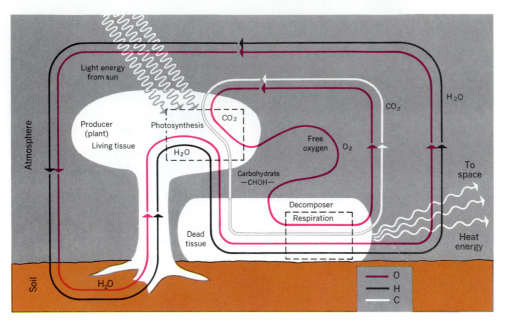

FIGURE 25.2 A simplified flow diagram of the essential components of photosynthesis and respiration through the biosphere.

now liberated. Here both carbon dioxide and water enter the atmosphere as gases.

An important concept emerges from this flow diagram. Energy passes through the system. It comes from the sun and returns eventually to outer space. On the other hand, the material components—hydrogen, oxygen, and carbon—are recycled within the total system. Of course, many other material components are recycled in the same way. These are plant nutrients, essential in the growth of plants. Nutrients are constantly recycled. Because the earth as a planet is a closed system, the material components never leave the total system; but they can be stored in other ways and forms where they are unavailable for use by plants for prolonged periods of geologic time. We shall develop this concept more fully later in this chapter.

NET PHOTOSYNTHESIS

Because both photosynthesis and respiration go on simultaneously in a plant, the amount of new carbohydrate placed in storage is less than the total carbohydrate being synthesized. We must thus distinguish between gross photosynthesis and net photosynthesis. *Gross photosynthesis* is the total amount of carbohydrate produced by photosynthesis; *net photosynthesis* is the amount of carbohydrate remaining after respiration has broken down sufficient carbohydrate to power the plant. Stated as an equation,

Net photosynthesis = Gross photosynthesis − Respiration.

Because both photosynthesis and respiration occur in the same cell, gross photosynthesis cannot be measured readily. Instead, we shall deal with net photosynthesis. In most cases, respiration will be held constant, so use of the net instead of the gross will show the same trends.

The rate of net photosynthesis is strongly dependent on the intensity of light energy available, up to a limit. Figure 25.3 shows this principle. The rate of net photosynthesis is indicated on the vertical axis by the rate at which a plant takes up carbon dioxide. On the horizontal axis, light intensity increases from left to right. At first, net photosynthesis rises rapidly as light intensity increases; but the rate then slows and reaches a maximum value, shown by the plateau in the curve. Above this maximum, the rate falls off because the incoming light is also causing heating. Increased temperature, in turn, increases the rate of respiration, which offsets gross production by photosynthesis.

Light intensity sufficient to allow maximum net photosynthesis is only 10 to 30 percent of full summer sunlight for most green plants. Additional light energy is simply ineffective. Duration of daylight then becomes the important factor in the rate at which products of photosynthesis accumulate as plant tissues. On this subject, you can draw on your knowledge of the seasons and the changing angle of the sun's rays with latitude. Figure

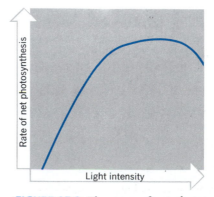

FIGURE 25.3 The curve of net photosynthesis shows a steep initial rise, then levels off as light intensity rises.

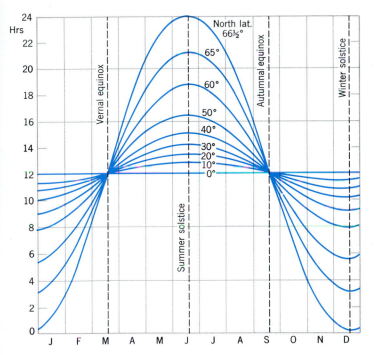

FIGURE 25.4 Duration of the daylight period at various latitudes throughout the year. The vertical scale gives the number of hours the sun is above the horizon.

25.4 shows the duration of the daylight period with changing seasons for a wide range of latitudes in the northern hemisphere. At low latitudes, days are not far from the average 12-hour length throughout the year; at high latitudes, days are short in winter but long in summer. The seasonal contrast in day length increases with latitude. In subarctic latitudes, photosynthesis can obviously go on in summer during most of the 24 hours, and this factor can compensate partly for the shortness of the season.

Photosynthesis also increases in rate as air temperature increases, up to a limit. Figure 25.5 shows the results of a laboratory experiment in which sphagnum moss was grown under constant illumination. Gross photosynthesis increased rapidly to a maximum at about 20°C (70°F), then leveled off. Respiration increased quite steadily to the limit of the experiment. Net photosynthesis, which is the difference between the values in the two curves, peaked at about 18°C (65°F), then fell off rapidly.

ENERGY FLOW ALONG THE FOOD CHAIN

The primary producers trap the sun's energy and make it available to support consuming organisms at higher levels in the food chain. But remember that energy is lost during each upward step in the chain, so the number of steps is limited. In general, anywhere from 10 to 50 percent of the energy stored in organic matter at one level can be passed up the chain to the next level. Four levels of consumers is about the normal limit.

Figure 25.6 is a bar graph showing the percentage of energy passed up the chain when only 10 percent moves from one level to the next. The horizontal scale is in powers of 10. In ecosystems of the lands, the mass of organic matter also decreases with each upward step in the chain; the number of individuals of the consuming animals also decreases with each upward step. In the food chain shown in Figure 25.1, there are only a few marsh hawks and owls in the third level of consumers, whereas there are countless individuals in the primary level. These facts serve as a clear message to humans; they tell us that when we raise beef cattle, swine, sheep, and fowl to eat, we are being quite wasteful of the world's total food resource. These animals, in storing food for humans to consume, have already wasted two-thirds or more of the energy they consumed as plant matter from the producing level. If we instead subsisted largely on grains and other food plants, such as legumes, grown on the available farm lands, our food resource could support a much larger population. Although this observation does not apply to sheep and cattle subsisting entirely on grazing lands that are unfit for cereal crops, it is all the same a potent concept.

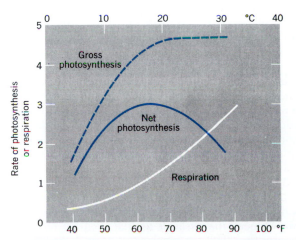

FIGURE 25.5 Respiration and gross and net photosynthesis vary with temperature. (Data of Stofelt, in A. C. Leopold, *Plant Growth and Development*, McGraw-Hill, New York.)

THE ENERGY FLOW SYSTEM OF A GREEN PLANT

The energy flow system of a green plant is shown in Diagram A of Figure 25.7. Solar shortwave energy falling on a green leaf is partly reflected and partly absorbed.

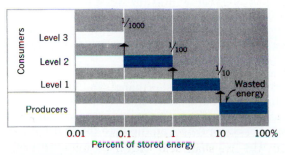

FIGURE 25.6 Percentage of energy passed up the steps of the food chain, assuming 90 percent is lost energy at each step.

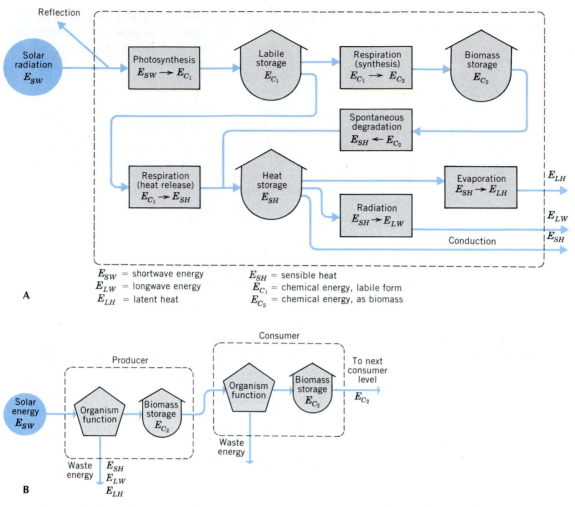

E_{SW} = shortwave energy E_{SH} = sensible heat
E_{LW} = longwave energy E_{C_1} = chemical energy, labile form
E_{LH} = latent heat E_{C_2} = chemical energy, as biomass

FIGURE 25.7 Schematic diagram of the energy flow systems of a green plant and a food chain.

Disregarding conversion to sensible heat within the leaf, we concentrate on the process of photosynthesis in which shortwave energy is converted into chemical energy, stored within molecules in plant cells. This initial form of chemical energy is referred to as *labile energy*, meaning that it is energy continually undergoing change by respiration and conversion into chemical energy stored in more complex molecules. Thus the respiration process transforms the labile energy into *biomass energy* that is stored in the tissues of the plant. Spontaneous degradation—an oxidation process—converts the biomass energy into sensible heat, which is held in temporary storage. Stored sensible heat is disposed of through evaporation, leaving the plant as latent heat, and by longwave radiation or direct conduction to the atmosphere.

Diagram B shows energy flow up the food chain. We have used a pentagon as a shorthand symbol for all the energy changes that are shown in Diagram A as leading to biomass storage. Stored biomass energy within the primary producer provides an input to the first level of consumer. These diagrams are repeated for higher levels in the food chain.

NET PRIMARY PRODUCTION

Plant ecologists measure the accumulated net production by photosynthesis in terms of the *biomass*, which is the dry weight of the organic matter. This quantity could, of course, be stated for a single plant or animal; but a more useful statement is made in terms of the biomass per unit of surface area within the ecosystem—the hectare, square meter, or square foot. Of all ecosystems, forests have the greatest biomass; that of grasslands and croplands is very small in comparison. For freshwater bodies and the oceans, the biomass is even smaller—on the order of one-hundredth that of the grasslands and croplands.

The annual rate of production of organic matter is the vital information we need, not the biomass itself. Granted that forests have a very large biomass and grasslands very little, what we want to know is, which ecosystem is the more productive? In other words, which ecosystem produces the greatest annual yield in terms of useful resources? For the answer we turn to figures on the *net primary production* of various ecosystems. Table

TABLE 25.1 Net Primary Production for Various Ecosystems

	Grams per Square Meter per Year	
	Average	Typical Range
Lands		
Rainforest of the equatorial zone	2000	1000–5000
Freshwater swamps and marshes	2500	800–4000
Midlatitude forest	1300	600–2500
Midlatitude grassland	500	150–1500
Agricultural land	650	100–4000
Lakes and streams	500	100–1500
Extreme desert	3	0–10
Oceans		
Algal beds and reefs	2000	1000–3000
Estuaries (tidal)	1800	500–4000
Continental shelf	360	300–600
Open ocean	125	1–400

25.1 gives this information in units of grams of dry organic matter produced annually from 1 sq m of surface. The figures are rough estimates, but they are nevertheless highly meaningful. Note that the highest values are in two quite unlike environments: forests and wetlands (estuaries). Agricultural land compares favorably with grassland; but the range is very large in agricultural land, reflecting many factors, such as availability of soil water, soil fertility, and use of fertilizers and machinery.

Productivity of the oceans is generally low. The deepwater oceanic zone is the least productive of the marine ecosystems. At the same time, however, it comprises about 90 percent of the world ocean area. Continental shelf areas are a good deal more productive and, in fact, support much of the world's fishing industry at the present time (Figure 25.8).

Upwelling of cold water from ocean depths, such as occurs in polar oceans, brings nutrients to the surface and greatly increases the growth of microscopic floating plants (phytoplankton). These, in turn, serve as food sources for marine animals in the food chain. Consequently, zones of upwelling near habitable coastlines are highly productive fisheries. An example is the Peru Current off the west coast of South America. Here, countless individuals of a single species of small fish, the anchoveta, provide food for larger fish and for birds. The birds, in turn, excrete their wastes on the mainland coast of Peru. The accumulated deposit, called guano, was a rich source of fertilizer.

Within the past decade or so, remote sensing has come into use as a tool for measuring and mapping primary productivity on a global scale. Figure 25.9 shows two polar views of the earth, displaying in image format satellite measurements of phytoplankton pigment concentration, which is an index of photosynthetic activity near the ocean surface. The red and yellow colors, marking high pigment concentrations, emphasize the importance of the polar seas in ocean productivity. The data were collected over a period of years, and show averages. It is important to realize that the productivity is highly seasonal, because it is limited by light. Thus polar productivity falls to near zero during the polar winter.

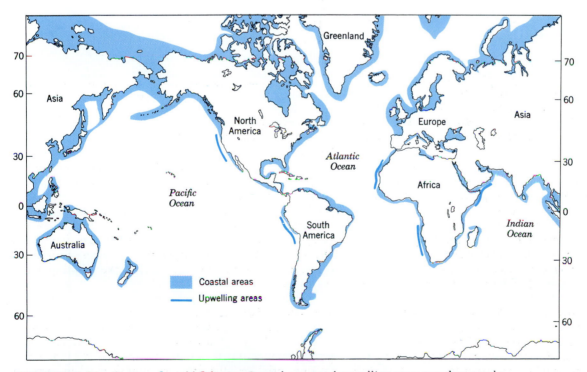

FIGURE 25.8 Distribution of world fisheries. Coastal areas and upwelling areas together supply over 99 percent of world production. (Compiled by the National Science Board, National Science Foundation.)

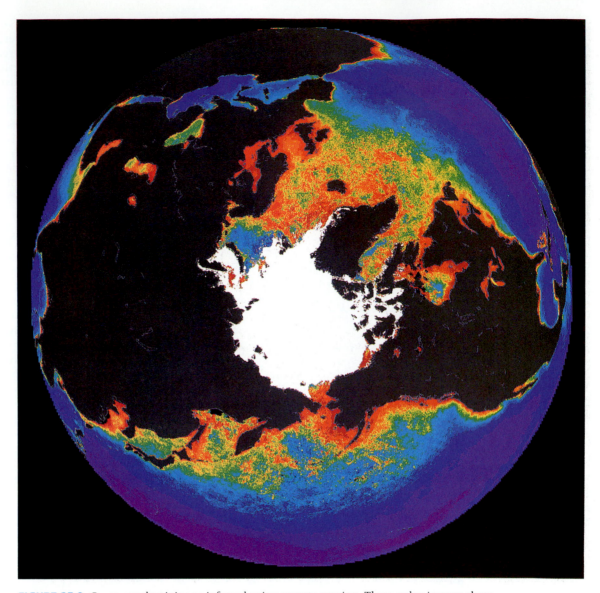

FIGURE 25.9 Ocean productivity as inferred using remote sensing. These polar images show phytoplankton pigment concentrations averaged over the period 1978 to 1986. They were made by processing data obtained from the Coastal Zone Color Scanner instrument on the *Nimbus-7* satellite. Pigment concentrations are indicated by colors: red (highest) through yellow, green, and blue, to purple (lowest). (NOAA/NESDIS/NCDC/SDSD.)

NET PRODUCTION AND CLIMATE

The geographer is interested in the climatic factors controlling organic productivity. Besides light intensity and temperature, an obvious factor will be availability of water. Soil-water shortage or surplus might be the best climatic factor to examine, but data are not available. Ecologists have related net annual primary production to mean annual precipitation, as shown in Figure 25.10. The production values are for plant structures above the ground surface. Although the productivity increases rap-

FIGURE 25.10 Net primary production increases rapidly with increasing precipitation, but levels off in the higher values. Observed values fall mostly within the shaded zone. (Data of Wittaker, 1970.)

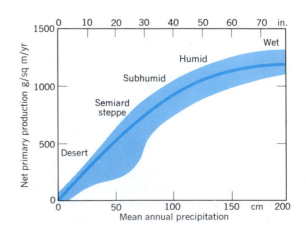

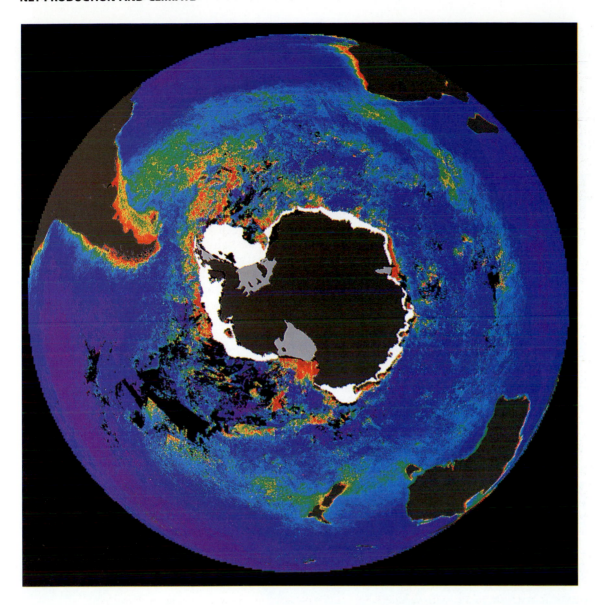

idly with precipitation in the lower range from desert through semiarid to subhumid climates, it seems to level off in the humid range. Apparently, a large soil-water surplus carries with it some counteractive influence, such as removal of plant nutrients by leaching.

Combining the effects of light intensity, temperature, and precipitation, you would guess that the maximum net productivity year-round would be in wet equatorial climates where soil water is adequate at all times. Productivity would be low in all desert climates and would decrease generally with increasing latitude. Productivity would be reduced to near zero in the arctic zone, where the combination of a short growing season and low temperatures would act together to slow the growth process.

These deductions are nicely displayed on a schematic diagram of net primary productivity on an imagined supercontinent (Figure 25.11). Comparing this diagram with the world climate diagram (Figure 8.8), we can assign rough values of productivity to each of the climates

(units are grams of carbon per square meter per year):

Highest (over 800)	Wet equatorial (1)
Very high (600–800)	Monsoon and trade-wind littoral (2) Wet-dry tropical (3)
High (400–600)	Wet-dry tropical (3) (Southeast Asia) Moist subtropical (6) Marine west coast (8)
Moderate (200–400)	Mediterranean (7) Moist continental (10)
Low (100–200)	Dry tropical, semiarid (4s) Dry midlatitude, semiarid (9s) Boreal forest (10)
Very low (0–100)	Dry tropical, desert (4d) Dry midlatitude, desert (9d) Boreal forest (10) Tundra (12)

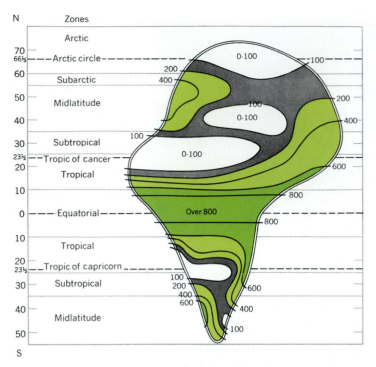

The wet-dry tropical climate (3) is in a narrow transition zone between very high productivity and low productivity, so some areas are in the "very high" rating and others are in the "high" rating. Transition zones such as this are highly sensitive to alternating cycles of drought and excessive rainfall. They pose the most difficult famine problem of all global regions because they carry dense populations, which can be fed adequately only under the best conditions.

Remote sensing can also be used to monitor net primary productivity of the earth's land surface. Figure 25.12 shows oceanic phytoplankton pigment concentration (as in Figure 25.9), in addition to which a land vegetation productivity index has been overlain on the continents. The dark-green colors are assigned to the most productive areas, while the reddish-purplish tones indicate the least productive. The world's deserts show clearly, and present an obvious contrast with equatorial rainforest regions. Tropical savanna and midlatitude grassland are intermediate in productivity, whereas boreal forest and tundra are low.

FIGURE 25.11 Schematic diagram of net primary production of organic matter on an imagined supercontinent. The units are grams of carbon per square meter per year. The quantities shown are estimates only. (Based on data of D. E. Reichle, 1970.)

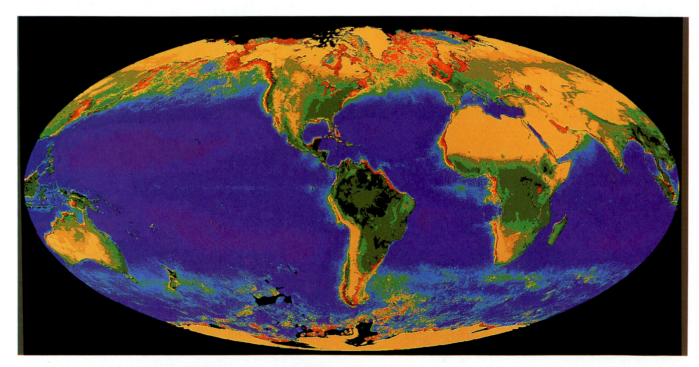

FIGURE 25.12 Global productivity as inferred using remote sensing. Oceans: From NOAA's Coastal Zone Color Scanner instrument as described for Figure 25.9. Lands: Displayed is a vegetation greenness index for the period April 1982 to March 1985, obtained from the Advanced Very High Resolution Radiometer (AVHRR) instrument aboard the *NOAA-7* Spacecraft. The most productive regions are shown in dark green tones, ranging through medium and light green tones to yellow-brown tones (least productive). (NOAA/NESDIS/NCDC/SDSD.)

BIOMASS ENERGY

Net primary production represents a source of renewable energy derived from the sun that can be exploited to fill human energy needs. The use of biomass as an energy source involves releasing solar energy that has been fixed in plant tissues through photosynthesis. This process can take place in a number of ways—the simplest is direct burning of plant matter as fuel, as in a campfire or a wood-burning stove. Other approaches involve the generation of intermediate fuels from plant matter—methane gas, charcoal, and alcohol, for example. Biomass energy conversion is not highly energy efficient; typical values for net annual primary production of plant communities range from 1 to 3 percent of available solar energy. However, the abundance of terrestrial biomass is so great that biomass utilization could provide the energy equivalent to 3 million barrels of oil per day for the United States by the year 2000.

One important use of biomass energy is the burning of firewood for cooking (and some space heating) in developing nations. The annual growth of wood in the forest of developing countries totals about half the world's energy production—plenty of firewood is thus available. However, fuelwood use exceeds production in many areas, creating local shortages and severe strains on some forest ecosystems. The forest–desert transition areas of thorntree, savanna, and desert scrub in central Africa south of the Sahara Desert are examples.

Even in closed stoves, wood burning is not very efficient, ranging from 10 to 15 percent for cooking. However, the conversion of wood to charcoal and/or gas can boost efficiencies to values as high as 70 to 80 percent with appropriate technology. In this process, termed pyrolysis, controlled partial burning in an oxygen-deficient environment reduces carbohydrate to free carbon (charcoal), and yields flammable gases such as carbon monoxide and hydrogen. Charcoal is more energy efficient than wood, burns more cleanly, and is easier to transport. As an added advantage, charcoal can be made from waste fibers and agricultural residues that would normally be discarded. Thus, charcoal is an efficient fuel that can help extend the firewood supply in areas where wood is in high demand.

A second method of extracting energy from biomass uses anaerobic digestion to produce *biogas*. In this process, animal and human wastes are fed into a closed digesting chamber, where anaerobic bacteria break down the waste to produce a gas that is a mixture of methane and carbon dioxide. The biogas can be easily burned for cooking or heating, or may be used to generate electric power. The digested residue is a sweet-smelling fertilizer. China now maintains a vigorous program of construction of biogas digesters for the use of small family units. The benefits include better sanitation and reduced air pollution, as well as more efficient fuel usage.

Another use of biomass of increasing importance is the conversion of agricultural wastes to alcohol. In this process, yeast microorganisms are used to convert the carbohydrate to alcohol through fermentation. With the rising price of motor fuels, alcohol achieved popularity as a substitute and extender for gasoline. Gasohol, a mixture of up to 10 percent alcohol in gasoline, can be burned in conventional engines without adjustment.

Brazil, a country without adequate petroleum production, has relied heavily on alcohol fuel derived from sugar cane. In 1988 alcohol provided 63 percent of Brazil's automotive fuel needs. Distillation of alcohol, however, requires heating, thus greatly reducing the net energy yield. Alcohol, charcoal, and firewood are all alternatives to fossil fuels that will become increasingly important as petroleum becomes scarcer and more costly in the coming decades.

Relying on biomass energy can also yield important benefits in reducing the carbon dioxide emissions that enhance the greenhouse effect and promote climate warming. Although CO_2 is released by combustion of biomass, it is CO_2 that would be released in a relatively short period of time through normal decay processes, in any event. Moreover, the power generated saves the burning of fossil fuels that would add permanently to atmospheric CO_2.

MATERIAL CYCLES IN ECOSYSTEMS

We have seen how energy of solar orgin flows through ecosystems, passing from one part of the food chain on to the next, until it is ultimately lost from the biosphere as energy radiated to space. Matter also moves through ecosystems, but because gravity keeps surface material earthbound, matter cannot be lost in the global ecosystem. As molecules are formed and reformed by chemical and biochemical reactions within an ecosystem, the atoms that compose them are not changed or lost. Thus matter is conserved within an ecosystem, and atoms and molecules can be used and reused, or cycled, within ecosystems.

Atoms and molecules move through ecosystems under the influence of both physical and biological processes. The pathways of a particular type of matter through the earth's ecosystem comprise a *material cycle* (sometimes referred to as a *biogeochemical cycle*, or *nutrient cycle*).

Ecologists recognize two types of material cycles—gaseous and sedimentary. In the *sedimentary cycle*, the compound or element is released from rock by weathering, then follows the movement of running water either in solution or as sediment to the sea. Eventually, by precipitation and sedimentation, these materials are converted into rock. When the rock is uplifted and exposed to weathering, the cycle is completed.

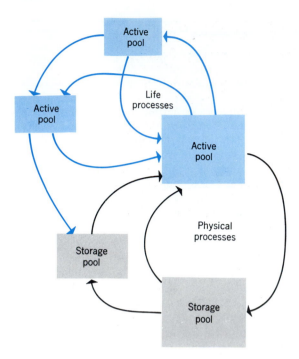

FIGURE 25.13 General features of a material cycle.

TABLE 25.2 Elements Comprising Global Living Matter, Taking 100 Percent as the Total of the 15 Most Abundant Elements

Basic Carbohydrate	
Hydrogen (M)	49.74
Carbon (M)	24.90
Oxygen (M)	24.83
	Subtotal 99.47
Other Nutrients	
Nitrogen (M)	0.272
Calcium (M)	0.072
Potassium (M)	0.044
Silicon	0.033
Magnesium (M)	0.031
Sulfur (M)	0.017
Aluminum	0.016
Phosphorus (M)	0.013
Chlorine	0.011
Sodium	0.006
Iron	0.005
Manganese	0.003
	M—macronutrient

Data source: E. S. Deevey, Jr., *Scientific American*, Vol. 223.

In the *gaseous cycle*, a shortcut is provided—the element or compound can be converted into a gaseous form. The gas diffuses throughout the atmosphere and thus arrives over land or sea, to be reused by the biosphere, in a much shorter time. The primary constituents of living matter—carbon, hydrogen, oxygen, and nitrogen—all move through gaseous cycles.

The major features of a material cycle are diagrammed in Figure 25.13. Any area or location of concentration of a material is a *pool*. There are two types of pools: *active pools*, where materials are in forms and places easily accessible to life processes, and *storage pools*, where materials are more or less inaccessible to life. A system of pathways of material flows connects the various active and storage pools within the cycle. Pathways between active pools are usually controlled by life processes, whereas pathways between storage pools are usually controlled by physical processes.

The magnitudes of the total storage and total active pools can be very different. In many cases, the active pools are much smaller than storage pools, and materials move more rapidly between active pools than between storage pools or in and out of storage. Taking an example from the carbon cycle, photosynthesis and respiration will cycle all the carbon dioxide in the atmosphere (active pool) through plants in about 10 years. But it may be many millions of years before the carbonate sediments (storage pool) now forming as rock will be uplifted and decomposed to release carbon dioxide.

NUTRIENT ELEMENTS IN THE BIOSPHERE

We will take the 15 most abundant elements in living matter as a whole and designate their total mass as 100 percent. Percentages of each of the elements are given in Table 25.2. The three principal components of carbohydrate—hydrogen, carbon, and oxygen—account for almost all living matter and are called *macronutrients*. The remaining one-half percent is divided among 12 elements. Six of these are also macronutrients: nitrogen, calcium, potassium, magnesium, sulfur, and phosphorus. The macronutrients are all required in substantial quantities for organic life to thrive. The first three macronutrients—hydrogen, carbon, oxygen—are materials whose pathways we have already followed in the photosynthesis-respiration circuits (Figure 25.2). We will undertake a more detailed analysis of the gaseous cycles of carbon and oxygen because they are influenced by our combustion of hydrocarbon compounds. We will also give special attention to nitrogen, the fourth most abundant element in the composition of living matter; it also moves in a gaseous cycle. Of the remaining macronutrients, three—calcium, potassium, and magnesium—are elements derived from silicate rocks through mineral weathering. Two other macronutrients derived from rock weathering are sulfur and phosphorus.

THE CARBON CYCLE

The movements of carbon through the life layer are of great importance because all life is composed of carbon compounds of one form or another. Of the total carbon available, most lies in storage pools as carbonate sediments below the earth's surface. Only about two-tenths of a percent are readily available to organisms as CO_2 or as decaying biomass in active pools.

Some details of the *carbon cycle* are shown in a schematic diagram, Figure 25.14. In the gaseous portion of its cycle, carbon moves as carbon dioxide (CO_2) as a free gas in the atmosphere and as a gas dissolved in fresh water of the lands and in salt water of the oceans. In

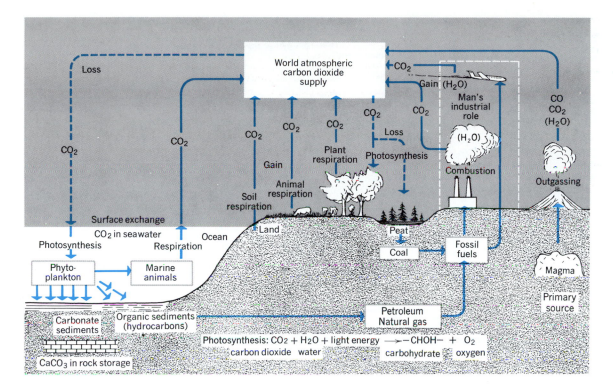

FIGURE 25.14 The carbon cycle. (Redrawn from A. N. Strahler, *Planet Earth: Its Physical Systems Through Geologic Time*, Harper & Row, Publishers. Copyright © by Arthur N. Strahler.)

the sedimentary portion of its cycle, carbon resides in carbohydrate molecules in organic matter, as hydrocarbon compounds in rock (petroleum, coal), and as mineral carbonate compounds such as calcium carbonate ($CaCO_3$). The world supply of atmospheric carbon dioxide is presented in Figure 25.14 by a box. It is a small portion of the carbon in active pools, constituting less than 2 percent. This atmospheric pool is supplied by respiration from plant and animal tissues in the oceans and on the lands. Under natural conditions, some new carbon enters the atmosphere each year from volcanoes by outgassing in the form of CO_2 and carbon monoxide (CO). Industry injects substantial amounts of carbon into the atmosphere through combustion of fossil fuels. This increment from fuel combustion and its probable effects on global air temperatures were discussed in Chapter 4.

Carbon dioxide leaves the atmospheric pool to enter the oceans, where it is used in photosynthesis by minute marine plants, the phytoplankton. These organisms are primary producers in the ocean ecosystem and are consumed by marine animals in the food chain. Phytoplankton also build skeletal structures of calcium carbonate. This mineral matter settles to the ocean floor to accumulate as sedimentary strata, an enormous storage pool not available to organisms until released later by rock weathering. Organic compounds synthesized by phytoplankton also settle to the ocean floor and eventually are transformed into the hydrocarbon compounds making up petroleum and natural gas. On the lands accumulating plant matter has, under geologically favorable circumstances, formed massive layers of peat, which were transformed into coal. Petroleum, natural gas, and coal

comprise the fossil fuels, and these represent huge storage pools of carbon.

THE OXYGEN CYCLE

Details of the *oxygen cycle* are shown in schematic form in Figure 25.15. The complete picture of the cycling of oxygen also includes its movements and storages when combined with carbon in carbon dioxide and in organic and inorganic compounds. These we have covered in the carbon cycle.

The world supply of atmospheric free oxygen is shown in Figure 25.15 by a box at the top of the diagram. Oxygen enters this storage pool through release in photosynthesis, both in the oceans and on the lands. Each year a small amount of new oxygen comes from volcanoes through outgassing, principally as CO_2 and H_2O (shown in Figure 25.14). Balancing the input to the atmospheric storage pool is loss through organic respiration and mineral oxidation. Adding to the withdrawal from the atmospheric oxygen pool is industrial activity through the combustion (oxidation) of wood and fossil fuels. Forest fires and grass fires (not shown) are another means of oxygen consumption. The oceans also contain a small active pool of dissolved gaseous oxygen. Some oxygen is continuously placed in storage in mineral carbonate form in ocean-floor sediments.

Human activity reduces the amount of oxygen in the air by (1) burning fossil fuels; (2) clearing and draining land, which speeds the oxidation of soils and soil organic matter; and (3) reducing photosynthesis by clearing for-

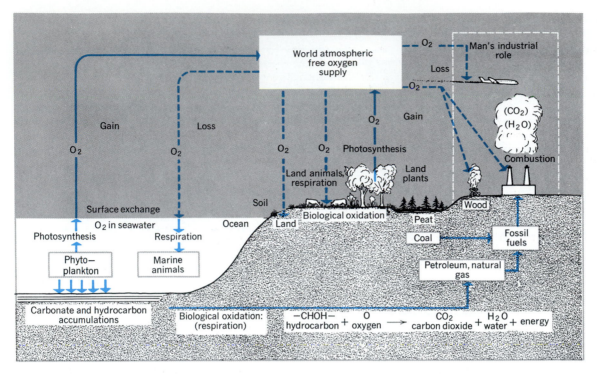

FIGURE 25.15 The oxygen cycle. (Redrawn from A. N. Strahler, *Planet Earth: Its Physical Systems Through Geologic Time*, Harper & Row, Publishers. Copyright © by Arthur N. Strahler.)

ests for agriculture and by paving and covering previously productive surfaces. The importance of urbanization can be appreciated by the fact that every 6 months a land area about the size of Rhode Island is covered by new construction in the United States alone. Fortunately, the oxygen pool is so large that the human impact or potential impact is very small, at least at this time.

THE NITROGEN CYCLE

Nitrogen moves through the biosphere in a gaseous *nitrogen cycle* in which the atmosphere, containing 78 percent nitrogen by volume, is a vast storage pool (Figure 25.16). Nitrogen in the atmosphere in the form of N_2 cannot be assimilated directly by plants or animals. Only certain

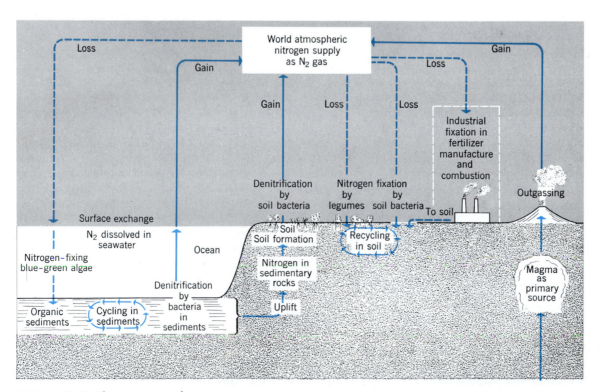

FIGURE 25.16 The nitrogen cycle.

microorganisms possess the ability to utilize N_2 directly, a process termed *nitrogen fixation*. One class of such microorganisms consists of certain species of free-living soil bacteria. Some blue-green algae can also fix nitrogen.

Another class consists of the symbiotic nitrogen fixers. In a symbiotic relationship, two species of organisms live in close physical contact, each contributing to the life processes or structures of the other. Symbiotic nitrogen fixers are bacteria of the genus *Rhizobium;* they are associated with some 190 species of trees and shrubs as well as almost all members of the legume family. Legumes important as agricultural crops are clover, alfalfa, soybeans, peas, beans, and peanuts. *Rhizobium* bacteria infect the root cells of these plants in root nodules produced jointly by action of the plant and the bacteria. The bacteria supply the nitrogen to the plant through nitrogen fixation, while the plant supplies nutrients and organic compounds needed by the bacteria. Crops of legumes are often planted in seasonal rotation with other food crops to ensure an adequate nitrogen supply in the soil. Both the action of nitrogen-fixing crops and of soil bacteria are shown in the nitrogen cycle diagram (Figure 25.16).

Nitrogen is lost to the biosphere by *denitrification*, a process in which certain soil bacteria convert nitrogen from usable forms back to N_2. This process is also shown in the diagram. Denitrification completes the organic portion of the nitrogen cycle, as nitrogen returns to the atmosphere.

At the present time, nitrogen fixation is far exceeding denitrification, and usable nitrogen is accumulating in the life layer. This excess of fixation is produced almost entirely by human activities. Human activity fixes nitrogen in the manufacture of nitrogen fertilizers and by oxidizing nitrogen in the combustion of fossil fuels. Widespread cultivation of legumes has also greatly increased worldwide nitrogen fixation. At present rates, nitrogen fixation attributable to human activity nearly equals all natural biological fixation.

Much of the excess nitrogen fixed by human activities is carried from the soil into rivers and lakes and ultimately reaches the ocean. Major water pollution problems can arise when nitrogen stimulates the growth of algae and phytoplankton, which can be detrimental to desirable forms of aquatic life. These problems will be accentuated in years to come because industrial fixation of nitrogen in fertilizer manufacture is doubling about every 6 years at present. Just what impact such large amounts of nitrogen reaching the sea will have on the earth's ecosystem remains uncertain.

SEDIMENTARY CYCLES

The oxygen, carbon, and nitrogen cycles are all referred to as gaseous cycles because they possess a gaseous phase in which the element involved is present in significant quantities in the atmosphere. Many other elements move in sedimentary cycles, that is, from the land to ocean in running water, returning after millions of years as uplifted terrestrial rock. These elements are not present in the atmosphere except in small quantities as blowing dust or condensation nuclei in precipitation.

Figure 25.17 shows how some important macronutrients move in sedimentary cycles. Within the large box representing the lithosphere are smaller compartments representing the parent matter of the soil and the soil itself. In the soil, nutrients are held as ions on the surfaces of soil colloids and are readily available to plants. (This was explained in Chapter 23.)

The nutrient elements are also held in enormous storage pools, where they are unavailable to organisms. These storage pools include seawater (unavailable to land organisms), sediments on the seafloor, and enor-

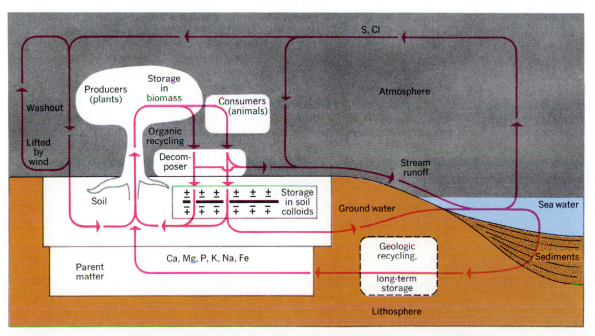

FIGURE 25.17 A flow diagram of the sedimentary cycle of materials in and out of the biosphere and within the inorganic realm of the lithosphere, hydrosphere, and atmosphere.

mous accumulations of sedimentary rock beneath both lands and oceans. Eventually, elements held in the geologic storage pools are released into the soil by weathering. Soil particles are lifted into the atmosphere by winds and fall back to earth or are washed down by precipitation. Chlorine and sulfur are shown as passing from the ocean into the atmosphere and entering the soil by the same mechanisms of fallout and washout.

The organic realm, or biosphere, is shown in three compartments: producers, consumers, and decomposers. Considerable element recycling occurs between organisms of these three classes and the soil. The elements used in the biosphere, however, are continually escaping to the sea as ions dissolved in stream runoff and groundwater flow.

AGRICULTURAL ECOSYSTEMS

The principles of energy use and flow in natural ecosystems also apply to *agricultural ecosystems*. Important differences exist, however, between natural ecosystems and those that are highly managed for agriculture. The first major difference is the reliance of agricultural ecosystems on inputs of energy that are ultimately derived from fossil fuels. The most obvious of these inputs is the fuel that runs the machinery used to plant, cultivate, and harvest crops. Another is the application of fertilizers and pesticides, which have required large expenditures of fuel to extract or synthesize and to transport. Fossil fuels also indirectly power such activities as the breeding of plants that have higher yields and are resistant to disease and the development of new chemicals to combat insect pests. Fuel is expended in transporting crops to distant sources of consumption, thus enabling large areas of similar climate and soils to be used for the same crop. In these and many other ways, the high yields obtained today are brought about only at the cost of a large energy input derived from fossil fuels.

A schematic flow diagram, Figure 25.18, shows how fuel-energy input enters into both the purchased inputs and the operations performed on the farm to raise and harvest crops. Solar energy of photosynthesis is, of course, a "free" input. But, to be delivered to the humans or animals that consume it, the raw food or feed product of this flow system requires the expenditure of fossil fuel energy.

Agricultural ecosystems, unlike natural ecosystems, are simple in structure and function. They often consist of one genetic strain of one species. Such ecosystems are overly sensitive to attacks by one or two well-adapted insects that can multiply rapidly to take advantage of an abundant food source. Thus pesticides are constantly needed to reduce insect populations. Weeds, too, are a problem, adapted as they are to rapid growth on disturbed soil in sunny environments. Weeds can divert much of the productivity to undesirable forms. Herbicides are often the immediate solution to these problems.

In natural ecosystems, nutrient elements are returned to the soil following the death of the plants that concentrate them. In agricultural ecosystems, this recycling is usually interrupted by harvesting the crop for consumption at a distant location. Continuous removal of nutrients in crop biomass introduces a new pathway in the nutrient cycle. To preserve soil fertility, still another pathway must be introduced—from the fertilizer plant to the farm. Keeping this new nutrient subcycle functioning requires considerable fossil fuel energy input.

PRODUCTIVITY AND EFFICIENCY OF AGRICULTURAL SYSTEMS

Human inputs of energy into managed ecosystems in the form of agricultural chemicals and fertilizers, as well as those in the form of the work of farm machinery, have acted to boost greatly the net primary productivity of the land. Recent studies show that the net primary productivity of agricultural ecosystems has increased more than five times over through the use of energy contained in fossil fuels.

Energy consumed in agriculture is subdivided into two major categories: (1) petroleum and electricity expended on farms, and (2) energy represented by materi-

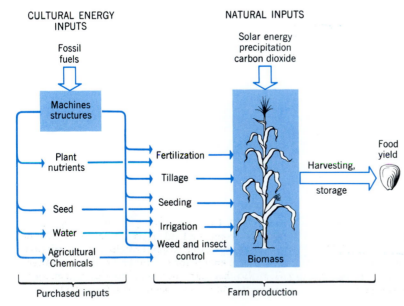

FIGURE 25.18 A schematic diagram of the inputs of cultural energy into various stages in the agriculture production system. (After G. H. Heichel, Agricultural production and energy resources, *American Scientist*, vol. 64, p. 65.)

TABLE 25.3 Energy Input to Grain Production, 1950–1985

Year	Energy Input, Barrels of Petroleum Per Tonne of Grain
1950	0.44
1960	0.65
1970	0.89
1980	1.13
1985	1.14

Data source: Lester R. Brown, 1988, *World Watch*, vol. 1, no. 2, p. 25.

als expended and equipment used on farms. Two important energy expenditures in the first category are fuel for tractors and electricity for water pumps in irrigation. The number of tractors in use in the world increased about fourfold—6 million to 24 million—between 1950 and 1985, and continues to grow each year. During the 1950s, 1960s, and 1970s, irrigated lands expanded greatly, increasing at a rate of about 4 percent per year. During the 1980s, this rate of expansion slowed to about 1 percent. Although much of the irrigated land was developed with gravity-fed water flows, extensive well irri-

gation, powered by fossil fuel, developed in the United States, India, and China, where millions of pumps are now in use.

Much of the energy consumed in modern agriculture is in the form of fertilizer, which requires large inputs of energy for its manufacture. About one-third of the total energy expended in agricultural production in the United States annually is in the use of fertilizers. Increasing world agricultural yields have been accompanied by much larger increases in fertilizer use. World grain production rose from 624 million tonnes (metric tons) in 1950 to 1660 million in 1986, an increase factor of about 2.7. In the same period, fertilizer use expanded nearly tenfold, from 14 million to 131 million tonnes per year. The fertilizer was used to make cropland more productive, with the result that yields increased at about 3 percent per year, while population grew at about 2 percent per year and grain area per capita fell about 1½ percent per year.

Although world food production is now highly dependent on energy input, the dependence has stabilized in recent years. Table 25.3 presents the energy required to produce a tonne of grain, expressed in barrels of oil, for the period 1950–1985. From 1950 to 1980, the energy

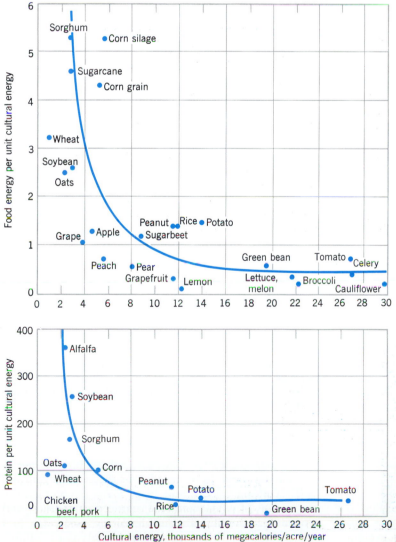

FIGURE 25.19 (upper graph) Cultural energy used to produce certain food crops in relation to yield of food energy. (lower graph) Protein yield in relation to cultural energy for several kinds of crops. (After G. H. Heichel, Agricultural production and energy resources, *American Scientist*, vol. 64, p. 66.)

input rose steadily, whereas between 1980 and 1985 it remained nearly constant.

Excluding the solar energy of photosynthesis, energy expended on the production of a raw food or feed crop can be referred to as *cultural energy*. Let us examine the cultural energy input needed to produce various feed and food crops. The upper graph in Figure 25.19 shows the relative efficiency of food production of 24 crops. The horizontal scale shows cultural energy expended in terms of thousands of megacalories per acre per year (mcal/acre/yr). The vertical scale shows the ratio of food energy produced to input of cultural energy; the higher the ratio, the greater the efficiency (and the lower the cultural input required). Notice that field crops, such as sorghum and corn, which are used largely as animal feeds, have the highest levels of efficiency, whereas foods consumed directly by humans have relatively low efficiencies. Garden crops require a very high cultural-energy input per unit area of land and also have low efficiencies.

The lower graph in Figure 25.19 shows, on the vertical scale, the protein derived from each unit of cultural energy. Alfalfa and soybeans rate very high on this scale. (Soybean protein is presently being processed in various forms for human diet as a meat substitute.) Notice that oats, wheat, and corn have intermediate values of protein yield, but that rice ranks very low. (Protein deficiency is a serious health problem for peoples subsisting largely on rice.) A small area of the graph at the lower left is labeled "chicken, beef, pork." This insertion serves to show that protein obtained from meat has an energy efficiency only about one-tenth as great as from soybeans. Indeed, the energy available from edible meat represents only about 10 percent of the energy expended in animal feed, a fact previously pointed out in our discussion of energy flow in the food chain.

The data thus indicate clearly that our food production system is not efficient in terms of cultural energy expended to furnish food to humans. The most highly coveted part of the American diet, meat and garden vegetables, is extremely wasteful of cultural energy as compared with, say, diets based largely on grain foods (bread, breakfast cereals) and soybean products.

Our discussion of agricultural ecosystems has emphasized their general nature and how they differ from natural ecosystems. We will give a fuller discussion of natural ecosystems in Chapters 26 and 27, which emphasize the diversity in structure of primary producers as related to climate and soils.

THE GREEN REVOLUTION—SUCCESS OR FAILURE?

The *green revolution* is a program of strategies and technologies designed to increase sharply the production of certain cereal crops—especially the widely cultivated staples, such as wheat, corn, and rice. These are the crops upon which survival hinges for hundreds of millions of inhabitants of Third World countries in the low latitudes. Barely able, at best, to keep pace with the demands imposed by rapid increase in population, hope for their improved food production has been promoted through development of new genetic strains of the cereals to the accompaniment of necessary changes in agricultural practices and technologies that include massive increases in the inputs of fertilizers and pesticides.

Countries affected by the green revolution lie mostly in the climates of low latitudes, Group I, and specifically in those with strongly developed contrasts between dry and wet seasons. Climate types include the wet-dry tropical climate (3) and the monsoon and trade-wind littoral climates (2) of Central America and Southeast Asia, including the Philippines. Also important are the tropical dry climates (4s, 4sd) of Mexico, Africa, Pakistan, and India, where rainfall is limited to a short period in the season of high sun and irrigation is practiced.

Closely associated with the green revolution is the name of Norman Ernest Borlaug, an agronomist who pioneered in the revolution in the 1940s and in 1970 was awarded the Nobel Peace Prize for seeing the program win a remarkable success. Together with a team of co-workers, Borlaug began his experimentation in new varieties of wheat in Mexico. His staff collected wheat varieties from many parts of the world and began to interbreed them to produce new strains of wheat. Improvements included increases in yield per acre, protein content, and resistance to fungus disease. Improvements also included the ability of the wheat plant to take in large amounts of fertilizer and to mature into a heavy ear of grain supported on a short stalk that would not easily fall over. The result was a strong-stemmed dwarf variety. After an initial crop failure, in which most of the new seed was lost to rust disease, the strain was further improved and a successful product emerged.

The new wheat moved Mexico into the ranks of a wheat-exporting country. Tried out in Pakistan and India, the new variety increased yields from a miserable 7 to 8 bushels per acre to 60 to 70 per acre. In India, the success was so astounding that armed guards had to be posted around fields at experiment stations to prevent theft of the new seed grain. Wheat production in the Punjab of India was doubled in the five years between 1966 and 1971. It was hoped that this phenomenal increase could sustain the growing needs of India's population, which at that time was increasing by some 17 million persons per year.

Meanwhile, the green revolution had led to improvements of rice yields, carried out at the International Rice Institute in the Philippines. Drawing upon more than 20,000 varieties of rice, new highly productive strains were developed and passed on to rice farmers. By 1970, the rice revolution in that country seemed to have been permanently won.

Not long thereafter, however, a disturbing theme in a minor key began to be heard. Was the green revolution really working? There had been many problems. The new strains of wheat and rice required substantial and dependable supplies of water, fertilizer, pesticides, and financial credit. These inputs were simply not available to a great number of poor farmers. By 1973, rice production in the Philippines had risen only from 20 to 25 percent for the nation as a whole and had leveled off. Then came the first great energy crunch, as the OPEC countries imposed oil embargoes and raised the price of crude oil. In north India, for example, the fuel shortage severely curtailed irrigation by pumping from wells. With the increase in oil prices came a major increase in the price of fertilizers, which require a large energy input in manufacturing. Wheat production there had already leveled off, and in 1974 was forced into sharp decline.

Meantime, human populations were increasing steadily in the developing nations. Tens of millions more mouths had appeared to consume the added production brought about by the green revolution. This trend came as no surprise to Dr. Borlaug, since he had said in 1970 that the green revolution simply "offers the possibility of buying 20 to 30 years of time . . . in which to bring population into balance with food production." In 1974, he was reported as concluding that time bought by the green revolution had been largely wasted.

Despite Borlaug's pessimistic appraisal, agricultural productivity in Asia generally increased thereafter to a point of self-sufficiency in food production. Although application of green revolution technology to Africa lagged behind that in Southeast Asia and Latin America, some important advances had been made there by the mid-1980s. For example, a drought-tolerant variety of sorghum was introduced into the Sudan with the capability of more than doubling yields of traditional varieties. Improved varieties of wheat introduced into Ethiopia were capable of doubling yields on small highland farms, if given adequate fertilizer. Unfortunately, genocidal warfare caused severe disruption of normal agricultural activity in these countries during the late 1980s, producing widespread famine.

One hazard of the green revolution lies in the requirement that the genetic strains bred for high yields must be used to the exclusion of a variety of native strains. Should the high-yield strain prove vulnerable to an epidemic plant disease, the entire crop of a whole nation could be wiped out in one season. Current research in crop breeding is now stressing the development of more individualized varieties that respond better to local conditions. These varieties are less subject to epidemic diseases and can still produce improved yields with less fertilizer input. A second hazard lies in the need to increase the size of fields by merging many small plots into a large one to allow mechanized agriculture to work most efficiently. In so doing, a variety of food crops is no longer grown, and dependence for survival comes to rest on the single crop. Under traditional practices, the Asiatic farmer planted several food crops to ensure that if some failed, others would yield enough to prevent starvation. The small farmer is therefore reluctant to move to more efficient, single-crop agriculture.

Insect predation has also presented a hazard to dependence on a single crop variety. An example comes from Indonesia, which by 1983 had succeeded in growing sufficient rice to feed its large population. That success was threatened in 1985 by a great rise in the numbers of the brown planthopper, an insect that laid waste to millions of acres of rice plants. Pesticides failed to halt this invasion, but the introduction of predator insects seems to have been successful.

It became obvious a decade ago that future increases in food crop yields would be achieved only by modifying the techniques of the green revolution to become less dependent on industrial technology practiced in middle latitudes in Europe and North America. New crop strains and methods of cultivation and fertilization will have to become more compatible with ancient agricultural systems and local culture patterns.

What about increasing the land area under cultivation? Soil scientists have pointed out that large areas of Vertisols remain to be placed under cultivation in the tropical wet-dry climate regions. The Vertisols are rich in nutrient bases, but will require machine cultivation to overcome difficult tillage. Primitive plows and hoes cannot work these soils. In Southeast Asia, most of the arable land of the monsoon environment has already been intensively developed by existing standards. To hope for major expansion of agriculture to poor Oxisols and to semidesert zones marginal to the tropical deserts is, at best, unrealistic.

At the start of the 1990s, much remains to be accomplished to bring self-sufficiency of staple foods to individual nations and continents. Mexico's agricultural trade balance switched from positive to negative in 1980 and has remained that way; its corn imports have continued to be required since 1970. Africa stands alone among the continents as continuing to suffer a declining food production—this despite the fact that it was self-sufficient until about 1960.

THE WORLD ECOSYSTEM IN REVIEW

As we have shown in this chapter, the organizing principles of energy flow and material cycling can greatly aid our understanding of the processes of the biosphere and their implications for human activity. Just as solar energy is the driving force for the circulation of global atmospheric and ocean fluids, so it also provides the power source for photosynthesis, on which all the world's organisms ultimately depend for sustenance. Humans are no different from other organisms. Without this input of solar energy and its conversion to consumable biomass, we would soon perish.

Of course, we are also dependent on the smooth functioning of material cycles. Without the tiny fraction of the earth's atmosphere that is carbon dioxide, no terrestrial photosynthesis would occur. If that component is enhanced by human activity through CO_2 release, it is sure to impact the productivity of the biosphere directly, as well as indirectly through climate change. Similarly, we influence the nitrogen and oxygen cycles at our peril.

As we approach the twenty-first century, our responsibility for the safe stewardship of our planet increases ever more with our numbers and our impact on the world's ecosystems. This stewardship will require a better global understanding of the earth and its life processes, and this chapter has provided an introduction to such an understanding.

CHAPTER 26

CONCEPTS OF BIOGEOGRAPHY

A major field within physical geography is *biogeography*, the study of the distribution patterns of plants and animals on the earth's surface and the processes that produce those patterns. We shall devote two chapters to the field of biogeography. This chapter deals with the processes that differentiate the world ecosystem into distinctive assemblages of plants and animals; the next chapter deals with the spatial distribution patterns and unique features of the world's vegetation. Throughout our two chapters, emphasis will be on terrestrial plants. Not only is vegetation the most obvious part of the biotic landscape, but it is also the most important ecologically because the plants provide the primary production on which the animals depend.

In considering how various factors of the physical environment influence plants and animals, two scale ranges can be treated. One is the global scale, which considers such climatic factors as the seasonal and latitudinal patterns of insolation, light and darkness, temperature, precipitation, and prevailing winds. The other scale of consideration is that of variations of the physical environment found within a relatively small area. Thus, within a generally humid region, a few small areas (such as a dune or cliff) that are extremely dry will be found. We may also find in a large desert some small places (such as a seep or spring) that are extremely wet most of the time.

Air temperatures and soil-water availability are the most important factors governing plant and animal distribution at both global and local scales. This chapter will begin by examining how plants and animals respond to the varying climatic factors of temperature and water availability.

THE BIOLOGICAL ROLE OF WATER

Water is probably the most important of the factors that determine the global distribution patterns of plants and animals. Water is important because, through evolution, plants and animals have become specialized, or adapted, to excesses or deficiencies in water availability.

The availability of water to terrestrial organisms at a particular point in time or space is determined by the balance among precipitation, evaporation, runoff, and infiltration. This balance is, in turn, affected by organisms—mainly the plant cover. Through transpiration, plants return much of the soil water to the atmosphere. By obstructing overland flow and increasing soil porosity, plants reduce runoff and increase infiltration. Although these movements of water are vital from the viewpoint of organic life, their effects are small compared to those of the physical processes that control the major features of the water cycle. Thus the major pattern of variation in water from place to place is still determined by the overall dynamics of the atmosphere and oceans. Because such a large portion of the earth's surface lies in areas where water shortages exist in at least some portion of the year, our discussion will emphasize the ways in which plants and animals are adapted to dry conditions.

ORGANISMS AND WATER NEEDS

Ecologists and biogeographers often classify plants according to their water requirements. Terms associated with the water factor are built on three simple prefixes

FIGURE 26.1 Prickly pear cactus (*Opuntia*) in the Sonoran Desert, Arizona. (Robert J. Ashworth/Photo Researchers.)

502

FIGURE 26.2 Hygrophytes are encroaching upon the borders of this small lake on the Gaspé Peninsula, Quebec. The leading edge of the bog vegetation, consisting of sedges, floats on the water. (Pierre Dansereau.)

of Greek roots: *xero-*, "dry"; *hygro-*, "wet"; and *meso-*, "intermediate." Plants that grow in dry habitats are *xerophytes;* those that grow in water or in wet habitats are *hygrophytes;* those of habitats of an intermediate degree of wetness and relatively uniform soil-water availability are *mesophytes.*

The xerophytes are highly tolerant of drought and can survive in habitats that dry quickly following rapid drainage of precipitation (e.g., on sand dunes, beaches, and bare rock surfaces). The plants typical of dry climates are also xerophytes; cactus is an example (Figure 26.1). The hygrophytes are tolerant of excessive water and may be found in shallow streams, lakes, marshes, swamps, and bogs (Figure 26.2). The mesophytes are found in upland habitats in regions of ample rainfall. Here the drainage of soil water is good and moisture penetrates deeply, where it can later be used by the plants.

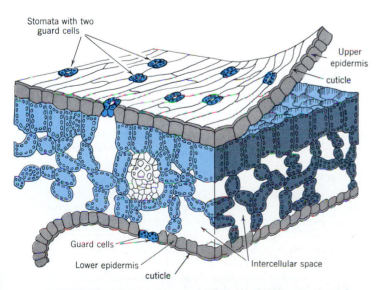

FIGURE 26.3 Cell structure of foliage leaf. (After W. W. Robbins and T. E. Weier, *Botany*. Copyright by John Wiley & Sons. Reprinted by permission of John Wiley & Sons, Inc.)

Water loss from plant tissues occurs through the process of transpiration, explained in Chapter 9. The rate of transpiration varies greatly according to the type of plant and the prevailing atmospheric conditions. High temperatures, low humidities, and winds favor high rates of transpiration. The plant structure, particularly of the leaf, largely determines the rate of water loss. Plants with foliage surfaces composed of broad, thin leaves have higher rates of loss than plants bearing needleleaves, spines, or small thick leaves. Under conditions of low water supply but high rates of evaporation, only those plants that minimize transpiration losses by their special leaf structures and their small size can survive.

The adaptation of plant structures to soil-water budgets with large water deficiencies is of particular interest to the plant geographer. Transpiration occurs largely from specialized leaf pores, called *stomata*, which are openings in the outer cell layer, and through the *cuticle*, the outermost protective layer of the leaf. Surrounding the openings of the stomata are *guard cells* that can open and close the openings and thus regulate the flow of water vapor and other gases (Figure 26.3). Although most of the transpiration occurs through the stomata, some water vapor may pass through the cuticle. This latter form of loss is reduced in some plants by thickening of the outer layers of cells or by the deposition of wax or waxlike material on or near the leaf surface. Thus many desert plants have thickened cuticle or wax-coated leaves, stems, or branches.

Other means of reducing transpiration are the development of stomata deeply sunken into the leaf surface to retard outward diffusion of water vapor into dry air and the restriction in location of stomata to the shaded undersurfaces of leaves. A plant may also adapt to a desert environment by greatly reducing the leaf area or by bearing no leaves at all. Thus needleleaves and spines representing leaves greatly reduce loss from transpiration. In cacti, the foliage leaf is not present and transpiration is limited to fleshy stems.

In addition to developing leaf structures that reduce water loss by transpiration, plants in a water-scarce en-

Labels in Figure 26.3: Stomata with two guard cells; Upper epidermis; cuticle; Guard cells; Lower epidermis; cuticle; Intercellular space

vironment improve their means of obtaining water and of storing it. Roots become greatly extended to reach soil moisture at increased depth. In cases where the roots reach to the ground-water table, a steady supply of water is assured. Plants drawing from such a source are termed *phreatophytes* and may be found along dry channels (draws) and alluvial valley floors in desert regions. Other desert plants produce a widespread but shallow root system enabling them to absorb the maximum quantity of water from sporadic desert downpours that saturate only the uppermost soil layer. Stems of desert plants are commonly greatly thickened by a spongy tissue in which much water can be stored. Plants employing this adaptation are termed *succulents*.

A quite different adaptation to extreme aridity is seen in many species of small desert plants that complete a very short cycle of germination, leafing, flowering, fruiting, and seed dispersal immediately following a desert downpour. Because they appear so briefly, these plants are referred to as *ephemeral annuals*.

Certain climates, such as the wet-dry tropical climate and the moist continental climate, have a yearly cycle with one season in which water is unavailable to plants because of lack of precipitation or because the soil water is frozen. This season alternates with one in which there is abundant water. Plants adapted to such regimes are called *tropophytes*, from the Greek word *trophos*, meaning "change," or "turn." Tropophytes meet the impact of the season of unavailable water by dropping their leaves and becoming dormant. When water is again available, they leaf out and grow at a rapid rate. Trees and shrubs that seasonally shed their leaves are *deciduous plants;* in distinction, *evergreen plants* retain most of their leaves in a green state throughout the year.

The Mediterranean climate also has a strong seasonal wet-dry alternation, the summers being dry and the winters wet. Plants in this climate adopt the habit of xerophytic plants and characteristically have hard, thick, leathery leaves. An example is the live oak, which holds most of its leaves throughout the dry season. Such hard-leaved evergreen trees and woody shrubs are called *sclerophylls*. The prefix *scler* is from the Greek word for "hard." Plants that hold their leaves throughout a dry or cold season have the advantage of being able to resume photosynthesis immediately when growing conditions become favorable, whereas the deciduous plants must grow a new set of leaves.

To cope with shortage of water, *xeric animals*, those adapted to dry conditions, have evolved methods that are somewhat similar to those used by the plants. Many of the invertebrates exhibit the same pattern as the ephemeral annuals—evading the dry period in dormant stages. When rain falls, they emerge to take advantage of the new and short-lived vegetation. For example, many species of birds regulate their behavior to nest only when the rains occur, the time of most abundant food for their offspring. The tiny brine shrimp of the Great Basin may wait many years in dormancy until normally dry lake beds fill with water, an event that occurs perhaps three or four times a century. The shrimp then emerge and complete their life cycles before the lake evaporates.

The mammals are by nature poorly adapted to desert environments, but many survive there by using a variety of mechanisms to avoid water loss. Just as plants reduce transpiration to conserve water, so many desert mammals do not sweat through skin glands; they rely instead on other methods of cooling. Many of the desert mammals conserve water by excreting highly concentrated urine and relatively dry feces. The desert mammals also evade the heat by nocturnal activity. In this respect, they are joined by most of the rest of the desert fauna, spending their days in cool burrows in the soil and their nights foraging for food.

ORGANISMS AND TEMPERATURE

The temperature of air and soil, another of the important climatic factors in ecology, acts directly on organisms through its influence on the rates at which the physiological processes take place. For plants, we can say that each species has an optimum temperature associated with each of its functions, such as photosynthesis, flowering, fruiting, and seed germination. Some overall optimum yearly temperature conditions exist for its growth in terms of size and numbers of individual plants. There are also limiting lower and upper temperatures for the individual functions of the plant and for its total survival.

Temperature also acts as an indirect factor. Higher air temperatures increase the water vapor capacity of the air, inducing greater transpiration as well as faster rates of direct evaporation of soil water.

In general, the colder the climate, the fewer are the species that are capable of surviving. A large number of tropical plant species cannot survive below-freezing temperatures. In the severely cold arctic and alpine environments of high latitudes and high altitudes, only a few species can survive. This principle explains why a forest in the equatorial zone has many species of trees, whereas a forest of the subarctic zone may be dominated by just a few. Tolerance to cold is closely tied to the ability of the plant to withstand the physical disruption that accompanies the freezing of water. If the plant has no means of disposing of the excess water in its tissues, the freezing of that water will damage the cell tissue.

The effects of temperature variations on animals are moderated by their physiology and by their ability to seek sheltered environments. Most animals lack a physiological mechanism for internal temperature regulation. These animals, including reptiles, invertebrates, fish, and amphibians, are *cold-blooded animals;* their body temperatures passively follow the environment. With a few exceptions (notably fish and some social insects), these animals are active only during the warmer parts of the year. They survive the cold weather of the midlatitude zone winter by becoming dormant. Some vertebrates enter a dormant state termed *hibernation*, in which metabolic processes virtually halt and body temperatures closely parallel those of the surroundings. Most hibernators seek out burrows, nests, or other environments where winter temperatures do not reach extremes or fluctuate rapidly. Because the annual range of soil temperatures is greatly reduced below the uppermost layers, soil burrows are particularly suited to hibernation.

Other animals maintain their tissues at a constant temperature by internal metabolism. This group includes the birds and mammals. These *warm-blooded animals* possess a variety of adaptations to maintain a constant internal temperature. Fur, hair, and feathers act as insulation by trapping dead air spaces next to the skin surface, reducing heat loss to the surrounding air or water. A thick layer of fat will also provide excellent insulation. Other adaptations are for cooling; for example, sweating or panting uses the high latent heat of vaporization of water to remove heat. Heat loss is also facilitated by exposing blood-circulating tissues to the cooler surroundings. The seal's flippers and bird's feet serve this function.

OTHER CLIMATIC FACTORS

The factor of light is also of importance in determining local plant distribution patterns. The amount of light available to a plant will depend in large part on the plant's position. Tree crowns in the upper layer of a forest receive maximum light, but correspondingly reduce the amount available to lower layers. In extreme cases, forest trees so effectively cut off light that the forest floor is almost free of shrubs and smaller plants. In certain deciduous forests of midlatitudes, the period of early spring, before the trees are in leaf, is one of high light intensity at ground level, permitting the smaller plants to go through a rapid growth cycle. In summer these plants will largely disappear as the leaf canopy is completed. Other low plants in the same habitat require shade and do not appear until later in the summer.

Treated on a global basis, the factor of light available for plant growth varies by latitude. Duration of daylight in summer increases rapidly with higher latitude and reaches its maximum poleward of the arctic and antarctic circles, where the sun may be above the horizon for 24 hours (see Figure 25.4). Thus, although the growing season for plants is greatly shortened at high latitudes by frost, the rate of plant growth in the short frost-free summer is greatly accelerated by the prolonged daylight.

In midlatitudes, where vegetation is of a deciduous type, the annual rhythm of increasing and decreasing periods of daylight determines the timing of budding, flowering, fruiting, leafshedding, and other vegetation activities. As to the importance of light intensity itself, even on overcast days there is sufficient light to permit most plants to carry out photosynthesis at their maximum rates.

Light also influences animal behavior. The day–night cycle controls the activity patterns of many animals. Birds, for example, are generally active during the day, whereas small foraging mammals, such as weasels, skunks, and chipmunks, are more active at night. Light also controls seasonal activity through *photoperiod*, or day length, in midlatitudes. As autumn days grow shorter and shorter, squirrels and other rodents hoard food for the coming winter season. Later, increasing photoperiod will trigger such activities as mating and reproduction in the spring.

Wind is also an important environmental factor in the structure of vegetation in highly exposed positions. Close to timberline in high mountains and along the northern limits of tree growth in the arctic zone, trees are deformed by wind in such a way that the branches project from the lee side of the trunk only (flag shape) (Figure 26.4). Some trees may show trunks and branches bent to near-horizontal attitude, facing away from the prevailing wind direction. The effect of wind is to cause excessive drying, damaging the exposed side of the plant. The tree limit on mountainsides thus varies in elevation with degree of exposure of the slope to strong prevailing winds and will extend higher on lee slopes and in sheltered pockets.

FIGURE 26.4 Alpine tundra near the summit of the Snowy Range, Wyoming. Flag-shaped spruce trees (left) represent the upper limit of tree growth. (A. N. Strahler.)

BIOCLIMATIC FRONTIERS

Taken separately or together, climatic factors of moisture, temperature, light, and wind can act to limit the distribution of plant and animal species. Biogeographers recognize that there is a critical level of climatic stress beyond which a species cannot survive and that there will be a geographic boundary marking the limits of the potential distribution of a species. Such a boundary is sometimes referred to as a *bioclimatic frontier*. Although the frontier is usually marked by a complex of climatic elements, it is sometimes possible to single out one climatic element related to soil water or temperature that coincides with it.

The distribution of ponderosa pine (*Pinus ponderosa*) in western North America provides an example (Figure 26.5). In this mountainous region, annual rainfall varies sharply with elevation. The 50 cm (20 in.) isohyet of annual total precipitation encloses most of the upland areas having the yellow pine. It is the parallelism of the isohyet with forest boundary, rather than actual degree of coincidence, that is significant. The sugar maple (*Acer saccharum*) is a somewhat more complex case (Figure 26.6). Here the boundaries on the north, west, and south coincide roughly with selected values of annual precipitation, mean annual minimum temperature, and mean annual snowfall.

Although bioclimatic limits must exist for all species,

no plant or animal need necessarily be found at its frontier. Many other factors may act to keep the spread of a species in check. A species may be limited by diseases or predators found in adjacent regions. In another example, a species (especially a plant species) may migrate slowly and may still be radiating outward from the location in which it evolved. Or, a species may be dependent on another species and therefore be limited by the latter's distribution.

INTERACTIONS AMONG SPECIES

Species interactions can also be important factors in determining the distribution patterns of plants and animals. Two species that are part of the same ecosystem can interact with one another in three ways: interaction may be negative to one or both species; or the two species may be neutral, not affecting each other; or interaction may be positive, benefiting at least one of the species. Negative interactions include competition, parasitism, predation, herbivory, and allelopathy; positive interac-

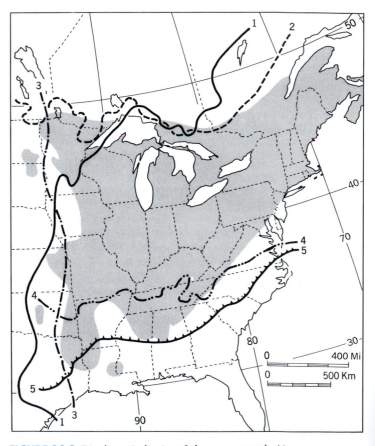

FIGURE 26.6 Bioclimatic limits of the sugar maple (*Acer saccharum*) in eastern North America. The shaded area represents the distribution of the sugar maple. Line 1, 76 cm (30 in.) annual precipitation. Line 2, −40°C (−40°F) mean annual minimum temperature. Line 3, eastern limit of yearly boundary between arid and humid climates. Line 4, 25 cm (10 in.) mean annual snowfall. Line 5, −10°C (16°F) mean annual minimum temperature. (Based on *Biogeography—An Ecological Perspective*, by Pierre Dansereau. Copyright © 1957, The Ronald Press Company, New York.)

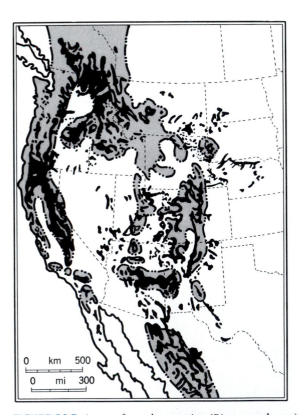

FIGURE 26.5 Areas of ponderosa pine (*Pinus ponderosa*) in western North America are shown in solid black. The edge of the shaded area represents the isohyet of 50 cm (20 in.) annual precipitation. (Based on *Biogeography—An Ecological Perspective*, by Pierre Dansereau. Copyright © 1957, The Ronald Press Company, New York.)

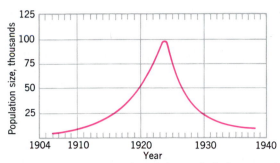

FIGURE 26.7 Rise and fall of the Kaibab deer population in the Kaibab National Forest, Arizona. (After D. I. Rasmussen, *Ecological Monographs*, vol. 11, p. 237.)

tions include commensalism, protocooperation, and mutualism.

Competition between species occurs whenever two species require a common resource that is in short supply. Because neither species has full use of the resource, both populations suffer, showing growth rates lower than those when only one of the species is present. Sometimes one species will win the competition and crowd out its competitor. At other times, the two species may remain in competition indefinitely.

Competition between species is an unstable situation—if a genetic strain within one of the populations emerges which uses a substitute resource for which there is no competition, its survival rate will be higher than that of the remaining strain, which still competes. The original strain may become extinct. In this way, evolutionary mechanisms tend to reduce competition among species.

Predation and *parasitism* are negative interactions in which one species feeds on the other. If the organism which gains energy is larger, the process is predation—

the individual gaining energy is the predator, and its food is the prey. If the organism gaining energy is smaller, the process is parasitism—the parasite gains energy, and the host serves as its energy source.

Although we tend to think of predation and parasitism as essentially negative processes which benefit one species at the expense of the other, it may well be that these interactions are really beneficial in the long term to the host or prey populations. A famous example is the growth of the deer herd on the Kaibab Plateau north of the Grand Canyon in Arizona (Figure 26.7). Initially at a population of about 4000, in an area of 283,000 hectares (700,000 acres), the herd grew to near 100,000 in the span from 1907 to 1924 in direct proportion to a government predator control and game protection program. Wolves became extinct in the area and populations of coyotes and mountain lions were greatly reduced. The huge deer population, however, proved too much for the land, and overgrazing led to a population crash. In one year, half the animals starved to death; by the late 1930s, the population had declined to a stable level near 10,000. Thus, predation maintained the deer population at levels which were in harmony with the supportive ability of the environment. In addition to maintenance of equilibrium population levels, predation and parasitism differentially remove the weaker individuals and can improve the genetic composition of the species.

Another type of predation is *herbivory*, in which grazing of plants by animals reduces the viability of the plant species population. Some species are well-adapted to grazing and can maintain themselves well in the face of grazing pressure. Others may be quite sensitive to this process. When overgrazing occurs, these differing sensitivities can produce significant changes in the structure and composition of plant communities (Figure 26.8).

FIGURE 26.8 This Landsat Thematic Mapper image of 1973 shows the southeast Mediterranean coast and includes a portion of the Sinai/Negev Desert. The Egypt/Israeli border stands out as a sharp diagonal line where it crosses a field of longitudinal and parabolic sand dunes. After the Six-Day War of 1967, the native pastoralists withdrew from the eastern (Negev) portion, and the border was fortified. There followed intensive grazing pressure on the western area (Sinai), which appears here as a much lighter area of sparser vegetation. Whereas in the Negev plant cover was maintained at 35 percent, in the Sinai it was only 10 percent. Irrigated fields of the Beersheeba region to the north show as red and pale green rectangles. Width of the area shown is about 60 km (35 mi). (GEOPIC™, Earth Satellite Corporation.)

A fourth type of negative interaction is *allelopathy*, a phenomenon of the plant kingdom in which chemical toxins produced by one species serve to inhibit the growth of others. As an example, sage, a common shrub species in the California chaparral, produces leaves rich in volatile toxins (cineole and camphor). As the leaves fall and accumulate in the soil, the allelopathic toxins build to a level sufficient to inhibit the growth of herbaceous plants, such as grasses, which are thus only found in adjacent areas. Still other chaparral shrubs produce water-soluble antibiotics that also inhibit the growth of nearby grasses. These chemical defenses, however, are broken down by periodic fires, which are events essential to the maintenance of the chaparral ecosystem. The fires destroy the toxins and also trigger the germination of seeds of many species of annual herbaceous plants by breaking the seed coats. The annuals then dominate the area until the shrubs grow and force them out by allelopathy, beginning the cycle anew.

The term *symbiosis* includes three types of positive interactions between species: commensalism, protocooperation, and mutualism. In commensalism, one of the species is benefited and the other is unaffected. Examples of commensals include the epiphytic plants—such as orchids or Spanish moss—which live on the branches of larger plants (Figure 27.3). These epiphytes depend on their hosts for physical support only. In the animal kingdom, small commensal crabs or fishes seek shelter in the burrows of sea worms; or the commensal remora fish attaches itself to a shark, feeding on bits of leftover food as its host dines.

When the relationship benefits both parties but is not essential to their existence, it is termed protocooperation. The attachment of a stinging coelenterate to a crab is an example of protocooperation. The crab gains camouflage and an additional measure of defense, while the coelenterate eats bits of stray food which the crab misses.

Where protocooperation has progressed to the point that one or both species cannot survive alone, the result is mutualism. A classic example is the association of the nitrogen-fixing bacterium *Rhizobium* with the root tissue of certain types of plants (legumes), in which the action of the bacteria converts nitrogen gas to a form directly usable by the plant. The association is mutualistic because *Rhizobium* cannot survive alone.

TERRESTRIAL ECOSYSTEMS—THE BIOMES

Because plants and animals have become adapted through evolution to varying environments, many different ecosystems exist, each adjusted to a different range of environmental conditions. Ecosystems fall into two major groups: aquatic and terrestrial. The *aquatic ecosystems* include life forms of the marine environments and the freshwater environments of the lands. Marine ecosystems include the open ocean, coastal estuaries, and coral reefs. Freshwater ecosystems include lakes, ponds, streams, marshes, and bogs. The *terrestrial ecosystems* comprise the assemblages of land plants and animals spread widely over the upland surfaces of the conti-

nents. The terrestrial ecosystems are determined largely by climate and soil and, in this way, are closely woven into the fabric of physical geography.

Within terrestrial ecosystems, the largest recognizable subdivision is the *biome*. Although the biome includes the total assemblage of plant and animal life interacting within the life layer, the green plants dominate the biome physically because of their enormous biomass, as compared with that of other organisms. Thus biogeographers classify the biomes by the characteristic life form of the green plants within the biome.

The following are the principal biomes, listed in order of availability of soil water and heat:

Forest	Ample soil water and heat.
Savanna	Transitional between forest and grassland.
Grassland	Moderate shortage of soil water; adequate heat.
Desert	Extreme shortage of soil water; adequate heat.
Tundra	Insufficient heat.

Biogeographers subdivide the biomes into smaller units, called *formation classes*, based on the size, shape, and structure of the plants. For example, at least four and perhaps as many as six kinds of forests are easily recognizable within the forest biome. At least two kinds of grasslands are easily recognizable. Deserts, too, span a wide range in terms of the abundance and life form of plants. Reviewing these formation classes and their unique features and adjustments to characteristic climates and soils will be our goal in the following chapter.

COMMUNITIES AND HABITATS

Although the distribution of terrestrial biomes is dependent on climate, there is much local variation in plants and animals. *Biotic communities* are local associations of plants and animals that are interdependent and often found together. The total biotic cover of a region is actually a mosaic of small community types that recur in different places on the landscape (see Figure A.II.9).

A primary influence on the distribution of biotic communities is the effect of varying landform and soil on vegetation. Landform refers to the configuration of the land surface, including such features as hills, valleys, ridges, or cliffs. Vegetation on an upland—relatively high ground with thick, well-drained soil—is quite different from that on an adjacent valley floor, where the water lies near the surface much of the time. Vegetation is often different on rocky ridges and steep cliffs, where water drains away rapidly and soil is thin or largely absent. Because the animals in an area are dependent on the vegetation for primary production, both plant and animal components of the biotic community will respond to these differences in the physical environment.

Habitat is the word used to refer to a type of physical environment that has a characteristic biotic community. For example, Figure 26.9 shows six habitats within the Canadian forest: upland, bog, bottomland, ridge, cliff,

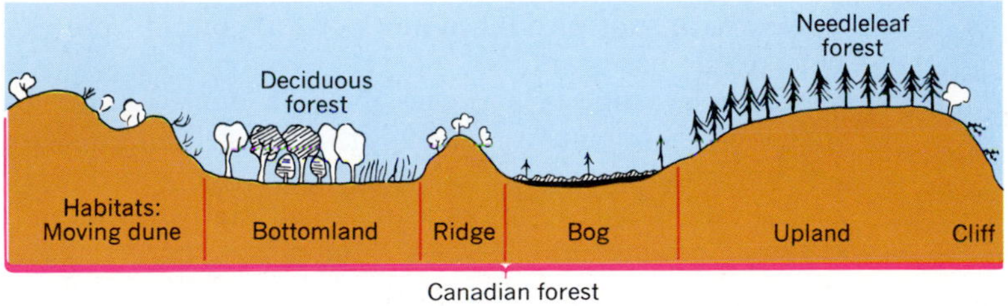

Needleleaf forest

Deciduous forest

Habitats: Moving dune | Bottomland | Ridge | Bog | Upland | Cliff

Canadian forest

FIGURE 26.9 Habitats within the Canadian forest. (Modified from Pierre Dansereau, *Ecology*, vol. 32.)

and active dune. Just where each habitat is located and how large an area it occupies depends largely on factors of soil and landform.

GEOMORPHIC FACTORS

Geomorphic factors (landform factors) influencing ecosystems are essentially the same as landform factors influencing the soil (Chapter 23). They include such elements as slope steepness (the angle that the ground surface makes with respect to the horizontal), slope aspect (the orientation of a sloping ground surface with respect to geographic north), and relief (the difference in elevation of divides and adjacent valley bottoms). In a much broader sense, geomorphic factors include the entire sculpturing of the landforms of a region by processes of erosion, transportation and deposition by streams, waves, wind, and ice, and by forces of volcanism and mountain building. These are topics covered in detail in Chapters 13 through 22.

Slope steepness acts indirectly by influencing the rate at which precipitation is drained from the surface. On steep slopes, surface runoff is rapid and soil-water recharge by infiltration is reduced. On gentle slopes, much of the precipitation can penetrate the soil and be retained. More rapid erosion on steep slopes may result in thin soil, whereas that on gentler slopes is thicker. Slope aspect has a direct influence on plants by increasing

or decreasing the exposure to sunlight and prevailing winds. Slopes facing the sun have a warmer, drier environment than slopes that face away from the sun and therefore lie in shade for much longer periods of the day. In midlatitudes, these slope-aspect contrasts may be so strong as to produce quite different biotic communities on north-facing and south-facing slopes (Figure 26.10).

Geomorphic factors are in part responsible for the dryness or wetness of the habitat within a region having essentially the same overall climate. Each community has its own microclimate. On divides, peaks, and ridge crests, the soil tends to dryness because of rapid drainage and because the surfaces are more exposed to sunlight and drying winds. By contrast, the valley floors are wetter because surface runoff over the ground and into streams causes water to converge there. In humid climates, the ground-water table in the valley floors may lie close to or at the ground surface to produce marshes, swamps, ponds, and bogs.

EDAPHIC FACTORS

Edaphic factors are those related to the soil. In Chapters 23 and 24, the principles of soil development (pedogenesis) were taken up systematically. In terms of biogeography, we can look at soils in two perspectives. One views the broad patterns of soil distribution as controlled by the climatic regimes. Patterns of soils and climates are closely correlated with the global patterns of the formation classes. These broad relationships will be taken up in Chapter 27.

A second viewpoint is in terms of habitats—the small-scale mosaic of place-to-place variations of the earth's surface. Among the edaphic factors important in differentiating the habitat are soil texture and structure, humus content; presence or absence of soil horizons, soil alkalinity, acidity, or salinity; and the activity of bacteria and animals in the soil.

Although this book treats the systematic principles of soil science ahead of those of natural ecosystems, a good argument might be made for reversing this order of treatment on the grounds that plants and animals play a leading role in the development of soil characteristics. Given a barren habitat, recently formed by some geologic event, such as the outpouring of lava or the emergence of a coastal zone from beneath the sea, the gradual evolution of a soil profile goes hand in hand with the occupance of the habitat by a succession of biotic communities. The plants profoundly alter the soil by such processes as contributing organic matter or producing acids that act on the mineral matter. Animal life, feeding

FIGURE 26.10 The two sides of this low hill in southern Idaho show a marked vegetation asymmetry in response to variation in microclimate. On the left (south-facing) side, dry conditions prevail and grasses are found. On the right (north-facing) side, conditions are more moist and support a young stand of pines. (Jerome Wycoff.)

on the plant life, also makes its contribution to physical and chemical processes of soil evolution. This change in communities through time forms our next topic.

ECOLOGICAL SUCCESSION

The phenomenon of change in plant communities through time is a familiar one. A drive in the country reveals patches of vegetation in many stages of development—from open, cultivated fields through grassy shrublands to forests. Clear lakes, gradually filled with sediment from the rivers that drain into them, become bogs. These kinds of changes, in which biotic communities succeed one another on the way to a stable end point, are referred to as *ecological succession*.

In general, succession leads to formation of the most complex community of organisms possible in an area, given its physical controlling factors of climate, soil, and water. The series of communities that follow one another on the way to the stable stage is called a *sere*. Each of the temporary communities is referred to as a *seral stage*. The stable community, which is the end point of succession, is the *climax*. If succession begins on a newly constructed deposit of mineral sediment, it is termed *primary succession*. If succession occurs on a previously vegetated area that has been recently disturbed by such agents as fire, flood, windstorm, or humans, it is referred to as *secondary succession*.

A new site on which primary succession occurs may have one of several origins: a sand dune, a sand beach, the surface of a new lava flow or freshly fallen layer of volcanic ash, or the deposits of silt on the inside of a river bend that is gradually shifting. Such a site will not have a true soil with horizons; rather, it may perhaps be little more than a deposit of coarse mineral fragments. In other cases, such as that of the floodplain silt deposits, the surface layer may represent redeposited soil endowed with substantial amounts of soil colloids and exchangeable base cations.

The first stage of a succession is a *pioneer stage;* it includes a few plant species unusually well adapted to adverse conditions of rapid water drainage and drying of soil and to excessive exposure to sunlight, wind, and extreme ground and lower-air temperatures. As these plants grow, their roots penetrate the soil; their subsequent decay adds humus to the soil. Fallen leaves and stems add an organic layer to the ground surface. Bacteria and animals begin to live in the soil in large numbers. Grazing mammals feed on the small plants. Birds forage the newly vegetated area for seeds and grubs.

Soon conditions are favorable for other species that invade the area and displace the pioneers. The new arrivals may be larger plant forms providing more extensive cover of foliage over the ground. In this case the climate near the ground, or *microclimate*, is considerably altered toward one of less extreme air and soil temperatures, high humidities, and less intense insolation. Now still other species can invade and thrive in the modified environment. When the succession has finally run its course, a climax community of plant and animal species in a more or less stable composition will exist.

The colonization of a sand dune provides an example of primary succession. Growing foredunes bordering the ocean or lakeshore present a sterile habitat. The dune sand—usually largely quartz, feldspar, and other common rock-forming minerals—lacks such important nutrients as nitrogen, calcium, and phosphorus, and its water-holding ability is very low. Under the intense solar radiation of the day, the dune surface is a hot, drying environment. At night, radiation cooling in the absence of moisture produces low surface temperatures. One of the first pioneers of this extreme environment is beachgrass (Figure 26.11, left). This plant reproduces vegetatively by sending out rhizomes (creeping underground stems), and the plant thus slowly spreads over

FIGURE 26.11 Stages in beach/dune succession at Sandy Hook, New Jersey. (left) Beachgrass is a pioneer on beach dunes and helps stabilize the dune against wind erosion. (right) The climax forest on dunes. Holly (*Ilex opaca*), seen on the left, is an important constituent of the climax forest. Note the leaves and decaying organic matter on the forest floor. (Alan H. Strahler.)

the dune. Beachgrass is well adapted to the eolian environment; it does not die when buried by moving sand, but instead puts up shoots to reach the new surface.

After colonization, the shoots of beachgrass act to form a baffle that suppresses movement of sand, and thus the dune becomes more stable. With increasing stabilization, plants that are more adapted to the dry, extreme environment but cannot withstand much burial begin to colonize the dune. Typically these are low, matlike woody shrubs, such as beach wormwood or false heather.

On older beach and dune ridges of the central Atlantic coastal plain, the species that follow matlike shrubs are typically larger woody plants and such trees as beach plum, bayberry, poison ivy, and choke cherry. These species all have one thing in common—their fruits are berries that are eaten by birds. The seeds from the berries are excreted as the birds forage among the low dune shrubs, thereby sowing the next stage of succession. As the scrubby bushes and small trees spread, they shade out the matlike shrubs and any remaining beachgrass. Pines may also enter at this stage.

At this point, the soil begins to accumulate a significant amount of organic matter. No longer dry and sterile, it now possesses organic compounds and nutrients, and has accumulated enough colloids to hold water for longer intervals. These soil conditions encourage the growth of such broadleaf species as red maple, hackberry, holly, and oaks, which shade out the existing shrubs and small trees (Figure 26.11, right). Once the forest is established, it tends to reproduce itself; the species of which it is composed are tolerant to shade, and their seeds can germinate on the organic forest floor. Thus the climax is reached. The stages through which

the ecosystem has developed constitute the sere, progressing from beachgrass to low shrubs to higher shrubs and small trees to forest.

Although this example has stressed the changes in plant cover, animal species are also changing as succession proceeds. Table 26.1 shows how some typical invertebrates appear and disappear through succession on the

Vegetation:

Rushes | Sedges | Sphagnum | Heaths

Mesophytic shrubs | Hygrophytic trees | Mesophytic trees

Peats:

Rush | Sedges | Sphagnum | Woody

FIGURE 26.12 Autogenic bog succession typical of the Laurentian shield area of Canada. (After Dansereau and Sagadas-Vianna, *Canadian Journal of Botany*, vol. 30.)

TABLE 26.1 Invertebrate Succession on the Lake Michigan Dunes

Invertebrate	Successional Stages				
	Beach grass–Cotton-wood	Jack Pine Forest	Black Oak Dry Forest	Oak and Oak–Hickory Moist Forest	Beech–Maple Forest Climax
White tiger beetle	x				
Sand spider	x				
Long-horn grasshopper	x	x			
Burrowing spider	x	x			
Bronze tiger beetle		x			
Migratory locust		x			
Ant lion			x		
Flatbug			x		
Wireworms			x	x	x
Snail			x	x	x
Green tiger beetle				x	x
Camel cricket				x	x
Sowbugs				x	x
Earthworms				x	x
Woodroaches				x	x
Grouse locust					x

Data source: V. E. Shelford, as presented in E. P. Odum, *Fundamentals of Ecology*, W. B Saunders Co., Philadelphia, p. 259.

Lake Michigan dunes. Note that the seral stages shown in the table for these inland dunes are somewhat different than those described for the coastal environment.

Another example of primary succession is that of *bog succession* (see Figure 26.2). Extensive land areas of North America and Europe have innumerable bogs. These are former glacial lake basins now filled with partially decomposed plant matter that we recognize as *freshwater peat*. The peat accumulates because decay of plant matter is slow in these cold climates. Plant matter that accumulates below the water level of a lake is in a continually saturated condition, with little oxygen available to promote the activity of decomposers.

Figure 26.12 shows, by a series of diagrams, the stages in the filling of lakes during bog succession. At the water's edge is a zone of sedges, followed by rushes. These construct a floating layer that encroaches upon the open water. There follows a zone of sphagnum (peat moss), which eventually completely fills the lake. Now the peat deposit supports hygrophytic trees (largely spruce), which produce a woody peat. The soil is now a Histosol. This community may in turn be replaced by mesophytic trees, marking the climax stage.

OLD-FIELD SUCCESSION

Where disturbance alters an existing community, secondary succession can occur. *Old-field succession*, taking place on abandoned farmland, is a good example of secondary succession. In the eastern United States, the first stages of the sere often depend on the last use of the land before abandonment. If row crops were last cultivated, one set of pioneers, usually annuals and biennials, will appear; if small grain crops were cultivated, the pioneers are often perennial herbs and grasses. If pasture is abandoned, those pioneers that were not grazed will have a head start. Where mineral soil was freshly exposed by plowing, pines are often important following the first stages of succession because pine seeds favor disturbed soil and strong sun for germination. Although slower-growing than other pioneers, the pines will eventually shade the others out and become dominant. Their dominance is only temporary, however, because their seeds cannot germinate in shade and litter on the forest floor. Seeds of hardwoods such as maples and oaks, however, can germinate under these conditions, and as the pines die, hardwood seedlings grow quickly to fill the holes

produced in the canopy. The climax, then, is the hardwood forest, which can reproduce itself. Figure 26.13 is a schematic diagram showing one example of old-field succession.

It is important to note that the changes of the sere result from the action of the plants and animals themselves; one set of inhabitants paves the way for the next. As long as nearby populations provide colonizers, the changes lead in an automatic fashion from old field to forest. This type of succession is often termed *autogenic* (self-producing) *succession*.

In many cases, however, autogenic succession does not run its full course. Environmental disturbances, such as wind, fire, flood, or clearing by humans, may occur often enough to permanently alter or divert the course of succession. In addition, habitat conditions, such as site exposure, unusual bedrock, impeded drainage, and the like can hold back the course of succession so successfully that the climax is never reached; instead, an earlier stage of the sere becomes more or less permanent and is as stable at that site as the climax may be on more favorable sites. Thus a mosaic of biotic communities will be the stable biogeographic pattern in an area with a diversity of habitats.

HUMAN IMPACT ON NATURAL ECOSYSTEMS

Some areas of natural vegetation appear to be untouched but are dominated by humans in a more subtle manner. For example, many of our national forests have been protected from fire for a number of decades, setting up an unnatural condition in terms of what might be expected of the natural ecosystem. When lightning starts a forest fire, the fire fighters parachute down and put out the flames as fast as possible. But periodic burning of forests and grasslands is a natural phenomenon, and it performs vital functions in the ecosystem. One such vital function is to release nutrients from storage in the biomass so that the soil can be revitalized. Another is to increase the areas covered by grasses and herbs on which grazing animals are dependent.

Humans have influenced ecosystems in yet another way: by moving plant and animal species from their indigenous habitats to foreign lands and foreign environments. The eucalyptus tree is a striking example. The various species of eucalypts have been transplanted from

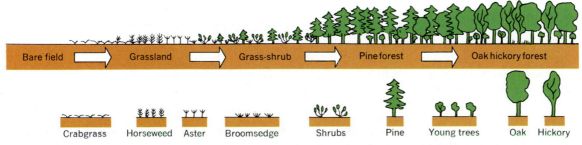

FIGURE 26.13 Old-field succession in the Piedmont region of the southeastern United States, following the abandonment of cornfields and cottonfields. This is a pictorial graph of continuously changing composition spanning about 150 years. (After E. P. Odum, *Fundamentals of Ecology*, W. B. Saunders Co., Philadelphia.)

Australia to such faroff places as California, North Africa, and India. Sometimes exported plants thrive like weeds, forcing out natural species and becoming a major nuisance. Transporting the jackrabbit to Australia resulted in a devastating rabbit population explosion, only recently checked by importing rabbit diseases as well. It is said that scarcely one of the kinds of grasses seen clothing the coast ranges of California is a native species, but the casual observer might think that these represent the native vegetation.

Inadvertent introduction of a disease from a foreign continent can cause the extinction of a particular plant or animal species. An example is the chestnut blight, which eliminated the American chestnut from the forests of the northeastern United States. Accidentally imported insects can also wipe out most of the mature individuals of a native plant when there are no native insect predators available to combat the invasion. These are just a few of the ways in which humans have interfered with natural ecosystems.

THE GREAT YELLOWSTONE FIRE—DISASTROUS OR BENEFICIAL?

Yellowstone is the largest and oldest American national park. Thanks in part to the magnificent photographs of its wondrous geysers and travertine terraces, taken by the pioneer photographer, William Henry Jackson, and brought back to Congress in Washington, it was declared a park in 1872. Almost a square in outline, its 900,000 hectares (2.2 million acres) occupy the northwest corner of Wyoming within the Rocky Mountain province. Volcanic rocks dominate the geology, and lava flows occupy much of the area, forming a high plateau. Because an ancient caldera underwent collapse here, geothermal activity continues to this day in many localities in the park (see Chapter 14 and Figure 14.13). Its rich, well-watered forests and parks support populations of bison, elk, and grizzly bear, along with many other mammal and bird species—a powerful attraction for millions of tourists. All in all, we have here a priceless forest ecosystem, little disturbed by natural catastrophe or human interference for at least the past two centuries.

Through August and September 1988, many forest fires burned out of control in Yellowstone Park, casting a smoke pall over large areas of surrounding states. Of the 45 fires, most were started by lightning strikes, and this number was not unusual for the area. It had, however, been the driest summer for more than a century of record, and temperatures were persistently high. Once started, the fires spread rapidly and uncontrollably under the driving force of strong winds that in one fire reached 160 km (100 mi) per hour. Firefighters were powerless to check the spread of the flames. Dense forests of lodgepole pine that had not burned for 250 years were killed by flames that raced through the forest canopy. Fires at ground level raced through areas of older forest with ample fuel supplies accumulated as fallen trees and branches. The fires also consumed patches of young forest that rarely ignite and have typically served well as natural firebreaks.

Contrary to greatly exaggerated stories in the news media, a later study of air photographs showed that only 20 percent of the park land had been burned over in some degree. Fears that massive extermination of wildlife had occurred also proved unfounded. Most of the large animals survived: only 9 bison of the herd of 2,500, and 350 elk of the herd of 30,000 perished in the fires. Mammals large and small, along with insects and birds, returned quickly to the burned areas, with the herbivores taking advantage of grasses that grew vigorously among the ashes of the forest floor (Figure 26.14).

For a full century since the establishment of this park in 1872, the National Park Service had practiced its "no burn" policy of not allowing any fire to spread unchecked. This policy was based on the concept that national parks must be maintained in a pristine condition, their plants being permitted to grow and die unaffected by any human activity or by fires. In 1972, this practice was reversed to a "let burn" policy under which naturally occurring fires were to be allowed to spread unchecked. As expected, the Park Service immediately came under strong criticism for having practiced their "let burn" policy by standing aside and allowing the July blazes to spread to mammoth proportions. By late July, park officials decided to suppress all fires, but a massive effort with national support was to no avail. In retrospect, however, it is questionable as to whether immediate action to put out the first fires could have made much difference over the vast area of the park. Another carefully drawn conclusion following study of the fire by forest ecologists was that the additional fuel accumulated in the 100-year "no burn" period made only a minor difference in the extent and intensity of the 1988 fire.

Ecologists view the occurrence of a season of great fires in Yellowstone Park about every 250 to 300 years as part of a natural cycle, or ecological succession. Following such a fire, saplings of lodgepole pine occupy the

FIGURE 26.14 This herd of elk is grazing among the ashes of burned out lodgepole pine forest in Yellowstone Park, Wyoming, during the fall season following the great summer fires of 1988. (Tom & Pat Leeson/Photo Researchers.)

burned area, growing tall and closely set to form a dense crown canopy, enduring between 150 and 250 years after the initial fire. It is during this phase that large canopy burns can occur, restarting the succession. If no fire occurs during this phase, the succession continues with a dying out of the pines and the growth of a new generation of other tree species, such as the spruce and fir. By this time—more than 300 years from the start—the forest floor becomes highly flammable and can be swept by fire to restart the succession.

Recent studies have established that the natural succession described above affects isolated patches of the entire forested region, resulting in a mosaic of patches in different stages of succession. In this mosaic, the younger patches serve as natural firebreaks to limit the spread of a fire occurring in adjacent older patches. Wildlife is able to escape such fires, and then to reinhabit the burned area, where new ground cover of grasses and shrubs can provide food for the grazing herbivores. The importance of this natural process of fire and renewal in shaping the mosaic pattern of vegetation has been summarized by two ecologists:

Fire itself is extremely heterogeneous in its impact. Many people imagine that almost half of Yellowstone has been completely blackened by the fire. Not so. It is a mosaic of destruction, a heterogeneity that will be reflected in the regrowth. We have to ensure that whatever policies we have for fires in parks, we maintain that heterogeneity. (Norman Christensen)[1]

It is well documented that fire is essential to the maintenance of many different types of habitat. Without fires, the tall grass prairie of East Kansas would be woodland, the pine forests of the South East would be hardwood forest, the conifer forests of the west would be completely transformed. (Ronald Myers)[1]

[1]From Roger Lewin, 1988, *Science*, vol. 241, p. 1762–63.

Additional data sources: Richard Monastersky, 1988, *Science News*, vol. 134, p. 314–17; William H. Romme and Don G. Despain, 1989, *Scientific American*, vol. 261, no. 5, p. 37–46.

BIOMASS BURNING AND ITS IMPACTS ON THE ATMOSPHERE

Biomass burning—the combustion of plant matter in fires occurring naturally as well as those set by humans—is widespread over the lands of the earth, but especially prevalent in the low latitude zones. The fuel consists of the trees and shrubs of forests and woodlands, of grasses and forbs of the savanna and prairie grasslands, and of cereal crops. Biomass burning is carried out to remove forests so that new areas can be turned into agricultural and pastoral lands. This conversion is largely permanent and irreversible because of the expanding food requirements of a burgeoning world population of humans. Annual burning of dead grasses along with seedling trees and shrubs is carried out over wide areas of the tropical grasslands prior to the annual rains to promote the growth of a dense crop of new grasses. Burning of savanna woodland and rainforest is a part of sustained programs of shifting agriculture. Waste products of agriculture are also burned on a large scale; for example, in the disposal of rice straw and sugar cane husks. Trees and woody shrubs are removed on a large scale for use as fuel for cooking fires in the Third World countries. Also included in the world total is disposal by burning of discarded wood and wood products used in building construction and paper made from wood pulp.

Recent estimates place the total carbon content of biomass released into the atmosphere annually by burning at between 2 and 5 billion tonnes (metric tons). This burning releases a variety of gases, along with particulate matter as smoke. By far the largest gaseous component is carbon dioxide (CO_2), carrying 90% of the released carbon. There follow in order of magnitude carbon monoxide (CO), methane (CH_4), nitrogen in several gaseous forms, hydrogen (H_2), various other hydrocarbons, and sulfur dioxide (SO_2). Much of the nitrogen is in the molecular form, N_2, produced by the high temperatures of the fires. As we have explained in earlier chapters, the greenhouse gases in the above list—carbon dioxide being the dominant one—make an important contribution to global warming.

Dust particles contained in the smoke of biomass combustion form a relatively small part of the total emission. Rising smoke plumes frequently induce spontaneous convection resulting in the production of rain, but this usually evaporates before reaching the ground. The finer smoke particles are eventually widely distributed through the global troposphere; some enter the stratosphere, where they remain suspended for long periods. In the stratosphere they serve to scatter incoming shortwave radiation, and can thus exert a cooling effect on the planetary temperature.

Smoke from fires contains the starting ingredients necessary for the production of ozone in the lower troposphere, and thus contributes (along with nitrogen oxides in the released gases) to a form of air pollution in the class of photochemical smog. Acid deposition is also a product of biomass burning and is responsible for low values of pH (on the order of 4.2) now commonly observed in low-latitude rainfall.

To summarize the contribution of greenhouse gases released by biomass burning to a global rise in temperature, it has been estimated that carbon dioxide from this source may be responsible for as much as 20 to 60 percent of the predicted greenhouse warming. Methane from burning also makes a proportionate contribution. Tropospheric smoke particles, by absorbing incoming shortwave radiation and outgoing longwave radiation, can be expected to contribute to global warming. Smoke particles also serve as nuclei of condensation, promoting cloud formation, which may lead to atmospheric warming or cooling, depending on the altitude and extent of the cloud cover (see Chapter 6).

Much remains to be learned from continued scientific investigation of the role that burning of the global biomass plays, along with the combustion of fossil fuels, in global climate change, but there seems to be no doubt that the role of this form of human intervention in natural processes is a major one.

Data sources: P. J. Crutzen and M. O. Andreae, 1990, *Science*, vol. 250, p. 1669–78; R. Monastersky, 1990, *Science News*, vol. 37, p. 196–97; J. S. Levine, 1990, *EOS*, vol. 71:37, p. 1075–77.

ECOSYSTEMS IN REVIEW

In looking back over this chapter, we can clearly see that two concepts of biogeography stand opposed but inseparable, like the two sides of a coin. One is the concept of the *natural ecosystem*, an ecosystem that attains its development without appreciable interference by humans and is subject to natural forces of modification and destruction, such as storms or fires. The other concept is that of the ecosystem sustained in a modified state by human activities. Extremes of both cases are found throughout the world.

Natural ecosystems can still be seen over vast areas of the wet equatorial climate where rainforests are as yet scarcely touched by humans. Much of the arctic tundra and the needleleaf forest of the subarctic zones is in a natural state. In contrast, much of the continental surface in midlatitudes is almost totally under our control through intensive agriculture, grazing, or urbanization. You can drive across an entire state, such as Ohio or Iowa, without seeing a vestige of the plant cover it bore before the coming of the European settlers. Only if you know where to look for them can you find a few small plots of undisturbed prairie or natural forest.

WORLD PATTERNS OF NATURAL VEGETATION

The distribution of natural vegetation and its relationship to soils and climate are of primary concern to geographers. To organize and classify the world's vegetation, biogeographers have adopted a number of systems. Most such systems are based on two criteria. First in importance is the structure of the vegetation in a region. Structure includes the growth forms of the dominant plants as well as their organization and arrangement in space. A second criterion is the flora, or list of plants found in a region. A biogeographer may wish to distinguish different vegetation regions, or floristic realms, on the basis of species composition. The world vegetation map described in this chapter is based on both structural and floristic criteria.

DESCRIBING THE STRUCTURE OF VEGETATION

The structural description of vegetation is based on the physical properties and outward forms of plants. Six categories of information have been included in a system of structural description set up by a distinguished Canadian biologist and ecologist, Pierre Dansereau. These six categories tell us about the growth form of the plants, their size and stratification, the degree to which they cover the ground, their time functions (periodicity), and their leaf forms.

1. Life-form. Plants can be classified according to their *life-form*. The first two forms, trees and shrubs, are erect, woody plants (Figure 27.1). A *tree* is a perennial plant having a single upright main trunk, often with few branches in the lower part, but branching in the upper part to a crown which, in the mature individual, contributes to one of the upper layers of vegetation (Figure 27.2). A *shrub* is a woody plant having several stems branching near the ground, so as to place the foliage in a mass starting close to ground level. Third are woody vines that climb on trees: the *lianas*. Although the term "liana" is usually associated with woody vines of tropical forests, the meaning may be broadened to include woody vines of midlatitude forests, such as the poison ivy and Virginia creeper. Thus life-form classes cut across the taxonomic categories of the plant world. For example, palms and tree ferns look much alike outwardly but are entirely different botanically.

Fourth of the life-forms is the plant group known as the *herbs*. These are usually small, tender plants lacking woody stems. The adjective *herbaceous* is applied to this

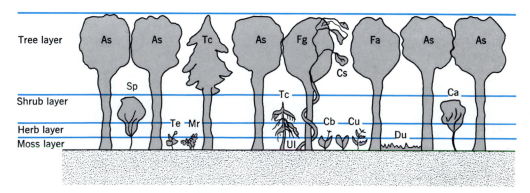

FIGURE 27.1 Schematic diagram of life forms in a beech-maple-hemlock forest. The tree layer consists of sugar maple (As), ash (Fa), beech (Fg), and hemlock (Tc) and includes a liana (Cs). The shrub layer includes elder (Sp), dogwood (Ca), and a young hemlock (Tc). An epiphyte (Ul) grows on the hemlock. Plants designated Te, Mr, Cb, and Cu form the herb layer. Moss (Du) forms the lowest layer. (From Pierre Dansereau, *Ecology*, vol. 32.)

FIGURE 27.2 Forest of the midlatitude zone is represented by this stand of beech and hemlock trees in Great Smoky Mountains National Park, Tennessee. This is a mixed deciduous–needleleaf forest. Lightning has shattered the tree at the left. The lower layer consists of scattered shrubs and seedling trees. (Donaldson Koons.)

life-form. Herbs occur in a wide variety of forms and leaf structures, including annuals and perennials, broad-leaved plants and grasses. Broadleaved herbs are termed *forbs* in distinction with grasses. The herbaceous layer will normally occupy a low position in a layered or stratified plant community. Smaller, and lying in close contact with the ground or attached to tree trunks, are the *bryoids*, a life-form that includes the mosses and liverworts.

Finally, among the life-forms are the *epiphytes*, plants that use other plants as supporting structures and thus live above ground level, out of contact with soil (Figure 27.3). Familiar to all are the tropical orchids that live high on tree limbs and are sometimes referred to as "air plants." Ferns also commonly exist as epiphytes.

Not included in our list of life-forms are lower forms of plants, grouped under the name of *thallophytes*. These include bacteria, algae, molds, and fungi—all of which

are plants lacking true roots, stems, and leaves. *Lichens*, plant forms in which algae and fungi live together to form a single plant structure, are included within the life-form classification of bryoids. Lichens occur as tough, leathery coatings or crusts and as leaflike masses attached to rocks and tree trunks (Figure 11.43). In arctic and alpine environments, lichens may grow in profusion and dominate the vegetation in the near absence of more conspicuous plant forms.

2. Size and stratification. Each of the life-forms just described may be classified according to size. The words "tall," "medium," and "low" may be given definite limits for each life-form. For example, a tree higher than 25 m (82 ft) is "tall"; from 10 to 25 m (33 to 82 ft) is "medium"; 8 to 10 m (26 to 33 ft) is "low." Such standardization of size enables the plant geographer to make precise descriptions of vegetation.

3. Coverage. The degree to which the foliage of individual plants of a given life-form cover the ground beneath them is the *coverage*. We may use four terms descriptive of the coverage: barren or very sparse, discontinuous, in tufts or groups, and continuous. For example, the trees may be of discontinuous coverage whereas the herb layer is continuous, or vice versa.

4. Periodicity. Of primary importance in the classification of forms of natural vegetation is the response of the plant foliage to the annual climatic cycle. *Deciduous plants* shed their leaves and become dormant in an unfavorable season, which is either too cold or too dry to permit growth. *Evergreen plants* retain green foliage year around, although in some cases they become almost dormant in a cold or dry season. Where the climate is equable (i.e., moist and not cold throughout the year), evergreen plants grow continuously. A third class, the

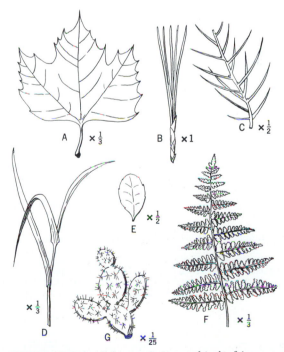

FIGURE 27.4 Leaf shapes. A, Large thin leaf (sycamore). B, Needleleaf (pine). C, Spine. D, Graminoid leaf (grass). E, Small leaf. F, Compound leaf (fern). G, Succulent stem (cactus).

FIGURE 27.3 This air-plant, or epiphyte, is a bromeliad, supported on a tree branch in subtropical evergreen rainforest. Everglades National Park, Florida. (A. N. Strahler.)

semideciduous plants, are those that shed their leaves at intervals not in phase with a season. Thus a semideciduous forest would not at any one time have all its individuals devoid of foliage. As a fourth class we recognize *evergreen-succulent plants*, those with very thick, fleshy leaves that retain their foliage year-round, and *evergreen-leafless plants*, those with fleshy stems but no functional leaves, such as the cacti.

5. Leaf shape and size. Recognition of the leaf form of a plant is an essential part of the structural description (Figure 27.4). One form is the *broadleaf,* familiar to us in such common trees as oak, maple, beach, and rhododendron. In contrast is the *needleleaf,* also familiar in the pine, spruce, fir, and hemlock. A similar form is the *spine,* which in some plants represents the transformation of

the entire leaf. The slender, tapering leaves of grass are *graminoid* in form. We may also recognize the *small leaf,* as in the birch or holly, and the *compound leaf,* as in the hickory and ash.

6. Leaf texture. Leaf textures range widely according to climate and habitat because of the different degrees to which the water loss from the leaf into the air must be controlled. Leaves of average thickness are described as *membranous;* those that are thin and delicate (as in the maidenhair fern) are described as *filmy.* Leaves that are hard, thick, and leathery are *sclerophyllous;* a forest dominated by trees and shrubs having such leaves is termed a *sclerophyll forest.* Very greatly thickened leaves, capable of holding much water in their spongy structure, are described as *succulent.*

TABLE 27.1 The Major Formation Classes

Formation Class and Map Symbol	Associated Climate Types	Associated Soil Orders or Suborders
Forest Biome		
Equatorial and tropical rainforest (Fe)	(1) Wet equatorial (2) Monsoon and trade-wind littoral	Oxisols, Ultisols
Montane forest (Fmt)	Highlands, climates (1) and (2)	Oxisols, Ultisols
Monsoon forest (Fmo)	(3) Wet-dry tropical (4s) Dry tropical, semiarid	Ultisols, Oxisols, Ustalfs, Vertisols
Broadleaf evergreen forest (laurel forest) (Fbe)	(6) Moist subtropical (8) Marine west-coast	Udults Udalfs
Midlatitude deciduous forest (Fd)	(8) Marine west-coast (10) Moist continental	Udalfs, Boralfs, Udolls
Needleleaf forest (Fcl, Fl, Fsp, Fbo, Fbl)	(8) Marine west-coast (North America) (10) Moist continental (North America) (11) Boreal forest	Spodosols, Boralfs Histosols, Cryaquepts
Sclerophyll forest (Fsm, Fss, Fsa)	(7) Mediterranean	Xeralfs, Xerolls, Xerults
Savanna Biome		
Savanna woodland (Sw)	(3) Wet-dry tropical	Ustalfs, Ultisols, Oxisols, Vertisols
Thorntree-tall grass savanna (Stg)	(4s, 4sd) Dry tropical, semiarid, semidesert (5s, 5sd) Dry subtropical, semiarid, semidesert	Ustalfs, Ultisols, Oxisols, Vertisols
Grassland Biome		
Prairie (tall-grass) (Gp)	(6) Moist subtropical (10) Moist continental	Udolls
Steppe (short-grass prairie) (Gs)	(9s) Dry midlatitude, semiarid (10sh) Moist continental, subhumid	Borolls, Ustolls, Xerolls, Aridisols
Desert Biome		
Thorntree semidesert (Dtw, Dtg)	(4sd, 4d) Dry tropical, semidesert, desert (5sd, 5d) Dry subtropical, semidesert, desert	Aridisols, Ustalfs
Semidesert (Dsd, Dss)	(4sd, 4d) Dry tropical, semidesert, desert (5sd, 5d) Dry subtropical, semidesert, desert (9sd, 9d) Dry midlatitude, semidesert, desert	Aridisols, Psamments
Dry desert (D, Dsp)	(4d) Dry tropical, desert (5d) Dry subtropical, desert (9d) Dry midlatitude, desert	Aridisols, Psamments
Tundra Biome		
Arctic tundra (T) Alpine tundra (Ta)	(12) Tundra Highland climate, alpine zone	Cryaquepts, Cryorthents

THE MAJOR BIOMES

The biome is a broad major grouping of natural ecosystems that includes animal life as well as plants (Chapter 26). The major biomes, however, are recognized primarily on the basis of their distinctive types of vegetation: forest, savanna, grassland, desert, and tundra.

The *forest biome* includes all regions of *forest*—an assemblage of trees growing close together, forming a layer of foliage that largely shades the ground. Forests often show stratification, with more than one layer. Shading of the ground gives a distinctly different microclimate than that found over open ground. Forests require a relatively large annual precipitation, but it does not need to be uniformly distributed throughout the year. In terms of the soil-water balance, forest is most closely associated with the humid and perhumid subtypes of the moist climates, as defined in Chapter 9. Soil-water storage usually remains high throughout most of the year, although it may be depleted during a dry season. Consequently,

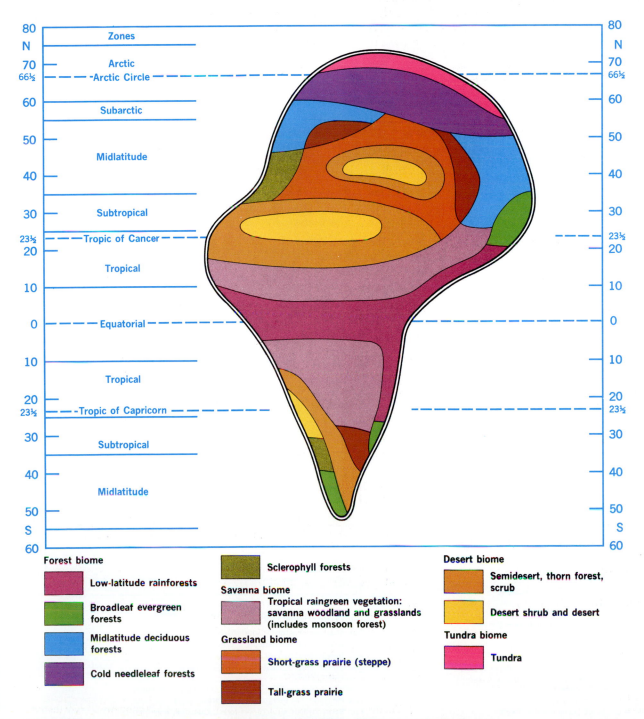

FIGURE 27.5 A schematic diagram of the vegetation formation classes on an idealized continent. Compare with the world vegetation map, Figure 27.6.

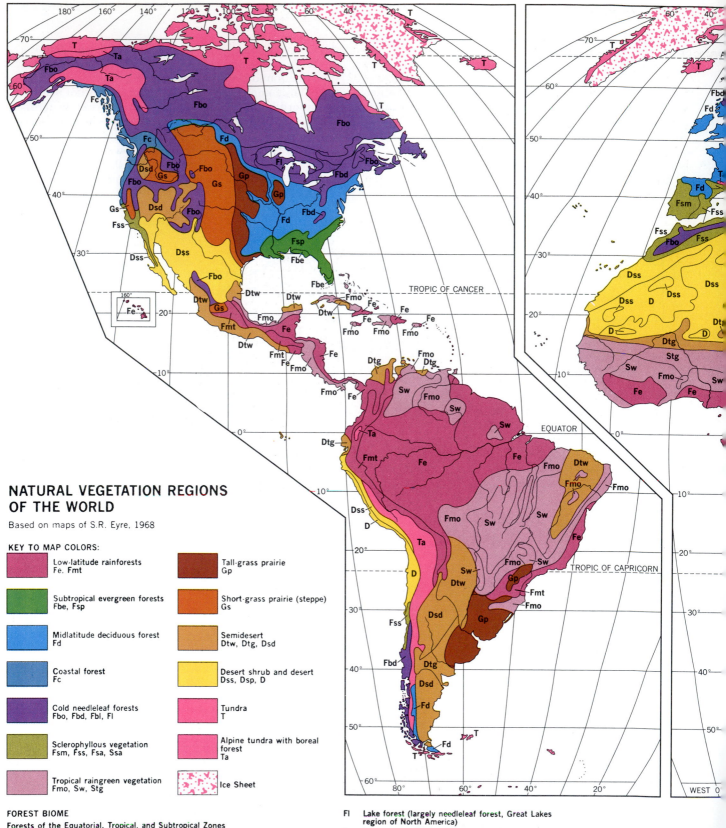

NATURAL VEGETATION REGIONS OF THE WORLD

Based on maps of S.R. Eyre, 1968

KEY TO MAP COLORS:

- Low-latitude rainforests — Fe, Fmt
- Subtropical evergreen forests — Fbe, Fsp
- Midlatitude deciduous forest — Fd
- Coastal forest — Fc
- Cold needleleaf forests — Fbo, Fbd, Fbl, Fl
- Sclerophyllous vegetation — Fsm, Fss, Fsa, Ssa
- Tropical raingreen vegetation — Fmo, Sw, Stg
- Tall-grass prairie — Gp
- Short-grass prairie (steppe) — Gs
- Semidesert — Dtw, Dtg, Dsd
- Desert shrub and desert — Dss, Dsp, D
- Tundra — T
- Alpine tundra with boreal forest — Ta
- Ice Sheet

FOREST BIOME

Forests of the Equatorial, Tropical, and Subtropical Zones

Fe   Equatorial and tropical rainforest (selva, broadleaf evergreen forest)

Fmt   Montane forest (may include conifers)

Fmo   Monsoon (raingreen) forest (tropical deciduous forest)

Fbe   Subtropical broadleaf evergreen forest (laurel forest; may include mixed broadleaf-needleleaf forest)

Forests of the Midlatitude and Subarctic Zones

Fd   Midlatitude deciduous (summergreen) forest

Fc   Coastal forest (largely needleleaf evergreen forest, west coast of North America)

Fl   Lake forest (largely needleleaf forest, Great Lakes region of North America)

Fsp   Southern pine forest (southeastern United States)

Fbd   Mixed boreal and deciduous forest

Fbo   Boreal forest (largely needleleaf evergreen forest)

Fbl   Boreal forest dominated by deciduous larch (Larix dahurica)

Sclerophyllous Forests of the Subtropical and Midlatitude Zones

Fsm   Mediterranean evergreen mixed forest

Fss   Sclerophyllous scrub (dwarf forest, chaparral; may be transitional to desert biome)

Fsa   Australian sclerophyll (Eucalyptus) forest

FIGURE 27.6 Natural vegetation of the world.

520

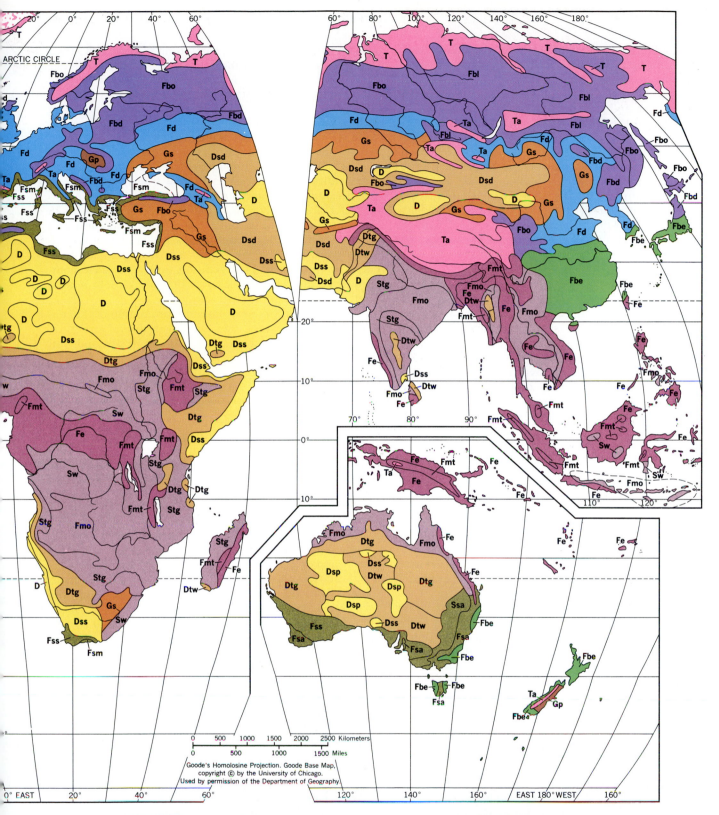

SAVANNA BIOME
- **Sw** Savanna woodland (broadleaf tree savanna)
- **Stg** Thorntree-tall grass savanna
- **Ssa** Australian sclerophyllous tree savanna

GRASSLAND BIOME
- **Gp** Tall-grass prairie
- **Gs** Short-grass prairie (steppe)

DESERT BIOME
- **Dtw** Thorn forest and thorn woodland (may be transitional to forest)
- **Dtg** Thorntree-desert grass savanna

- **Dsd** Semidesert scrub and woodland
- **Dss** Semidesert scrub
- **Dsp** Desert alternating with porcupine grass semidesert
- **D** Desert

TUNDRA BIOME
- **T** Arctic tundra
- **Ta** Alpine tundra (includes boreal forest)

Data source: S. R. Eyre, Vegetation and soils; a world picture, Second Edition, Aldine Publishing Company, copyright © 1968 by S. R. Eyre. See Appendix I, Maps 1–10. Map boundaries and classes have been simplified and modified by the authors with permission of S. R. Eyre, Edward Arnold (Publishers) Ltd., and The Aldine Publishing Company.

521

the forest biome spans a wide range of climates and latitudes, from the wet equatorial climate (1) to the boreal forest climate (11).

Vegetation of the *savanna biome* consists of a combination of trees and grassland in various proportions. The appearance of the vegetation can be described as park-like, with trees spaced singly or in small groups and surrounded by, or interspersed with, surfaces covered by grasses or by some other plant life form, such as shrubs or annuals in a low layer. Savanna is closely associated with warm climates having alternate wet and dry seasons. The most widespread areas of savanna are identified with the wet-dry tropical climate (3) and with the semiarid subtype (4s) of the dry tropical climate. The latter climate shows a brief, but often intense, rainy season.

Vegetation of the *grassland biome* consists largely or entirely of herbs, which may include grasses, grasslike plants, and forbs (broadleaf herbs). Degree of coverage may range from continuous to discontinuous and there may be stratification. Grassland can include some trees in moister habitats of valley floors and along stream courses where ground water is within reach of tree roots. The grassland biome is closely associated with both the semiarid subtype of the dry midlatitude climate (9s) and the subhumid subtypes of the moist subtropical climate (6sh) and the moist continental climate (10sh). Soil-water deficit ranges from zero to 20 cm or greater, and there is little or no water surplus. Thus soil-water storage is well below the storage capacity throughout most of the year.

The *desert biome*, associated with climates of extreme aridity, has thinly dispersed plants. A large proportion of the surface is bare ground, exposed to direct insolation and to the forces of wind and water erosion or to freeze–thaw action. Although essentially treeless, desert may have scattered woody plants, grasses, and perennial herbs.

Vegetation of the *tundra biome* consists of grasses and grasslike plants and an abundance of flowering herbs, dwarf shrubs, mosses, and lichens. Ground cover can range from complete to spotty. Tundra climate is cold throughout most of the year, with severe frost possible during all the warmer months. Tundra is encountered not only in the arctic zone, but also at high elevations as alpine tundra.

FORMATION CLASSES

Biogeographers subdivide the biomes into formation classes, smaller vegetation units based largely on vegetation structure. The formation classes we introduce are major widespread types clearly associated with 12 climate types and their soil-water budgets. Association with major soil types will also be important. In this way we survey the global scope of plant geography as a synthesis of climate with organic and physical processes of the life layer. Table 27.1 lists 16 formation classes and their approximate associations with climate types and soil types. Figure 27.5 is a schematic diagram to show how the vegetation formation classes would be arranged

on an idealized continent, combining Eurasia and Africa into a single supercontinent.

On world vegetation maps, it is possible to recognize many more than 16 formation classes. When further subdivisions are made, floristic composition becomes of equal importance to vegetation structure. Thus we can define lesser units to distinguish between vegetation types that are structurally very similar but consist of greatly differing assemblages of species. Some subordinate units are also necessary to show transitions or mixes between major vegetation types.

Our world vegetation map, Figure 27.6, recognizes 28 distinctive vegetation units, using both structural and floristic criteria. It is based on a world system of vegetation types prepared by Professor S. R. Eyre. The Eyre system recognizes 34 vegetation units. We have combined certain of Eyre's classes that are closely related but of relatively minor occurrence.

We now turn to a systematic examination of the major formation classes, their composition and structure, and their relationships with climate types and soil orders.

FOREST BIOME

Equatorial and Tropical Rainforests

Equatorial rainforest (Fe) consists of tall, closely set trees. Their crowns form a continuous canopy of foliage and provide dense shade for the ground and lower layers (see Figure 10.8). The trees are characteristically smooth barked and unbranched in the lower two-thirds (Figure 27.7). Trunks commonly are buttressed at the base by radiating, wall-like roots (Figure 27.8). Tree leaves are large and evergreen; from this characteristic the equatorial rainforest is often described as "broadleaf evergreen forest." Crowns of the trees tend to form into two or three layers, or strata, of which the highest layer consists of scattered emergent crowns rising to 40 m (130 ft) and protruding conspicuously above a second layer, 15 to 30 m (50 to 100 ft), which is continuous (Figure 27.9). A third, lower layer consists of small, slender trees 5 to 15 m (15 to 50 ft) high with narrow crowns.

Typical of the equatorial rainforest are thick, woody lianas, supported by the trunks and branches of tall trees. Some are slender, like ropes; others reach thicknesses of 20 cm (8 in.). They rise to heights of the upper tree levels where light is available and may have profusely branched crowns. Many lianas have tendrils or suckers to cling to a growing tree. Others rise by winding about the tree trunk.

Epiphytes are numerous in the equatorial rainforest. These plants are attached to the trunk, branches, and foliage of trees and lianas, using the "host" solely as a means of physical support. Epiphytes are of many plant classes and include ferns, orchids, mosses, and lichens. Some epiphytes are *stranglers*. These woody vines send down their roots to the soil and may eventually surround the tree, perhaps ultimately replacing it. The strangling fig (*Ficus*) is an example (Figure 27.10). Other stranglers begin as lianas.

FIGURE 27.7 Interior of lowland rainforest, Baiyer River, New Guinea. The trees are smooth-trunked and support thick lianas (left). (Tom McHugh/Photo Researchers.)

FIGURE 27.8 Buttress roots at the base of a large rainforest tree. From a forest preserve near Abidjan, Côte d'Ivoire, West Africa. (Alan H. Strahler.)

A particularly important botanical characteristic of the equatorial rainforest is the large number of species of trees that coexist. As many as 300 species may be found in 1 hectare. Individuals of a species are often widely separated. Consequently, if a particular tree species is to be extracted from the forest for commercial uses, considerable labor is involved in seeking out the trees and transporting them from their isolated positions.

The floor of the equatorial rainforest is usually so densely shaded that plant foliage is sparse close to the ground. The forest has an open aspect, making it easy to

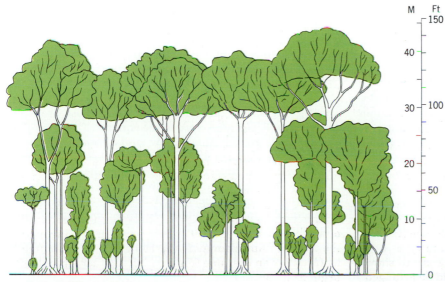

FIGURE 27.9 Diagram of the structure of tropical rainforest in Trinidad, British West Indies. A representative species of the tallest trees is *Mora exelsa.* (After J. S. Beard, *The Natural Vegetation of Trinidad*, Clarendon Press, Oxford.)

FIGURE 27.10 This strangler fig has surrounded the trunk of a living oak tree. Everglades National Park, Florida. (A. N. Strahler.)

FIGURE 27.11 A coastal mangrove forest growing in muddy water. Everglades National Park, Florida. (A. N. Strahler.)

traverse. The ground surface is covered only by a thin litter of leaves. Rapid consumption of dead plant matter by bacterial action results in the absence of humus on the soil surface and within the soil profile.

Equatorial rainforest is a response to an equable climatic regime that is continuously warm, frost free, and has abundant precipitation in all months of the year (or, at most, only one or two dry months). The equatorial rainforest is thus closely correlated in extent with the wet equatorial climate (1) and the monsoon and trade-wind littoral climate (2). A large water surplus characterizes the annual water budget, so soil water is adequate at all times. An important climatic factor is the extreme uniformity of monthly mean air temperature, which is typically between 26° and 29°C (79° and 84°F) for stations at low elevation in the equatorial zone. In Chapter 10 we illustrated the equatorial thermal regime by graphs of minimum and maximum daily temperatures for 2 months in the city of Panama, lat. 9°N (Figure 10.3). Recall that the daily range greatly exceeds the annual range. In the absence of a cold or dry season, plant growth goes on continuously throughout the year. Individual species have their own seasons of leaf shedding, possibly caused by slight changes in day length.

Soil orders of the equatorial rainforest are typically Oxisols and Ultisols on well-drained upland surfaces. Re-

call from Chapters 23 and 24 that soils of these orders have low base status. The meager supply of nutrient base cations is held in a thin surface layer, where it is effectively recycled by rainforest plants.

Variations in the equatorial rainforest structure are found in specialized habitats and where humans have disturbed the vegetation. Where the forest has been cleared by cutting and burning (as for small-plot agriculture), the returning plant growth is low and dense and may be described as *jungle*. Jungle can consist of lianas, bamboo scrub, thorny palms, and thickly branching shrubs. This tangled growth can form an impenetrable barrier to travel, in contrast to the open floor of the climax rainforest. Eventually, rainforest species will return, unless disturbance is chronic.

Coastal vegetation in areas of equatorial rainforest is highly specialized. Coasts that receive suspended sediment (mud) from river mouths, and where water depths are shallow, typically have *mangrove swamp forest*, consisting of stilted trees (Figure 27.11). The mangrove prop roots trap sediment from ebb and flood tidal currents, so the land is gradually extended seaward. Mangroves commonly consist of several shoreward belts of the red (*Rhizophora*), the black (*Avicennia*), and the white (*Laguncularia*) mangrove. Another common coastal salt-marsh plant of the humid low-latitude zone is the screw pine (*Pandanus*). Typical of recently formed coastal deposits, such as sandy beaches, are belts of palms, of which the coconut palm (*Cocos nucifera*) is an example (Figure 10.10).

World distribution of equatorial rainforest is shown on the vegetation map, Figure 27.6, by those areas of rainforest (Fe) located close to the equator. The principal world areas are the Amazon lowland of South America; the Congo lowland of Africa and a coastal zone extending westward from Nigeria to Guinea; and the East Indian region, from Sumatra on the west to the islands of the western Pacific on the east. Poleward of these areas, equatorial rainforest regions are transitional into tropical rainforest of higher latitudes.

See Chapter 10 for a description of the rainforest fauna as well as the plant products and food resources.

Tropical rainforest (Fe) is quite similar in structure to

the equatorial variety, but extends through the tropical zone (lat. 10°–25°N and S) along coasts windward to the trades. The tradewind littoral climate in which the tropical rainforest thrives has a short dry season, but not intense enough to deplete the soilwater. There is also a marked annual temperature cycle resulting from the variations in height of the sun's path in the sky in tropical latitudes. The cooler temperatures, coinciding approximately with the period of reduced rainfall, impose some stress on the plants. As a result, there are fewer species and fewer lianas. Epiphytes are abundant, however, because of continued exposure to humid air and cloudiness of the maritime tropical air masses that impinge on the coastal hill and mountain slopes. Soils of the tropical rainforest are largely Oxisols and Utisols.

In global distribution, the tropical rainforest is represented on the world map (Figure 27.6) by those areas of rainforest, designated Fe, that lie between lat. 10° and 25°. The Caribbean lands represent one important area of tropical rainforest, although the rainforest is predominantly limited to windward locations. The Everglades of southernmost Florida contains small isolated communities of tropical rainforest, termed "hammocks," in which one finds the mahogany tree and strangler fig, along with abundant epiphytes. The coast of the same area has extensive mangrove swamp forest.

In southern and southeastern Asia, tropical rainforest is extensive in coastal zones and highlands that have heavy monsoon rainfall and a very short dry season. The Western Ghats of India and the coastal zone of Burma have tropical rainforest supported by orographic rains of the southwest monsoon. In the zone of combined northeast trades and Asiatic summer monsoon are rainforests of the coasts of Vietnam and the Philippine Islands. In the southern hemisphere, belts of tropical rainforest extend down the eastern Brazilian coast, the Madagascar coast, and the coast of northeastern Australia.

Within the regions of equatorial and tropical rainforest are many islandlike highlands of cooler climate in which precipitation is increased by the orographic effect. Here rainforest extends upward on the rising mountain slopes. Between 1000 and 2000 m (3000 and 6000 ft), the rainforest gradually changes in structure, becoming *montane forest* (Fmt). Several areas of montane forest are shown on the vegetation map—for example, in Central America and along the eastern slopes of the Andes mountains of South America. Montane forest is lower in tree height and less dense in canopy than equatorial and tropical rainforest of adjacent lowlands; with increasing elevation, forest height becomes even lower. Tree ferns and bamboos are numerous; epiphytes are particularly abundant. Heavy accumulations of mosses attached to tree branches become conspicuous at higher altitudes. In these *mossy forests* (cloud forests), enshrouded much of the time in clouds, mist and fog are persistent. Relative humidity is high, and condensed moisture keeps the plant surfaces wet. Near its upper limit, in *elfin forest*, the forest trees are dwarfed and densely festooned with mosses. Above the forest limit, which occurs at about 3600 m (12,000 ft) in the equatorial zone, forest gives way to a treeless vegetation that may be alpine tundra, scrub, grassland, or heath, depending on altitude and available moisture.

EXPLOITATION OF THE LOW-LATITUDE RAINFOREST ECOSYSTEM

In the past, humans farmed in low-latitude rainforests by the *slash-and-burn* method—cutting down all the vegetation in a small area, then burning it (Figure 27.12). Most of the nutrients in a rainforest ecosystem are tied up in the biomass rather than in the soil. Burning the slash on the site releases the trapped nutrients, returning a portion of them to the soil. The supply of nutrients derived from the original biomass is small, however, and the harvesting of crops rapidly depletes the nutrients. After a few seasons of cultivation, the soil loses much of its productivity. A new field is then cleared in another area, and the old field is abandoned. Plants of the rainforest are able to reinvade the abandoned area because fruiting species and animal seed carriers are close by, and so the rainforest soon returns to its original state. In this way the primitive slash-and-burn agriculture is fully compatible with the maintenance of the rainforest ecosystem.

On the other hand, modern intensive agriculture uses large areas of land and is not compatible with the rainforest ecosystem. When such lands are abandoned, seed sources are so far away that the forest species cannot take hold. Instead, secondary species dominate, often accompanied by species from other vegetation types. The

FIGURE 27.12 In this scene in the Guiana Highlands of Venezuela, the Panare Indians—inhabitants of this rainforest—have cleared a small area of rainforest and burned the vegetation in place. This area will be planted with subsistence crops, then abandoned to revert to new rainforest. (Jacques Jangoux/Peter Arnold, Inc.)

FIGURE 27.13 The Trans–Amazon Highway in construction cuts a great swath through the rainforest and requires extensive earth moving to grade the roadbed and provide drainage. (Toby Molenaar/Woodfin Camp.)

dominance of these secondary species is permanent, at least on the human time scale. Thus the rainforest ecosystem is, in this sense, a nonrenewable genetic resource of many, many species of plants and animals. Once displaced by large-scale cultivation, those species can never return to reoccupy the area. Ecologists warn that the disappearance of thousands of species of organisms from the rainforest environment would mean the loss of millions of years of evolution, together with the destruction of the most complex ecosystem on the globe.

Transformation of large areas of the rainforest of Amazonia into agricultural land has made use of heavy machinery to carve out major highways, such as the Trans-Amazon Highway in Brazil, and innumerable secondary roads and trails; to cut down or push down the great broadleaved rainforest trees, removing the trunks for commercial lumber; to burn the unwanted plant matter in place; and to replace the forest with clear fields for cattle pasture or commercial crops. Figures 27.13, 27.14,

and 27.15 are scenes typical of this transformation process.

Now we can take up a question arising from destruction of the rainforest: Can the rainforest environment be exploited as a new source of food? Some agricultural specialists were at first highly optimistic; but others, specializing in ecology, have been pessimistic, or at best dubious. What we can do here is to point out restraints on such expansion when it relates to raising such staples as rice, soybeans, corn, or sugar cane, which are field crops harvested seasonally in other environments.

First, the low content of nutrients in the Oxisols and Ultisols requires massive and repeated applications of fertilizers. These applied nutrients are not held in storage in substantial quantities in the soil and are quickly exported from the land in runoff because of the large water surplus. Second, there is no dry season in which crops can reach maturity and be harvested under dry conditions, as is the case in the wet-dry tropical climates and the subhumid and semiarid continental climates of midlatitudes. Third, the Oxisols are capable, in some areas, of becoming lithified—literally turning to rock—when denuded of plant cover and exposed to the atmosphere. This has actually happened on agricultural test farms in the rainforest of Brazil.

It is easy to say that agricultural technology can overcome these and other difficulties. Even if this conclusion is valid, no one questions the fact that the cost will be enormous in terms of energy input through all parts of the agricultural system. Fertilizers, pesticides, machinery, and fuel are only part of that energy input. Large amounts of capital are needed to set up efficient management and marketing systems. In view of the pending destruction of the rainforest ecosystem as a consequence of agricultural expansion, we can justifiably raise the question as to whether benefits outweigh losses.

According to a study published in 1990 by the World Resources Institute of Washington, D.C., the global total of low-latitude rainforest removed in 1989 was between 16 to 20 million hectares. At the same time, figures re-

FIGURE 27.14 Trunks of the rainforest trees are prepared and loaded for export as commercial lumber. (Abril Imagens/Andre Penner.)

FIGURE 27.15 This 1989 aerial view of the Brazilian rainforest near the Rio Branco in the state of Acre shows the pristine rainforest (left) in process of being cleared and burned (center and right); burned and largely cleared (lower left); and finally transformed into green pastureland (lower right). (R. Funari/Imagens da Terra/Impact Visuals.)

leased by the United Nations put the annual destruction rate at 17 million hectares, an area about that of the state of Washington. These figures are roughly twice as large as estimates made in 1980 and are probably more reliable because of improvements in quality of satellite imagery. The greatest loss of rainforest in 1989 was in Brazil—about 8 million hectares—but that number is down from the peak year of 1987, probably because certain Brazilian laws that encouraged forest removal were rescinded in 1988. Increases in deforestation rates in the decade 1980–89 took place in India, Indonesia, and Burma, where the rise was dramatic, and in Cameroon, Costa Rica, the Philippines, and Vietnam. Also experiencing increases were Colombia, Ivory Coast, Laos, and Nigeria.

A computer model study completed in 1990 by scientists of the University of Maryland and the Brazilian Space Research Institute indicated that when Amazon rainforest is entirely removed and replaced with pasture, surface and soil temperatures will be increased 1 to 3 Celsius degrees, while precipitation will decline by 26 percent and evaporation by 30 percent. Deforestation will greatly reduce the quantity of water vapor entering the Amazon basin from outside sources. In areas where a marked dry season occurs, that season will be lengthened. Although such models make simplifications and are subject to error, results confirm the pessimistic conclusion that once large-scale deforestation has occurred, artificial restoration of a rainforest comparable to the original one will be impossible to achieve—i.e., that deforestation is irreversible.

In 1985 the United Nations Food and Agriculture Organization (FAO) established the Tropical Forestry Action Plan (TFAP)—its purpose, to work with individual nations to protect remaining rainforests and to promote a sustainable use program for those forest resources. TFAP works in cooperation with the World Bank and the World Resources Institute. Initial success of the TFAP effort to bring about reforestation in Nepal has been widely publicized, but other plans of the TFAP have come under unfavorable criticism by conservation groups as having actually increased deforestation.

Quite different from the TFAP is The International Tropical Timber Organisation (ITTO), a trade group representing both timber users and suppliers. Their stated goal has been to achieve sustainable development of the low-latitude forests. A commission was appointed by the ITTO to investigate logging practices in Sarawak, a province of Malaysia and former British colony occupying the northwestern region of the Island of Borneo. Much of Sarawak's timber goes to Japan and Great Britain. The commission reported in 1990 that logging there is not being conducted in a sustainable manner; that it is destroying the rainforest environment; and that if the practice continues, all of Sarawak's primary forests will be destroyed within a decade. A related development of major importance in Sarawak, as in several other rainforest countries, was the entry of local tribal peoples into conflict with the central government. Representatives of two Sarawak tribes went on a world tour to publicize the devastation being wrought on their lands that threatens their livelihood as forest dwellers.

What seems to be emerging from such protests by natives of the rainforests and their supporters on the outside is a call for protecting and strengthening the economic base available to native inhabitants whose use of forest resources is fitted to the natural environment. In Brazil, for example, tapping of latex from isolated rubber trees that grow naturally throughout the forest is a lasting source of income for one group of natives. Many food products—nuts and fruits, for example—and medicinal plants of the rainforest can be harvested and processed to provide a living for the indigenous populations. A bold step in this direction was taken by Colombia, whose president announced in 1990 that 50 million hectares of Amazonian rainforest has been handed over to local Indian tribes along with the legal rights for them to live in their traditional manner and, in so doing, protect the rainforest. Programs of this kind require that large areas of existing rainforest be placed in permanent preserves under government protection from loggers, gold seekers, prospective cattle ranchers, and would-be crop farmers. Resistance to such programs comes not only from the commercial exploiters, but also from national governments beset by problems arising from vast migrations of poverty-stricken rural peoples into the large cities, where employment is lacking. An easy solution is to encourage the unemployed to pioneer in the rainforests.

Data sources: R. Repetto, *Scientific American*, 1990, vol. 262, no. 4, p. 36–42. R. Monastersky, *Science News*, 1990, vol. 137, p. 164; vol. 138, p. 40–42. *New Scientist*, 1990, vol. 125/1699, p. 42–45; vol. 126/1713, p. 22; vol. 126/1715, p. 46–49; vol. 126/1719, p. 25; vol. 126/1726, p. 29; vol. 127/1724, p. 45–48; vol. 127/1733, p. 32; vol. 127/1734, p. 61; vol. 128/1745, p. 28–29.

Monsoon Forest

Monsoon forest (Fmo) presents a more open tree growth than the equatorial and tropical rainforests. Consequently there is less competition among trees for light but a greater development of vegetation in the lower layers. Maximum tree heights range from 12 to 35 m (40 to 100 ft), which is less than in the equatorial rainforest. Many tree species are present and may number 30 to 40 in a small tract. Tree trunks are massive; the bark is often thick and rough. Branching starts at a comparatively low level and produces large round crowns (Figure 27.16).

Perhaps the most important feature of the monsoon forest is the deciduousness of most of the tree species present. The shedding of leaves results from the stress of a long dry season that occurs at time of low sun and cooler temperatures. Thus the forest in the dry season has somewhat the dormant winter aspect of deciduous forests of midlatitudes. A representative example of a monsoon forest tree is the teakwood tree (*Tectona grandis*).

Lianas and epiphytes are locally abundant in monsoon rainforest but are fewer and smaller than in the equatorial rainforest. Undergrowth is often a dense shrub thicket. Where second-growth vegetation has

FIGURE 27.16 Monsoon woodland in the Bandipur Wild Animal Sanctuary in the Nilgiri Hills of southern India. The scene is taken in the rainy season, with trees in full leaf. (John S. Shelton.)

formed, it is typically jungle. Clumps of bamboo are an important part of the vegetation in climax teakwood forest.

Monsoon forest is a response to a wet-dry tropical climate in which a rainy season with a water surplus alternates with a dry, rather cool season having a soil-water deficit. These conditions are most strongly developed in the Asiatic monsoon climate, but are not limited to that area. Typical regions of monsoon forest are in India, Burma, Thailand, and Cambodia. Deciduous tropical forest also occurs in Africa, Central and South America, and northern Australia.

Monsoon forest is closely identified with Ultisols and Oxisols in Southeast Asia and South America. These are soils of low base status and may have developed in periods of moister climate in the Pleistocene Epoch. According to our world map of soils, Figure 24.2, several areas of monsoon forest coincide with areas of Ustalfs, the ustic-regime suborder of Alfisols associated with a climate having a dry season. Base status of Alfisols is high, in contrast to the Ultisols and Oxisols.

Broadleaf Evergreen Forest

Broadleaf evergreen forest (Fbe), also referred to as *temperate rainforest* or *laurel forest*, differs from the equatorial and tropical rainforests in having relatively few species of trees. Thus there are large populations of individuals of a species. Trees are not as tall as in the low-latitude rainforests; the leaves tend to be smaller and more leathery, the leaf canopy less dense. Among the trees commonly found in the broadleaf evergreen forests of southern Japan and the southeastern United States are evergreen oaks (such as *Quercus virginiana*) and members of the laurel and magnolia families (such as *Magnolia grandiflora*) (Figure 11.7). A quite different evergreen forest flora is found in New Zealand and consists of large tree ferns, large conifers such as the kauri tree (*Agathis*

FIGURE 27.17 This lush rainforest of Rotorua, North Island of New Zealand, is one type of evergreen forest. It is known as the Smoking Forest because of the natural steam emissions of geothermal origin. Note the tree fern in the foreground. (F. Kenneth Hare.)

FIGURE 27.18 A winter scene along the Blue Ridge Parkway of Virginia, showing deciduous trees in their leafless phase. The evergreen pines in the foreground provide a bit of color in the drab landscape. (Alan H. Strahler.)

australis), podocarp trees (*Podocarpus*), and small-leaved southern beeches (*Nothofagus*) (Figure 27.17). Another important type of broadleaf evergreen forest, found in the Azores and Canary island groups, is the Canary laurel forest, which formerly covered Europe in the Miocene geologic epoch.

Broadleaf evergreen forests tend to have a well-developed lower stratum of vegetation that, in different places, may include tree ferns, small palms, bamboos,

shrubs, and herbaceous plants. Lianas and epiphytes are abundant. An example of conspicuous epiphyte accumulation is the Spanish "moss" (*Tillandsia usneoides*) that festoons the Evangeline oak and bald cypress trees of the Gulf Coast of the southeastern United States (Figure 11.7).

Broadleaf evergreen forest represents a response to a moist climate strongly influenced by maritime air masses. The annual range of temperature is small or

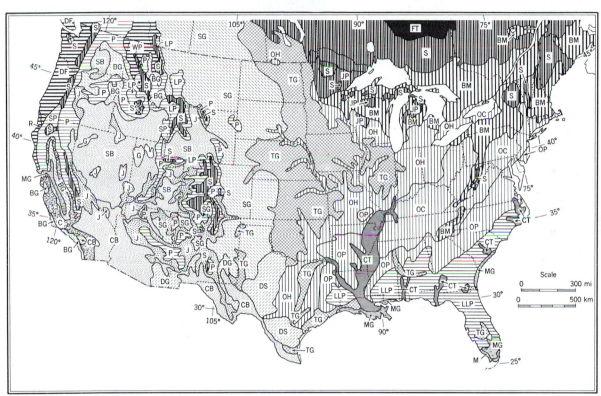

EASTERN FOREST VEGETATION

FT   Subarctic forest-tundra transition (Canada)
S   Spruce-fir (Northern coniferous forest)
JP   Jack, red, and white pines (Northeastern pine forest)
BM   Birch-beech-maple-hemlock (Northeastern hardwoods)
Oak forest (Southern hardwood forest):
        OC   Chestnut-chestnut oak-yellow poplar
        OH   Oak-hickory
        OP   Oak-pine
CT   Cypress-tupelo-red gum (River bottom forest)
LLP   Longleaf-loblolly-slash pines (Southeastern pine forest)
M   Mangrove (Subtropical forest)

WESTERN FOREST VEGETATION

S   Spruce-fir (Northern coniferous forest)
Cedar-hemlock (Northwestern coniferous forest):
        WP   Western larch-western white pine
        DF   Pacific Douglas fir
        R   Redwood

Yellow pine-Douglas fir (Western pine forest)
        SP   Yellow pine-sugar pine
        P   Yellow pine-Douglas fir
        LP   Lodgepole pine
J   Pinon-Juniper (Southwestern coniferous woodland)
C   Chaparral (Southwestern broad-leaved woodland)

DESERT SHRUB VEGETATION

SB   Sagebrush (Northern desert shrub)
CB   Creosote bush (Southern desert shrub)
G   Greasewood (Salt desert shrub)

GRASS VEGETATION

TG   Tall grass (Prairie grassland)
SG   Short grass (Plains grassland)
DG   Mesquite-grass (Desert grassland)
DS   Mesquite and desert grass savanna (Desert savanna)
BG   Bunch grass (Pacific grassland)
MG   Marsh grass (Marsh grassland)
Alpine meadow (Not shown)

FIGURE 27.19 Vegetation of the contiguous 48 United States and Southern Canada. (Modified and simplified from maps of H. L. Shantz and Raphael Zon, in *Atlas of American Agriculture*, and Canada Department of Forestry, Bulletin 123.)

moderate, and rainfall is abundant and well distributed throughout the year. Broadleaf evergreen forest of North America and eastern Asia occurs in the moist subtropical climate (6); that of New Zealand and Tasmania is in the marine west-coast climate (8).

Broadleaf evergreen forest of the southeastern United States and southern China is closely identified with the Udults, the udic-regime suborder of the Ultisols. Low base status is a characteristic of this soil order. Nutrient bases are concentrated close to the surface, where they are recycled by forest plants. Broadleaf evergreen forest of southeastern coastal Australia coincides with an area of Udalfs, the udic-regime suborder of Alfisols.

Midlatitude Deciduous Forest

Midlatitude deciduous forest (Fd), sometimes called *summergreen deciduous forest*, is familiar to inhabitants of eastern North America and western Europe as a native forest type. It is dominated by tall, broadleaf trees that provide a continuous and dense canopy in summer (see Figure 27.2) but shed their leaves completely in the winter (Figure 27.18). Lower layers of small trees and shrubs are weakly developed. In spring a luxuriant low layer of herbs develops quickly, but is greatly reduced after the trees have reached full foliage and have shaded the ground.

The deciduous forest is almost entirely limited to the midlatitude landmasses of the northern hemisphere. Figure 27.19, a map of vegetation of the United States and southern Canada, shows forest subdivisions emphasizing floristic composition. Common trees of the deciduous forests of eastern North America, central and eastern Europe, and eastern Asia (all in the moist continental climate) are oak (*Quercus*), beech (*Fagus*), birch (*Betula*), hickory (*Carya*), walnut (*Juglans*), maple (*Acer*), basswood (*Tilia*), elm (*Ulmus*), ash (*Fraxinus*), tulip tree (*Liriodendron*), chestnut (*Castanea*), and hornbeam (*Carpinus*). The needleleaved conifers pine (*Pinus*), hemlock (*Tsuga*), and balsam fir (*Abies*) are often interspersed with these broadleaved species. In western Europe, under a marine west-coast climate, dominant trees are mostly oak and ash, with beech (*Fagus sylvatica*) in cooler and moister areas.

In poorly drained habitats, the deciduous forest consists of such trees as alder, willow, ash, and elm and many hygrophytic shrubs. Where the deciduous forests have been cleared by windthrow or landslide, pines readily develop as second-growth vegetation on the freshly exposed mineral soil.

Because the regions of midlatitude deciduous forest have for centuries been the areas of dense populations, few remnants of a primeval forest have survived. Instead, most existing forests are modified by practices of tree farming. Large areas are completely and permanently removed from forest growth by farming, urban development, and roadways.

The midlatitude deciduous forest represents a response to a continental climate in which precipitation is adequate throughout the growth season. There is a strong annual temperature cycle with a cold winter season and a warm summer. Precipitation is markedly

FIGURE 27.20 Needleleaf forest of red and black spruce in southern Ontario. (Alan H. Strahler.)

greater in the summer months (especially so in eastern Asia) and thus increases at the time of year when evapotranspiration is great and the soil-water demand is high. Only a small water deficit is incurred in the summer, whereas a large surplus normally develops in spring. The winter is exceptionally dry in eastern Asia but this factor is compensated for by cold.

Soils most closely identified with the midlatitude deciduous forests are Udalfs and Boralfs, suborders of the Alfisols. Some forest areas in the subhumid climate subtype are underlain by Udolls, the udic-regime suborder of the Mollisols. All these suborders are soils of high base status. Favorable parent materials of glacial origin, including till plains, lake plains, and loess covers underlie wide areas where the midlatitude deciduous forest occurs.

Needleleaf Forests

Needleleaf forest (Fc, Fl, Fsp, Fbo, Fbl) is composed largely of straight-trunked, conical trees with relatively short branches and small, narrow, needlelike leaves. These trees are conifers (Figure 27.20). Where evergreen,

FIGURE 27.21 The Siberian forest of deciduous larch trees is seen here in its fall colors. (Wolfgang Kaehler.)

FIGURE 27.22 Two kinds of needleleaf forest of the western United States. (left) Open forest of western yellow pine (ponderosa pine), in the Kaibab National Forest, Arizona. (A. N. Strahler.) (right) A grove of great redwood trees (*Sequoia*) in Humboldt State Park, California. (Alan H. Strahler.)

the needleleaf forest provides continuous and deep shade to the ground so that lower layers of vegetation are sparse or absent, except for a thick carpet of mosses in many places. Species are few, and large tracts of forest consist almost entirely of but one or two species. The world vegetation map (Figure 27.6) includes five units within the needleleaf forest. These subdivide the needleleaf forest by floristic composition.

Boreal forest (*Fbo*, *Fbl*) predominates in two great continental belts, one in North America and one in Eurasia, which span the landmasses from west to east in latitudes 45° to 75°. The boreal forest (Fbo) of North America, Europe, and western Siberia is composed of evergreen conifers, such as spruce (*Picea*), fir (*Abies*), and pine (*Pinus*), whereas that of north-central and eastern Siberia is boreal forest dominated by larch (Fbl). The larch (*Larix*) sheds its needles in winter and is thus a deciduous forest (Figure 27.21). Associated with the needleleaf trees is mountain ash. Aspen and balsam poplar (*Populus*), willow, and birch tend to take over rapidly in areas of needleleaf forest that have been burned over, or they can be found bordering streams and in open places.

Needleleaf evergreen forest (*Fbo*) extends into lower latitudes wherever mountain ranges and high plateaus exist. Thus in western North America, this formation class extends southward into the United States on the Sierra Nevada and Rocky Mountain ranges and over parts of the higher plateaus of the southwestern states (Figure 27.22). In Europe, needleleaf evergreen forests flourish on all the higher mountain ranges.

The *coastal needleleaf evergreen forest* (*Fc*) of coastal southeastern Alaska, British Columbia, Washington, Oregon, and California is of particular interest. Under a regime of heavy orographic precipitation and prevailing high humidities, those areas have perhaps the densest of all coniferous forests (Figure 11.20). Forests of coastal redwood (*Sequoia sempervirens*) (Figure 27.22), big tree (*Sequoiadendron giganteum*), and Douglas fir (*Pseudotsuga menziesii*) are particularly remarkable (Figure 11.20). Individuals of redwood and big tree attain heights of over 100 m (325 ft) and girths of over 20 m (65 ft).

Lake forest (*Fl*), another needleleaf evergreen forest type, occurs in the Great Lakes area of North America. It is dominated by white pine (*Pinus strobus*), red pine (*P. resinosa*), and eastern hemlock (*Tsuga canadensis*). Little of this commercially valuable forest remains.

Another needleleaf forest type unique to North America is the *southern pine forest* (*Fsp*), consisting of a number of different pine species. It is found in sandy soils of the southeastern coastal plain, where it appears to be a specialized vegetation type dependent on fast-draining sandy soils and frequent fires for its preservation (Figure 27.23).

In the northernmost range, the boreal needleleaf forest grades into *cold woodland*. This form of vegetation is limited to the very cold subarctic climate. Trees are low in height and spaced well apart; a shrub layer may be well developed. The ground cover of lichens and mosses is distinctive (Figure 11.38). Cold woodland is often re-

FIGURE 27.23 This plantation of longleaf pine grows on sandy soil of the Georgia coastal plain. The trees are used as pulpwood for paper production. The blackened trunks show that fires periodically sweep through the area, consuming the undergrowth. (Charles R. Belinky/Photo Researchers.)

ferred to as *taiga*. It is transitional into the treeless tundra and arctic heath at its northern fringe.

Because much of the area of needleleaf evergreen forest in North America and Europe was subjected to glaciation in the Wisconsinan stage of the Pleistocene Epoch, lakes and poorly drained depressions abound. These bear the hygrophytic vegetation, described in Chapter 26 as forming a bog succession and leading to large, thick peat accumulations known in Canada as *muskeg*.

Boreal forest (Fbo) is closely identified with the boreal forest climate (11) throughout North America and Eurasia. Both northern and southern boundaries of the formation class coincide rather well with the climatic boundaries, defined in terms of annual water need. Much of the boreal forest region is underlain by Spodosols, acid soils of low base status. In interior areas of western Canada, Russia, and Siberia, the soils are Boralfs, the cold-regime suborder of the Alfisols. Also important are large areas of Histosols and Cryaquepts. The Cryaquepts are a cold-climate great soil group within the Aquepts, which is a suborder of the Inceptisols characterized by poor drainage and saturated soil. Lake forest (Fl) and coastal forest (Fcl) are closely associated with Spodosols. The southern pine forest (Fsp), which lies in the moist subtropical climate (6), is in a region of Udults and Aquults. (The Aquults, a suborder of the Ultisols characterized by poor drainage, occupy much of the coastal fringe of the southeastern United States.)

Sclerophyll Forest

Sclerophyll forest (*Fsm, Fss, Fsa*) consists of low trees with small, hard, leathery leaves. Typically the trees are low-branched and gnarled, with thick bark. The formation class includes much *woodland*, an open forest in which the canopy coverage is only 25 to 60 percent. Also included are extensive areas of *scrub*, a plant formation type consisting of shrubs having a canopy coverage of perhaps 50 percent. The trees and shrubs are evergreen, their thickened leaves being retained despite a severe annual drought. There is little stratification in the sclerophyll forest and scrub, although there may be a spring herb layer.

Sclerophyll forest is closely associated with the Mediterranean climate (7) and quite narrowly limited in geographic extent—primarily to west coasts between 30° and 40° or 45° latitude (Figure 27.6). In the Mediterranean lands the sclerophyll forest forms a narrow peripheral coastal belt. Here the *Mediterranean evergreen mixed forest* (*Fsm*) consists of such trees as cork oak (*Quercus suber*), live oak (*Quercus ilex*), Aleppo pine (*Pinus halepensis*), stone pine (*Pinus pinea*), and olive (*Olea europaea*) (Figure 11.15). What may have once been luxuriant forests of such trees were greatly disturbed by humans over the centuries and reduced to woodland or entirely destroyed. Today, large areas consist of dense scrub, locally called *maquis*, which includes many species, some of them very spiny.

The other northern hemisphere region of sclerophyll forest is that of the California coast ranges. Some of this is a woodland of live oak (*Quercus agrifolia*) and white oak (*Quercus lobata*) (Figure 11.16). Much of the vegetation is *sclerophyllous scrub* (*Fss*) or "dwarf forest," known as *chaparral* (Figure 27.24). It varies in composition with elevation and exposure. Chaparral may contain wild lilac (*Ceanothus*), manzanita (*Arctostaphylos*), mountain mahogany (*Cercocarpus*), "poison oak" (*Rhus diversiloba*), and live oak.

The sclerophyll forest is represented in central Chile and in the Cape region of South Africa by a scrub vegetation that is of quite different flora from scrub of the northern hemisphere. Important areas of sclerophyll forest, woodland, and scrub are found in southeast, south-central, and southwest Australia, including several species of eucalypts and acacias (Figure 27.25). This formation type is designated *Australian sclerophyll forest* (*Fsa*) on the world climate map.

FIGURE 27.24 Chaparral, often called "dwarf forest," in the San Gabriel Mountains of southern California. (left) Firebreaks are kept clear along the ridge crests, while narrow winding fire roads allow limited access. (right) A close-up shows a rain gauge, mounted in a tilted position, placed on a small patch of open ground. (A. N. Strahler.)

FIGURE 27.25 A eucalyptus woodland in northern New South Wales, Australia. With its light trunk and clusters of drooping leaves, the eucalyptus presents a distinctive appearance. (Alan H. Strahler.)

The Mediterranean regime of the sclerophyll forest is one of great environmental stress because the severe drought season coincides with high air temperatures. Thus a large water deficit occurs in the summer. The wet, mild winter is, by contrast, highly favorable to rapid plant growth.

The soils of the sclerophyll forest, woodland, and scrub are closely associated with the Xeralfs, a xeric-regime suborder of the Alfisols. These soils have high base status. Also important locally are Xerolls and Xerults, xeric-regime suborders of the Mollisols and Ultisols, respectively.

FORESTS AND GLOBAL WARMING

A young, rapidly growing forest withdraws carbon dioxide (CO_2) from the atmosphere and places that carbon in storage within its biomass. Very old forests, on the other hand, increase their biomass very little, or not at all. From these statements, it is easy to conclude that we might be able to slow down the impending global warming by replacing our old forests with new ones. That course of action was put forward in the late 1980s by representatives of the logging industry as a viable program to follow in managing the Douglas fir and redwood forests of northern California and the Pacific Northwest. Naturally, this raised an outcry from the conservationist side, and has escalated a "timber war" widely publicized in the news media. Hostility throughout the general public to the loggers' plan has focused on the stands of old-growth redwoods as yet uncut, but located outside of sequestered state and national parks. A few activist extremists have sought to sabotage clear-cutting operations in these and Douglas fir forests.

More soberly-minded ecologists set out to examine this conclusion more closely. It rests in part on the assumption that when a mature forest is cleared, its wood goes mostly into long-term storage in buildings constructed of its finished lumber products. However, research scientists of the College of Forestry of Oregon State University in 1990 published a report showing that the conversion of old-growth forests to young, fast-growing forests will not significantly decrease the atmospheric CO_2. First, they calculated that while 42 percent of the harvested timber goes into comparatively long-term storage (greater than 5 years) in building structures, much of the remainder is directly discarded on the logging site where it is burned or rapidly decomposes. In addition, some becomes waste in factory processing of the lumber, where sawdust and scrap are burned as fuel. Similarly, the manufacture of paper also results in short-term conversion of a large proportion of the harvested trees to CO_2. In sum, harvesting of old-growth forests as now practiced actually contributes substantially to atmospheric CO_2.

Turning to the other side of the equation, that of introducing new forests in place of the old, there seems to be no question that this rapid growth will remove CO_2 from the atmosphere, but it is doubtful that the amount of that withdrawal will compensate for the amount injected during harvesting of old-growth forests and preparation of lumber and paper from the trees. Moreover, the withdrawal to storage of CO_2 by the growth of the young forest (or successive forests) requires some two centuries to equal that initially removed.

The concept of planting trees on land now largely barren of plants remains viable as a means of withdrawing CO_2 from the atmosphere. A large percentage of the 48 contiguous United States was completely deforested in the early years of occupation by European settlers, and most remains that way. Reforestation has been increasing in recent decades in certain parts of the country. A bonus in the equation comes from the realization that as the atmospheric CO_2 concentration rises, photosynthesis by green plants will be intensified, accelerating the growth rate of new forests and of food crops generally. (This reinforcing effect has been called the "fertilization theory" of CO_2.) And, since photosynthesis removes CO_2 from the air, its enhancement will reduce atmospheric CO_2. This phenomenon is an example of negative feedback—a process that helps maintain the status quo.

A number of grandiose tree planting schemes were put forward in the late 1980s. To remove the quantity of CO_2 now annually entering the earth's atmosphere (6 billion tons) would require some 7 million square kilometers of new closed-crown broadleaved deciduous forest—an area about the size of Australia. To absorb only the net increase would require about half that area. In 1988 the American Forestry Association announced its program, called Global Releaf, that would require Americans to plant 100 million new trees over a 4-year period. Ultimate disposal of the mature trees would again require a reversal of the CO_2 flow back into the atmosphere, but in the meantime some decades of reduction in the warming rate would allow more time to adjust to global shifts in zones of forest and agriculture.

Adjustment to global warming through geographic changes in forest regions in the middle latitudes has been anticipated. For example, warming might be accompanied by a northward shift in the loblolly pine range in the southeastern states, and in the Pacific Northwest, a shift of the Douglas fir forest environment upward into higher elevation zones where cool, moist conditions would remain in place. These and many other speculations will continue to be advanced and evaluated as the next decade or two reveals more about trends in global temperatures.

Data sources: M. E. Harmon, W. K. Ferrell, and J. E. Franklin, 1990, *Science*, vol. 247, p. 699–702; R. Monastersky, 1990, *Science News*, vol. 137, p. 85.

SAVANNA BIOME

Savanna Formation Classes

Savanna woodland (*Sw*) consists of trees spaced rather widely apart, permitting development of a dense lower layer, which usually consists of grasses. The woodland has an open, parklike appearance. Although plant formations of this general description can be found in a wide range of latitudes, most geographers associate savanna woodland with the wet-dry tropical climate (3) of Africa and South America. There the soil-water regime shows a seasonal decline in soil-water storage to an amount too small to sustain a closed-canopy forest. Savanna woodland typically forms a belt adjacent to equatorial rainforest in which soil-water storage remains high throughout the year.

In the tropical savanna woodland of Africa, the trees are of medium height, the crowns are flattened or umbrella-shaped, and the trunks have thick, rough bark (Figure 10.14). Some species of trees are xerophytic forms with small leaves and thorns; others are broad-leaved deciduous species that shed their leaves in the dry season. In this respect, savanna woodland is akin to monsoon forest, into which it grades in some places.

Fire is a frequent occurrence in the savanna woodland during the dry season. The trees of the savanna are of species that are particularly resistant to fire. Many biogeographers hold the view that periodic burning of the savanna grasses is responsible for the maintenance of the grassland against the invasion of forest. Fire does not kill the underground parts of grass plants, but it limits tree growth to a few individuals of fire-resistant species. The browsing of animals, which kills many young trees, is also a factor in maintaining grassland at the expense of forest. Many rainforest tree species that might otherwise grow in the wet-dry climate regime are prevented by fires from invading.

In Africa, the savanna woodland grades into a belt of *thorntree-tall grass savanna* (*Stg*), a formation class transitional into the desert biome. The trees, largely of thorny species, are more widely scattered and the open grassland is more extensive than in the savanna woodland. One characteristic tree is the flat-topped acacia (*Acacia*), another is the grotesque baobab (*Adansonia digitata*), with its barrel-shaped water-storing trunk. Elephant grass (*Pennisetum purpureum*) is a common species; it may grow to a height of 5 m (16 ft) to form a thicket impenetrable to humans. Because of the dominance of grassland over woodland, this formation class may equally well be placed in the grassland biome in terms of structure. However, since it is geographically and climatically intergradational with savanna woodland, inclusion in the savanna biome seems the more desirable alternative.

The thorntree-tall grass savanna is closely identified with the semiarid subtype of the dry tropical and subtropical climates (4s, 5s). In Africa and India, the formation class includes areas of the semidesert climate subtype (4sd, 5sd). In the semiarid climate, soil-water storage is very low in 10 out of the 12 months. Only during the brief rainy season is soil water adequate for the needs of plants. Onset of the rains is quickly followed by greening of the trees and grasses. For this reason, vegetation of the savanna biome is described as "rain-green," an adjective that applies also to the monsoon forest.

Soils of the savanna biome include Ustalfs, Ultisols, Oxisols, and Vertisols. The distribution of these orders varies according to parent matter of the soil and past climatic history. Ustalfs of the savanna biome in West Africa are believed to owe their high base status to a rain of fine dust particles carried from the Sahara Desert on the north. These dusts replenish primary minerals rich in base cations. The same process may explain the occurrence of Ustalfs in savanna regions of other parts of Africa, and in South America and India, where savannas are downwind from deserts.

The savanna biome in South America is exemplified in the *campo cerrado* of the interior Brazilian Highlands.

FIGURE 27.26 Sawgrass savanna with scattered pines in the dry season (winter). Everglades National Park, Florida. Clumps of tropical forest (left) are known as hammocks. (A. N. Strahler.)

Here the trees are largely broadleaved and evergreen. They are deep-rooted and capable of tapping lower levels of soil water not available to the grasses during the dry season.

A different expression of the savanna biome is found in eastern Australia. Here the *Australian sclerophyllous tree savanna (Ssa)* forms a north–south belt in which trees are evergreen species of *Eucalyptus*. The climate is a semidesert subtype of the dry subtropical climate (5sd). According to the world soils map, Figure 24.2, the soils here are Vertisols.

Of particular interest to American geographers is a small patch of rather specialized tropical savanna in southern Florida—the Everglades (Figure 27.26). Underlain by a limestone formation, the extremely low, flat plain of the Everglades is flooded by a shallow layer of runoff from summer rains and becomes a swamp. In winter the area becomes extremely dry. Coarse saw grass (*Cladium effusum*) covers much of the plain. Scattered trees are represented by palms and, on higher ground, by pines.

DROUGHT AND LAND DEGRADATION IN THE AFRICAN SAHEL

The tropical savanna environment is subject to devastating drought years as well as to years of abnormally high rainfall that can result in severe floods. *Drought* is the occurrence of lower-than-average precipitation in that season in which ample precipitation usually occurs. The term is usually used with respect to the growing season of plants that provide food for humans and their domesticated animals. Rainfall of the wet season of the savanna environment can be expected to vary greatly from one year to the next. Typically, climate records show that two or three successive years of abnormally low rainfall (a drought) alternate with several successive years of average or higher-than-average rainfall (see Figure 10.25).

The drier savanna zone of North Africa, including the semidesert zone adjoining it on the north, provides a lesson in human impact on a delicate ecological system. Countries of this perilous belt, called the Sahel, or Sahelian zone, are shown in Figure 27.27. All these countries were struck a severe blow by a drought that extended from 1968 through 1974. Both nomadic cattle herders and grain farmers share this zone. During the drought,

grain crops failed and cattle could find no forage. In the worst stages of the Sahel drought, nomads were forced to sell the remaining cattle that were their sole means of subsistence. Some 5 million cattle perished and it has been estimated that 100,000 people died of starvation and disease in 1973 alone (Figure 27.28).

Associated with the Sahelian drought of 1968–1974 is a special phenomenon, given the name *desertification*—the transformation of the land surface by human activities superimposed on a natural drought situation. That term has now been abandoned in favor of *land degrada-*

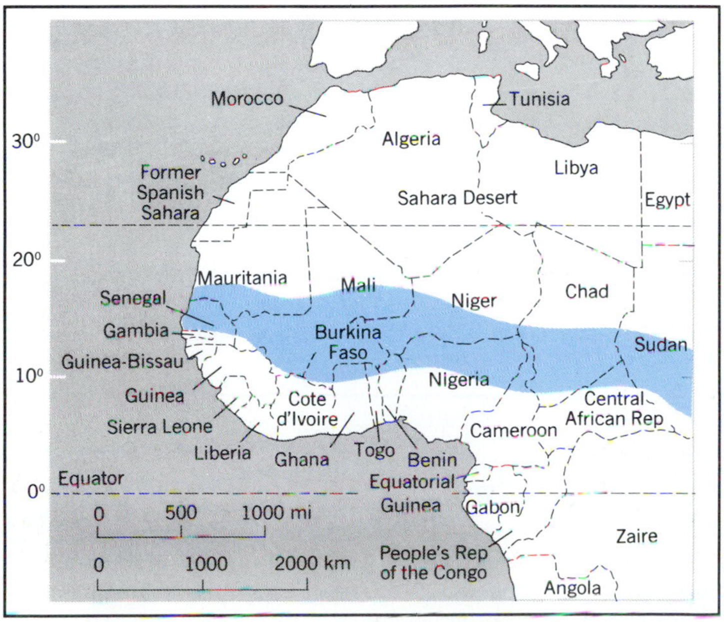

FIGURE 27.28 At the height of the Sahelian drought vast numbers of cattle had perished and even the goats were hard pressed to survive. Trampling of the dry ground prepared the region for devastating soil erosion in rains that eventually ended the drought. (Alain Nogues/Sygma.)

FIGURE 27.27 The Sahel, or Sahelian Zone, shown in color, lies south of the great Sahara Desert of North Africa.

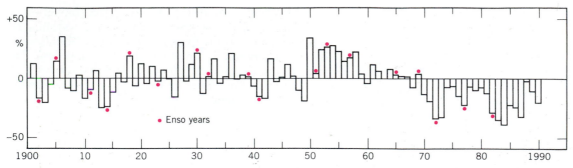

FIGURE 27.29 Rainfall fluctuations for stations in the western Sahel, 1901 through 1990, expressed as percent departure from the long-term mean. ENSO years are marked by dots. (Courtesy of Sharon E. Nicholson, Department of Meteorology, Florida State University, Tallahassee.)

tion, which consists of a change in appearance of the land surface toward closer resemblance to a desert, largely through the destruction of grasses, shrubs, and trees that previously existed. Also visible are the effects of accelerated soil erosion, such as rilling and gullying of slopes, and accumulations of sediment in stream channels. Deflation, too, is intensified and may be seen in scour of soil surfaces and accumulations of soil drifts in the lee of obstacles.

Land degradation in the African savanna environment can be attributed to greatly increased numbers of humans and their cattle. European managers of African colonies contributed to population increases by supplying food in time of famine, by reducing the death rate from disease, and by developing ground-water supplies for crop irrigation and livestock. With each succeeding drought, a heavier impact was made on the vegetation and soil by the increased population. Ultimately, land degradation was too sustained and too severe to permit recovery of the plant cover in the rainy season or over a moister period of ample rainfall seasons. Thus, as long as the population pressure persists, desertification is permanent, and takes on the outward appearances of a change toward a desert climate.

Periodic droughts throughout past decades are well documented in the Sahel, as they are in other world regions of the tropical savanna with its wet-dry climate. Figure 27.29 is a graph showing the percentage departures from the long-term mean of each year's rainfall in the western Sahel from 1901 through 1990. Since about 1950, the durations of periods of continuous departures both above and below the mean seem to have increased substantially over those of earlier decades. The period of sustained high rainfall years in the 1950s contrasts sharply with the severe drought years of 1970–73. There

followed several years of lessened rainfall deficiencies, but severe drought returned in 1982 and continued through 1984, a period in which the effects were particularly severe in Ethiopia. The period of continuous deficiencies spanned 20 years, and was broken in 1988 by near average rainfall, continued through 1989. The Sudan experienced exceptionally heavy rains in 1988 that caused disastrous flooding of the Nile River. A drought year followed in 1990. There is nothing in the modern record to confirm fears that changes in the land surface under the land degradation process have produced a permanent climate change in the direction of aridity.

Going much further back in the record, periods of rainfall deficiency and excess can be recognized: 1820–1840, below normal; 1870–1895, above normal; 1895–1920, below normal. Evidence of these periods is preserved in the record of changing shorelines of Lake Chad, which lies within the Sahelian zone.

Some climatologists are now searching for links between periods of Sahelian drought and ENSO (El Niño and the Southern Oscillation). In Figure 27.29, the bars for years coinciding with ENSO events are marked with dots and these appear to be about equally divided between wet and dry years.

The effect of an accelerated global warming in coming decades on climate of the Sahelian zone is difficult to predict. The Sahelian zone may be in for greater swings between drought and surplus precipitation. At the same time, the location of that zone may shift northward, along with the desert zone, as a result of intensification of the Hadley cell circulation. These suggested changes are at present highly speculative.

Data source: S. E. Nicholson, 1989, *Geophysical Monograph* 52, American Geophysical Union, p. 79–100.

GRASSLAND BIOME

Prairie (Gp) consists of tall grasses, comprising the dominant herbs, and of subdominant forbs (Figure 27.30). Trees and shrubs are almost totally absent but may occur in the same region as forest or woodland patches in valley bottoms. The grasses are deeply rooted and form

a continuous and dense sward. The grasses flower in spring and early summer; the forbs, in late summer. In Iowa, a representative region of tall-grass prairie, typical grasses are big bluestem (*Andropogon gerardi*) and little bluestem (*A. scoparius*); a typical forb is black-eyed susan (*Rudbeckia nitida*).

The largest areas of tall-grass prairie, those of North

FIGURE 27.30 Close-up view of virgin tall-grass prairie preserved in Kalsow Prairie, a State Botanical Monument in Iowa. Among the tall leaves of grass are forbs in flower. (Gene Ramsay, Webb Publishing Company.)

FIGURE 27.31 Cattle grazing on short-grass prairie in the northern Great Plains. (Charlton Photos.)

and South America, are closely associated with the subhumid subtype of the moist continental climate (10sh) and moist subtropical climate (6sh). There the soil-water budget shows a moderate soil-water deficiency, ranging from 0 to 15 cm. The tall-grass prairie also extends far into the area of the humid subtype of the same climates (10h, 6h). Details of the distribution of tall-grass prairie in North America are shown in Figure 27.19. Areas of deciduous forest are interfingered with areas of prairie over a broad transition belt. In Uruguay and eastern Argentina, the prairie plains region is known as the *Pampa*. A small but important European tall-grass prairie region is the *Puszta* of Hungary.

Tall-grass prairie is closely identified with Udolls, the udic-regime suborder of the Mollisols. Ustolls, the ustic-regime suborder of the Mollisols, underlie areas of prairie near its western limit in North America. These soils are exceptionally rich in base cations important to grasses.

Steppe (Gs), also called *short-grass prairie*, is a formation class consisting of short grasses occurring in sparsely distributed clumps or bunches (Figure 27.31). Scattered shrubs and low trees may also be found in the steppe. Ground coverage is small and much bare soil is exposed. Many species of grasses and other herbs occur. A typical grass of the American steppe is buffalo grass (*Buchloe dactyloides*); other typical plants are the sunflower (*Helianthus rigidus*) and loco weed (*Oxytropis lambertii*). Steppe grades into semidesert in dry environments and into prairie where rainfall is higher.

The world vegetation map (Figure 27.6) shows that steppe grassland is largely concentrated in vast midlatitude areas of North America and Eurasia. The only southern hemisphere occurrence shown on the map is the *veldt* of South Africa, formed on a highland surface in Orange Free State and Transvaal. We have excluded from the steppe formation class many occurrences of grasslands in low latitudes; these are, however, included in formation classes of transitional nature within the savanna and desert biomes.

The steppe formation class coincides quite closely with the semiarid subtype of the dry midlatitude climate (9s). The soil-water budget shows a substantial soil-water deficiency, and there is no water surplus. Soil-water storage is far below the soil storage capacity in all months. During a spring period of soil-water recharge, a substantial amount of water is made available to grasses and results in rapid growth into early summer. By midsummer the grasses are usually dormant, although occasional summer rainstorms cause periods of revived growth.

Soils of the steppe grasslands are largely within the order of Mollisols. Borolls are found in the colder northerly areas, Ustolls in the warmer southerly areas. Xerolls underlie the steppe grasslands west of the Rocky Mountains. Aridisols are represented in some of the more arid steppe lands, transitional into the desert biome.

DESERT BIOME

The desert biome includes several formation classes that are transitional from grassland and savanna biomes into vegetation of the more severe desert environment. We recognize two basic transitional formation classes: thorntree semidesert and semidesert. The first of these, *thorntree semidesert*, is found in low latitudes and is associated with tropical and subtropical climates. On the world vegetation map, Figure 27.6, thorntree semidesert includes two lesser formation classes: *thorn forest and thorn woodland (Dtw)* and *thorntree-desert grass savanna (Dtg)*.

Thorntree semidesert consists of xerophytic trees and shrubs adapted to a climate with a very long, hot dry season and only a very brief but intense rainy season. These conditions are found in the semidesert and desert subtypes of the dry tropical and dry subtropical climates (4sd, 4d, 5sd, 5d). The thorny trees and shrubs are locally known as *thorn forest, thornbush*, and *thornwoods*. These are deciduous forms that shed their leaves in the dry season (Figure 10.23). The plants may be closely intergrown to form impenetrable thickets. Cactus plants are present in some localities. A lower layer of herbs may consist of annuals, which largely disappear in the dry season, or grasses that become dormant in the dry season. Examples of thorntree semidesert are the *caatinga* of northeastern Brazil and the *dornveldt* of South Africa. Aridisols are the dominant soil order of the thorntree semidesert, but Ustalfs are present in some areas.

Semidesert is a transitional formation class of very wide latitudinal distribution; it ranges from the tropical zone to the midlatitude zone and is identified with the semidesert and desert subtypes of all three dry climates. On the world vegetation map, Figure 27.6, semidesert includes two secondary formation classes: *semidesert scrub and woodland (Dsd)* and *semidesert scrub (Dss)*.

Semidesert is a xerophytic shrub vegetation with a poorly developed herbaceous lower layer. An example is the sagebrush (*Artemisia tridentata*) vegetation of the middle and southern Rocky Mountain region and Colorado Plateau (Figure 27.32). Semidesert shrub vegetation seems recently to have expanded widely into areas of the western United States that were formerly steppe grasslands as a result of overgrazing and trampling by livestock. Soils of the semidesert are largely Aridisols. Also present are many areas of Entisols, particularly the Psamments developed on dune sands and sandy alluvium.

Dry desert is a formation class of xerophytic plants widely dispersed and providing no important degree of ground cover. On the world vegetation map, Figure 27.6, dry desert is represented by two formation classes: *desert (D)* and *desert alternating with porcupine grass semidesert (Dsp)*. The latter type is limited in occurrence to Australia.

The visible vegetation of dry desert consists of small hardleaved or spiny shrubs, succulent plants (cacti), or hard grasses. Many species of small annuals may be present, but appear only after a rare but heavy desert downpour.

Desert floras differ greatly from one part of the world to another. In the Mojave–Sonoran deserts of the southwestern United States, plants are often large and in places give a near-woodland appearance (Figure 27.33). Well known are the treelike saguaro cactus (*Carnegiea gigantea*), the prickly-pear cactus (*Opuntia imbricata*), the ocotillo (*Fouquiera splendens*), the creosote bush (*Larrea tridentata*) and the smoke tree (*Dalea spinosa*).

FIGURE 27.32 Sagebrush semidesert along the base of the White Cliffs in southern Utah. The reddish soil is an Aridisol. The classic automobile is an air-cooled Franklin, ca. 1934. (Donald L. Babenroth.)

FIGURE 27.33 A desert scene near Phoenix, Arizona. The tall, columnar plant is saguaro cactus; the delicate wandlike plant is ocotillo. Small clumps of prickly pear cactus are seen between groups of hardleaved shrubs. (Alan H. Strahler.)

In the Sahara Desert (most of it very much drier than the American desert), a typical plant is a hard grass (*Stipa*); another, found along the dry beds of watercourses, is tamarisk (*Tamarix*).

Much of the world map area assigned to desert vegetation has no plants of visible dimensions because the surface consists of shifting dune sands or sterile salt flats. Soils of the dry desert belong to the order of Aridisols or to Entisols, suborder Psamments.

The soil-water budget of the dry desert vegetation usually shows a greater quantity of water need than precipitation in every one of the twelve months. The annual total soil-water shortage is therefore very large. When rain falls in rare, torrential rainstorms, the soil water is used to maximum advantage by the plants and is rapidly depleted. During long periods when soil-water storage is near zero, the plants must maintain a nearly dormant state to survive.

TUNDRA BIOME

Arctic tundra (T) and *alpine tundra (Ta)* are formation classes of the tundra climate (12). Tundra of the arctic regions flourishes under a regime of long summer days during which the ground ice melts only in a shallow surface layer (Figure 11.39). The frozen ground beneath (permafrost) remains impermeable and meltwater cannot readily escape. Consequently, in summer a marshy condition prevails for at least a short time over wide areas. Humus accumulates in well-developed layers because bacterial action is very slow. Size of plants is partly limited by the mechanical rupture of roots during freeze and thaw of the surface layer of soil, producing shallow-rooted plants. In winter, drying winds and mechanical abrasion by wind-driven snow tend to reduce any portions of a plant that project above the snow.

Plants of the arctic tundra are low and mostly herbaceous, although dwarf willow (*Salix herbacea*) occurs in places. Sedges, grasses, mosses, and lichens dominate the tundra in a low layer (Figure 11.42). Typical species are ridge sedge (*Carex bigelowii*), arctic meadow grass (*Poa arctica*), cotton grasses (*Eriophorum*), and snow lichen (*Cetraria nivalis*) (Figure 11.43). There are also many species of forbs that flower brightly in the summer. Tundra composition varies greatly because soils range from wet to well drained. One form of tundra consists of sturdy hummocks of plants with low, water-covered ground between. Some areas of arctic scrub vegetation composed of willows and birches are also found in tundra.

Traced southward the vegetation changes into birch–lichen woodland, then into needleleaf forest. In some places a distinct tree line separates the forest and tundra. Coinciding approximately with the 10°C (50°F) isotherm of the warmest month, this line has been used by geographers as a boundary between boreal forest and tundra.

Vegetation is scarce on dry, exposed slopes and summits. The rocky pavement of these areas gives them the name *fell-field*, the Danish term meaning "rock desert."

In all latitudes where altitude is sufficiently high, alpine tundra is developed above the limit of tree growth and below the vegetation-free zone of barren rock and perpetual snow. Alpine tundra resembles arctic tundra in many physical respects (Figure 26.4).

Soils of the arctic tundra include representatives from the Inceptisols, Entisols, and Histosols. Large areas of arctic soils are classified as Cryaquepts, a great soil group within the suborder of Aquepts, order Inceptisols. Cryaquepts are poorly drained mineral soils. Other soil areas are classified as Cryorthents, a great soil group within the suborder of Orthents, order Entisols. Histosols are found in filled lake basins.

The number of animal species in the tundra ecosystem is small, but the abundance of individuals is high. Vast herds of caribou in North America or reindeer (their Eurasian relatives) roam the tundra, lightly grazing the lichens and plants and moving constantly. A smaller number of musk-oxen are also primary consumers of the tundra vegetation. Wolves and wolverines, arctic foxes, and polar bears are predators. Among the smaller mammals, snowshoe rabbits and lemmings are important

herbivores. Invertebrates are scarce in the tundra, except for a small number of insect species. Blackflies, deerflies, mosquitoes, and no-see-ums (tiny biting midges) are all abundant and can make July on the tundra most uncomfortable for humans and other mammals. Reptiles and amphibians are also rare. The boggy tundra, however, offers an ideal summer environment for many migratory birds, such as waterfowl, sandpipers, and plovers.

The food chain of the tundra ecosystem is simple and direct. The important producer is "reindeer moss," a lichen (see Figure 11.43). In addition to the caribou and reindeer, lemmings, ptarmigan (arctic grouse), and snowshoe rabbits are important lichen grazers. The important predators are the foxes, wolves, lynx, and humans, although all these animals may feed directly on plants as well. During the summer, the abundant insects help support the migratory waterfowl populations. The directness of the tundra food chain makes it particularly vulnerable to fluctuations in the population of a few species.

ALTITUDE ZONES OF VEGETATION

In earlier chapters we described the effects of increasing elevation on climatic factors, particularly air temperature and precipitation, and on soils. Vegetation also responds to an increase in elevation, as we showed in the case of the transition of rainforest into montane forest in low latitudes. To round out this concept, we turn to the dry midlatitude climate of the southwestern United States because there altitude zonation of vegetation is particularly striking.

Figure 27.34 shows the vegetation zones of the Colorado Plateau region of northern Arizona and adjacent states. Zone names, elevations, dominant forest trees, and annual precipitation data are given in the figure. Ecologists have set up a series of life zones, whose names suggest the similarities of these zones to latitude zones encountered in poleward travel on a meridian. The Hudsonian zone, 2900 to 3500 m (9500 to 11,500 ft), bears a needleleaf forest essentially similar to needleleaf boreal forest of the subarctic zone. Here Spodosols lie beneath

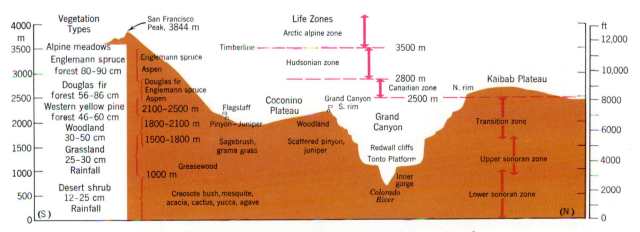

FIGURE 27.34 Altitude zoning of vegetation in the arid southwestern United States. Grand Canyon–San Francisco Mountain district of northern Arizona. (After G. A. Pearson, C. H. Merriam, and A. N. Strahler.)

S   Equatorial   Savanna   Savanna   Tropical scrub   Tropical   N
    rainforest   woodland   grassland   desert

Profile from equator to tropic of cancer, Africa

S   Tropical   Subtropical   Sclerophyll   Midlatitude   Subarctic   Subarctic   Arctic   N
    desert   steppe   forest   deciduous forest   needleleaf forest woodland (taiga)   tundra

Profile from tropic of cancer to Arctic circle, Africa-Eurasia

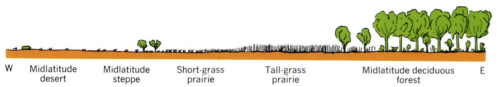

W   Midlatitude   Midlatitude   Short-grass   Tall-grass   Midlatitude deciduous   E
    desert   steppe   prairie   prairie   forest

Profile across United States, 40°N, Nevada to Ohio

FIGURE 27.35 Three schematic profiles showing the succession of plant formation classes across climatic gradients. (Drawn by A. N. Strahler.)

the forest. As the limit of forest, or tree line, is approached, the coniferous trees take on a stunted appearance and decrease in height to low shrublike forms (Figure 26.4).

A zone of alpine tundra lies above tree line. The snow line is encountered at about 3500 to 4000 m (11,500 to 13,000 ft) in midlatitudes, which is of course much lower than at the equator (see Figure 10.28).

GLOBAL VEGETATION IN REVIEW

The leading theme of this chapter has been that the structure of the natural plant cover is a response to environments dictated by climate. Each of the life forms favored by green plants has a capability for survival and a limit to survival. Forests exist where the environment is most favorable to net productivity of the biomass be-

cause of the abundance of heat and soil water during a long growth season. Where soil water is in short supply but heat remains adequate, forest gives way to the savanna structure in which shrubs and herbs have the upper hand over trees (Figure 27.35). The savanna woodland grades into grassland and this, in turn, grades into desert, where only those plants capable of living through long drought periods can survive. Traced into colder, higher latitudes, where heat supply becomes reduced below the optimum level, the forest gives way to arctic tundra. There plants contend with the effects of prolonged freezing of soil water and have only a short, cold annual period of growth.

We have found a significant correspondence of vegetation structure with soil type because soil-forming processes are also strongly dominated by climate. But organic processes also shape soil character, so the close associations of vegetation formation classes and soil types are partly a result of interaction.

EPILOGUE

Our survey of physical geography has brought together four major ingredients that define the physical environment: climate, soils, vegetation, and landforms. Climate provides the basic input of heat and water into the life layer where soil is formed and plants develop their distinctive formation classes.

Landforms add another dimension of variation within the broad climatic regions. Geologic processes play an important, independent role in landform evolution through tectonic and volcanic activities that have shaped the continents and the major landform units within the continents. Climate is also a powerful factor in shaping landforms. In this way, each of the global environments has some distinctive quality of landform instilled by climate through the same inputs of heat and water that vary the quality of soil and plant cover.

Perhaps the impression most deeply implanted by this survey is the enormous range in environmental quality, judged in terms of plant resources and the potential for production of plant food for human consumption. It may be a bit disconcerting to consider that vast areas of the continents are almost total deserts, either too dry or too cold to support human populations. In the desert wastes of the Sahara or the snow-covered plateau of Antarctica, the average American family might stay alive for at most a few days without a supply of water or heat, or both, brought in from more favorable environments.

Another distinct impression our survey might leave is that the human family has already done a remarkably thorough job of exploiting the favorable environments of the globe. For certain environments we have made an assessment of the prospects for increased food production by using massive inputs of irrigation water, fertilizers, fuels, and pesticides in concert with the newest in genetic strains of staple crops and the best in management techniques.

Interaction of humans with the natural environment has been a persistent theme of our study of physical geography. Here we are in the heartland of geography as a science—a study in human ecology and the spatial distribution of the varied ingredients. Many forms of interaction appear harmful to the natural environment and pose threats to the very sources of water and food on which humans and lesser life-forms depend. In closing this survey of physical geography we point out that physical geography can provide an important part of the science background needed to analyze environmental problems from a sound perspective.

With the availability of enormous supplies of energy and a high level of technology, twentieth-century civilization has a greatly broadened choice of environments that it can occupy. No finer example of this principle can be found on a vast scale than in the Los Angeles basin of southern California. With no flowing rivers of fresh water close at hand, no coal to speak of, only modest local reserves of petroleum and natural gas, and no good natural harbor, a great metropolis has arisen. Water importation by aqueduct systems on a vast scale solved the water problem; ships and pipelines solved the fuel problem; rail and truck transport solved the food problem; construction machinery and dredges solved the harbor problem.

A more extreme example of this type is the maintenance of human habitation on the ice shelves and ice sheet of Antarctica. If a nation chooses to pay the high cost of sending by ship and plane all food, fuel, and structural materials needed to keep a group of humans alive and well, even the most severe of global environments is not barred from occupation. As the cost of energy rises steadily, the allocation of national resources to such projects is given increasing attention in terms of benefits to be derived. To overcome severe environmental shortcomings by sheer force of energy and technology we must pay an increasingly higher price. Would it not be wise, instead, to optimize the benefits and resources of those naturally favorable environments in which the sun provides an unending source of energy and both heat and water are abundantly received by the soil under a favorable climate?

Finally, we must take note of a factor entering with increasing force into decisions regarding what courses of human activity are possible and most probable. Throughout the Industrial Revolution and its extension into European colonial expansion and growth of modern industrialized nations, human activity was dominated by self-serving motives—a better way of life and the acquisition of wealth and power. Destruction of the natural environment and its ecosystems proceeded unchecked well into the twentieth century with only a few voices raised in warning or protest.

Today, however, great concern is loudly expressed by many citizens for preservation of the remaining fragments of undamaged ecosystems. Many courses of action that were once viable possibilities have been ruled out or have come under severe criticism. The setting up of new national parks and wilderness areas is an example of a series of free choices guided by the philosophy of environmental preservation. Another, more recent example is the criticism leveled by ecologists at Brazil's plans to destroy the rainforest of Amazonia and replace it with farmland, cultured forest, and towns. This criticism led to a decision by the Brazilian government to save a part of the nation's rainforest. Perhaps, in time (and if not too late), forces of environmental preservation will prevail over those who would allow removal of the rainforest biome from the face of the earth.

MAP PROJECTIONS

As we stated in Chapter 1, a fundamental geographic problem is to find satisfactory ways to transfer the spherical network of parallels and meridians shown on a globe to the flat surface of a map; i.e., to devise useful *map projections*. Cartographers have struggled with this problem for centuries, driven by the need of the early navigators to have usable charts in their discovery voyages. Despite all the efforts of the finest mathematicians and cartographers, the quest for a perfect chart showing parallels and meridians of an entire globe has ended in frustration. It is simply impossible to transform a spherical surface to a flat (planar) surface without some violation of the true surface as a result of cutting, stretching, or otherwise distorting the information that lies on the sphere.

The problem of map projection must be faced squarely. As a geographer, you will find it essential to know which types of networks of parallels and meridians are best suited to the illustration of various portions and properties of the earth's surface. Keep in mind that no map projection will ever substitute fully for a globe to show general world relationships. We recommend the use of a globe in conjuction with flat maps.

DEVELOPABLE GEOMETRIC SURFACES

Certain geometric surfaces are said to be developable because, by cutting along certain lines, they can be made to unroll or unfold to make a flat sheet. Two such forms are the cone and the cylinder (Figure A.I.1). Were the earth conical or cylindrical, the map-projection problem would be solved once and for all by using the developed surface. No distortion of surface shapes or areas would occur, although it is true that the surface would be cut apart along certain lines. The earth, being spherical, belongs to a group of geometric forms said to be undevelopable because, no matter how they are cut, they cannot be unrolled or unfolded to lie flat. It is possible to draw a true straight line in one or more directions on the surface of a developable solid, but this cannot be done anywhere on an undevelopable form, such as a spherical surface. To make the parts of a spherical surface lie perfectly flat, the surface must be stretched—more in some

places than in others. So it is impossible to make a perfect map projection.

When a map is made of a very small part of the earth's surface, for example, an area 10 km across, the map-projection problem can be ignored. If the meridians and parallels are drawn as straight lines, intersecting at right angles and spaced apart correctly, the actual error is probably so small as to fall within the width of the lines drawn and is not worth correcting. As the area included on the map is increased, however, the problem gains in importance. When we attempt to show the whole globe, serious trouble develops. Only by some compromise can the distortion be reduced to a reasonable degree over important parts of the earth's surface.

SCALE OF GLOBES AND MAPS

All globes and maps depict the earth's features in much smaller size than the true features they represent. Globes are intended in principle to be perfect models of the earth itself, differing from the earth only in size, but not in shape. The *scale* of a globe is therefore the ratio between the size of the globe and the size of the earth, where size is expressed by some measure of length or distance (but not area or volume). Take, for example, a globe 20 cm in diameter representing the earth, whose diameter is about 13,000 km. The scale of the globe is

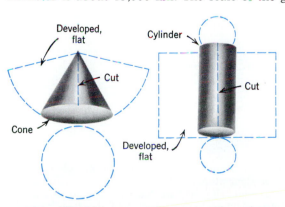

FIGURE A.I.1 The cone and the cylinder are developable geometric forms.

therefore the ratio between 20 cm and 13,000 km. Dividing 13,000 by 20, this ratio reduces to a scale stated as follows: 1 cm represents 650 km. This relationship holds true for distances between any two points on the globe.

Scale is more usefully stated as a simple fraction, termed the *fractional scale,* or *representative fraction (RF),* which can be obtained by reducing both earth and globe distances to the same unit of measure, which in this case is centimeters:

$$\frac{1 \text{ cm on globe}}{650 \text{ km on earth}} = \frac{1 \text{ cm}}{650 \times 100,000} = \frac{1}{65,000,000}$$

This fraction may be written as 1:65,000,000 for convenience in printing. The advantage of the representative fraction is that it is entirely free of any specified units of measure, such as the foot, mile, meter, or kilometer. Persons of any nationality can understand the fraction, regardless of their language or units of measure.

A globe is a true-scale model of the earth in that the same scale of kilometers or miles applies to any distance on the globe, regardless of the latitude or longitude and regardless of the compass direction of the line whose distance is being considered (Figure A.I.2A). Thus the scale remains constant over the entire globe. Map projections, however, cannot have the uniform-scale property of a globe, no matter how cleverly devised. In flattening the curved surface of the sphere to conform to a plane surface, all map projections stretch the earth's surface

in a nonuniform manner so that the map scale changes from place to place. Thus we cannot say about a map of the world, "the scale of this map is 1:65,000,000" because the statement is false for any form of projection.

It is quite possible, however, to have the scale of a flat map remain true, or constant, in certain specified directions. For example, one type of projection preserves constant scale along all parallels, but not along the meridians. This condition is illustrated in Figure A.I.2B. Another type of projection keeps a constant scale along all meridians, but not along parallels, as shown in Figure A.I.2C. Still other projections have changing scale along both meridians and parallels, as illustrated in Figure A.I.2D.

PRESERVING AREAS ON MAP PROJECTIONS

Because a globe is a true-scale model of the earth, given areas of the earth's surface are shown to correct relative size everywhere over its surface. If we should take a small wire ring, say 2 cm in diameter, and place it anywhere on the surface of the 20-cm globe, the area enclosed by the ring will represent the same area of earth's surface. But a similar procedure would not enclose constant areas on all parts of most map projections, only on one having the special property of being an *equal-area projection.*

At this point a good question arises. If, as we have stated, no projection preserves a true, or constant, scale of distances in all directions over the projection, how can circles of equal diameter placed on the map enclose equal amounts of earth area? The answer is suggested in Figure A.I.3. The square, 1 km on a side, encloses 1 sq km between two meridians and two parallels. The square can be deformed into rectangles of different shapes, but if the dimensions are changed in an inverse manner, each will enclose 1 sq km. The scale has been changed in one direction to compensate for change in another in just the right way to preserve equal areas of map between corresponding parts of intersecting meridians and parallels. On such a map, any small square or circle moved about over the map surface will enclose a piece of the map representing a constant quantity of area of the earth's surface. Projections shown in Figures A.I.11 and A.I.12 have the equal-area property, but it is also obvious that these networks have strong distortions of shape, particularly near the outer edges of the map.

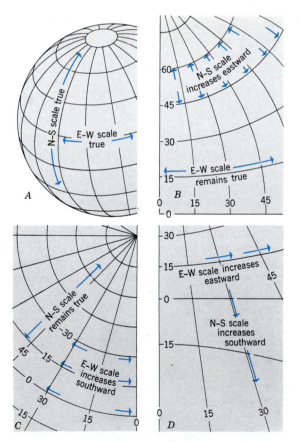

FIGURE A.I.2 A, Scale true in all directions on a globe.
B, Scale true along all parallels but not along all meridians.
C, Scale true along all meridians but not along all parallels.
D, Scale changes along both parallels and meridians.

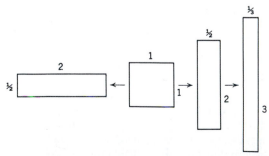

FIGURE A.I.3 Areas can be preserved even though scales and shapes change radically.

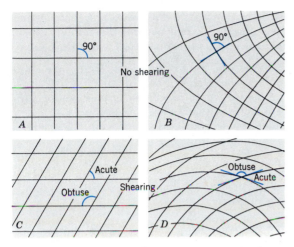

FIGURE A.I.4 Shearing of areas on map projections.

PRESERVING SHAPES ON MAP PROJECTIONS

A projection is said to be a *conformal projection* when any small piece of the earth's surface has the same shape on the map as it does on a globe. Thus the appearance of small islands or countries is faithfully preserved by a conformal map. One characteristic of a conformal projection is that parallels and meridians cross each other at right angles everywhere on the map, just as they do on the globe. (Not all projections whose parallels and meridians cross at right angles are conformal.)

Another way of saying that parallels and meridians intersect at right angles is that no shearing of areas occurs. Figure A.I.4 illustrates the meaning of *shearing*. For those projections consisting of straight parallels and meridians, shearing gives parallelograms formed of acute and obtuse angles. For projections with curved meridians and parallels, straight lines are drawn tangent to the curves at the point of intersection. If these tangent lines

cross at right angles, the projection is not sheared; but if the tangents form obtuse and acute angles, shearing is present. Although conformal maps are without shearing, not all maps without shearing are conformal. A conformal map cannot also have equal-area properties, so some areas are greatly enlarged at the expense of others. Generally speaking, areas near the margin of a conformal map have a much larger scale than areas near the center.

Whether we should select a conformal projection or an equal-area projection depends on what is to be shown. When the surface extent of something, such as grain crops or forest-covered lands, is to be shown, an equal-area projection is needed. For most general purposes, a conformal type is preferable because physical features most nearly resemble their true shapes on the globe. Many map projections are neither perfectly conformal nor equal area, but represent a compromise between the two. This compromise may achieve a map of more all-around usefulness. In some cases, the projection has some special property that makes it essential for a specific use, such as for navigation.

CLASSES OF MAP PROJECTIONS

Three important classes of map projections are the following: (1) zenithal, (2) conic, and (3) cylindric. Many unique types, not falling into these groups, have also been invented.

Zenithal projections are centered about a point and have a radial, or wheel-like, symmetry. Some zenithal projections can actually be demonstrated by use of a wire replica of the earth, in which the wires represent parallels and meridians (Figure A.I.5). A tiny light source, such as a flashlight bulb or an arc light, is placed at the center of the wire globe, or at any one of several prescribed positions, shown by letters A, B, C, D, in Figure A.I.5. In a darkened room the shadow of the wire globe is cast upon a screen, a wall, or the ceiling. This shadow is a true geometric projection.

All zenithal projections are characterized by the following properties (Figure A.I.6):

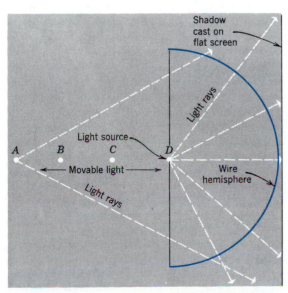

FIGURE A.I.5 Principle of the zenithal projections.

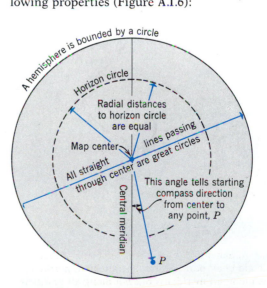

FIGURE A.I.6 Properties of zenithal projections.

1. A line drawn from the center point of the map to any other point gives the true compass direction taken by a great circle as it leaves the center point, headed for the outer point.
2. When a complete globe or hemisphere is shown, the map is circular in outline. (Because any map can be trimmed down to have a circular outline, this feature is not a reliable criterion of the zenithal class.)
3. The map possesses radial symmetry, in which all its properties are arrayed in wheel-like fashion around a center point. All changes of scale and distortion of shapes occur uniformly (concentrically) outward from this center.
4. All points equidistant from the center lie on a circle, known as a *horizon circle*. When the entire globe is shown by a zenithal map, the circular edge of the map represents the opposite point, or *antipode*, on the globe. When a hemisphere is shown, the outer edge of the map represents a great circle, everywhere equidistant from the point on which the projection is centered.
5. All great circles that pass through the center point of the projection appear as straight lines on the map. Likewise, all straight lines drawn through the center point of the map are true great circles.

Zenithal projections appear in three positions, or orientations: (1) polar; (2) equatorial; and (3) oblique; or tilted, as illustrated in Figure A.I.9. In the polar position, the center of the projection coincides with the north or south pole; in the equatorial position, the center is somewhere on the equator; and in the oblique position, the center is at some point intermediate between equator and poles. Although the equatorial and oblique types may not look radially symmetrical, they possess, just as truly as the polar type, the five characteristics of all zenithal projections.

Conic projections are based on the principle of transferring the geographic grid from a globe to a cone, then developing the cone to a flat map. This principle, too, can be demonstrated with the wire globe and a point source of light (Figure A.I.7). Instead of a vertical flat screen, however, a clear plastic cone is seated on the wire globe, much as a lampshade is seated on a lamp. The shadow of the wires cast upon the conical shade gives a

conic projection. If this shadow were traced in pencil and the cone unrolled, a true conic projection would result. Conic projections possess the following features:

1. All meridians are straight lines, converging to a common point at the north (or south) pole.
2. All parallels are arcs of concentric circles, whose common center lies at the north (or south) pole.
3. A complete conic projection is a sector of a circle, never a complete circle.
4. A conic projection cannot show the whole globe and usually shows little more than the northern (or southern) hemisphere.

Cylindric projections are based on the principle of transferring the geographic grid first to a cylinder wrapped about the earth, then unrolling the cylinder to make a flat map (Figure A.I.8). Simple cylindric projections are easy to draw because they consist of intersecting horizontal and vertical lines (see Figure A.I.10). The completed map is rectangular in outline, and the whole circumference of the globe can be shown. When the cylinder is tangent to the equator (as in Figure A.I.8), meridians are equally spaced vertical lines. Parallels are spaced in various ways, according to the particular projection desired.

Many other kinds of map projections have been invented, each based on some unique principle. We have selected several important map projections to illustrate useful types spanning a wide range of properties. Included are both conformal and equal-area types.

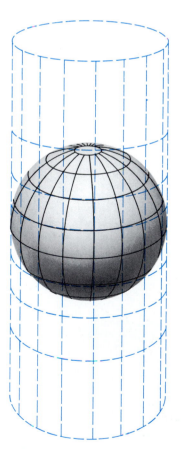

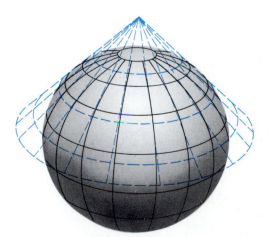

FIGURE A.I.7 Principle of the conic projections.

FIGURE A.I.8 Principle of the cylindric projections.

STEREOGRAPHIC PROJECTION

The *stereographic projection* is a conformal projection belonging to the zenithal class. Figure A.I.9 shows how an accurate stereographic net can be constructed by using a compass, a straightedge, and a protractor. The construction lines, or rays, emanate from a point diametrically opposite to the point where the tangent plane touches the globe. The stereographic net can show much more than one hemisphere, although it cannot show the whole globe. Parallels and meridians show closer spacing near the center and increasingly wider spacing toward the margins. On any stereographic projection the parallels and meridians are either straight lines or true arcs of circles. Because the stereographic projection is truly conformal, all lines that are circles on the globe are shown as circles on the map. The map scale, however, always increases from the map center toward the periphery.

With the enormous growth in importance of polar regions in the age of scientific discovery and long-range missile and aircraft operation, the polar stereographic projection has assumed great importance.

World Aeronautical Charts, issued by the U.S. National Ocean Survey, are based on a polar stereographic projection for lat. 80° to 90°. The data of scientific observations of the Arctic Ocean and Antarctic continent are usually shown on this projection. The National Weather Service weather maps use a polar stereographic projection.

MERCATOR PROJECTION

Perhaps the best known of all map projections is the *Mercator projection*, which was devised by Gerardus Mercator in 1569 (Figure A.I.10). It is based on a mathemati-

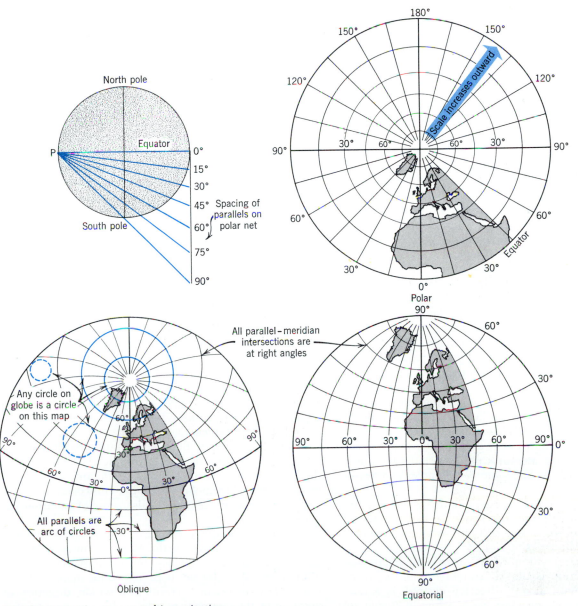

FIGURE A.I.9 The stereographic projection.

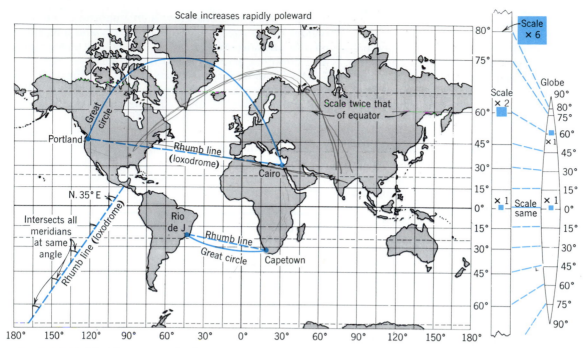

FIGURE A.I.10 The Mercator projection.

cal formula. The principle can, however, be explained without mathematics. On any cylindrical projection in which the meridians are straight vertical lines, equidistantly spaced, the meridians have had to be spread apart (see right side of Figure A.I.10). Only along the equator are meridians the same distance apart as on a globe of the same equatorial scale. To maintain them as parallel lines, the normally converging meridians have had to be spread apart in a greater and greater ratio as the poles are approached. At lat. 60°N and S, the meridians are spread apart twice as far as originally because at that place a degree of longitude is only half what it is at the equator. At the poles, the spreading is infinitely greater because the poles themselves are infinitely tiny points. Now, to maintain the map as a truly conformal map, we must space the parallels increasingly far apart toward the poles, using the same ratio of increase that resulted when the meridians were spread to make vertical lines. For example, near the 60th parallel north, parallels must be spread twice as far apart as on the globe because, as already explained, the meridians here are also spread twice as far apart. Thus, at lat. 60°N, the map scale is double that at the equator. At lat. 80° the map scale is enlarged almost six times. Near the poles the spacing of parallels increases enormously and rapidly approaches infinity. Therefore the Mercator map is usually trimmed off at about lat. 80° or 85°N and S. The poles can never be shown.

The Mercator chart is a true conformal projection. Any small island or country is shown in its true shape. The scale of the map, however, becomes enormously greater toward the poles.

The really important, unique feature of an equatorial Mercator projection is that a straight line drawn anywhere on the map, in any direction desired, is a line of constant compass bearing. Such a line is known to navigators as a *rhumb line*, or *loxodrome* (Figure A.I.10). If this line is followed, the ship's (or plane's) compass will show that the course is always at a constant angle with respect to geographic north. Once the proper compass bearing is determined, the ship is kept on the same bearing throughout the voyage, if the rhumb line is to be followed. The Mercator chart is the only one of all known projections on which all rhumb lines are true straight lines, and vice versa. A protractor can be used with reference to any meridian on the map, and the compass bearing of any straight line can be measured off directly.

The relation of great-circle navigation routes to rhumb lines is also shown in Figure A.I.10. The great-circle course is actually the shortest surface distance between two selected points, although on the Mercator map the rhumb line usually appears to be shorter. Along the equator and all meridians (but only on these lines), rhumb lines and great circles are identical and are straight lines on both charts.

Although indispensable for navigational uses, the Mercator projection has serious shortcomings for use as a world map to show geographic information dealing with areas of distribution. Except for equatorial regions (for which it provides an excellent grid), distortions of scale are very serious. Because of infinite stretching toward the poles, this map fails completely to show how the land areas of North America, Asia, and Europe are grouped around the polar sea. In the mind of an inexperienced user, it may enhance a false sense of isolation between these lands.

On the other hand, certain forms of geographic information are best shown on the Mercator projection. Because of its accurate depiction of the compass directions of lines, the Mercator net is preferred for maps of direction of flow of ocean currents and winds, direction of pointing of the compass needle, or lines of equal value

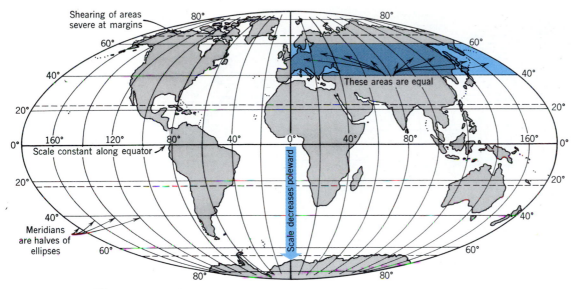

FIGURE A.I.11 The homolographic projection.

of air pressure and air temperature. Examples of such uses of the Mercator projection will be seen in later chapters. This projection is often supplemented by two stereographic projections, one for each polar region.

HOMOGRAPHIC PROJECTION

A projection rather widely used by geographers to show the entire globe is the *homolographic projection* (Figure A.I.11). "Homolographic" is a word often used to mean "equal area," a property this projection has. One hemisphere is outlined by a circle; the other hemisphere, divided into two parts, is included within an ellipse. All meridians except the straight central meridian and the hemisphere circle, are halves of ellipses. The equator is twice as long as the central meridian, which is also true

on a globe. Parallels are straight, horizontal lines, becoming more closely spaced toward the poles.

The homolographic projection has distinct advantages and disadvantages. Its equal-area property makes it valuable for showing the global distribution of geographic properties that cover areas, such as crops, soil types, or political units. It is an excellent grid for low-latitude regions. For example, it is often used for maps of Africa. Severe distortion in the polar regions, however, has hindered its wider use.

SINUSOIDAL PROJECTION

In some ways the *sinusoidal projection* is similar to the homolographic projection. It is an equal-area projection with straight central meridian and horizonal straight

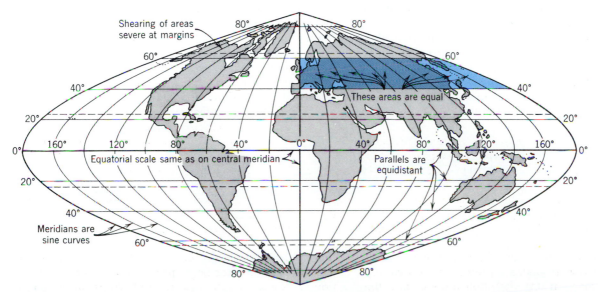

FIGURE A.I.12 The sinusoidal projection.

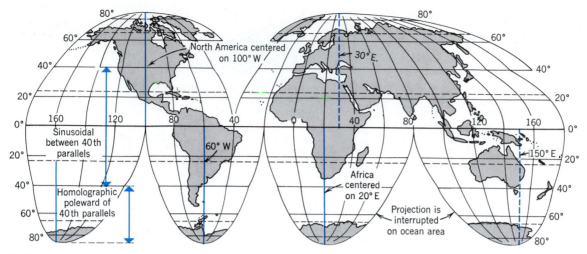

FIGURE A.I.13 Goode's interrupted homolosine projection. (Based on Goode Base Map. Copyright by the University of Chicago. Used by permission of the Geography Department.)

parallels (Figure A.I.12). The difference lies in the type of curve used in meridians. Whereas the homolographic net uses ellipses, the sinusoidal net uses *sine curves*. The parallels are uniformly spaced apart (equidistant). For this reason, more space is available for the high-latitude zones than on the homolographic projection. Even so, shearing at high latitudes is severe.

HOMOLOSINE PROJECTION

The *homolosine projection*, invented by Dr. J. Paul Goode, is a combination of the homolographic and sinusoidal types. The sinusoidal projection is used between lat. 40°N and 40°S, the homolographic for the remaining poleward parts. The interrupted form of the homolosine projection, Figure A.I.13, shows North America, Eurasia, South America, Africa, and Australia, each based on the best-suited meridian. In another variation of this map, Eurasia is split down the meridian of 60°E, greatly improving the presentation of eastern Asia.

Finally, as a means of showing information about land areas to greatest possible advantage on the limited space of a book page, the interrupted homolosine projection can be compressed by carving out intervening ocean areas and pulling the continents together (see, for example, our world map of climate, Figure 8.9).

SELECTING A MAP PROJECTION

Select a map projection best suited to the class of information it must show. Use an equal-area projection to show the distribution of geographic entities that occupy surface area—soils, vegetation types, climates, or political units. Always choose a conformal projection to show systems of lines whose compass directions are important and should not be distorted—lines of equal temperature or pressure, winds, ocean currents. Use a polar projection for high latitudes.

Certain environmental properties of special interest in physical geography change systematically from the equator toward either pole. These properties are strongly dependent on latitude. Examples are world temperatures, climates, soils, and vegetation types. For such subjects, always choose a projection with horizontal, straight parallels so the eye can easily follow a given latitude zone across the entire map.

Before you begin to analyze data and draw conclusions from a world map, be sure you know the qualities of the projection, whether equal-area, conformal, or neither. If in doubt, compare two or more projections of the same regions and see what possible erroneous concepts each might give. Consult a good globe whenever points or areas are to be related in distance and direction. There is no fully satisfactory substitute for a true-scale model of the earth's surface features.

REMOTE SENSING FOR PHYSICAL GEOGRAPHY

In various branches of geography and the earth sciences, a new technical discipline called *remote sensing* has expanded rapidly within the past decade and is adding greatly to our ability to perceive and analyze the physical, chemical, biological, and cultural character of the earth's surface. In its broadest sense, remote sensing is the measurement of some property of an object using means other than direct contact. Hearing and seeing are remote sensing activities of organisms and depend on reception of wave forms of energy transmitted from the object to the observer. In its normal operational meaning, remote sensing refers to gathering information from great distances and over broad areas, usually through instruments mounted on aircraft or orbiting space vehicles. All substances, whether naturally occurring or synthetic, are capable of reflecting, absorbing, and emitting energy in forms that can be detected by instruments known collectively as remote sensors.

TWO KINDS OF SENSING SYSTEMS

Two classes of electromagnetic sensor systems are recognized: passive systems and active systems. *Passive systems* measure radiant energy reflected or emitted by an object. Mostly, this energy falls in the visible light and near-infrared regions (reflected) and the thermal infrared region (emitted). The most familiar instrument of this class is the camera, using film sensitive to reflected energy at wavelengths in the visible range. *Active systems* use a beam of wave energy as a source, sending the beam toward an object or surface. A part of the energy is reflected back to the source, where it is recorded by a detector. A simple analogy would be the use of a spotlight on a dark night to illuminate a target and reflect light back to the eye. Active systems used in remote sensing operate mostly in the radar region.

MICROWAVES AND RADAR

At longer wavelengths beyond the visible to infrared portions of the electromagnetic spectrum lies a form of radiation known as *microwaves* (see Figure 3.3). Shifting from units of micrometers, used in the visible light spectrum, to centimeters, we can place the microwave region as between about 0.03 and about 100 cm. Most persons are familiar with microwaves as the form of energy used in the microwave oven for the rapid heating or cooking of foods. Microwaves are also used in direct-line transmission of messages from one tower to another across country. Within the microwave region is the *radar* region, beginning at about 0.1 cm and extending through to about 100 cm (1 m). (The word "radar" was originally an acronym for "RAdio Detection And Ranging," but now is recognized as a word in its own right.) Radar systems are active microwave sensor systems. The radar system emits a short pulse of microwave radiation and then "listens" for a returning microwave echo. Radar sensors are described more fully in a following section. Frequencies at which radar systems operate grade into television and radio frequencies, the latter extending into wavelengths exceeding 300 m.

ABSORPTION OF ENERGY BY THE ATMOSPHERE

Because remote sensing usually involves imaging the earth from spacecraft or aircraft, the energy reflected from or emitted by an object must travel through the atmosphere before it reaches the sensor. Figure 3.4 shows how different wavelengths of electromagnetic energy are transmitted and absorbed by the atmosphere. Oxygen and ozone absorb ultraviolet radiation with wavelengths shorter than about 0.3 micrometers. In the visible region (0.4–0.7 micrometers), the atmosphere absorbs little radiation. Between 0.7 and about 2.5 micrometers, in the near- and middle-infrared regions, the atmosphere is also relatively transparent, except for absorption bands at about 1.4 and 1.9 micrometers that are produced by water vapor. The thermal-infrared region extends from about 3.5 to 100 micrometers. There are also regions in the thermal infrared, at about 5, 10, and 20 micrometers, where the atmosphere absorbs little radiation. Thus, a large portion of the electromagnetic spectrum in the visi-

DIGITAL IMAGES

FIGURE A.II.1 (left) This infrared color photograph was taken with a large-format camera carried on the NASA Space Shuttle. It gives an extremely detailed picture of the terrain and is suitable for precision mapping. The area shown, about 125 km in width, includes the Sulaiman Range in West Pakistan (upper left). The Indus and Sutlej rivers, fed by snowmelt from the distant Hindu Kush and Himalaya ranges, cross the scene flowing from northeast to southwest. The larger of the two, the Indus, is crossed by a diversion dam (upper right) that feeds water into irrigation canals. A mottled pattern of green fields (red) interspersed with barren patches of saline soil (white) covers much of the lower fourth of the area. Small streams arising in deeply eroded anticlines of sedimentary rock (upper left) have built a series of alluvial fans (white and pale blue) that bear clusters of irrigated fields. (NASA No. 57-20380-4107-1783.)

ble, near- and middle-infrared, and thermal-infrared regions is usable for remote sensing from high-flying aircraft or spacecraft.

In the microwave region, the atmosphere is also relatively transparent. At longer wavelengths, microwaves do not interact with the water droplets that constitute fog and clouds. Thus, many radar systems can penetrate clouds to provide images of the earth's surface in any weather. At short wavelengths, however, microwaves can be scattered by water droplets and produce a return signal sensed by the radar apparatus. This effect is the basis for weather radars, which can detect rain and hail and are used in local weather forecasting.

AERIAL PHOTOGRAPHY

Of the passive sensor systems, camera photography in the visible portion of the spectrum is most familiar to the average person. Black-and-white aerial photographs, taken by cameras from aircraft, have been in wide use by geographers and other environmental scientists since before World War II. Commonly, the field of one photograph overlaps the next along the plane's flight path, so that the photographs can be viewed stereoscopically for a three-dimensional effect. Color film can be used to increase the level of information on the aerial photographs. Because of its high resolution (degree of sharpness), aerial photography remains one of the most valuable of the older remote-sensing techniques.

Photography has been extended to greater distances through the use of cameras operated by astronauts in orbiting space vehicles. Most persons are familiar with the striking color photographs obtained during the Gemini missions of the early 1960s. A recent Space Shuttle flight included a specially constructed large-format camera, designed to produce very large (23 × 46 cm, 9 × 18 in.), very detailed transparencies of the earth's surface suitable for precise topographic mapping (Figure A.II.1).

Photography using reflected electromagnetic radiation also extends into the ultraviolet and infrared wavelengths. Conventional cameras equipped with suitable filters and film can be used in the near-ultraviolet region between 0.3 and 0.4 micrometers. However, atmospheric absorption and scattering of light in these wavelengths limits the use of ultraviolet photography to objects within a limited distance range. Within the infrared

spectrum, there is a region in the near-infrared, immediately adjacent to the visible red region, in which reflected rays can be recorded by cameras with suitable film and filter combinations. Because the atmosphere is very clear in this portion of the spectrum, direct infrared photographs taken from high-flying aircraft are extremely sharp and render a great deal of information on vegetation, soil surface conditions, and land use.

COLOR INFRARED PHOTOGRAPHY

Color infrared film is often used in aerial photography (Figures A.II.2, A.II.3, A.II.4). In this type of film, the red color is produced as a response to infrared light, the green color is produced by red light, and the blue color by green light. Because healthy, growing vegetation reflects strongly in the infrared, vegetation has a characteristic red appearance on color infrared film. Thus, agricultural crops appear as color shades ranging from pink to orange-red to deep red. Mature crops and dried vegetation (dormant grasses) appear yellow or brown. Urbanized areas typically appear in tones of blue and gray. Shallow water areas appear blue; deep water appears dark blue to blue-black. Color infrared photography is particularly useful in geographic interpretation and land-use planning, and thus has found wide application since it was first developed during World War II.

DIGITAL IMAGES

Since 1975, the science of remote sensing made great strides. This progress has largely been due to the use of computers to process remotely-sensed data. Computer processing of pictorial data, however, requires that images be in digital format (i.e., as collections of numbers on which a computer can operate). Figure A.II.5 illustrates the concept of a *digital image*. The picture can be thought of as consisting of a very large number of grid cells, each of which records a brightness value. The cells are referred to as *pixels*, a term that arises as a contraction of the term "picture element." Normally, low numbers code for darkness (low reflectance) and high numbers code for light (high reflectance). The numbers are typically stored on magnetic media (disks or tapes) by the computer, and so the image does not exist in a viewable form. To create an image that is visible, as, for example, on a television monitor, the brightness values are fed to a special electronic device that generates a corresponding television signal. Or, the digital image may be sent to a film-writing device that exposes film to light, a tiny spot at a time, in proportion to the brightness values within the digital image. The result is a film negative or transparency that can be printed or viewed directly.

The importance of the digital image is that it allows the numbers that constitute the image to be modified by the computer. This activity is referred to as *image processing*. By manipulating these numbers in the computer, it is easy to enhance the image—for example, to modify the contrast selectively within certain areas of the image, or to emphasize edges or boundaries within the image (Figure A.II.6).

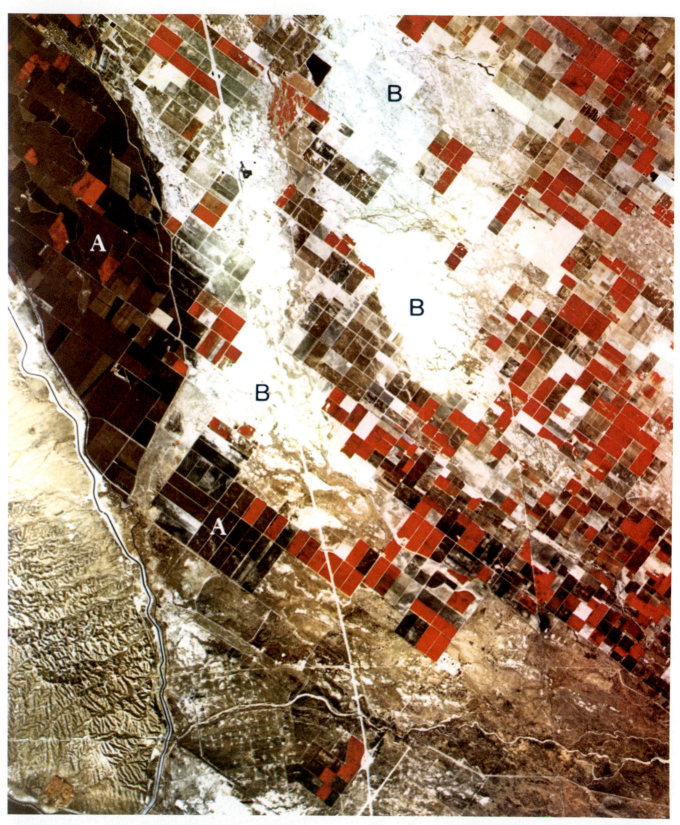

FIGURE A.II.2 High-altitude infrared photograph of an area near Bakersfield, California, in the southern San Joaquin Valley, taken by a NASA U-2 aircraft flying at approximately 18 km (60,000 ft). The original photo scale was about 1:120,000. Photos such as this can be used to study problems associated with agriculture. Problems affecting crops yields arise in (A) perched ground water areas, which appear dark, and (B) areas of high soil salinity, which appear light. The various red tones are associated with the different types of crops that are being grown. By knowing which crops are typically grown in this area and their growing cycle, the total areal extent of individual crops being cultivated can be determined. The areal estimates can then be combined with other data on weather conditions and soils to predict crop yields. (Photo by NASA, compiled and annotated by John E. Estes and Leslie W. Senger.)

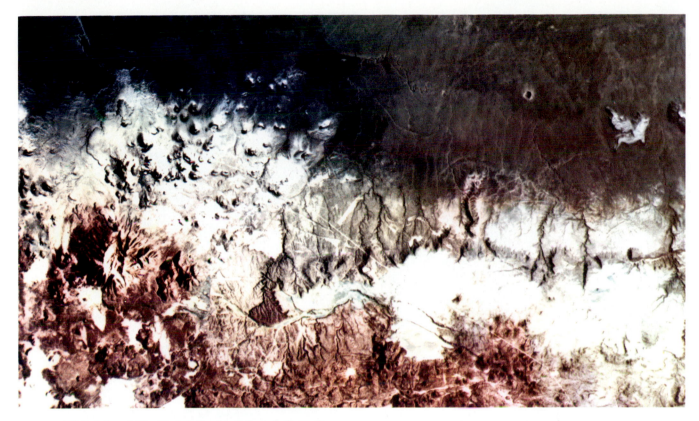

FIGURE A.II.3 (above) The Skylab 4 color infrared photos in this figure and Figure A.II.4 were imaged by the Earth Resources Experiment Package (EREP) S-1908 (13-cm terrain camera). This photo of the Flagstaff, Arizona, area shows Humphrey's Peak, the Sunset Crater volcanic field, and Flagstaff at the left. Meteor Crater, seen as a small circle in the right center of the photo, may be used for scale comparison; the crater is approximately 1.6 km (1 mi) in diameter. Living vegetation appears red; snow, white; water, blue. Photos such as this are used for snow mapping, which helps in local water resource assessment. (Photo by NASA. Annotated by John E. Estes and Leslie W. Senger.)

FIGURE A.II.4 (below) This color infrared photo of the eastern coast of Sicily was taken during the Skylab 3 mission. It shows Mt. Etna, the highest volcano in Europe—still active, as evident by the thin plume of smoke emanating from its crest. On Etna's flanks recent lava flows appear black, while older lava flows and volcanic debris, which now support vegetation, appear in shades of red. Photos such as this can be used not only to study the localized patterns of geologic phenomena, such as volcanic activity or faulting, but regional or global patterns and distributions as well. (Photo by NASA. Annotated by John E. Estes and Leslie W. Senger.)

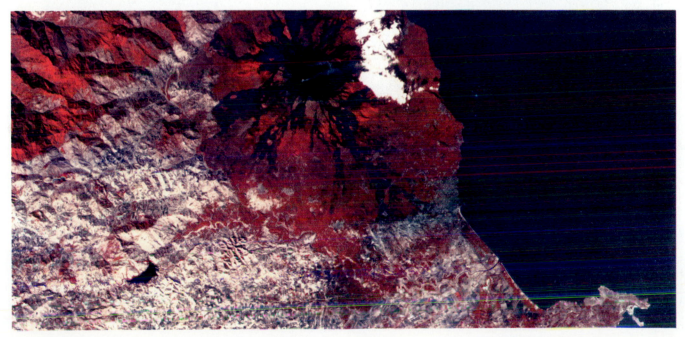

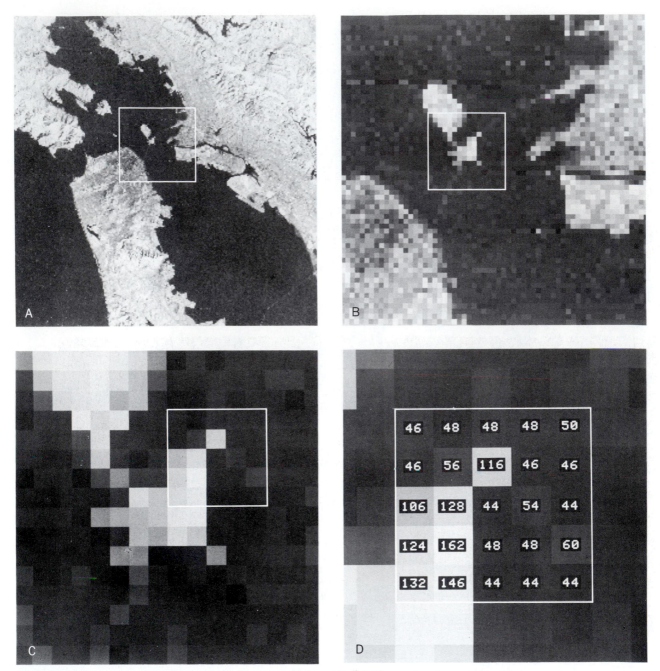

FIGURE A.II.5 Four Landsat images of San Francisco (infrared band) illustrating the concept of the digital image. (A–C) Progressively smaller subimages zooming in on the Bay Bridge and Yerba Buena Island in San Francisco Bay. (D) Actual brightness values for a small array of 25 pixels, scaled to range from 0 (darkest) to 255 (lightest). (Alan H. Strahler.)

SCANNING SYSTEMS

Digital images are usually generated by *scanning systems*. Scanning is the process of receiving information instantaneously from only a very small portion of the area under surveillance (Figure A.II.7). The scanning instrument projects a very small field of view that runs rapidly across the ground scene. Light from the field of view is focused on a detector that measures its intensity, and eventually a string of digital brightness values is output. As the process is repeated, information along a set of closely-spaced parallel lines is obtained. In this way, a digital image is built up by the scanning system. An important characteristic of the image is the *resolution cell size*—the size of the area on the ground associated with each digital measurement. This value is determined primarily by the angular field of view of the scanner and its altitude above the earth.

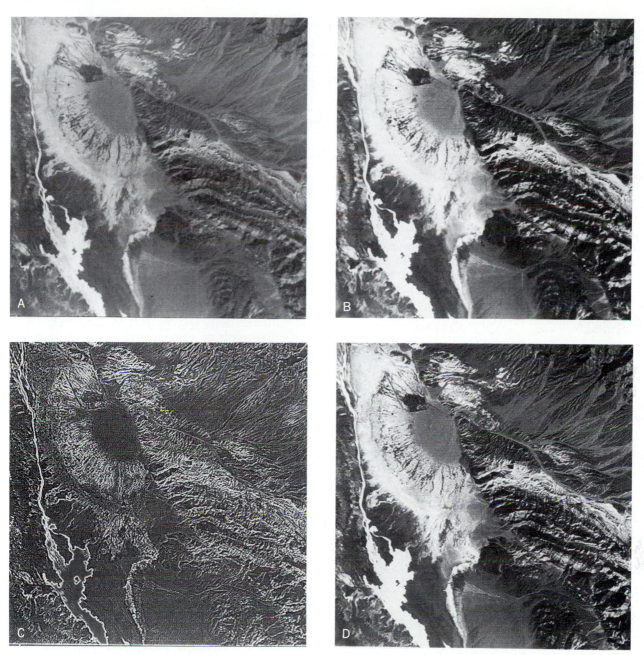

FIGURE A.II.6 (above) Thematic Mapper image of Death Valley (red band) showing how digital images can be enhanced. (A) Original image. (B) Contrast-enhanced image. (C) Image produced by applying edge detection procedure. (D) Edge-enhanced image created by adding images B and C together. (Alan H. Strahler.)

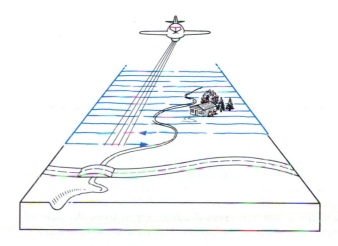

FIGURE A.II.7 (right) Multispectral scanning from aircraft. As the aircraft flies forward, the scanner sweeps side to side. The result is a digital image covering the overflight area.

MULTISPECTRAL SCANNERS

Most scanning systems in common use are *multispectral scanners*. These devices measure brightness in several wavelength regions simultaneously. An example is the Multispectral Scanning System (MSS) used aboard the Landsat series of earth-observing satellites. This instrument simultaneously collects reflectance data in four spectral bands (numbered 4, 5, 6, and 7) with ranges of 0.5–0.6 (green), 0.6–0.7 (red), 0.7–0.8 (infrared), and 0.8–1.1 (infrared) micrometers (Figure A.II.8). The resolution cell size for the Landsat MSS is 56 by 79 meters. A successor to this system, the Landsat Thematic Mapper (TM), collects data in six spectral bands, ranging from 0.45 (blue) to 2.35 (middle infrared) micrometers. Data obtained by the TM are of much finer resolution than data from the MSS, for the resolution cell size of this instrument is 30 × 30 m. The TM also senses in a thermal infrared band in the wavelength region 10.4–12.5 micrometers. However, the resolution cell size of this band is larger than the others: 120 × 120 m.

MULTISPECTRAL IMAGERY

Although it is possible to produce black and white photos or TV signals from an image within a single spectral band, multispectral data are normally viewed as *multispectral images*. Since the human eye can respond to three primary colors at once, it is possible to assign one spectral band to each color and view the result as a color image. The colors produced will then not resemble true colors unless, of course, the primary colors of red, green, and blue are assigned to the red, green, and blue bands of the sensor. For example, Landsat MSS data are often presented as a color composite image with the red color intensity controlled by one of the infrared bands (usually band 7), the green color controlled by the red band, and the blue color controlled by the green band (Figure A.II.9). This presentation of the data conforms to that used in color-infrared film, discussed above, and illustrated in Figure A.II.2; it is therefore quite familiar to photointerpreters.

SPECTRAL SIGNATURES

As shown in Figure A.II.8, various types of earth surfaces have varying degrees of reflectivity over the wavelength ranges covered by the Landsat MSS and TM instruments. Water surfaces, although always relatively dark, show their highest reflectivity in the blue-green region. There is often variation produced by the amount of sediment and algae found within the upper layers of the water surface. Vegetation appears dark green to the human eye—which means that it reflects moderately well in the green portion of the spectrum, but reflects less well in the blue and red regions. In the infrared region, healthy growing vegetation reflects very strongly and looks quite bright. Although soils vary in their spectral response, a typical curve such as that presented in Figure A.II.8 shows soils to be generally dark in the visible and becoming gradually lighter in the infrared. In geologic applications of remote sensing, some minerals also have distinctive patterns of absorption that influence their brightness within the multispectral bands that are sensed by instruments such as the TM. Thus, it is often possible to distinguish an object by its *spectral signature*—the pattern of relative brightness within spectral bands that characterizes the object. Multispectral signatures are important because they allow the use of statistical techniques to classify the image. In this process of *image classification*, the analyst identifies a type of object within the image (e.g., a cornfield), and then asks the computer to identify all those pixels with a similar spectral signature. These pixels then presumably also correspond to similar objects (cornfields). In this way, a large region may be mapped quickly, easily, and accurately using digital satellite data.

THERMAL INFRARED SENSING

Electromagnetic radiation is emitted by all objects because they possess sensible heat. At the range of temperatures encountered on the earth's surface, this radiation is within the *thermal infrared* portion of the spectrum. Thermal infrared radiation is usually sampled by a scanning system. However, the detector that absorbs the radiation so that it can be measured and digitized is composed of a material sensitive to infrared wavelengths rather than to light. Since warmer objects emit more infrared radiation than do cooler ones, the former will appear lighter on thermal infrared imagery. Notice also that objects will emit thermal infrared radiation as long as they possess sensible heat, and therefore thermal infrared imagery can be collected during both day and night.

Besides absolute temperature, the intensity of infrared emission depends on an intrinsic property of the object, known as *infrared emissivity*. Whereas the black body, mentioned earlier in Chapter 3, is an ideal perfect

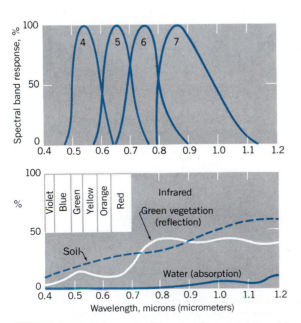

FIGURE A.II.8 Schematic diagram of sensitivities of four filters in the multispectral scanner on Landsat, compared with reflectivity of green vegetation, water, and soil surfaces.

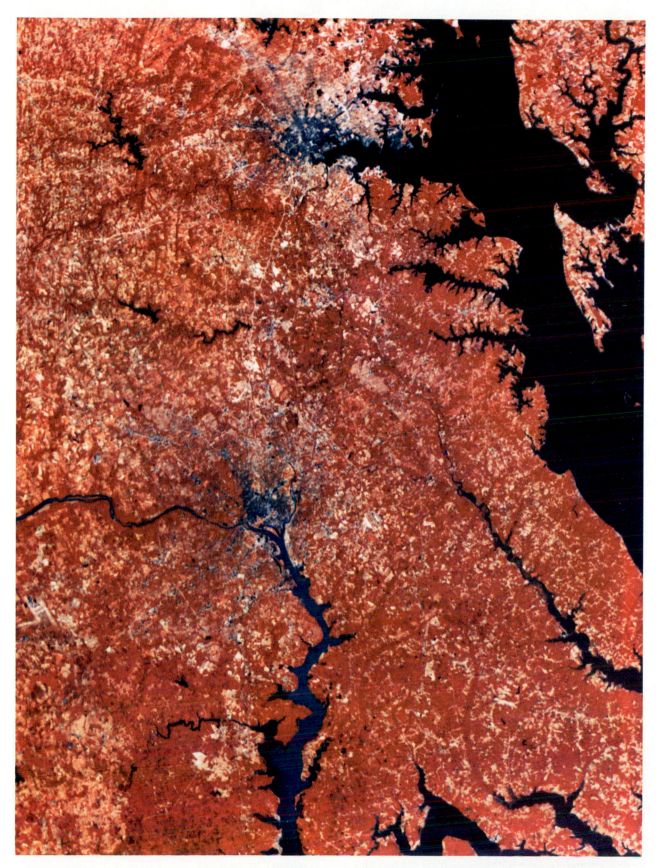

FIGURE A.II.9 Portion of a Landsat color infrared composite (Multispectral Scanner Bands 4, 5, and 7) image of the Washington–Baltimore region. The Chesapeake Bay occupies the upper right portion. Urban patterns and transportation networks, including beltways girdling the two metropolitan regions, can be readily interpreted. Vegetated areas (red) stand out clearly. This type of composite false-color image has applications for land use analysis, sediment studies, and general resource surveys. (Photo by NASA and Hughes Aircraft Company. Annotated by John E. Estes and Leslie W. Senger.)

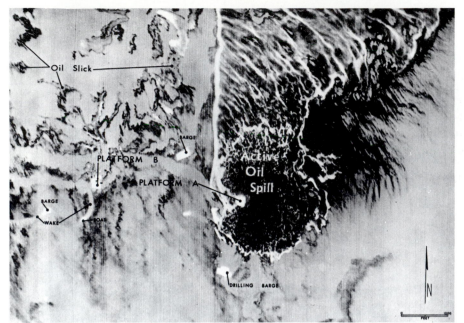

FIGURE A.II.10 Thermal infrared image of the source area of the Santa Barbara Oil Spill, taken in early February 1969. This image shows the patterns of emitted thermal infrared energy from the sea surface. Imagery such as this can be acquired day or night. (Photo by courtesy of North American Rockwell Corp.; labels by Geography Remote Sensing Unit, University of California, Santa Barbara. Annotated by John E. Estes and Leslie W. Senger.)

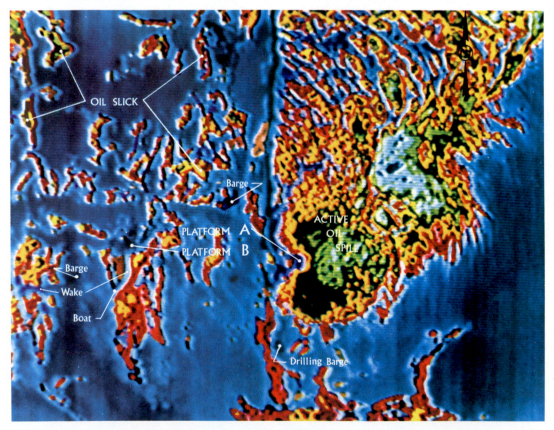

FIGURE A.II.11 An electronic enhancement of Figure A.II.10. Image enhancement systems, such as the one which generated this image, expand the image interpreter's ability to study the meaning of subtle variations in an image's optical density. This effect is achieved by assigning colors to specific gray levels on the original image. In this instance, researchers were attempting to ascertain whether or not the patterns of thermal energy emitted from the sea surface were related to variations in thickness of the oil film. (Photo by courtesy of Spatial Data Systems Inc.; labels by Geography Remote Sensing Unit, University of California, Santa Barbara. Annotated by John E. Estes and Leslie W. Senger.)

radiator of energy, all substances are imperfect radiators. An imperfect radiator is referred to as a *gray body*. Infrared emissivity is the ratio of emission of a gray body to that of a black body at the same absolute temperature; it ranges from very low for a body with almost no emission to unity (100%) for a black body. For most natural terrestrial surfaces, the infrared emissivity is comparatively high—in the range of 85 to 99 percent. Differences in emissivity can be important in determining the patterns of the infrared image. For example, two objects may be at the same temperature, but if the emissivity of one is greater, it will appear lighter than the other in a thermal infrared image.

INFRARED IMAGERY

An example of infrared imagery acquired at night is shown in Figure 3.12. Keep in mind while examining this image that the differences in tone by means of which objects are delineated are caused by differences in the level of emission of infrared energy. In turn, the level of emission of objects is controlled by their temperatures and emissivities. For this scene, however, differences in temperature are the most important variable in causing the differences in tone, and in general the lighter tones indicate warmer temperatures. In the figure, an image taken during the early morning hours (between 2:00 and 4:00 A.M.), pavement and water emit infrared radiation more strongly and appear light in tone. Trees lining many of the roadways also appear light in tone. Agricul-

tural areas, moist soil surfaces, and buildings are "cooler" and appear darker. Figures A.II.10 and A.II.11 show thermal imagery of an oil spill in the Santa Barbara Channel. The thermal response is related to the depth of oil on the ocean surface.

RADAR-SENSING SYSTEMS

We turn next to the active mode of remote sensing within the radar portion of the electromagnetic spectrum (0.1–100 cm). As noted earlier, the active sensing system uses pulses of energy emitted by transmitters mounted on aircraft or spacecraft. A beam of such pulses is directed at the ground, typically at an angle of 30 to 60° down from the horizon. A portion of the energy is returned as an echo signal. The strength of the return signal will depend partly on the nature of the surface. A smooth surface will act like a mirror, scattering the pulse forward and away from the sensor; it will therefore look relatively dark on a radar image. A rough surface will contain many facets or projections that scatter part of the pulse back toward the sensor, and thus will appear brighter. The distance between the sensor and the surface is also quite important. Since time is required for the pulse to travel to the surface and return, the farther away is the surface, the later the return signal will arrive at the sensor. Distance will also affect signal strength—the further away is the surface from the sensor, the weaker will be the signal returned. The electronics inside radar sensors are designed to utilize these principles, amplifying signals

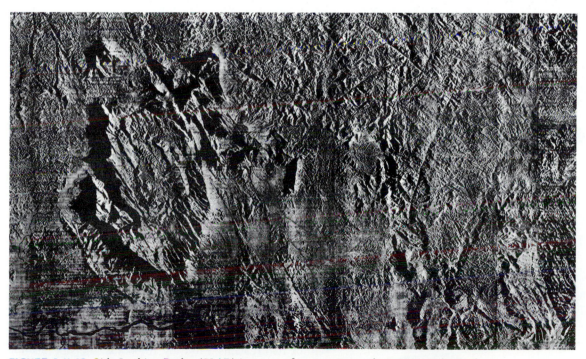

FIGURE A.II.12 Side-Looking Radar (SLAR) imagery of an area in southern Venezuela, in the headwater region of the Orinoco River, lat. 3° to 4°N. The entire region is clothed in dense equatorial rainforest. A massive synclinal remnant of sedimentary rock forms a highland at left. Closely-set fractures, trending northwest–southeast, control the trends of minor stream valleys carved into the highland escarpment. Other intersecting lineaments in the right-hand portion of the area suggest fault control of major drainage lines. Horizontal dimension of the area shown is about 160 km (100 mi). (Courtesy of Goodyear Aerospace Corporation and Aero Service Corporation.)

more strongly when they arrive later, and also displaying these signals as returning from farther away.

Radar-sensing systems are used effectively on aircraft or spacecraft. The type most often used to produce imagery is known as *side-looking airborne radar* (SLAR). A SLAR system sends its impulses toward either side of the vehicle, producing a long swath of imagery as the aircraft or spacecraft flies forward. SLAR images show terrain features with remarkable sharpness and contrast (Figure A.II.12). Surfaces oriented most nearly at right angles to the radar beam will return the strongest echo and therefore appear lightest in tone. In contrast, those surfaces facing away from the beam will appear darkest. The effect is to produce an image resembling a relief map using plastic shading (see Figure 19.33). Various types of surfaces, such as forest, rangeland, and agricultural fields, can also be identified by variations in image tone and pattern.

ORBITING EARTH SATELLITES

It is only since the advent of orbiting earth satellites that remote sensing has burgeoned into a major branch of geographic research, going far beyond the limitations of conventional aerial photography. One of the reasons for this development has already been mentioned—the fact that most satellite remote sensors provide digital images that can be processed and enhanced by computers. The other reason is the ability of orbiting satellites to monitor nearly all of the earth's surface.

In determining the orbit of a satellite, two factors are important. First, the orbit must allow the satellite to image as much as possible of the globe. Second, the orbit should be chosen so that the plane of the satellite's orbit remains fixed with respect to the sun. The latter ensures that the solar illumination of the earth under the satellite's track will remain constant in direction and intensity. The only orbit that allows a satellite to image the

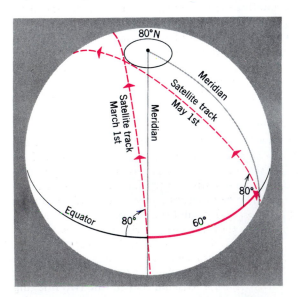

FIGURE A.II.13 Earth-track of a sun-synchronous satellite. In the period from March 1 to May 1 the orbit has shifted eastward about 60° with respect to space coordinates. (Copyright © by Arthur N. Strahler.)

entire globe is a polar orbit, in which the satellite crosses both poles. However, a perfectly polar orbit will hold the plane of the satellite's motion fixed with respect to the stars, not to the sun, and thus illumination conditions will change with the seasons.

The solution to this problem is a *sun-synchronous orbit*, illustrated in Figure A.II.13. The plane of this orbit makes an angle of about 80° with the plane of the earth's equator, and the satellite's earth track makes a tangent contact with the corresponding parallels of latitude north and south. With this orbit, the torque exerted by the earth's equatorial bulge upon the satellite's motion is just sufficient to shift the orbit westward at a rate matching the shift in angle between the sun and the stars. Therefore, the satellite will always pass overhead at the same time of day, no matter what the season.

In the example illustrated in Figure A.II.13, the satellite completes about 12.6 orbits around the earth in the mean solar day of 24 hours. At this rate, 1 orbit is completed in about 1 hour and 54 minutes (114 min). During this time, the earth is rotating at a rate of 4° per minute, and thus the satellite will cross the equator at a point about 28.5° longitude to the west of its previous crossing. This distance is 3167 km (1968 mi). Therefore, in a single day the satellite will image about 13 swaths that are about 3000 km apart at the equator.

THE LANDSAT PROGRAM

A major step forward in remote sensing of the environment was taken by the National Aeronautics and Space Administration (NASA) in July 1972 with the launching of the first of the earth-observing satellites known as Landsats. Since then, five Landsats have been launched. The first three, known as Landsats-1, -2, and -3, carried the MSS instrument described in an earlier section. The latter two, Landsats-4 and -5, also carried the Thematic Mapper. Landsats-6 and -7 are planned to continue the program in coming years. Landsat orbits are of two slightly different types. Landsats-1, -2, and -3 were inserted into a sun-synchronous orbit with an altitude of 917 km (570 mi) and a period of about 1 hour and 43 minutes. The orbit was designed to repeat every 18 days, allowing imaging of the same location about every 2½ weeks. The MSS instrument images a strip 185 km (115 mi) wide, with a 14 percent overlap between strips at the equator. This overlap increases with latitude as the orbit paths converge at about 80°. The orbit of Landsats-4 and -5, which carry both the TM and MSS, is somewhat lower: 705 km (438 mi). The repeat period is also slightly shorter: 16 days.

The Landsat orbits require the satellite to descend across the daylight portion of the globe between about 9:00 and 10:00 A.M. local time. This time of day was chosen for two reasons. First, by 10:00 A.M. the sun has not reached the zenith and thus casts strong shadows that serve to emphasize three-dimensional relief features. This timing aids geologic interpretation. Second, by this time of the morning, cloud cover created by convection from unequal solar heating of the surface has not usually begun to develop. Thus, images acquired at this time of the day will tend to have fewer clouds.

RECEIVING AND STORING LANDSAT IMAGERY

Imagery obtained by the MSS and TM instruments is converted to digital form by electronics on board the satellite. The values are then transmitted by radio to ground receiving stations operating in the United States and 18 other nations. Landsats-1, -2, and -3 also carried on-board tape recorders to store MSS images when the satellite was out of range of receiving stations. These tapes could then be played back later in the orbit, when the satellite was in range of a ground receiving station. Because TM images are more detailed, and further, seven bands are imaged as compared to four with the MSS, they consist of many times more digital values than MSS images. Transmitting all this information requires much higher-capacity radio transmitting and receiving equipment. Rather than tape record these images when out of range, they are relayed through NASA's geostationary communications satellite network, which is designed to handle data from these types of satellites as well as data generated by the flights of the Space Shuttles. After relay, the radio signals are beamed to a receiving station at White Sands, New Mexico, and then relayed by domestic communications satellites to Goddard Space Flight Center in Maryland. At Goddard, the data are processed into a computer-readable format. They are then archived at the EROS Data Center, in Sioux Falls, South Dakota. Users may order digital data directly, or may have the data converted to several photographic formats.

Under a contract with the U.S. Government, Landsat data are now acquired and distributed by Eosat Corporation, a special corporation formed by Hughes Aircraft and RCA in 1985 to market and sell Landsat data. The contract requires Eosat to build and launch two more satellites in the Landsat series: Landsats-6 and -7. Both of these spacecraft will include TM and MSS scanners in order to provide continuity of data into the 1990s. Other new imaging devices may be carried by these satellites as well.

In anticipation of the fact that the Landsats have a limited lifetime, Eosat, with significant government support, is now constructing Landsat-6 for a launch in May 1992. This new satellite will feature the Enhanced Thematic Mapper Sensor, which will provide identical imagery to the TM aboard Landsats-4 and -5. It will also provide new 15 m imagery in a single, broad panchromatic band (0.5–0.9 micrometers). The Landsat-6 spacecraft will be equipped with two tape recorders, like the earlier Landsats-1, -2, and -3. This will allows better global coverage without relying on NASA communications satellites.

SPOT SATELLITES

Although NASA dominated the field of earth-orbiting satellites in the early days of remote sensing, several other nations have launched or plan new satellite systems. The first such system was developed by France, and includes the SPOT-1 and SPOT-2 satellites, launched in 1986 and 1990. The name, SPOT, is an acronym for the French title *Système Probatoire d'Observation de la Terre*, trans-

lated as "Probatory System for Earth Observation." SPOT satellites have two modes of operation. They can image in three spectral bands (green, red, and near-infrared) at a resolution of 20 m, or in a single broad panchromatic band at a 10 m resolution. Figure A.II.14 shows a color SPOT image of New York harbor. The resolution and color are sufficient to reveal many fine details, including the street grid pattern and many individual buildings. Boats in the harbor and their wakes can also be distinguished.

SPOT satellites are also pointable; that is, they can look from side to side to image areas not directly below the orbital path. This feature allows them to image the same area on the ground at intervals of 1 or 2 days, instead of 16 or 18 days, as in the case of the Landsat satellites. Of course, the images will be sensed from different angles (i.e., from different positions in the sky). This can actually be an advantage if the two images are used as a stereo pair, in which case it is possible to produce a full stereo image from space.

The French government is committed to supporting the SPOT system through the turn of the century, and maintains a SPOT-3 instrument, identical to SPOTs-1 and -2, for launch as needed to provide data continuity. SPOT-4, scheduled for launch in 1995, will feature simultaneous 10 m red and 20 m multispectral imagery, and will include a 10 m middle infrared band (1.58–1.75 micrometers). Technical improvements will extend the expected life of the instrument from 2 years to 5 years.

Like the Landsat system, SPOT utilizes tape recorders to store images when the satellite is not within range of a ground receiving station. In 1991, 11 ground stations in 10 nations were operating, with 8 more in the construction or planning stages.

OTHER SATELLITE SYSTEMS

In February 1987, Japan launched MOS-1, the first in a series of satellites designed specifically for marine studies. It included a multispectral scanner for ocean color with 50 m resolution and bands similar to the Landsat MSS; a visible and thermal-infrared radiometer with 0.9 km and 2.7 km resolution for ocean surface temperature; and a third instrument to measure water vapor content of the atmosphere. MOS-1 was succeeded by MOS-2 in 1990.

The Soviet Union has an extensive remote sensing program, with a number of satellites in orbit that are similar to Landsats. However, in 1988, the Soviet Soyuzkarta agency began to sell photographic imagery collected from space with a large-format camera in low Earth orbit. The photographs have a ground resolution varying between 1.3 m and 6 m, depending on the contrast of the objects on the ground. Because of the security implications of the fine resolution, Soyuzkarta will sell only images of a purchaser's home country.

Also in 1988, India launched its first Remote Sensing Satellite, IRS-1A, on a Russian rocket. The two imaging instruments aboard this satellite have moderate resolution and image in four spectral bands in the visible and near-infrared regions. They are similar in design to the SPOT instruments, using a technology that allows a wide area to be imaged by thousands of detectors operating simultaneously without the need for a scanning mirror to sweep across the scene.

ORBITAL RADAR SENSING

The NASA program of earth observations has also included a number of radar missions. In 1978, Seasat, the first NASA satellite devoted to oceanic observations, was launched. The instrument package included five different sensors, one of which was an L-band (23 cm wavelength) radar system designed to measure wave heights. Although the satellite functioned for only a little more than 3 months, the radar system sent back many useful radar images of the land surface. However, the design of the system, which looked to the side at an average angle of only 23.6° below the horizontal plane, was not well

FIGURE A.II.14 Digital image of New York Harbor, taken by SPOT-1 Satellite, May 1, 1986. This image was acquired in multispectral mode at 20 meter resolution and later merged with 10 meter panchromatic SPOT data in a digital process that yields very high resolution images. The tip of Manhattan is at the top center; the blocky, dark structures at the tip are Wall Street skyscrapers and their shadows. Just below the tip is Governor's Island, a Coast Guard military reservation, and immediately to the west (left) is small Liberty Island, home of the Statue of Liberty. The long container-loading piers of Port Newark are visible on the left, as are piers on the Brooklyn shoreline to the right. At the center bottom, the Verrazano-Narrows bridge connects Brooklyn with Staten Island. The fine detail of the street grid pattern, and even of the wakes of ships in the harbor, shows this satellite to have the finest resolution of any unclassified satellite system. (© 1991 CNES. Provided by SPOT Image Corporation)

FIGURE A.II.15 Digital image composite of Shuttle Imaging Radar-A (SIR-A) and Seasat radar data, California coast near Santa Barbara. The SIR-A values are used to produce the color in the image and the Seasat values (central portion) produce the intensity (brightness). The area shown is about 90 km wide. The image demonstrates how data from different sensors can be merged digitally. (Photo courtesy of Charles Elachi, Jet Propulsion Laboratory, Pasadena.)

FIGURE A.II.16 Color composite of Seasat and Landsat data for a portion of Orange County, California. In the image, the red color is assigned to Landsat Band 7 (infrared), the blue to Landsat Band 4 (green), and the green to Seasat radar response (central area). The combination of two types of data produces much more detailed information about surface cover than either does alone. (Photo courtesy of Jerry Clarke, Jet Propulsion Laboratory, Pasadena.)

suited to land observations. At such a low angle, much of the surface is hidden from view or imaged improperly in mountainous terrain. Thus, NASA launched a nearly identical sensor system—SIR-A (Shuttle Imaging Radar-A)—on one of the first operational Space Shuttle flights, in November 1981. In contrast to Seasat, SIR-A looked down to the side at an average angle of 50°. Although data were collected for only 8 hours of the 2½-day flight, many useful images were received. An example is shown in Figure A.II.15.

Figure A.II.16 presents an image made by combining Seasat and SIR-A radar data. In this image, SIR-A data are used to derive the color and Seasat data are used to modulate the intensity of the color presented. The low incidence angle of Seasat accentuates the relief very strongly. The color variations produced by SIR-A reveal differences in underlying rock types that are not visible with Seasat. In 1984, NASA flew SIR-B, a similar instrument, on another shuttle flight. Much useful information at a range of look angles was acquired.

Although the Challenger disaster halted NASA's program of Shuttle-based observations for more than 2 years, NASA's commitment to orbital radar has continued. Missions are planned for 1993, 1994, and 1996 utilizing two radars together in the Shuttle's cargo bay—SIR-C, built by NASA, and X-SAR built by the German Aerospace Research Establishment. Operating simultaneously, the two instruments will acquire data from as many as three frequency bands at 25 m and 40 m resolution. The objective of the missions is to provide data for studies of vegetation biomass and extent, soil moisture, snow properties, recent climatic change, tectonic activity, and ocean wave spectra.

The 1990s will also see three free-flying radar imaging instruments on three satellite platforms: ERS-1, the European Remote Sensing Satellite launched by ESA, the European Space Agency (planned for 1991); JERS-1, the Earth Resources Satellite launched by Japan (1992); and Radarsat, from the Canadian Space Agency (1994). These radars will provide extensive coverage of high-latitude regions, allowing the monitoring of sea ice, snow cover, glaciers, permafrost, and vegetation area and extent. Besides the direct commercial use of this information, it will be important in monitoring global climatic change, which is expected to impact high latitudes more strongly.

AVHRR IMAGERY

Another important remote sensing satellite system is the province of NOAA. Since 1978, NOAA has maintained a series of sun-synchronous, polar-orbiting satellites launched annually that collect data primarily for studying the atmosphere. Although the instrument complement of each platform varies somewhat, each includes an Advanced Very High Resolution Radiometer (AVHRR). This scanner has a resolution of about 1 km, and images a very wide field of view, thus allowing full global coverage on a 2-day basis. Data are returned in either the full resolution mode (1 km) or in a compressed 4-km mode. The instrument has bands in the red, near-infrared and three thermal-infrared regions, and is used

by NOAA primarily for monitoring the temperatures of clouds, sea surface, and land.

The AVHRR sensing system is particularly useful for monitoring global vegetation. In this process, the red and near-infrared bands are used to calculate a vegetation index that responds to the amount of green vegetation cover on the land. Although the AVHRR can collect images every other day, a problem arises with cloud cover. To solve this problem, a week's worth of images are combined (composited) at a time by computer—that is, only those pixels that are least influenced by clouds appear in the final image. The result is a type of composite image that shows the weekly extent of vegetation development over all of the earth's vegetated surface, as if the sky were completely clear.

Figure A.II.17 presents four global AVHRR-derived maps from four intervals during the year. The values displayed are also composited, but they are composited from weekly data over a period of 5 weeks. Thus, we can assume that nearly every value has been derived from a cloud-free period somewhere within the 5 weeks that constitute each image. In these maps, the brightness values are displayed as colors, ranging from brown and yellow (lowest brightness), through green, red, white, and shades of violet (greatest brightness). In the computer processing method used, brightness values increase as green vegetation cover increases. The seasonal cycle in the midlatitude zones is conspicuous. In April–May, green vegetation is well-developed only in the southeastern United States, eastern China, and in Spain, France, and the British Isles in Europe. By June–July, the green wave has swept eastern North America and penetrated far across Europe to central and northeast Asia. By September–October, vegetation development has proceeded to far northern Canada, northern Alaska, and Siberia, and is beginning to wane somewhat. In December–January, only the eastern and western continental fringes in North America, Europe, and Asia remain green. Of course, the situation is just reversed in South America, southern Africa, and Australia. Note also that the equatorial rainforest of Brazil, central Africa, and Malaysia remains very well vegetated, although there are some variations throughout the seasons. The Asian monsoon is quite obvious, too, and can be seen especially well in the seasonal changes on the Indian subcontinent.

Weekly and monthly composite images of this nature can be particularly helpful in monitoring the severity and extent of drought in the Sahelian zone of Africa. Scientists from NASA, NOAA, and the United Nations are currently working on ways of using the AVHRR data directly in drought assessment and relief planning.

HIGH SPECTRAL RESOLUTION REMOTE SENSING

One of the most exciting new concepts for instruments that NASA is currently developing involves high spectral resolution remote sensing. In this concept, a large number of very narrow spectral bands are imaged. Thus, where the Landsat MSS once imaged a single spectral band between 0.8 and 1.1 micrometers, a high spectral

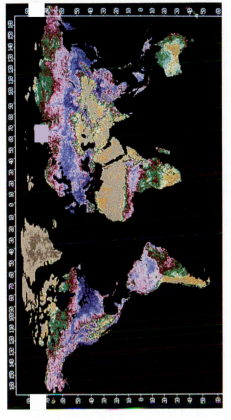

A April 1982

B July 1982

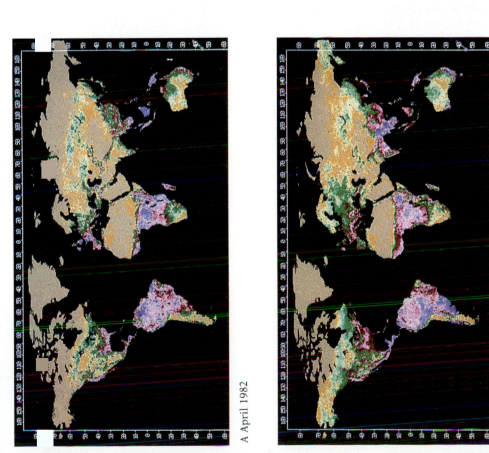

C October 1982

D January 1983

FIGURE A.II.17 Satellite-derived maps of vegetation index for the earth at four seasons of the year. Color indicates the intensity of vegetation cover, ranging from brown and yellow (leafless or barren of vegetation) through green, red, and white, to shades of violet (densest green canopy). The progressive poleward spread of green vegetation with the onset of the growing season is clearly shown for midlatitude climates and tropical wet-dry (monsoon) climates. (Photos courtesy of C. J. Tucker, NASA-Goddard Space Flight Center.)

resolution instrument might image 30 bands. An instrument built for aircraft use images 224 spectral bands simultaneously in the 0.4 to 2.4 micrometer spectral region. Why are so many bands an advantage? Because they will permit identification of specific absorption signatures of important minerals that can indicate rock composition or the presence of certain types of ore deposits.

In addition to its utility in geologic applications, high spectral resolution remote sensing may be able to detect subtle changes in vegetation produced by stresses such as acid rain or smog pollution. Identification of some distinctive plant species may also be possible. Further, the ability to select a few very narrow bands may help scientists who study the earth's surface and atmosphere for many other diverse purposes. Thus, high spectral resolution remote sensing will most likely prove to be a very important technique for studying the earth's surface from space.

THE EOS PROGRAM

Beginning in 1998, an international program of remotely sensed data acquisition will begin to provide orbital measurements and imagery that will be of an order of magnitude greater than heretofore acquired. This program, the Earth Observing System (EOS), will include five large polar-orbiting platforms, each with many sensing instruments simultaneously observing the earth. Two platforms will be supplied by NASA, two by the European Space Agency, and one by Japan. Further, each platform will be built in three copies launched 5 years apart, thus providing at least 15 years of continuous earth observation. Accompanying the polar platforms will be a number of other free-fliers, some with different orbits designed to sample tropical and equatorial zones over the full diurnal cycle, others with special instruments, such as the EOS–SAR imaging radar that is due to be launched just after the turn of the century.

The data obtained by EOS will be used by an international consortium of scientists from the United States, Canada, Japan, Europe, and other nations to study global environmental variables and how they change through time. Some of the variables and processes to be studied include cloud properties; surface temperature; energy exchange between earth, sun, and space; structure, composition, and dynamics of the atmosphere; accumulation and ablation of snow; biological activity on land and in ocean-surface waters; circulation of the oceans; exchange of energy, momentum, and gases between the earth's surface and atmosphere; structure and motion of sea ice and glaciers; mineral composition of exposed soils and rocks; and changes in stress and surface elevation around geologic faults.

NASA's part of the EOS program is that agency's contribution to a larger effort, the Global Change Research Program, which involves the National Science Foundation, NOAA, the U.S. Departments of Interior and Agriculture, the Environmental Protection Agency, and the Department of Energy. The goal of the program is to gain a predictive understanding of the interactive physical, geological, chemical, biological, and social processes that regulate the total earth system. This understanding is to establish the scientific basis for national and international policy formulation as well as decisions relating to natural and human-induced changes in the global environment and regional impacts of such changes.

As we enter the twenty-first century, EOS and other international environmental monitoring programs will mark the dawn of a new era in human understanding of our planet. As this knowledge is accumulated, let us hope that human institutions will be able to respond in a timely fashion and wisely manage the earth's surface, oceans and atmosphere for the benefit of all of the earth's biological inhabitants.

CLIMATE DEFINITIONS AND BOUNDARIES

The following table summarizes the definitions and boundaries of climates and climate subtypes based on the soil-water balance, as described in Chapter 9 and shown on the world climate map, Figure 8.9. All definitions and boundaries are provisional. (See Chapter 9 for definitions of symbols used.)

Group I Low-Latitude Climates

1. Wet equatorial climate
   Ep $\geq$ 10 cm in every month, and
   S $\geq$ 20 cm in 10 or more months.

2. Monsoon and trade-wind littoral climate
   Ep $\geq$ 4 cm in every month, or
   Ep $>$ 130 cm annual total, or both, and
   S $\geq$ 20 cm in 6, 7, 8, or 9 consecutive months, or, if
   S $>$ 20 cm in 10 or more months, then Ep $\leq$ 10 cm in 5 or more consecutive months.

3. Wet-dry tropical climate
   D $\geq$ 20 cm, and
   R $\geq$ 10 cm, and
   Ep $\geq$ 130 cm annual total, or Ep $\geq$ 4 cm in every month, or both, and
   S $\geq$ 20 cm in 5 months or fewer, or minimum monthly S $<$ 3 cm.

4. Dry tropical climate
   D $\geq$ 15 cm, and
   R $=$ 0, and
   Ep $\geq$ 130 cm annual total, or Ep $\geq$ 4 cm in every month, or both.
   Subtypes of dry climates (4, 5, 7, and 9)
   s Semiarid subtype (Steppe subtype)
   At least 2 months with S $\geq$ 6 cm.
   sd Semidesert subtype (Steppe-desert transition)
   Fewer than 2 months with S $>$ 6 cm, and at least 1 month with S $>$ 2 cm.
   d Desert subtype
   No month with S $>$ 2 cm.

Group II Midlatitude Climates

5. Dry subtropical climate
   D $\geq$ 15 cm, and

R $=$ 0, and Ep $<$ 130 cm annual total, and
Ep $\geq$ 0.8 cm in every month, and
Ep $<$ 4 cm in 1 month.
(Subtypes 5s, 5sd, 5d, defined under 4.)

6. Moist subtropical climate
   D $<$ 15 cm when R $=$ 0, and
   Ep $<$ 4 cm in at least 1 month, and
   Ep $\geq$ 0.8 cm in every month.

   Subtypes of moist climates (6, 8, 10, 11, 12)
   sh Subhumid subtype
   15 cm $>$ D $>$ 0 when R $=$ 0, or
   D $>$ R when R $>$ 0.
   h Humid subtype
   R $\geq$ 0.1 cm, and
   R $>$ D, and R $<$ 60 cm.
   p Perhumid subtype R $>$ 60 cm.

7. Mediterranean climate
   D $\geq$ 15 cm, and
   R $\geq$ 0, and
   Ep $\geq$ 0.8 cm in every month, and storage index $>$ 75%, or P/Ea $\times$ 100 $<$ 40%.

   Subtypes 7s, 7sd, defined under 4, and 7h Humid subtype R $>$ 15 cm.

8. Marine west-coast climate
   D $<$ 15 cm, and
   Ep $<$ 80 cm annual total, and
   Ep $\geq$ 0.8 cm in every month.
   (Subtypes 8sh, 8h, 8p, defined under 6.)

9. Dry midlatitude climate
   D $\geq$ 15 cm, and
   R $=$ 0, and
   Ep $\leq$ 0.7 cm in at least 1 month, and
   Ep $>$ 52.5 cm annual total.
   (Subtypes 9s, 9sd, 9d, defined under 4.)

10. Moist continental climate
    D $<$ 15 cm when R $=$ 0, and
    Ep $\leq$ 0.7 cm in at least 1 month, and
    Ep $>$ 52.5 cm annual total.
    (Subtypes 10sh, 10h, 10p, defined under 6.)

Group III High-Latitude Climates

11. Boreal forest climate
    52.5 cm > Ep > 35 cm annual total, and
    Ep = 0 in fewer than 8 consecutive months.
    Subtypes
       11s Dry subtype D > 15 cm.
       (Moist subtypes 11sh, 11h, defined under 6.)

12. Tundra climate
    Ep < 35 cm annual total, and
    Ep = 0 in 8 or more consecutive months.

Subtypes
    11s Dry subtype D > 15 cm.
    (Moist subtypes 12sh, 12h, defined under 6.)

13. Ice-sheet climate
    Ep = 0 in all months.

TOPOGRAPHIC MAP READING

Several methods are used to show accurately the configuration of the land surface on *topographic maps:* plastic shading, altitude tints, hachures, and contours (see Figure A.IV.2). The first three methods give a strong visual effect of three dimensions so that most persons can easily grasp the essential character of the landscape features.

These methods of showing relief are, however, inadequate because they do not tell the reader the elevation above sea level of all points on the map or how steep the slopes are. Topographic contours give this information and make the most useful type of topographic map.

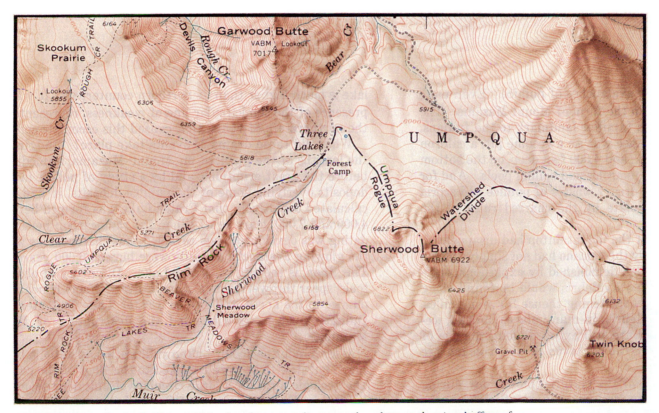

FIGURE A.IV.1 Plastic shading combined with contour lines greatly enhances the visual effect of relief. (Portion of U.S. Geological Survey, Crater Lake National Park and Vicinity topographic map. Scale 1:62,500.)

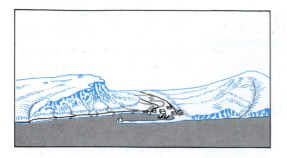

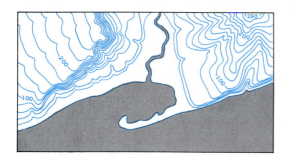

FIGURE A.IV.2 Various ways in which relief can be shown are, from top to bottom: (1) perspective diagram or terrain sketch, (2) hachures, (3) contours, (4) hachures and contours combined, and (5) plastic shading and contours combined. (Drawn by E. Raisz.)

PLASTIC SHADING

Maps using *plastic shading* to show relief look very much like photographs taken looking down on a plastic relief model of the land surface (Figure A.IV.1). The effect of relief is produced by gray or brown tones applied according to the oblique illumination method. Light rays are imagined as coming from a point in the northwestern sky somewhere intermediate between the horizon and zenith. Thus all slopes facing southeast receive the heaviest shades and are darkest where the slopes are steepest (Figure A.IV.2). Maps with plastic shading look much like photographs (Figure A.IV.3).

ALTITUDE TINTS

In its simplest form, *altitude tinting* consists of assigning a certain color, or a certain depth of tone of a color, to all areas on the map lying within a specified range of elevation. Maps in atlases and wall maps commonly show low areas in green, intermediate ranges of elevations in successive shades of buff and light brown, and high mountain elevations in darker shades of brown, red, or violet. Each shade of color is assigned a precise elevation range. The method is effective for small-scale maps that are viewed at some distance.

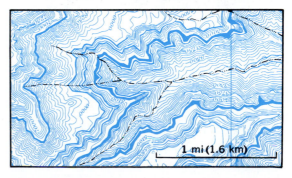

FIGURE A.IV.3 A vertical air photograph (above) is a kind of topographic map, showing all relief details but lacking elevation information. The contour topographic map (below) covers the same area and shows a small side canyon of the Grand Canyon. North is toward the bottom of these maps to give the proper effect of relief in the photograph. (U.S. Forest Service and U.S. Geological Survey.)

HACHURES

Hachures are very fine, short lines arranged side by side into roughly parallel rows. Each hachure line lies along the direction of the steepest slope and represents the direction that would be taken by water flowing down the surface.

A precise system of hachures, adapted to representation of detailed topographic features on accurate, large-scale maps, was invented by J. G. Lehmann and was widely used on military maps of European countries throughout the nineteenth century. In the Lehmann system, steepness of slope is indicated by thickness of the hachure line (Figure A.IV.4).

Because hachures do not tell the elevation of surface points, it is necessary to print numerous elevation figures on hilltops, road intersections, towns, and other strategic locations. These numbers are known as *spot heights*. Without them a hachure map would be of little practical value.

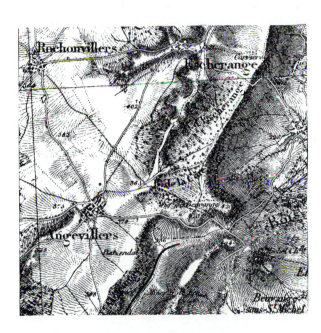

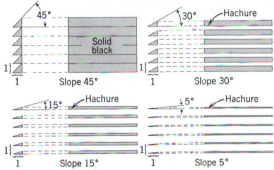

FIGURE A.IV.4 A portion of the Metz sheet is shown above to correct scale. This map is one of the French 1:80,000 topographic series using black hachures and spot heights. The Lehmann system of hachuring, shown below, varies the thickness of hachure line according to ground slope.

CONTOUR LINES

A *contour line* is an imaginary line on the ground, every point of which is at the same elevation above sea level. Contour lines on a map are simply the graphic representations of ground contours, drawn for each of a series of specified elevation, such as 10, 20, 30, 40, 50 meters or feet above sea level or any other chosen base, known as a *datum plane*. The resulting line pattern not only gives a visual impression of topography to the experienced map reader, but also supplies accurate information about elevations and slopes.

To clarify the contour principle, imagine a small island, shown in Figure A.IV.5. The shoreline is a natural contour line because it is a line connecting all points having zero elevation. Suppose that sea level could be made to rise exactly 10 meters (or that the island could be made to sink exactly 10 meters); the water would come to rest along the line labeled "10." This new water line would be the 10 meter contour because it connects all points on the island that are exactly 10 meters higher than the original shoreline. By successive rises in water level, each exactly 10 meters more than the last, the positions of the remaining contours would be fixed.

CONTOUR INTERVAL AND SLOPE

Contour interval is the vertical distance separating successive contours. The interval remains constant over the entire map, except in special cases where two or more intervals are used on the map sheet.

Because the vertical contour interval is fixed, horizontal spacing of contours on a map varies with changes in land slope. The rule is that close crowding of contour lines represents a steep slope; wide spacing represents a gentle slope. Figure A.IV.6 shows a small island, one side of which is a steep, clifflike slope. From the summit point B to the cliff base at A, the contours are crossed within a short horizontal distance and therefore appear closely spaced on the map. To travel from B to the shore at C requires the same total vertical descent; but, because the slope is gentle, the horizontal distance traveled is much

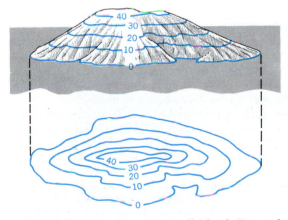

FIGURE A.IV.5 Contours on a small island. (Drawn by A. N. Strahler.)

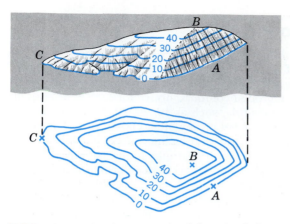

FIGURE A.IV.6 On the steep side of this island, the contours appear more closely spaced. (Drawn by A. N. Strahler.)

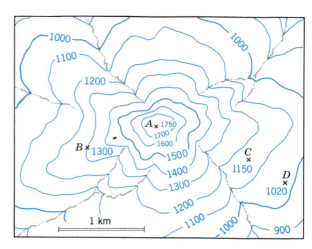

FIGURE A.IV.7 Stream valleys produce V-shaped indentations of the contours. Elevations are given in meters.

greater and the contours between B and C are widely spaced on the map.

Selection of a contour interval depends both on relief of the land and on scale of the map. Topographic maps showing regions of strong relief require a large interval, such as 10, 25, or 50 m (50, 100, or 200 ft); regions of moderate relief require such intervals as 2 or 5 m (10 or 25 ft).

Because much of the earth's land surface is sculptured by streams flowing in valleys, special note should be made of how contours change direction when crossing a stream valley. Figure A.IV.7 is a small contour sketch map illustrating some stream valleys. Notice that each contour is bent into a V whose apex lies on the stream and points in an upstream direction.

DETERMINING ELEVATIONS BY MEANS OF CONTOURS

Figure A.IV.7 can be used to illustrate the determination of elevations. Point B is easy to determine because it lies exactly on the 1300 m contour. Point C requires interpolation. Because it lies midway between the 1100 and 1200 m lines, its elevation is estimated as the midvalue of the vertical interval, or 1150 m. Point D lies about one-fifth of the distance from the 1000 m to the 1100 m contour. One-fifth of the contour interval is 20 m. Thus we estimate that point D has an approximate elevation of 1020 m. If the ground is not highly irregular, the error of estimate will probably be small. Determination of the summit elevation, point A, involves more uncertainty. It is certain that the summit point is more than 1700 m and less than 1800 m because the 1700 m contour is the highest one shown. Because a fairly large area is included within the 1,700 m contour, we may suppose that the summit rises appreciably higher than 1700 m. A guess would place the actual elevation at about 1750 m.

On many topographic maps the elevations of hilltops, road intersections, bridges, streams, and lakes are printed on the map to the nearest meter or foot. These spot heights do away with the need for estimating elevations at key points.

Government agencies, such as the U.S. Geological Survey, determine the precise elevation and position of convenient reference points. These points are known as *bench marks*. On the map they are designated by the letters B.M., together with the elevation stated to the nearest meter or foot.

DEPRESSION CONTOURS

A special type of contour is used where the land surface has basinlike hollows, or *closed depressions*, which would make small lakes if they could be filled with water. This contour line is the *hachured contour*, or *depression contour*. Figure A.IV.8 is a sketch of a depression in a gently undulating plain. Below the sketch is the corresponding contour map. Hachured contours have the same elevations and contour intervals as regular contours on the same map.

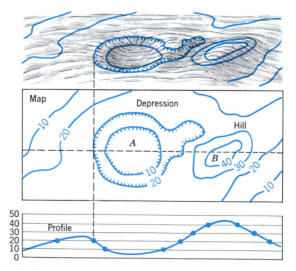

FIGURE A.IV.8 Contours that close in a circular pattern show either closed depressions or hills.

MAP SCALE

Distance between points shown on a map depends on the scale of the map: the ratio between map distances and the actual ground distances that the map represents (Appendix I). The fractional scale (representative fraction, or R.F.) is converted to conventional units of length. For example, a map scale of 1:100,000 can be read as "one centimeter represents one kilometer." An ordinary centimeter scale can thus be used to measure map distances.

Most maps of small areas carry a *graphic scale* printed on the map margin. This device is a length of line divided off into numbered segments (Figure A.IV.9). The units are in conventional terms of measurement, such as meters and kilometers or feet, yards, and miles. To use the graphic scale, hold the edge of a piece of paper along the line to be measured on the map and mark the distance on the edge of the paper. Then place the paper along the graphic scale and read the length of the line directly.

TOPOGRAPHIC PROFILES

To visualize the relief of a land surface, *topographic profiles* can be drawn. These are lines that show the rise and fall of the surface along a selected line crossing the map. Figure A.IV.10 shows how to construct a profile. Draw a line, XY, across the map at the desired location. Place a piece of paper, ruled with horizontal lines, so that its top edge lies along the line XY.

Each horizontal line represents a contour level and is so numbered along the left-hand side. Starting at the left, drop a perpendicular from the point a where the

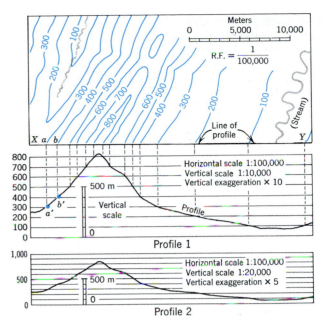

FIGURE A.IV.10 Construction of a topographic profile.

map contour intersects the profile line, XY, down to the corresponding horizontal level. Mark the point a' on the horizontal line. Next, repeat the procedure for the 400 m contour at point b, and so on, until all points have been plotted. Then draw a smooth line through all the points, completing the profile. Where contours are widely spaced, some judgment is required in drawing the profile.

Figure A.IV.10 shows two profiles, both of which are

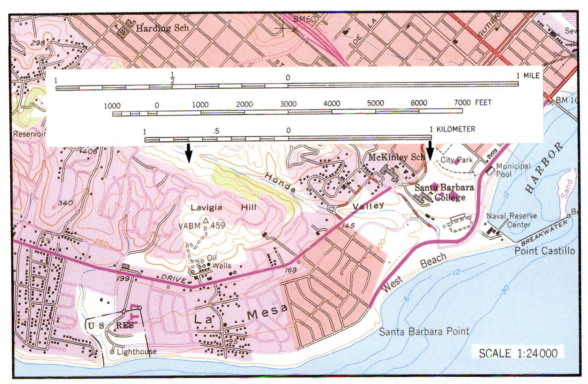

FIGURE A.IV.9 Portion of a modern, large-scale map for which three graphic scales have been provided. (U.S. Geological Survey.)

drawn along the same line XY. The difference is one of degree of exaggeration of the vertical scale. In this illustration, horizontal map scale is 1 cm to 1000 m, or 1:100,000, whereas the vertical scale of the upper profile is 1 cm to 100 m, or 1:10,000. The vertical scale is thus 10 times as large as the horizontal map scale, and the profile is said to have a *vertical exaggeration* of 10 times. In the lower profile the horizontal scale remains the same, of course, but the vertical scale is 1 cm to 200 m, or 1:20,000. The vertical exaggeration is therefore 5 times. Some degree of vertical exaggeration is usually needed to bring out the details of the topography.

LARGE-SCALE AND SMALL-SCALE MAPS

The relative size of two different scales is determined according to which fraction (R.F.) is the larger quantity. For example, a scale of 1:10,000 is twice as large as a scale of 1:20,000. Maps with scales ranging from 1:600,000 down to 1:100,000,000 or smaller are known as small-scale maps; those of scale 1:600,000 to 1:75,000 are medium-scale maps, and those of scale greater than 1:75,000 are large-scale maps.

For representing details of the earth's surface, large-scale maps are needed and the area of land surface shown by an individual map sheet must necessarily be small. A topographic sheet measuring 40 cm by 50 cm, on a scale of 1:100,000 (1 cm = 1 km), would of course include an area 40 km by 50 km, or 2000 sq km. Of the common sets of topographic maps published by national governments for general distribution, most fall within the scale range of 1:20,000 to 1:250,000.

RELATIONSHIP BETWEEN SCALES AND AREAS

Assuming that two maps, each on a different scale, have the same dimensions, what is the relationship between the ground areas shown? Figure A.IV.11 shows three maps, each having the same dimensions but representing scales of 1:20,000, 1:10,000, and 1:5,000, respectively, from left to right. Although map B is on twice the scale of map A, it shows a ground area only one-fourth as great. Map C is on four times as large a scale as map A, yet it covers a ground area only one-sixteenth as much. Thus the ground area that is represented by a map of given dimensions varies inversely with the square of the change in scale. For example, if the scale is reduced to

one-third its original value, the area that can be shown on a map of fixed dimensions increases to nine times the original value.

MAP ORIENTATION AND DECLINATION OF THE COMPASS

By convention, large-scale topographic maps are oriented so that north is in a direction toward the top of the map and south is toward the bottom of the map.

The geographic north pole, to which all meridians converge, forms a reference point for *true north*, or *geographic north*. There is, however, another point on the earth, the *magnetic north pole*, to which magnetic compasses point (Figure A.IV.12). The magnetic north pole in 1980 was located in the Northwest Territories of Canada, just south of King Christian Island at about lat. 77½°N, long. 102°W. It is moving poleward at an average speed of about 24 km (15 mi) per year. Most large-scale maps have printed on the margin two arrows stemming from a common point. One arrow designates true north, the other *magnetic north*. The angular distance between the two directions is known as the *magnetic declination*.

Magnetic declination varies greatly in different parts of the world, depending principally on one's position relative to the geographic and magnetic poles. Lines on a map drawn through all places having the same compass declination are known as *isogonic lines* (Figure A.IV.13). The line of zero declination (agonic line) runs through eastern North America. Anywhere along this line, the compass points to true geographic north and no adjustment is required.

Magnetic declination changes appreciably with the passage of years. The amount of annual change in declination is usually stated on the margin of the map.

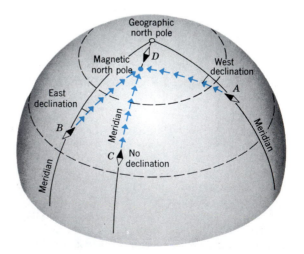

FIGURE A.IV.12 Whether declination is east or west depends on the observer's global position with respect to the magnetic and geographic north poles. (From A. N. Strahler, *The Earth Sciences*, 2nd ed., Harper & Row, Publishers. Copyright © by Arthur N. Strahler.)

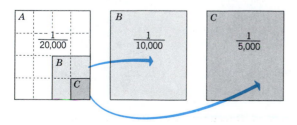

FIGURE A.IV.11 Map area decreases as scale increases.

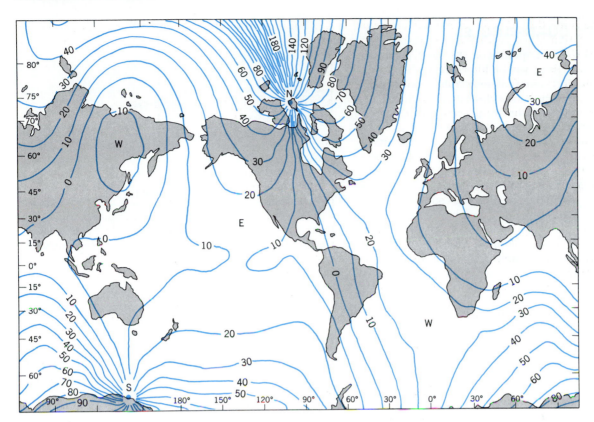

FIGURE A.IV.13 On this world isogonic map, declination is given in intervals of 10°. (Data of U.S. Naval Oceanographic Office. From A. N. Strahler, *The Earth Sciences*, 2nd ed., Harper & Row, Publishers. Copyright © by Arthur N. Strahler.)

BEARINGS AND AZIMUTHS

When a map is used in the field, it is often necessary to state the direction followed by a road or stream or to describe the direction that can be taken to locate a particular object with respect to some known reference point. For this purpose, the observer determines the horizontal angle between the line to the objective and a north–south line. The common unit of angular measurement is the degree, 360 of which comprise a complete circle. Other systems of angular measurement, such as the mil (of which there are 6400 in a complete circle), are sometimes preferable for special applications.

Two systems are used to state direction with respect to north. *Compass quadrant bearings* are angles measured eastward or westward of either north or south, whichever happens to be the closer. Examples are shown in Figure A.IV.14A. The direction from a given point to some object on the map is written as "N49°E" or "S70°W." All bearings range between 0 and 90°. Compass bearings may be magnetic bearings, related to magnetic north, or true bearings, related to geographic north. Unless otherwise stated, a bearing should be assumed to be a true bearing.

Azimuths are used by military services and in air and marine navigation generally. As shown in Figure A.IV.14B, all azimuthus are read in a clockwise direction from north and range between 0° and 360°. Azimuths are measured from either magnetic north or true north, referred to as *magnetic azimuth* and *true azimuth*, respectively.

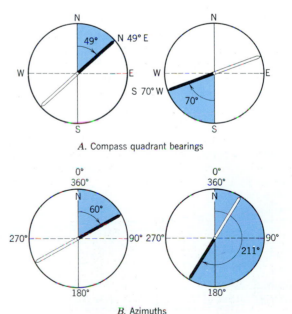

A. Compass quadrant bearings

B. Azimuths

FIGURE A.IV.14 Directions are expressed as bearings or azimuths.

TOPOGRAPHIC QUADRANGLES

Any system used to locate points on the earth's surface with reference to a fixed set of intersecting lines is a *coordinate system*. The system of geographic coordinates using parallels and meridians is described in Chapter 1. Two other coordinate systems are the military grid and the township grid of the U.S. Land Office Survey.

Most published sets of large-scale topographic maps use the geographic grid to determine the position and size of individual map sheets in a series. A single map sheet, or *quadrangle*, is bounded on the right- and left-hand margins by meridians, and on the top and bottom by parallels, which are a specified number of minutes or degrees apart.

Seven standard scales comprise the U.S. National Topographic Map Series. English units of length continue to be used, but a change to metric units has begun.

Series	R.F.	Unit Equivalents
7.5-minute	1:24,000	1 in. to 2000 ft
7.5-minute	1:31,680	1 in. to ½ mi
15-minute	1:62,500	1 in. to about 1 mi
Alaska	1:63,360	1 in. to 1 mi
30-minute	1:125,000	1 in. to about 2 mi
1:250,000	1:250,000	1 in. to about 4 mi
1:1,000,000	1:1,000,000	1 in. to about 16 mi
		(1 cm to 10 km)

Figures A.IV.15 and A.IV.16 compare the coverages and sizes of standard quadrangles on several of these series.

TOPOGRAPHIC MAP SYMBOLS

Large-scale topographic maps of the U.S. Geological Survey use a standard set of symbols to show many kinds of features that cannot be represented to true scale. Reproduced on the back end paper of this book is the set of symbols used by the U.S. Geological Survey on its recent

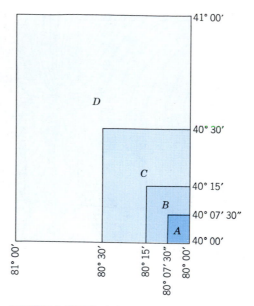

FIGURE A.IV.16 If the four quadrangles of Figure A.IV.15 were reduced to the same scale, their areas would compare as shown here.

series of large-scale topographic maps, together with a representative example of a map on the scale of 1:24,000.

In general, it is conventional to show relief features in brown, hydrographic (water) features in blue, vegetation in green, and cultural (human-made) features in black or red.

THE U.S. LAND OFFICE SURVEY

Topographic maps of the central and western United States show the civil divisions of land according to the U.S. Land Office Survey.

In 1785 Congress authorized a survey of the territory lying north and west of the Ohio River. To avoid the irregular and unsystematic type of land subdivision that had grown up in seaboard states during colonial times,

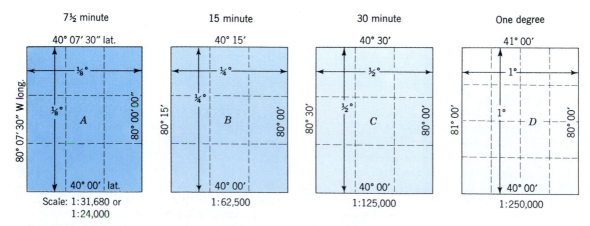

FIGURE A.IV.15 Large-scale maps of the U.S. Geological Survey are bounded by parallels and meridians to form quadrangles. The four quadrangles shown here represent the scales and areas commonly used in the United States, exclusive of Alaska.

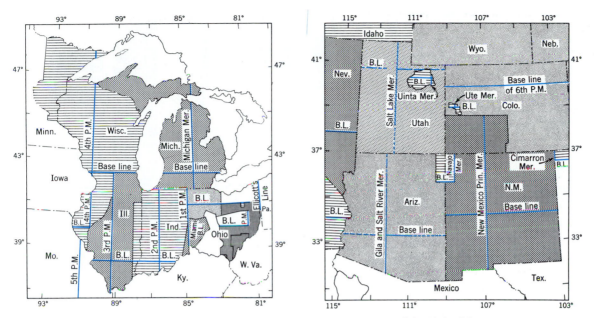

FIGURE A.IV.17 Base lines and principal meridians of these two portions of the United States are representative of the system used by the U.S. Land Office. (After U.S. Department of Interior, General Land Office Map of the United States.)

Congress specified that the new lands should be divided into 6-mile squares, now called *congressional townships,* and that the grid of townships should be based on a carefully surveyed east–west base line, designated the "geographer's line." Meridians and parallels laid off at 6-mile intervals from the base line were to form the boundaries of the townships. This general plan, believed to have been proposed by Thomas Jefferson, was subsequently carried out to cover the balance of the central and western states.

The *principal meridians* and *base lines,* from which rows of townships were laid off, are shown in Figure A.IV.17. Principal meridians run north or south, or both, from selected points whose latitude and longitude were originally calculated by astronomical means. Thirty-two principal meridians have been surveyed. Westward from

the Ohio–Pennsylvania boundary, these are numbered from 1 through 6, beyond which they are designated by names.

Through the initial point selected for starting the principal meridian, an east–west base line was run, corresponding to a parallel of latitude through that point. North and south from the base line, horizontal tiers of townships were laid off and numbered accordingly. Vertical rows of townships, called *ranges,* were laid off to the right and left of the principal meridians and were numbered accordingly (Figure A.IV.18).

The area governed by one principal meridian and its base line is restricted to a particular section of country, usually about as large as one or two states. Where two systems of townships meet, they do not correspond because each system was built up independently of the others.

Because the range lines, or eastern and western boundaries of townships, are meridians converging slightly as they are extended northward, the width of townships is progressively diminished in a northward direction. To avoid a considerable reduction in township widths in the more northerly tiers, new base lines, known as *standard parallels,* have been surveyed for every four tiers of townships. They are designated 1st, 2nd, 3rd standard parallel north, and so on (Figure A.IV.18). The ranges will be found to offset at the standard parallels; consequently, roads that follow range lines make an offset or jog when crossing standard parallels.

Subdivisions of the township are square-mile *sections* of which there are 36 to the township. These are usually numbered as illustrated in Figure A.IV.19. Each section may be subdivided into halves, quarters, and halfquarters, or even smaller units. These divisions, together with the number of acres contained in each, are illustrated in Figure A.IV.20.

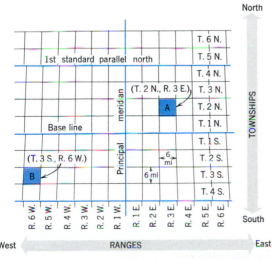

FIGURE A.IV.18 Designation of townships and ranges.

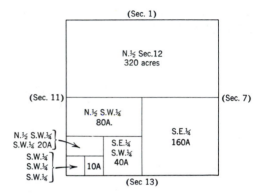

FIGURE A.IV.19 A township is divided into 36 sections, each one a square mile.

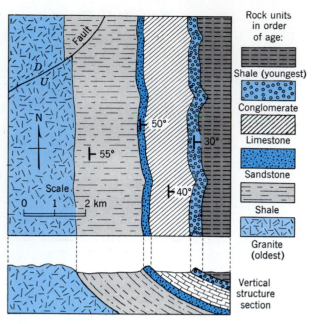

FIGURE A.IV.21 A geologic map shows the surface distribution of rocks and structures. The structure section shows rocks at depth.

FIGURE A.IV.20 A section may be subdivided into many units.

GEOLOGIC MAPS AND STRUCTURE SECTIONS

The *areal geologic map* shows, by means of colors or patterns, the surface distribution of each rock unit, with special emphasis on the lines of contact of rocks unlike in variety or age. Faults are shown as lines. Strikes and dips of strata are added by special symbols.

Figure A.IV.21 is a simple geologic map of the same area shown by a perspective diagram in Figure 19.1.

When map reproduction is limited to black and white, patterns are applied to differentiate the rock unit. Shorthand letter combinations may be added as a code to set apart formations of different ages. Small T-shaped symbols, seen on the map, tell strike and dip. The long bar gives direction of strike; the short bar that abuts it at right angles shows direction of dip. Amount of dip in degrees is given by a figure beside the symbol. A small fault, cutting across the northwest corner of the map, is shown by a solid line. The letters D and U indicate which side moved downward, which upward.

To show the internal geologic structure of an area, a *structure section* is used. It is an imaginary vertical slice through the rocks. A structure section is illustrated in the lower part of Figure A.IV.21. The upper line of the section is a topographic profile.

SUBORDERS OF THE SOIL TAXONOMY

There are 47 suborders within the Soil Taxonomy. Names of the suborders combine a special set of formative elements, listed in Table A.V.1, with formative elements of the soil orders (see Table 24.1). The suborders within each order are listed below with a brief description of each.*

*Much of the description of soil suborders in the remainder of this appendix is selectively quoted or paraphrased from one of the following two sources: (1) Guy D. Smith, *The Soil Orders of the Classification Used in the U.S.A.*, Romanian Geological Institute, Technological and Economic Bulletin, Series C—Pedology, no. 18, pp. 509–31, 1970. Dr. Guy D. Smith is the former Director, Soil Survey Investigations, Soil Conservation Service, U.S. Department of Agriculture. (2) Roy W. Simonson and Darrell L. Gallup, *Soils of the World*, unpublished report, 1972. This material has been used with permission of the authors.

ENTISOLS

Aquents (E1) are Entisols of wet places (aquic regime). They are dominantly gray throughout and stratified or are saturated with water at all times.

Arents are Entisols with recognizable fragments of pedogenic horizons that have been mixed by deep plowing, spading, or other disturbance by humans.

Fluvents are Entisols formed from recent alluvium, rarely saturated with water, usually stratified, and brownish to reddish throughout—at least in the upper 50 cm (20 in.)—and with loamy or clayey texture in some part of the upper 1 m (40 in.).

Orthents (E2) are Entisols on recent erosional surfaces. They have a shallow depth to consolidated rock or

TABLE A.V.1 Formative Elements in Names of Soil Suborders

Formative Element	Derivation	Memory Word (Mnemonicon)	Connotation
Alb	L. *albus*, white	albino	Presence of albic horizon
And	Modified from *andosol*		Resembling Ando soils of Indonesia, formed from volcanic ash
Aqu	L. *aqua*, water	aquarium	Aquic soil-water regime
Ar	L. *arare*, to plow	arable	Horizons mixed by plowing
Arg	L. *argilla*, white clay	argillaceous	Presence of argillic horizon
Bor	Gr. *boreas*, northern	boreal	Formed in cold climate
Ferr	L. *ferrum*, iron	ferrous	Presence of iron
Fibr	L. *fibra*, fiber	fibrous	Least decomposed stage of organic matter
Fluv	L. *fluvius*, river	fluvial	Formed on floodplain deposits
Fol	L. *folia*, leaf	foliage	Layer of forest litter
Hem	Gr. *hemi*, half	hemisphere	Intermediate stage of decomposition of organic matter
Hum	L. *humus*, earth	humus	Presence of organic matter
Ochr	Gr. *ochros*, pale	ocher	Presence of ochric epipedon
Orth	Gr. *orthos*, true	orthodox	The common suborder
Plagg	Ger. *plaggen*, sod		Presence of plaggen epipedon
Psamm	Gr. *psammos*, sand		Sand texture
Rend	From *Redzina*, a soil order in 1938 USDA system.		High carbonate content
Sapr	Gr. *sapros*, rotten	saprolite	Most decomposed stage of organic matter
Torr	L. *torridus*, hot and dry	torrid	Torric soil-water regime
Ud	L. *udus*, humid	udder	Udic soil-water regime
Umbr	L. *umbra*, shade	umbrella	Presence of umbric epipedon
Ust	L. *ustus*, burnt	combustion	Ustic soil-water regime
Xer	Gr. *xeros*, dry	xerophyte	Xeric soil-water regime

have unconsolidated materials, such as loess, and have skeletal, loamy, or clayey texture in some part of the upper 1 m (40 in.).

Psamments (E3) are Entisols that have sandy texture (sands and loamy sands) throughout the upper 1 m (40 in.). Parent material is usually dune sand or beach sand.

INCEPTISOLS

Andepts commonly have a dark surface horizon, are rich in humus and amorphous silica (allophane), and have a high CEC. Andepts are formed chiefly from recent volcanic ash deposits. An example is the occurrence of Andepts on the island of Hawaii, where volcanic activity is in progress.

Aquepts are Inceptisols of wet places (aquic regime). They have a gray to black mollic or histic epipedon, a gray subsoil, and are seasonally saturated with water.

Ochrepts have a brownish horizon (ochric epipedon) of altered materials at or close to the surface. Ochrepts are found in midlatitude and high-latitude climates.

Plaggepts are Inceptisols with a plaggen epipedon—a thick surface horizon of materials added by humans through long habitation or long-continued manuring. Most are in Europe and the British Isles.

Tropepts are Inceptisols of low latitudes. They have a brownish or reddish horizon of altered material (ochric epipedon) at the surface.

Umbrepts are Inceptisols with a dark acid surface horizon (umbric or mollic epipedon) more than 25 cm (10 in.) thick. They contain crystalline clay minerals and are brownish in color. They occur in moist midlatitude and high-latitude climates.

HISTOSOLS

Fibrists are Histosols composed mostly of *Sphagnum* moss or are saturated with water most of the year or artificially drained and consist mainly of recognizable plant remains so little decomposed that rubbing does not destroy the bulk of the fibers.

Folists are Histosols formed with free drainage and consist of forest litter resting on bedrock or rock rubble.

Hemists are Histosols saturated with water most of the year or artificially drained, with between 10 and 40 percent of the volume remaining as fibers after rubbing.

Saprists are Histosols saturated with water most of the year or artificially drained, and so decomposed that after rubbing less than 10 percent of the volume remains as fibers.

OXISOLS

Aquox are Oxisols of wet places (aquic regime). They are dominantly gray, or mottled dark red and gray, and are seasonally saturated with water.

Humox are Oxisols of relatively cool moist regions, with very large accumulations of organic carbon. Humox occur on high plateaus.

Orthox are Oxisols of warm, humid regions with short dry seasons or no dry season (perudic regime). These soils are never dry in any horizon for as long as 60 consecutive days. Orthox are widespread in the equatorial zone.

Torrox are Oxisols of the aridic or torric soil-water regime. They are dry in all horizons for more than 6 months of the year, are never continuously moist for as long as 3 consecutive months, and are not appreciably darkened by humus. Although occurring in dry climates, Torrox may have originated under a formerly moist climate.

Ustox are Oxisols of the ustic soil-water regime. They are moist in some horizon more than 6 months (cumulative) or more than 3 consecutive months, but are dry in some horizon for more than 60 consecutive days (roughly equivalent to a 3-month dry season). Ustox are closely associated with the wet-dry tropical climate.

ULTISOLS

Aquults (U1) are Ultisols of wet places (aquic regime). They are seasonally saturated with water and are dominantly gray in color throughout. A few have black epipedons on gray subsoils.

Humults (U2) are Ultisols with large accumulations of organic matter formed under relatively high, well-distributed rainfall in middle and low-latitude zones.

Udults (U3) are Ultisols of the udic and perudic soil-water regimes. They have moderate to small amounts of organic matter, reddish or yellowish B horizons, and no periods or only short periods when some part of the soil is dry.

Ustults (U4) are Ultisols of the ustic soil-water regime, in which the soil is dry during a pronounced dry season. The soil-temperature regime is usually thermic. Ustults are brownish to reddish throughout.

Xerults are Ultisols of Mediterranean climate (7) with cool moist winters and rainless summers. The xeric soil-water regime prevails. Xerults are brownish to reddish throughout.

VERTISOLS

Torrerts are Vertisols of the torric regime. They have cracks that remain open throughout the year in most years.

Uderts (V1) are Vertisols of the udic regime. They have cracks that remain open for only short periods in most years, less than 90 cumulative days and less than 60 consecutive days.

Usterts (V2) are Vertisols of the ustic regime with distinct wet and dry seasons. They are widespread in areas of wet-dry tropical climate. The cracks remain open more than 90 cumulative days in most years.

Xererts are Vertisols of the xeric regime in the Mediterranean climate. The soil cracks close in winter but open in summer every year for more than 60 consecutive days.

ALFISOLS

Aqualfs are Alfisols of wet places (aquic regime). They are dominantly gray throughout and are seasonally saturated with water.

Boralfs (A1) are Alfisols of boreal forests or high mountains. They have a gray surface horizon, a brownish subsoil, and are associated with a mean annual soil temperature less than 8°C (47°F).

Udalfs (A2) are brownish Alfisols of the udic soil-water regime. They occur in regions having no periods or only short periods (fewer than 90 days in most years) when part or all of the soil is dry. A deciduous forest was the typical vegetation on Udalfs, but today these soils are intensively farmed.

Ustalfs (A3) are Alfisols of the ustic soil-water regime, with long periods (more than 90 cumulative days in most years) when the soil is dry. Ustalfs are brownish to reddish throughout.

Xeralfs (A4) are Alfisols of the xeric soil-water regime, under a Mediterranean climate with cool, moist winters and rainless summers. Xeralfs are brownish to reddish throughout. A typical occurrence is in coastal and inland valleys of central and southern California.

SPODOSOLS

Aquods (S2) are Spodosols of wet places (aquic regime), seasonally saturated with water. The surface horizon may be either black or white and lacks free iron oxides. The B horizon is black, brownish, or reddish and may be firm or cemented.

Ferrods are freely drained Spodosols having B horizons of accumulation of free iron oxides. This suborder is of rare occurrence.

Humods (S3) are freely drained Spodosols in which at least the upper part of the B horizon has accumulated humus and aluminum, but not iron.

Orthods (S4) are freely drained Spodosols with B horizons in which iron, aluminum, and humus have accumulated. Orthods are widespread in North America and northern Europe.

MOLLISOLS

Albolls (M1) are Mollisols with a bleached gray to white horizon (albic horizon) above a slowly permeable horizon of clay accumulation, saturated with water in at least the surface horizons at some season.

Aquolls are Mollisols of wet places (aquic regime). They have a nearly black mollic epipedon resting on a mottled grayish B horizon and are seasonally saturated with water.

Borolls (M2) are Mollisols of cold-winter semiarid plains (steppes) or high mountains. They have a mean annual soil temperature lower than 8°C (47°F).

Rendolls (M3) are Mollisols formed on highly calcareous parent materials and with more than 40 percent calcium carbonate in some horizon within 50 cm (20 in.) of the surface, but without a horizon of accumulation of calcium carbonate. Rendolls are mostly formed in moist climates under forest vegetation.

Udolls (M4) are Mollisols of the udic soil-water regime. They have brownish hues throughout and no horizon of accumulation of calcium carbonate in soft powdery forms.

Ustolls (M5) are Mollisols of the ustic soil-water regime, with long periods (more than 90 cumulative days in most years) in which the soil is dry. Most Ustolls have a horizon of accumulation of soft, powdery calcium carbonate starting at a depth of about 50 to 100 cm (20 to 40 in.). A petrocalcic horizon may be formed within 1 m (3 ft) of the soil surface. In the southwestern United States, this material is known as *caliche*.

Xerolls (M6) are Mollisols of the xeric regime under a Mediterranean climate with cool, moist winters and rainless summers.

ARIDISOLS

Argids (D2) are Aridisols with an argillic horizon. Argids have formed largely on surfaces of late-Pleistocene (Wisconsin) age or older and, consequently, have developed throughout one or more of the Pleistocene pluvial (moist) periods. Illuviation of the B horizon is thought to have taken place during these moist periods.

Orthids are soils without an argillic horizon but with one or more pedogenic horizons that extend at least 25 cm (10 in.) below the surface, so some horizon persists in part if the soil is plowed. One variety of Orthids (Salorthids) is characterized by having a salic horizon (horizon of salt accumulation).

REFERENCES FOR FURTHER STUDY

CHAPTER 1
OUR ROTATING PLANET

Carrigan, Richard A., Jr. (1978), Decimal time, *American Scientist*, vol. 66, pp. 305–13.

Bartky, I. R., and E. Harrison (1979), Standard and daylight-saving time, *Scientific American*, vol. 240, no. 5, pp. 46–53.

Moyer, Gordon (1982), The Gregorian calendar, *Scientific American*, vol. 246, no. 5, pp. 144–51.

The World Almanac and Book of Facts (issued annually), Pharos Books, A Scripps Howard Company, New York. See Astronomy, Calendar, International Time Zones (map).

CHAPTER 2
THE EARTH'S ATMOSPHERE AND OCEANS

Menard, H. W., ed. (1977), *Ocean Science*, Readings from *Scientific American*, W. H. Freeman & Company, San Francisco.

Broecker, Wallace S. (1983), The ocean, *Scientific American*, vol. 249, no. 3, pp. 146–60.

Akasofu, Syun-Ichi (1989), The dynamic aurora, *Scientific American*, vol. 260, no. 5, pp. 90–97.

National Research Council (1989), *Ozone depletion, greenhouse gases, and climate change*, National Academy Press, Washington, DC.

Rowland, F. Sherwood (1989), Chlorofluorocarbons and the depletion of stratospheric ozone, *American Scientist*, vol. 77, pp. 36–45.

CHAPTER 3
THE EARTH'S RADIATION BALANCE

Sellers, W. D. (1965), *Physical climatology*, University of Chicago Press.

Kreith, Frank, and Richard T. Meyer (1983), Large-scale use of solar energy with central receivers, *American Scientist*, vol. 71, pp. 598–605.

Hamakawa, Yoshihiro (1987), Photovoltaic power, *Scientific American*, vol. 256, no. 4, pp. 87–92.

CHAPTER 4
HEAT AND COLD AT THE EARTH'S SURFACE

Hansen, J. E., and T. Takahashi, eds. (1984), *Climate processes and climate sensitivity*, Geophysical Monograph 29, American Geophysical Union, Washington, DC.

Berger, A., R. E. Dickinson, and J. W. Kidson, eds. (1989), *Understanding climate change*, Geophysical Monograph 52, American Geophysical Union, Washington, DC.

Berner, Robert A., and Antonio C. Lasaga (1989), Modeling the geochemical carbon cycle. *Scientific American*, vol. 260, no. 3, pp. 74–81.

Houghton, Richard A., and George M. Woodwell (1989), Global climatic change, *Scientific American*, vol. 260, no. 4, pp. 36–44.

Schneider, Stephen H. (1989), The changing climate, *Scientific American*, vol. 261, no. 3, pp. 70–79.

CHAPTER 5
WINDS AND THE GLOBAL CIRCULATION

Forrester, Frank H. (1982), Winds of the world, *Weatherwise*, vol. 35, no. 5, pp. 204–10.

Moretti, Peter M., and Louis V. Divone (1986), Modern windmills, *Scientific American*, vol. 254, no. 6, pp. 110–18.

Lessard, Arthur G. (1988), The Santa Ana wind of Southern California, *Weatherwise*, vol. 41, no. 2, pp. 100–104.

Clarke, Alexi (1989), How green is the wind? *New Scientist*, vol. 122, no. 1666, pp. 62–65.

CHAPTER 6
ATMOSPHERIC MOISTURE AND PRECIPITATION

Ludlam, F. H. (1980), *Clouds and storms*, Pennsylvania State University Press, University Park.

Hallett, John (1984), How snow crystals grow, *American Scientist*, vol. 72, pp. 582–89.

Graedel, Thomas E., and Paul J. Crutzen (1989), The changing atmosphere, *Scientific American*, vol. 261, no. 3, pp. 58–68.

Smith, Kirk R. (1989), Air pollution—Assessing total exposure in developing countries, *Environment*, vol. 30, no. 10, pp. 16–20, 28–35.

National Research Council (1986), *Acid deposition: Long-term trends*, National Academy Press, Washington, DC.

Schwartz, Stephen E. (1989), Acid deposition: Unraveling a regional phenomenon, *Science*, vol. 243, pp. 753–62.

CHAPTER 7
AIR MASSES AND CYCLONIC STORMS

Dunn, G. E., and B. I. Miller (1964), *Atlantic hurricanes*, Louisiana State University Press, Baton Rouge.

Simpson, R. H., and H. Riehl (1981), *The hurricane and its impact*, Louisiana State University Press, Baton Rouge.

Anthes, R. A. (1982), *Tropical cyclones: Their evolution, structure and effects*, Meteorological Monographs, vol. 119, no. 41, American Meteorological Society, Boston.

Woods Hole Oceanographic Institution (1984), The 1982–83 El Niño, *Oceanus*, vol. 27, no. 2, pp. 1–62.

Rasmusson, E. M. (1985), El Niño variations in climate, *American Scientist*, vol. 73, pp. 168–77.

Emanuel, Kerry A. (1988), Toward a general theory of hurricanes, *American Scientist*, vol. 76, pp. 371–79.

Smith, W. L., et al. (1986), The meteorological satellite: overview of 25 years of operation, *Science*, vol. 231, pp. 455–70.

Philander, George (1989), El Niño and La Niña, *American Scientist*, vol. 77, pp. 451–59.

CHAPTER 8
GLOBAL CLIMATE SYSTEMS

Trewartha, Glenn T. (1961), *The earth's problem climates*, University of Wisconsin Press, Madison, Methuen & Company, London.

Wilcock, A. A. (1968), Köppen after fifty years, *Annals, Association of American Geographers*, vol. 58, pp. 12–18.

Oliver, John (1973), *Climate and man's environment: An introduction to applied climatology*, John Wiley & Sons, New York.

Mather, John R. (1974), *Climatology: Fundamentals and applications*, McGraw-Hill Book Company, New York.

CHAPTER 9
THE SOIL–WATER BALANCE AND CLIMATE

Thornthwaite, C. W. (1948), An approach toward a rational classification of climate, *Geographical Review*, vol. 38, pp. 55–94.

Thornthwaite, C. W., and J. R. Mather (1955), The water balance, Drexel Institute of Technology, Laboratory of Climatology, *Publications in Climatology*, vol. 8, no. 1, Centerton, NJ.

Sellers, W. D. (1965), *Physical climatology*, University of Chicago Press. See Chapter 7.

Miller, David H. (1977), *Water at the surface of the earth*, Academic Press. See Chapters 6–14.

CHAPTERS 10 AND 11
LOW-LATITUDE, MIDLATITUDE, AND HIGH-LATITUDE CLIMATES

Love, R. Merton (1970), The rangelands of the western U.S., *Scientific American*, vol. 222, no. 2, pp. 89–96.

Bell, Richard H. V. (1971), A grazing system in the Serengeti, *Scientific American*, vol. 225, no. 1, pp. 86–93.

Hare, F. Kenneth, and J. C. Ritchie (1972), The boreal bioclimates, *Geographical Review*, vol. 62, pp. 333–65.

Anderson, Myrdene (1985), The Saami Reindeer-Breeders of Norwegian Lapland, *American Scientist*, vol. 73, pp. 524–32.

Colinvaux, Paul A. (1989), The past and future Amazon, *Scientific American*, vol. 260, no. 5, pp. 102–108.

Prance, Ghillean (1990), Fruits of the rainforest, *New Scientist*, vol. 125, no. 1699, pp. 42–45.

CHAPTER 12
MATERIALS OF THE EARTH'S CRUST

Ernst, W. G. (1969), *Earth materials*, Prentice-Hall, Englewood Cliffs, NJ.

Simpson, Brian (1966), *Rocks and minerals*, Pergamon Press, Oxford and London.

Hurlbut, C. S., and C. Klein (1977), *Manual of mineralogy (after James D. Dana)*, nineteenth edition, John Wiley & Sons, New York.

Nockolds, S. R., R. W. O'B. Knox, and G. A. Chinner, 1978, *Petrology for students*, Cambridge University Press.

Friedman, G. M., and J. E. Sanders (1978), *Principles of sedimentology*, John Wiley & Sons, New York.

CHAPTER 13
THE LITHOSPHERE AND PLATE TECTONICS

Hallam, A. (1973), *A revolution in the earth sciences*, Clarendon Press and Oxford University Press, London.

Sullivan, Walter (1974), *Continents in motion: The new earth debate*, McGraw-Hill Book Company, New York.

Schwartzbach, Martin (1985), *Alfred Wegener: The father of continental drift*, Science Tech., Madison, WI.

Wood, Robert M. (1985), *The dark side of the earth*, Allen & Unwin, London.

Cox, Allan, and Robert B. Hart (1986), *Plate tectonics: How it works*, Blackwell Scientific Publications, Palo Alto, CA.

CHAPTER 14
VOLCANIC AND TECTONIC LANDFORMS

Dutton, Clarence E. (1890), reprinted 1979, *The Charleston earthquake of August 31, 1886*, U.S. Geological Survey, U.S. Government Printing Office, Washington, DC.

Macdonald, G. A., and A. T. Abbott (1970), *Volcanoes in the sea: The geology of Hawaii*, University of Hawaii Press, Honolulu.

Bullard, Fred M. (1976) *Volcanoes of the earth*, revised edition, University of Texas Press, Austin.

Bolt, Bruce A. (1978), *Earthquakes, a primer*, W. H. Freeman & Company, San Francisco.

Lipman, P. W., and D. R. Mullineaux, eds. (1981), *The 1980 eruptions of Mount St. Helens, Washington*, U.S. Geological Survey Professional Paper 1250, U.S. Government Printing Office, Washington, DC.

Bolt, Bruce A. (1982), *Inside the earth: Evidence from earthquakes*, W. H. Freeman & Company, San Francisco.

Simkin, T., and R. S. Fiske (1983), *Krakatoa 1883: The volcano and its effects*, Smithsonian Institution Press, Washington, DC.

CHAPTER 15
LANDFORMS OF WEATHERING AND MASS WASTING

Sharpe, C. F. S. (1938), *Landslides and related phenomena*, Columbia University Press, New York.

Highway Research Board (1958), *Landslides and engineering practice*, Special Report 29, NAS-NRC Publication 544, Washington, DC.

Ritter, Dale F. (1978), *Process geomorphology*, William C. Brown, Dubuque, IA.

Embleton, C., and J. Thornes (1979), *Processes in geomorphology*, Halstead Press, John Wiley & Sons, New York.

CHAPTER 16
RUNOFF, STREAMS, AND GROUND WATER

Linsley, R. K, Jr., M. A. Kohler, and J. L. H. Paulhus (1975), *Hydrology for engineers*, second edition, McGraw-Hill Book Company, New York.

Dunne, T., and L. B. Leopold (1978), *Water in environmental planning*, W. H. Freeman & Company, San Francisco.

Ward, R. C. (1978), *Floods: A geographic perspective*, Halstead Press, John Wiley & Sons, New York.

Dingman, S. L. (1984), *Fluvial hydrology*, W. H. Freeman & Company, New York.

Todd, D. K. (1980), *Ground water hydrology*, second edition, John Wiley & Sons, New York.

Issar, Arie (1985), Fossil water under the Sinai–Negev Peninsula, *Scientific American*, vol. 253, no. 1, pp. 104–10.

CHAPTER 17
LANDFORMS MADE BY RUNNING WATER

Leopold, L. B., M. G. Wolman, and J. P. Miller (1964), *Fluvial processes in geomorphology*, W. H. Freeman & Company, San Francisco.

Morisawa, M. (1986), *Streams: Their dynamics and morphology*, McGraw-Hill Book Company, New York.

Schumm, Stanley A. (1977), *The fluvial system*, John Wiley & Sons, New York.

Chorley, R. J., S. A. Schumm, and D. E. Sugden (1985), *Geomorphology*, Methuen, New York.

CHAPTER 18
DENUDATION AND CLIMATE

Davis, William Morris (1909, 1954), *Geographical essays*, Dover Publications, New York. See pp. 249–78, 196–322, 381–412.

Carson, M. A., and M. J. Kirkby (1972), *Hillslope process and form*, John Wiley & Sons, New York. See Chapters 5, 6, 8, 11.

Young, A. (1972), *Slopes*, Oliver & Boyd, Edinburgh.

Derbyshire, E., ed. (1973), *Climatic geomorphology*, Harper & Row, New York.

Thomas, M. F. (1974), *Tropical geomorphology*, Macmillan Press, London.

Douglas, I. (1977), *Humid landforms*, M.I.T. Press, Cambridge, MA.

Mabutt, J. A. (1977), *Desert landforms*, M.I.T. Press, Cambridge, MA.

Kiewiet de Jonge, C. (1984), Büdel's geomorphology, *Progress in Physical Geography*, vol. 8, pp. 218–48, 365–97.

CHAPTER 19
LANDFORMS AND ROCK STRUCTURE

Davis, William Morris (1909, 1954), *Geographical essays*, reprinted by Dover Publications, New York. See pp. 413–84, 725–72.

Johnson, Douglas (1931), *Stream sculpture on the Atlantic slope*, Columbia University Press, New York.

Shelton, John S. (1966), *Geology illustrated*, W. H. Freeman & Company, San Francisco.

Jennings, J. N. (1971), *Karst*, M.I.T. Press, Cambridge, MA.

Giusti, Ennio V. (1978), *Hydrogeology of the karst of Puerto Rico*, U.S. Geological Survey Professional Paper 1012, U.S. Government Printing Office, Washington, DC.

CHAPTER 20
LANDFORMS MADE BY WAVES AND CURRENTS

Johnson, D. W. (1919, 1965), *Shore processes and shoreline development*, John Wiley & Sons, New York, reprinted by Hafner Publishing Co., New York.

Strahler, Arthur N. (1966), *A geologist's view of Cape Cod*, Natural History Press (Doubleday & Co.), Garden City, NY. Reprinted 1988 by Parnassus Imprints, Orleans, MA.

Bird, E. C. F. (1969), *Coasts*, M.I.T. Press, Cambridge, MA.

King, C. A. M. (1972), *Beaches and coasts*, second edition, St. Martin's Press, New York.

Fisher, J. S., and R. Dolan (1977), *Beach processes and coastal hydrodynamics*, Dowden, Hutchinson & Ross, Stroudsburg, PA.

CHAPTER 21
LANDFORMS MADE BY WIND

Bagnold, R. A. (1941), *The physics of blown sand and desert dunes*, Methuen & Company, London.

McKee, E. D., ed. (1979), *A study of global sand seas*, U.S. Geological Survey Professional Paper 1052, U.S. Government Printing Office, Washington, DC.

Péwé, T. L., ed. (1981), *Desert dust: Origin, characteristics, and effect on man*, Special Paper 186, Geological Society of America, Boulder, CO.

Nielson, Jamie, and Gary Kocurek (1987), Surface processes, deposits, and development of star dunes, *Bulletin, Geological Society of America*, vol. 99, pp. 177–86.

Blount, Grady, and Nicholas Lancaster (1990), Development of the Gran Desierto sand sea, northwestern Mexico, *Geology*, vol. 18, pp. 724–28.

CHAPTER 22
GLACIAL LANDFORMS AND THE ICE AGE

Hough, J. L. (1958), *Geology of the Great Lakes*, University of Illinois Press, Urbana.

Paterson, W. S. B. (1969), *The physics of glaciers*, Pergamon Press, Oxford, U.K.

Flint, Richard F. (1971), *Glacial and quaternary geology*, John Wiley & Sons, New York.

Embleton, C., and C. A. M. King (1975), *Glacial geomorphology*, Edward Arnold, London.

Frakes, L. A. (1979), *Climates throughout geologic time*, Elsevier, Amsterdam.

Lockwood, J. G. (1979), *Causes of climate*, John Wiley & Sons, New York. See Chapter 5.

Broecker, Wallace S., and George H. Denton (1990), What drives glacial cycles? *Scientific American*, vol. 262, no. 1, pp. 49–56.

Ruddiman, William F., and John E. Kutzbach (1991), Plateau uplift and climatic change, *Scientific American*, vol. 264, no. 3, pp. 66–75.

CHAPTER 23
THE SOIL LAYER

Loughnan, F. C. (1969) *Chemical weathering of the silicate minerals*, American Elsevier Publishing Company. See Chapters 2, 3.

Birkeland, P. W. (1974), *Pedology, weathering, and geomorphological research*, Oxford University Press, New York.

Soil Survey Staff (1975), *Soil taxonomy*, U.S. Department of Agriculture, Agriculture Handbook No. 436, Government Printing Office, Washington, DC. See Chapters 1–6.

Foth, Henry D. (1984), *Fundamentals of soil science*, seventh edition, John Wiley & Sons, New York.

CHAPTER 24
WORLD SOILS

Cruikshank, J. G. (1972), *Soil geography*, David and Charles Publishers, Newton Abbot, Devon, England.

Soil Survey Staff (1975), *Soil taxonomy: A basic system of soil classification for making and interpreting soil surveys*, Soil Conservation Service, U.S. Department of Agriculture, Agriculture Handbook No. 436, Government Printing Office, Washington, DC.

Steila, Donald (1976), *The geography of soils*, Prentice-Hall, Englewood Cliffs, NJ.

Gersmehl, P. J. (1977), Soil taxonomy and mapping, *Annals, Association of American Geographers*, vol. 67, pp. 419–28.

Buol, S. W., F. D. Hole, and R. J. McCracken (1980), *Soil genesis and soil classification*, second edition, Iowa State University Press, Ames.

Lal, Rattan (1987), Managing the soils of Sub-Saharan Africa, *Science*, vol. 236, pp. 1069–76.

CHAPTER 25
ENERGY FLOWS AND MATERIAL CYCLES IN THE BIOSPHERE

Odum, E. P. (1971), *Fundamentals of ecology*, third edition, W. B. Saunders Company, Philadelphia.

Odum, H. T. (1971), *Environment, power, and society*, John Wiley & Sons, New York.

Strahler, A. N., and A. H. Strahler (1974), *Introduction to environmental science*, Hamilton Publishing Co., Santa Barbara, Calif. (John Wiley & Sons, New York). See Chapters 20, 21.

Pimentel, D., and M. Pimentel (1979), *Food, energy, and society*, Halstead Press, John Wiley & Sons, New York.

Clark, William C., and R. E. Munn, eds. (1986), *Sustainable development of the biosphere*, Cambridge University Press, Cambridge, England.

Scientific American (1988) *Managing Planet Earth*, W. H. Freeman & Company, New York.

CHAPTER 26
CONCEPTS OF BIOGEOGRAPHY

Dansercau, Pierre (1957), *Biogeography: An ecological perspective*, Ronald Press, New York.

Strahler, A. N., and A. H. Strahler (1974), *Introduction to environmental science*, Hamilton Publishing Company, Santa Barbara, Calif. (John Wiley & Sons, New York). See Chapters 22–24.

Barbour, M. G., J. H. Burk, and W. D. Pitts (1980), *Terrestrial Plant Ecology*, Benjamin/Cummings Publishing Company, Menlo Park, CA.

Brown, J. H., and A. C. Gibson (1983), *Biogeography*, C. V. Mosby Company, St. Louis.

CHAPTER 27
WORLD PATTERNS OF NATURAL VEGETATION

Küchler, A. W. (1964), *Potential natural vegetation of the coterminous United States* (map and manual), Special Publication No. 36. American Geographical Society, New York.

Eyre, S. R. (1968), *Vegetation and soils: A world picture*, second edition, Aldine Publishing Company, Chicago.

Glanz, M. H., ed. (1977), *Desertification: Environmental degradation in and around arid lands*, Westview Press, Boulder, CO.

Walter, Heinrich (1983), *Vegetation of the Earth*, third edition, Springer-Verlag, Berlin.

APPENDIX I
MAP PROJECTIONS

Ricardus, P., and R. K. Adler (1972), *Map projections: An introduction*, American Elsevier Publishing Company, New York.

Snyder, John P. (1987), *Map projections—a working manual*, U.S. Geological Survey Professional Paper 1395, U.S. Government Printing Office, Washington, DC.

Snyder, John P., and P. M. Voxland (1989), *An album of map projections*, U.S. Geological Survey Professional Paper 1453, U.S. Government Printing Office, Washington, DC.

APPENDIX II
REMOTE SENSING FOR PHYSICAL GEOGRAPHY

Short, N. M., P. D. Lowman, Jr., S. C. Freden, and W. A. Finch (1976), *Mission to earth: Landsat views the world*, NASA SP-360, National Aeronautics and Space Administration, Washington, DC.

Siegal, B. S., and A. R. Gillespie, eds. (1980), *Remote sensing in geology*, John Wiley & Sons, New York.

National Geographic Society (1985), *Atlas of North America: Space age portrait of a continent*, National Geographic Society, Washington, DC.

Jensen, J. R. (1986), *Introductory digital image processing*, Prentice-Hall, Englewood Cliffs, NJ.

Short, Nicholas M., and Robert W. Blair, eds. (1986), *Geomorphology from space: A global overview of regional landforms*, NASA Scientific and Technical Information Branch, National Aeronautics and Space Administration, Washington, DC.

Lillesand, T. W., and R. Keifer (1987), *Remote sensing and image interpretation*, second edition, John Wiley & Sons, New York.

GLOSSARY

This Glossary contains definitions of italicized terms in the text. For terms not listed in the Glossary, refer to the Index, where definitions can be located by italicized page numbers.

ablation   Wastage of glacial ice by both melting and evaporation.

ablational till   Glacial till deposited as ice melts away, leaving the enclosed rock particles behind.

abrasion   Erosion of bedrock of a stream channel by impact of particles carried in a stream and by rolling of larger rock fragments over the stream bed. Abrasion is also an activity of glacial ice, waves, and wind.

abrasion platform   Seaward sloping, nearly flat bedrock surface extending out from foot of marine cliff under shallow water of breaker zone.

absolute temperature   Temperature measured by the Kelvin scale on which absolute zero is equivalent to −273°C.

absorption of radiation   Transfer of energy of electromagnetic radiation into heat energy within a gas or liquid through which the radiation is passing.

abyssal plain   Large expanse of very smooth, flat ocean floor found at depths of 4600 to 5500 m.

accelerated erosion   Soil erosion occurring at a rate much faster than soil horizons can be formed from the parent regolith.

accretion of lithosphere   Production of new oceanic lithosphere at an active spreading plate boundary by the rise and solidification of magma of basaltic composition.

accretionary prism   Mass of deformed trench sediments and ocean floor sediments accumulated in wedgelike slices on the underside of the overlying plate above a plate undergoing subduction.

acid (acidity)   A chemical condition of the soil solution caused by the presence of readily exchangeable acid-generating cations, comprising from 5 to 60 percent of the total cation exchange capacity (CEC).

acid deposition   (See *acid rain*.)

acid-generating cations   Cations, mostly cations of aluminum and hydrogen, whose presence in large numbers in the soil solution produces an acid condition of the soil.

acid mine drainage   Sulfuric acid effluent from coal mines, mine tailings, or spoil ridges made by strip mining of coal.

acid rain   Rainwater having an abnormally high content of the sulfate ion and showing a pH between 2 and 5 as a result of air pollution by combustion products of fuels having high sulfur content.

active continental margins   Continental margins that coincide with tectonically active plate boundaries. (See also *continental margins, passive continental margins*.)

active layer   Shallow surface layer subject to seasonal thawing in permafrost regions.

active pool   Type of pool in the material cycle in which the materials are in forms and places easily accessible to life processes. (See also *storage pools*.)

active systems   Remote sensor systems that emit a human-made beam of wave energy at a source and measure the intensity of that energy reflected back to the source.

actual evapotranspiration (water use)   Actual rate of evapotranspiration at a given time and place.

adiabatic lapse rate   (See *dry adiabatic lapse rate, wet adiabatic lapse rate*.)

adiabatic process   Change of sensible temperature within a gas because of compression or expansion, without gain or loss of heat from the outside.

advection fog   Fog produced by condensation within a moist basal air layer moving over a cold land or water surface.

advection process   The horizontal mixing of cold and warm air in midlatitude cyclones and anticyclones, resulting in poleward meridional transport of heat to reach higher latitudes.

agate   Banded variety of chalcedony.

aggradation   Upbuilding of the floor of a stream channel by continued deposition of bed load.

agric horizon   Illuvial soil horizon formed under cultivation and containing significant amounts of illuvial silt, clay, and humus.

agricultural ecosystems   Ecosystem modified and managed by human beings for agricultural purposes.

A horizon   Mineral horizon of the soil solum, usually rich in organic matter, overlying the E horizon.

air mass   Extensive body of air within which upward gradients of temperature and moisture are fairly uniform over a large area.

albedo   Percentage of electromagnetic radiation reflected from a surface.

albic horizon   Pale, often sandy soil horizon from which clay and free iron oxides have been removed. Found in the profile of the Spodosols.

Albolls   Suborder of the Mollisols with a bleached gray to white horizon

(albic horizon) above a slowly permeable horizon of clay accumulation, saturated with water in at least the surface horizons at some season.

Aleutian low   Persistent center of low atmospheric pressure located in the area of the Aleutian Islands and strongly intensified in winter.

Alfisols   Soil order within the Soil Taxonomy consisting of soils of humid and subhumid climates, with high base status and a dense clay (argillic) horizon.

alkaline soil   Condition of the soil solution present when the large majority of cations held by soil colloids are base cations.

allelopathy   Form of negative interaction among plant species in which the roots of one species secrete substances in the soil that are toxic to other species and thus inhibit their growth.

alluvial fan   Low, gently sloping, conical accumulation of coarse alluvium deposited by a braided stream undergoing aggradation below the point of emergence of the channel from a narrow canyon.

alluvial meanders   Sinuous bends of a graded stream flowing in the alluvial deposit of a floodplain.

alluvial river   River (stream) of low gradient flowing on thick deposits of alluvium and experiencing approximately annual overbank flooding of the adjacent floodplain.

alluvial terrace   Terrace carved in alluvium by a stream during degradation.

alluvium   Any stream-laid sediment deposit found in a stream channel and in low parts of a stream valley subject to flooding.

alpine chains   High mountain ranges that are narrow tectonic belts severely deformed by folding and thrusting in comparatively recent geologic time.

alpine debris avalanche   Rapid downvalley movement of a debris tongue consisting of a mixture of rock waste and glacial ice originating on high steep slopes of glaciated mountains.

alpine glacier   Long narrow mountain glacier on a steep downgrade, occupying the floor of a troughlike valley.

Alpine system   Global system of alpine chains, mostly of Cenozoic tectonic activity, but including some adjacent inactive belts affected by Mesozoic tectonic activity.

alpine tundra   Plant formation class within the tundra biome, found at high altitudes above the limit of tree growth.

altitude tinting Applications of varied color tints to altitude zones on a map.

altocumulus Cloud of middle height range formed into individual small masses within a single layer and fitted closely together in a geometric pattern.

altostratus Cloud of middle height range with blanketlike form.

aluminosilicates Silicate minerals containing aluminum as an essential element. Example: feldspars.

amphibole group (amphiboles) Group of complex aluminosilicate minerals rich in calcium, magnesium, and iron, dark in color and classified as mafic minerals. Example: hornblende.

Andepts Suborder of the Inceptisols commonly having a dark surface horizon and rich in humus and amorphous silica (allophane); usually formed from alteration of volcanic ash.

andesite Extrusive igneous rock of diorite composition, dominated by plagioclase feldspar of intermediate composition and with amphibole and pyroxene as important constituents; the extrusive equivalent of diorite.

anemometer Weather instrument used to indicate wind speed.

aneroid barometer Barometer using a mechanism consisting of a partially evacuated air chamber and a flexible diaphragm.

angle of repose Natural surface inclination (dip) of a slope consisting of loose, coarse, well-sorted rock or mineral fragments; for example, the slip face of a sand dune, a talus slope, or the sides of a cinder cone.

angstrom Wavelength unit equal to 0.000,000,01 cm (10^{-8} cm).

anhydrite Evaporite mineral, composition calcium sulfate.

anion Negatively charged ion.

annual temperature range Difference between mean monthly temperatures of the warmest and coldest months of the year.

annular drainage pattern A stream network dominated by concentric (ringlike) major subsequent streams.

antarctic air mass Cold air mass with source region over Antarctica.

antarctic circle Parallel of latitude at 66½°S.

antarctic circumpolar current Ocean current system flowing eastward at high latitudes in the zone of prevailing westerly winds of the southern hemisphere.

antarctic zone Latitude zone in the latitude range 60° to 75°S (more or less), centered about on the antarctic circle, and lying between the subantarctic zone and the polar zone.

antecedent stream A stream that has maintained its course across a rising rock barrier, such as an anticlinal fold or a fault block.

anthracite Grade of coal very high in fixed carbon content, with little volatile matter, and of metamorphic development; also called "hard coal."

anticlinal mountain Long, narrow ridge or mountain formed on an anticline of resistant strata.

anticlinal valley Valley eroded in weak strata along the central line or axis of an eroded anticline.

anticline Upfold of strata or other layered rocks in an archlike structure; a class of folds. (See also *syncline*.)

anticyclone Center of high atmospheric pressure.

antipode That point which lies diametrically opposite to a given point on the surface of a globe or the earth.

anvil top Flattened top of a cumulonimbus cloud, produced as ice particles are carried downwind at high levels.

aphelion Point on the earth's elliptical orbit at which the earth is farthest from the sun.

Aqualfs Suborder of Alfisols found in wet places (aquic regime), dominantly gray throughout and seasonally saturated with water.

aquatic ecosystems Ecosystems consisting of life forms of the marine environments and the freshwater environments of the lands.

Aquents Suborder of Entisols found in wet places (aquic regime), dominantly gray throughout and stratified or saturated with water at all times.

Aquepts Suborder of Inceptisols found in wet places (aquic regime) and having a gray to black mollic or histic epipedon, a gray subsoil, and seasonally saturated with water.

aquiclude Rock mass or layer of low permeability that impedes or largely prevents the gravitational movement of ground water.

aquic regime Soil-moisture regime in which the soil is saturated with water most of the time because the ground-water table lies at or close to the surface throughout most of the year.

aquifer Rock mass or layer, of both high porosity and high permeability, that readily transmits and holds ground water.

Aquods Suborder of the Spodosols found in wet places (aquic regime) and seasonally saturated with water.

Aquolls Suborder of the Mollisols found in wet places (aquic regime) and having a nearly black mollic epipedon resting on a mottled grayish B horizon; they are seasonally saturated with water.

Aquox Suborder of the Oxisols found in wet places (aquic regime), dominantly gray, or mottled dark red and gray, and seasonally saturated with water.

Aquults Suborder of the Ultisols found in wet places (aquic regime), seasonally saturated with water and dominantly gray in color throughout.

aragonite Carbonate mineral, composition calcium carbonate.

arc–continent collision Collision of a volcanic arc with continental lithosphere along a subduction boundary.

arch (marine) Arch of bedrock remaining when a rock promontory of a marine cliff has been cut through from two sides by wave action.

arctic air mass Cold air mass developed over a source region in the Arctic Ocean and fringing lands.

arctic circle Parallel of latitude at 66½°N.

arctic front Frontal zone of interaction between arctic air masses and polar air masses.

arctic front zone Belt in the subarctic zone and arctic zone in which the fluctuating arctic front is found.

arctic tundra Plant formation class within the tundra biome, consisting of low, mostly herbaceous plants, but with some small stunted trees, associated with the tundra climate.

arctic zone Latitude zone in the latitude range 60° to 75°N (more or less), centered about on the arctic circle, and lying between the subarctic zone and the polar zone.

arcuate delta Variety of delta with shoreline curved convexly outward from the land.

areal geologic map Map that depicts the surface exposure and boundaries of various rock units and other geologic features, such as faults.

area strip mining Form of strip mining practiced in regions where a horizontal coal seam lies beneath a land surface that is approximately horizontal.

Arents Suborder of the Entisols with recognizable fragments of pedogenic horizons that have been mixed by deep plowing, spading, or other disturbance by humans.

arête Sharp, knifelike divide or crest formed between two cirques by alpine glaciation.

Argids Suborder of the Aridisols having an argillic horizon, formed largely on surfaces of late Pleistocene (Wisconsinan) age and older, with illuviation of the B horizon having taken place during pluvial periods.

argillans Cutans that are formed of clay; they can be found in argillic horizons.

argillic horizon Illuvial soil horizon, usually the B horizon, in which lattice-layer clay minerals have accumulated by illuviation.

aridic regime Soil-water regime for warm soils in which the soil is never moist in some or all parts for as long as 90 consecutive days. (Also called *torric regime*.)

Aridisols Soil order within the Soil Taxonomy consisting of soils of dry climates, with or without argillic horizons, and with accumulations of carbonates or soluble salts.

artesian flow Spontaneous rise of water in a well or fracture zone above the level of the surrounding water table.

artesian well Drilled well in which water rises under hydraulic pressure above the level of the surrounding water table and may reach the surface.

artificial levee Earth embankment built parallel with an alluvial river channel, usually on the crest of the natural levee, to contain the river in flood stage. (Also called a *dike*.)

aspect Compass orientation of a slope as an inclined element of the ground surface.

asthenosphere Soft layer of the upper mantle, beneath the rigid lithosphere. Rock of the asthenosphere is close to its melting point and has low strength.

astronomical hypothesis Explanation for glaciations and interglaciations making use of cyclic variations in the form of the earth's orbit and the angle of inclination of the earth's axis as controls of cyclic variations in the intensity of solar energy received at the earth's surface.

astronomical seasons Spring, summer, autumn, and winter; subdivisions of the tropical year defined by the successive occurrences of vernal equinox, summer solstice, autumnal equinox, and winter solstice.

Atlantic stage Warm climate stage of the Holocene Epoch, spanning a period about from −8000 to −5000 years. (See also *climatic optimum*.)

atmosphere Envelope of gases surrounding the earth, held by gravity.

atmospheric pressure Pressure exerted by the atmosphere because of the force of gravity acting on the overlying column of air.

atoll Circular or closed-loop coral reef enclosing a lagoon of open water with no central island.

aurora australis Southern hemisphere counterpart of the *aurora borealis*.

aurora borealis Shifting light bands, rays, or draperies, emanating from the ionosphere and appearing at night in the northern sky.

Australian sclerophyll forest Formation class of vegetation, a subtype of the sclerophyll forest found in Australia, and including forest, woodland, and scrub dominated by several species of eucalypts and acacias.

Australian sclerophyllous tree savanna Formation class subtype of the savanna trees, widespread in eastern Australia.

autogenic succession Form of ecological succession that is self-producing, i.e., the result of the action of the plants and animals themselves. (See also *old-field succession*.)

autumnal equinox Equinox occurring on September 22 or 23, when the sun's declination is 0°.

available landmass Landmass above base level that can be consumed by fluvial denudation.

axial rift Narrow, trenchlike depression situated along the center line of the midoceanic ridge and identified with active seafloor spreading.

azimuth Direction referred to a circular scale of degrees read clockwise and ranging from 0° to 360°. (See also *magnetic azimuth*, *true azimuth*.)

Azores high Persistent cell of high atmospheric pressure located over the subtropical zone in the North Atlantic Ocean. (Sometimes also referred to as the *Bermuda high*.)

backarc basin Comparatively small ocean basin, underlain by oceanic crust and lithosphere, lying between an island arc and the continental mainland or between two island arcs. Example: Sea of Japan, Bering Sea.

backrush (See *backwash*.)

backswamp Area of low, swampy ground on the floodplain of an alluvial river between the natural levee and the bluffs.

backwash Return flow of swash water of a breaker on a beach under the influence of gravity.

badlands Rugged land surface of steep slopes, resembling miniature mountains, developed on weak clay formations or clay-rich regolith by fluvial erosion too rapid to permit plant growth and soil formation.

bajada Graded slope that is an alluvial fan or a pediment, extending from mountain base to playa in a mountainous desert region.

bank caving Incorporation of masses of alluvium or other weak bank materials into a stream channel because of undermining, usually in high flow stages.

bank-full stage Stream stage corresponding with flood stage, at which time discharge is contained entirely within the limits of the stream channel.

bar (See *baymouth bar*, *cuspate bar*, *offshore bar*.)

bar (pressure) Unit of pressure or capillary tension equal to 1 million dynes per sq cm; it is approximately equal to the pressure of the earth's atmosphere at sea level.

bar-and-swale Assemblage of floodplain landforms consisting of alternate low ridges (bars) and low troughs (swales), formed as point-bar deposits on the outsides of alluvial meanders.

barchan dune Sand dune of crescentic base outline with sharp crest and steep lee slip face, with crescent points (horns) pointing downwind.

barometer Instrument for measurement of atmospheric pressure.

barometric pressure (See *atmospheric pressure*.)

barrier island Long narrow island, built largely of beach sand and dune sand, parallel with the mainland and separated from it by a lagoon.

barrier-island coast Coast with broad zone of shallow water offshore (a lagoon) shut off from the ocean by a barrier island.

barrier reef Coral reef separated from mainland shoreline by a lagoon.

basalt Extrusive igneous rock of gabbro composition; occurs as lava flows, dikes, shield volcanoes, and cinder cones.

base cations Certain cations in the soil solution that are also plant nutrients; the most important are cations of calcium, magnesium, potassium, and sodium.

base flow That portion of the discharge of a stream contributed by ground-water seepage and interflow.

base level Lower limiting surface or level that can ultimately be attained by a stream under conditions of stability of the earth's crust and sea level; an imaginary surface equivalent to sea level projected inland.

base line Reference parallel used in the U.S. Land Office Survey.

base status of soils Quality of a soil as measured by the percentage base saturation (PBS); soils with PBS greater than 35 percent are soils of high base status; those with PBS less than 35 percent are soils of low base status.

batholith Large, deep-seated pluton (body of intrusive igneous rock) with an area of exposure greater than 100 sq km, and usually consisting of coarse-grained granitic rock.

bauxite Mixture of several residual clay minerals, largely aluminum oxides and hydroxides, with various impurities; a principal ore of aluminum.

baymouth bar A low ridge of sand built above water level across the mouth of a bay by littoral drift and wave action; a variety of beach.

beach Wedge-shaped accumulation of sand, gravel, or cobbles in the foreshore and offshore zones of breaking and shoaling waves. (See also *pocket beach*.)

beach drift Transport of sand in the foreshore zone of a beach parallel with the shoreline by an alternating succession of landward and seaward water movements at times when the swash approaches obliquely.

beach ridges Low coastal sand ridges, representing berms produced in succession during progradation of a beach.

bearing (See *compass quadrant bearings.*)

bedded chert Uniform layers of chert, usually of marine origin, separated by thin dark bands of shale.

bedding planes Planes of separation between individual strata in a sequence of sedimentary rocks.

bed load That portion of the stream load moving close to the stream bed by rolling, sliding, and low leaps.

bedrock Solid rock in place with respect to the surrounding and underlying rock and relatively unaltered or softened by weathering processes.

bench mark A surface point inscribed on a permanent monument to provide a fixed reference of elevation above the datum plane.

berm Low embankment or ridge on a sand beach constructed by swash of breaking waves. (See also *summer berm, winter berm.*)

Bermuda high (See *Azores high.*)

B horizon Mineral soil horizon located beneath the A horizon, and usually characterized by a gain of mineral matter (such as clay minerals and oxides of aluminum and iron) and organic matter (humus).

bioclimatic frontier Geographic boundary corresponding with a critical limiting level of climatic stress, beyond which a species cannot survive.

biogas Mixture of methane and carbon dioxide generated by action of anaerobic bacteria in animal and human wastes enclosed in a digesting chamber.

biogenic sediments (See *organically derived sediment.*)

biogeochemical cycle (See *material cycle.*)

biogeography Study of the distribution patterns of plants and animals on the earth's surface and the processes that produce those patterns.

biomass Dry weight of living organic matter in an ecosystem within a designated surface area; units are grams of organic matter per square meter.

biomass energy Energy derived for human use from natural biomass sources, such as combustible plant tissues or the fermentation of plant matter to produce alcohol or methane gas.

biome Largest recognizable subdivision of the terrestrial ecosystems, including the total assemblage of plant and animal life interacting within the life layer.

biosphere All living organisms of the earth and the environments with which they interact.

biotic communities Local associations of plants and animals that are interdependent and often found together.

bird-foot delta Delta with long projecting distributary fingers extending out into open water.

bitumen Combustible mixture of hydrocarbons that is highly viscous and will flow only when heated; considered a form of petroleum. Enclosed in sand, the mixture is known as *bituminous sand* or *tar sand.*

bituminous coal Grade of coal with substantial content of volatiles; also called "soft coal."

bituminous sand (See *bitumen.*)

black body Ideal object or surface that is a perfect radiator of energy and will absorb all radiation falling on it.

block separation Separation of individual joint blocks during the process of physical weathering.

blowout Shallow depression produced by continued deflation; also called a *deflation hollow.*

bluffs Steeply rising slopes or scarps, marking the outer limits of a floodplain.

bog succession Ecological succession typical of a shallow glacial lake, in which the lake basin is eventually filled with freshwater peat.

bora Strong, cold wind occurring during the winter along the Adriatic coastal region when a strong pressure gradient exists.

Boralfs Suborder of the Alfisols occurring in boreal forests or high mountains. They have a gray surface horizon, a brownish subsoil, and are associated with a mean annual soil temperature less than 8°C (47°F).

boreal forest Variety of needleleaf forest found in the boreal forest climate regions of North America and Eurasia.

boreal forest climate Cold climate of the subarctic zone in the northern hemisphere with long, extremely severe winters and several consecutive months of zero potential evapotranspiration (water need).

Boreal stage First (earliest) climate stage of the Holocene Epoch.

bornhardt Prominent knob of massive granite or similar plutonic rock with rounded summit and often showing exfoliation shells.

Borolls Suborder of the Mollisols found in cold-winter steppes (semiarid plains) or high mountains, where mean annual soil temperature is lower than 8°C (47°F).

braided channel (See *braided stream.*)

braided stream Stream with shallow channel in coarse alluvium carrying multiple deeper threads of faster flow, subdividing and rejoining repeatedly and continually shifting in position.

breaker Sudden collapse of a steepened water wave as it approaches the shoreline.

breccia General term for a rock consisting of angular rock fragments in a matrix of finer particles. (See *sedimentary breccia, volcanic breccia.*)

broadleaf Leaf form that is wide in relation to length, thin, and comparatively large.

broadleaf evergreen forest Formation class in the forest biome consisting of broadleaf evergreen trees and found in the moist subtropical climate and in parts of the marine west-coast climate. (Also known as *temperate rainforest.*)

broadleaf evergreen rainforest General term for equatorial and tropical (low-latitude) rainforests. (See *equatorial rainforest, tropical rainforest.*)

bryoids Group of low, small plants that lie close to the ground or are attached to tree trunks; most are mosses and liverworts.

bush-fallow farming Agricultural system practiced in the African savanna woodland in which trees are cut and burned to provide cultivation plots.

butte Prominent, steep-sided flat-topped hill or mountain, often representing the final erosional remnant of a resistant rock layer in a region of horizontal strata.

caatinga Region of thorntree semidesert in northeastern Brazil.

calcic horizon Soil horizon of accumulation of calcium carbonate or magnesium carbonate.

calcification Accumulation of calcium carbonate in a soil, usually occurring in the B horizon or in the C horizon below the soil solum.

calcite Mineral of calcium carbonate composition with perfect rhombohedral cleavage, the principal constituent of limestone.

calcrete Rocklike layer rich in calcium carbonate, formed below the soil solum in the C horizon. (See also *caliche, petrocalcic horizon.*)

caldera Large, steep-sided circular depression resulting from the terminal explosion and subsidence of a large composite volcano.

caliche Name applied in the southwestern United States to the petrocalcic horizon of the soil, associated with Aridisols. (See also *calcrete.*)

calorie Standard energy unit. Gram calorie is the quantity of heat needed to raise the temperature of one gram of water by 1°C. Kilocalorie, or kilogram calorie (large calorie), equals 1000 gram calories. (See also *joule.*)

calving Breaking off of blocks of glacial ice from a glacier terminus that extends into tidewater; the process by which icebergs are formed.

cambic horizon Altered soil horizon that has lost sesquioxides or bases, including carbonates, through leaching.

campo cerrado Region of savanna biome in the interior Brazilian highlands of South America.

Canadian high Region of above-average atmospheric pressure centered over central North America during the winter.

canyon (See *gorge*.)

capable fault Fault that has exhibited movement at or near the ground surface at least once within the past 35,000 years or movement of a recurring nature within the past 500,000 years.

capillary fringe Saturated layer lying at the top of the ground water zone and consisting of water held by capillary tension above the level that would represent a surface of hydrostatic equilibrium under gravity alone. (See also *capillary water*.)

capillary water Water in the unsaturated zone and soil-water zone held in the form of capillary films attached to mineral particles.

carbohydrate Class of organic compounds consisting of the elements carbon, hydrogen, and oxygen.

carbonates (carbonate minerals) Minerals that are carbonate compounds of calcium or magnesium or both, i.e., calcite (calcium carbonate) or dolomite (magnesium carbonate).

carbonation Chemical reaction of carbonic acid in rainwater, soil water, and ground water with minerals; most strongly affects carbonate minerals and rocks, such as limestone and marble; an activity of chemical weathering.

carbon cycle Material cycle in which carbon moves through the biosphere; includes both gaseous cycles and sedimentary cycles.

carbonic acid Weak acid formed by the solution of atmospheric carbon dioxide gas in surface water and ground water.

cation Positively charged ion.

cation exchange Replacement of certain cations by other cations on the surfaces of clay mineral particles of colloidal dimensions, following a replacement order.

cation exchange capacity (CEC) Capacity of a given quantity of soil to hold and to exchange cations.

caverns Subterranean systems of interconnected large cavities formed in limestone by the carbonic acid action (carbonation) of circulating ground water.

Celsius scale Temperature scale in which the freezing point of water is 0°, the boiling point 100°.

Cenozoic Era Last (youngest) of the eras of geologic time.

central eye Cloud-free central vortex of a tropical cyclone.

chalcedony Variety of mineral quartz composition silica (silicon dioxide), microcrystalline in structure.

chalk Variety of limestone that is soft, earthy, and white, formed of hard parts of various planktonic marine organisms (foraminifera, algae).

channel flow (See *stream flow*.)

chaparral Sclerophyllous scrub and dwarf forest plant formation class found throughout the coastal mountain ranges and hills of central and southern California.

chatter marks Curved impact fractures produced by ice pressure on the surface of bedrock subjected to abrasion by moving glacial ice. The impacts are made by rock masses held in the ice.

chemical energy Energy stored within an organic molecule and capable of being transformed into heat during metabolism.

chemical pollutants Gases introduced into the atmosphere from industrial activities, fuel combustion, and other human activities; not included are the normal gaseous constituents of pure dry air.

chemical precipitate Sediment consisting of mineral matter chemically precipitated from a water solution in which the matter has been transported in the dissolved state as ions.

chemical weathering Chemical change in rock-forming minerals through exposure to atmospheric conditions in the presence of water; mainly involving oxidation, hydrolysis, carbonic acid reaction, and direct solution.

chert Variety of sedimentary rock composed largely of chalcedony and various impurities, in form of nodules and layers. (See also *bedded chert*.)

chinook winds One of a group of local winds, similar to the foehn, occurring at certain times to the lee of the Rocky Mountains; a very dry wind with a high capacity to evaporate snow.

C horizon Soil horizon lying beneath the soil solum (A and B horizons); it is a layer of sediment or regolith that is the parent material of the solum.

cinder cone Conical hill built of coarse tephra ejected from a narrow volcanic vent; a type of volcano.

circle of illumination Great circle that divides the globe at all times into a sunlit hemisphere and a shadowed hemisphere.

circum-Pacific belt Chain of island arcs and mountain arcs surrounding the Pacific Ocean basin and formed by volcanic and tectonic activity in Cenozoic time.

cirque Bowl-shaped bedrock depression holding the collecting ground and firn of an alpine glacier.

cirrocumulus High cloud formed of ice particles and shaped into small patches arranged in geometrical patterns.

cirrostratus High clouds formed of ice particles appearing as a whitish veil and producing a halo around the sun or moon.

cirrus High cloud formed of ice and shaped into delicate white filaments, streaks, or narrow bands.

clastic sediment Sediment consisting of particles broken away physically from a parent rock source.

clay Sediment size grade of particles smaller than 0.004 mm diameter. (May also be used to mean regolith or sediment consisting of clay minerals.)

clay minerals Class of minerals, produced by alteration of silicate minerals, having lattice-layer atomic structure and plastic properties when moist.

claystone Sedimentary rock formed by lithification of clay and lacking fissile structure.

cleavage of loess Natural vertical partings typical of thick deposits of loess.

cliff Sheer, near-vertical rock wall formed from flat-lying resistant layered rocks, usually sandstone, limestone, or lava flows. The term also may refer to any near-vertical rock wall. (See also *marine cliff*.)

climate Generalized statement of the prevailing weather conditions at a given place, based on statistics of a long period of record and including mean values, deviations from those means, and the probabilities associated with those deviations.

climate-process systems Concept of a number of geomorphic systems, each of which contains a unique combination of levels of intensity of several basic denudation processes.

climate types (climates) Varieties of climate recognized under a system of climate classification.

climatic optimum Past period of climate warmer than the present climate; usually in reference to the *Atlantic stage*.

climatology Science of climate.

climax Stable biotic community of plants and animals reached at the end point of ecological succession; end point of a *sere*.

climogenetic regions Concept of a classification of global land surfaces into several basic landscape types, each having a set of landforms uniquely dictated by climate. (Same as *morphogenetic regions*.)

climograph A graph on which two or more climatic variables, such as monthly mean temperature and monthly mean precipitation, are plotted for each month of the year.

clod Variety of ped (natural soil aggregate) caused by breakage during plowing.

closed depression Area of ground that is upwardly concave with all of its enclosing surface sloping inward.

closed system Flow system that is completely self-contained within a boundary through which no matter or energy is exchanged with the external environment. (Only the closed material system is capable of existence in the global realm.) (See also *open system*.)

cloud families Groups of cloud varieties defined in terms of height range or degree of vertical development.

cloud reflection Reflection of incoming shortwave radiation from the upper surfaces of clouds to space.

clouds Dense atmospheric concentrations of suspended water or ice particles in diameter range 20 to 50 microns.

cloud seeding Fall of ice crystals from the anvil top of a cumulonimbus cloud, serving as nuclei of condensation at lower levels. (Seeding may also be carried out artificially.)

coal Form of rock consisting of hydrocarbon compounds, formed of compacted, lithified, and altered accumulations of plant remains (peat).

coal measures Sequences of strata consisting of coal seams interbedded with layers of shale, sandstone, and limestone.

coal seam Miner's term for a single bed of coal.

coastal blowout dune High sand dune of the parabolic dune class formed adjacent to a beach, usually with a deep deflation hollow enclosed within the dune ridge.

coastal needleleaf evergreen forest Subtype of needleleaf evergreen forest found in the humid coastal zone of the northwestern United States and western Canada.

coastal plain Coastal belt, emerged from beneath the sea as a former continental shelf, underlain by strata with gentle dip seaward.

coastline (coast) Zone in which coastal processes operate or have a strong influence.

cockpit karst Variety of karst landscape consisting of numerous closely set hills, between which are numerous sinkholes.

col Natural pass or low notch in an *arête* between opposed cirques.

colatitude Angular value in degrees of arc equal to 90° minus the latitude.

cold-blooded animal Animal whose body temperature passively follows the temperature of the environment.

cold-core ring Circular eddy of cold water, surrounded by warm water and lying adjacent to a warm, poleward moving ocean current, such as the Gulf Stream. (See also *warm-core ring*.)

cold front Moving weather front

along which a cold air mass is forcing itself between a warm air mass, causing the latter to be lifted.

cold woodland Plant formation class consisting of woodland with low, widely spaced trees and a ground cover of lichens and mosses, found along the northern fringes of the region of boreal forest climate; also called *taiga*.

colloids (mineral) Mineral particles of extremely small size, capable of remaining indefinitely in suspension in water; typically in the form of thin plates or scales.

colluvium Deposit of sediment or rock particles accumulating from overland flow at the base of a slope and originating from higher slopes where sheet erosion is in progress. (See also *alluvium*.)

compass quadrant bearings System of direction measurement using four divisions of the compass referred to the four cardinal points.

competition (in ecosystem) Form of interaction between plant or animal species in which both draw resources from the same pool. Example: two animal species that feed from the same sources.

composite volcano Volcano constructed of alternate layers of lava and tephra (volcanic ash).

compound leaf Leaf form in which a single leaf consists of three or more separated leaflets, each joining a single main stem.

Comprehensive Soil Classification System (CSCS) Scheme of soil classification developed through the 1950s and 1960s by soil scientists of the U.S. Department of Agriculture and other scientists; formerly known as the Seventh Approximation. (See also *Soil Taxonomy*.)

condensation Process of change of matter in the gaseous state (water vapor) to the liquid state (liquid water) or solid state (ice).

conduction of heat Transmission of *sensible heat* through matter by transfer of energy from one atom or molecule to the next in the direction of decreasing temperature.

cone of depression Conical configuration of the lowered water table around a well from which water is being rapidly withdrawn.

confined aquifer Aquifer above which lies an impermeable rock layer (an aquiclude) commonly of clay or shale in a sedimentary sequence.

conformal projection Map projection that preserves without shearing the true shape or outline of any small surface feature of the earth.

congressional township Unit of land area, 6 miles on a side, used in the U.S. Land Office survey.

conic projections A group of map

projections in which the geographic grid is transformed to lie on the surface of a developed cone.

consequent stream Stream that takes its course down the slope of an initial landform, such as a newly emerged coastal plain or a volcano.

consumers Animals in the food chain that live on organic matter formed by primary producers or by other consumers. (See also *primary consumers*, *secondary consumers*.)

consumption of plate Destruction or disappearance of a subducting lithospheric plate in the asthenosphere, in part by melting of the upper surface, but largely by softening because of heating to the temperature of the surrounding mantle rock.

continental air mass Air mass developed over a continental source region.

continental collision Event in plate tectonics in which subduction brings two segments of lithosphere bearing continental crust into head-on contact, closing the intervening ocean basin and causing suturing with the formation of a continental suture.

continental crust Crust of the continents, of felsic composition in the upper part, thicker and less dense than oceanic crust.

continental drift Hypothesis introduced by Alfred Wegener and others early in the 1900s of the breakup of a parent continent, Pangaea, starting in the middle of the Mesozoic Era, and by drifting apart of the fragments resulting in the present arrangement of continents and intervening ocean basins.

continental glacier (See *ice sheet*.)

continentality Tendency of large land areas in midlatitudes and high latitudes to impose a large annual temperature range on the air temperature cycle.

continental lithosphere Lithosphere bearing continental crust. (See also *oceanic lithosphere*.)

continental margins (1) Topographic: one of three major divisions of the ocean basins, being the zones directly adjacent to the continent and including the continental shelf, continental slope, and continental rise. (2) Tectonic: marginal belt of continental crust and lithosphere that is in contact with oceanic crust and lithosphere, with or without an active plate boundary being present at the contact. (See also *active continental margins*, *passive continental margins*.)

continental rise Gently sloping seafloor lying at the foot of the continental slope and leading gradually into the abyssal plain of the deep ocean floor.

continental rupture Extension accompanied by up-doming affecting the continental lithosphere and its crust so as to cause a rift-valley system to appear and to widen, eventually creating a new belt of oceanic lithosphere and a new ocean basin.

continental shelf   Shallow, gently sloping belt of seafloor adjacent to the continental shoreline and terminating at its outer edge in the continental slope.

continental shelf wedge   A thick, wedge-shaped body of sediments formed by deposition on a subsiding passive continental margin, in shallow water of the continental shelf.

continental shields   Areas of continental crust underlain largely by deformed felsic igneous (granitic) and metamorphic rocks and mostly of Precambrian age.

continental slope   Steeply descending belt of seafloor between the continental shelf and the continental rise.

continental suture   Long, narrow zone of crustal deformation, including underthrusting and intense folding with nappes, produced by a continental collision. Examples: Himalayan Range, European Alps.

continent–continent collision   Collision between two large masses of continental lithosphere along a subduction plate boundary, resulting in a continental suture.

continents   High-standing areas of continental lithosphere capped by continental crust. (See also *ocean basins.*)

contour interval   Vertical distance represented by two adjacent contour lines on a topographic map.

contour line   Line drawn on a topographic map to pass through all surface points having the same altitude or height above a datum plane.

contour strip mining   Form of strip mining practiced in hilly or mountainous regions where coal seams form natural outcrops along the contour of the hillslopes.

convection (atmospheric)   Air motion consisting of strong updrafts taking place within a convection cell. General term for any overturning motions within fluids. Also applied to slow rising and sinking motions within the asthenosphere and deeper mantle.

convection cell   Individual column of strong updrafts produced by atmospheric convection, usually associated with a thunderstorm.

converging plate boundary   Boundary along which two lithospheric plates are coming together, requiring one plate to pass beneath the other by subduction; same as *subduction boundary.*

coordinate system   In cartography, any system of fixed intersecting lines used to designate the positions of points. Example: geographic coordinates of the geographic grid.

coral reef   Rocklike accumulation of carbonate mineral matter secreted by corals and algae in shallow water along a marine shoreline.

coral-reef coast   Coast built outward and upward by accumulations of limestone in coral reefs.

corange lines   Lines on a map (isopleths) drawn through all points having the same annual temperature range.

core of earth   Spherical central mass of the earth composed largely of iron and consisting of an outer liquid zone and an interior solid zone; it is surrounded by the mantle.

core of jet stream   Line of greatest air speed in a jet stream.

Coriolis effect   Effect of the earth's rotation tending to turn the direction of motion of any object or fluid toward the right in the northern hemisphere and to the left in the southern hemisphere.

corrosion   Erosion of bedrock of a stream channel (or other rock surfaces) by chemical reactions between solutions in stream water and mineral surfaces.

counterradiation   Longwave radiation of atmosphere directed downward to the earth's surface.

covered shields   Areas of continental shields in which the ancient rocks are covered beneath a veneer of younger sedimentary strata that show little disturbance by tectonic activity.

crater (volcanic)   Central summit depression associated with the principal vent of an active volcano.

creep   In mass wasting, the extremely slow downslope motion of soil (soil creep) or of rock fragments (rock creep) under gravity because of continual or seasonal agitation of the particles. (See *soil creep.*)

crescent dune   (See *barchan dune.*)

crescentic gouge   Form of curved fracture on a bedrock surface affected by glacial abrasion; it is convex toward the downstream direction of ice motion.

crevasse   Gaping crack in the brittle surface ice of a glacier.

crevice (marine)   Narrow, cleftlike cavity eroded into a marine cliff by the action of breaking storm waves.

crude oil   Liquid form of petroleum, as distinguished from natural gas and bitumen or asphalt.

crust of earth   Outermost solid shell or layer of earth, composed largely of silicate minerals and ranging in rock composition from felsic rocks in the upper continental crust to mafic rocks in the oceanic crust. (See also *continental crust, oceanic crust.*)

cryic regime   Soil-temperature regime in which mean annual soil temperature is greater than 0°C, but less than 8°C.

crystal lattice   The three-dimensional internal structural arrangement of atoms or ions that forms a crystalline mineral.

cuesta   Erosional landform developed on resistant strata having low to moderate dip and taking the form of an asymmetrical low ridge or hill belt with one side a steep scarp and the other a gentle slope; usually associated with a coastal plain.

cultural energy   Energy in forms exclusive of solar energy of photosynthesis that is expended on the production of raw food or feed crops in agricultural ecosystems.

cumuliform cloud   Cloud of globular shape, often with extended vertical development.

cumulonimbus cloud   Large, dense cumuliform cloud yielding precipitation.

cumulus   Cloud type consisting of low-lying, white cloud masses of globular shape well separated from one another.

current meter   A device to measure the velocity of flow of water at a given point in a stream.

cuspate bar   Low coastal sand ridge projecting seaward in a toothlike form; a form of beach.

cuspate delta   Delta with a shoreline sharply pointed toward the open water.

cuspate foreland   Accumulation of beach ridges projecting seaward in a toothlike or arcuate form, formed by progradation.

cutan   Thin film or coating on a soil ped or an individual coarse mineral grain; commonly such cutans are clay skins (*argillans*).

cuticle   Outermost protective cell layer of a leaf.

cutoff of meander   Cutting-through of a meander neck so as to bypass the stream flow in an alluvial meander and cause it to be abandoned.

cut-off high   Upper-air anticyclone formed by occlusion of an upper-air wave, or Rossby wave.

cut-off low   Upper-air cyclone formed by occlusion of the trough of an upper-air wave, or Rossby wave.

cycle of denudation   Concept of an orderly sequence of evolutionary stages of fluvial denudation in which relief of the landmass declines with time to reach a late stage, when the landscape becomes a peneplain.

cycle of rock transformation   (See *rock transformation cycle.*)

cyclone   Center of low atmospheric pressure. (See also *tropical cyclone, wave cyclone.*)

cyclone family   Succession of wave cyclones tracking eastward along the polar front while developing from open stage to occluded stage.

cyclonic storm Intense weather disturbance within a moving cyclone generating strong winds, cloudiness, and precipitation.

cylindric projections Group of map projections in which the geographic grid is transformed to lie on the surface of a developed cylinder.

datum plane Horizontal plane of reference or zero altitude base, usually mean sea level, applied to a topographic map.

daylight saving time Time system under which time is advanced by one hour with respect to the standard time of the prevailing standard meridian.

debris avalanche (See *alpine debris avalanche*.)

debris flood Streamlike flow of muddy water heavily charged with sediment of a wide range of size grades, including boulders, generated by sporadic torrential rains on steep mountain watersheds.

decalcification Removal of calcium carbonate from a soil horizon or soil solum as carbonic acid reacts with carbonate mineral matter during time of soil water surplus.

deciduous plant Tree or shrub that sheds its leaves seasonally, i.e., a *tropophyte*.

declination (See *magnetic declination*.)

declination of sun Terrestrial latitude of the subsolar point, or point at which sun's rays are perpendicular at noon. (Also defined as sun's angular distance north or south of the celestial equator on the ecliptic circle in the celestial sphere.)

décollement Detachment and extensive sliding of a rock layer, usually sedimentary, over a near-horizontal basal rock surface; a special form of low-angle thrust faulting.

decomposers Organisms that feed on dead organisms from all levels of the food chain; most are microorganisms and bacteria that feed on decaying organic matter.

deep environment Environment of high pressure and high temperature to which rock is subjected deep within the earth's crust.

deflation Lifting and transport in turbulent suspension by wind of loose particles of soil or regolith from dry ground surfaces.

deflation hollow (See *blowout*.)

deglaciation Widespread recession of ice sheets during a period of warming global climate, leading to an interglaciation. (See also *glaciation, interglaciation*.)

degradation Downcutting by a stream, causing the stream channel to be lowered in altitude and resulting in trenching of alluvial deposits or cutting of a bedrock gorge.

delta Sediment deposit built by a stream entering a body of standing water and formed of the stream's load.

delta coast Coast bordered by a delta.

delta kame Flat-topped hill of stratified drift representing a glacial delta constructed adjacent to an ice sheet in a marginal glacial lake.

dendritic drainage pattern Drainage pattern of treelike branched form, in which the smaller streams take a wide variety of directions and show no parallelism or dominant trend.

denitrification Biochemical process in which nitrogen in forms usable to plants is converted into molecular nitrogen in the gaseous form and returned to the atmosphere—a process that is part of the nitrogen cycle.

density Quantity of mass per unit of volume, stated in g/cc.

denudation Total action of all processes of weathering, mass wasting, and erosion whereby the exposed rocks of the continents are worn down and the resulting sediments are transported to the sea by the fluid agents.

deposition (See *stream deposition*.)

depositional landforms Sequential landforms created by the deposition of sediment by a fluid agent of denudation.

depression contour (See *hachured contour*.)

depression storage (See *surface detention*.)

desalinization Natural process of removal of soluble salts from a soil.

desert biome Biome of the dry climates consisting of thinly dispersed plants that may be shrubs, grasses, or perennial herbs, but lacking in trees.

desert climate subtype Subtype of the dry climates in which no month has soil-water storage exceeding 2 cm.

desertification (See *land degradation*.)

desert pavement Surface layer of closely fitted pebbles or coarse sand from which finer particles have been removed by deflation.

desert varnish Dark iridescent coating found on rock surfaces in a desert climate.

detrital sediment Sediment consisting of mineral fragments derived by weathering of preexisting rock and transported to places of accumulation by currents of air, water, or ice.

dew-point temperature Temperature of saturated air.

diagenesis (See *lithification*.)

diagnostic horizons Certain soil horizons, rigorously defined, that are used as diagnostic criteria in classifying soils within the Soil Taxonomy.

diffuse reflection Form of scattering in which solar rays are deflected or reflected by minute dust particles or cloud particles.

digital image Numeric representation of a picture, consisting of a collection of numeric brightness values (*pixels*) arrayed in a fine grid pattern.

dike (on floodplain) (See *artificial levee*.)

dike (igneous) Thin layer of intrusive igneous rock, often nearly vertical or with steep dip, occupying a widened fracture in the surrounding older rock and typically cutting across the older rock planes.

diorite Intrusive igneous rock consisting dominantly of intermediate plagioclase feldspar and pyroxene, with some amphibole and biotite; a felsic igneous rock, it occurs as a pluton.

dip Acute angle between an inclined natural rock plane or surface and an imaginary horizontal plane of reference; always measured perpendicular to the strike. (Also a verb, meaning "to incline toward.")

discharge Volume of water flow moving through a given cross section of a stream in a given unit of time; commonly given in cubic meters per second.

distributary channels Branching stream channels of a delta, terminating in open water of a lake or ocean.

doldrums Belt of calms and variable winds occurring at times along the equatorial trough.

dolomite Carbonate mineral or sedimentary rock having the composition calcium magnesium carbonate.

dome (See *sedimentary dome*.)

dornveldt Local name for a region of thorntree semidesert in South Africa.

down-scatter Scattered shortwave radiation directed earthward within the atmosphere.

downvalley sweep Slow downvalley migration of meander loops on the floodplain of an alluvial river.

drainage basin Total land surface occupied by a drainage system, bounded by a drainage divide or watershed.

drainage system A branched network of stream channels and adjacent ground slopes, bounded by a drainage divide and converging to a single channel at the outlet.

drainage winds Winds, usually cold, that flow from higher to lower regions under the direct influence of gravity; synonymous with *katabatic winds*.

drawdown Difference in height between base of cone of depression and original water table surface.

drifts (in mines) Horizontal tunnels or passageways in mines.

dripstone Mineral deposit in limestone cave formed by dripping water. (See also *stalactite, stalagmite*.)

drought Occurrence of substantially lower-than-average precipitation in a season that normally has ample precipitation for the support of food-producing plants.

drumlin Hill of glacial till, oval or elliptical in basal outline, and with smoothly rounded summit, formed by plastering of till beneath moving, debris-ladened glacial ice.

dry adiabatic lapse rate Rate at which rising air is cooled by expansion when no condensation is occurring; 1.0 C°/100 m (5.5 F°/1000 ft).

dry climate Climate in which the total annual soil-water shortage is 15 cm (5.9 in.) or greater, and there is no water surplus.

dry desert Plant formation class in the desert biome consisting of widely dispersed xerophytic plants, including hardleaved or spiny shrubs, succulent plants (cacti), or hard grasses.

dry mid-latitude climate Dry climate of the midlatitude zone with a strong annual cycle of potential evapotranspiration (water need) and cold winters.

dry subtropical climate Dry climate of the subtropical zone, transitional between the dry tropical climate and the dry midlatitude climate.

dry tropical climate Dry climate of the tropical zone with large total annual potential evapotranspiration (water need).

dune (See *sand dune.*)

dunite Ultramafic igneous rock consisting almost entirely of olivine.

duripan Dense, hard subsurface soil horizon cemented by silica that does not soften during prolonged soaking.

dust storm Heavy concentration of dust in a turbulent air mass, often associated with a cold front.

earthflow Moderately rapid downhill flowage of masses of water-saturated regolith, clay, or weak shale, typically forming a steplike terrace at the top and a bulging toe at the base.

earthquake Trembling or shaking of the ground produced by the passage of seismic waves.

earthquake epicenter Ground surface point directly above the earthquake focus.

earthquake focus Point within the earth at which the energy of an earthquake is first released by rupture and from which seismic waves emanate.

earthquake intensity scale Scale of numbers indicating the intensity of earth shaking felt at a particular place during an earthquake. (See *modified Mercalli scale.*)

earthquake magnitude Strain energy released as kinetic energy by a single earthquake at its focus, in units of joules or ergs. (See also *Richter scale.*)

earth rotation Spinning of the earth on its axis.

easterly wave Weak, slowly moving trough of low pressure within the belt of tropical easterlies; it causes a weather disturbance with rain showers.

ebb current Oceanward flow of tidal current in a bay or tidal stream.

ecological succession Time-succession (sequence) of distinctive plant and animal communities occurring within a given area of newly formed land or land cleared of plant cover by burning, clear cutting, or other agents.

ecology Science of interactions between life forms and their environment; the science of ecosystems.

ecosystem Group of organisms and the environment with which the organisms interact.

edaphic factors Factors relating to soil that influence a terrestrial ecosystem.

effluent stream Stream that receives an increment of flow by out-seepage of ground water.

E horizon Mineral horizon of the soil solum lying below the A horizon and characterized by the loss of clay minerals and oxides of iron and aluminum; it may show a concentration of quartz grains and is often pale in color.

electromagnetic radiation Wavelike form of energy radiated by any substance possessing heat; it travels through space at the speed of light.

elfin forest High-altitude variety of forest found in the tropical zone and equatorial zone; it consists of dwarfed trees festooned with mosses.

El Niño Episodic cessation of the typical upwelling of cold deep water off the coast of Peru; literally, "The Christ Child," for its occurrence in the Christmas season, once every few years.

eluviation Pedogenic process consisting of the downward transport of fine particles, particularly the colloids (both mineral and organic), carrying them out of an upper soil horizon.

emergence (coastal) Exposure of submarine landforms by a lowering of sea level or a rise of the crust, or both.

end moraine (See *terminal moraine.*)

energy deficit Condition in which rate of outgoing radiant energy exceeds rate of incoming radiant energy for a unit surface area at a given time and place.

energy flow system Open system that receives an input of energy, undergoes internal energy flow, energy transformation, and energy storage, and has an energy output.

energy subsystem Energy system completely contained within a larger energy system.

energy surplus Condition in which rate of incoming radiant energy exceeds the rate of outgoing radiant energy for a unit surface area at a given time and place.

ENSO Acronym for "El Niño" and "Southern Oscillation."

Entisols Soil order in the Soil Taxonomy, consisting of mineral soils lacking soil horizons that would persist after normal plowing.

entrainment Drawing in of surrounding air by a rapidly rising, bubblelike air body within a thunderstorm cell.

entrenched meanders Winding, sinuous valley produced by degradation of a stream with trenching into the bedrock by downcutting.

environmental temperature lapse rate Rate of temperature decrease upward through the troposphere; standard value is 6.4 C°/km (3½ F°/1000 ft).

epeirogenic movement (epeirogeny) Slow rising or sinking of the crust over a large area, without appreciable faulting or folding.

ephemeral annuals Small desert plants that complete a life cycle very rapidly following a desert downpour.

epicenter (See *earthquake epicenter.*)

epilimnion Uppermost, warm water layer of a lake.

epipedon Soil horizon that forms at the surface.

epiphytes Plants that live above ground level out of contact with the soil, usually growing on the limbs of trees or shrubs; also called "air plants."

epoch of geologic time Geologic time unit, a subdivision of a period.

equal-area projections Class of map projections on which any given area of the earth's surface is shown to correct relative areal extent, regardless of position on the globe.

equatorial air mass Warm, moist air mass with source region over an ocean in the equatorial zone.

equatorial countercurrent Narrow ocean current, flowing west to east between the two equatorial currents.

equatorial current West-flowing ocean current in the belt of the trade winds.

equatorial easterlies Upper-level easterly air flow over the equatorial zone.

equatorial rainforest Plant formation class within the forest biome, consisting of tall, closely set broadleaf trees of evergreen or semideciduous habit.

equatorial trough Atmospheric low-pressure trough centered more or less over the equator and situated between the two belts of trade winds.

equatorial zone Latitude zone lying between lat. 10°S and 10°N (more or less) and centered on the equator.

equinox Instant in time when the

subsolar point falls on the earth's equator and the circle of illumination passes through both poles. Vernal equinox occurs on March 20 or 21; autumnal equinox on September 22 or 23.

era Major subdivision of geologic time consisting of a number of geologic periods. The three eras following *Precambrian time* are *Paleozoic, Mesozoic,* and *Cenozoic.*

erg Large expanse of active dunes in the Sahara Desert of North Africa.

erosion General term for the removal of mineral particles from exposed surfaces of bedrock or regolith by impact of a fluid—water, air, or ice—and by impact of solid particles carried by a fluid (*abrasion*). See also (*stream erosion, wind abrasion.*)

erosional landforms Class of the sequential landforms shaped by the removal of regolith or bedrock by agents of erosion. Examples: canyon, glacial cirque, *marine cliff.*

erratic boulder Boulder carried by glacial ice far from the place of its origin from a bedrock outcrop.

esker Narrow, often sinuous embankment of coarse gravel and boulders deposited in the bed of a meltwater stream enclosed in a tunnel within stagnant ice of an ice sheet.

estuarine delta Delta built into the lower part of an estuary.

Eurasian-Indonesian belt Major mountain arc system extending from southern Europe across southern Asia and Indonesia.

eutrophication Excessive growth of algae and other primary producers in a stream or lake as a result of the input of large amounts of nutrient ions, especially phosphate and nitrate.

evaporating pan Straight-sided, circular container in which changes of level in water surface are measured to determine quantity of evaporation.

evaporation Process in which a substance (such as water) in the liquid state or solid state passes into the vapor state.

evaporites Chemically precipitated sediments and sedimentary rocks composed of soluble salts deposited from evaporating saltwater bodies. Example: halite.

evapotranspiration Combined water loss to the atmosphere by evaporation from the soil and transpiration from plants.

evergreen leafless plants Evergreen plants with fleshy stems but no functional leaves. Example: the cacti.

evergreen plant Tree or shrub that holds most of its green leaves throughout the year.

evergreen-succulent plant Evergreen plant with thick, fleshy leaves or stems.

exfoliation Development of sheets, shells, or scales of rock on exposed bedrock outcrops or boulders, usually in concentric spheroidal arrangements. (See also *exfoliation dome, spheroidal weathering.*)

exfoliation dome Smoothly rounded rock knob or hilltop bearing rock sheets or shells produced by spontaneous expansion accompanying unloading.

exotic river Stream that flows across a region of dry climate and derives its discharge from adjacent uplands where a water surplus exists.

exponential decay Program of decrease in a variable quantity at a rate such that the quantity is halved in a constant time interval called the *half-life.*

exponential decay system Any physical system in which an initial quantity diminishes with time according to a negative exponential function.

exposed shields Areas of continental shields in which the ancient basement rock, usually of Precambrian age, is exposed to the surface.

external magnetic field Lines of force of the earth's magnetic field surrounding the earth's solid surface and extending far out into space.

extrusive igneous rock Rock produced by the solidification of lava or ejected igneous fragments (tephra). (See also *intrusive igneous rocks.*)

Fahrenheit scale Temperature scale in which the freezing point of water is 32°, the boiling point 212°.

fallout Gravity fall of atmospheric pollutant particulate matter, reaching the ground.

fault Sharp break in crustal rock with displacement (slippage) of crustal block on one side with respect to the adjacent block. (See *normal fault, overthrust fault, transcurrent fault, transform fault.*)

fault block Blocklike crustal mass lying between two parallel normal faults. (See also *graben, horst.*)

fault-block mountains Elevated, blocklike crustal masses bounded by normal faults; i.e., a succession of lifted blocks (horsts) or tilted blocks.

fault coast Coastline formed when the shoreline comes to rest against a fault scarp.

fault creep More or less continuous slippage on a fault plane, relieving some of the accumulated strain.

fault line Surface trace of a fault.

fault-line scarp Erosion scarp developed upon an inactive fault line.

fault plane Surface of slippage between two crustal masses moving relative to one another during faulting.

fault scarp Clifflike surface feature produced by faulting and exposing the fault plane; commonly associated with

a normal fault. (See also *fault-line scarp.*)

feldspar Aluminosilicate mineral group containing one or two of the metals potassium, sodium, or calcium. (See *plagioclase feldspar, potash feldspar.*)

fell-field In regions of tundra climate, a ground surface littered with rock fragments that may be formed into stone polygons.

felsenmeer Expanse of large blocks of rock produced by joint block separation and shattering by frost action at high altitudes or in high latitudes; also called a *boulder field.*

felsic igneous rock Igneous rock dominantly composed of felsic minerals.

felsic minerals, felsic mineral group Quartz and feldspars treated as a silicate mineral group showing pale colors and relatively low density. (See also *mafic minerals.*)

fenlands Diked salt marsh that has been converted to a freshwater environment, as in coastal southeast England.

ferricrete Rocklike surface layer rich in sesquioxide of iron (hematite); essentially the same as an exposed layer of laterite.

Ferrods Suborder of Spodosols that are freely drained and have B horizons of accumulation of free iron oxides.

fibric soil materials Diagnostic soil materials consisting of organic fibers readily identifiable as to botanical origin. Example: *Sphagnum* peat from bogs in cold climates.

Fibrists Suborder of Histosols that are composed mostly of *Sphagnum* moss or are saturated with water most of the year and consist of fibric soil materials.

field capacity (See *storage capacity.*)

filmy leaves Leaves that are thin and delicate, as compared with leaves described as membranous.

finger lake Long, narrow lake occupying part of the floor of a glacial trough; a variety of trough lake.

fiord Narrow, deep ocean embayment partially filling a glacial trough.

fiord coast Deeply embayed, rugged coastline formed by partial submergence of glacial troughs.

firn Granular old snow forming a surface layer in the zone of accumulation of a glacier.

firn field Field of glacier in which firn accumulates.

fissile Adjective describing a rock, usually shale, that readily splits up into small flakes or scales.

fixed dunes Sand dunes covered by protective vegetation and no longer in the active state.

flash flood Sudden and short-lived overbank stream discharge following torrential thunderstorm rain within a small watershed.

flocculation The clotting together of colloidal mineral particles to form larger particles; it occurs when a colloidal suspension in fresh water mixes with seawater.

flood Stream flow at a stage so high that it cannot be accommodated within the stream channel and must spread over the banks to inundate the adjacent floodplain.

flood basalts Large-scale outpourings of basalt lava to produce thick accumulations of basalt layers over a large area.

flood crest Maximum stage (surface elevation) attained during the passage of a flood.

flood current Landward flow of a tidal current.

floodplain Belt of low, flat ground, present on one or both sides of a graded stream channel, subject to inundation by a flood about once annually and underlain by alluvium.

flood stage Designated stream stage for a particular point on a stream, higher than which overbank flooding may be expected.

flood wave Time sequence of rising and falling stream stage during the passage of a flood.

flowstone Deposit of calcium carbonate left by flowing water on the walls or floor of a limestone cave.

fluid Substance that flows readily when subjected to unbalanced stresses; may exist as a gas or a liquid.

fluid agents Fluids that erode, transport, and deposit mineral and organic matter; they are running water, waves and coastal currents, glacial ice, and wind.

Fluvents Suborder of Entisols formed from recent alluvium, rarely saturated with water, usually stratified, and brownish to reddish throughout.

fluvial landforms Landforms shaped by running water.

fluvial processes Geomorphic processes in which running water is the dominant fluid agent, acting as overland flow and stream flow.

focus (See *earthquake focus*.)

foehn (föhn) Warm, dry wind produced to the lee of a mountain range as air descends and is adiabatically warmed; essentially the same phenomenon as chinook winds.

fog Cloud layer in contact with land or sea surface, or very close to that surface. (See *advection fog, radiation fog*.)

folding Process by which folds are produced; a form of tectonic activity.

folds Wavelike corrugations of strata (or other layered rock masses) as a result of crustal compression; a product

of tectonic activity. (See also *anticline, syncline*.)

foliation in rock Crude layering structure or parting imposed of rock during metamorphism; typical of *schist*.

Folists Suborder of Histosols formed with free drainage and consisting of forest litter resting on bedrock or rock rubble.

food chain Organization of an ecosystem into steps or levels through which energy flows as the organisms at each level consume energy stored in the bodies of organisms of the next lower level.

forb Broadleaved herb, as distinguished from the grasses.

foredunes Ridge of irregular sand dunes typically found adjacent to beaches on low-lying coasts and bearing a partial cover of plants.

foreland folds Folds produced on the continental foreland zone by arc-continent and continent-continent collisions.

foreshore Sloping face of a beach that is within the zone of swash and backwash. (See also *offshore*.)

forest Assemblage of trees growing close together, their crowns forming a layer of foliage that largely shades the ground.

forest biome Biome that includes all regions of forest over the lands of the earth.

formation classes Subdivisions within a biome based on the size, shape, and structure of the plants that dominate the vegetation.

formative elements Syllables used to form the names of soil orders, suborders, and great soil groups in the Soil Taxonomy.

fossil fuels Collective term for coal, petroleum, and natural gas capable of being utilized by combustion as energy sources.

Foucault pendulum Pendulum so designed as to be perfectly free to undergo a steady change in direction of swing as the earth rotates on its axis.

fractional scale Ratio of distance between two points on a map or a globe to the actual distance between the same two points on the earth's surface.

fracture zone (oceanic) Linear ocean-floor scarps or ridges offsetting the mid-oceanic ridge and its axial rift. Most of these features are now recognized as active *transform faults* or *transform scars*.

fragipan Dense, moderately brittle layer in the soil.

freezing Change from liquid state to solid state accompanied by release of latent heat of fusion to become sensible heat.

freshwater peat Form of peat produced by bog succession; often identi-

fied as the organic soil horizon of a Histosol.

frigid regime Soil temperature regime in which the mean annual soil temperature is below 8°C and the difference between means of warm season and cold season soil temperatures is greater than 5C°.

fringing reef Coral reef directly attached to land with no intervening lagoon of open water.

front Surface of contact between two unlike air masses. (See *cold front, occluded front, polar front, warm front*.)

frost (See *killing frost*.)

frost action Disintegration of rock or regolith by forces accompanying the freezing of water in pore spaces and fractures.

fumarole Vent in earth's surface emitting volcanic gases (mostly steam) at high temperature.

funnel cloud Long, narrow cloud of tubular or funnel shape hanging from the base of a cumulonimbus cloud; it represents a tornado or a waterspout.

gabbro Intrusive igneous rock consisting largely of pyroxene and calcic plagioclase feldspar, with variable amounts of olivine; a mafic igneous rock, occurring as a pluton.

gamma rays High-energy form of radiation at the extreme short wavelength (high-frequency) end of the electromagnetic spectrum.

gas, gaseous state Fluid of very low density (as compared with a liquid of the same chemical composition) that expands to fill uniformly any small container and is readily compressed.

gaseous cycle Type of material cycle in which an element or compound is converted into gaseous form, diffuses through the atmosphere, and passes rapidly over land or sea where it is reused in the biosphere.

geographic grid Complete network of parallels and meridians on the surface of the globe, used to fix the locations of surface points.

geographic north (See *true north*.)

geologic map Map showing the surface extent and boundaries of rock units of different ages or rock varieties.

geologic norm Stable natural condition in a moist climate in which slow soil erosion is paced by maintenance of soil horizons bearing a plant community in an equilibrium state.

geology Science of the solid earth, including the earth's origin and history, materials comprising the earth, and the processes acting within the earth and upon its surface.

geomorphology Science of landforms, including their history and processes of origin.

geopressurized zone Zone deep within the strata of a coastal plain where abnormally high fluid pressures exist in heated ground water.

geostationary orbit Satellite orbit that holds a fixed position over a selected point on the earth's equator.

geostrophic wind Wind at high levels above the earth's surface blowing parallel with a system of straight, parallel isobars.

geothermal energy Heat energy of igneous origin drawn from steam, hot water, or dry hot rock beneath the earth's surface.

geothermal gradient The rate at which temperature increases with increasing depth beneath the earth's solid surface.

geothermal locality Place where geothermal heat reaches the earth's surface, emanating from a deep-seated magma body in the crust.

geyser Periodic jetlike emission of hot water and steam from a narrow vent at a geothermal locality.

gilgae Small surface relief features of the soil that may be knobs and basins or narrow ridges with valleys between; typical of the Vertisols.

glacial abrasion Abrasion by a moving glacier of the bedrock floor beneath it.

glacial delta Delta built by meltwater streams of a glacier into standing water of a marginal glacial lake.

glacial drift General term for all varieties and forms of rock debris deposited in close association with Pleistocene ice sheets.

glacial plucking Removal of masses or fragments of bedrock from the rock floor beneath a glacier as the ice moves forward suddenly.

glacial trough Deep, steep-sided rock trench of U-shaped cross section formed by alpine glacier erosion.

glaciation (1) General term for the total process of growth of glaciers and the landform modifications they make. (2) Single episode or time period in which ice sheets form and spread widely, as contrasted with an interglaciation.

glacier Large natural accumulation of land ice affected by present or past flowage. (See also *alpine glacier, outlet glacier.*)

glacier terminus Lower end of an alpine glacier.

glaciofluvial sediment Sediment accumulation formed by meltwater streams issuing from a glacier terminus.

glaciolacustrine sediment Sediment that has accumulated on the floor of a marginal glacial lake.

glaze Ice layer accumulated on solid surfaces by the freezing of falling rain or drizzle.

global water balance Balance among the three hydrologic components—precipitation, evaporation, and runoff—for the earth as a whole.

gneiss Variety of metamorphic rock showing banding and commonly rich in quartz and feldspar.

gorge (canyon) Steep-sided bedrock valley with a narrow floor limited to the width of a stream channel.

graben Trenchlike depression representing the surface of a fault block dropped down between two opposed, infacing normal faults. (See also *rift valley.*)

graded profile Smoothly descending longitudinal profile displayed by a graded stream.

graded stream Stream (or stream channel) with stream gradient so adjusted as to achieve a balanced state in which average bed load transport is matched to average bed load input; an average condition over periods of many years' duration.

graminoid Having long, narrow leaves; for example, the leaves of grasses.

granite Intrusive igneous rock consisting largely of quartz, potash feldspar, and plagioclase feldspar, with minor amounts of biotite and hornblende; a felsic igneous rock, occurring as a large pluton (batholith or stock).

granitic rock General term for rock of the upper layer of the continental crust, composed largely of felsic igneous and metamorphic rock; rock of composition similar to that of granite.

granular disintegration Grain-by-grain breakup of the outer surface of coarse-grained rock, yielding sand and gravel and leaving behind rounded boulders.

graphic scale Map scale shown by a line marked off into length units.

grassland biome Biome consisting largely or entirely of herbs, which may include grasses, grasslike plants, and forbs.

gravitation Mutual attraction between any two masses.

gravity Gravitational attraction of the earth on any small mass near the earth's surface; also applies to the moon. In practical usage of geophysics, it is the acceleration of gravity (g) at the earth's surface.

gravity percolation Downward movement of water under the force of gravity through the soil-water belt and intermediate belt, eventually arriving at the water table.

gray body An object that does not behave like a perfect thermal radiator (*black body*), but instead radiates less energy than would be expected given the temperature of the body. Since a perfect thermal radiator exists only in theory, all real objects can be regarded as gray bodies.

great circle Circle formed by passing a plane through the exact center of a perfect sphere; the largest circle that can be drawn on the surface of a sphere.

great groups Third level of classification in the Soil Taxonomy.

greenhouse effect Accumulation of heat in the lower atmosphere through the absorption of longwave radiation from the earth's surface.

green revolution Major advance in securing increased agricultural production in the developing nations through improved genetic strains of wheat and rice and the application of cultural energy.

Greenwich meridian Meridian passing through the Royal Observatory at Greenwich, England, universally accepted as the prime meridian of zero longitude.

groin Human-made wall or embankment built out into the water at right angles to the shoreline to control beach erosion.

gross photosynthesis Total amount of carbohydrate produced by photosynthesis by a given organism or group of organisms in a given unit of time.

ground moraine Moraine formed of till distributed beneath a large expanse of land surface covered at one time by an ice sheet.

ground radiation Longwave radiation emitted by land or water surfaces and passing upward into the overlying atmosphere.

ground water Subsurface water occupying the saturated zone and moving under the force of gravity.

ground water recharge Replenishment of ground water by downward movement of water through the unsaturated zone, or from stream channels, or through recharge wells.

grus Loose gravel formed by granular disintegration of a coarse-grained felsic igneous rock, such as granite, in a dry climate.

guard cells Cells surrounding the openings of the stomata that can open and close those openings to regulate the flow of water vapor and other gases.

gullies Deep, V-shaped trenches carved by newly formed streams in rapid headward growth during advanced stages of accelerated soil erosion.

gypsic horizon Soil horizon of accumulation of hydrous calcium sulfate, the mineral gypsum.

gypsum Evaporite mineral, composition hydrous calcium sulfate.

gyre Large circular ocean current system centered on an oceanic subtropical high-pressure cell.

habitat Subdivision of the plant environment having a certain combination of slope, drainage, soil type, and other controlling physical factors.

hachured contour Contour line with attached hachures (ticks) used to denote a closed depression on a topographic map.

hachures Minute, short lines drawn on a topographic map to show direction and steepness of slope.

Hadley cell Atmospheric circulation cell in low latitudes involving rising air over an equatorial trough and sinking air over subtropical high-pressure belts.

hail Form of precipitation consisting of pellets or spheres of ice with a concentric layered structure.

hairpin dune Type of parabolic dune of highly elongate form.

half-life Time required for an initial quantity at time-zero to be reduced by one-half in an exponential decay system or program.

halite Evaporite mineral, composition sodium chloride; commonly known as rock salt.

halocarbons Synthetic compounds containing carbon, fluorine, and chlorine atoms; used as aerosol propellants and refrigerants.

hanging trough Tributary glacial trough with floor high above the floor of the main trough that it joins.

hanging valley Stream valley that has been truncated by marine erosion so as to appear in cross section in a marine cliff, or truncated by glacial erosion so as to appear in cross section in the upper wall of a glacial trough.

Hawaiian high Persistent cell of high atmospheric pressure located in the subtropical zone of the central and eastern North Pacific Ocean.

haze Minor concentration of pollutants or natural forms of particulate matter in the atmosphere causing a reduction in visibility.

heat (See *latent heat, sensible heat*.)

heat engine Mechanical system in which kinetic energy of motion is derived from heat energy.

heat island Persistent region of higher air temperatures centered over a city.

heavy minerals Group of minerals having exceptionally high density, usually 4 g/cc and greater, typically occurring in clastic waterlaid sediments.

heavy oil Asphalt-base crude oil of relatively high density.

hematite Mineral, composition sesquioxide of iron, a common product of chemical weathering of mafic minerals present in regolith.

hemic soil materials Organic matter of intermediate composition between fibric materials (peat) and sapric materials (decomposed organic matter).

Hemists Suborder of Histosols that

are saturated with water most of the year or artificially drained, with between 10 and 40 percent of the volume remaining as fibers after rubbing.

herb Tender plant, lacking woody stems, usually small or low; it may be annual or perennial, broadleaf (forb), or graminoid (grass).

herbaceous plant Plants that are herbs.

herbivory Form of interaction between species in which an animal (herbivore) grazes on a herbaceous plant.

hertz Wave frequency of one cycle per second.

hibernation Dormant state of some vertebrate animals during the winter season.

high base status (See *base status of soils*.)

high-latitude climates Group of climates in the subarctic zone, arctic zone, and polar zone, dominated by arctic air masses and polar air masses.

high water Highest water level reached by the ocean tide in a given tidal cycle.

hillslope Sloping land surface between a drainage divide and a stream channel on a landmass undergoing fluvial denudation.

hillslope profile Profile of a hillslope plotted along the path followed by overland flow from a drainage divide to the nearest stream channel.

histic epipedon Thin surface soil horizon (epipedon) consisting of peat.

Histosols Soil order within the Soil Taxonomy, consisting of soils with a thick upper layer of organic matter.

hogback Sharp-crested, often sawtooth ridge formed of the upturned edge of a resistant layer of sandstone, limestone, or lava.

Holocene Epoch Last epoch of geologic time, commencing about 10,000 years ago; it followed the Pleistocene Epoch and includes the present.

homolographic projection An equal-area map projection consisting of straight horizontal parallels and elliptical meridians.

homolosine projection Composite map projection consisting of the sinusoidal projection between lat. 40°N and 40°S and the homolographic projection poleward of those latitudes.

horizonation Degree of development of soil horizons as a result of complex combinations of the pedogenic processes.

horizon circle Circle representing the positions of all points equidistant from the center point of a zenithal projection.

horizontal strata Strata lying in approximately horizontal attitude, as typical of the covered continental shields.

horn Sharp, steep-sided mountain peak formed by the intersection of headwalls of glacial cirques.

horse latitudes Subtropical high-pressure belt of the North Atlantic Ocean, coincident with the central region of the Azores high; a belt of weak, variable winds and frequent calms.

horst Fault block uplifted between two normal faults.

hot spot Center of persistent volcanic activity usually located within a lithospheric plate and not related to volcanism of either a subduction boundary (volcanic arc) or an oceanic spreading boundary; postulated to be formed over a rising mantle plume. Examples: Hawaii, Yellowstone Park.

hot springs Springs discharging heated ground water at a temperature close to the boiling point; found in geothermal areas and thought to be related to a magma body at depth.

humid climate subtype Subtype of the moist climate in which the annual water surplus is 1 mm or greater but less than 60 cm, with annual water surplus always greater than the annual soil-water shortage.

humidity General term for the amount of water vapor present in the air. (See also *relative humidity, specific humidity*.)

humification Pedogenic process of transformation of plant tissues into humus.

Humods Suborder of the Spodosols that are freely drained and in which at least the upper part of the B horizon has accumulated humus and aluminum, but not iron.

Humox Suborder of Oxisols of relatively cool moist regions, with very large accumulations of organic carbon.

Humults Suborder of the Ultisols with large accumulations of organic matter formed under relatively high, well-distributed rainfall in middle and low latitudes.

humus Dark brown to black organic matter on or in the soil, consisting of fragmented plant tissues partly oxidized by soil organisms.

hurricane Tropical cyclone of the western North Atlantic and Caribbean Sea.

hydraulic action Stream erosion by impact force of the flowing water on alluvium or regolith exposed in the bed and banks of the stream channel.

hydraulic head Difference in level of the water table between one point and another, setting up a pressure difference and causing the flow of ground water.

hydrogenic sediments Sediments that are chemical precipitates from an aqueous solution.

hydrograph Graphic presentation of the variation in stream discharge with elapsed time, based on data of stream gauging at a given station on a stream.

hydrologic cycle   Total plan of movement, exchange, and storage of the earth's free water in gaseous, liquid, and solid states.

hydrology   Science of the earth's water and its motions through the hydrologic cycle.

hydrolysis   Chemical union of water molecules with minerals, to form different, usually more stable mineral compounds, classed as hydroxides.

hydrosphere   Total water realm of the earth's surface zone, including the oceans, surface waters of the lands, ground water, and water held in the atmosphere.

hydrostatic pressure surface   Imaginary surface that represents the level to which ground water will rise if free to do so in a well or tube penetrating the ground water body.

hygrograph   Recording hygrometer; it produces a continuous record of relative humidity.

hygrometer   Instrument that measures the water vapor content of the atmosphere; some types measure relative humidity directly.

hygrophytes   Plants adapted to a wet environment on the lands.

hygroscopic nuclei   Nuclei of condensation having a high affinity for water; usually salt particles.

hyperthermic regime   Soil temperature regime in which mean annual soil temperature is greater than 22°C.

hypolimnion   Deep, cold water layer of a lake, found below the thermocline.

ice age   Span of geologic time, usually on the order of 1 to 3 million years, or longer, in which glaciations alternate with interglaciations repeatedly in rhythm with cyclic global climate changes. (See also *glaciation, interglaciation*.)

iceberg   Mass of glacial ice floating in the ocean, derived by calving from a glacier that extends into tidal water.

icecap   Platelike mass of glacial ice limited to the high summit region of a mountain range or plateau; a variety of glacier.

ice fall   Abrupt steepening of the downvalley gradient of a valley glacier, causing the surface ice to break apart in deep crevasses.

ice floes   Individual patches of pack ice detached by the forces of wind and ocean currents.

ice island   Large, thick body of floating freshwater ice, derived from a mass of landfast ice.

Icelandic low   Persistent center of low atmospheric pressure located over the North Atlantic Ocean and strongly intensified in winter.

ice lobes (glacial lobes)   Broad

tonguelike extensions of an ice sheet resulting from more rapid ice motion where terrain is more favorable.

ice sheet   Large thick plate of glacial ice moving outward in all directions from a central region of accumulation; also called a *continental glacier*. Example: Greenland Ice Sheet.

ice sheet climate   Severely cold climate, found on the Greenland and Antarctic ice sheets, with potential evapotranspiration (water need) effectively zero throughout the year.

ice shelf   Thick plate of partially floating glacial ice attached to an ice sheet and fed by the ice sheet and by snow accumulation.

ice storm   Occurrence of heavy glaze of ice as falling rain or drizzle freezes on solid surfaces.

ice wedge   Vertical, wall-like body of ground ice, often tapering downward, occupying a shrinkage crack in silt of permafrost areas.

ice-wedge polygons   Polygonal networks of ice wedges.

igneous rock   Rock solidified from a high-temperature molten state; rock formed by cooling of magma. (See *extrusive igneous rock, felsic igneous rock, intrusive igneous rocks, mafic igneous rocks, ultramafic igneous rocks*.)

illite group   Clay mineral group derived by chemical weathering from such silicate materials as feldspar and muscovite mica.

illuviation   Accumulation in a lower soil horizon (typically, the B horizon) of materials brought down from a higher horizon; a pedogenic process.

image classification   Digital *image processing* technique in which each *pixel* is assigned to a particular class. An example is the computer classification of a Landsat *digital image* of an agricultural region into crop types, such as corn, soybeans, alfalfa, or small grains, to produce an agricultural map.

image processing   Mathematical manipulation of *digital images*, for example, to enhance contrast or edges.

imagery   General term in remote sensing for the graphic form of presentation of data obtained by scanning systems.

Inceptisols   Soil order in the Soil Taxonomy, consisting of soils having weakly developed soil horizons and containing weatherable minerals.

infiltration   Absorption and downward movement of precipitation into the soil and regolith.

infiltration capacity   Limiting rate at which falling rain or melting snow can be absorbed by a soil surface in the process of infiltration.

influent stream   Stream that loses discharge by seepage through the channel

floor to recharge the ground water body beneath.

infrared emissivity   Ratio of infrared radiation emission of a *gray body* to that of a *black body* at the same absolute temperature.

infrared imagery   Graphic image, resembling a photograph taken in ordinary light, recording the intensity of emission of invisible infrared radiation.

infrared rays (radiation)   Electromagnetic radiation in the wavelength range of 0.7 to about 200 microns.

initial landforms   Landforms produced directly by internal earth processes of volcanism and tectonic activity. Examples: volcano, fault scarp. (See also *sequential landforms*.)

inner lowland   On a coastal plain, a shallow valley lying between the first *cuesta* and the area of older rock (*oldland*).

inselberg   Small, islandlike hill or mountain rising sharply above a surrounding *pediment* or *alluvial fan*.

insolation   Interception of solar energy (shortwave radiation) by an exposed surface.

interception   Catching and holding of precipitation above the ground surface on leaves and stems of plants.

interflow   Movement of soil water through a permeable soil horizon or other shallow permeable layer in a downslope direction parallel with the ground surface; same as *throughflow*.

interglaciation   Within an ice age, a time interval of mild global climate in which continental ice sheets were largely absent or were limited to the Greenland and Antarctic ice sheets; the interval between two glaciations. (See also *deglaciation, glaciation*.)

interlobate moraine   Moraine formed between two adjacent lobes of an ice sheet.

intermediate belt   Zone below the soil-water belt too deep to supply capillary water to plants; i.e., too deep to be reached by plant roots.

International Date Line   The 180° meridian of longitude, together with deviations east and west of that meridian, forming the time boundary between adjacent standard time zones that are 12 hours fast and 12 hours slow with respect to Greenwich standard time.

intertropical convergence zone (ITC)   Zone of convergence of air masses of tropical easterlies along the axis of the equatorial trough.

intrusive igneous rocks   Igneous rocks formed by the solidification of magma beneath the surface in contact with older rock (country rock). (See also *extrusive igneous rock*.)

inversion   (See *temperature inversion*.)

inversion lid   Top surface of a low-level temperature inversion, resisting mixing of colder air below with warmer air above.

ion   Atom or group of atoms bearing an electrical charge as the result of a gain or loss of one or more electrons. (See also *cation, anion*.)

ionizing radiation   Very short wavelength electromagnetic radiation, such as that of X rays and radioactivity, capable of causing ionization of atoms exposed to the radiation.

ionosphere   A region 80 to 400 km above the earth where a dense concentration of positive ions and negative electrons occurs in a series of layers.

island arc   Chain of islands paralleling a subduction boundary and formed of volcanic rocks or rocks of accretionary prisms. (See also *tectonic arc, volcanic arc*.)

isobar   Line on a map passing through all points having the same atmospheric pressure.

isobaric surface   Surface of equal atmospheric pressure.

isogonic line   Line drawn on a map to pass through all points having the same magnetic declination.

isohyet   Line on a map drawn through all points having the same numerical value of precipitation.

isoseismal   Line on a map drawn through all points having equal value of earthquake intensity, usually given in numbers according to the modified Mercalli scale.

isostasy   Equilibrium state, resembling hydrostatic flotation, in which crustal and lithospheric masses stand at relative levels determined by their thickness and density, equilibrium being achieved by plastic flowage of denser mantle rock of the underlying asthenosphere.

isostatic compensation   Crustal rise or sinking in response to unloading by denudation or loading by sediment deposition, following the principle of isostasy.

isotherm   Line drawn on a map to pass through all points having the same air temperature.

isotope–ratio stages   Time units of the geologic past determined by cyclic fluctuations in the oxygen–isotope ratio; they have been numbered from 1 through 41 to cover the past 2 million years of record based on plankton from deep-sea cores.

jet stream   High-speed air flow in narrow tubelike zones within the upper-air westerlies and in certain other global latitude belts at high levels.

joint blocks   Blocklike masses of bedrock bounded by joint surfaces.

joints   Internal bedrock fracture surfaces of hairline thickness along which no slippage has occurred.

joule   Unit of work or energy in the SI system, equal to 10^7 ergs.

jungle   Type of forest that is low and dense, consisting partly of lianas, bamboo scrub, thorny palms, or thickly branching shrubs, usually formed where tropical rainforest has been disturbed or destroyed.

kame   Hill composed of sorted coarse water-laid glacial drift, largely sand and gravel, built into an impounded water body within stagnant ice or against the margin of an ice sheet.

kame terrace   Kame taking the form of a flat-topped terrace built between a body of stagnant glacial ice and a rising valley wall.

kaolinite   Clay mineral, composition hydrous aluminum oxide, typically formed by hydrolysis from potash feldspar and other aluminosilicate minerals.

karst   Landscape or class of topography dominated by surface features of limestone solution and underlain by limestone cavern systems.

katabatic winds   (See *drainage winds*.)

Kelvin scale (°K)   Temperature scale on which the starting point is absolute zero, equivalent to -273°C. Kelvin degrees are the same increments as Celsius degrees.

kerogen   Waxy substance of hydrocarbon composition held in oil shale and capable of yielding petroleum on distillation by heating.

killing frost   Occurrence of below-freezing temperature in the air layer near the ground, capable of damaging frost-sensitive plants.

knob-and-kettle   Terrain of numerous small knobs of glacial drift and intervening deep depressions usually situated within the moraine of a former ice sheet.

knot   Speed of 1 nautical mile per hour.

labile energy   Chemical energy temporarily stored as carbohydrate in plant biomass and readily broken down during metabolism.

lagoon   Shallow body of open water lying between a *barrier island* or a *barrier reef* and the mainland.

lag time   Interval of time between occurrence of precipitation and peak discharge of a stream.

lahar   Rapid down-slope or down-valley movement of a tonguelike mass of water-saturated *tephra* (volcanic ash) originating high up on a steep-sided volcanic cone; a variety of mudflow.

lake   Terrestrial body of standing water surrounded by land or glacial ice.

lake forest   Variety of *needleleaf* evergreen forest occurring in the region of the Great Lakes of North America and dominated by white pine, red pine, and hemlock.

land breeze   Local wind blowing from land to sea during the night.

land degradation   Degradation of the quality of plant cover and soil as a result of overuse by humans and their domesticated animals, especially during periods of drought.

landform (pedology)   General term for the configuration of the ground surface as a factor in soil formation; it includes slope steepness and aspect, as well as relief.

landforms   Configurations of the land surface taking distinctive forms and produced by natural processes. Examples: hill, valley, plateau. (See *depositional landforms, erosional landforms, initial landforms, sequential landforms*.)

landmass   Large area of continental crust lying above sea level (base level) and thus subject to removal by denudation. (See also *available landmass*.)

landmass rejuvenation   (See *rejuvenation of landmass*.)

landslide   Rapid sliding of large masses of bedrock on steep mountain slopes or from high cliffs.

La Niña   Counterpart of El Niño in which the subtropical high of the Southern Pacific becomes strongly developed and sea-surface temperatures in the central and eastern Pacific Ocean fall to lower levels.

lapse rate   (See *environmental temperature lapse rate, dry adiabatic lapse rate, wet (saturation) adiabatic lapse rate*.)

latent heat   Heat absorbed and held in storage in a gas or liquid during the processes of evaporation or melting, respectively; distinguished from sensible heat.

latent heat of fusion   Latent heat absorbed during melting or released during freezing.

latent heat of vaporization   Latent heat released during condensation from gaseous state to liquid state or absorbed during evaporation.

lateral cutting   Sidewise shifting of a stream channel caused by undercutting of steep banks on the outsides of bends.

lateral moraine   Moraine forming an embankment between the ice of an alpine glacier and the adjacent valley wall.

laterite   Rocklike layer rich in sesquioxide of aluminum and iron, including bauxite and limonite, found in low latitudes in association with the Ultisols and Oxisols.

latitude   Arc of a meridian between the equator and a given point on the globe.

lattice layers   Geometric structure of certain minerals such as the clay minerals, in which extremely thin layers of strongly bonded atoms and ions are separated from adjacent layers by weak bonds.

laurel forest   Variety of broadleaf evergreen forest dominated by species of the laurel family (*Lauraceae*.)

lava Magma emerging on the earth's solid surface, exposed to air or water.

lava flow Outpouring of molten lava upon the earth's surface, congealing to form extrusive igneous rock.

layer silicates Silicate minerals having lattice layer structure. Examples: mica, clay minerals.

leachate Solution of various ions and dissolved compounds carried down from the waste body of a sanitary landfill site to the ground water system below.

leaching Pedogenic process in which material is lost from the soil by downward washing out and removal by percolating surplus soil water.

leads Narrow strips of open ocean water between ice floes.

liana Woody vine supported on the trunk or branches of a tree.

lichens Plant forms in which algae and fungi live together (in a symbiotic relationship) to create a single structure; they typically form tough, leathery coatings or crusts attached to rocks and tree trunks.

life-form Characteristic physical structure, size, and shape of a plant or of an assemblage of plants.

life layer Shallow surface zone containing the biosphere; a zone of interaction between atmosphere and land surface, and between atmosphere and ocean surface.

lignite Low grade of coal, intermediate in development between peat and coal; also called "brown coal."

lime Soil-conditioning material that may be either calcium oxide or calcium carbonate (limestone), applied for the purpose of reducing the acidity of the soil.

limestone Nonclastic sedimentary rock in which calcite is the dominant mineral, and with varying minor amounts of magnesium carbonate, silica, or other minerals, and clay.

limestone caverns (See *caverns*.)

limonite Mineral or mineral group consisting largely of hydrous ion oxide, produced by chemical weathering of other iron-bearing minerals.

liquefaction (See *spontaneous liquefaction*.)

liquid Fluid that maintains a free upper surface and is only very slightly compressible, as compared with a gas.

liquid state Fluid state of matter having the properties of a liquid.

lithification Process of formation of sedimentary rock from sediment by volume compression during exclusion of water, or by cementation of grains with mineral matter.

lithosphere (1) General term for the entire solid earth realm. (2) In plate tectonics, the strong brittle outermost earth shell, lying above the asthenosphere.

lithospheric plate Large segment of the lithosphere moving as a unit in contact with adjacent lithospheric plates along plate boundaries.

Little Ice Age Climatic episode of below-normal temperatures from about −550 to −150 years (A.D. 1450–1850), during which alpine glaciers enlarged and advanced to lower levels.

littoral cell Coastal sediment flow system in which fluviatile sediment provides the basic input while discharge of beach sediment through a submarine canyon to the deep ocean floor provides the output.

littoral drift Transport of sediment parallel with the shoreline by the combined action of beach drift and longshore current transport.

live dunes Sand dunes in an active state of change as sand is moved in saltation.

loam Soil-texture class in which no one of the three size grades (sand, silt, clay) dominates over the other two.

local time For a given point on the globe, the mean solar time based on the local meridian.

local winds General term for winds generated as direct or immediate effects of the local terrain.

lodgment till Glacial till formed beneath moving glacial ice; a densely compacted form of till, often with a high clay content.

loess Yellowish to buff-colored, fine-grained sediment, largely of silt grade, deposited on upland surfaces after transport by wind in a dust storm.

longitude Arc of a parallel between the prime meridian and a given point on the globe.

longitudinal dunes Class of sand dunes in which the dune ridges are oriented parallel with the prevailing wind.

longitudinal profile Graphic representation of the descending course of a stream or stream channel; altitude is plotted on the vertical scale, downstream distance on the horizontal scale.

longshore current Current in the breaker zone, set up by the oblique approach of waves and running parallel with the shoreline.

longshore drift Littoral drift caused by action of a *longshore current*.

longwave radiation Electromagnetic radiation emitted by the earth, largely in the range from 3 to 50 microns.

low-angle overthrust fault Overthrust fault in which the fault plane or fault surface has a low angle of dip or may be horizontal.

low base status (See *base status of soils*.)

lowland Broad, open valley between two cuestas of a coastal plain. (The term may refer to any relatively low area of land surface.)

low-latitude climates Group of climates of the equatorial zone and tropical zone dominated by the subtropical high-pressure belt and the equatorial trough.

low-latitude rainforest Broadleaf evergreen rainforest of the equatorial and tropical zones.

low-latitude rainforest environment Environment of the low-latitude broadleaf evergreen rainforest.

low-level temperature inversion Reversal of normal environment temperature lapse rate in an air layer near the ground.

low water Point in cycle of ocean tide when ocean water reaches its lowest point.

loxodrome Line of constant compass bearing drawn on a map or navigational chart; also known as a *rhumb line*.

macronutrients Nine elements required in abundance for organic growth, including primary production by green plants. (See also *micronutrients*.)

mafic igneous rocks Igneous rocks composed dominantly of mafic minerals.

mafic minerals, mafic mineral group Rock-forming minerals, largely silicates, rich in magnesium and iron, dark in color, and of relatively high density. (See also *felsic minerals*.)

magma Mobile, high-temperature molten state of rock, usually of silicate mineral composition and with dissolved gases and other volatiles.

magnetic azimuth Azimuth system referred to magnetic north.

magnetic declination Horizontal angle between geographic north and magnetic north.

magnetic field (See *external magnetic field*.)

magnetic north Direction from a given point on the earth's surface following a great circle toward the *magnetic north pole*.

magnetic north pole Surface point toward which the north-seeking point of a magnetic compass is directed.

magnetic storm Intense disturbance of the field of the magnetosphere caused by an influx of energetic particles from a solar flare.

magnetopause Outer boundary surface of the magnetosphere.

magnetosphere External portion of the earth's magnetic field, shaped by pressure of the solar wind.

mangrove swamp forest Coastal vegetation of mangrove plants found in shallow muddy-water environments in the equatorial zone and tropical zone.

mantle Rock layer or shell of the earth beneath the crust and surrounding the core, composed of ultramafic rock of silicate mineral composition.

mantle plume A hypothetical columnlike or stalklike rising of heated mantle rock, thought to be the cause of a hot spot in the overlying lithospheric plate.

map projection Any orderly system of parallels and meridians drawn on a flat surface to represent the earth's surface.

maquis Form of dense sclerophyllous scrub found throughout the Mediterranean region.

marble Variety of metamorphic rock derived from limestone or dolomite by recrystallization under pressure.

marginal glacial lake Lake impounded between an ice front and rising ground slopes.

marine cliff Rock cliff shaped and maintained by the undermining action of breaking waves.

marine scarp Steep seaward slope in poorly consolidated alluvium or other forms of regolith, produced along a coast by the undermining action of waves.

marine terrace Former abrasion platform elevated to become a benchlike coastal landform.

marine west-coast climate Cool, moist climate of west coasts in the midlatitude zone, usually with a substantial annual water surplus and a distinct winter precipitation maximum.

maritime air mass Moist air mass developed over an ocean source region.

mass wasting Spontaneous downward movement of soil, regolith, and bedrock under the influence of gravity; does not include the action of fluid agents.

material cycle Total system of pathways by which a particular type of matter (a given element, compound, or ion, for example) moves through the earth's ecosystem or biosphere; also called biogeochemical cycle or nutrient cycle.

material flow system System of interconnected flowpaths of material (matter) that may constitute a closed system or an open system.

material subsystem A secondary system of flow of material within a larger material system.

maximum-minimum thermometer Pair of thermometers recording the maximum and minimum air temperatures since last reset.

mean annual temperature Mean of daily air temperature means for a given year or succession of years.

mean daily temperature Sum of daily maximum and minimum air temperature readings divided by two.

meanders (See *alluvial meanders*.)

mean monthly temperature Mean of daily air temperature means for a given calendar month.

mean velocity Mean or average speed of flow of water through the entire cross section of a stream.

mechanical weathering (See *physical weathering*.)

medial moraine Moraine riding on the surface of an alpine glacier and composed of debris carried downvalley from the point of junction of two ice streams.

Mediterranean climate Climate type of the subtropical zone, characterized by the alternation of a very dry summer and a mild, rainy winter.

Mediterranean evergreen mixed forest Variety of sclerophyll forest consisting of various species of oak and pine, once common in lands bordering the Mediterranean Sea.

megahertz Wave frequency of 1 million cycles per second.

melting Change from solid state to liquid state, accompanied by absorption of sensible heat to become latent heat.

membranous leaves Leaves of the broadleaf type that are of normal thickness.

Mercator projection Conformal map projection with horizontal parallels and vertical meridians and with map scale rapidly increasing with increase in latitude.

mercurial barometer Barometer using the Torricelli principle, in which atmospheric pressure counterbalances a column of mercury in a tube.

meridian of longitude North–south line on the surface of the global oblate ellipsoid, connecting the north and south poles.

meridional transport Flow of energy (heat) or matter (water) across the parallels of latitude, either poleward or equatorward.

meridional winds Winds moving across the parallels of latitude, in a north–south, or south–north direction, along the meridians of longitude.

mesa Table-topped plateau of comparatively small extent bounded by cliffs, occurring in a region of horizontal strata and capped by a resistant formation.

mesic regime Soil temperature regime in which mean annual soil temperature is greater than 8°C but less than 15°C, while the difference between mean temperatures of warm and cold seasons is greater than 5°C.

mesopause Upper limit of the mesosphere.

mesophytes Plants adapted to a habitat of intermediate degree of wetness and to uniform soil water availability.

mesosphere Atmospheric layer of upwardly diminishing temperature, situated above the stratopause and below the mesopause.

Mesozoic Era Second of three geologic eras following Precambrian time.

metamorphic rocks Rocks altered in physical structure and/or chemical (mineral) composition while in the solid state (without melting) by action of heat, pressure, and shearing action, or by infusion or loss of elements, taking place at substantial depth below the earth's surface.

meteorology Science of the atmosphere; particularly the physics of the lower or inner atmosphere.

mica group Aluminosilicate mineral group of complex chemical formula having perfect cleavage into thin sheets.

microburst Brief onset of intense winds close to the ground beneath the downdraft zone of a thunderstorm cell.

microclimate Climate of a shallow layer of air near the ground, and including the soil surface and plant community within which it is in contact.

microcontinent Fragment of continental crust and its lithosphere of subcontinental dimensions that is embedded in an expanse of oceanic lithosphere.

micron Length unit; one micron equals 0.0001 cm.

micronutrients Elements essential to organic growth, but only in very small amounts. (See also *macronutrients*.)

microwaves Waves of the electromagnetic radiation spectrum in the wavelength band from about 0.03 cm to about 1 cm.

midlatitude climates Group of climates of the midlatitude zone and subtropical zone, located in the polar front zone and dominated by both tropical air masses and polar air masses.

midlatitude deciduous forest Formation class within the forest biome dominated by tall, broadleaf deciduous trees, found mostly in the moist continental climate and marine west-coast climate. (Also called *summergreen deciduous forest*.)

midlatitude zones Latitude zones occupying the latitude range 35° to 55° N and S (more or less) and lying between the subtropical zones and the subarctic (subantarctic) zones.

midnight meridian Imaginary time meridian opposite to the noon meridian, marking the occurrence of midnight and traveling westward around the globe at 15° per hour.

mid-oceanic ridge One of three major topographic divisions of the ocean basins, the central belt of submarine mountain topography with a characteristic axial rift that marks a spreading plate boundary.

Milankovitch curve Curve showing

the fluctuations in insolation at a given latitude resulting from the combined effects of variations in the form and precession of the earth's orbit and tilt of the earth's axis, for a period extending back to a half million years or more before present.

millibar (mb) Unit of atmospheric pressure; one-thousandth of a bar. Bar is a force of 1 million dynes per square centimeter.

milliequivalent Cation-exchange capacity measured as the ratio of weight of cations to weight of soil.

mineral Naturally occurring homogeneous inorganic solid substance, having either a definite chemical composition and an orderly atomic structure or one that is variable between stated limits. (See also *felsic minerals, mafic minerals, silicate minerals*.)

mineral alteration Chemical change of minerals to more stable compounds on exposure to atmospheric conditions; essentially synonymous with *chemical weathering*.

mineral horizons Soil horizons designated by the letters A and B, having less than 20 percent organic matter when no clay is present and less than 30 percent organic matter when the mineral fraction consists of 50 percent or more clay.

mistral Local drainage wind of cold air affecting the Rhône Valley of southern France.

modified Mercalli scale Earthquake intensity scale, modified in 1956 by C. F. Richter, using 12 intensity levels and relating each to phenomena observed during an earthquake.

Moho (Mohorovičić discontinuity, M-discontinuity) Contact surface between the earth's crust and mantle; named for A. Mohorovičić, the seismologist who discovered the discontinuity.

moist climate Climate in which the annual soil-water shortage is less than 15 cm (5.9 in.).

moist continental climate Moist climate of the midlatitude zone with strongly defined winter and summer seasons, adequate precipitation throughout the year, and a substantial annual water surplus.

moist subtropical climate Moist climate of the subtropical zone, characterized by a moderate to large annual water surplus and a strongly seasonal cycle of potential evapotranspiration (water need).

mollic epipedon Relatively thick, dark-covered surface soil horizon, or epipedon, containing substantial amounts of organic matter (humus) and usually rich in base cations.

Mollisols Soil order within the Soil Taxonomy consisting of soils with a mollic epipedon and high base status.

monadnock Prominent, isolated mountain or large hill rising conspicuously above a surrounding peneplain and composed of a rock more resistant than that underlying the peneplain; a landform of denudation in moist climates. (See also *bornhardt, inselberg*.)

monsoon forest Plant formation class within the forest biome consisting in part of deciduous trees adapted to a long dry season in the wet-dry tropical climate.

monsoon system System of low-level winds blowing into a continent in summer and out of it in winter, controlled by atmospheric pressure systems developed seasonally over the continent.

monsoon and trade-wind littoral climate Moist climate of low latitudes showing a strong rainfall peak in the season of high sun and a short period of reduced rainfall.

montane forest Plant formation class of the forest biome found in cool upland environments of the tropical zone and equatorial zone.

montmorillonite Hydrous aluminosilicate clay mineral derived by chemical alteration of silicate minerals in various igneous rocks; expands greatly when it adsorbs water.

moraine Accumulation of rock debris previously carried by an alpine glacier or an ice sheet and deposited by the ice to become a depositional landform. (See *ground moraine, interlobate moraine, lateral moraine, medial moraine, recessional moraine, terminal moraine*.)

morphogenetic regions (See *climogenetic regions*.)

mossy forest Form of upland forest (montane forest) in highlands of the tropical zone and equatorial zone, characterized by massive accumulations of mosses attached to tree branches.

mountain arc Curved (arcuate) segment of an alpine mountain chain, usually of complex geologic structure and associated with either a subduction boundary or a continental suture.

mountain roots Erosional remnants of deep portions of ancient orogens that were once alpine chains.

mountain winds Daytime movements of air up the gradient of valleys and mountain slopes; alternating with nocturnal valley winds.

mud Sediment consisting of a mixture of clay and silt with water, often with minor amounts of sand and sometimes with organic matter.

mud flat Nearly flat expanse of mud, rich in organic matter; deposited in a tidal estuary or bay.

mudflow Rapid flowage of a mud stream down a canyon floor and spreading out on a plain at the foot of a mountain range; often contributing to the building of an alluvial fan.

mudstone Sedimentary rock formed by the lithification of mud.

multiband spectral photography Photography using a combination of several narrow spectral bands within the visible light region and the near-infrared region.

multispectral image Image consisting of two or more images, each of which is taken from a different portion of the spectrum (e.g., blue, green, red, infrared).

multispectral scanner Remote sensing instrument, flown on an aircraft or spacecraft, that simultaneously collects multiple *digital images* (*multispectral images*) of the ground. Typically, images are collected in four to eight spectral bands.

muskeg Thick water-saturated accumulations of peat produced by bog succession in glaciated regions of Canada.

nappe Large, sheetlike recumbent fold or thrust sheet that has moved horizontally or on a low grade for many kilometers; an orogenic structure typical of collision tectonics.

natric horizon Soil horizon having prismatic structure and a high content of the sodium ion.

natural bridge Natural rock arch spanning a stream channel, formed by cutoff of an entrenched meander bend.

natural ecosystem Ecosystem that attains its development without appreciable interference by humans and is subject to natural forces of modification and destruction. (See also *agricultural ecosystems*.)

natural gas Naturally occurring mixture of hydrocarbon compounds (principally methane) in the gaseous state held within porous reservoir rocks.

natural levee Belt of higher ground paralleling a meandering alluvial river on both sides of the channel and built up by deposition of fine sediment during periods of overbank flooding.

nautical mile Unit of distance measurement approximately equal to the average length of 1 minute of latitude, or of 1 minute of longitude measured along the earth's equator; 1 nautical mile equals 1.85 km (1.15 statute mi), approximately.

near-infrared region Narrow wavelength band within infrared region, immediately adjacent to the red band of the visible light region.

needleleaf Leaf form that is very narrow in relation to length, as illustrated by needles of the spruce, pine, or fir.

needleleaf evergreen forest Needleleaf forest composed of evergreen tree species, such as spruce, fir, and pine.

needleleaf forest Plant formation class within the forest biome, consisting largely of needleleaf trees.

net photosynthesis Carbohydrate production remaining in an organism after respiration has broken down sufficient carbohydrate to power the metabolism of the organism.

net primary production Rate at which carbohydrate is accumulated in the tissues of plants within a given ecosystem; units are grams of dry organic matter per year per square meter of surface area.

net radiation Difference in intensity between all incoming energy (positive quantity) and all outgoing energy (negative quantity) carried by both shortwave radiation and longwave radiation.

névé (See *firn*.)

nimbostratus Low, dense stratiform cloud from which rain or snow is falling.

nitrogen cycle Material cycle in which nitrogen moves through the biosphere by the processes of nitrogen fixation and denitrification.

nitrogen fixation Chemical process of conversion of gaseous molecular nitrogen of the atmosphere into compounds or ions that can be directly utilized by plants; a process carried out within the nitrogen cycle by certain microorganisms.

nonclastic sediments Class of sediments formed of mineral compounds precipitated from chemical solution or from organic activity.

noon (See *solar noon*.)

noon altitude of sun Vertical angle or arc between the horizon and the sun in its noon position over the local meridian.

normal fault Variety of fault in which the fault plane inclines (dips) toward the downthrown block and a major component of the motion is vertical.

nuclei (atmospheric) Minute particles of solid matter suspended in the atmosphere and serving as cores for condensation of water or ice.

nuée ardente Glowing volcanic avalanche; a cloud of incandescent dust and gas that travels rapidly down the steep side of a volcano.

nutrient cycle (See *material cycle*.)

oblate ellipsoid Geometric solid resembling a flattened sphere, with polar axis shorter than the equatorial diameter, elliptical in any polar cross section but circular in any cross section taken at right angles to the polar axis.

oblateness Ratio of difference between length of polar axis and length of equatorial diameter of an oblate ellipsoid to its equatorial diameter, expressed as a simple fraction; also known as the flattening of the poles.

obsidian Black or dark-colored volcanic glass of rhyolite composition.

occluded front Weather front along which a moving cold front has overtaken a warm front, forcing the warm air mass aloft.

ocean-basin floors Major topographic division of the ocean basins, consisting of the deep portions, including abyssal plains and low hills.

ocean basins Deep depressions of subglobal dimension, underlain largely by oceanic lithosphere and crust and holding the water of the oceans. (See also *continents*.)

ocean current Persistent, dominantly horizontal flow of ocean water.

oceanic crust Crust of basaltic composition beneath the ocean floors, capping oceanic lithosphere. (See also *continental crust*.)

oceanic lithosphere Lithosphere bearing oceanic crust. (See also *continental lithosphere*.)

oceanic trench Long, narrow, deep, troughlike depression in the ocean floor representing the line of subduction of oceanic lithosphere beneath the margin of *continental lithosphere*.

oceanography (See *physical oceanography*.)

oceans Standing bodies of salt water occupying the ocean basins of the earth. (See also *world ocean*.)

ocean tide Periodic rise and fall of the ocean level induced by gravitational attraction between the earth and moon in combination with earth rotation.

Ochrepts Suborder of the Inceptisols having a brownish horizon (ochric epipedon) of altered materials at or close to the surface.

ochric epipedon Surface soil horizon (epipedon) that is light in color and contains less than 1 percent organic matter.

offshore That part of the beach profile lying in the zone of shoaling waves and below the level of low tide. (See also *foreshore*.)

offshore bar A low bar of sand in the offshore region of a beach.

oil sand (See *bituminous sand*.)

oil shale Shale containing dispersed hydrocation compounds, capable of yielding petroleum by distillation when heated.

old-field succession Form of secondary succession typical of an abandoned field, such as might be found in the moist continental climate or moist subtropical climate of the eastern and central United States; a form of autogenic succession.

oldland As used in reference to a coastal plain, the region of exposed older basement rock bordering the innermost limit of the coastal plain strata.

olivine Mafic mineral, a silicate of magnesium and iron without aluminum; olive-green, grayish-green, or brown in color; a dominant mineral in the ultramafic igneous rocks.

open folds Folds of strata having moderate to low dips of the fold flanks.

open system System of interconnected flowpaths of energy or material with a boundary through which energy or matter can enter and leave the system. (See also *closed system*.)

orbit Path followed by a planet revolving around the sun, or by a planetary satellite revolving around a planet.

organically derived sediment Class of sediments consisting of the remains of nonliving plants or animals, or of mineral matter produced by the activities of plants or animals. (Also known as *biogenic sediments*.)

organic horizon Soil horizon, designated as the O horizon, overlying the mineral horizons and formed of accumulated organic matter derived from plants and animals.

orogen The mass of tectonically deformed rocks and related igneous rocks produced during an orogeny.

orogeny Major episode of severe tectonic deformation with attendant igneous intrusion occurring in a long, relatively narrow belt of the continental margin as a result of formation of a new subduction boundary on the adjacent ocean basin or by continental collision.

orographic precipitation Precipitation induced by the forced rise of moist air over a mountain barrier.

Orthents Suborder of Entisols formed on recent erosional surfaces and having shallow depth to consolidated rock (bedrock) or unconsolidated materials (regolith), such as loess.

Orthids Suborder of Aridisols lacking an argillic horizon but with one or more pedogenic horizons that extend at least 25 cm (10 in.) below the surface.

Orthods Suborder of Spodosols that are freely drained and have a B horizon in which iron, aluminum, and humus have accumulated.

Orthox Suborder of Oxisols found in warm humid regions with a short dry season or no dry season (perudic regime).

oscillatory waves Water waves in which particles move in vertical orbits, completing one orbital circle with the passage of one wave.

outcrop Surface exposure of bedrock.

outgassing Process of exudation of volatiles as gases from the earth's crust, largely through volcanic activity, to become a part of the earth's hydrosphere and atmosphere.

outlet glacier Tonguelike ice stream,

resembling an alpine glacier, fed by ice from the margin of an ice sheet.

outwash deposit　Accumulation of layers of sand and gravel deposited by meltwater streams near the margin of a stagnant ice sheet or alpine glacier.

outwash plain　Flat, gently sloping plain built up of sand and gravel by the aggradation of meltwater streams in front of the margin of an ice sheet.

overbank flooding　Rise of flood waters in an alluvial river so as to spill over the banks and inundate the floodplain.

overland flow　Gravity flow of a surface layer of water over a sloping ground surface at times when the infiltration rate is exceeded by the precipitation rate; a form of runoff.

overthrust fault　Fault characterized by the overriding of one crustal mass over the adjacent mass on an inclined fault plane, a result of crustal compression during orogeny. (See also *low-angle overthrust fault.*)

overwash　Movement of storm swash entirely across a barrier island or barrier beach to reach the lagoon or salt marsh on the inland side.

oxbow lake, oxbow swamp　Crescent-shaped lake or swamp representing the abandoned channel left by the cutoff of an alluvial meander.

oxic horizon　Highly weathered soil horizon rich in clay minerals and sesquioxides of low cation exchange capacity (CEC).

oxidation　Chemical weathering process in which free oxygen unites with metallic ions of minerals to produce mineral oxides.

Oxisols　Soil order in the Soil Taxonomy consisting of very old, highly weathered soils of low latitudes, with an oxic horizon and low cation exchange capacity (CEC).

oxygen cycle　Material cycle in which oxygen moves through the biosphere in both gaseous and sedimentary forms.

oxygen–isotope ratio　Ratio of oxygen-18 to oxygen-16 as determined from foraminifera tests of deep-ocean cores and from ice layers of the Greenland Ice Sheet. (See also *isotope-ratio stages.*)

ozone　Gas molecule consisting of three atoms of oxygen, O_3.

ozone layer　Layer in the stratosphere, mostly in the altitude range 20 to 35 km (12 to 31 mi), in which a concentration of ozone is produced by the action of solar ultraviolet rays.

pack ice　Floating sea ice that completely covers the sea surface.

paleoglaciation curve　Curve of oxygen–isotope ratios plotted back into time so as to reveal fluctuations interpreted as synchronous with fluctuations in the total volume of glacier

ice present on the globe at any given time. (See also *isotope-ratio stages.*)

Paleozoic Era　First of three geologic eras comprising all geologic time younger than Precambrian time.

Pampa　Geographic region of plains in Uruguay and eastern Argentina, characterized by the dominance of tall-grass prairie as the native plant formation class.

Pangaea　Hypothetical parent continent, enduring from late Paleozoic time to late in the Mesozoic Era, consisting of the continental shields of Laurasia and Gondwana joined into a single unit. (See also *continental drift.*)

parabolic blowout dune　Type of parabolic dune formed to the lee of a shallow deflation hollow, usually found on interior plains in a dry climate.

parabolic dunes　Isolated low sand dunes of parabolic outline, with points directed into the prevailing wind.

parallel of latitude　Imaginary east–west circle on the earth's surface, lying in a plane parallel with the equator and at right angles to the axis of rotation.

parallel retreat　Program of hillslope profile evolution in which the angle of slope remains essentially constant as the slope retreats.

parasitism　Form of negative interaction between species in which a small species (parasite) feeds on a larger one (host) without necessarily killing it.

particulate matter　Solid and liquid particles capable of being suspended for long periods in the atmosphere.

passive continental margins　Continental margins lacking active plate boundaries at the contact of continental crust with oceanic crust. A passive margin thus lies within a single lithospheric plate. Example: Atlantic continental margin of North America. (See also *continental margins, active continental margins.*)

passive systems　Electromagnetic remote sensor systems that measure radiant energy reflected or emitted by an object or surface.

patterned ground　General term for a ground surface that bears polygonal or ringlike features, including stone polygons, and ice-wedge polygons.

peat　Partially decomposed, compacted accumulation of plant remains forming in a saturated environment that may be either a freshwater bog or a salt marsh. (See also *freshwater peat.*)

ped　Individual natural soil aggregate.

pediment　Gently sloping, rock-floored surface found at the base of a mountain mass or cliff in an arid region.

pediplain　Desert land surface of low relief composed in part of pediment surfaces and in part of alluvial fan and playa surfaces.

pedogenic processes　Group of recognized basic soil-forming processes, mostly involving the gain, loss, translocation, or transformation of materials within the soil body.

pedology　Science of the soil as a natural surface layer capable of supporting living plants; synonymous with *soil science.*

pedon　Soil column extending down from the surface to reach a lower limit in some form of regolith or bedrock.

peneplain　Land surface of low elevation and slight relief produced in the late stage of landmass denudation.

percentage base saturation (PBS)　Percentage of exchangeable base cations with respect to the total cation exchange capacity (CEC) of a given soil.

perched water table　Water table of a layer or lens of ground water formed over an aquiclude and occupying a position above the main water table.

pergelic regime　Soil temperature regime in which mean annual soil temperature is below 0°C.

perhumid climate subtype　Subtype of the moist climate in which the soil-water surplus is 60 cm or greater.

peridotite　Igneous rock consisting largely of olivine and pyroxene; an ultramafic igneous rock occurring as a pluton, also thought to compose much of the upper mantle.

periglacial environment　The physical environment in close proximity to the margin of a continental ice sheet.

perihelion　Point on the earth's elliptical orbit at which the earth is nearest to the sun.

period of geologic time　Time subdivision of the era, each ranging in duration between about 35 and 70 million years.

permafrost　Condition of permanently frozen water in the regolith and bedrock in cold climates of subarctic and arctic regions.

permeability　Property of relative ease of movement of ground water through rock or regolith when unequal pressure exists.

petrocalcic horizon　Hardened calcic horizon in the soil that does not break apart when soaked in water. (See also *caliche, calcrete.*)

petroleum　Natural mixture of many complex hydrocarbon compounds occurring in rock; it includes natural gas, crude oil, and bitumens, but in common usage the term is synonymous with crude oil.

pH　Measure of the concentration of hydrogen ions in a solution; the number represents the logarithm to the base 10 of the reciprocal of the weight in grams of hydrogen ions per liter of water.

photochemical reactions Chemical reactions occurring in polluted air through the action of sunlight on pollutant gases to synthesize new toxic compounds or gases.

photoperiod Duration of daylight on a given day at a given latitude.

photosynthesis Production of carbohydrate by the union of water with carbon dioxide while absorbing light energy. (See *gross photosynthesis, net photosynthesis*.)

phreatophytes Plants that draw water from the ground-water table beneath alluvium of dry stream channels and valley floors in desert regions.

physical geography The study and synthesis of selected subject areas from the natural sciences—especially atmospheric science, hydrology, physical oceanography, geology, geomorphology, soil science, and plant ecology—in order to gain a complete picture of the physical environment of humans and to examine the interactions of humans with that environment.

physical oceanography Physical science of the oceans, as distinguished from biological oceanography.

physical weathering Breakup of massive rock into smaller particles through the action of physical stresses at or near the earth's surface; also called *mechanical weathering*. (See also *chemical weathering, weathering*.)

phytogenic dunes Class of sand dunes formed under partial cover of plants that influence the building and shaping of the dunes.

pingo Conspicuous conical mound or circular hill, having a core of ice, found on plains of the arctic tundra where permafrost is present.

pioneer stage First stage of an ecological succession.

pixel Individual brightness value within a *digital image*.

plaggen epipedon Human-made surface soil horizon (epipedon) produced by long-continued manuring, incorporating sod or other livestock bedding materials into the soil.

Plaggepts Suborder of the Inceptisols having a plaggen epipedon (thick surface horizon of materials added by humans through long habitation or long-continued manuring); most are in Europe or the British Isles.

plagioclase feldspar Aluminosilicate mineral group with sodium or calcium or both; of the felsic mineral class.

plain Surface of low inclination and low relief underlain by soil, regolith, marine sediments, or bedrock and covered by air or water. (See also *abyssal plain, coastal plain, floodplain, outwash plain, peneplain, pediplain*.)

plane of the ecliptic Imaginary plane in which the earth's orbit lies.

plastic shading Method of showing relief on topographic maps by use of color shades varied according to aspect of the slope.

plateau Upland surface, more or less flat and horizontal, upheld by a resistant bed or formation of sedimentary rock or by lava flows and bounded by a steep cliff. (In common usage, any upland surface bounded by steeper descending slopes.)

plate tectonics Theory of tectonic activity of the earth's lithospheric plates, their present and past interactions, and the influence of this activity upon all aspects of geology.

playa Flat land surface underlain by fine sediment or evaporite minerals deposited from shallow lake waters in a dry climate in the floor of a closed topographic depression.

Pleistocene Epoch Epoch of the Cenozoic Era; it followed the Pliocene Epoch and preceded the Holocene Epoch.

plinthite Iron-rich mineral concentration, usually in the form of dark red mottles, present in deeper soil horizons of certain Oxisols and Ultisols and capable of hardening into rocklike material (laterite) with repeated wetting and drying.

plucking (See *glacial plucking*.)

plume (See *mantle plume*.)

plunging fold Fold of strata with a descending or rising crest or fold axis.

pluton Body of intrusive igneous rocks. Examples: batholith, dike.

pluvial lake Lake that reached full development during past climatic periods of relatively high ratio of precipitation to evaporation and is presently extinct or a small remnant of its former extent. The term applies specifically to Pleistocene lakes that formerly occupied closed depressions in the floors of basins in the Basin-and-Range Province. Example: Lake Bonneville.

pocket beach Small beach, usually crescentic in plan, formed at the head of a coastal embayment and often consisting of pebbles or cobbles (shingle).

point-bar deposit Deposit of coarse bed-load alluvium accumulated on the inside of a growing alluvial meander.

polar air Air originating in high latitudes and having the characteristics of a polar air mass.

polar air mass Cold air mass with source region over continents and oceans in lat. 50° or 60° N and S.

polar easterlies System of easterly surface winds at high latitude, best developed in the southern hemisphere, over Antarctica.

polar front Front lying between a cold polar air mass and a warm tropical air mass, often situated along a jet stream within the upper-air westerlies.

polar front jet stream Jet stream formed along the polar front, where

cold polar air and warm tropical air are in contact.

polar front zone Broad atmospheric zone in midlatitudes and high latitudes, occupied by the shifting polar front.

polar high Persistent low-level center of high atmospheric pressure located over the polar zone of Antarctica.

polar low Persistent center of low atmospheric pressure over high latitudes in the upper atmosphere.

polar outbreak Tongue of cold polar air, preceded by a cold front, penetrating far into the tropical zone and often reaching the equatorial zone; it brings rain squalls and unusual cold.

polar zones Latitude zones lying between 75° and 90° N and S.

polders Diked salt marshes and tidal mudflats converted to a freshwater environment and used for agriculture, as in coastal Netherlands.

pollutants In air pollution context, foreign matter injected by humans into the lower atmosphere as particulate matter or as chemical pollutants.

pollution dome Broad, low dome-shaped layer of polluted air, formed over an urban area at times when winds are weak or calm prevails.

pollution plume (1) The trace or path of leachate or other pollutant substances, moving along the flow paths of ground water. (2) Trail of polluted air carried downwind from a pollution source by strong winds.

polypedon Smallest distinctive geographic unit of the soil of a given area; it consists of pedons.

pool of materials Area or location of concentration of a given material in the biogeochemical material cycle; two types are active pools and storage pools.

porosity Total volume of pore space within a given volume of rock; a ratio, expressed as a percentage.

postglacial rebound Spontaneous rise of continental crust following melting and disappearance of an ice sheet, tending to restore isostatic equilibrium.

potash feldspar Class of feldspars in which potassium is dominant, also with some sodium.

potential evapotranspiration (water need) Ideal or hypothetical rate of evapotranspiration estimated to occur from a complete canopy of green foliage of growing plants continuously supplied with all the soil water they can use; a real condition reached in those situations where precipitation is sufficiently great or irrigation water is supplied in sufficient amounts.

pothole Cylindrical cavity in hard bedrock of a stream channel produced by abrasion of a spherical or discus-shaped rock fragment rotating within the cavity.

prairie Plant formation class of the grassland biome, consisting of dominant tall grasses and subdominant forbs, widespread in subhumid continental climate regions of the subtropical zone and midlatitude zone.

Precambrian time All of geologic time older than the beginning of the Cambrian Period, i.e., older than about 600 million years.

precipitation Particles of liquid water or ice that fall through the atmosphere and may reach the ground.

predation Form of negative interaction between animal species in which one species (predator) kills and consumes the other (prey).

pressure cell Center of high or low atmospheric pressure, identified with a cyclone or an anticyclone.

pressure gradient Change of atmospheric pressure measured along a line at right angles to the isobars.

pressure-gradient force Force acting horizontally, tending to move air in the direction of lower atmospheric pressure.

prevailing westerly winds (westerlies) Surface winds blowing from a generally southwesterly direction in the midlatitude zone, but varying greatly in direction and intensity.

primary consumers Animals that live by feeding on producers.

primary minerals In soil science, the original, unaltered silicate minerals of the igneous rocks and metamorphic rocks.

primary producers Organisms that use light energy to convert carbon dioxide and water to carbohydrates through the process of photosynthesis.

primary production (See *net primary production*.)

primary succession Ecological succession that begins on a newly constructed deposit of mineral sediment.

prime meridian Reference meridian of zero longitude; universally accepted as the Greenwich meridian.

principal meridian Reference meridian used in the U.S. Land Office Survey.

producers (See *primary producers*.)

progradation Shoreward building of a beach, bar, or sandpit by addition of coarse detrital sediment carried by littoral drift or brought from deeper water offshore. (See also *retrogradation*.)

Psamments Suborder of Entisols that have sandy texture (sands and loamy sands) throughout the upper 1 m (40 in.).

psychrometer (See *sling-psychrometer*.)

pumice Glassy, light-colored scoria of very low density, resembling solidified foam; it commonly has the composition of rhyolite.

Puszta Lowland region in Hungary, formerly covered by tall-grass prairie.

pyranometer Instrument that measures the intensity of insolation (short-wave radiation), including both direct solar beam and indirect sky radiation from down-scatter.

pyroclastic sediment Sediment consisting of particles thrown into the air by volcanic explosion in the form of *tephra* or volcanic ash.

pyroxenes, pyroxene group Group of complex aluminosilicate minerals rich in calcium, magnesium, and iron, dark in color, relatively high in density, classed as mafic minerals. Example: augite.

quadrangle Map area within designated boundary meridians and parallels.

quartz Mineral, composition silicon dioxide (silica); an essential constituent of the felsic igneous rocks and commonly the major constituent of sand and sandstone.

quartzite Metamorphic rock consisting largely of quartz. (Some forms of quartzite are recognized as sedimentary rocks.)

quick clays Clay layers that spontaneously change from a solid condition to a near-liquid condition when disturbed, a process called spontaneous liquefaction.

radar Wavelength band within the microwave region, beginning at about 0.1 cm and extending to about 100 cm (1 m).

radial drainage pattern Stream pattern consisting of streams, radiating outward from a central peak or highland, such as a sedimentary dome or a volcano.

radiation balance Condition of balance between incoming energy of solar shortwave radiation and outgoing longwave radiation emitted by the earth into space.

radiation fog Fog produced by radiational cooling of the basal air layer.

radiocarbon dating method Method of absolute age determination by analysis of the ratio of carbon-14 to ordinary carbon in organic materials.

rain Form of precipitation consisting of falling water drops, usually 0.5 mm or large in diameter.

rainfall variability Long-term percentage of departure of yearly rainfall quantities from the average or mean of record.

rain gauge Instrument used to measure the amount of rain that has fallen.

raingreen vegetation Vegetation that puts out green foliage in the wet season, but becomes largely dormant in the dry season; found in the tropical zone, it includes the savanna biome and monsoon forest.

rainshadow desert Belt of arid climate to lee of a mountain barrier, produced as a result of adiabatic warming of descending air.

raised shoreline Former shoreline, lifted above the limit of wave action; also called an elevated shoreline.

range of tide Difference in height between low water and high water of the tide in a single tidal cycle or as an average value.

ranges Vertical rows of congressional townships in the U.S. Land Office Survey.

rapids Steep-gradient reaches of a stream channel in which stream velocity is high.

recessional moraine Moraine produced at the ice margin during a temporary halt in the recessional phase of deglaciation.

recharge (See *ground water recharge*.)

recharge well A tube well used to provide artificial ground water recharge.

reclining retreat Program of hillslope profile evolution in which the angle of slope decreases with time.

recording thermometer Thermometer equipped with a mechanism for making a continuous record of temperature; same as *thermograph*.

reef rock Limestone consisting of skeletal structures secreted by corals and algae in coral reefs, and accumulations of detrital fragments of reefs.

reg Desert surface armored with pebble layer, resulting from long-continued deflation; found in the Sahara Desert of North Africa.

regolith Layer of mineral particles overlying the bedrock; it may be derived by weathering of underlying bedrock (residual regolith) or by transportation and deposition from other locations by fluid agents (transported regolith). Geological usage includes the soil solum in the term regolith. (See also *residual regolith, transported regolith*.)

rejuvenation of landmass Episode of rapid fluvial denudation set off by a rapid crustal rise, increasing the available landmass. (See also *rejuvenation of stream*.)

rejuvenation of stream Episode of rapid degradation (downcutting) by a stream in an attempt to reestablish grade at lower level following crustal uplift or a falling of sea level. (See also *rejuvenation of landmass*.)

relative humidity Ratio of water vapor present in the air to the maximum quantity possible for saturated air at the same temperature.

relief Measure of the average elevation difference between adjacent high and low points in a land surface, as, for example, hilltops and valley bottoms.

remote sensing   Measurement of some property of an object or surface by means other than direct contact; usually refers to the gathering of scientific information about the earth's surface from great heights and over broad areas, using instruments mounted on aircraft or orbiting space vehicles.

remote sensors   Collective term for detection instruments or devices used in remote sensing.

Rendolls   Suborder of the Mollisols formed on highly calcareous parent materials and with more than 40 percent calcium carbonate in some horizon within 50 cm (20 in.) of the surface, but without a horizon of accumulation of calcium carbonate.

representative fraction (R.F.)   Fractional scale stated as a simple fraction.

reservoir rock   A porous rock mass capable of holding a large concentration of petroleum.

reservoir trap   Arrangement of rock layers or structures capable of trapping and holding petroleum or natural gas. Examples: anticline, sedimentary dome, salt dome.

residual regolith   Regolith formed in place by alteration of the bedrock directly beneath it. (See also *transported regolith*.)

resolution cell size   The rectangular area on the ground associated with each *pixel* in a *digital image*.

respiration   Metabolic process in which organic compounds are oxidized within living cells to yield biochemical energy and waste heat.

retrogradation   Cutting back (retreat) of a shoreline, beach, marine cliff, or marine scarp by wave action. (See also *progradation*.)

reverse fault   Type of fault in which one fault block rides up over the other on a steep fault plane.

revolution   Motion of a planet in its orbit around the sun, or of a planetary satellite around a planet.

rhumb line   (See *loxodrome*.)

rhyolite   Extrusive igneous rock of granite composition; it occurs as lava or tephra.

ria coast   Deeply embayed coast formed by partial submergence of a landmass previously shaped by fluvial denudation.

Richter scale   Logarithmic scale of magnitude numbers stating the relative quantity of energy released by an earthquake.

rift valley   Trenchlike valley with steep, parallel sides; essentially a graben between two normal faults; associated with crustal spreading. (See also *axial rift*.)

rill erosion   Form of accelerated erosion in which numerous, closely spaced miniature channels (rills) are scored into the surface of exposed soil or regolith.

riverine environment   Zone of influence of an alluvial river, including the floodplain.

roches moutonnées   Knobs of bedrock formed by glacial abrasion.

rock   Natural aggregate of minerals in the solid state; typically hard, it consists of one or more mineral varieties.

rock basin   Overdeepened section of bedrock floor of a cirque or glacial trough forming a depression, typically holding a tarn.

rock creep   (See *creep*.)

rock-defended terrace   Alluvial terrace protected from further removal by undercutting by the presence of a bedrock outcrop.

rockfall   Free fall of particles or masses of bedrock from a steep cliff face, typically accumulating at the cliff base in the form of talus.

rock glacier   A tonguelike body of angular rock fragments extending down from the foot of a steep talus slope in high mountains of the alpine zone and showing evidence of present or past motion by rock creep.

rockslide   Form of landslide consisting of the slippage of a bedrock mass on a dipping rock plane.

rock step   Abrupt downstep in the rock floor of a glacial trough.

rock terrace   Terrace carved in bedrock during the degradation of a stream channel induced by the crustal rise or a fall of the sea level. (See also *alluvial terrace, marine terrace*.)

rock texture   Physical property of rock related to the size, shape, and arrangement of mineral particles constituting the rock.

rock transformation cycle   Total cycle of changes in which rock of any one of the three major rock classes—igneous, sedimentary, metamorphic—is transformed into rock of one of the other classes.

Rossby waves   Horizontal undulations in the flow path of the upper-air westerlies; also known as *upper-air waves*.

rotation of earth   (See *earth rotation*.)

runoff   Flow of water from continents to oceans by way of stream flow and ground water flow; a term in the water balance of the hydrologic cycle. In a more restricted sense, runoff refers to surface flow by overland flow and channel flow.

salic horizon   Soil horizon enriched by soluble salts.

salinization   Precipitation of soluble salts within the soil.

saltation   Leaping, impacting, and rebounding of spherical sand grains transported over a sand or pebble surface by wind.

salt-crystal growth   A form of weathering in which rock is disintegrated by the expansive pressure of growing salt crystals during dry weather periods when evaporation is rapid.

salt marsh   Peat-covered expanse of sediment built up to the level of high tide over a previously formed tidal flat.

saltwater intrusion   Contamination of fresh ground water by landward movement of the interface between fresh ground water and salt ground water as a result of excessive pumping from wells.

sand   Grade size of sediment particles between 0.06 mm and 2 mm in diameter.

sand dune   Hill or ridge of loose, well-sorted sand shaped by wind and usually capable of slow downwind motion.

sand-ridge desert   Area dominated by longitudinal dunes.

sand sea   Field of transverse sand dunes.

sandspit   Narrow, fingerlike embankment of sand constructed by littoral drift into the open water of a bay.

sandstone   Variety of sedimentary rock consisting dominantly of mineral particles of sand grade size.

sand storm   Dense, low cloud of sand grains traveling in saltation over the surface of a sand dune or beach.

sanitary landfill   Disposal of solid wastes by burial beneath a cover of soil or regolith.

Santa Ana   Easterly wind, often hot and dry, that blows from the interior desert region of southern California and passes over the coastal mountain ranges to reach the Pacific Ocean.

sapric soil materials   Diagnostic soil material consisting of decomposed organic matter, denser than fibric and hemic materials, with little content of identifiable fibers.

Saprists   Suborder of Histosols that are saturated with water most of the year or artificially drained, and so decomposed that after rubbing less than 10 percent of the volume remains as fibers.

saprolite   Surface layer of deeply decayed igneous rock or metamorphic rock, rich in clay minerals, typically formed in warm, humid climates; a form of residual regolith.

saturated air   Air holding the maximum possible quantity of water vapor at a given temperature and pressure.

saturated zone   Subsurface water zone in which all pores of the bedrock or regolith are filled with ground water, which moves under the force of gravity.

savanna biome   Biome that consists of a combination of widely spaced trees and grassland in various proportions.

savanna environment Environment associated with the savanna biome of the wet-dry tropical climate.

savanna woodland Plant formation class of the savanna biome consisting of a woodland of widely spaced trees and a grass layer, found throughout the wet-dry tropical climate regions in a belt adjacent to the monsoon forest and equatorial rainforest.

scale of globe Ratio of size of a globe to size of the earth, where size is expressed by a measure of length or distance.

scale of map Ratio of distance between two points on a map and the same two points on the ground. (See *fractional scale, graphic scale*.)

scanning systems Remote sensing systems that make use of a scanning beam to generate images over the frame of surveillance.

scarification General environmental impact term for human-made excavations and other land disturbances produced for purposes of extracting or processing mineral resources.

scarp (See *fault scarp, marine scarp*.)

scattering Turning aside by reflection of solar shortwave radiation by gas molecules of the atmosphere.

schist Foliated metamorphic rock in which mica flakes are typically found oriented parallel with the foliation surfaces.

sclerophyll forest Plant formation class of the forest biome, consisting of low sclerophyllous trees, and often including sclerophyllous woodland or scrub, associated with regions of Mediterranean climate.

sclerophyllous leaves Leaves of sclerophylls; they are hard, thick, and leathery.

sclerophylls Hard-leaved evergreen trees and shrubs capable of enduring a long, dry summer.

scoria Lava or tephra containing numerous cavities produced by expanding gases during cooling. (See also *pumice*.)

scree slope (See *talus slope*.)

scrub Plant formation class or subclass consisting of shrubs and having a canopy coverage of about 50 percent.

sea breeze Local wind blowing from sea to land during the day.

sea cave Cave near the base of a marine cliff, eroded by breaking waves.

sea cliff (See *marine cliff*.)

seafloor spreading Pulling apart of the oceanic crust along the axial rift of the mid-oceanic ridge, and representing the continual separation of two lithospheric plates made up of oceanic lithosphere.

sea ice Floating ice of the oceans formed by direct freezing of ocean water.

seasons (See *astronomical seasons*.)

secondary consumers Animals that feed on primary consumers.

secondary minerals In soil science, minerals that are stable in the surface environment, derived by mineral alteration of the primary minerals

secondary succession Ecological succession beginning on a previously vegetated area that has been recently disturbed by such agents as fire, flood, windstorm, or humans.

sections Units of land, 1 mile square, used in the U.S. Land Office Survey.

sediment Finely divided mineral matter, derived directly or indirectly from preexisting rock, and organic matter produced by life processes, usually transported and deposited by a fluid agent. (See also *clastic sediment, organically derived sediment*.)

sedimentary breccia Class of detrital sedimentary rock consisting of angular rock fragments in a matrix of finer fragments.

sedimentary cycle Type of material cycle in which the compound or element is released from rock by weathering, follows the movement of running water either in solution or as sediment to reach the sea, and is eventually converted into rock.

sedimentary dome Uparched strata forming a circular structure with domed summit and flanks with moderate to steep outward dip.

sedimentary rock Rock formed from accumulations of sediment by the process of lithification (diagenesis).

sediment yield Quantity of sediment removed by overland flow from a land surface of given unit area in a given unit of time.

seif dune Large, tapering sand ridge with sharp crest rising and falling in alternate peaks and saddles; also known as sword dune.

seismic sea wave (tsunami) Train of sea waves set off by an earthquake or other seafloor disturbance and traveling over the ocean surface with a wave velocity proportional to the square root of the ocean depth.

seismic waves Waves sent out during an earthquake by faulting or other form of crustal rupture from an earthquake focus and propagated through the solid earth.

semiarid (steppe) climate subtype Subtype of the dry climate in which soilwater storage equals or exceeds 6 cm in at least 2 months of the year.

semideciduous plants Plants that shed their leaves at intervals not in phase with a season. (See also *deciduous plant*.)

semidesert Plant formation class of the desert biome, consisting of xerophytic shrub vegetation with a poorly developed herbaceous lower layer; subtypes are semidesert scrub and woodland and semidesert scrub.

semidesert climate subtype Subtype of the dry climate in which soil-water storage exceeds 6 cm in fewer than 2 months, but is greater than 2 cm in at least 1 month.

semidesert scrub Subtype of the semidesert plant formation class.

semidesert scrub and woodland Subtype of the semidesert plant formation class.

sensible heat Heat measurable by a thermometer; an indication of the intensity of kinetic energy of molecular motion within a substance.

sequential landforms Landforms produced by external earth processes in the total activity of denudation. Examples: canyon, alluvial fan, floodplain.

seral stage Stage in a *sere*.

sere In an ecological succession, the series of biotic communities that follow one another on the way to the stable stage, or climax.

sesquioxides Oxides of aluminum or iron with a ratio of two atoms of aluminum or iron to three atoms of oxygen.

shale Fissile sedimentary rock of mud or clay composition, showing lamination.

shattering Form of physical weathering in which fresh rock fractures are produced by strong mechanical stresses.

shearing of map Distortion on a map projection resulting in the intersection of parallels and meridians in acute and obtuse angles.

sheet erosion Form of rapid soil erosion in which thin layers of soil are removed by overland flow.

sheet flow Overland flow taking the form of a continuous thin film of water over a smooth surface of soil, regolith, or rock.

sheeting structure Thick, subparallel layers of massive bedrock formed by spontaneous expansion accompanying unloading.

shield volcano Domelike accumulation of basalt lava flows emerging from radial fissures.

shoreline Shifting line of contact between land and the water of a lake or ocean.

short-grass prairie (See *steppe*.)

shortwave radiation Electromagnetic radiation in the range from 0.2 to 3 microns, including most of the energy in the solar radiation spectrum.

shrub Woody perennial plant, usually small or low, with several low-branching stems and a foliage mass close to the ground.

Siberian high Center of high atmospheric pressure located over north-central Asia in winter.

side-looking airborne radar (SLAR) Remote sensing using radar sensor systems that send their impulses toward either side of an aircraft.

silcrete Rocklike surface layer cemented largely with silica, widespread in the tropical zone.

silica Silicon dioxide in any of several mineral forms, including the crystalline form as quartz and partly crystalline or amorphous forms such as allophane and chalcedony.

silicate magma Magma from which silicate minerals are formed.

silicates, silicate minerals Minerals containing silicon and oxygen atoms linked in crystal space lattice units of four oxygen atoms surrounding a central silicon atom.

silication Pedogenic process in which the proportion of silica in a soil horizon is increased because the silica remains behind while other materials are removed.

sill Pluton taking the form of plate, representing magma forced into a natural parting in the crustal rock, such as a bedding surface in a sequence of sedimentary rocks.

silt Sediment particles between 0.004 mm and 0.06 mm in diameter.

siltstone Detrital sedimentary rock similar to sandstone but composed of silt grade particles.

sine curve Curve on a graph on which the sine of an angle is plotted against the value of that angle from 0° to 360°.

sinkhole Surface depression or cavity in limestone, leading down into limestone caverns.

sinusoidal projection Conformal map projection consisting of straight horizontal parallels and of meridians that are sine curves.

skeletal minerals Mineral particles, mostly of sand and silt grades, that make up the chemically inactive fraction of the soil, as distinguished from the clay minerals and related weathering products.

slash-and-burn Agricultural system, practiced in the low-latitude rainforest, in which small areas are cleared and the trees burned, forming plots that can be cultivated for brief periods.

slate Fine-textured metamorphic rock derived from shale and showing well-developed cleavage.

sleet Form of precipitation consisting of ice pellets, which may be frozen raindrops.

slickensides Soil surfaces with grooves or striations made by movement of one mass of soil against the other while the soil is in a moist, plastic state; typical of the *Vertisols.*

sling psychrometer Form of hygrometer consisting of a wet-bulb thermometer and a dry-bulb thermometer.

slip face Steep face of an active sand dune, receiving sand by saltation over the dune crest and repeatedly sliding because of oversteepening; it assumes the angle of repose of noncohesive mineral particles on a free slope.

slope (1) Degree of inclination from the horizontal of an element of ground surface, analogous to dip in the geologic sense. (2) Any portion or element of the earth's solid surface, as in hillslope. (3) Verb "to incline."

slump Form of landslide in which a single large block of bedrock moves downward with backward rotation on an upwardly concave fracture surface.

slump block A block of bedrock that moves to a lower position of rest on a curved slip surface. (See also *slump.*)

small circle Circle produced when a plane is passed through a sphere, but does not pass through the center of the sphere. (See also *great circle.*)

small leaf-form Leaves that are thin, flat, and comparatively wide, but of small overall dimensions as compared with the broadleaf.

smog Mixture of particulate matter and chemical pollutants in the lower atmosphere over urban areas.

snow Form of precipitation consisting of ice particles formed by sublimation.

snowline Altitude above which snowbanks remain throughout the entire year on high mountains.

soft layer of mantle A layer within the mantle in which the temperature is close to the melting point, causing the mantle rock to be weak; synonym for *asthenosphere.*

soil Natural terrestrial surface layer containing living matter and supporting or capable of supporting plants.

soil consistence Quality of stickiness of wet soil, the plasticity of moist soil, and the degree of coherence or hardness of soil when it holds small amounts of moisture or is in the dry state.

soil creep Extremely slow downhill movement of regolith as a result of continued agitation and disturbance of the particles by such activities as frost action, temperature changes, or wetting and drying of the soil.

soil enrichment Additions of materials to the soil body; one of the pedogenic processes.

soil erosion Erosional removal of material from the soil surface by action of rainbeat and overland flow.

soil horizon Distinctive layer of the soil, more or less horizontal, set apart from other soil zones or layers by differences in physical and chemical composition, organic content, structure, or a combination of those properties, produced by soil-forming processes (*pedogenic processes*).

soil moisture (See *soil water.*)

soil orders Those 10 soil classes forming the highest category in the Soil Taxonomy within the Comprehensive Soil Classification System of the U.S. Department of Agriculture.

soil profile Display of soil horizons on the face of a pedon, or on any freshly cut vertical exposure through the soil.

soil science (See *pedology.*)

soil solum That part of the soil made up of the A and B soil horizons; the soil zone in which living plant roots exert a control on the soil horizons.

soil solution Aqueous solution held in the soil as soil water and containing dissolved atmospheric gases and ions.

soil structure Presence, size, and form of aggregations (lumps or clusters) of soil particles.

soil suborders Second level of classification of soils in the Soil Taxonomy.

Soil Taxonomy Taxonomic system of classifying soils; part of the Comprehensive Soil Classification System.

soil-temperature regime Characteristic annual cycle of soil temperature defined in terms of mean annual soil temperature and the difference between mean temperatures of the warm and cold seasons.

soil-texture classes Classes of the mineral portion of the soil based on varying proportions of sand, silt, and clay, expressed as percentages.

soil water Water held in the soil and available to plants through their root systems; a form of subsurface water. (Synonymous with *soil moisture.*)

soil-water balance Balance among the component terms of the soil-water budget: precipitation, evapotranspiration, change in soil-water storage, and water surplus.

soil-water belt Soil layer from which plants draw soil water.

soil-water budget Accounting system evaluating the daily, monthly, or yearly amounts of precipitation, evapotranspiration, soil-water storage, water deficit, and water surplus.

soil-water control section Depth zone in the soil recognized for the purpose of establishing the soil-water regime and defined in terms of the depth to which a given amount of applied water will moisten a dry soil.

soil-water recharge Restoring of depleted soil water by infiltration of precipitation.

soil-water regime Characteristic type of annual soil-water budget adapted to purposes of soil classification and defined in terms of conditions prevailing in the soil-water control section.

soil-water shortage Difference between potential evapotranspiration (water need) and actual evapotranspiration (water use), representing the quantity of irrigation water that would be required to sustain maximum plant growth.

soil-water storage Actual quantity of water held in the soil-water belt at any given instant; usually applied to a soil layer of given depth, such as 300 cm (12 in.).

solar collectors Mechanical devices for absorbing direct solar energy and allowing the energy to be transported for use or storage.

solar constant Intensity of solar radiation falling on a unit area of surface held at right angles to the sun's rays at a point outside the earth's atmosphere; equal to about 1400 watts per square meter (1400 W/m²) or about 2 gram-calories per square centimeter per minute (2 cal/cm²/min).

solar energy Energy arriving as electromagnetic radiation from the sun, including such energy stored as heat in air, soil, or water, and chemical energy stored in the biomass of plants through photosynthesis.

solar noon Instant at which the sun crosses the celestial meridian of a given point on the earth; instant at which the sun's shadow points exactly due north or due south.

solar wind Flow of electrons and protons emanating from the sun and traveling outward in all directions through the solar system.

solids Substances in the solid state that resist changes in shape and volume, are usually capable of withstanding large unbalanced forces without yielding, but will ultimately yield by sudden breakage.

solid state State of matter that is dense, has strength to resist flowage, shows elastic properties, and may be a crystalline solid or an amorphous (noncrystalline) substance. Examples: crystalline mineral rock, volcanic glass.

solifluction Arctic tundra variety of earthflow in which the saturated thawed layer over permafrost flows slowly downhill to produce multiple terraces and lobes.

solifluction lobe Bulging mass of saturated regolith with steep curved front moved downhill by solifluction.

solum (See *soil solum*.)

sorting Separation of one size grade of sediment particles from another by the action of currents of air or water.

source region Extensive land or ocean surface over which an air mass derives its temperature and moisture characteristics.

Southern Oscillation Episodic reversal of prevailing barometric pressure differences between two regions, one centered on Darwin, Australia, and the other on Tahiti in the eastern Pacific Ocean; a precursor to the occurrence of an El Niño event.

southern pine forest Subtype of nee-

dleleaf forest dominated by pines and occurring in the moist subtropical climate.

spalling Form of weathering in which concentric, curved rock shells are produced; a small-scale form of exfoliation.

specific humidity Mass of water vapor contained in a unit mass of air.

spectral signature The unique spectral reflectance of a surface or object by which it may be possible to identify it. To the eye, this is the "color" of the surface or object.

spectrum of electromagnetic radiation Total range of wavelengths and frequencies from gamma rays to radio waves.

spheroidal weathering Formation of thin, soft concentric shells of decomposed rock as chemical weathering penetrates joints in bedrock under a protective cover of regolith.

spine Leaf form that is a hard, sharp-pointed spike.

spit (See *sandspit*.)

splash erosion Soil erosion caused by direct impact of falling raindrops on a wet surface of soil or regolith.

spodic horizon Soil horizon containing precipitated amorphous materials composed of organic matter and sesquioxides of aluminum, with or without iron.

Spodosols Soil order within the Soil Taxonomy consisting of soils with a spodic horizon, an albic horizon, and with low cation exchange capacity (CEC), and lacking in carbonate materials.

spontaneous liquefaction In quick clays, the sudden loss of internal strength when shearing occurs during initial stages of flowage or shaking by an earthquake.

spot heights Numerals printed on a map to show the altitude of selected points.

spreading plate boundary Lithospheric plate boundary along which two plates of oceanic lithosphere are undergoing separation, while at the same time new lithosphere is being formed by accretion. (See also *converging plate boundary, transform plate boundary*.)

spring Discharge of ground water from a point on the land surface, on the floor of a stream or lake, or at a shoreline.

stable air mass Air mass in which the environmental temperature lapse rate is less than the dry adiabatic lapse rate and thus resists being lifted.

stack (marine) Isolated columnar mass of bedrock left standing in front of a retreating marine cliff.

staff gauge Graduated vertical scale used to measure stream stage.

stalactite Vertical, conical, or tubular

mineral deposit, usually of calcium carbonate, attached to the ceiling of a limestone cave and built by the action of dripping water.

stalagmite Upright conical or cylindrical mineral deposit, usually of calcium carbonate, built upward from the floor of a limestone cave by action of dripping water.

standard meridians Standard time meridians separated by 15° of longitude and having values that are multiples of 15°. (In some cases meridians are used that are multiples of 7½°.)

standard parallels Parallels of latitude, spaced 24 miles apart, used as secondary base lines in the U.S. Land Office Survey.

standard time Time system based upon the local time of a standard meridian and applied to belts of longitude extending 7½° (more or less) on either side of that meridian.

star dune Large, isolated sand dune with radial ridges culminating in a peaked summit; found in the deserts of North Africa and the Arabian Peninsula.

Stefan-Boltzmann law Total energy radiated by a unit area of surface per unit of time varies as the fourth power of the absolute temperature (°K).

steppe Plant formation class in the grassland biome consisting of short grasses sparsely distributed in clumps and bunches and some shrubs, widespread in areas of semiarid climate in continental interiors of North America and Eurasia; also called *short-grass prairie*.

steppe climate (See *semiarid climate subtype*.)

stereographic projection Conformal zenithal projection on which any arc of a circle on the globe is shown as an arc of a circle on the map.

stomata Specialized leaf pores that are openings in the outer cell layer through which transpiration occurs.

stone polygons Ringlike networks of coarse rock fragments found on barren land surfaces in the arctic environment.

storage capacity Maximum capacity of soil to hold water against the pull of gravity; same as *field capacity*.

storage pools Type of pool in the material cycle in which materials are more or less inaccessible to life. (See also *active pool*.)

storage recharge Restoration of stored soil water during periods when precipitation exceeds potential evapotranspiration (water need).

storage withdrawal Depletion of stored soil water during periods when evapotranspiration exceeds precipitation, calculated as the difference between actual evapotranspiration (water use) and precipitation.

storm surge   Rapid rise of coastal water level accompanying the onshore arrival of a tropical cyclone.

strangler   Twining plant or liana that surrounds a tree trunk, eventually killing and replacing the tree.

strata   Layers of sediment or sedimentary rock separated from one another by stratification planes (bedding planes).

stratified drift   Glacial drift made up of sorted and layered clay, silt, sand, or gravel, deposited from meltwater in stream channels or in marginal lakes close to the ice front.

stratiform clouds   Clouds of layered, blanketlike form.

stratocumulus   Cloud type of the low-height family consisting of a layer of individual dense cloud masses.

stratopause   Upper limit of the stratosphere, transitional upward into the mesosphere.

stratosphere   Layer of atmosphere lying directly above the troposphere.

stratus   Cloud type of the low-height family formed into a dense, dark-gray layer.

stream   Long, narrow body of flowing water occupying a stream channel and moving to lower levels under the force of gravity. (See *consequent stream, graded stream, subsequent stream.*)

stream capacity   Maximum load of solid matter that can be carried by a stream for a given discharge.

stream channel   Long, narrow, troughlike depression occupied and shaped by a stream moving to progressively lower levels.

stream deposition   Accumulation of transported rock particles on a stream bed, on the adjacent floodplain, or as a delta in a body of standing water.

stream erosion   Progressive removal of mineral particles from the floor or sides of a stream channel by drag force of the moving water, or by abrasion, or by chemical reaction with ions in stream water.

stream flow   Water flow in a stream channel; also known as *channel flow.*

stream gauging   Measurement of stream discharge, mean velocity, and depth continuously or at intervals over a long period of time at a selected station.

stream gradient   Rate of descent to lower elevations along the profile of a stream or stream channel, stated in m/km, degrees, or percent.

stream load   Solid matter carried by a stream in dissolved form, in turbulent suspension, and as bed load. (See also *bed load, suspended load.*)

stream stage   Height or elevation of the surface of a stream at any given instant.

stream transportation   Downvalley movement of eroded particles in a stream, in solution, in turbulent suspension, or in traction as bed load.

stream velocity   Speed of flow (m/sec) of water in a stream, measured in the downstream direction at a given point above the bed or averaged for the entire stream cross section.

striations, glacial   Scratches made by glacier abrasion on bedrock outcrops or the surfaces of rock fragments carried by the ice.

strike   Compass direction of the line of intersection of an inclined rock plane and a horizontal plane of reference. (See also *dip.*)

strike-slip fault   (See *transcurrent fault.*)

strip mining   Mining method in which overburden is first removed from a seam of coal, or a sedimentary ore, allowing the coal or ore to be extracted.

structure section   Graphic presentation of a vertical cross section from the surface to a given depth, showing the arrangement of rock units and other geologic features.

subantarctic low-pressure belt   Persistent belt of low atmospheric pressure centered about at lat. 65°S over the Southern Ocean.

subantarctic zone   Latitude zone lying between lat. 55° and 60° S (more or less) and occupying a region between the midlatitude zone and the antarctic zone.

subarctic zone   Latitude zone between lat. 55° and 60° N (more or less), occupying a region between the midlatitude zone and the arctic zone.

Subboreal climatic stage   Climate stage of the Holocene Epoch, with below-average temperatures, spanning the period of about −5000 to −2000 years.

subduction   Descent of the downbent edge of a lithospheric plate into the asthenosphere so as to pass beneath the edge of the adjoining plate along an active plate boundary.

subduction boundary   (See *converging plate boundary.*)

subhumid climate subtype   Subtype of the moist climate in which annual soil-water shortage is greater than zero but less than 15 cm.

sublimation   Process of change of water vapor (gaseous state) to ice (solid state) or vice versa. Examples: formation of hoar frost, evaporation of snow.

submarine canyon   Narrow, V-shaped submarine valley cut into the continental slope, usually attributed to erosion by turbidity currents.

submarine fan (cone)   Submarine accumulation of coarse-textured sediment carried by turbidity currents to form a large fan-shaped deposit on the deep ocean floor, usually situated at the lower end of a submarine canyon system leading down the outer slopes of a major delta on the continental shelf.

submergence, marine   Inundation or partial drowning of a former land surface by a rise of sea level or a sinking of the crust or both.

subsequent stream   Stream that develops its course of erosion along a zone or belt of weaker rock, such as the crushed zone of a fault line.

subsidence of air   Downsinking of a large mass of air in the central region of an anticyclone.

subsidence theory   Hypothesis advanced by Charles Darwin to explain the formation of an atoll by subsidence of the ocean floor with continued upbuilding of coral reefs upon the summit of a drowned submarine volcano (seamount).

subsolar point   Point at which solar rays are perpendicular to the earth's ellipsoidal surface of reference.

subsurface water   Water of the lands held below the surface in regolith or bedrock.

subtropical high-pressure belts   Belts of persistent high atmospheric pressure trending east-west and centered about on lat. 30° N and S.

subtropical jet stream   Jet stream of westerly winds forming at the tropopause, just above the Hadley cell.

subtropical zones   Latitude zones occupying the region of lat. 25° and 35° N and S (more or less) and lying between the tropical zones and the midlatitude zones.

succulent   Adjective applied to thickened, spongy leaves or stems capable of holding much water.

succulents   Plants adapted to resist water loss (xerophytes) by means of thickened spongy tissue in which water is stored.

summer berm   Berm produced by weak waves during the summer season in midlatitudes. (See also *winter berm.*)

summergreen deciduous forest   (See *midlatitude deciduous forest.*)

summer monsoon   Inflow of maritime air at low levels from the Indian Ocean toward the Asiatic low pressure center in the season of high sun; associated with the rainy season of the wet-dry tropical climate and the Asiatic monsoon climate.

summer solstice   Solstice occurring on June 21 or 22, when the sun's declination is 23½°N.

sun's declination   (See *declination of sun.*)

sun's noon altitude   (See *noon altitude of sun.*)

sun-synchronous orbit   Satellite orbit in which the orbital plane remains fixed in position with respect to the sun.

supercooled water Water existing in the liquid state at a temperature lower than the normal freezing point.

surface detention Temporary holding of precipitation in minor surface depressions; same as *depression storage.*

surface environment Environment of low pressure and low temperature to which rock is exposed near the earth's surface.

surface water Water of the lands flowing exposed as streams or impounded as ponds, lakes, or marshes.

surge of glacier Episode of very rapid downvalley movement within an alpine glacier.

suspended load That part of the stream load carried in turbulent suspension.

suture (See *continental suture.*)

swales Long, narrow depressions lying between parallel beach ridges, or between bars of a point-bar deposit on a river floodplain.

swash Surge of water up the beach slope (landward) following collapse of a breaker.

symbiosis Form of positive interaction between species that is beneficial to one of the species and does not harm the other.

synclinal mountain Steep-sided ridge or elongate mountain developed by erosion on a syncline.

synclinal valley Valley eroded on weak strata along the central trough or axis of a syncline.

syncline Downfold of strata or other layered rock in a troughlike structure; a type of fold. (See also *anticline.*)

system boundary Bounding surface, real or imagined, limiting the extent of a flow system of matter or energy.

taiga (See *cold woodland.*)

tall-grass prairie (See *prairie.*)

talus Accumulation of loose rock fragments (slide rock) derived by rockfall from a cliff.

talus cone Accumulation of talus in the form of a partial cone with apex at top, heading in a ravine or gully.

talus slope Slope formed of slide rock; the surface of a talus cone.

tarn Small lake occupying a rock basin in a cirque or glacial trough.

tectonic Relating to tectonic activity and tectonics.

tectonic activity Crustal processes of bending (folding) and breaking (faulting), usually concentrated on or near active lithospheric plate boundaries. (See also *plate tectonics, tectonic, tectonics.*)

tectonic arc Long, narrow chain of islands or mountains or a narrow submarine ridge adjacent to a subduction boundary and its trench, formed by tectonic processes, such as the con-

struction and rise of an accretionary prism.

tectonic crest Ridgelike summit line of a tectonic arc associated with an accretionary prism.

tectonic erosion Removal of masses of rock from the lower edge of a lithospheric plate by downdrag exerted by a subducting plate passing beneath it.

tectonics Branch of geology relating to tectonic activity and the features it produces. (See also *plate tectonics, tectonic activity.*)

temperate rainforest (See *broadleaf evergreen forest, laurel forest.*)

temperature inversion Upward reversal of the normal environmental temperature lapse rate, so that the air temperature increases upward. (See *low-level temperature inversion, upper-level temperature inversion.*)

tephra Collective term for all size grades of particles of solidified magma blown out under gas pressure from a volcanic vent.

terminal moraine Moraine deposited as an embankment at the terminus of an alpine glacier or at the leading edge of an ice sheet.

terrane Continental crustal rock unit having a distinctive set of lithologic properties, reflecting its geologic history, that distinguish it from adjacent or surrounding continental crust.

terrestrial ecosystems Ecosystems of land plants and animals found on upland surfaces of the continents.

texture of rock (See *rock texture.*)

thallophytes Low forms of plant life lacking in true roots, stems, and leaves; they include bacteria, molds, and fungi.

thermal environment Total influence of heat and cold on living organisms in the life layer.

thermal erosion In regions of permafrost, the physical disruption of the land surface by melting of ground ice, brought about by removal of a protective organic layer.

thermal infrared Electromagnetic radiation in the infrared radiation wavelength band, approximately from 1 to 20 microns.

thermal pollution Form of water pollution in which heated water is discharged into a stream or lake from the cooling system of a power plant or other industrial heat source.

thermal regimes Global set of air-temperature regimes, based on distinctive annual cycles of mean monthly air temperature.

thermic regime Soil temperature regime in which mean annual soil temperature is greater than 15°C but less than 22°C and the difference between mean temperatures of warm season and cold season is greater than 5C°.

thermocline Water layer of a lake or

the ocean in which temperature changes rapidly in the vertical direction.

thermograph (See *recording thermometer.*)

thermometer Instrument measuring temperature. (See also *maximum–minimum thermometer; recording thermometer.*)

thermosphere Atmospheric layer of upwardly increasing temperature, lying above the mesopause.

thornbush Term used locally for occurrences of thorntree semidesert vegetation.

thorn forest and thorn woodland Subtype of the thorntree semidesert formation class.

thorntree-desert grass savanna Subtype of the thorntree semidesert plant formation class.

thorntree semidesert Formation class within the desert biome, transitional from grassland biome to savanna biome and consisting of xerophytic trees and shrubs.

thorntree-tall grass savanna Plant formation class, transitional between the savanna biome and the grassland biome, consisting of widely scattered trees in an open grassland.

thornwoods Term used locally for occurrences of thorntree semidesert vegetation.

throughflow (See *interflow.*)

thrust sheet Sheetlike mass of rock moving over a low-angle overthrust fault.

thunderstorm Intense, local convectional storm associated with a cumulonimbus cloud and yielding heavy precipitation, along with lightning and thunder, and sometimes the fall of hail.

tidal current Current set in motion by the ocean tide.

tidal inlet Narrow opening in a barrier island or baymouth bar through which tidal currents flow.

tidal power Power derived from tidal currents moving through constricted coastal passages, usually modified by dam structures.

tide (See *ocean tide.*)

tide curve Graphical presentation of the rhythmic rise and fall of ocean surface level because of ocean tides.

tide range (See *range of tide.*)

till Heterogeneous mixture of rock fragments ranging in size from clay to boulders, deposited beneath moving glacier ice or directly from the melting in place of stagnant glacier ice.

till plain Undulating, plainlike land surface underlain by glacial till.

tombolo Sand bar or narrow beach connecting an island with the mainland.

topographic map Map showing surface configuration or topography by means of contours, plastic shading, hachures, or other graphic devices.

topographic profile Graph with elevation (altitude) on the vertical scale and distance on the horizontal scale, on which the relief of the land surface along a given traverse line is shown by a rising and falling line.

tor Group of boulders or joint blocks forming a small but conspicuous hill.

tornado Small, very intense wind vortex with extremely low air pressure in center, formed beneath a dense cumulonimbus cloud in proximity to a cold front.

Torrerts Suborder of *Vertisols* found in the torric regime and having cracks that remain open throughout the year in most years.

torric regime (See *aridic regime.*)

Torrox Suborder of *Oxisols* of the aridic or torric soil-water regime, dry in all horizons for more than 6 months of the year and never continuously moist for as long as 3 consecutive months.

tower karst Variety of karst landscape consisting of tall, steep-sided limestone hills separated by a plainlike surface.

township (See *congressional township.*)

trace gases Atmospheric gases present in relatively small concentrations (principally, methane, nitrous oxide, ozone, and chlorofluorocarbons) that absorb outgoing longwave radiation.

trade winds (trades) Persistent easterly surface winds in low latitudes, representing the low-level air flow within the tropical easterlies.

transcurrent fault Fault on which the relative motion is dominantly horizontal, in the direction of the strike of the fault; also called a *strike-slip fault.*

transform fault Special case of a transcurrent fault making up the boundary of two moving lithospheric plates; most commonly found transverse to the mid-oceanic ridge where it connects the offset segments of a spreading plate boundary.

transform plate boundary Lithospheric plate boundary along which two plates are in contact on a transform fault; the relative motion is that of a transcurrent fault.

transform scar Linear topographic feature of the ocean floor taking the form of an irregular scarp or ridge and originating at the offset axial rift of the mid-oceanic ridge; it represents a former transform fault but is no longer a plate boundary.

transpiration Evaporative loss of water to the atmosphere from leaf pores of plants.

transportation (See *stream transportation.*)

transported regolith Regolith formed of mineral matter carried by fluid agents from a distant source and deposited on the bedrock or on older regolith. Examples: floodplain silt, lake clay, beach sand. (See also *residual regolith.*)

transverse dunes Field of wavelike sand dunes with crests trending at right angles to the direction of the prevailing wind.

travertine Carbonate mineral matter, usually calcite, accumulating on limestone cavern surfaces situated in the unsaturated zone.

tree Large erect woody perennial plant typically having a single main trunk, few branches in the lower part, and a branching crown.

trellis drainage pattern Drainage pattern characterized by a dominant parallel set of major subsequent streams, joined at right angles by numerous short tributaries; typical of coastal plains and belts of eroded folds.

trench (See *oceanic trench.*)

triangular facets Steeply inclined bedrock surfaces of triangular outline occurring between canyons carved into the fault scarp near the base of a fault block mountain.

Tropepts Suborder of Inceptisols found in low latitudes. They have a brownish or reddish horizon of altered material (ochric epipedon) at the surface.

tropical air Air originating in the subtropical zone and tropical zone, having the characteristics of a tropical air mass.

tropical air mass Warm air mass with source region over continents and oceans in lat. 20° to 35° N and S.

tropical cyclone Intense traveling cyclone of tropical and subtropical latitudes, accompanied by high winds and heavy rainfall.

tropical easterlies Low-latitude wind system of persistent tropospheric air flow from east to west between the two subtropical high-pressure belts.

tropical easterly jet stream Upper-air jet stream of seasonal occurrence, running east to west at very high altitudes over Southeast Asia.

tropical rainforest Plant formation class similar in structure to the equatorial rainforest, but extending into the tropical zone along coasts windward to the trade winds.

tropical year Year defined as the time elapsing between one vernal equinox and the next.

tropical zones Latitude zones centered on the tropic of cancer and the tropic of capricorn, within the latitude ranges 10° to 25°N and 10° to 25°S, respectively.

tropic of cancer Parallel of latitude at 23½°N.

tropic of capricorn Parallel of latitude at 23½°S.

tropopause Boundary between troposphere and stratosphere.

tropophyte Plant that sheds its leaves and enters a dormant state during a dry or cold season when little soil water is available.

troposphere Lowermost layer of the atmosphere in which air temperature falls steadily with increasing altitude.

trough (See *glacial trough, hanging trough.*)

trough lake Lake occupying part of the floor of a glacial trough.

true azimuth Azimuth system referred to true north (geographic north).

true north Direction from a point on the earth's surface to the north geographic pole, following a meridian through the point; same as *geographic north.*

truncated spur Valley spur that has been beveled off by glacial abrasion to become part of the sidewall of a glacial trough.

tsunami (See *seismic sea wave.*)

tube well Drilled water well that makes use of a metal casing.

tuff Consolidated or lithified deposit of volcanic ash; a form of pyroclastic sediment or rock.

tundra biome Biome of the cold regions of arctic tundra and alpine tundra, consisting of grasses, grasslike plants, flowering herbs, dwarf shrubs, mosses, and lichens.

tundra climate Cold climate of the arctic zone with 8 or more consecutive months of zero potential evapotranspiration (water need).

turbidity current Rapid streamlike flow of turbid (muddy) seawater close to the seabed, often confined within a submarine canyon on the continental slope or flowing down the inner wall of an oceanic trench.

turbulence (See *turbulent flow.*)

turbulent flow Mode of fluid flow in which individual fluid particles (molecules) move in complex eddies, superimposed on the average downstream flow path.

turbulent suspension Stream transportation in which particles of sediment are held in the body of the stream by turbulent eddies.

typhoon Tropical cyclone of the western North Pacific and coastal waters of Southeast Asia.

Udalfs Brownish suborder of the *Alfisols* formed under the udic soil-water regime in regions having no periods or only short periods when part or all of the soil is dry.

Uderts Suborder of the Vertisols of the udic soil-water regime. They have cracks that remain open only for short periods in most years.

udic regime   Soil-water regime of moist climates in which the soil is not dry in any part of the control section as long as 90 days per year and there is a seasonal water surplus that causes water to move through the soil at some part of the year.

Udolls   Suborder of the *Mollisols* of the udic soil-water regime. They have brownish hues throughout and no horizon of accumulation of calcium carbonate in soft powdery forms.

Udults   Suborder of *Ultisols* of the udic and perudic soil-water regimes. They have moderate to small amounts of organic matter, reddish or yellowish B horizons, and no periods or only short periods when some part of the soil is dry.

Ultisols   Soil order in the *Soil Taxonomy*, consisting of soils of mesic and warmer soil-temperature regimes with an argillic horizon and low base status.

ultramafic igneous rocks   Igneous rocks composed almost entirely of mafic minerals, usually olivine or pyroxene.

ultraviolet rays   Electromagnetic radiation in the wavelength range of 0.2 to 0.4 microns.

Umbrepts   Suborder of *Inceptisols* with a dark acid surface horizon (umbric or mollic epipedon) more than 25 cm (10 in.) thick.

umbric epipedon   Dark surface soil horizon (epipedon) resembling the mollic epipedon, but with percent base saturation (PBS) less than 50.

unconfined aquifer   Aquifer in which the water table is free to receive recharge from above and to fluctuate up or down.

unloading   Process of removal of overlying rock load from bedrock by denudation, accompanied by spontaneous expansion and often leading to the development of sheeting structure.

unsaturated zone   Subsurface water zone in which pores are not fully saturated, except at times when infiltration is very rapid; it lies above the saturated zone.

unstable air mass   Air mass with substantial content of water vapor, capable of breaking into spontaneous convectional activity leading to the development of heavy showers and thunderstorms.

upper-air waves   (See *Rossby waves*.)

upper-air westerlies   System of westerly winds in the upper atmosphere over middle and high latitudes.

upper-level temperature inversion   Temperature inversion produced by subsidence of air within an anticyclone and occurring at the top of a cool, stable air layer.

uprush   (See *swash*.)

Ustalfs   Suborder of the *Alfisols* formed under the ustic soil-water re-

gime, with long periods when the soil is dry.

Usterts   Suborder of the *Vertisols* formed in the ustic soil-water regime with distinct wet and dry seasons; widespread in areas of the wet-dry tropical climate.

ustic regime   Soil-water regime in which the soil has a moderate quantity of water in storage during the season when conditions are favorable to plant growth and the soil water is not frozen, but the soil is dry for 90 or more cumulative days in most years.

Ustolls   Suborder of the *Mollisols* formed in the ustic soil-water regime, with long periods in which the soil is dry. •

Ustox   Suborder of the *Oxisols* formed under the ustic soil-water regime.

Ustults   Suborder of the *Ultisols* formed under the ustic soil-water regime.

valley train   Deposit of alluvium extending downvalley from a melting glacier.

valley winds   Air movement at night down the gradient of valleys and the enclosing mountainsides; alternating with daytime mountain winds.

Van Allen radiation belt   Doughnut-shaped belt of intense ionizing radiation surrounding the earth, within the inner magnetosphere.

varves   Annual bands in the form of alternate light and dark layers found in fine-textured glaciolacustrine sediments.

veldt   Region of steppe grassland in Orange Free State and the Transvaal of South Africa.

vermiculite   Group of clay minerals consisting of hydrous aluminosilicates rich in iron and magnesium; a common product of the weathering of mafic volcanic rocks.

vernal equinox   Equinox occurring on March 20 or 21, when the sun's declination is 0°.

vertical exaggeration   Ratio by which the vertical scale on a topographic profile is greater than the horizontal scale.

Vertisols   Soil order of the *Soil Taxonomy*, consisting of soils of the subtropical zone and the tropical zone with high clay content, developing deep, wide cracks when dry, and showing evidence of movement between aggregates.

visible light   Electromagnetic radiation in the wavelength range of 0.4 to 0.7 microns.

volatiles   Elements and compounds normally existing in the gaseous state under atmospheric conditions, dissolved in magma. (See also *outgassing*.)

volcanic arc   Long, narrow chain of composite volcanoes on an active con-

tinental margin or island arc paralleling an active subduction boundary.

volcanic ash   Finely divided extrusive igneous rock blown under gas pressure from a volcanic vent; a form of *tephra*.

volcanic breccia   Breccia consisting of pyroclastic sediment.

volcanic neck   Isolated, narrow steep-sided peak formed by erosion of igneous rock previously solidified in the feeder pipe of an extinct volcano.

volcanism   General term for volcano building and related forms of extrusive igneous activity.

volcano   Conical, peaklike hill or mountain built by accumulations of lava flows and tephra, including volcanic ash. (See *composite volcano, shield volcano*.)

volcano coast   Coastline formed against volcanoes and lava flows built partly below and partly above sea level.

warm-blooded animal   Animal that possesses one or more adaptations to maintain a constant internal temperature despite fluctuations in the environmental temperature.

warm-core ring   Circular eddy of warm water, surrounded by cold water and lying adjacent to a warm, poleward moving ocean current, such as the Gulf Stream. (See also *cold-core ring*.)

warm front   Moving weather front along which a warm air mass is sliding up and over a cold air mass, leading to production of stratiform clouds and precipitation.

washout   Downsweeping of atmospheric particulate matter by precipitation.

water deficit   Difference between soil water present in the soil (actual soil-water storage) and the storage capacity of the soil.

waterfall   Free-fall descent of a stream over an abrupt, clifflike bedrock downstep in the stream channel.

watergap   Narrow transverse gorge cut across a narrow ridge by a stream, usually in a region of eroded folds.

waterlogging   Rise of a water table in alluvium to bring the zone of saturation into the root zone of plants.

water need   (See *potential evapotranspiration*.)

waterspout   Intense perpendicular vortex of air and water droplets formed beneath a cumulonimbus cloud over water; visible as a funnel cloud.

water surplus   Water disposed of by runoff, interflow, or percolation to the ground water zone after the storage capacity of the soil is full.

water table   Upper boundary surface of the saturated zone; the upper limit of the ground water body. (See also *perched water table*.)

water use　(See *actual evapotranspiration.*)

water vapor　Gaseous state of water.

watt　Unit of power equal to the quantity of work done at the rate of one joule per second, or 10^7 ergs per second.

wave amplitude　In wave motion, the width of space between crests and troughs, measured at right angles to the direction of wave travel.

wave-cut notch　Rock recess at the base of a marine cliff where wave impact is concentrated.

wave cyclone　Traveling, vortexlike cyclone involving interaction of cold and warm air masses along sharply defined fronts.

wave frequency　Number of waves passing a fixed point in a given unit of time.

wavelength　Distance separating one wave crest from the next in any uniform succession of traveling waves.

wave refraction　Bending of a wave front as it travels in shallow water, caused by changes in depth of the bottom.

wave theory　Ideal model of development of the wave cyclone as put forward by J. Bjerknes.

weak equatorial low　Weak, slowly moving low-pressure center (cyclone) accompanied by numerous convectional showers and thunderstorms; it forms close to the ITC in the rainy season, or summer monsoon.

weather　Physical state of the atmosphere at a given time and place.

weathering　All processes acting at or near the earth's surface to cause physical disruption and chemical decomposition of rock and regolith. (See *chemical weathering, physical weathering.*)

westerlies　(See *prevailing westerly winds, upper-air westerlies.*)

west-wind drift　Surface drift of ocean water moving eastward in the zone of prevailing westerlies.

wet (saturation) adiabatic lapse rate　Reduced adiabatic lapse rate when condensation is taking place in rising air; value ranges from 3 to 6C°/1000 m (2 to 3F°/1000 ft).

wet-dry tropical climate　Climate of the tropical zone characterized by a very wet season alternating with a very dry season.

wet equatorial climate　Moist climate of the equatorial zone with a large annual water surplus, and with uniformly warm temperatures and high values of soil-water storage throughout the year.

wetted perimeter　Length of the line of contact between the water of a stream and its channel.

wilting point　Quantity of stored soil water, less than which the foliage of plants not adapted to drought will wilt.

wind　Air motion, dominantly horizontal relative to the earth's surface. (See *geostrophic wind, meridional winds, prevailing westerly winds, upper-air westerlies.*)

wind abrasion　Mechanical wearing action by wind-driven sand grains striking exposed rock surfaces.

windows　Certain wavelength bands of the electromagnetic radiation spectrum within which energy is radiated through the atmosphere to escape into outer space.

wind vane　Weather instrument used to indicate wind direction.

winter berm　Berm produced by large storm waves during the winter in midlatitudes. (See also *summer berm.*)

winter monsoon　Outflow of continental air at low levels from the Siberian high passing over Southeast Asia as a dry, cool northerly wind.

winter solstice　Solstice occurring on December 21 or 22, when the sun's declination is 23½°S.

Wisconsinan Stage (Glaciation)　Last glacial stage (glaciation) of the Pleistocene Epoch.

woodland　Plant formation class, transitional between forest biome and savanna biome, consisting of widely spaced trees with canopy coverage between 25 and 60 percent.

world ocean　Collective term for combined oceans of the globe.

Xeralfs　Suborder of the *Alfisols* formed under the xeric soil-water regime.

Xererts　Suborder of the *Vertisols* formed under a xeric soil-water regime.

xeric animals　Animals adapted to dry conditions typical of a desert climate.

xeric regime　Soil-water regime applying to areas of Mediterranean climate with long, dry summer and rainy winter.

Xerolls　Suborder of the *Mollisols* formed under the xeric soil–water regime.

xerophytes　Plants adapted to a dry environment.

Xerults　Suborder of *Ultisols* formed under a xeric soil-water regime.

X rays　High-energy form of radiation at the extreme short wavelength (high-frequency) end of the electromagnetic spectrum.

yardang　Long, narrow ridge of semiconsolidated silty alluvium or former dune deposit, shaped by wind abrasion into a streamlined form with conspicuous upwind prow and tapering tail.

yazoo stream　Stream that enters on the floodplain of a larger alluvial river and is forced by the presence of the natural levees to flow far downvalley before making a junction with the larger stream.

year　Period of time required for one complete revolution of a planet in its orbit around the sun. (See also *tropical year.*)

zenithal projection　One of a class of map projections in which the geographic grid is centered on a point and has perfect radial symmetry of its geometrical properties.

zone of ablation　Lower portion of glacier in which ablation exceeds gain of mass by snowfall; the zone of wastage of a glacier.

zone of accumulation　Upper portion of a glacier in which the firn becomes transformed into glacial ice; the zone of nourishment of a glacier.

INDEX

Numbers in *italics* refer to definitions or explanations of terms.

TOPOGRAPHIC MAP SYMBOLS

VARIATIONS WILL BE FOUND ON OLDER MAPS

Hard surface, heavy duty road, four or more lanes . . . . . .

Hard surface, heavy duty road, two or three lanes . . . .

Hard surface, medium duty road, four or more lanes . . . .

Hard surface, medium duty road, two or three lanes . . . . .

Improved light duty road . . . .

Unimproved dirt road and trail . . . .

Dual highway, dividing strip 25 feet or less . . . .

Dual highway, dividing strip exceeding 25 feet . . . .

Road under construction . . . .

Railroad, single track and multiple track . . . .

Railroads in juxtaposition . . . .

Narrow gage, single track and multiple track . . . .

Railroad in street and carline . . . .

Bridge road and railroad . . . .

Drawbridge, road and railroad . . . .

Footbridge . . . .

Tunnel, road and railroad . . . .

Overpass and underpass . . . .

Important small masonry or earth dam . . . .

Dam with lock . . . .

Dam with road . . . .

Canal with lock . . . .

Buildings (dwelling, place of employment, etc.) . . . .

School, church, and cemetery . . . . Cem

Buildings (barn, warehouse, etc.) . . . .

Power transmission line . . . .

Telephone line, pipeline, etc. (labeled as to type) . . . .

Wells other than water (labeled as to type) . . . . o Oil . . . . o Gas

Tanks; oil, water, etc. (labeled as to type) . . . . ● ● ● Water

Located or landmark object; windmill . . . . o

Open pit, mine, or quarry; prospect . . . . x x

Shaft and tunnel entrance . . . . ▪ . . . . Y

Horizontal and vertical control station:

   Tablet, spirit level elevation . . . . BM△ 5653

   Other recoverable mark, spirit level elevation . . . . △ 5455

Horizontal control station: tablet, vertical angle elevation VABM △ 9519

   Any recoverable mark, vertical angle or checked elevation △3775

Vertical control station: tablet, spirit level elevation . . . . . BM × 957

   Other recoverable mark, spirit level elevation . . . . × 954

Checked spot elevation . . . . ×4675

Unchecked spot elevation and water elevation . . . . ×5657 870

Boundary, national . . . .

   State . . . .

   County, parish, municipio . . . .

   Civil township, precinct, town, barrio . . . .

   Incorporated city, village, town, hamlet . . . .

   Reservation, national or state . . . .

   Small park, cemetery, airport, etc. . . . .

   Land grant . . . .

Township or range line, United States land survey . . . . . .

Township or range line, approximate location . . . .

Section line, United States land survey . . . .

Section line, approximate location . . . .

Township line, not United States land survey . . . .

Section line, not United States land survey . . . .

Section corner, found and indicated . . . . + ... +

Boundary monument: land grant and other . . . . ▫ . . . . ▫

United States mineral or location monument . . . . ▲

Index contour . . . . Intermediate contour . .

Supplementary contour Depression contours . .

Fill . . . . Cut . . . .

Levee . . . . Levee with road . . . . .

Mine dump . . . . Wash . . . .

Tailings . . . . Tailings pond . . . .

Strip mine . . . . Distorted surface . . . . .

Sand area . . . . Gravel beach . . . .

Perennial streams . . . . Intermittent streams . .

Elevated aqueduct . . . . Aqueduct tunnel . . . . .

Water well and spring . o Disappearing stream . .

Small rapids . . . . Small falls . . . .

Large rapids . . . . Large falls . . . .

Intermittent lake . . . . . Dry lake . . . .

Foreshore flat . . . . Rock or coral reef . . . .

Sounding, depth curve . 10 Piling or dolphin . . . . .

Exposed wreck . . . . . . Sunken wreck . . . .

Rock, bare or awash; dangerous to navigation . . . . . ✻ ⊛

Marsh (swamp) . . . . Submerged marsh . . . .

Wooded marsh . . . . Mangrove . . . .

Woods or brushwood . . Orchard . . . .

Vineyard . . . . Scrub . . . .

Inundation area . . . . Urban area . . . .